GEOMETRIC STRUCTURES OF PHASE SPACE IN MULTIDIMENSIONAL CHAOS

A SPECIAL VOLUME OF ADVANCES IN CHEMICAL PHYSICS
VOLUME 130

PART B

EDITORIAL BOARD

BRUCE J. BERNE, Department of Chemistry, Columbia University, New York, New York, U.S.A.
KURT BINDER, Institut für Physik, Johannes Gutenberg-Universität Mainz, Mainz, Germany
A. WELFORD CASTLEMAN, JR., Department of Chemistry, The Pennsylvania State University, University Park, Pennsylvania, U.S.A.
DAVID CHANDLER, Department of Chemistry, University of California, Berkeley, California, U.S.A.
M. S. CHILD, Department of Theoretical Chemistry, University of Oxford, Oxford, U.K.
WILLIAM T. COFFEY, Department of Microelectronics and Electrical Engineering, Trinity College, University of Dublin, Dublin, Ireland
F. FLEMING CRIM, Department of Chemistry, University of Wisconsin, Madison, Wisconsin, U.S.A.
ERNEST R. DAVIDSON, Department of Chemistry, Indiana University, Bloomington, Indiana, U.S.A.
GRAHAM R. FLEMING, Department of Chemistry, University of California, Berkeley, California, U.S.A.
KARL F. FREED, The James Franck Institute, The University of Chicago, Chicago, Illinois, U.S.A.
PIERRE GASPARD, Center for Nonlinear Phenomena and Complex Systems, Brussels, Belgium
ERIC J. HELLER, Institute for Theoretical Atomic and Molecular Physics, Harvard-Smithsonian Center for Astrophysics, Cambridge, Massachusetts, U.S.A.
ROBIN M. HOCHSTRASSER, Department of Chemistry, The University of Pennsylvania, Philadelphia, Pennsylvania, U.S.A.
R. KOSLOFF, The Fritz Haber Research Center for Molecular Dynamics and Department of Physical Chemistry, The Hebrew University of Jerusalem, Jerusalem, Israel
RUDOLPH A. MARCUS, Department of Chemistry, California Institute of Technology, Pasadena, California, U.S.A.
G. NICOLIS, Center for Nonlinear Phenomena and Complex Systems, Université Libre de Bruxelles, Brussels, Belgium
THOMAS P. RUSSELL, Department of Polymer Science, University of Massachusetts, Amherst, Massachusetts, U.S.A.
DONALD G. TRUHLAR, Department of Chemistry, University of Minnesota, Minneapolis, Minnesota, U.S.A.
JOHN D. WEEKS, Institute for Physical Science and Technology and Department of Chemistry, University of Maryland, College Park, Maryland, U.S.A.
PETER G. WOLYNES, Department of Chemistry, University of California, San Diego, California, U.S.A.

GEOMETRIC STRUCTURES OF PHASE SPACE IN MULTIDIMENSIONAL CHAOS
APPLICATIONS TO CHEMICAL REACTION DYNAMICS IN COMPLEX SYSTEMS

ADVANCES IN CHEMICAL PHYSICS
VOLUME 130

PART B

Edited by

M. TODA, T. KOMATSUZAKI, T. KONISHI,
R. S. BERRY, and S. A. RICE

Series Editor

STUART A. RICE

Department of Chemistry
and
The James Franck Institute
The University of Chicago
Chicago, Illinois

AN INTERSCIENCE PUBLICATION
JOHN WILEY & SONS, INC.

Copyright © 2005 by John Wiley & Sons, Inc. All rights reserved.

Published by John Wiley & Sons, Inc., Hoboken, New Jersey.
Published simultaneously in Canada.

No part of this publication may be reproduced, stored in a retrieval system, or transmitted in any form or by any means, electronic, mechanical, photocopying, recording, scanning, or otherwise, except as permitted under Section 107 or 108 of the 1976 United States Copyright Act, without either the prior written permission of the Publisher, or authorization through payment of the appropriate per-copy fee to the Copyright Clearance Center, Inc., 222 Rosewood Drive, Danvers, MA 01923, 978-750-8400, fax 978-646-8600, or on the web at www.copyright.com. Requests to the Publisher for permission should be addressed to the Permissions Department, John Wiley & Sons, Inc., 111 River Street, Hoboken, NJ 07030, (201) 748-6011, fax (201) 748-6008.

Limit of Liability/Disclaimer of Warranty: While the publisher and author have used their best efforts in preparing this book, they make no representations or warranties with respect to the accuracy or completeness of the contents of this book and specifically disclaim any implied warranties of merchantability or fitness for a particular purpose. No warranty may be created or extended by sales representatives or written sales materials. The advice and strategies contained herein may not be suitable for your situation. You should consult with a professional where appropriate. Neither the publisher nor author shall be liable for any loss of profit or any other commercial damages, including but not limited to special, incidental, consequential, or other damages.

For general information on our other products and services please contact our Customer Care Department within the U.S. at 877-762-2974, outside the U.S. at 317-572-3993 or fax 317-572-4002.

Wiley also publishes its books in a variety of electronic formats. Some content that appears in print, however, may not be available in electronic format.

Library of Congress Catalog Number: 58:9935

ISBN 0-471-70527-6 (Part A)
ISBN 0-471-71157-8 (Part B)
ISBN 0-471-71158-6 (Set)

Printed in the United States of America

10 9 8 7 6 5 4 3 2 1

CONTRIBUTORS TO VOLUME 130

YOJI AIZAWA, Department of Applied Physics, Faculty of Science and Engineering, Waseda University, Tokyo, 169-8555, Japan

R. STEPHEN BERRY, Department of Chemistry, The University of Chicago, Chicago, Illinois 60637, USA

JENS BREDENBECK, Max-Planck-Institüt für Strömungsforschung, D-37073 Göttingen, Germany. *Present address*: Physikalisch-Chemisches Institüt, Universität Zürich, CH-8057 Zürich, Switzerland

LINTAO BU, Department of Chemistry, Boston University, Boston, Massachusetts, 02215, USA

MASSIMO CENCINI, Dipartimento di Fisica, Università di Roma "la Sapienza" and Center for Statistical Mechanics and Complexity INFM UdR Roma 1 Piazzale Aldo Moro 5, I-00185 Roma, Italy

STAVROS C. FARANTOS, Institute of Electronic Structure and Laser Foundation for Research and Technology, Hellas, Greece; and Department of Chemistry, University of Crete, Iraklion 711 10, Crete, Greece

HIROSHI FUJISAKI, Department of Chemistry, Boston University, Boston, Massachusetts, 02215, USA

JIANGBIN GONG, Department of Chemistry and The James Franck Institute, The University of Chicago, Chicago, Illinois 60637 USA

SERGY YU. GREBENSHCHIKOV, Max-Planck-Institüt für Strömungsforschung, D-37073 Göttingen, Germany

HIROSHI H. HASEGAWA, Department of Mathematical Sciences, Ibaraki University, Mito, 310-8512, Japan; and Center for Studies in Statistical Mechanics and Complex Systems, The University of Texas at Austin, Austin, Texas 78712, USA

SEIICHIRO HONJO, Department of Basic Science, Graduate School of Arts and Sciences, University of Tokyo, Komaba, Meguro-ku, Tokyo, 153-8902, Japan

KYOKO HOSHINO, Nonlinear Science Laboratory, Department of Earth and Planetary Sciences, Faculty of Science, Kobe University, Nada, Kobe, 657-8501, Japan

KENSUKE S. IKEDA, Department of Physical Sciences, Faculty of Science and Engineering, Ritsumeikan University, Kusatsu, 525-8577, Japan

CHARLES JAFFÉ, Department of Chemistry, West Virginia University, Morgantown, West Virginia 26506-6045, USA

MARC JOYEUX, Laboratoire de Spectrométrie Physique (CNRS UMR 5588), Université Joseph Fourier, Grenoble 1, F-38402 St. Martin d'Hères Cedex, France

KUNIHIKO KANEKO, Department of Basic Science, College of Arts and Sciences, University of Tokyo, Komaba, Meguro-ku, Tokyo, 153-8902, Japan

SHINNOSUKE KAWAI, Department of Chemistry, Graduate School of Science, Kyoto University, Kyoto, 606-8502, Japan

TAIZO KOBAYASHI, Department of Physical Sciences, Ritsumeikan University, Kusatsu, 525-8577, Japan

TAMIKI KOMATSUZAKI, Nonlinear Science Laboratory, Department of Earth and Planetary Sciences, Faculty of Science, Kobe University, Nada, Kobe, 657-8501, Japan

TETSURO KONISHI, Department of Physics, Nagoya University, Nagoya, 464-8602, Japan

DAVID M. LEITNER, Department of Chemistry and Chemical Physics Program, University of Nevada, Reno, Nevada 89557, USA

YASUHIRO MATSUNAGA, Nonlinear Science Laboratory, Department of Earth and Planetary Sciences, Faculty of Science, Kobe University, Nada, Kobe, 657-8501, Japan

TAKAYUKI MIYADERA, Department of Information Sciences, Tokyo University of Science, Noda City, 278-8510, Japan

TERUAKI OKUSHIMA, Department of Physics, Tokyo Metropolitan University, Minami-Ohsawa, Hachioji, Tokyo, 192-0397, Japan

YOSHIKAZU OHTAKI, Department of Mathematical Sciences, Ibaraki University, Mito, 310-8512, Japan

YOSHITSUGU OONO, Department of Physics, University of Illinois at Urbana-Champaign, Urbana, Illinois, 61801-3080, USA

JESÚS PALACIÁN, Departamento de Matemática e Informática, Universidad Pública de Navarra, 31006 Pamplona, Spain

STUART A. RICE, Department of Chemistry and The James Franck Institute, The University of Chicago, Chicago, Illinois 60637 USA

SHINJI SAITO, Department of Chemistry, Nagoya University, Furo-cho, Chikusa-ku, Nagoya, 464-8602, Japan

MITSUSADA M. SANO, Graduate School of Human and Environmental Studies, Kyoto University, Sakyo, Kyoto, 606-8501, Japan

SHIN'ICHI SAWADA, School of Science and Technology, Kwansei Gakuin University, Sanda, 669-1337, Japan

REINHARD SCHINKE, Max-Planck-Institüt für Strömungsforschung, D-37073 Göttingen, Germany

NORIHIRO SHIDA, Omohi College, Graduate School of Engineering, Nagoya Institute of Technology, Gokiso-cho, Showa-ku, Nagoya, 466-8555, Japan

YASUSHI SHIMIZU, Department of Physical Sciences, Ritsumeikan University, Kusatsu, 525-8577, Japan

AKIRA SHUDO, Department of Physics, Tokyo Metropolitan University, Minami-Ohsawa, Hachioji, Tokyo, 192-0397, Japan

JOHN E. STRAUB, Department of Chemistry, Boston University, Boston, Massachusetts, 02215, USA

Y-H. TAGUCHI, Department of Physics, Faculty of Science and Technology, Chuo University, Bunkyo-ku, Tokyo, 112-8551, Japan; and Institute for Science and Technology, Chuo University, Bunkyo-ku, Tokyo, 112-8551, Japan

KIN'YA TAKAHASHI, The Physics Laboratories, Kyushu Institute of Technology, Iizuka, 820-8502, Japan

TOSHIYA TAKAMI, Institute for Molecular Science, Okazaki, 444-8585, Japan

KAZUO TAKATSUKA, Department of Basic Science, Graduate School of Arts and Sciences, University of Tokyo, Komaba, 153-8902, Tokyo, Japan

MIKITO TODA, Physics Department, Nara Women's University, Nara, 630-8506, Japan

TURGAY UZER, Center for Nonlinear Science, School of Physics, Georgia Institute of Technology, Atlanta, GA 30332-0430, USA

DAVIDE VERGNI, Istituto Applicazioni del Calcolo, CNR Viale del Policlinico 137, I-00161 Roma, Italy

ANGELO VULPIANI, Dipartimento di Fisica, Università di Roma "la Sapienza" and Center for Statistical Mechanics and Complexity INFM UdR Roma 1 Piazzale Aldo Moro 5, I-00185 Roma, Italy

LAURENT WIESENFELD, Laboratoire d'Astrophysique, Observatoire de Grenoble, Université Joseph-Fourier, BP 53, F-38041 Grenoble Cédex 9, France

YOSHIYUKI Y. YAMAGUCHI, Department of Applied Mathematics and Physics, Kyoto University, 606-8501, Kyoto, Japan

TOMOHIRO YANAO, Department of Complex Systems Science, Graduate School of Information Science, Nagoya University, 464-8601, Nagoya, Japan

PATRICIA YANGUAS, Departamento de Matemática e Informática, Universidad Pública de Navarra, 31006 Pamplona, Spain

MEISHAN ZHAO, Department of Chemistry and The James Franck Institute, The University of Chicago, Chicago, Illinois, 60637 USA

INTRODUCTION

Few of us can any longer keep up with the flood of scientific literature, even in specialized subfields. Any attempt to do more and be broadly educated with respect to a large domain of science has the appearance of tilting at windmills. Yet the synthesis of ideas drawn from different subjects into new, powerful, general concepts is as valuable as ever, and the desire to remain educated persists in all scientists. This series, *Advances in Chemical Physics*, is devoted to helping the reader obtain general information about a wide variety of topics in chemical physics, a field that we interpret very broadly. Our intent is to have experts present comprehensive analyses of subjects of interest and to encourage the expression of individual points of view. We hope that this approach to the presentation of an overview of a subject will both stimulate new research and serve as a personalized learning text for beginners in a field.

STUART A. RICE

PREFACE

The study of chemical reactions covers a variety of phenomena, ranging from the microscopic mechanisms of reaction processes through structural changes involving macromolecules such as proteins, to biochemical networks within cells. One common question concerning these seemingly diverse phenomena is how we can understand the temporal development of the system based on its dynamics.

At the microscopic level, chemical reactions are dynamical phenomena in which nonlinear vibrational motions are strongly coupled with each other. Therefore, deterministic chaos in dynamical systems plays a crucial role in understanding chemical reactions. In particular, the dynamical origin of statistical behavior and the possibility of controlling reactions require analyses of chaotic behavior in multidimensional phase space.

In contrast, conventional reaction rate theory replaces the dynamics within the potential well by fluctuations at equilibrium. This replacement is made possible by the assumption of local equilibrium, in which the characteristic time scale of vibrational relaxation is supposed to be much shorter than that of reaction. Furthermore, it is supposed that the phase space within the potential well is uniformly covered by chaotic motions. Thus, only information concerning the saddle regions of the potential is taken into account in considering the reaction dynamics. This approach is called the transition state theory.

Recently, however, experimental studies have cast a doubt on this assumption (see Ref. 1 for a review). For example, spectroscopic studies reveal hierarchical structures in the spectra of vibrationally highly excited molecules [2]. Such structures in the spectra imply the existence of bottlenecks to intramolecular vibrational energy redistribution (IVR). Reactions involving radicals also exhibit bottlenecks to IVR [3]. Moreover, time-resolved measurements of highly excited molecules in the liquid phase show that some reactions take place before the molecules relax to equilibrium [4]. Therefore, the assumption that local equilibrium exists prior to reaction should be questioned. We seek understanding of reaction processes where the assumption does not hold.

The problem requires analyses of phase-space structures in systems with many degrees of freedom. In particular, appreciating the global structure of the phase space becomes essential for our understanding of reactions under nonequilibrium conditions. In order to make this point clear, we briefly summarize the present status of the study.

Since the 1980s, concepts and results from nonlinear physics have been incorporated into studies of unimolecular reactions. (For a review, see Rice and co-workers' contribution in this volume.) In particular, concepts established for systems with two degrees of freedom have played an important role in defining the reaction rate based on dynamics [5]. The concept of transition state has been examined from the standpoint of dynamical system theory, and reformulated in terms of normally hyperbolic invariant manifolds (NHIMs). While transition states in the conventional sense are situated in configuration space, NHIMs corresponding to saddles are structures in phase space. In order to formulate transition states as dividing surfaces, we have to resort to NHIMs and their stable and unstable manifolds. These phase space structures enable us to avoid the so-called recrossing problem. Moreover, Lie perturbation theory makes it possible to calculate the dividing surfaces at least locally near the NHIMs (see Ref. 6 for a review).

However, in systems with more than two degrees of freedom, the dividing surfaces do not generally exist globally in phase space [7,8]. Thus, the attempt to define the reaction rate based on dynamics has not been successful for systems with many degrees of freedom. Instead, global features of the phase space, such as the network of reaction paths, emerge as crucial ingredients in studying reactions from the dynamical point of view.

The reason why the dividing surfaces do not generally exist globally is because intersections between the stable and unstable manifolds of NHIMs sometimes involve tangency. This tangency reveals that branching structures exist in the network of reaction paths. Moreover, combining these branching structures with the Arnold web in the potential well, the global aspects of the phase space offer rich possibilities for nonergodic behavior for reactions in systems with many degrees of freedom. Implications of this possibility are to be sought in reactions under nonequilibrium conditions.

Thus, we shift our attention from quantities related to local equilibrium, notably reaction rate constants, to nonequilibrium aspects of reaction processes. In particular, we list the following three closely related questions as most important.

First, do dynamical correlations exist in processes involving multiple saddles, such as structural changes of macromolecules in clusters and proteins? In the conventional theory, it is supposed that consecutive processes of going over saddles take place independent of one another. In other words, the system loses its memory of the past immediately, since the vibrational relaxation within a well is assumed to be much faster than the escape from it and multistep processes are conventionally assumed to be Markov processes. To the contrary, when the characteristic time scale of IVR is comparable to that of the reaction, the system can keep dynamical correlations as it goes over successive saddles.

These correlations result in (a) acceleration of reactions for some initial conditions and (b) deceleration for others. This approach will shed new light on problems such as why reactions proceed on multibasin energy landscapes without being trapped in deep minima [9], why proteins fold so effectively, how enzymes help specific reactions to take place, and so on.

Second, how we can characterize nonequilibrium reactions using a dynamical viewpoint? Since the conventional concepts are not sufficient here, we need new ideas that relate measurable quantities to reaction dynamics. In particular, for reactions involving structural changes of macromolecules, collective variables will be necessary to describe processes, and the degrees of freedom that compose collective variables will change as the reaction proceeds over multiple saddles. Furthermore, dynamical correlations are likely to play important roles. Then, we need methods that answer the following questions: What degrees of freedom are necessary to describe reaction dynamics, in what way do they evolve and vary during the processes, and how we can extract information on their dynamics from measurements?

Third, what is the dynamical origin of Maxwell's demon? As is well known since the work of Maxwell, Szilard, and Brillouin, nonequilibrium conditions are necessary for systems to do information processing. Therefore, in studying biochemical reactions, we are interested in how nonequilibrium conditions are maintained at the molecular level. From the viewpoint of dynamics, in particular, the following problem stands out as crucial: Does any intrinsic mechanism of dynamics exist which helps to maintain nonequilibrium conditions in reaction processes? In other words, are there any reactions in which nonergodicity plays an essential role for systems to exhibit functional behavior?

Keeping these subjects in perspective, we organized a conference entitled "Geometrical Structures of Phase Space in Multidimensional Chaos— Applications to Chemical Reaction Dynamics in Complex Systems" from 26th October to 1st November, 2003, at the Yukawa Institute for Theoretical Physics, Kyoto University, Kyoto, Japan. A pre-conference was also held at Kobe University from 20th to 25th October.

This conference was interdisciplinary, where researchers from physics (including astrophysics), biophysics, physical chemistry, and nonlinear science gathered to discuss a wide range of problems in reaction dynamics with the common theme that chaos in dynamical systems plays a crucial role in studying chemical reactions. Furthermore, we argue that reactions involving macromolecules such as clusters, liquids, and proteins are important examples of dynamical systems with many degrees of freedom. Thus, we expect that studies of these reactions from a dynamics point of view will shed new light on phenomena such as phase transitions in clusters, slow relaxation in liquids, and the efficiency of protein folding, as well as in seeking the possibility of manipulating these reactions.

In particular, in the Conference we focused our attention on the following topics.

1. Transition state theory revisited from the dynamical point of view, including a historical perspective of the study.
2. Phase-space structure of Hamiltonian systems with multiple degrees of freedom—in particular, normally hyperbolic invariant manifolds (NHIMs), intersections between their stable and unstable manifolds, and the Arnold web.
3. Analyses of reaction processes based on the phase space structure of the system.
4. Quantum aspects of chaos and how we can control them.
5. Nonstatistical properties, such as nonstationary behavior and multiple scales of time and distance for evolution, in systems of many degrees of freedom.
6. Dynamical understanding of reaction processes in macromolecules and liquids, such as phase transitions, fast alloying, energy redistribution, and structural changes in clusters and proteins.
7. Data mining to extract information on dynamics from time series data from experiments and simulations of molecular dynamics.
8. Dynamical insights into reactions at the macroscopic level, including chemical networks in cells and their evolution.

Here, in this volume, we have collected contributions from the invited speakers, from poster presentations that received the best poster awards (Yanao, Honjo, and Okushima), and from poster presentations chosen to cover topics that were not treated by the invited speakers. The best poster awards were decided based on a jury vote by the invited speakers and a popular vote by all the participants. Note, however, that there were many other posters that also deserved inclusion here.

In the following, we give a brief overview of the content of this volume. The volume consists of the following three parts:

I. Phase-space geometry of multidimensional dynamical systems and reaction processes.
II. Complex dynamical behavior in clusters and proteins, and data mining to extract information on dynamics.
III. New directions in multidimensional chaos and evolutionary reactions.

In the first part, our aim is to discuss how we can apply concepts drawn from dynamical systems theory to reaction processes, especially unimolecular reactions of few-body systems. In conventional reaction rate theory, dynamical aspects are replaced by equilibrium statistical concepts. However, from the standpoint of chaos, the applicability of statistical concepts itself is problematic. The contribution of Rice's group gives us detailed analyses of this problem from the standpoint of chaos, and it presents a new approach toward unimolecular reaction rate theory.

In statistical reaction rate theory, the concept of transition state plays a key role. Transition states are supposed to be the boundaries between reactants and products. However, the precise formulation of the transition state as a dividing surface is only possible when we consider "transition states" in phase space. This is the place where the concepts of normally hyperbolic invariant manifolds (NHIMs) and their stable and unstable manifolds come into play.

The contributions of Komatsuzaki and Berry, and of Uzer's group, discuss these manifolds, and they present their calculations using Lie perturbation theory methods. The contribution of Wiesenfeld discusses these manifolds in reaction processes involving angular momenta, and the contribution by Joyeux et al. shows applications of the perturbation theory method to reactions involving Fermi resonance. The contribution of Sano discusses invariant manifolds in the Coulomb three-body problem.

The importance of NHIMs, and their stable and unstable manifolds is shared strikingly between chemical reactions and astrophysics. Therefore in the conference at Kyoto, Koon, from Caltech, discussed controlling an orbiter in astrophysics, and Uzer presented his study of asteroids near Jupiter, where analyses of these manifolds were essential.

In reaction processes for which there is no local equilibrium within the potential well, global aspects of the phase space structure become crucial. This is the topic treated in the contribution of Toda. This work stresses the consequences of a variety of intersections between the stable and unstable manifolds of NHIMs in systems with many degrees of freedom. In particular, "tangency" of intersections is a feature newly recognized in the phase space structure. It is a manifestation of the multidimensionality of the system, where reaction paths form a network with branches.

Here, we also include the contributions related to quantum mechanics: The chapter by Takami et al. discusses control of quantum chaos using coarse-grained laser fields, and the contribution of Takahashi and Ikeda deals with tunneling phenomena involving chaos. Both discuss how chaos in classical behavior manifests itself in the quantum counterpart, and what role it will play in reaction dynamics.

In the second part, we collect contributions concerning dynamical processes in complex systems such as clusters and proteins. Here, we also include those ideas related to data mining, since this topic is an indispensable part of the studies on dynamics of macromolecules.

The contribution of Berry presents an overview of the study of clusters as vehicles for investigating complex systems. The study of clusters has given birth to a variety of new ideas which turned out to be fruitful in other complex systems such as proteins. The contribution of Takatsuka discusses dynamical and statistical aspects of phase transitions in clusters, and the contribution of Yanao and Takatsuka studies the gauge structure arising from the dynamics of floppy molecules. Shida's contribution presents an important issue related to saddles of index of two or more, and shows their role in the phase transitions of clusters. Another interesting phenomenon of clusters is fast alloying, discussed in the contribution of Shimizu et al. from the standpoint of reaction dynamics.

Liquids and proteins are complex systems for which the study of dynamical systems has wide applicability. In the conference, relaxation in liquids (ε-entropy by Douglas at the National Institute of Standards and Technology, nonlinear optics by Saito, and energy bottlenecks by Shudo and Saito), energy redistribution in proteins (Leitner and Straub et al.), structural changes in proteins (Kidera at Yokohama City University), and a new formulation of the Nosé-Hoover chain (Ezra at Cornell University) were discussed. Kidera's talk discussed time series analyses in molecular dynamics, and it is closely related to the problem of data mining. In the second part of the volume, we collect the contributions by Leitner and by Straub's group, and the one by Shudo and Saito in the third part.

The contribution by Komatsuzaki's group bridges the two research fields—that is, dynamics in complex systems and data mining. They apply to a model of proteins the methods of embedding and Allan variance, both of which have been developed in dynamical system theory. Their results reveal, using the Allan variance, nonstationary behavior in protein dynamics, and they show, by embedding, how many degrees of freedom are necessary to describe this dynamics. Thus, this contribution indicates a crucial role for the methods of data mining in the study of processes involving macromolecules.

Therefore, contributions to methods of data mining are included here. It is uncommon to discuss this topic in the context of reaction processes. However, as we have already discussed, data mining becomes ever more important in analyzing experiments and simulations. In conventional data analyses, the concepts of equilibrium statistical physics have been routinely applied. To the contrary, in situations in which local equilibrium breaks down, established methods do not exist to analyze experiments and simulations. Thus, data mining to extract information on dynamics is crucial here. In the conference, several methods were discussed (Broomhead at Manchester University on embedding, Vulpiani on finite-size Lyapunov exponents, Taguchi on nonmetric methods, and Hasegawa on inductive thermodynamics approach from time series). Here we include the contributions by Taguchi and Oono and by Hasegawa and Ohtaki.

In the third part, those contributions are collected which discuss nonergodic and nonstationary behavior in systems with many degrees of freedom, and seek new possibilities to describe complex reactions, including even the evolution of living cells.

Conventional theory supposes that statistical ideas would be more applicable to systems of many degrees of freedom than to few-body systems. To the contrary, in these systems, new kinds of behavior such as multiergodicity, nonstationarity, and an anomalous approach to equilibrium can emerge. Consequently, their implications for reaction dynamics should be explored, especially in those cases where biological functions are involved.

Thus, the contribution of Shudo and Saito starts by presenting the problem concerning the relation between nonergodicity and $1/f$ noise. For systems with two degrees of freedom, the dynamical origin of $1/f$ noise is attributed to the hierarchical structures of resonant tori (Aizawa). However, for systems with many degrees of freedom, this relationship is not well understood. This discussion goes on to systems with a gap in the spectrum of characteristic time scales and nonergodic behavior, based on the studies of the Italian group (Benettin et al.). The contributions of Aizawa and of Yamaguchi also discuss these problems in the context of cluster formation (Aizawa) and of an approach to equilibrium (Yamaguchi). These features will become important in understanding reaction processes in complex systems such as protein folding and slow relaxation in complex liquids.

Nonlinear resonances are important factors in reaction processes of systems with many degrees of freedom. The contributions of Konishi and of Honjo and Kaneko discuss this problem. Konishi analyzes, by elaborate numerical calculations, the so-called Arnold diffusion, a slow movement along a single resonance under the influence of other resonances. Here, he casts doubt on the usage of the term "diffusion." In other words, "Arnold diffusion" is a dynamics completely different from random behavior in fully chaotic regions where most of the invariant structures are lost. Hence, understanding "Arnold diffusion" is essential when we go beyond the conventional statistical theory of reaction dynamics. The contribution of Honjo and Kaneko discusses dynamics on the network of nonlinear resonances (i.e., the Arnold web), and stresses the importance of resonance intersections since they play the role of the hub there.

Here we also include the contribution of Okushima, in which the concept of the Lyapunov exponents is extended to orbits of finite duration. The mathematical definition of the Lyapunov exponents requires ergodicity to ensure convergence of the definition. On the other hand, various attempts have been made to extend this concept to finite time and space, to make it applicable to nonergodic systems. Okushima's idea is one of them, and it will find applications in nonstationary reaction processes.

The contributions of Vulpiani's group and of Kaneko deal with reactions at the macroscopic level. The contribution of Vulpiani's group discusses asymptotic analyses to macroscopic reactions involving flows, by presenting the mechanism of front formation in reactive systems. The contribution of Kaneko deals with the network of reactions within a cell, and it discusses the possibility of evolution and differentiation in terms of that network. In particular, he points out that molecules that exist only in small numbers can play the role of a switch in the network, and that these molecules control evolutionary processes of the network. This point demonstrates a limitation of the conventional statistical quantities such as density, which are obtained by coarse-graining microscopic quantities. In other words, new concepts will be required which go beyond the hierarchy in the levels of description such as micro and macro.

We hope that the contributions collected in this volume convey the stimulating and interdisciplinary atmosphere of the conference. We also expect that the results and discussions in these contributions form a first and decisive step toward understanding reaction processes from the standpoint of dynamics.

The conference was supported by the following grants and institute. We greatly appreciate these organizations for their financial support.

- Japan Society for Promotion of Science, Japan–U.S. Cooperative Science Program.
- The Inoue Foundation for Science.
- Yukawa Institute for Fundamental Physics, Kyoto University.
- Grant-in-Aid for Scientific Research on Priority Areas "Control of Molecules in Intense Laser Fields" from the Ministry of Education, Science, Sports, and Culture.

References

1. M. Toda, *Adv. Chem. Phys.* **123**, 3643 (2000).
2. K. Yamanouchi, N. Ikeda, S. Tsuchiya, D. M. Jonas, J. K. Lundberg, G. W. Adamson, and R. W. Field, *J. Chem. Phys.* **95**, 6330 (1991).
3. T. Shibata, H. Lai, H. Katayanagi, and T. Suzuki, *J. Phys. Chem.* **A102**, 3643 (1998).

4. S. L. Schultz, J. Qian, and J. M. Jean, *J. Phys. Chem.* **A101**, 1000 (1997).
5. M. J. Davis and S. K. Gray, *J. Chem. Phys.* **84**, 5389 (1986).
6. T. Komatsuzaki and R. S. Berry, *Adv. Chem. Phys.* **123**, 79 (2002).
7. R. E. Gillilan and G. S. Ezra, *J. Chem. Phys.* **94**, 2648 (1991).
8. S. Wiggins, *Physica* **D44**, 471 (1990).
9. L. Sun, K. Song, and W. L. Hase, *Science* **296**, 875 (2002).

Spring 2004

M. Toda
T. Komatsuzaki
T. Konishi
R. S. Berry
S. A. Rice

YITP International Symposium on Geometrical Structures of Phase Space in Multidimensional Chaos—Applications to Chemical Reaction Dynamics in Complex System, October 26–November 1, 2003, at the Yukawa Institute for Theoretical Physics, Kyoto University, Kyoto, Japan.

(1) Tamiki Komatsuzaki (2) Gregory Ezra (3) R. Stephen Berry (4) Charles Jaffé (5) Angelo Vulpiani (6) Yoji Aizawa (7) Mikito Toda (8) Masanori Shimono (9) Dave F. Broomhead (10) Shinichiro Goto (11) Yoshihiro Taguchi (12) Shinji Saito (13) David M. Leitner (14) Laurent Wiesenfeld (15) Koji Hotta (16) Seiichiro Honjo (17) Wang Sang Koon (18) Stuart A. Rice (19) Akinori Kidera (20) Toshiya Takami (21) Kyoko Hoshino (22) Ayako Nozaki (23) Yoko K. Ueno (24) Kazuo Takatsuka (25) Yasushi Shimizu (26) Kin'ya Takahashi (27) Mitsusada M. Sano (28) Hiroshi H. Hasegawa (29) Koichi Fujimoto (30) Turgay Uzer (31) Tetsuro Konishi (32) Hidetoshi Morita (33) Yoshiyuki Y. Yamaguchi (34) John E. Straub (35) Hiroshi Fujisaki (36) Mitsunori Takano (37) Sotaro Fuchigami (38) Jack F. Douglas (39) Kazuo Kuwata (40) Taku Mizukami (41) Teruaki Okushima (42) Kim Kyeon-deuk (43) Norihiro Shida (44) Akira Shudo (45) Takefumi Yamashita (46) Kunihiko Kaneko (47) Youhei Koyama (48) Marc Joyeux (49) Lintao Bu (50) Statue of Hideki Yukawa (Nobel Prize Laureate (Physics) 1949)

xxi

CONTENTS PART B

PART II COMPLEX DYNAMICAL BEHAVIOR IN CLUSTERS AND PROTEINS, AND DATA MINING TO EXTRACT INFORMATION ON DYNAMICS   1

CHAPTER 10 ATOMIC CLUSTERS: POWERFUL TOOLS TO PROBE COMPLEX DYNAMICS   3
By R. Stephen Berry

CHAPTER 11 TEMPERATURE, GEOMETRY, AND VARIATIONAL STRUCTURE IN MICROCANONICAL ENSEMBLE FOR STRUCTURAL ISOMERIZATION DYNAMICS OF CLUSTERS: A MULTICHANNEL CHEMICAL REACTION BEYOND THE TRANSITION-STATE CONCEPT   25
By Kazuo Takatsuka

CHAPTER 12 EFFECTS OF AN INTRINSIC METRIC OF MOLECULAR INTERNAL SPACE ON CHEMICAL REACTION DYNAMICS   87
By Tomohiro Yanao and Kazuo Takatsuka

CHAPTER 13 ONSET DYNAMICS OF PHASE TRANSITION IN Ar_7 129
By Norihiro Shida

CHAPTER 14 RAPID ALLOYING IN BINARY CLUSTERS: MICROCLUSTER AS A DYNAMIC MATERIAL   155
By Yasushi Shimizu, Taizo Kobayashi, Kensuke S. Ikeda, and Shin'ichi Sawada

CHAPTER 15 VIBRATIONAL ENERGY RELAXATION (VER) OF A CD STRETCHING MODE IN CYTOCHROME c   179
By Hiroshi Fujisaki, Lintao Bu, and John E. Straub

CHAPTER 16 HEAT TRANSPORT IN MOLECULES AND REACTION KINETICS: THE ROLE OF QUANTUM ENERGY FLOW AND LOCALIZATION   205
By David M. Leitner

CHAPTER 17 REGULARITY IN CHAOTIC TRANSITIONS ON MULTIBASIN LANDSCAPES   257
By Tamiki Komatsuzaki, Kyoko Hoshino, and Yasuhiro Matsunaga

CHAPTER 18 NONMETRIC MULTIDIMENSIONAL SCALING AS A
DATA-MINING TOOL: NEW ALGORITHM AND NEW TARGETS   315
By Y-H. Taguchi and Yoshitsugu Oono

CHAPTER 19 GENERALIZATION OF THE FLUCTUATION–DISSIPATION
THEOREM FOR EXCESS HEAT PRODUCTION   353
By Hiroshi H. Hasegawa and Yoshikazu Ohtaki

PART III NEW DIRECTIONS IN MULTIDIMENSIONAL CHAOS AND EVOLUTIONARY REACTIONS 373

CHAPTER 20 SLOW RELAXATION IN HAMILTONIAN SYSTEMS WITH
INTERNAL DEGREES OF FREEDOM   375
By Akira Shudo and Shinji Saito

CHAPTER 21 SLOW DYNAMICS IN MULTIDIMENSIONAL PHASE
SPACE: ARNOLD MODEL REVISITED   423
By Tetsuro Konishi

CHAPTER 22 STRUCTURE OF RESONANCES AND TRANSPORT IN
MULTIDIMENSIONAL HAMILTONIAN DYNAMICAL SYSTEMS   437
By Seiichiro Honjo and Kunihiko Kaneko

CHAPTER 23 MULTIERGODICITY AND NONSTATIONARITY IN
GENERIC HAMILTONIAN DYNAMICS   465
By Yoji Aizawa

CHAPTER 24 RELAXATION AND DIFFUSION IN A GLOBALLY
COUPLED HAMILTONIAN SYSTEM   477
By Yoshiyuki Y. Yamaguchi

CHAPTER 25 FINITE-TIME LYAPUNOV EXPONENTS IN
MANY-DIMENSIONAL DYNAMICAL SYSTEMS   501
By Teruaki Okushima

CHAPTER 26 THE ROLE OF CHAOS FOR INERT AND
REACTING TRANSPORT   519
By Massimo Cencini, Angelo Vulpiani, and Davide Vergni

CHAPTER 27 ON RECURSIVE PRODUCTION AND EVOLVABILITY
OF CELLS: CATALYTIC REACTION NETWORK APPROACH   543
By Kunihiko Kaneko

AUTHOR INDEX   599

SUBJECT INDEX   627

CONTENTS PART A

PART I PHASE-SPACE GEOMETRY OF MULTIDIMENSIONAL
DYNAMICAL SYSTEMS AND REACTION PROCESSES   1

CHAPTER 1   CLASSICAL, SEMICLASSICAL, AND QUANTUM MECHANICAL
UNIMOLECULAR REACTION RATE THEORY   3
By Meishan Zhao, Jiangbin Gong, and Stuart A. Rice

CHAPTER 2   REGULARITY IN CHAOTIC TRANSITIONS ON TWO-BASIN
LANDSCAPES   143
By Tamiki Komatsuzaki and R. Stephen Berry

CHAPTER 3   A NEW LOOK AT THE TRANSITION STATE: WIGNER'S
DYNAMICAL PERSPECTIVE REVISITED   171
By Charles Jaffé, Shinnosuke Kawai, Jesús Palacián,
Patricia Yanguas, and Turgay Uzer

CHAPTER 4   GEOMETRY OF PHASE-SPACE TRANSITION STATES:
MANY DIMENSIONS, ANGULAR MOMENTUM   217
By Laurent Wiesenfeld

CHAPTER 5   INTRAMOLECULAR DYNAMICS ALONG ISOMERIZATION
AND DISSOCIATION PATHWAYS   267
By Marc Joyeux, Sergy Yu. Grebenshchikov, Jens Bredenbeck,
Reinhard Schinke, and Stavros C. Farantos

CHAPTER 6   CLASSICAL COULOMB THREE-BODY PROBLEM   305
By Mitsusada M. Sano

CHAPTER 7   GLOBAL ASPECTS OF CHEMICAL REACTIONS IN
MULTIDIMENSIONAL PHASE SPACE   337
By Mikito Toda

CHAPTER 8   CLASSICAL MECHANISM OF MULTIDIMENSIONAL
BARRIER TUNNELING   401
By Kin'ya Takahashi and Kensuke S. Ikeda

CHAPTER 9 COARSE-GRAINED PICTURE FOR CONTROLLING
QUANTUM CHAOS   435
    By Toshiya Takami, Hiroshi Fujisaki, and Takayuki Miyadera

AUTHOR INDEX   459

SUBJECT INDEX   487

PART II

COMPLEX DYNAMICAL BEHAVIOR IN CLUSTERS AND PROTEINS, AND DATA MINING TO EXTRACT INFORMATION ON DYNAMICS

CHAPTER 10

ATOMIC CLUSTERS: POWERFUL TOOLS TO PROBE COMPLEX DYNAMICS

R. STEPHEN BERRY

*Department of Chemistry, The University of Chicago
Chicago, Illinois 60637, USA*

CONTENTS

I. Introduction
II. Distributions of Level Spacings
III. Power Spectra, Phase-Space Dimensions, Liapunov Exponents, and Kolmogorov Entropy
IV. Local Characteristics of Regularity, Chaos, and Ergodicity
V. The Next Step: The Magic Transformation and Its Consequences
VI. Conclusion
Acknowledgments
References

I. INTRODUCTION

We may look at atomic clusters as particularly apt and useful models to study virtually every aspect of what we call, diffusely, "complexity." Simulations are particularly powerful means to carry out such studies. For example, we can follow trajectories for very long times with molecular dynamics and thereby evaluate the global means of the exponential rates of divergence of neighboring trajectories. This is the most common way to evaluate those exponents, the Liapunov exponents. The sum of these is the Kolmogorov entropy, one gross measure of the volume of phase space that the system explores, and hence one

---

Geometric Structures of Phase Space in Multidimensional Chaos: A Special Volume of Advances in Chemical Physics, Part B, Volume 130, edited by M. Toda, T Komatsuzaki, T. Konishi, R.S. Berry, and S.A. Rice. Series editor Stuart A. Rice.
ISBN 0-471-71157-8 Copyright © 2005 John Wiley & Sons, Inc.

kind of entropy. We may also examine the distribution of intervals between the eigenvalues of the Hamiltonian, the intervals of the energy spectrum; the pattern of level spacings is an indication of whether the system has regularity in the form of "extra" constants of motion. Still another variety of probes has been less frequently used than these, but proves very powerful for the study of systems of several bodies, namely the examination of regularity and chaos in localized regions of space or time. We shall see how all of these have been used to provide new insights into the behavior of the terribly complex systems composed of collections of atoms, notably atomic clusters. The extensions of the ideas here to other systems will be apparent, but will not be discussed in this presentation.

II. DISTRIBUTIONS OF LEVEL SPACINGS

The distributions of spacings of energy levels of a quantized system give a clear indication of whether the system's states can be characterized by some kind of symmetry or constant of motion beyond those of energy, momentum, total angular momentum, and mass. If the system has any such symmetries and constants of motion, then there is no constraint on how close the energy levels may be. This is because states of different symmetry exhibit no "repulsions of levels." That is, there are no nonzero off-diagonal matrix elements of the Hamiltonian to couple nearby states of differing symmetry. Such nonzero elements do occur between approximate states of the same symmetry, and cause their more accurate forms, corresponding to a diagonalized Hamiltonian, to yield energy spectra with spacings wider than those of the approximate system. However, there are no constants of motion apart from energy, mass, total momentum, and total angular momentum for a closed, chaotic Hamiltonian system. Hence there are very few small level spacings of a system with no symmetry, and no essential degeneracies at all.

If the system has some degree of regularity and "repelling" energy levels, the distribution of level spacings can be expected to be Poissonian, but if the system is chaotic, with strongly interacting levels, then it should shift to a Gaussian orthogonal ensemble (GOE) or Wigner distribution [1]. Analyses of the distributions of nearest-neighbor level spacings for H_3^+ [2] and for the Ar_3 cluster [3] show clearly that these systems have nearest-neighbor level spacings that follow Wigner distributions. Figure 1 illustrates this, for the first 133 vibrational energy levels of the argon trimer that have the same symmetry as the ground vibrational state. It is indeed representative of the other symmetries as well. The conclusion is easy to draw: This system, a simple three-body cluster, becomes chaotic if its temperature brings it into a range where the accessible

ATOMIC CLUSTERS: POWERFUL TOOLS TO PROBE COMPLEX DYNAMICS 5

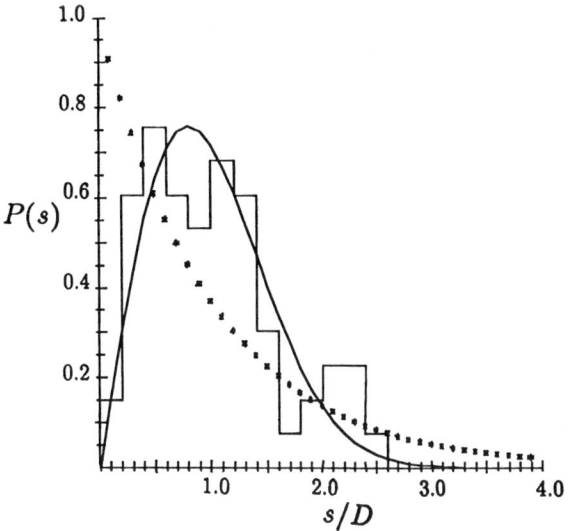

Figure 1. Distribution of spacings of nearest-neighbor vibrational energy levels of the vibrational states of Ar_3 with A'_1 symmetry, that of the ground vibrational state. The histogram shows the computed level spacings. The solid curve is that of a fitted Wigner distribution, and the dotted curve is that of a Poisson distribution. [Reprinted with permission from D. M. Leitner, R. S. Berry, and R. M. Whitnell, *J. Chem. Phys.* **91**, 3470 (1989). Copyright © 1989, American Institute of Physics.]

level distribution fits the Wigner distribution. From molecular dynamics simulations, we can infer that this happens at about 27 K [4].

III. POWER SPECTRA, PHASE-SPACE DIMENSIONS, LIAPUNOV EXPONENTS, AND KOLMOGOROV ENTROPY

The argon trimer, even with its Wigner distribution of vibrational energy level spacings, is, in some respects, not a very wildly chaotic system. This became apparent when Beck et al. determined the energy dependence of the power spectrum, the effective dimensionality of its phase-space trajectory, and its complete set of classical Liapunov exponents (4). The phase-space dimension, the Hausdorff dimension, is a measure of the ergodicity of the system; the latter two indicate the extent of its chaotic behavior. The Ar_3 cluster might be expected to show a smooth increase in the extent of its chaotic behavior with increasing energy or temperature, and to be reasonably ergodic when it populates vibrational levels in the range where anharmonicity is not negligible. Larger clusters of rare-gas atoms, notably

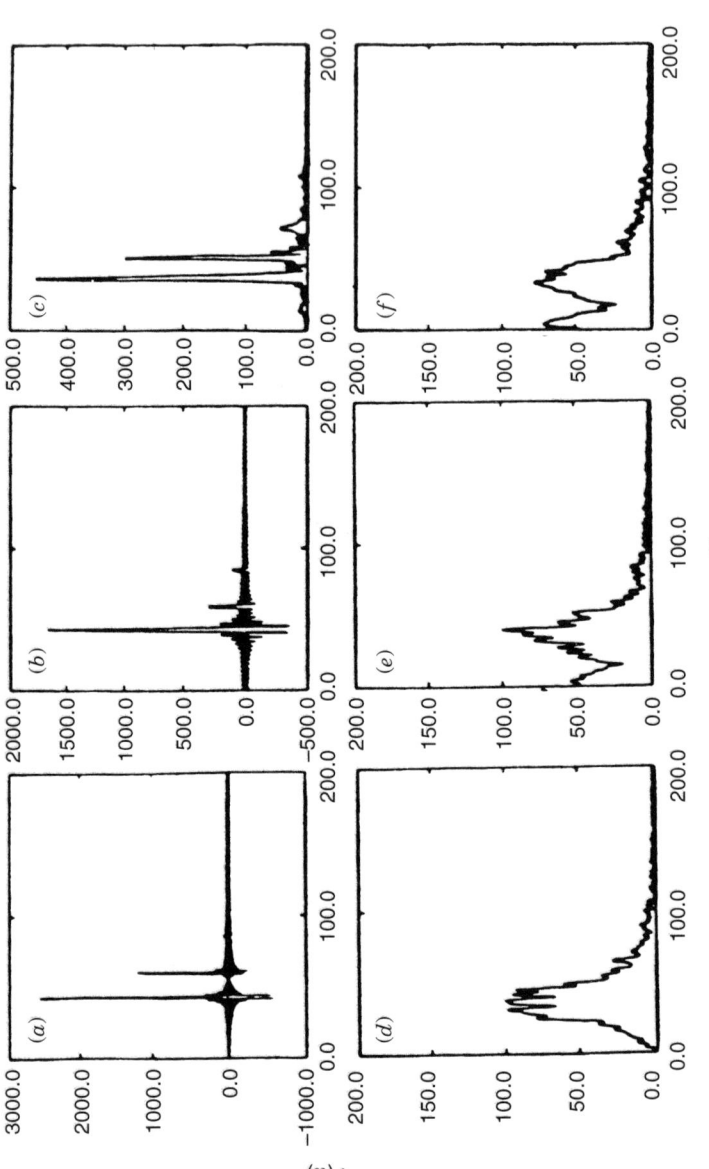

Figure 2. Power spectra of Ar_3 at energies of (a) -1.641×10^{-14} erg/atom (equivalent to 2.12 K), (b) -1.613×10^{-14} erg/atom (equivalent to 4.15 K), (c) -1.394×10^{-14} erg/atom (equivalent to 18.19 K), (d) -1.164×10^{-14} erg/atom (equivalent to 28.44 K), (e) -0.939×10^{-14} erg/atom (equivalent to 30.54 K), and (f) -0.792×10^{-14} erg/atom (equivalent to 36.51 K). [Reprinted with permission from T. L. Beck, D. M. Leitner, and R. S. Berry, *J. Chem. Phys.* **89**, 1681 (1988). Copyright © 1988, American Institute of Physics.]

with seven or more atoms, might be expected to show a sharper kind of change of behavior because they show clear transitions from solid-like to liquid-like behavior. By contrast, the only faintly comparable behavior of the equilateral-triangular Ar_3 is its passage over the potential energy saddle at its linear configuration. However, something interesting does happen there that has led to new insights.

The power spectra of Ar_3, shown in Fig. 2, make it clear that there is a distinct change between mean energies of -1.4×10^{-14} erg/atom (equivalent to 18.19 K) and -1.16×10^{-14} erg/atom (equivalent to 28.44 K). From sharp, distinct vibrations, the system has transformed to one with a continuous spectrum of available classical energies.

At the equivalent of 18.19 K (Fig. 2c), the system vibrates through a region in which the modes are "softer" than at the bottom of the potential, but the three modes are clear and distinguishable. At the equivalent of 28.44 K in (Fig. 2d), the system no longer has clearly separated modes, but it does not have enough energy to pass over the linear saddle. Only when the energy reaches the equivalent of 30.54 K in (Fig. 2e) can it pass through that saddle, a phenomenon revealed by the density of very low-frequency modes at this and higher energies. This behavior might be considered a primitive "melting" [5].

The dimensionality of the phase space in which the system moves is a measure of the extent of separability of the modes, and thus of the extent to which the system sweeps out its phase space—that is, the extent of its ergodicity. A system of N separable harmonic oscillators has $3N - 6$ normal modes and hence moves in a space of that number of dimensions. A system of N particles with no constraints on the interactions and exchanges among its $6N$ dimensions in phase space has only to preserve its three center-of-mass coordinates, three center-of-mass momenta, three components of angular momentum and its total energy, 10 in all. Hence such a system may move in a phase space of $6N - 10$ dimensions. For the argon trimer, these two extreme cases correspond to $3 \times 3 - 6 = 3$ and $3 \times 6 - 10 = 8$, respectively. We might, therefore, expect to see an increase in that dimensionality as the energy of a several-body system increases.

There is one paradoxical subtlety here, insofar as the dimensionality one computes will usually be based on some finite-time sampling. Consequently if a system is truly ergodic but requires a very long time interval to exhibit its ergodicity, the dimensionality one obtains can be no better than a lower bound. We shall return to this topic later, when we examine local properties of several-body systems. Meanwhile, we just quote the results of computations of the dimensionality of the phase space for Ar_3, as a function of energy [4]. The calculations were carried out by the method introduced by Grassberger and Procaccia [6,7]. Values of the dimension, specifically the "correlation

TABLE I
Effective Temperatures, Energies, and Correlation Dimensions for the Argon Trimer[a]

Temperature (K)	Energy (ε)	D_2
2.12	−0.98	3.1
4.15	−0.97	3.3
10.31	−0.91	3.5
18.19	−0.83	5.9
28.44	−0.70	2.5, 7.6
30.54	−0.56	2.6, 5.3
36.51	−0.47	6.2

[a] Where two values are given, the system exhibits separated populations in two regions of phase space. [Constructed with information from T. L. Beck, D. M. Leitner, and R. S. Berry, *J. Chem. Phys.* **89**, 1681 (1988).]

dimension" D_2, are those given in Table I; for sufficiently long trajectories, this quantity equals the fractal or Hausdorff dimension D_0.

The point to be made here is the obvious one that this dimensionality increases with the energy of the system, but that this is not a simple monotonic increase. There is a symptom here of something we will explore more and more extensively. The inference we draw from these results is becoming clear when we compare these numbers with the patterns of the power spectra in Fig. 2. The system becomes more ergodic and chaotic as its energy increases, until it has enough energy to enter the saddle region. When this occurs, some trajectories remain in the basin region around the energy minimum corresponding to an equilateral triangular structure; these trajectories are associated with high kinetic energies and hence with high dimensionalities. Other trajectories go into the saddle region, using most of their energy to climb the potential "hill" so that they have very low kinetic energy in the saddle region, and hence behave rather regularly and remain for significant time intervals there. These two kinds of trajectories are distinguishable on the time scale used to estimate the dimensionalities in these calculations. This phenomenon is clear when we examine the distributions of "short-time" averages (500 time steps) of the kinetic energies, expressed as effective temperatures, for two values of the total energy. Figure 3 shows these distributions for the two energies for which two distinguishable values of D_2 appear (Table I), 28.44 K and 30.54 K. The number of excursions into the saddle region is quite low at the lower total energy; the system can enter the saddle region, but hasn't enough energy to reach the saddle.

Next, we turn to the Liapunov exponents and the Kolmogorov entropy (K-entropy). The systems concerning us here are Hamiltonian, so the Liapunov

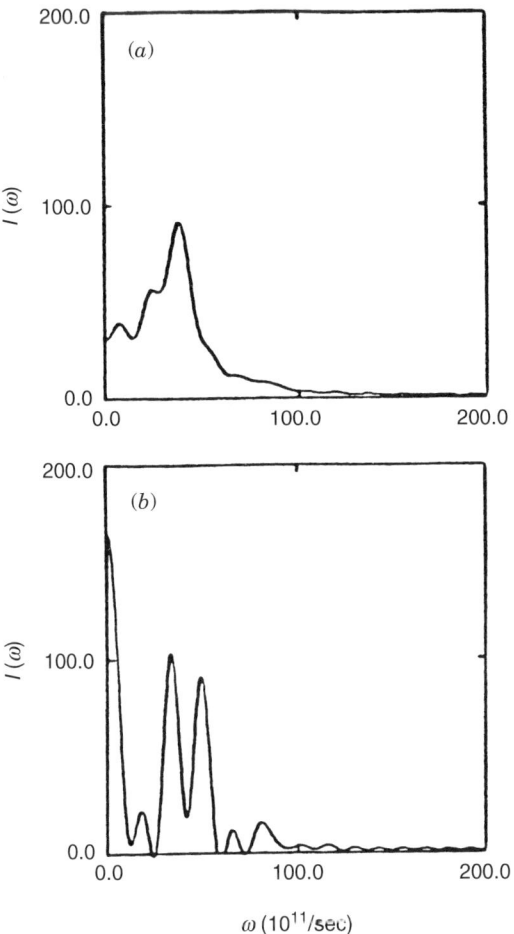

Figure 3. Distributions of "short-term" kinetic energies, expressed as effective temperatures, for Ar$_3$, at total energies corresponding to (a) 28.44 K and (b) 30.54 K. The low kinetic energies correspond to trajectory segments in the saddle region; the high kinetic energy parts of the distribution are associated with motion above the deep well of the equilibrium geometry. [Reprinted with permission from T. L. Beck, D. M. Leitner, and R. S. Berry, *J. Chem. Phys.* **89**, 1681 (1988). Copyright © 1988, American Institute of Physics.]

exponents come in pairs with equal magnitude and opposite sign. We focus specifically on the positive exponents, which measure the (exponential) rate at which trajectories close at some initial point diverge. The sum of the positive Liapunov exponents is the K-entropy. In the case of Ar$_3$, there are only two positive and two negative Liapunov exponents. These quantities were evaluated

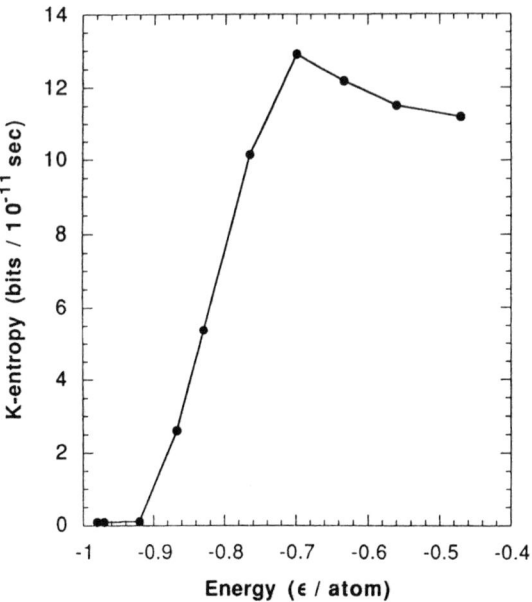

Figure 4. The K-entropy of the Lennard-Jones three-particle cluster, simulating Ar_3, as a function of the cluster's energy. [Reprinted with permission from R. J. Hinde, R. S. Berry, and D. J. Wales, *J. Chem. Phys.* **96**, 1376 (1992). Copyright © 1992, American Institute of Physics.]

roughly for Ar_3 by Beck et al. in the work just cited, and then they were evaluated more accurately for Lennard-Jones clusters modeling Ar_3 and Ar_7 by Hinde et al. [8]. Here we find a striking difference between these two clusters. We can focus our discussion on the K-entropies as functions of the total energies. Figure 4 shows the variation of the K-entropy for the three-particle Lennard-Jones cluster as a function of energy. The particularly striking aspect to this function is the appearance of a maximum just above -0.7 ε/atom (where ε is the energy parameter in the Lennard-Jones potential function), corresponding to the energy at which the system can just cross the saddle.

The behavior of the same function for the seven-particle Lennard-Jones cluster, simulating Ar_7, as shown in Fig. 5, is strikingly different. Here we see no maximum, just a smooth, monotonically increasing function. This is particularly striking because this seven-particle cluster displays distinct solid-like and liquid-like forms, coexisting within a significant energy band well within the range of the scale of this figure. The origin of this difference between the three-particle and seven-particle cluster became clear in the next step of our studies [9].

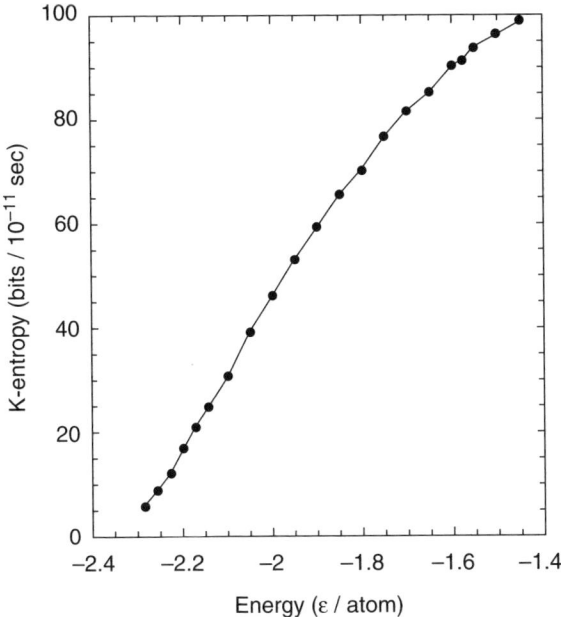

Figure 5. The K-entropy of the seven-particle Lennard-Jones cluster, a smooth, monotonic increasing function with no indication of any maximum. [Reprinted with permission from R. J. Hinde, R. S. Berry, and D. J. Wales, *J. Chem. Phys.* **96**, 1376 (1992). Copyright © 1992, American Institute of Physics.]

IV. LOCAL CHARACTERISTICS OF REGULARITY, CHAOS, AND ERGODICITY

In order to go to the next stage, we turn back to considering properties such as K-entropy and ergodicity in a manner quite different from their traditional mathematical formulations. The rigorous mathematical definitions of Liapunov exponents, for example, are based on the limits of arbitrarily long trajectories, trajectories that pass through all parts of the phase space that the system can reach. To understand the behavior of real physical systems, we must depart from this approach and find ways to interpret how a system behaves *in different regions of its accessible phase space*. This necessitates inventing ways, albeit not rigorous in the mathematical sense, to characterize local regions in terms of quantities similar to, and related to, their counterparts from mathematics. In other words, we must soften and extend concepts such as Liapunov exponent,

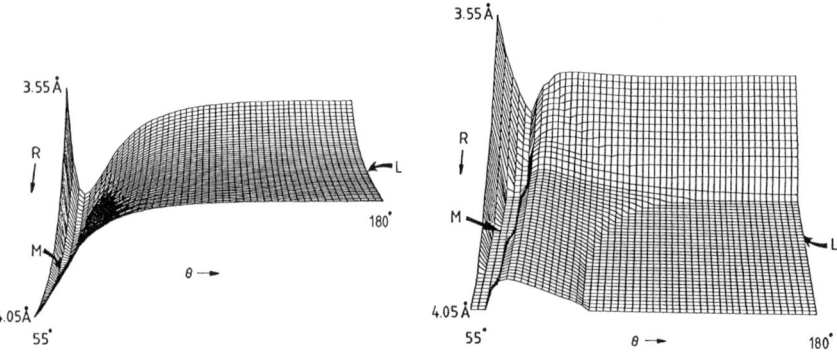

Figure 6. Effective potential energy (**left**) and local K-entropy (**right**) for the argon trimer, as functions of the bending angle (left to right axis) and the (equal) lengths of the two Ar–Ar bonds (back to front axes, 3.05 Å at back to 4.05 Å at front. [Reprinted with permission from D. J. Wales and R. S. Berry, *J. Phys. B* **24**, L351 (1991). Copyright © 1991, Institute of Physics.]

K-entropy, and ergodicity to enable us to differentiate behavior in different time periods, different time scales, and different regions and scales of phase space.

We can begin by examining the "local" Kolmogorov entropy for Ar_3, a quantity based on trajectories of selected finite length [10]. By evaluating the approximate *local* values of the Liapunov exponents with a sufficiently fine integration grid, one can construct maps of the effective K-entropy over the same range of configuration space for which one can construct an effective potential. Figure 6 compares the potential surface and the local K-entropy for Ar_3 as functions of the bending angle θ and the bond length R, for the situation in which the bonds have the same length. It is clear from this figure that the two multidimensional functions have similar form. The local K-entropy is lowest in regions where the potential is flat. Of course the local K-entropy is very low and constant in both the trough of the potential well and in the region of large internuclear distances of the linear or obtuse-triangular structure (lower right in the figure). The local K-entropy is highest in the regions of highest slope and curvature of the potential. These are the regions in which the forces drive the particles hardest, in ways that depend most sensitively on precisely where the particles are. Hence neighboring trajectories depart from one another rapidly in such regions. And this is precisely the condition for a high level of chaotic behavior.

Ergodicity is also amenable to a kind of local and temporal scrutiny. Here we can look at some of the results obtained until now. However, the whole subject of how ergodicity develops in time, and how that development is related to the kinds of interactions among the component particles of a system and thus to the effective potential surface, is a vast, open, and tantalizing subject. First, we can

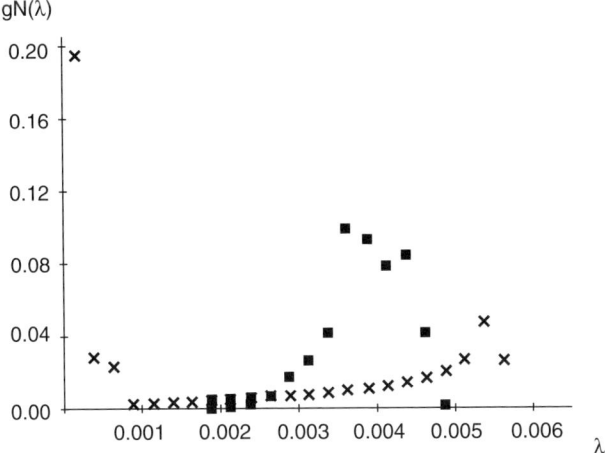

Figure 7. Two distributions of sample Liapunov exponents for a cluster of three Lennard-Jones particles simulating Ar_3 at energy equivalent to 4.15 K. The only difference between the two calculations is in the initial conditions. [Reprinted with permission from C. Amitrano, and R. S. Berry, *Phys. Rev. Lett.* **68**, 729 (1992). Copyright © 1992, American Physical Society.]

see something obvious, that the systems simply cannot achieve ergodicity in short trajectories. Figure 7 shows two distributions of short-trajectory (256-step) sample approximations to Liapunov exponents, for two different initial conditions for the Ar_3 cluster at a constant energy equivalent to 4.15 K. The difference is unambiguous evidence that the system does not achieve ergodicity in this time interval, even in the very limited subspace accessible to it at such a low energy [11].

Now we turn to more specific indicators of the differences of properties in different local regions on potential surfaces [9]. We begin with Ar_3 and then go on to larger clusters. Four properties reveal most of the relevant behavior: the local K-entropy, the extent of mode-mode coupling of the vibrations, the mean kinetic energy (expressed here as an effective temperature), and the magnitude of the negative curvature of the potential in each region. For the purpose of this analysis, we divide the potential surface of Ar_3 into four regions: the uppermost distributions in the following figures correspond to the region of the well around the global minimum. The second describes behavior in a region approaching but not yet entering the saddle zone. The third rows describe behavior in the saddle region, and the last, in the region beyond the saddle, which, in symmetrical cases, is equivalent to the second row. Figure 8 shows the distributions of these properties for the argon trimer at energy at which the system can readily enter the saddle region, equivalent to roughly 30 K. These distributions make quite

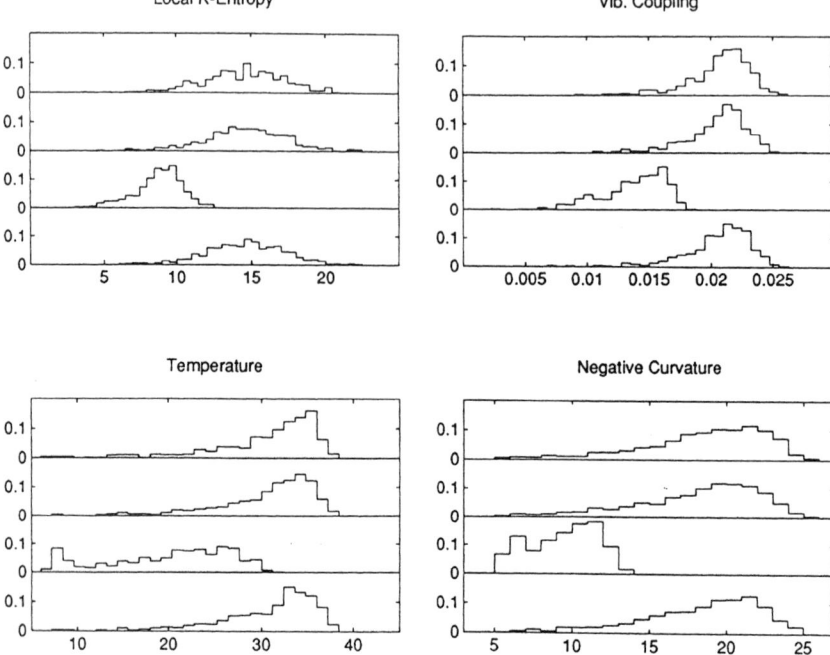

Figure 8. Distributions of local K-entropy, vibrational mode-mode coupling, mean kinetic energy and negative curvature of the potential for the three-particle Lennard-Jones cluster simulating Ar_3 at constant energy. The energy of the cluster corresponds to a temperature of approximately 30 K. [Reprinted with permission from R. J. Hinde and R. S. Berry, *J. Chem. Phys.* **99**, 2942 (1993). Copyright © 1993, American Institute of Physics.]

clear how different is the behavior of the cluster, in terms of all four of these related properties, in the saddle region from the other three regions.

The next two figures reveal how these overall cluster properties vary with the size of the cluster, even for quite small clusters. Figure 9 shows the same kind of distributions as Fig. 8, but for the four-particle Lennard-Jones cluster, and Figs. 10 and 11 do the same for the 5-particle cluster, a system with two kinds of saddles (but only one locally stable structure), so two sets of distributions are shown there. Only the distributions over the higher-energy saddle show any detectable differences from the distributions elsewhere on the surface. With still larger clusters, the distinctions between saddle regions and the other parts of the surface essentially disappear.

Finally, in this particular investigation, we look at the same distributions for the six-particle Lennard-Jones cluster, the first that has two geometrically different locally stable structures. The lower-energy structure is a regular

ATOMIC CLUSTERS: POWERFUL TOOLS TO PROBE COMPLEX DYNAMICS   15

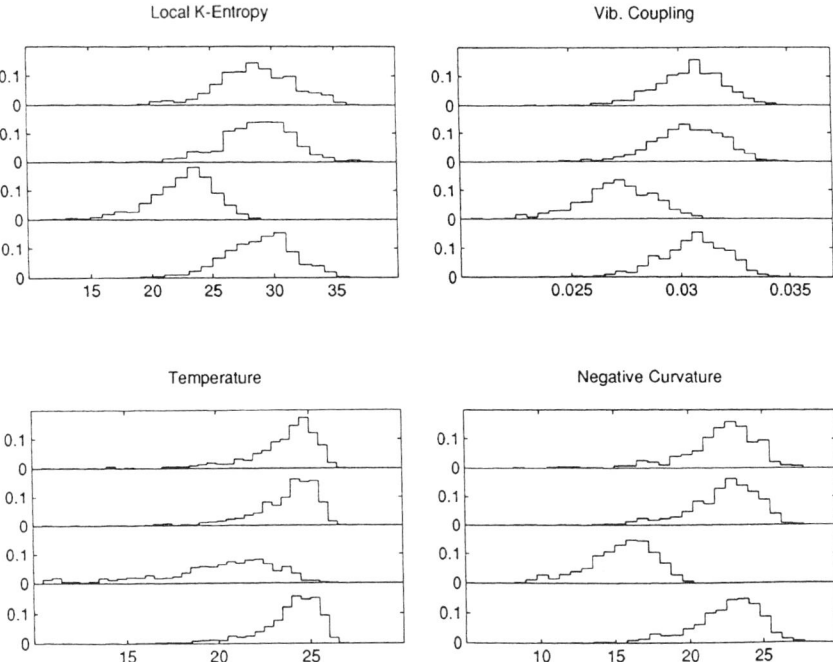

Figure 9. Distributions as shown in Fig. 8, but for a four-particle Lennard-Jones cluster. The saddle region is clearly different from the others. [Reprinted with permission from R. J. Hinde and R. S. Berry, *J. Chem. Phys.* **99**, 2942 (1993). Copyright © 1993, American Institute of Physics.]

octahedron; the higher is a capped trigonal bipyramid. This system also has two kinds of saddles: One joins the octahedron to the capped bipyramid, whereas the other joins two bipyramidal structures. In Fig. 12, we show the same four properties as above, for five regions. From top to bottom, these are: the capped trigonal bipyramidal well, the transition to the bipyramid-to-octahedron saddle, the saddle, the exit region toward the octahedron, and the octahedral well. Clearly the saddle region is just not different here in the way it is for the smaller clusters. There must be some phenomenon that can show itself in very small but still complex systems that becomes hidden when one examines the overall properties of even slightly larger systems. There is a difference between something with three or six degrees of freedom that is well on its way to disappearing (or, more precisely, hiding) if the system has nine or more degrees of freedom.

Next we can pursue this line of thought a bit further to examine in more detail what happens in the saddle region of the small systems. We do this by

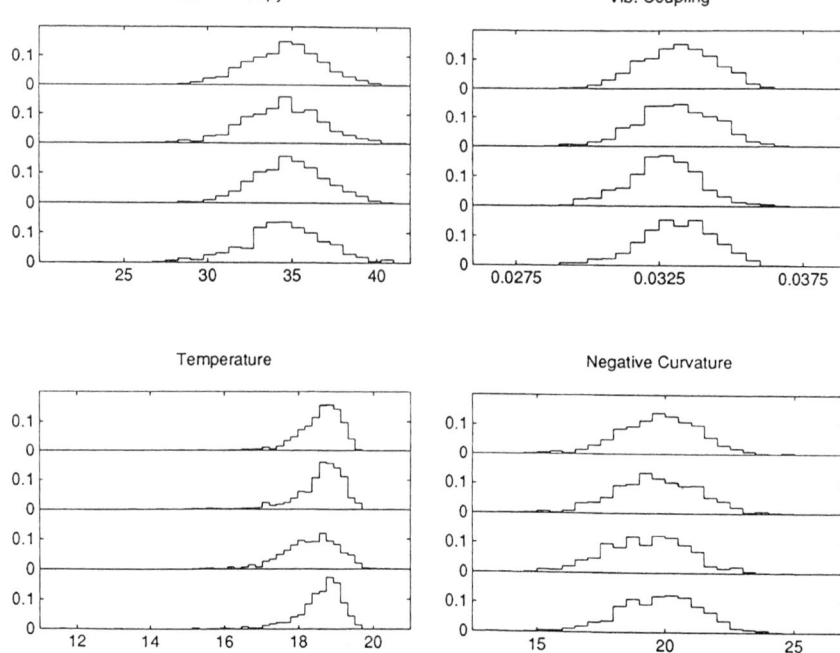

Figure 10. Distributions as shown in Fig. 8 but for a five-particle Lennard-Jones cluster, going over its lower-energy ("diamond-square-diamond") saddle. No significant distinction makes the saddle region different from other regions. [Reprinted with permission from R. J. Hinde and R. S. Berry, *J. Chem. Phys.* **99**, 2942 (1993). Copyright © 1993, American Institute of Physics.]

examining the time dependence of the accumulation of action along pathways that take the system through its saddle, first for the three-particle system and then for the four-particle system. What we see for these is simply not apparent in comparable analyses of systems of five or more particles.

We look now at Fig. 13, in which the accumulation of action is shown, first in a rather long segment of trajectory and then in a close-up of the most relevant part of that trajectory [9]. The greater part of the path shows action accumulating in a sort of "devil's staircase," with random jumps of assorted sizes. However, when the system enters the saddle region, this changes dramatically. In the saddle region, the action accumulates in very distinct, periodic steps, precisely as one expects for a *regular* system. This is in marked contrast to the other parts of the trajectory, which give every evidence of being chaotic by this criterion as well as by the others we have examined.

Then, Fig. 14 shows the same kind of behavior for the four-particle Lennard-Jones cluster, as it passes through its saddle, on a single, typical trajectory [9].

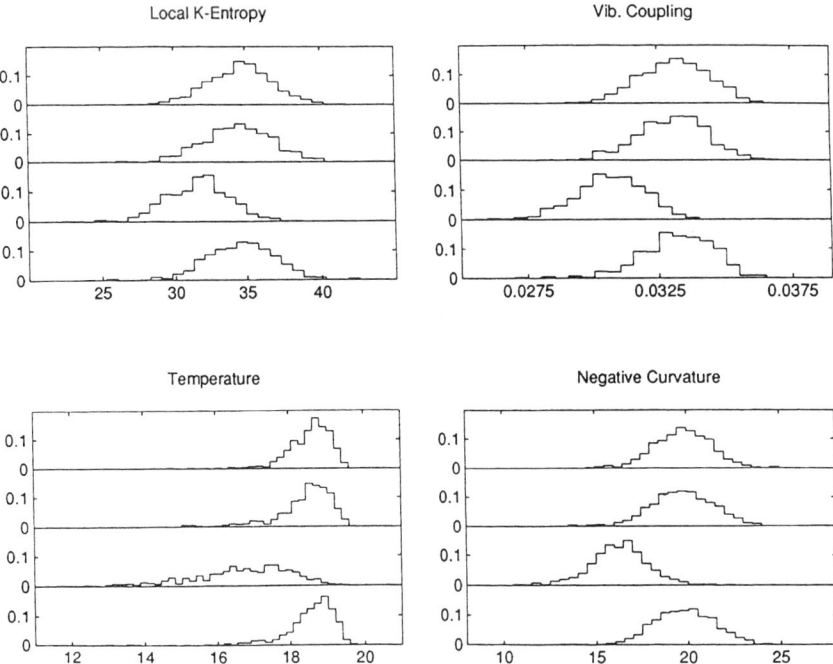

Figure 11. Distributions as shown in Fig. 8, for a five-particle Lennard-Jones cluster passing over its higher-energy ("edge-bridging") saddle. This saddle region is distinguishable from other regions of the potential. [Reprinted with permission from R. J. Hinde and R. S. Berry, *J. Chem. Phys.* **99**, 2942 (1993). Copyright © 1993, American Institute of Physics.]

This system also shows very regular behavior in the saddle region, according to this criterion, but not as simply periodic as for the three-body system in Fig. 13.

Finally, in this section we examine one other kind of locality, to which we referred in the context of the distribution of sample values of Liapunov exponents. Many kinds of clusters of six or seven or more particles exhibit distinguishable phase-like forms existing in dynamic equilibrium with one another, within well-defined bands of energy or temperature [12]. In these ranges, it has proved possible to distinguish time-scale separations, dividing the short-time period in which a system is only locally ergodic, from the long-time region in which it is globally ergodic. This is apparent in the forms of the distributions of sample values of the Liapunov exponents for trajectories of different lengths [11,13]. This behavior is apparent in systems too large to show the distinctions of the three- and four-body clusters, because those are so small that they cannot really behave like liquids or multiphase forms of solids. In

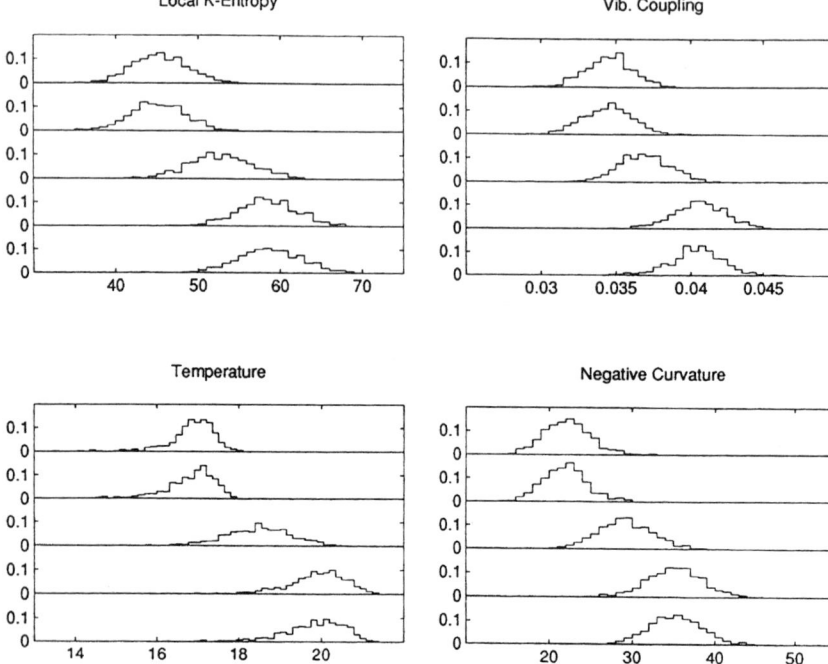

Figure 12. The same four distributions as in Fig. 8, now for a six-particle system. These are shown for five stages along the pathway from the capped trigonal bipyramid (top band in each panel) through an approach to the bipyramid-to-octahedron saddle, to the saddle region itself (middle band), then to the exit region toward the octahedron and, finally (at bottom), the behavior in the well around the octahedral global minimum structure. [Reprinted with permission from R. J. Hinde and R. S. Berry, *J. Chem. Phys.* **99**, 2942 (1993). Copyright © 1993, American Institute of Physics.]

Fig. 15, we see how the distributions of sample values change from bimodal to unimodal (in Fig. 15c) when the energy lies in the range of solid–liquid coexistence. This figure is constructed from simulations of the seven-particle Lennard-Jones cluster, the smallest Lennard-Jones cluster that exhibits such two-phase equilibrium. The short trajectories give bimodal distributions, and the long trajectories give only a single maximum. The three examples at energies below and above the coexistence region show essentially only unimodal distributions.

Thus we have seen some ways to explore the *local* aspects of chaotic and ergodic behavior. However, at this stage of progress, a significant challenge loomed, which was only resolved when one of the organizers of the 2003 Kyoto Conference appeared, some five years after the work here was carried out. The

ATOMIC CLUSTERS: POWERFUL TOOLS TO PROBE COMPLEX DYNAMICS 19

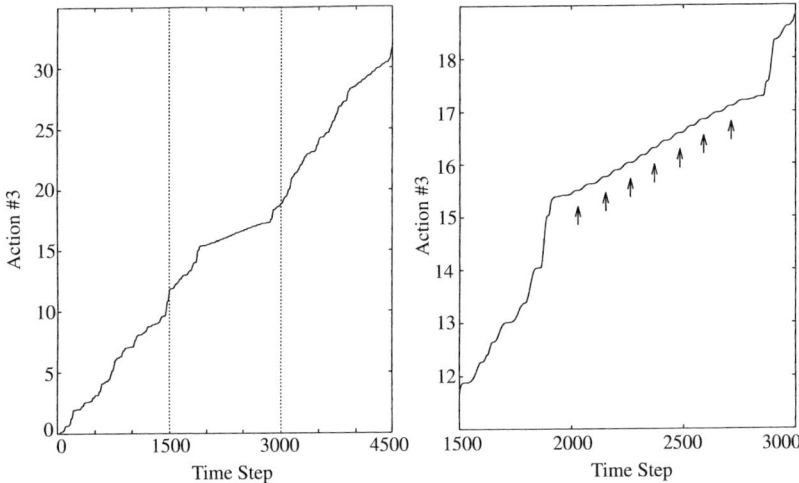

Figure 13. The accumulated action, as a function of time, for a typical saddle-crossing trajectory of the three-particle Lennard-Jones cluster. The right panel is a close-up view of the central part of the longer trajectory on the left, with arrows indicating the individual oscillations. [Reprinted with permission from R. J. Hinde and R. S. Berry, *J. Chem. Phys.* **99**, 2942 (1993). Copyright © 1993, American Institute of Physics.]

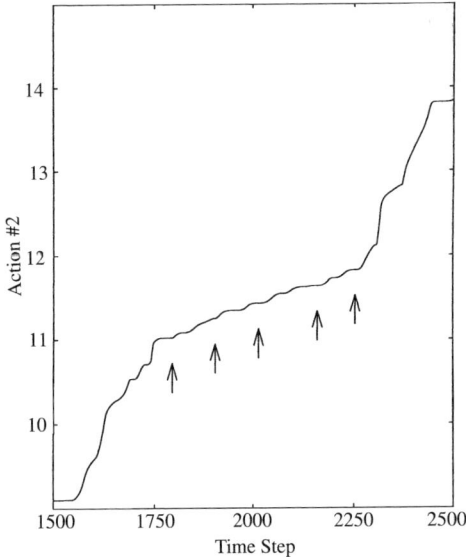

Figure 14. The accumulated action as a function of time as a four-particle Lennard-Jones particle passes through its saddle region. The periodicity is not as sharp and clearly marked as with the three-particle cluster in the previous figure. [Reprinted with permission from R. J. Hinde and R. S. Berry, *J. Chem. Phys.* **99**, 2942 (1993). Copyright © 1993, American Institute of Physics.]

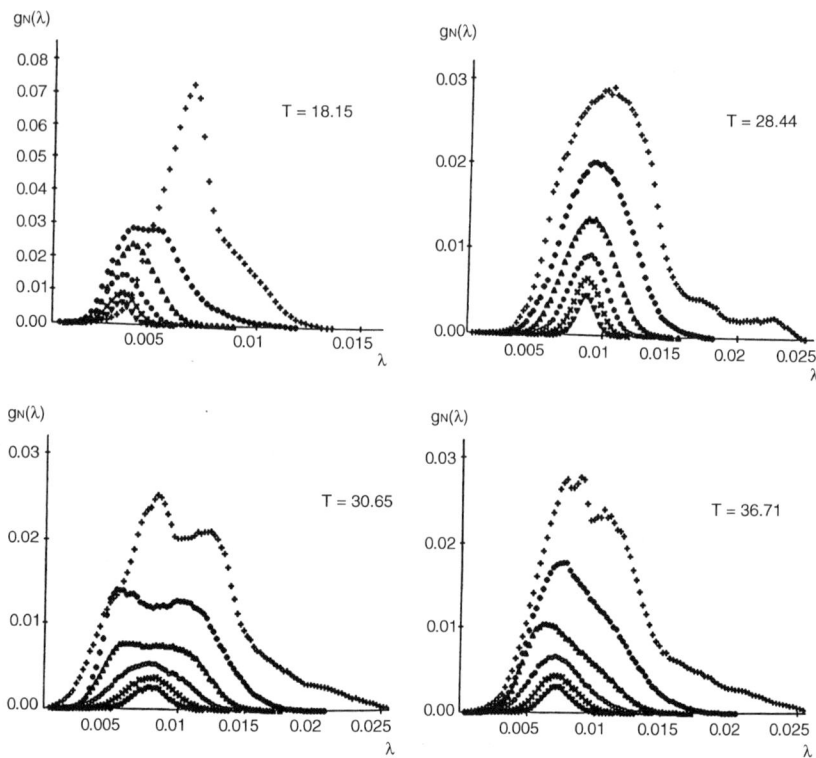

Figure 15. Distributions of sample values of Liapunov exponents for the seven-particle Lennard-Jones cluster at energies corresponding to (a) 18.15 K, (b) 28.44 K, (c) 30.65 K, and (d) 36.71 K. The trajectories from top to bottom in each figure correspond to 256, 512, 1024, 2048, 4096, and 8192 steps. Only at the energy of part c does two-phase equilibrium occur. [Reprinted with permission from C. Amitrano and R. S. Berry, *Phys. Rev. E* **47**, 3158 (1993). Copyright © 1993, American Physical Society.]

specific challenge, recognized by the people whose work has just been described, was finding a way to transform the representation to a form that would separate the reaction coordinate from all the others in a way that would make it possible to recognize and study the kind of regularity that the three-body and four-body systems show. The presumption, based on the results illustrated by Figs. 8–14, is that the multitude of degrees of freedom in the larger systems simply conceal the regularity that the reactive degree of freedom shows in the small systems and presumably also carries, when the system is near a saddle, in larger systems. Hence the problem, easy to identify but not to solve, was finding that magic transformation.

V. THE NEXT STEP: THE MAGIC TRANSFORMATION AND ITS CONSEQUENCES

This author was approached by a young theorist from Japan who had studied the work described above and who had, he felt, made a significant step to break through the problem of finding the transformation. With M. Nagaoka, Tamiki Komatsuzaki had developed a means to expand the potential around a saddle point in a power series and use this expansion to build a *nonlinear but canonical transformation in phase space* that apparently separated the reaction coordinate in a manner that preserved its regularity as much as possible [14,15]. The method, basing the transformation on Lie canonical perturbation theory, looked very encouraging on the basis of the relatively simple system they studied, the isomerization of malonaldehyde. The results were so promising that Komatsuzaki and this writer soon began working on a more complex system, the six-atom Lennard-Jones cluster. This system has, thus far, been the workhorse of our studies. Here we will see only a very brief summary of what has emerged from that work. The methods can be found in the original articles [16–18].

Like many of the studies presented here, a major difference with much of traditional theoretical work in chemical kinetics is the use of phase space for transformations, rather than just the coordinate space. This may seem conceptually more difficult than working only with coordinates, but it proves far more powerful and more appropriate.

The first characteristic of the transformed trajectory to recognize here is the way it maintains regularity in the reaction coordinate's degree of freedom, precisely the quality we hoped to find. Figure 16 shows the action in the reaction coordinate of the six-particle Lennard-Jones cluster as the system passes through the saddle between the two stable structures. The three dominant curves show the action when the coordinate is constructed in zeroth order, first order, and second order of perturbation theory. The energy is high, half again the barrier height. The forest of faint curves in the background are actions of some of the other modes. Only the reaction coordinate shows any regularity at this high energy; all the other degrees of freedom are very chaotic. And, it is quite apparent, the reaction coordinate's degree of freedom is indeed very regular in the saddle region. This is completely in accord with the expectation based on Hernandez and Miller's recognition that the reaction coordinate must separate from the others somewhere in the vicinity of the saddle point, simply because the frequency of that mode must be imaginary [19].

The second point is that the new phase-space representation permits the definition of a true dividing surface *in phase space* which truly separates the reactant and product sides of a reaction. Traditional transition state theory of chemical reactions, based simply on coordinate-space definitions of the degrees of freedom, required an empirical "correction factor," the "transmission

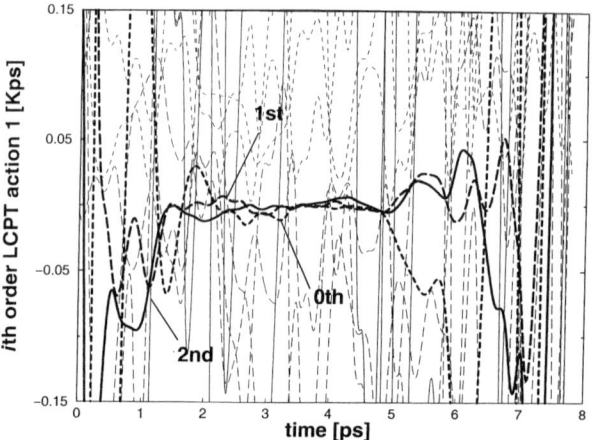

Figure 16. The action in the reaction coordinate's degree of freedom of the six-particle Lennard-Jones cluster, as it passes through the saddle between octahedral and capped trigonal bipyramidal structures. The numbers labeling the curves refer to the order of Lie canonical perturbation theory in which the curves were computed. The faint, wildly oscillating curves in the background are actions in other degrees of freedom. The trajectory was carried out at energy of 1.5 times the height of the barrier above the global minimum—that is, 50% higher than the barrier. The constancy of the action in the saddle region is indisputable evidence of local regularity in this specific mode. [Reprinted with permission from T. Komatsuzaki and R. S. Berry, *J. Chem. Phys.* **110**, 9160 (1999). Copyright © 1999, American Institute of Physics.]

coefficient," to take into account that trajectories could re-cross even the most precisely defined dividing surface between reactant and product. With the phase-space representation, the transmission coefficient is essentially unity [20,21].

One issue opens with the separability and regularity of the reaction coordinate. It seems that the range of regularity becomes narrower as the energy of the system increases above the top of the saddle. At 50% higher than the barrier, the regularity is very apparent, as Fig. 16 shows, but at twice the barrier height—that is, a full barrier height above the barrier itself; this regularity has nearly disappeared. The qualitative reason for this is not difficult to recognize: With increasing energy, the mode-mode coupling becomes stronger and stronger, and the range of separability of the reaction coordinate from the other degrees of freedom grows narrower and narrower. When the width of the region of separability corresponds to only a single period of vibration or less, the separability and regularity simply become meaningless. However, this concept has yet to be put into quantitative terms.

VI. CONCLUSION

We have reviewed the analysis of regularity and chaos in chemical reactions, primarily from the viewpoint of the use of local characteristics and finite-time analyses. These help us see how the reaction process itself occurs and how traditional formulations of chemical kinetics can be refined to achieve the clarity initially hoped for them. Many challenges are now apparent. For example, integrating the formulations reviewed here with more formal, mathematical descriptions of kinetic processes in phase space such as those described by Jaffé, Uzer, Wiggins, and others is an obvious next step. Relating the behavior of reactive systems to concepts in chaos theory, concepts such as the relation of reactive trajectories to Arnold diffusion and Levy flights, of resonance scattering to attractors, such questions lie ahead. Clear as the challenges are, many of the problems are still very imprecisely formulated and need careful articulation before they can be addressed properly—but that is always the most challenging stage of a science!

Acknowledgments

The author wishes to thank the former students and collaborators whose work forms the basis of this review. He also acknowledges the very important support of Grants from the National Science Foundation during the course of the research.

References

1. M. V. Berry, *Proc. Roy. Soc. London Ser. A* **413**, 183 (1987).
2. R. M. Whitnell and J. C. Light, *J. Chem. Phys.* **90**, 1774 (1989).
3. D. M. Leitner, R. S. Berry, and R. M. Whitnell, *J. Chem. Phys.* **91**, 3470 (1989).
4. T. L. Beck, D. M. Leitner, and R. S. Berry, *J. Chem. Phys.* **89**, 1681 (1988).
5. I. Benjamin, Y. Alhassid, and R. D. Levine, *Chem. Phys. Lett.* **115**, 113 (1985).
6. P. Grassberger and I. Procaccia, *Phys. Rev. A* **28**, 2591 (1983).
7. P. Grassberger and I. Procaccia, *Phys. Rev. Lett.* **50**, 346 (1983).
8. R. J. Hinde, R. S. Berry, and D. J. Wales, *J. Chem. Phys.* **96**, 1376 (1992).
9. R. J. Hinde and R. S. Berry, *J. Chem. Phys.* **99**, 2942 (1993).
10. D. J. Wales and R. S. Berry, *J. Phys. B* **24**, L351 (1991).
11. C. Amitrano and R. S. Berry, *Phys. Rev. Lett.* **68**, 729 (1992).
12. R. S. Berry, *Compt. Rend. Phys.* **3**, 1 (2002).
13. C. Amitrano and R. S. Berry, *Phys. Rev. E* **47**, 3158 (1993).
14. T. Komatsuzaki and M. Nagaoka, *Chem. Phys. Lett.* **265**, 91 (1997).
15. T. Komatsuzaki and M. Nagaoka, *J. Chem. Phys.* **105**, 10838 (1996).
16. T. Komatsuzaki and R. S. Berry, *J. Chem. Phys.* **110**, 9160 (1999).

17. T. Komatsuzaki and R. S. Berry, *Phys. Chem. Chem. Phys.* **1**, 1387 (1999).
18. T. Komatsuzaki and R. S. Berry, *Adv. Chem. Phys.* **123**, 79 (2002).
19. R. Hernandez and W. H. Miller, *Chem. Phys. Lett.* **214**, 129 (1993).
20. T. Komatsuzaki and R. S. Berry, *J. Mol. Struct. (Theochem)* **506**, 55 (2000).
21. T. Komatsuzaki and R. S. Berry, *J. Chem. Phys.* **115**, 4105 (2001).

CHAPTER 11

TEMPERATURE, GEOMETRY, AND VARIATIONAL STRUCTURE IN MICROCANONICAL ENSEMBLE FOR STRUCTURAL ISOMERIZATION DYNAMICS OF CLUSTERS: A MULTICHANNEL CHEMICAL REACTION BEYOND THE TRANSITION-STATE CONCEPT

KAZUO TAKATSUKA

Department of Basic Science, Graduate School of Arts and Sciences, University of Tokyo, Komaba, 153-8902, Tokyo, Japan

CONTENTS

I. Introduction
II. Isomerization on Multi-Basin Potential
   A. M_7-like System; Description of the System
   B. Solid–Liquid Transition
   C. Anomalous Time Series of Structural Transitions
III. Strange Behavior of the Lifetime of Isomers in Liquid-like State
   A. Unimolecular Dissociation via a Transition State—A Preliminary
   B. Uniformity of the Average Lifetime for Passing Through a Basin
      1. Accumulated Residence Time and Ergodicity
      2. Passage Time and Uniformity
      3. The Uniformity in an Exponential Decay Expression
   Appendix A: A Short Detour to Non-RRKM Behaviors
IV. Geometry Behind the Memory-Losing Dynamics; Inter-Basin Mixing
   A. The Concept of Mixing Is Not Enough to Account for the Markow-Type Appearance of Isomers
   B. Bifurcation of Reaction Tubes
   C. Time Scale to Realize Inter-Basin Mixing
   D. Fractal Distribution of Turning Points

Geometric Structures of Phase Space in Multidimensional Chaos: A Special Volume of Advances in Chemical Physics, Part B, Volume 130, edited by M. Toda, T Komatsuzaki, T. Konishi, R.S. Berry, and S.A. Rice. Series editor Stuart A. Rice.
ISBN 0-471-71157-8 Copyright © 2005 John Wiley & Sons, Inc.

V. Microcanonical Temperature and an Arrhenius Relation with the Lifetime of Isomers
   A. Another Law for the Average Lifetimes of Isomers
   B. Evaluation of Classical Density of States
   C. Microcanonical Temperature
      1. Definition
      2. Local Microcanonical Temperatures
      3. Numerical Observation of an Arrhenius-like Relation
   D. An Exponential Relation Between the Microcanonical Temperature and Average Lifetimes
      1. Multiexponential Form
      2. Single Exponential Form
      3. Case Study on M_7 in Terms of the Single Exponential Form
      Appendix B: On Ergodicity and Nonergodicity of the Liquid-like Dynamics
      Appendix C: Canonical Temperature
VI. Linear Surprisal Revisited: A Theoretical Foundation
   A. Exploring Temperature in Chemical Reaction Dynamics
   B. Linear Surprisal
   C. Variational Structure in Microcanonical Ensemble to Determine the Final Energy Disposals of Chemical Reactions and Associated Exponential Distributions
      1. Chemical Reaction as a Nonequilibrium Stationary Flow
      2. Linear Surprisal as the Most Probable State
      3. The Meaning of the Prior Distribution
      4. Some New Features of the Surprisal Theory
   D. Study of the Surprisal Should be Resumed
VII. General Conclusions
Acknowledgments
References

I. INTRODUCTION

Isomerization dynamics of clusters composed of identical atoms exhibits many characteristic yet universal features. Take an explicit example from M_7 cluster, where M can be various metal atoms and rare gas atoms. In a low energy region, it simply behaves like a solid body undergoing a small vibration around a given molecular "structure." As the total energy is raised, a drastic change of molecular shapes (structural isomerization) can happen. If the energy is raised high enough, the cluster begins to undergo a continuous change of isomerization and ultimately behaves like a liquid droplet, which finally ends up with dissociation (evaporation). These rather drastic changes of the "phases" can be regarded as a microcanonical analog of the first-order solid–liquid and liquid–gas phase transitions. Furthermore, there is a very interesting state in between solid-like and liquid-like phases, which may represent a glassy and slow dynamics on a complicated potential function having a huge number of local minima [1]. On the other hand, the liquid-like droplet phase can be regarded as a chemical reaction (isomerization reaction) and large-amplitude vibrational motion.

A focus of the present review is placed on the isomerization reaction in the liquid-like phase and its relevant law as a statistical chemical reaction theory. The story of this review spins around a strange behavior of the lifetimes of isomers in the shape-changing dynamics. It is "strange" in that the reaction cannot be understood within the scheme of the transition-state concept. Yet, it shows a beautiful statistical behavior that can be comprehended in terms of a novel statistical theory. The present chapter is composed of three main subjects related to this strange cluster dynamics and of additional topic on the theoretical foundation of the linear surprisal theory [2–7], which is a natural outcome of our study on "temperature" in an isolated (microcanonical) system in characterizing the isomerization dynamics. After presenting the very basic energetics of the potential surface of a Lennard-Jones M_7 cluster, the four different potential basins of which can support four different isomers, and the outline of its dynamics in Section II, we first discuss the following three topics:

1. In Section III, we begin with our finding in our numerical studies III—that is, the uniformity of the average lifetimes (passage-times) of the isomers irrespective of the next visited potential basins in spite of the fact that the heights and geometries of the transition states to be surmounted are different among the different channels [8]. That is, the average passage time is characterized only by a basin through which a trajectory is currently passing and hence does not depend on the next visiting basins. Therefore, it is conceived that as soon as a classical path enters a potential basin, it becomes involved into a chaotic zone in which many paths leading to different channels are entangled among each other, and it effectively (in the statistical sense) loses its memory about which basin it came from and where it should visit next time. This idea is verified by confirming that the distributions of the lifetime of transition from one basin to others are expressed basically in exponential functions that have virtually a common exponent (time constant). The inverse of this exponent is essentially proportional to the average lifetime for a trajectory to pass through this basin. This sets a foundation for the multichannel generalization of the RRKM theory [9], with a very important exception that the present dynamics do not care about the transition state. A non-RRKM behavior in this context is also mentioned briefly.

2. A geometry that can actually materialize such a memory-losing dynamics is studied in Section IV. We here propose a notion of inter-basin mixing that is responsible for the Markov-type stochastic appearance of molecular structures in the above memory-losing isomerization dynamics. An extension of the Liapunov exponent to quantify the time scale to reach inter-basin mixing is also proposed [10].

3. It turns out that the average lifetime, the key quantity to characterize the isomerization dynamics of a liquid-droplet cluster, satisfies an Arrhenius-like exponential law with what we call microcanonical temperature (Section V). This quantity has been introduced to characterize a variational structure of a

phase-space distribution on a constant energy plane. We emphasize the fact that even if a phase-space distribution function satisfying the principle of equal *a priori* distribution, it can become sharply localized having a single peak if projected onto the coordinate of a potential (or the kinetic) energy. The microcanonical temperature is defined as a kinetic energy at which this projected distribution takes the maximum value. Then the most probable statistical events would be found in the vicinity of the peak, provided that the projected distribution is singly and sharply peaked and that the associated dynamics is ergodic. Moreover, the microcanonical temperature can be determined in the individual potential basins, giving rise to a different temperature depending on each isomer. It is highlighted that (the inverse of) the lifetime of an isomer bears an Arrhenius-like relation with thus defined local microcanonical temperature assigned to the corresponding potential basin [11]. We also present a theoretical analysis of how the Arrhenius relation can arise, which establishes new features of the isomerization dynamics of a liquid droplet.

The above three parts constitute a theoretical foundation of a novel chemical reaction, which we call the multichannel chemical reaction, with the above isomerization dynamics being its prototype. The variational statistical theory applied there highlighting the microcanonical temperature is so general that we revisit the linear surprisal theory with respect to final energy disposals to a given vibrational (rotational or translational) mode in chemical reaction dynamics under a fixed total energy. As is well known, the linear surprisal theory (actually an empirical law rather than a theory) deduces many different temperatures by comparing experimental values of product distributions with those estimated in terms of a primitive statistical theory. To our best knowledge, the theoretical foundation of this really remarkable finding due to Bernstein, Levein, and Kinsey [2–7] has not been well established, except for information theoretic reasoning or the maximum entropy principle [12–14]. However, we do not feel comfortable with explaining an elementary process of chemical reaction in terms of the information-scientific logic on human inference. We therefore try to give a new account for the linear surprisal as an outcome of our study of the cluster isomerization dynamics [15].

II. ISOMERIZATION ON MULTI-BASIN POTENTIAL

A. M_7-like System; Description of the System

Our interest is in dynamics arising from the following classical Hamilton:

$$H = \sum_{i=1}^{7} \frac{p_i^2}{2m} + 4\epsilon \sum_{i<j} \left\{ \left(\frac{\sigma}{r_{ij}}\right)^{12} - \left(\frac{\sigma}{r_{ij}}\right)^{6} \right\} \qquad (1)$$

With the following transformations $Q_{ij} = r_{ij}/\sigma$, $\tau = (\sqrt{\epsilon}t)/(\sigma\sqrt{m})$, and $P_i = dQ_i/d\tau$, the Hamiltonian turns into a unique form having no arbitrary parameter:

$$\frac{H}{\epsilon} = \frac{1}{2}\sum_{i=1}^{7} P_i^2 + 4\sum_{i<j}\left\{\left(\frac{1}{Q_{ij}}\right)^{12} - \left(\frac{1}{Q_{ij}}\right)^{6}\right\} \quad (2)$$

It is an exciting experience to realize that such a simple and well-studied system is still pregnant with very rich science to be revealed.

As is well known [1,16], this potential support has four local minima supporting different locally stable isomers. Note, however, that permutations among the atoms actually bring about more isomers by $\sim 7! = 5040$ times. In this review, we disregard the permutationally distinctive isomers, and they are regarded as the same species. The stable structures and their energies in units of ϵ are summarized in Fig. 1. Figure 1 also shows the energy of the lowest saddle

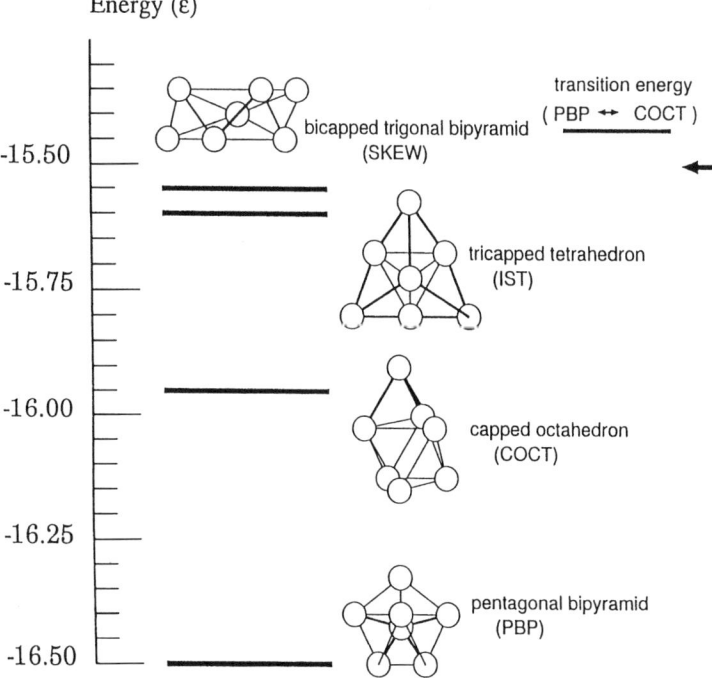

Figure 1. The four local minima of the Ar$_7$-like cluster. The minimum energies of the four structures -16.505ϵ for PBP, -15.935ϵ for COCT, -15.593ϵ for IST, and -15.533ϵ for SKEW. The lowest transition-state energy (-15.45ϵ) is also depicted, which is in between PBP and COCT structures. (Reproduced from Ref. 19 with permission.)

(about -15.45ϵ) connecting the bipyramid (PBP) and capped octahedron (COCT) structures. The highest saddle ever found is -14.5960ϵ, which is located in between PBP and the bicapped trigonal bipyramid (SKEW) isomers. The height of a permutational isomerization from COCT back to COCT amounts to -14.5482ϵ, which is also among the highest. There are a huge number of transition states among the structures [17].

Identification of the potential basins to which any point on a classical trajectory belongs is made with use of the so-called quenching technique due to Stillinger and Weber [18],

$$\dot{q} = -\nabla V \qquad (3)$$

We study this simple-looking system in classical dynamics. The initial conditions for running the classical trajectories are systematically set as follows. All the initial geometry is selected so as to be located in the basin of PBP structure and deformed slightly and randomly from the minimum energy structure, at which a trajectory begins to run with the zero initial momenta so that the velocity of the center of mass and the total angular momentum are all taken to be zero. The relevant technical details are found in Ref. 19.

B. Solid–Liquid Transition

The microcanonical analog of the macroscopic phases is quite often defined by means of the energy dependence of Lindemann index. This quantity is designed to detect the stiffness of a molecule by measuring the deviation of the bond lengths from their averaged values as

$$\delta = \frac{2}{n(n-1)} \sum_{i<j} \frac{(\langle r_{ij}^2 \rangle_t - \langle r_{ij} \rangle_t^2)^{1/2}}{\langle r_{ij} \rangle_t} \qquad (4)$$

where n is a number of particles, r_{ij} is a distance between ith and jth particles, and $\langle \ \rangle_t$ means a time average for an interval t.

The behavior of the Lindemann index for our M_7 as a function of the total energy is shown in Fig. 2, in which three stages are observed: the "solid-like" phase below about -14.5ϵ which is called "freezing" energy, the "liquid-like" phase above about -13.0ϵ called "melting" energy, and the coexistence region in between. The steep rise of the Lindemann index suggests that the present "melting" is similar to that of the first-order phase transition. However, unlike the latter, the former does not undergo an abrupt change, and thus the freezing and melting points do not coincide with each other. Besides, they are obviously not clear-cut values.

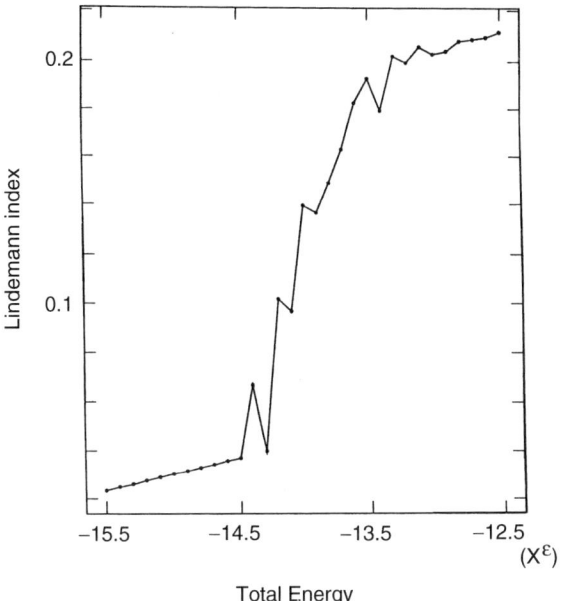

Figure 2. The Lindemann index (δ) versus the total energy. Only a single but very long trajectory has been sampled for each energy. (Reproduced from Ref. 19 with permission.)

In a similar context, Nayak and Ramaswamy have shown that the maximum Liapunov exponent (MLE) rises very steeply just as the Lindemann index and thereby can detect the aforementioned transition very well [20]. Since MLE is well established to measure the exponential divergence of the distance between nearby trajectories in phase space [21,22], their numerical results seem to suggest that the "phase-transition" could be a consequence of strong chaos behind the dynamics. We henceforth examine the resultant phenomena, the geometry on which the dynamics is characterized, and a statistical law of associated isomerization reaction.

C. Anomalous Time Series of Structural Transitions

As described in the preceding section, our model molecule has four local minima [16]. (See Fig. 1.) As the energy is increased from the solid-like phase, trajectories begin to get out of the potential basin of PBP structure and travel around the other basins. We would like to explore how the dynamics of structural change proceeds with time. To this end, we define a simple indicator that can detect the structural transitions with high sensitivity. Let $\vec{r}_{\alpha i}$ be a position vector from the αth to ith particles. Suppose a triangle plane that is expanded by two vectors $\vec{r}_{\alpha i}$ and $\vec{r}_{\alpha j}$. A vector normal to this plane is written as

$$\vec{n}_{ij} = \frac{\vec{r}_{\alpha i} \wedge \vec{r}_{\alpha j}}{|\vec{r}_{\alpha i}||\vec{r}_{\alpha j}|} \quad (5)$$

the absolute magnitude of which is proportional to the area of the triangle. Similarly, another vector $\vec{n}_{kl}$ is made up for the other set of two vectors $\vec{r}_{\alpha k}$ and $\vec{r}_{\alpha l}$ as

$$\vec{n}_{kl} = \frac{\vec{r}_{\alpha k} \wedge \vec{r}_{\alpha l}}{|\vec{r}_{\alpha k}||\vec{r}_{\alpha l}|} \quad (6)$$

Note that these triangle planes are orientable—that is, the direction can be expressed in terms of the sign of the normal vector $\vec{n}_{ij}$. This property is particularly sensitive to the permutation of the particles. For example, if the positions of i and j are replaced with each other, $\vec{n}_{ij}$ turns to $-n_{ij}$. Thus one can observe a part of the effect of permutational isomerization using this quantity. Now the geometrical change between the two triangles are conveniently detected in terms of a simple inner product between $\vec{n}_{ij}$ and $\vec{n}_{kl}$ in such a way that

$$K_{ijkl} = \frac{\vec{n}_{ij} \cdot \vec{n}_{kl}}{|\vec{n}_{ij}||\vec{n}_{kl}|} \quad (7)$$

The index K_{ijkl} depends on the position of the origin α. Of course, not all kinds of the molecular deformation are detected with a single selection of the set $\{i,j,k,l\}$. We first label each atom from 1 to 7 as in Fig. 3, and the origin α is set on the first atom. K_{2334} monitors the angle between 123-plane and 134-plane. Obviously, K_{2334} takes 1 or -1 when these planes are mutually parallel, and it becomes 0 when vertical. K_{ijkl} does not vary appreciably when the molecule vibrates only slightly. On the other hand, in the case where a shape of the molecule changes drastically, K_{2334} varies significantly. Figure 3 depicts an example that the transition from PBP structure to COCT results in a drastic change of K_{2334}.

Figure 4 shows three typical patterns of the time-series of K_{2334} taken from (a) solid-like phase (-15.505ϵ), (b) coexistence region (-13.505ϵ), and (c) liquid-like phase (-11.505ϵ). As expected, the structural transition takes place more frequently as the energy is increased. Nonetheless, the patterns can be distinguished qualitatively. In the solid-like energy region, an isomerization did not happen in the displayed time range, while the liquid-like phase exhibits a chaotic or almost random pattern. On the other hand, the time series in the panel (b) of Fig. 4 has evidenced very clearly that the motion in the coexistence range looks intermittent; that is, sudden transitions occur with no periodicity nor predictability. This intermittent behavior is quite similar to the so-called phase-space large-amplitude motion, which is essentially a weakly chaotic motion in a thin quasi-separatrix [23,24]. A trajectory in a thin separatrix can randomly wander around several tori that correspond to different vibrational modes. As long as a trajectory stays close to each torus, the corresponding mode can be

TEMPERATURE, GEOMETRY, AND VARIATIONAL STRUCTURE 33

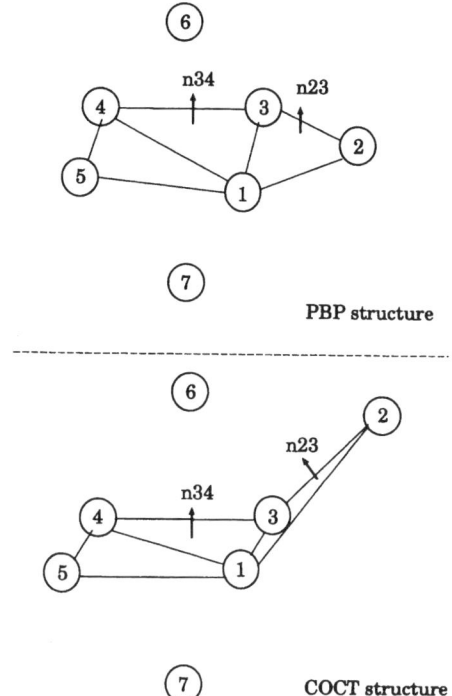

Figure 3. Definition of the vectors normal to the triangle planes. When an isomerization occurs from the PBP structure to the COCT, the angle between $\vec{n}_{34}$ and $\vec{n}_{23}$, namely K_{2334}, undergoes a large change. (Reproduced from Ref. 19 with permission.)

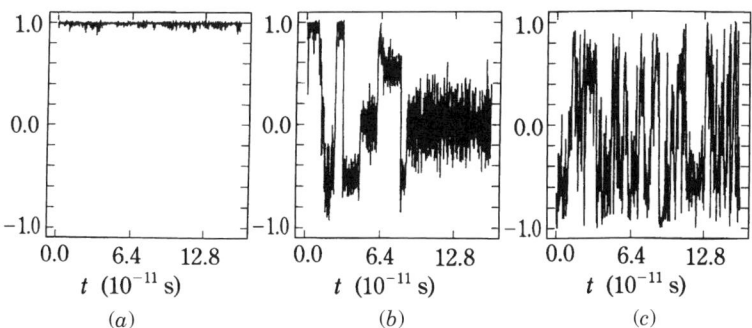

Figure 4. The time series of K_{2334} at three typical energy regions. (a) -15.505ϵ (solid-like phase), (b) -13.505ϵ (coexistence region), and (c) -11.505ϵ (liquid-like phase). (Reproduced from Ref. 19 with permission.)

clearly identified. Nevertheless, the trajectory undergoes an unpredictable transitions among the distinctive vibrational modes. An extensive study on the quantum mechanical manifestation of the phase-space large-amplitude motion has also been performed [25]. Although these two patterns of transition are very similar, it is not yet clear whether or not the motion in the coexistence region is indeed a high-dimensional version of the separatrix motion in phase space. However, this important problem is out of the scope of this review.

III. STRANGE BEHAVIOR OF THE LIFETIME OF ISOMERS IN LIQUID-LIKE STATE

A. Unimolecular Dissociation via a Transition State—A Preliminary

Suppose we have a simple unimolecular dissociation embedded in a microcanonical ensemble in phase space, in which only one dissociating channel is available. The Rice–Ramsperger–Kassel–Marcus (RRKM) rate constant is given as [9]

$$k(E) = \frac{N^{\ddagger}(E)}{hV(E)} \qquad (8)$$

where $V(E)$ is the phase-space volume of the basin (or the density of states) at an energy E, while $N^{\ddagger}(E)$ is the number of states in an appropriately defined dividing surface (usually termed simply as the transition state) which is transversal to the reaction coordinate. h is the Planck constant. The essential premise of the RRKM theory is a rapid mixing in phase space due to chaos that is supposed to be completed before the dissociation. If it happens, the phase-space population of the composite state (or resonant state) should decay in an exponential form like

$$P(t) = \frac{1}{\tau} \exp\left(-\frac{t}{\tau}\right) \qquad (9)$$

The rate is characterized by the inverse of τ, which is an average time for a trajectory to spend in the basin area before dissociation. The canonical rate constant is given by a Laplace transform, or the Boltzmann average over $k(E)$ as

$$k(T) = \int dE \exp\left(-\frac{E}{k_B T}\right) k(E) \qquad (10)$$

where k_B is the Boltzmann constant.

The main focus in the application of this beautiful statistical theory is usually placed on the identification of the transition state, the number of which is usually supposed to be one for a single reaction channel, and the evaluation of the relevant partition functions and related quantities. On the other hand, people are often interested in the theoretical foundation of this statistical assumption and in a possible deviation from the statistical behavior, which is termed the non-RRKM behavior.

Isomerization reaction or structural transition in clusters is quite different from the ordinary collision and dissociation dynamics in that isomerization occurs repeatedly many times in a bound state. Hence it is not trivial to define reaction rate or quantities related to that for isomerization reactions. Moreover, there are many different reactive channels involved in a high-energy region, since isomerization can take place among many possible isomers; actually four in our studied M_7 Lennard-Jones cluster. Therefore the possible theoretical form of the reaction rate in such a repeating multichannel process could be entirely different from that of the RRKM theory. Under this recognition, we study below the isomerization reaction of M_7 and try to extract a quantity that can best characterize its dynamical process. We will find out then that the average lifetimes of the individual isomers in the liquid-like phase have surprising features, which are not understandable within the theoretical realm of the transition-state (or RRKM) concept.

B. Uniformity of the Average Lifetime for Passing Through a Basin

We survey some basic quantities relevant to the rate of chemical reactions and then find a remarkable property in the lifetimes of the isomers.

1. Accumulated Residence Time and Ergodicity

First of all we show, in Table I, the accumulated residence time for trajectories to reside in the four possible basins (denoted by $t^{(a)}$, $a =$ PBP, COCT, IST, and SKEW) at three typical energies, -13.505ϵ and -11.505ϵ, which are sampled from the coexistence and liquid-like regions, respectively. The residence times have been obtained by applying periodically the so-called quenching technique [18] with which to search for the local minima of the potential basins.

TABLE I
The Accumulated Residence times ($t^{(a)}$) for the Four Possible Basins in Two Classical Trajectories (in nanoseconds)

Energy	PBP[a]	COCT[b]	IST[c]	SKEW[d]
-13.505ϵ	11.5	2.80	0.58	1.15
	(71.7%)[e]	(17.5%)	(3.6%)	(7.2%)
-11.505ϵ	5.49	3.58	1.92	5.01
	(34.3%)[e]	(22.4%)	(12.0%)	(31.3%)

[a]Pentagonal bipyramid.
[b]Capped octahedron.
[c]Tricapped tetrahedron.
[d]Bicapped trigonal bipyramid.
[e]The ratio.

The accumulated residence time is not necessarily relevant directly to the rate of isomerization, but instead should have a certain statistical implication if ergodicity holds. In order to check an ergodicity directly, we carried out numerical calculations of the 30-dimensional microcanonical ensembles in phase space and assigned them to the individual potential basins (structures) [19]. If the ergodicity holds, the ratio of the accumulated residence times listed in Table I should be proportional to the phase-space volumes. It turned out that in the liquid-like phase the ergodicity holds quite well, but considerable deviation has been observed in the coexistence region. An analysis of these distributions, particularly in conjunction with statistical inference, has been presented before [19]. We note again that the inverse of each residence time does not represent a rate for a trajectory to pass through the corresponding basin, simply because information about the frequencies of transitions from basin to basin is lacking. Rather, the law of large numbers suggests that the ratio of the accumulated residence times (or the phase-space volumes assigned to the basins) should be proportional to that between the numbers of those isomers to be found in a large ensemble of clusters.

2. Passage Time and Uniformity

We here pursue the fate of a single classical trajectory for a very long time, particularly by looking at the timing of the structural transitions. The sampled trajectories are selected randomly but carefully lest they should be pathological so as to stick to a particular basin. The occurrence of the isomerization is monitored using the quenching technique periodically with an interval 0.31 ps, which gives a discrete time series of the events. During this single interval, the molecules can undergo a global vibration as many as 5 to 6 times. As it will turn out later, the averaged passage times in the coexistence region are much longer, and hence the statistical error is hard to be expected. On the other hand, those in the liquid phase are sometimes as short as 0.5 ps. In such a case, we have shortened the time interval of quenching.

To see the dynamical characteristics of the present system, we first look at the (absolute) frequencies of structural transitions as well, which is tabulated in Table II. Let $n_{a \to b}$ be the number of transitions from a basin a to another basin b during a given running time. Nothing very special is found in this table except that it looks almost a symmetric matrix reflecting the microscopic reversibility [26].

It is natural then to define the relative frequency for a trajectory to make a transition from a basin to the others by

$$f_{a \to b} = \frac{n_{a \to b}}{t^{(a)}} \qquad (11)$$

which is an average frequency of the transition in a unit time during the total residence time in a basin a. The relative frequency is certainly a standard

TABLE II
The Absolute Frequencies of Transitions $n_{a \to b}$ from Basin a (Row) to Basin b (Column) During the Time Intervals of Table I

	PBP	COCT	IST	SKEW
		Energy -13.505ϵ (Coexistence Region)		
PBP		736	38	73
COCT	759		401	618
IST	30	394		47
SKEW	59	645	33	
		Energy -11.505ϵ (Liquid Region)		
PBP		2670	897	2110
COCT	2550		1619	3276
IST	895	1570		1278
SKEW	2204	3214	1248	

quantity to measure a reaction rate such as $k(E)$ in Eq. (8). Table III lists the relative frequencies thus calculated on the basis of the data of Tables I and II. Unfortunately again, nothing very characteristic happens in the calculated $f_{a \to b}$. They are distributed with a large variance, and therefore it is not easy to find a simple rule or law behind these distributions.

In contrast to experiments in reaction dynamics, the individual passage time $t_i^{a \to b}$ can be readily obtained in theoretical calculations, which is a time for a path to spend in a basin a on a single passage to another basin b. Each passing occasion is identified with the suffix i. A precise piece of information is of course

TABLE III
The Relative Frequencies of Isomerization $f_{a \to b}$ from Basin a (Row) to Basin b (Column) (in reciprocal nanoseconds)[a]

	PBP	COCT	IST	SKEW
		Energy $-13.505\,\epsilon$ (Coexistence Region)		
PBP		64	3.3	6.3
COCT	271		143	221
IST	52	679		81
SKEW	51	561	29	
		Energy $-11.505\,\epsilon$ (Liquid Region)		
PBP		486	163	384
COCT	712		452	915
IST	466	818		666
SKEW	440	642	249	

[a] The entries in each row in Table II are divided by the corresponding residence times of Table I.

lost when they are summed up to reproduce the accumulated residence time

$$t^{(a)} = \sum_b \sum_i t_i^{a \to b} \qquad (12)$$

which has been used back in Eq. (11). We further define an average passage time as

$$\langle t_{a \to b} \rangle = \frac{\sum_i t_i^{a \to b}}{n_{a \to b}} \qquad (13)$$

which is our main quantity. Note again that the structural transitions are monitored with the quenching of a fixed time interval. Thus the individual $t_i^{a \to b}$ are allotted to the bins of unit interval (0.31 ps). The average passage time is thus calculated as

$$\langle t_{a \to b} \rangle = \frac{(\text{width of the bins})}{n_{a \to b}} \times \sum_k (\text{number of sampling points in the } k\text{th bin}) \qquad (14)$$

We tabulate the average passage times $\langle t_{a \to b} \rangle$ in Table IV. On comparing them with the relative frequencies in Table III, surprising uniformity in the average

TABLE IV
The Average Passage Time from an Isomer to Others (in picoseconds)

	PBP	COCT	IST	SKEW
		$-13.505\epsilon^a$		
PBP		13.5	11.5	11.6
COCT	1.35		1.50	1.45
IST	0.86	1.08		1.24
SKEW	1.37	1.40	1.53	
		$-12.505\epsilon^a$		
PBP		2.00	1.95	2.06
COCT	0.50		0.48	0.48
IST	0.44	0.50		0.49
SKEW	0.70	0.69	0.75	
		$-11.505\epsilon^a$		
PBP		0.77	0.82	0.76
COCT	0.31		0.32	0.31
IST	0.34	0.34		0.34
SKEW	0.56	0.56	0.56	

[a]The total energy.

TEMPERATURE, GEOMETRY, AND VARIATIONAL STRUCTURE   39

passage time at the two high energies (-12.505ϵ and -11.505ϵ) is noticed (see the table in a fixed row scanning the columns). More specifically, $\langle t_{a \to b} \rangle$ scarcely depends on the next visiting channels b, despite the fact that the energy and geometry of the transition state between the basins a and b depends strongly on b. That is,

$$\langle t_{a \to b} \rangle \approx \langle t \rangle_a \qquad (15)$$

It has been confirmed numerically that this uniformity holds well in the energy range higher than about -12.505ϵ. On the other hand, considerable deviation is observed at -13.505ϵ, which is in the middle of the coexistence phase. Despite the large fluctuation in the relative frequency seen in Table III, such a clear uniformity in the average passage time $\langle t_{a \to b} \rangle$ has been observed and this phenomenon is robust.

The relative frequency and the average passage time are mutually interconnected through a simple relation

$$f_{a \to b} = \frac{n_{a \to b}}{\sum_c n_{a \to c} \langle t_{a \to c} \rangle} \qquad (16)$$

This relation can be readily confirmed with use of Tables III and IV. Since $\langle t_{a \to b} \rangle$ can always reproduce $f_{a \to b}$ and because of the clear uniformity, it seems sensible to investigate the passage time to characterize this kind of continuous dynamics. It is this uniformity that we would like investigate deeply as a measure of the mixing property in chaotic dynamics.

When the uniformity of the average passage times holds, Eq. (16) is led to a simpler relation

$$\sum_b f_{a \to b} = \frac{1}{\langle t_{a \to b} \rangle} \equiv \frac{1}{\langle t \rangle_a} \qquad (17)$$

Thus the sum of the relative frequencies over the next visiting basins is nearly equal to the inverse of the average passage time. But, the inverse of the individual average passage time for the channels $a \to b$ does not give the corresponding rate. This is not quite strange, since the time necessary for a trajectory to pass through a basin has nothing direct to do with the flux getting out of it.

3. The Uniformity in an Exponential Decay Expression

To understand how the uniformity arises, we make a working hypothesis as follows: First of all, it is already known that the present system in the liquid-like phase is strongly chaotic [19,20]. As soon as a classical path enters a basin, it becomes involved with a chaotic region in which many paths having different channels are entangled among each other, and it effectively loses its memory about which basin it came from and where it should visit next time in the statistical sense. This is just an ideal case of the Markov process. Hence, the time scale for a trajectory to get out of the tangled region should depend exclusively on the size of this region but virtually not on the exit channels.

This hypothesis can be verified phenomenologically as follows. The mixing, if it really occurs, should be associated with an exponential form in the population dynamics as in Eq. (9). Let $N_{a \to b}(t)$ be the number density of events for a trajectory to move from a basin a to another one b in a short time interval $[t, t + \Delta t]$. Under the present situation, it may be written as

$$N_{a \to b}(t) = \frac{1}{\langle t \rangle_a} n_{a \to b} \exp\left(-\frac{t}{\langle t_{a \to b} \rangle}\right) \tag{18}$$

which in turn should be rewritten as

$$N_{a \to b}(t) = \frac{1}{\langle t \rangle_a} n_{a \to b} \exp\left(-\frac{t}{\langle t \rangle_a}\right) \tag{19}$$

due to the uniformity. Here we make sure that the time is reset to zero as soon as a trajectory gets in a new basin. We check 12 such possible combinations of $\log N_{a \to b}(t)$ versus t at the energy -11.505ϵ. As Fig. 5 demonstrates very clearly, all the lines belonging to the common basins are nearly straight and parallel with each other. This supports that a very efficient mixing is materialized in the dynamics.

Once the above exponential form is established, it is further used to reconfirm the uniformity as follows. First, the average passage time is calculated in terms of the distribution function such that

$$\langle t_{a \to b} \rangle = \frac{1}{n_{a \to b}} \int_0^\infty t N_{a \to b}(t) dt \cong \langle t \rangle_a \tag{20}$$

which is identical to Eq. (13). Recall here that

$$n_{a \to b} = \int_0^\infty N_{a \to b}(t) \, dt \tag{21}$$

On the other hand, defining

$$n_a = \sum_b n_{a \to b} = \frac{t^{(a)}}{\langle t \rangle_a} \tag{22}$$

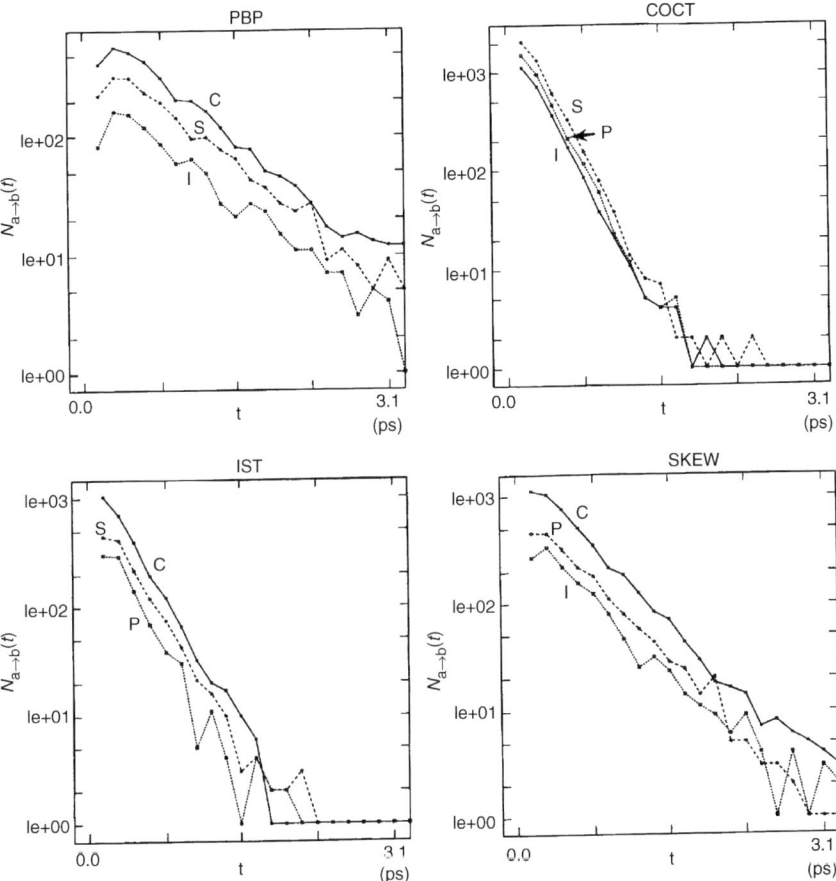

Figure 5. $\log N_{a \to b}(t)$ versus t at the energy of -11.505ϵ. The total running time is 16.09 ns. The quenching interval is 0.31 ps. The symbols P, C, I, and S stand for the next visiting basins PBP, COCT, IST, and SKEW, respectively. (Reproduced from Ref. 8 with permission.)

and using the relation of Eq. (17), one can rewrite Eq. (19) as

$$N_{a \to b}(t) = n_a f_{a \to b} \exp\left(-\frac{t}{\langle t \rangle_a}\right) \tag{23}$$

Thus a probability density for a path to get out of a basin a during $[t, t + \Delta t]$ is naturally defined as

$$P_a(t) = \frac{1}{n_a} \sum_b N_{a \to b}(t) = \frac{1}{\langle t \rangle_a} \exp\left(-\frac{t}{\langle t \rangle_a}\right) = \sum_b f_{a \to b} \exp\left(-\sum_c f_{a \to c} t\right) \tag{24}$$

TABLE V
The Passage Times (in picoseconds) from Basins to Others Estimated from the Slopes of the Linear Parts in Fig. 5[a]

	PBP	COCT	IST	SKEW
PBP		0.66	0.76	0.72
COCT	0.26		0.30	0.28
IST	0.41	0.33		0.38
SKEW	0.53	0.53	0.47	

[a]The total energy is -11.505ϵ.

This brings us back to Eq. (9) with the multichannel extension. The role of the relative frequency $f_{a \to b}$ defined by Amar and Berry [26] in isomerization dynamics has become clearer now.

Let us then compare $\langle t_{a \to b} \rangle$ obtained in the numerical average in Eq. (14) and that estimated from the slope of the above "exponential" functions in Fig. 5. The curves in Fig. 5 have been fitted to straight lines to estimate the inverse of $\langle t_{a \to b} \rangle$. Table V lists such values for the energy -11.505ϵ, which should be compared with the last entries of Table IV. The two kinds of the averaged passage time agree fairly well. However, the uniformity in Table V is not as good as that in Table IV. This is partly because the numbers of the sampling points in Fig. 5 are not large enough. Nevertheless, the absolute values are very similar to each other, although the values based on the slope of Fig. 5 tend to be slightly smaller than the simply averaged values by Eq. (13). (Later we consider the physical origin of this deviation in greater detail.) Thus it turns out that the exponential form of $N_{a \to b}(t)$ given in Eq. (19) is consistent with the uniformity of the average passage time.

Appendix A: A Short Detour to Non-RRKM Behaviors

Before concluding this section to proceed to the next topic about a geometry that can bring the statistical nature into the system, we slightly touch upon the nonexponential behavior, which is a key to consider the nonstatistical behavior of chemical reaction. This is interesting from the viewpoint of the onset of statistical mechanics and also from the study of the role of chaos in mechanics.

Short-Time Behavior in the Liquid-like Phase. It has been confirmed that the passage times in the high-energy region is mostly dominated by the mixing property that produces the exponential distributions. There are in general two regions where we can expect the exponential behavior to be possibly violated. One is in the long passage-time region. Indeed, the slope in the long-time region looks much more flat than that of the exponential part. This can be partly due to the small number of the sampling points. Fortunately though, this region should not make a large contribution to the average passage time anyway because of the very small populations. On the other hand, the short-time behavior can be important for the opposite reason. In fact, Dumont and Brumer [27] have pointed out the existence of the so-called induction time for the mixing property to dominate the dynamics before the dissociation occurs. Berblinger and Schlier [28] have actually shown that there are many direct paths in H_3^+ that lead to dissociation before the statistical property comes into effect. Because of these large number of unstatistical paths, the RRKM theory should always overestimate (underestimate) the average passage time (reaction rate). In other words, the straight line given by $\log N_{a \to b}(t)$ should be corrected so as to increase the short-time components.

Figure 6 shows a plot of $\log N_{a \to b}(t)$ at the energy of -13.505ε in the coexistence region. A precise look at this figure shows a significant reduction of the population in PBP and COCT, and it shows SKEW structures in the shortest time regions available. (Note that this shortest time is still longer than the periods of the molecular vibrations.) The populations in the short time region are pushed down rather than lifted up, and thus make holes. This tendency is in a marked contrast to the contribution of the direct paths. (Remember that the direct paths lift up the population in the short-time region.) Therefore the average passage times evaluated with Eq. (20) tend to become longer than those that are estimated by the exponential components only.

It is natural to conceive that this short-time behavior should be due to some time interval for a trajectory to spend to look for exit ways to the next basins in the complicated structure of phase space. In the next section, we will propose a geometrical view that shows what this complexity is. Hence we consider that the hole of $N_{a \to b}(t)$ in the short-time region should be a reflection of chaos, which is just opposite to the behavior arising from nonchaotic direct paths as observed in H_3^+ dynamics. The present effect is therefore expected to be more significant as the molecular size increases or the potential surface and corresponding phase-space structure become more complicated. Another important aspect of the hole in $N_{a \to b}(t)$ is an induction time for a transport of the flow of trajectories in phase space. It is of no doubt that the RRKM theory does not take account of a finite speed for the transport of nonequilibrium phase flow from the mid-area of a basin to the "transition states." Berblinger and Schlier [28] removed the contribution from the direct paths and equate the statistical part only to the RRKM rate. One should be able to do the same procedure to factor out the effect of the induction time due to transport. We believe that the transport in phase space is essentially important in a nonequilibrium rate theory and have reported a diffusion model to treat them [29].

Finally, we note that there is another pattern in the short-time behavior in $N_{a \to b}(t)$ observed in our M_7 dynamics, in which a direct-path contribution is embedded in the hole structure. These direct paths represent a motion that

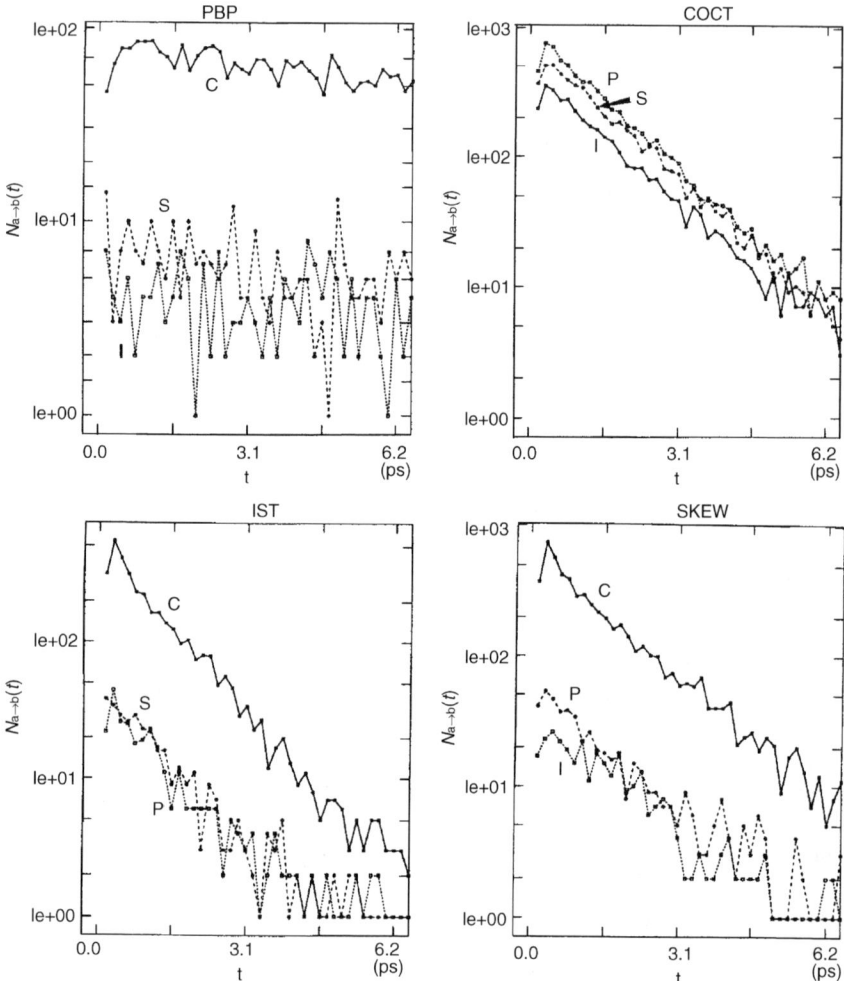

Figure 6. $\log N_{a\to b}(t)$ versus t at the energy of -13.505ϵ. The total running time is 134.84 ns. The quenching interval is 0.155 ps. (Reproduced from Ref. 8 with permission.)

does not penetrate deep into a basin after entering and immediately get out of it [30].

Nonstatistical Behavior in the Low Energy. Comparing $\log N_{a\to b}(t)$ in Fig. 6 with Fig. 5, we immediately notice (see Figs. 6b to 6d) that the behaviors of COCT, IST, and SKEW are not very strange in that (i) the curves are mostly linear, (ii) they are parallel to each other within the same structure, except in the long-time regions, and (iii) the holes in the short-time region are clearly identified, although they are not very large. So, coming back to Table IV, we can conclude that the deviations from the RRKM behavior in these three isomers are not very large. On the other hand, the PBP structure exhibits very peculiar distributions in $\log N_{a \to b}(t)$ (see Fig. 6a). The fluctuation of the curves are so large that even the seemingly huge hole in the short-time region for PBP to COCT is not well identified. It thus has turned out that the major deviation from the RRKM behavior arises from the PBP structure. However, the other three isomers would basically undergo the similar behavior as the energy becomes even lower. As for the PBP structure, we cannot identify which factor dominates the non-RRKM behavior. Thus M_7 at -13.505ϵ consists of three basins (COCT, IST, and SKEW) of nearly statistical nature and one mechanically dominated basin (PBP). These two categories of dynamics should be treated independently. It is thus suggested that the width in the sinusoidal increase of the Lindemann index and the maximum Lyapunov exponent [19,20,31] in the coexistence region could have been born by the mixture of the basins of different natures.

IV. GEOMETRY BEHIND THE MEMORY-LOSING DYNAMICS; INTER-BASIN MIXING

We now explore a geometrical structure that brings about the "memory-losing" dynamics found in the preceding section. Chaotic dynamics in two-dimensional space (four-dimensional phase space) has been beautifully geometrized in terms of the scenario of Poincaré–Birkhoff–Smale [21]. Its extension to a system of even three degrees of freedom in terms of the notions of invariant manifolds, heteroclinic crossings, and so on, is already extremely difficult—especially our 15-dimensional (30-dimensional in phase space) M_7 dynamics, which is simply impossible. We therefore employ a "chemical" way to characterize and quantify the geometry behind the memory-losing dynamics, although it may be rather intuitive and less rigorous.

A. The Concept of Mixing Is Not Enough to Account for the Markow-Type Appearance of Isomers

For a concrete description of our problem, let us take an example again from the classical dynamics of an M_7 cluster composed of seven identical atoms. The scaled Morse Hamiltonian for this system has the following canonical form:

$$\frac{H}{\epsilon} = \frac{1}{2} \sum_{i=1}^{7} \left[\left(\frac{d\bar{x}_i}{d\tau}\right)^2 + \left(\frac{d\bar{y}_i}{d\tau}\right)^2 + \left(\frac{d\bar{z}_i}{d\tau}\right)^2 \right] + \sum_{i<j} [\exp(-2(\rho_{ij} - \rho_0)) - 2\exp(-(\rho_{ij} - \rho_0))] \quad (25)$$

where $(\tilde{x}_i, \tilde{y}_i, \tilde{z}_i)$ are the scaled coordinates with ρ_{ij} representing the distance between atoms i and j. τ is the (dimensionless) time and the energy is measured in units of ϵ. Only one parameter ρ_0 can determine the essential feature of the potential function. We set $\rho_0 = 6.0$ throughout the present communication, which is similar to the 12-6 Lennard-Jones potential. This system has again four locally stable structures, which are named PBP (pentagonal bipyramid), COCT (capped octahedron), IST (tricapped tetrahedron), and SKEW (bicapped trigonal bipyramid), the potential minima of which are -16.208ϵ, -15.564ϵ, -15.248ϵ, and -15.216ϵ, respectively. Let us consider again the isomerization dynamics in the liquid-like phase of M_7 with a time series in which a cluster changes its shape from one isomer to another in a Markovian way. Suppose that we are pursuing a history of a classical trajectory for a some long time, in which a sequence of the isomers the trajectory visits looks, for instance, like $\{a, c, d, b, a, d, c, a, c, b, \ldots\} = \{\#\}$, where a–d are the names of the isomers like PBP, COCT, and so on.

It is well known that mixing is a key notion in considering statistical properties. On a multidimensional anharmonic potential, classical dynamics can be strongly chaotic even in a single-basin system. This would lead to a mixing, which is specified here as the *global mixing* in this communication to distinguish *inter-basin mixing* we are proposing. A system is said to be globally mixing if the following holds:

$$\lim_{t \to \infty} \mu[(\phi_t A) \cap B] = \mu(A)\mu(B) \quad (26)$$

where $\mu(A)$ is the measure of a set A in phase space, and $\phi_t A$ is a set mapped by a dynamics ϕ during time t [32]. The global mixing is a time-independent concept. On the other hand, chaos arising from an exponential separation of nearby orbits is usually measured with the eigenvalues of a stability (Jacobian) matrix $[\partial Z_f / \partial Z_i] = [\partial (q_f, p_f) / \partial (q_i, p_i)]$, where Z_i and Z_f are, respectively, the initial and final points in phase space connected by a trajectory. The positive stability exponents (or their sum) are frequently adopted as the positive Liapunov exponents [22], which can be viewed as a rate for this chaotic system to be relaxed to a mixing state. To quantify the rate for a system to approach the mixing state of Eq. (26), we here take the following naive method. Let us first pick an arbitrary trajectory as a reference path in phase space. Prepare a sphere of a short radius (r_0) in phase space, the center of which is located on the trajectory at a given time t_1. Sample many points in a random fashion from this spherical region. These phase space points are carried by their own classical trajectories and thereby represent the deformation of the initial sphere. At time $t_2(> t_1)$, monitor the number of trajectories $N(t_2 - t_1)$ that are found within the sphere that has been carried by the reference path for time $t_2 - t_1$ with the same radius r_0. One can numerically confirm that $N(t_2 - t_1)$ decreases exponentially in the liquid-lake phase,

$$\mu[(\phi_{t_2-t_1} V(t_1)) \cap V(t_2)] \propto N(t_2 - t_1) \approx N(t_1)^2 \exp[-\alpha(t_2 - t_1)] \quad (27)$$

where $V(t_1)$ and $V(t_2)$ are the spheres defined above [29]. In short, the global mixing is concerned with a situation in which trajectories initially prepared in a small region will be eventually "diffused" so that any portions in phase space of the relevant energy are to have the uniform density of those trajectories.

If one could look at the entire history of an ergodic trajectory embedded in the globally mixing state, the ratio of the numbers of conversion of, say, $a \to b$, $a \to c$, and $a \to d$ in the time series should converge to $n_{a \to b} : n_{a \to c} : n_{a \to d}$ after all, even though it may take an infinite time, where $n_{a \to b}$ etc. are the absolute frequencies for a given observing time [recall Eq. (11)]. This ratio must be realized by an ensemble of initial points randomly sampled in phase space a for trajectories whose next visiting basins are b, c, and d. The ratio $n_{a \to b} : n_{a \to c} : n_{a \to d}$ should be essentially determined by the size (number of states) of the accessible boundary surrounding the basin a. However, this convergence does not necessarily warrant the stochastic and Markov-type appearance of isomerization that is supposed to materialize the exponential expression of Eq. (19) within a relatively short time. Suppose on the other hand that we sample an arbitrary small phase-space volume in a basin a, and let many classical trajectories start to run from this sampling volume. Count the number of trajectories whose first visiting basin is b, and let this number be $m_{a \to b}$. (Note that we do not follow the fate of the trajectories after the first isomerization.) A simple-minded idea suggests that if the system is globally mixing, a ratio $m_{a \to b} : m_{a \to c} : m_{a \to d}$ should take a common value everywhere on a given energy plane in the basin a; furthermore, it should hold that $m_{a \to b} : m_{a \to c} : m_{a \to d} = n_{a \to b} : n_{a \to c} : n_{a \to d}$. However, this is not generally the case (we show a counterexample later), since the ratio $m_{a \to b} : m_{a \to c} : m_{a \to d}$ depends on where the small volume is sampled in phase space and how small its size is. We will show the trick behind this possible inequality. By the equality $m_{a \to b} : m_{a \to c} : m_{a \to d} = n_{a \to b} : n_{a \to c} : n_{a \to d}$ we define the inter-basin mixing. If this condition is satisfied at most of the points in a basin, trajectories passing this basin should behave as though they lose their memories, and the globally statistical appearance of the structural isomers with the uniform lifetime (passage-time) must be realized. We identify the inter-basin mixing more clearly in what follows by looking at the phase space structure geometry. The following examples are all made for the PBP basin at the liquid-like phase with an energy -11.208ϵ.

B. Bifurcation of Reaction Tubes

The need for the concept of inter-basin mixing may be best explained in terms of the so-called reaction tube, which has been established by De Leon et al. in the

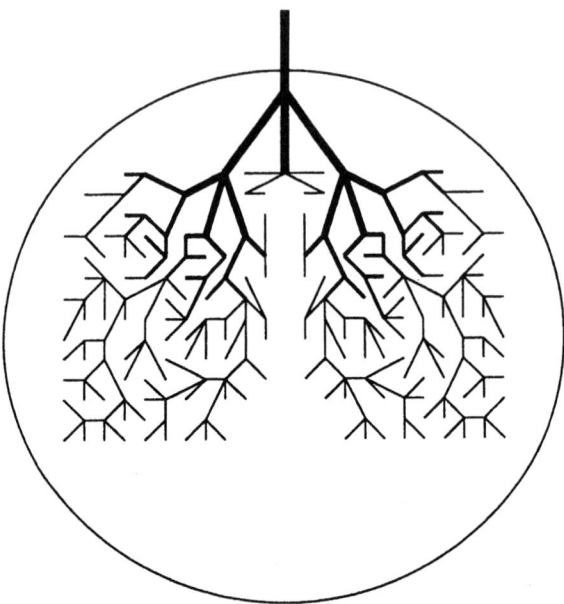

Figure 7. Schematic view of the bifurcation of a reaction tube.

studies of chemical reaction dynamics [33]. Pick an arbitrary trajectory that is moving from a basin b into a across the associated transition region. This is our reference path. It is then quite natural to assume that there should be a large bundle of similar trajectories flowing from b to a around the transition region, which includes the reference trajectory. These bundles are called the reaction tube. The reaction tubes under global mixing must generally undergo series of bifurcations before getting out from the basin a, since trajectories initially sharing a common tube can eventually go to different basins (see Fig. 7). In chaotic phase space a reaction tube can be broken apart to many branches before a single occurrence of isomerization, the diameters of which should become smaller and smaller exponentially as the bifurcations continue. We try to measure its rate as follows.

Here again, we take a classical trajectory as a reference path in a reaction tube that passes across the transition region between two basins a and b with the flow direction $b \to a$. Set the time origin $t = 0$ at just the moment of transition. At a given time t, we take a sphere of a radius r_t in 30-dimensional phase space, the center of which proceeds along the reference trajectory. Pick random points in this sphere, and let them run backward in time. Some of them will go back to the basin b if the sphere still lies inside the same tube, and the others will move to some other basins if this sphere is already out of the tube. Should the latter happen, a similar procedure is to be redone with a smaller radius r_t. Repeating

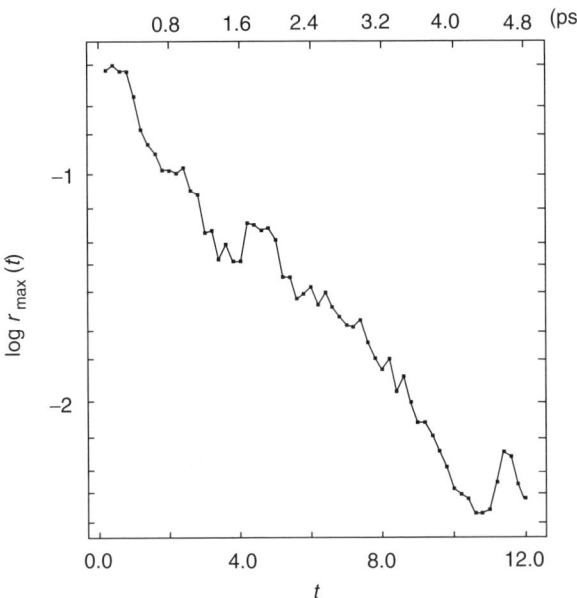

Figure 8. Globally exponential decay to inter-basin mixing at the energy of -11.208ϵ. (Reproduced from Ref. 10 with permission.)

this procedure, one can find the maximum sphere from which all the points come back to the basin b. The largest possible radius r_t would give a rough estimate of the diameter of the reaction tube $r_{\max}(t)$ at this particular time t. Since the tube is expected to become finer as time passes, $r_{\max}(t)$ should accordingly become smaller. In Fig. 8, we show an example of $\log r_{\max}(t)$ versus t, which indeed has evidenced the exponential decay of a reaction tube, although it is not smooth [10]. Since the present calculation is time-consuming, we do not have many examples. However, the overall feature of the $r_{\max}(t)$ indicates an exponential decay form as

$$r_{\max}(t) \approx A \exp(-\beta t) \qquad (28)$$

The constant β can be regarded as a Liapunov exponent, the inverse of which gives a time scale for a system to reach the state of inter-basin mixing. In Fig. 8, we have $\beta \approx 0.17$ or $\beta^{-1} \approx 6.0$. And, at time $t = 6$, it is observed in Fig. 9 that $r_{\max}(6) = 0.05$. This means $r_{\max}(6) = 0.37 \times r_{\max}(0)$, but since the dimensionality of the cross section of a phase space is 29, the area of the cross section of this reaction tube is already as small as roughly $r_{\max}(6)^{29} \sim 3 \times 10^{-13} \times r_{\max}(0)^{29}$.

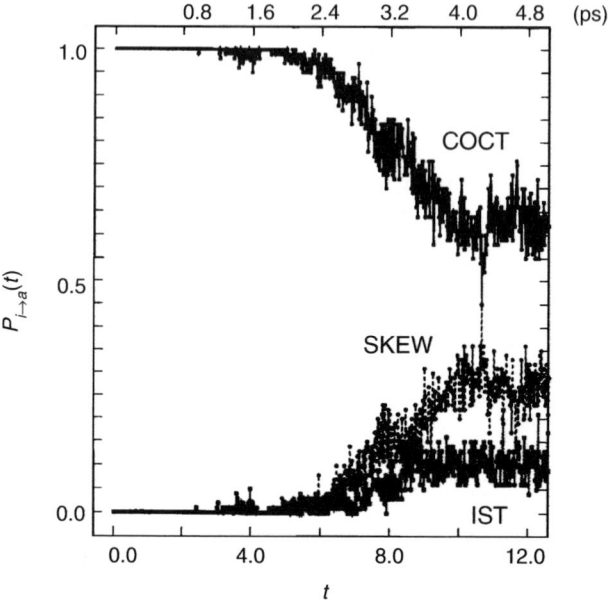

Figure 9. Time-dependent behavior of a reaction tube in terms of $P_{i \to a}$, where a = PBP, and COCT, IST, and SKEW are at the energy of -11.208ϵ. $P_{i \to a}$ has been measured with use of a sphere of the radius $r_{\text{fixed}} = 0.05$. Since the reference trajectory runs from COCT to PBP, $P_{\text{COCT} \to \text{PBP}} = 1.0$ in the initial stage. (Reproduced from Ref. 10 with permission.)

C. Time Scale to Realize Inter-Basin Mixing

We next show how a small vicinity in phase space proceeding along the reference trajectory is contaminated by trajectories that used to lie outside the reference tube. The computational procedure is as follows. A reference trajectory starts from a transition point of $b \to a$ at $t = 0$. A sphere with a fixed radius is prepared in phase space, with the center proceeding along the reference trajectory. Here we set $r_{\text{fixed}} = 0.05$ in view of the time constant discussed above [recall $r_{\max}(6) = 0.05$]. At a given time t, we pick many random points from this sphere, and their corresponding trajectories are traced backward in time to identify which basins they came from just before entering the present basin a. Count the normalized ratio of the trajectories thus characterized in terms of the preceding basins, which are denoted by $P_{i \to a}$ ($i = b, c, d$). Performing this calculation at different t, one can see the time evolution of the composition of the close vicinity around the reference path. In Fig. 9, we have plotted an example of $P_{i \to a}$ ($i = b, c, d$) for, b = COCT, c = IST, and d = SKEW as a function of time. For this choice of $r_{\text{fixed}} = 0.05$, the reference sphere keeps occupied by a single component until

$t \approx 6$; that is, $P_{\text{COCT} \to \text{PBP}} = 1.0$, $P_{\text{IST} \to \text{PBP}} = 0.0$, and $P_{\text{SKEW} \to \text{PBP}} = 0.0$. This is because the diameter of the tube happens to be larger than $r_{\text{fixed}} = 0.05$ by $t \approx 6$ as will be shown above. After $t \approx 6$, the reaction tube becomes finer than r_{fixed} and the sphere around the reference path begins to be contaminated by trajectories coming from the other basins. Moreover, the composition in this sphere becomes uniform because many such fine tubes are expected to entangle with each other. In fact, the ratio $P_{b \to a} : P_{c \to a} : P_{d \to a}$ in the case of Fig. 9 actually has approached the ratios $n_{a \to b} : n_{a \to c} : n_{a \to d}$, which is the state of inter-basin mixing [10]. This is not an exceptional situation to a special trajectory, but the converged ratio $P_{b \to a} : P_{c \to a} : P_{d \to a}$ in fact does not depend much on the selection of the reference trajectories as far as the converged cases are concerned, which also ensures the inter-basin mixing.

The above view in phase space can be inverted in time, making use of the time reversal symmetry in classical mechanics. For instance, the bifurcated tubes are rebundled (merge) around the reference trajectory, which is now proceeding to get out of the basin PBP. Accordingly, the composition in the small sphere around the reference trajectory should be $m_{\text{PBP} \to \text{COCT}} = 1.0$, $m_{\text{PBP} \to \text{IST}} = 0.0$, and $m_{\text{PBP} \to \text{SKEW}} = 0.0$ in the vicinity of the transition area from PBP to COCT. Obviously therefore the equality $m_{a \to b} : m_{a \to c} : m_{a \to d} = n_{a \to b} : n_{a \to c} : n_{a \to d}$ does not hold at all in this region. Deep in the basin by $t \approx -6$ before the trajectory gets out the basin PBP, $m_{a \to b} : m_{a \to c} : m_{a \to d} = n_{a \to b} : n_{a \to c} : n_{a \to d}$ holds, where the reaction tubes are fine enough. Similarly, if the rate of bifurcation of the tubes in a given basin is so slow (due to weaker chaos, for instance) that the diameters of the tubes do not become short enough before the trajectories leave the basin, $m_{a \to b} : m_{a \to c} : m_{a \to d}$ sampled across this tube will deviate from the statistical ratio $n_{a \to b} : n_{a \to c} : n_{a \to d}$. Thus, even if a system is globally mixing, inter-basin mixing will not necessarily follow. What is essential here is competition between the time scale of the bifurcation (or the decrease of the diameter) of the tubes and the average time for trajectories to pass through a basin. For the same reason, it can happen that some basins are inter-basin mixing while others may not be.

D. Fractal Distribution of Turning Points

It turns out that the bifurcation of reaction tubes well characterizes a geometrical structure behind the inter-basin mixing. Let us go back to Fig. 7, which gives a schematic view of a bifurcated reaction tube. By letting time flow backwards, we regard this dynamics as one in which many tubes merge successively to form a large trunk that is flowing out of this basin. In this figure we notice that there is another way to quantify the branching ratio of a tube. Suppose a classical path that is running near these bifurcated branches. If it is not included in one of the branches, this trajectory should turn back deep into the basin. If the branching (merging) does not occur frequently and thereby if the trunk is thick, the number of such returning trajectories around this tree (reaction tube) should be fewer. Such crotches of the trees may be partly detected as the so-called turning points of trajectories [30]. Furthermore, one can easily imagine that the spatial distribution of those crotches should have a fractal distribution.

With this observation, we calculated a correlation function [30]

$$C(r) = \frac{2}{n(n-1)} \sum_{i \neq j}^{n} \Theta(r - |\mathbf{X}_i - \mathbf{X}_j|) \tag{29}$$

where $\Theta(x)$ is the Heaviside step function, and $\mathbf{X}_i$ are the coordinate of the turning points in configuration space. Figure 10 shows a log–log plot of $C(r)$ for two different cases having $\rho_0 = 3.0$ and 6.0 in Eq. (25), the former has only two isomers in contrast to four isomers in the Morse potential of $\rho_0 = 6.0$. Therefore $\rho_0 = 6.0$ should give more complicated dynamics, and this is really the case in many aspects. As seen in the figure, the spatial distribution of the turning points certainly shows a fractal structure. It is understandable that the fractal dimension for the dynamics of $\rho_0 = 6.0$ is larger than that of $\rho_0 = 3.0$. Although it is

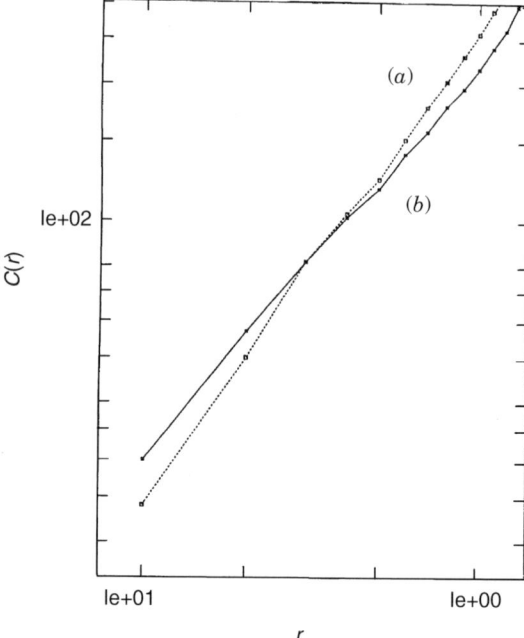

Figure 10. A log–log plot of the correlation function $C(r)$ versus r. (a) Curve representing the dynamics for $\rho_0 = 6.0$, and its slope is 1.04. (b) Curve representing the dynamics of $\rho_0 = 3.0$, and its slope is 0.89. (Reproduced from Ref. 30 with permission.)

interesting to study the relationship between this fractal dimension and the exponent β in the expression of Eq. (28), nothing is yet known about it.

V. MICROCANONICAL TEMPERATURE AND AN ARRHENIUS RELATION WITH THE LIFETIME OF ISOMERS

A. Another Law for the Average Lifetimes of Isomers

In Section III, we have observed the uniformity of the lifetimes of the isomers in the liquid-like phase. This fact suggests that the present dynamics does not care about the particular transition state (the lowest first-rank saddle), and we will confirm it in another manner later in this section. The transition state theory and its extensions including the variational version and quantum mechanical description are of fundamental importance. Nevertheless, the dynamics we are interested in this review lies outside the concept the transition state theory that had dominated the theory of chemical reaction dynamics (or reaction rate) of the twentieth century. Note that the present dynamics is beyond the transition state concept not because it is not statistical. It is highly statistical as our geometrical study in the preceding section has illustrated. Besides, we will find another novel statistical law of chemical reaction that these uniform averaged lifetimes satisfy. Briefly it is summarized as an Arrhenius-like relation between the inverse of the average lifetime and a "temperature" that we call the microcanonical temperature. Recall that we are discussing the dynamics of an isolated cluster, and the microcanonical temperature is not the so-called canonical temperature deduced from a canonical ensemble. This section is dedicated to a presentation of the definition of microcanonical temperature and the Arrhenius relation associated with it. We hope in the following discussions that the new Arrhenius like relation should not be confused with the well-known rate expression of the transition state theory in canonical ensemble.

When studying cluster dynamics theoretically or experimentally, one is tempted to rethink about "What is temperature?" This is simply because clusters are anticipated to bridge a gap between small molecules and bulk states that are composed of atoms of Avogadro's number. Here again, we should note that a cluster we consider is isolated and of a finite size. In the literature of simulation of molecular dynamics it is a usual practice that the time average and ensemble average of the kinetic energy are adopted as a temperature. However, this temperature is not necessarily a good quantity to characterize intramolecular dynamical processes. A clear counter example can be seen in classic articles of Jellinek, Beck, and Berry [31,34], in which they have shown remarkable examples of time dependence of the short-time average of the kinetic energy arising from isomerization dynamics of Ar clusters (see, for instance, Fig. 11). This isolated cluster varies its kinetic energy in an

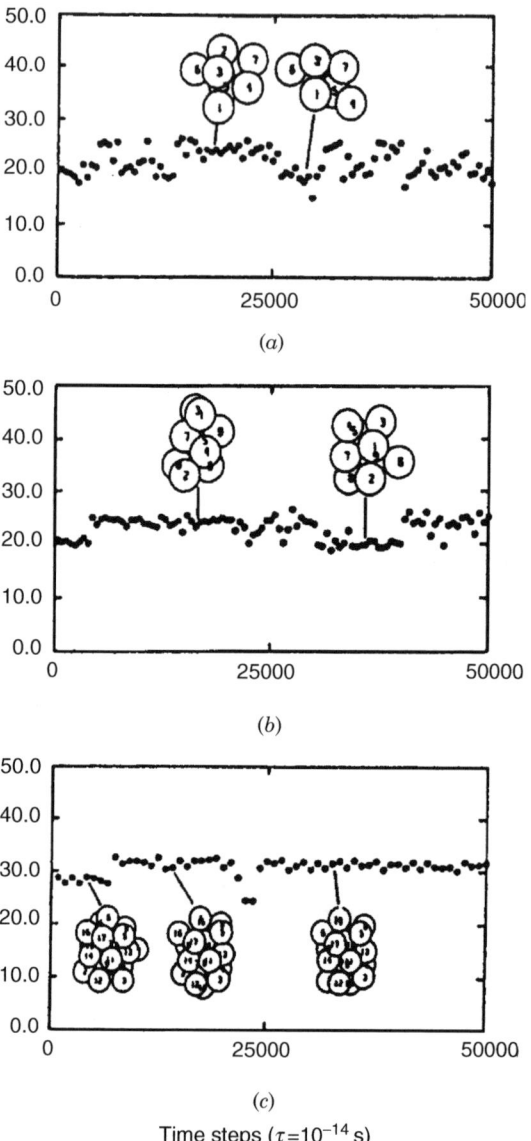

Figure 11. Time series of the short-time average of the total kinetic energy. (a) Ar_7, (b) Ar_9, and (c) Ar_{17}, due to Beck at al. (Reproduced from Ref. 34 with permission.)

intermittent and stair-like manner depending on the possible shapes it can take. It is obvious that this time series reflects very important information about the structural change, which should not be smoothed away by averaging. We hence formulate a new temperature that can be assigned to an individual potential basin of an isomer.

To do so, we first investigate the phase-space volume (classical density of states) and note that even a phase-space distribution on a constant energy plane can have a sharply biased (usually single-peaked) distribution if projected onto the potential energy coordinate (or equivalently onto the total kinetic energy coordinate). This is because the density of states in configuration space generally tends to increase as a potential energy is raised (with the total energy fixed), while that in momentum space becomes smaller, and hence these two conflicting factors form a single peak in their convolution to reproduce the total density of states. A kinetic energy maximizing this projected distribution is defined as the microcanonical temperature, since the most probable statistical phenomena should be dominated by those populations that are in vicinity of the microcanonical temperature. One can further determine the microcanonical temperature in the individual basins, which we call the local microcanonical temperature. The main aim of this section is to present a numerical fact such that each local microcanonical temperature bears an Arrhenius-like exponential relation with the lifetime in its potential basin. We then try to set a theoretical foundation of the Arrhenius-like relation, through which the characteristics and generality of the microcanonical temperature are considered.

B. Evaluation of Classical Density of States

Statistical mechanics begins with the statistical entropy in terms of the total number of states up to a given total energy E (see, for instance, Pearson, Halicioglu, and Tiller [35]),

$$v_{st}(E) = \int d\vec{Q} \, d\vec{P} \, \Theta(E - H(\vec{Q}, \vec{P})) \tag{30}$$

where Θ is the step function and $\vec{P}$ is a momentum vector conjugate to $\vec{Q}$ in an N-dimensional space. The Boltzmann entropy is given as usual as

$$S = k_B \ln v_{st}(E) \tag{31}$$

and the (canonical) temperature is

$$k_B T = \left(\frac{\partial S}{\partial E}\right)^{-1} = \frac{2}{N} \langle K \rangle \tag{32}$$

The temperature is thus the ensemble average of the kinetic energy $\langle K \rangle$ over the distribution function in Eq. (30). For the the later convenience, we redefine the thermodynamical temperature T simply as

$$T = \langle K \rangle \qquad (33)$$

This temperature is absolutely suitable to measure a direction and extent of heat transfer between two contacting subsystems [32]. (See also Appendix C.) However, as stated above, averaging of the kinetic energy can lose a very important piece of information [31] in an attempt to characterize multiple-basin dynamics on a single molecular basis such as our isomerization dynamics.

A classical density of states, which is slightly different from Eq. (30), is

$$V_{st}(E) = \int d\vec{Q}\, d\vec{P}\, \delta(E - H(\vec{Q}, \vec{P})) \qquad (34)$$

The phase-space volume at a given energy E considered in Eq. (34) has been based on this density of states. $V_{st}(E)$ is often rewritten in the Thomas–Fermi form [36]:

$$V_{st}(E) = C_N \int d\vec{Q}(E - V(\vec{Q}))^{N-2/2} \qquad (35)$$

with

$$C_N = \frac{(2\pi)^{\frac{N}{2}}}{\Gamma\left(\frac{N}{2}\right)} \qquad (36)$$

$V_{st}(E)$ in Eq. (35) may be evaluated directly with a random sampling technique, which is outlined below. The integration of Eq. (35) is first transformed to a sum form,

$$V_{st}(E) \sim \sum_j (E - V(\vec{Q}_j))^{\frac{N-2}{2}} \qquad (37)$$

where $\vec{Q}_j$ is the jth randomly sampled vector. A single vector $\vec{Q}_j$ specifies a molecular geometry with use of the quenching technique. The total number of the sampled vectors should be large enough to attain a sufficient accuracy. A very practical method of random generation of molecular geometries has been described elsewhere [19] (see also Ref. 37). Hence it is easy to assign a structure $\vec{Q}_j$ to one of the four basins. The entire set of $\vec{Q}_j$ is thus regrouped into four classes and each is characterized in terms of $\vec{Q}_j^a$, where the superfix a denotes the basins, $a = 1, 2, 3, 4$, and j is the sequential number in each basin.

C. Microcanonical Temperature

1. Definition

Since we have confirmed that the phase space volume can be practically calculated as above (see Appendix B for an illustrated example of such a numerical example), we now make use of it to define a microcanonical temperature. A useful alternative definition of $V_{st}(E)$ is a convolution type

$$V_{st}(E) = \int_0^E d\varepsilon\, \Omega_Q(\varepsilon) \Omega_P(E - \varepsilon) \tag{38}$$

where $\Omega_Q(\varepsilon)$ is the density of states in configuration space

$$\Omega_Q(\varepsilon) = \int d\vec{Q}\, \delta\left(\varepsilon - V(\vec{Q})\right) \tag{39}$$

at a given potential energy ε, while $\Omega_P(E - \varepsilon)$ is that in momentum space

$$\Omega_P(E - \varepsilon) = \int d\vec{P}\, \delta\left(E - \varepsilon - \tfrac{1}{2} \sum_i^N p_i^2\right)$$

$$= \int \delta\left(E - \varepsilon - \tfrac{1}{2} P^2\right) P^{N-1} dP\, d(\text{angular part}) = C_N (E - \varepsilon)^{\frac{N-2}{2}} \tag{40}$$

with C_N being the value of Eq. (36). Equivalence among the above three forms of the density of states, Eqs. (34), (35), and (38), is readily proved.

The volume of each basin in configuration space at a given potential energy ε should be proportional to the number of the sampling points thus assigned. We therefore write

$$\Omega_Q^{(a)}(\varepsilon) \Delta\varepsilon = \text{constant} \times (\text{number of } \vec{Q}_j^a \text{ in the bin of } [\varepsilon, \varepsilon + \Delta\varepsilon]) \tag{41}$$

where the above constant is common to all the basins and hence does not have to be considered here. The volume of the phase space is also partitioned and assigned to the individual basins.

A component of $V_{st}(E)$ on the potential energy coordinate ε is given as

$$\Xi(\varepsilon) = \Omega_Q(\varepsilon) \Omega_P(E - \varepsilon) \tag{42}$$

as seen in Eq. (38). In the case where a potential function can dominate dynamics and thereby bias $\Xi(\varepsilon)$, it is quite crucial to take account of this effect in any statistical theory. $\Xi(\varepsilon)$ thus defined is a convolution of two opposing quantities with respect to the potential energy ε: Usually $\Omega_Q(\varepsilon)$ ($\Omega_P(E - \varepsilon)$) increases

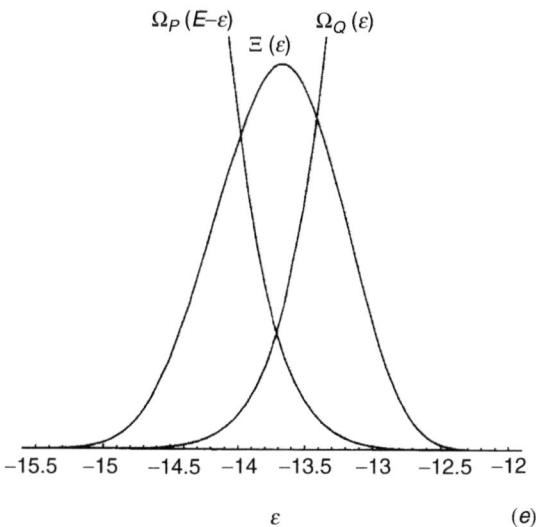

Figure 12. The density of states (DS) on the potential energy (ε) space for an M_7 cluster. (*a*) DS in configuration space $\Omega_Q(\varepsilon)$ calculated for the Lennard-Jones potential, increasing with ε. (*b*) DS in momentum space $\Omega_P(E - \varepsilon)$, decreasing with ε. (*c*) A product $\Xi(\varepsilon) = \Omega_Q(\varepsilon)\Omega_P(E - \varepsilon)$ giving the total density of state at the total energy E after the convoluting integral over ε. $\Xi(\varepsilon)$ has a single sharp peak. (Reproduced from Ref. 11 with permission.)

(decreases) as ε (the kinetic energy $E - \varepsilon$) increases (decreases). Therefore $\Xi(\varepsilon)$ is usually expected to be a singly peaked function, the actual shape of which depends on a potential field. See Fig. 12 for the M_7 case with the Lennard-Jones potential. (See also the kinetic energy distributions in Fig. 13 of Ref. 30, which have been sampled from classical trajectories for an M_7 cluster having the Morse potential.) The peak position is found at a potential energy at which the distributions in configuration and momentum spaces are compromised to have $\Xi(\varepsilon)$ maximized. This situation is mathematically (but not physically) very analogous to a case in which the canonical temperature is defined in two physically contacting subsystems [32]. Here in our intramolecular dynamics, a classical path is evolved in time continuously exchanging a part of its kinetic and potential energies. Hence it is natural to assume that most probable dynamical events should be overwhelmingly dominated by the major population that lies in vicinity of the peak. Although extremely simple, this idea is nontrivial in that we abandon the simple picture of equal *a priori* distribution in phase space, which assumes that any point in this space should contribute with an equal chance.

Thus we may define an entropy to quantify such a biased population as

$$S_{\text{ps}} = \ln \Xi(\varepsilon) = \ln \Omega_Q(\varepsilon)\Omega_P(E - \varepsilon) \qquad (43)$$

TEMPERATURE, GEOMETRY, AND VARIATIONAL STRUCTURE 59

the maximum of which represents the most probable state to be observed. It is straightforward to locate the maximum point of $\Xi(\varepsilon)$ by setting

$$\frac{\partial}{\partial \varepsilon} \ln \Xi(\varepsilon) = 0 \qquad (44)$$

or

$$\frac{\partial}{\partial \varepsilon} \ln \Omega_Q(\varepsilon) = \frac{\partial}{\partial (E - \varepsilon)} \ln \Omega_P(E - \varepsilon) \equiv \frac{\kappa_N}{T_m} \quad (\text{at } \varepsilon = \varepsilon^*) \qquad (45)$$

where ε^* is the potential energy giving the maximum. This expression defines the microcanonical temperature. Other statistical quantities such as an analog to the heat capacity can be defined as well to characterize $\Xi(\varepsilon)$. With the help of Eq. (40), we have

$$\frac{\partial}{\partial (E - \varepsilon)} \ln \Omega_P(E - \varepsilon) = \frac{N - 2}{2} \frac{1}{E - \varepsilon^*} \quad (\text{at } \varepsilon = \varepsilon^*) \qquad (46)$$

and

$$T_m = E - \varepsilon^* \qquad (47)$$

with

$$\kappa_N = \frac{N - 2}{2} \qquad (48)$$

T_m is thus the total kinetic energy at which the maximum density is realized. (Note that neither T_m nor T in Eq. (33) is divided by the number of degrees of freedom N.) Implicitly, therefore, the positions (actually a manifold) in configuration space satisfying $V(\vec{Q}) = \varepsilon^*$ are expected to have a large contribution to the density of states. At this temperature, mutual conversion between the kinetic and potential energies is supposed to be equilibrated. T_m generally tends to be closer to T of Eq. (33) in a larger system. Furthermore, if $\Xi(\varepsilon)$ happens to be of a single-peaked and symmetric form (see Fig. 12), T_m should exactly coincide with the average kinetic energy in the statistical sense:

$$T = \frac{\int d\varepsilon \Xi(\varepsilon)(E - \varepsilon)}{\int d\varepsilon \Xi(\varepsilon)} = \frac{\int d\varepsilon (E - \varepsilon) \Omega_Q(\varepsilon) \Omega_P(E - \varepsilon)}{V_{\text{st}}(E)} \qquad (49)$$

One can readily estimate the first-order densities of states both in configuration and momentum spaces individually around the ε^*. In particular, with the help of Eq. (45), we have

$$\Omega_P(E - \varepsilon) = C_N(E - \varepsilon)^{\frac{N-2}{2}} \cong \Omega_P(E - \varepsilon^*) \exp\left[-\frac{\kappa_N}{T_m}(\varepsilon-\varepsilon^*)\right] \quad (50)$$

which is crucial and will be applied later. This first-order approximation is valid only when $\Xi(\varepsilon)$ is sharply and singly peaked. This is really the case in our numerical study of the M_7 molecule (see also Refs. 19 and 30).

The second or higher-order effect can be readily taken into account by a straightforward extension. For instance,

$$\Omega_P(E - \varepsilon) \cong \Omega_P(E - \varepsilon^*) \exp\left[-\frac{\kappa_N}{T_m}(\varepsilon-\varepsilon^*) + \text{const.} \times (\varepsilon-\varepsilon^*)^2\right] \quad (51)$$

This formula can be vital if a deviation from the simple exponential form is large. The higher-order terms should represent an anisotropy of the distribution function $\Xi(\varepsilon)$ and large fluctuation in dynamics.

2. Local Microcanonical Temperatures

The distribution function $\Xi(\varepsilon)$ can be partitioned and assigned to the individual basins, since $\Omega_Q(\varepsilon) = \sum_a \Omega_Q^{(a)}(\varepsilon)$. We again assume that the distributions of configuration and momentum spaces are equilibrated locally in the basins. Thus, the local microcanonical temperature that maximizes

$$\Omega_Q^{(a)}(\varepsilon)\, \Omega_P(E - \varepsilon) \quad (52)$$

can be determined as well in the individual basins, and ε^* in Eq. (47) should be replaced with ε_a^* as

$$T_m^{(a)} = E - \varepsilon_a^* \quad (53)$$

These quantities are anticipated to be suitable for describing dynamics separately in the individual basins. The stair-like variation of the kinetic energy in Ar_{13}, and so on (Fig. 11), has now been idealized to be a statistical quantity as the local microcanonical temperatures.

3. Numerical Observation of an Arrhenius-like Relation

It is quite natural to try to plot $\log\langle t\rangle_a$ versus $1/T_m^{(a)}$. Figure 13 displays a very clear numerical fact: An Arrhenius-like exponential relation is observed between the local microcanonical temperature and the inverse of the average lifetime in each basin. Here we have used the numerical data studied in Refs. 8 and 19. The local microcanonical temperatures in Fig. 13 (and Fig. 14) cover the range of the total energy $[-13.5\epsilon, -11.0\epsilon]$. -13.5ϵ is close to the lowest end of the liquid-like

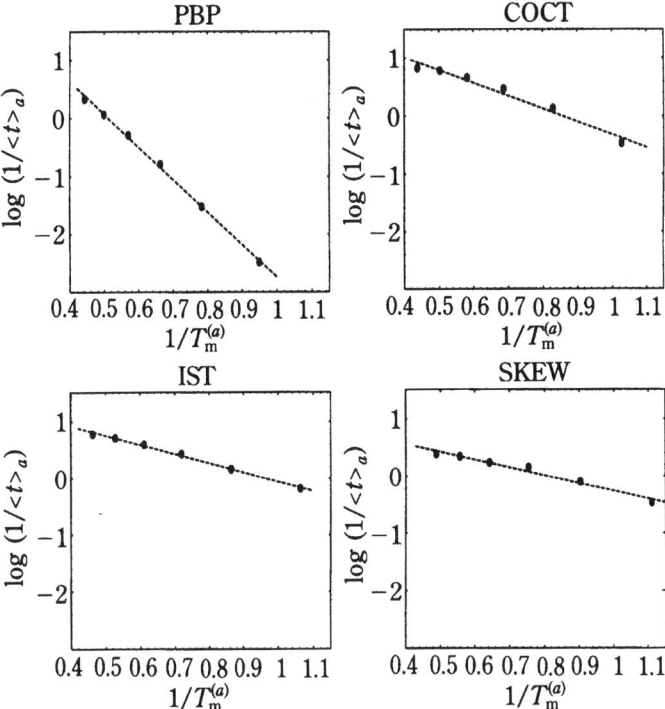

Figure 13. Plots of $\ln 1/\langle t \rangle_a$ versus $1/T_m^{(a)}$: An Arrhenius relation between the inverse of the lifetimes ($\langle t \rangle_a$) of the four isomers and their associated local microcanonical temperatures ($T_m^{(a)}$). (Reproduced from Ref. 11 with permission.)

phase, and the plot is expected to deviate from the exponential relation below this energy. This energy range is never narrow, since a substitution of 1 eV to 1ϵ, as an illustrative example, gives rise to about 1.16×10^4 Kelvin. Furthermore, although our numerical calculation has been truncated at -11.0ϵ, the exponential relations should hold in the higher energy as well up to the dissociation limit $\sim -9.0\epsilon$.

These numerical facts can be approximately expressed in an Arrhenius form as

$$\langle t \rangle_a^{-1} \simeq A_a \exp\left(-\frac{\Delta_a}{T_m^{(a)}}\right) \quad (54)$$

Here again this relation does not depend on the next visited isomers b. We maintain a possibility, however, that the preexponential factor A_a can be a polynomial function of $T_m^{(a)}$ in a low order.

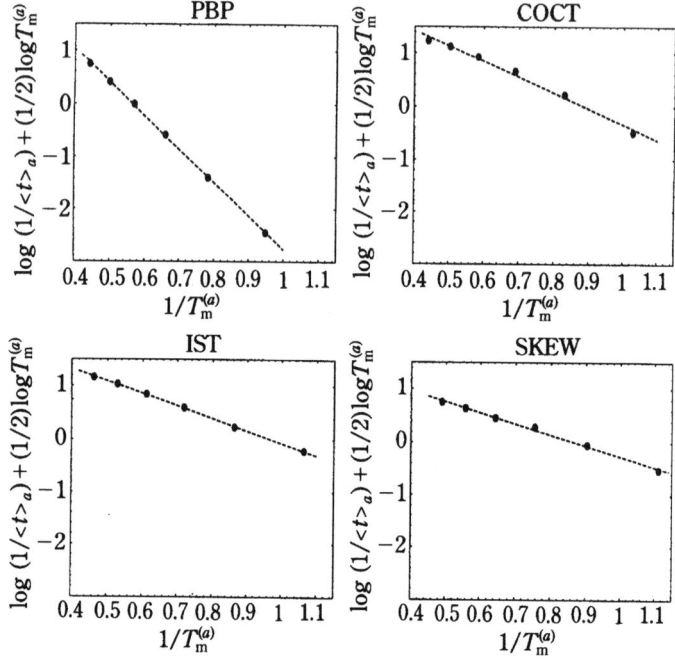

Figure 14. $\ln 1/\langle t \rangle_a + \frac{1}{2}\ln T_m^{(a)}$ versus $1/T_m^{(a)}$ is plotted to confirm the relation $1/\langle t \rangle_a \approx A_a[T_m^{(a)}]^{-1/2}\exp[-\Delta_a/T_m^{(a)}]$. (Reproduced from Ref. 11 with permission.)

D. An Exponential Relation Between the Microcanonical Temperature and Average Lifetimes

The rest of this section is devoted to the study of a possible origin of the Arrhenius relation in microcanonical dynamics. Since the content of this section is rather technical, readers may want to skip to Eq. (60) for a conclusion of the formulation.

1. Multiexponential Form

The average lifetime is defined by Eqs. (13), (15), (20) and (22). According to a general prescription in statistical reaction theory or the phase-space theory due to Light [38], the lifetime of an isomer in a basin a should be given by

$$\frac{1}{\langle t \rangle_a} = \frac{n_a}{t^{(a)}} = \frac{F^{(a)}}{V^{(a)}}$$

$$= \frac{\int_0^E d\varepsilon \, \Omega_Q^{\partial a}(\varepsilon) \int_0^{p_a^x} dp_a \Omega_P(E - \varepsilon - \frac{1}{2}p_a^2, \frac{1}{2}p_a^2)p_a}{\int_0^E d\varepsilon \, \Omega_Q^{(a)}(\varepsilon) \, \Omega_P(E - \varepsilon)} \quad (55)$$

where the denominator represents the volume of phase space for basin a, while the numerator is the net flux flowing out of this basin. The notations in this expression are:

$F^{(a)}$: Flux out from basin a.

$\Omega_Q^{\partial a}(\varepsilon)$: Density of states in configuration space at the boundary (∂a) of basin a having a potential energy ε. It should be noted that ∂a does not have to be a transition state, and it constitutes an $(N-1)$-dimensional manifold enclosing the basin a.

p_a: A momentum transversal to the boundary ∂a.

$\Omega_P(E - \varepsilon - \frac{1}{2}p_a^2, \frac{1}{2}p_a^2)$: Density of states in momentum space that is a convolution of the components orthogonal and parallel to the direction of p_a. $E - \varepsilon - \frac{1}{2}p_a^2$ is the kinetic energy allowed in the orthogonal directions.

p_a^x: Maximum value possible for p_a, that is, $\frac{1}{2}(p_a^x)^2 = E - \varepsilon$.

After a lengthy derivation [11], it turns out that $1/\langle t \rangle_a$ can be represented as a product of two terms, one that depends weakly on the temperature in the low power of it and the other which is sensitive to the temperature as in the exponential form. The temperature dependence can be roughly estimated as

$$\frac{1}{\langle t \rangle_a} \sim \sum_i (E - V(\vec{Q}_i^a))[T_m^{(a)}]^{-1/2} (Det.V''(\vec{Q}_i^a))^{-1/2} \exp\left[-\frac{\kappa_N}{T_m^{(a)}}(V(\vec{Q}_i^a) - \varepsilon_a^0)\right]$$
(56)

In this expression, ε_a^0 is the energy at the bottom of the basin a. $\vec{Q}_i^a$ are the stationary points satisfying $V'(\vec{Q}_i^a) = 0$, which have arisen from the steepest descent evaluation of the numerator of Eq. (55). The Hessian $V''(\vec{Q}_i^a)$ is taken only within the space of the boundary ∂a. Furthermore, only those stationary points are supposed to appear in Eq. (56) whose $N-1$ eigenvalues are all positive. Thus, the stationary points taken into account of this estimate are actually the first-rank saddles, and it is assumed that the boundary ∂a can be topologically built up by connecting such first-rank saddles by ridge lines [actually $(N-1)$-dimensional manifolds]. (Note that these particular roles of the first-rank saddle is just a matter of the steepest decent estimate. Indeed trajectories actually pass many different places, too. See the next subsubsection and Fig. 15.)

The physical meanings of the individual terms of Eq. (56) are as follows: It is obvious that $V(\vec{Q}_i^a) - \varepsilon_a^0$ is a relative barrier energy at $\vec{Q}_i^a$, which is measured from the bottom of basin a. The geometrical meaning of $Det.V''(\vec{Q}_i^a)$ is as follows: The larger value of $V''(\vec{Q}_i^a)$ indicates the narrower diameter (gate-size) of a valley at $\vec{Q}_i^a$ that is open to a product channel (transversal to the reaction coordinates). This is also a product of the frequencies of the normal modes at the critical points, and therefore $(Det.V''(\vec{Q}_i^a))^{-1/2}$ is proportional to the density of states based on the local normal modes.

The minus factor in the power $[T_m^{(a)}]^{-1/2}$ arises because the phase-space volume of basin a increases as the temperature becomes higher. On the other hand, the factor $E - V(\vec{Q}_i^a)$ is the total kinetic energy at a critical point $\vec{Q}_i^a$, which may be regarded as an approximate temperature at the surface of the boundary. In the case where $E - V(\vec{Q}_i^a)$ happens to be close to the value of $T_m^{(a)}$, which should be realized in a high E, it is expected that

$$(E - V(\vec{Q}_i^a))[T_m^{(a)}]^{-1/2} \sim [T_m^{(a)}]^{1/2} \tag{57}$$

2. Single Exponential Form

For a single-channel problem, only one term in the sum in Eq. (56) is to be considered. On the other hand, it is not unusual for multichannel chemical reactions to have a multiexponential form of reaction probability. Reduction to a single exponential form can depend on individual cases under study. For our cluster system, in which various isotropic properties dominate due to the identical particle composition (see Ref. 39), it is quite likely that there are very many critical points that are topologically equivalent or energetically similar to each other. In these cases, averaging should work. In the two familiar means over variables $\{X_i | i = 1, 2, \ldots, M\}$

$$\frac{1}{M}\sum_i^M X_i \geq \left(\prod_{i=1}^M X_i\right)^{1/M} \tag{58}$$

we may assume that they must be close to each other, since many critical points are topologically and energetically similar to each other. Regarding the sum of Eq. (56) as an arithmetic mean and furthermore approximating it with the geometric mean over such dominant terms, we write

$$\frac{1}{\langle t \rangle_a} \sim M_a [T_m^{(a)}]^{-1/2} \left(\prod_{i=1}^{M_a}(E - V(\vec{Q}_i^a))(Det.V''(\vec{Q}_i^a))^{-1/2}\right)^{1/M_a} \\ \times \exp\left[-\frac{\kappa_N}{T_m^{(a)}}\frac{1}{M_a}\sum_i^{M_a}\{V(\vec{Q}_i^a) - \varepsilon_a^0\}\right] \tag{59}$$

where M_a is the number of contributing critical points, which is also regarded as the number of leading reaction channels. This expression is further simplified as

$$\frac{1}{\langle t \rangle_a} \sim \text{Av.}[T_m^{(a)}]^\gamma \exp\left[-\frac{\kappa_N}{T_m^{(a)}}(\overline{V}_a - \varepsilon_a^0)\right] \quad (60)$$

where $\overline{V}_a$ is designated as the average energy of the potentials at the critical points and Av. is defined in comparison of Eq. (60) with Eq. (59). The power γ in the polynomial factor should be

$$-\tfrac{1}{2} \leq \gamma \leq \tfrac{1}{2} \quad (61)$$

depending on the situation as described above. It is expected from Eq. (57) that $\gamma \sim \tfrac{1}{2}$ for a higher-energy region.

To find the value of γ in our M_7 system, we plot $\ln\langle t \rangle_a^{-1} + \tfrac{1}{2}\ln T_m^{(a)}$ versus $1/T_m^{(a)}$ in Fig. 14, confirming

$$\langle t \rangle_a^{-1} \simeq A_a[T_m^{(a)}]^{-1/2} \exp\left[-\frac{\Delta_a}{T_m^{(a)}}\right] \quad (62)$$

The linearity is significantly improved, thereby suggesting $\gamma \sim -\tfrac{1}{2}$ in the wide temperature range.

3. Case Study on M_7 in Terms of the Single Exponential Form

Table VI lists A_a and Δ_a, which are estimated from Fig. 14 [$\gamma = -1/2$, as in Eq. (62)]. We first notice that A_a of PBP is considerably larger than those of the other isomers. Equation (59) suggests one of the possible reasons: An average value of $(Det.V''(\vec{Q}_i^a))$ may be particularly small for PBP. The geometrical meaning of this quantity has been described above. According to it, one may say that the channel gates from the PBP basin through which to pass to the neighboring basins are wide open. Another reason can be that the PBP basin may be associated with more channels connected to the other basins [see M_a in Eq. (59)]. Although the high symmetry of the PBP structure seems to make this situation possible, we have no clear evidence to support the above possibilities at this moment.

TABLE VI
The Parameters in the Rate Expression with $\gamma = -\tfrac{1}{2}$

	PBP	COCT	IST	SKEW
$A_a(\text{ps}^{-1}\epsilon^{\frac{1}{2}})$	36.2	13.4	9.48	5.85
$\Delta_a(\epsilon)$	6.36	2.93	2.32	2.03

Let us next examine the energy term in Table VI. Since $\Delta_a = \kappa_N(\overline{V}_a - \varepsilon_a^0)$ and $\kappa_N = 6.5$ [see Eq. (48) and $N = 15$], the effective barrier height $\overline{V}_a - \varepsilon_a^0$ for the PBP basin, for instance, is estimated to be about 0.98ϵ. Comparing with the lowest transition-state energy bridging to the COCT basin—that is, 1.05ϵ (see Fig. 1)—one sees that the estimated average barrier height is not very bad. But it is not very good either, since the Lindemann index indicates that isomerization practically begins to take place at the total energy as high as 2.0ϵ measured from the bottom of the PBP basin [19]. This is a little puzzling. One possible way to comprehend this situation would be as follows. Our statistical treatment has assumed that all the internal degrees of freedom $(N = 15)$ are involved in the fully mixed statistical state. This assumption therefore brings about $\kappa_N = N - 2/2 = 6.5$ in Eq. (60). If, on the other hand, the dimensionality of the statistical space (chaotic space) is smaller, we expect a smaller κ_N and a larger effective barrier height $\overline{V}_a - \varepsilon_a^0$. For instance, the dimensionality 8 gives $\kappa_N = 3.0$ and $\overline{V}_a - \varepsilon_a^0 = 2.12\epsilon$, which is consistent with the melting energy detected by the steep rise of the Lindemann index. Therefore it can happen that the dimensionality of the fully chaotic space is actually limited just as the notion of effective oscillators in the RRKM theory. Since our system is composed of identical atoms, the chaotic subspace(s) can change from time to time.

We now show that the dimensionality of the chaotic sub-space is indeed limited and considerably smaller than that of the entire 15-dimensional space. To observe it, the so-called local K entropy due to Hinde, Berry, and Wales [40,41] provides a convenient quantity, which measures the local and instantaneous instability of a classical trajectory;

$$K = \frac{1}{\ln 2} \sum_{\omega_j^2 < 0} |\omega_j^2|^{1/2} \qquad (63)$$

where ω_j^2 are negative eigenvalues of the Hessian matrix of a potential. K is to be calculated along a trajectory, which thereby exhibits the instantaneous magnitude of chaos. Accordingly, the dimensionality of an instantaneous chaotic subspace can be given by the number of such negative eigenvalues. We therefore plot, in Fig. 15, a frequency distribution of the number of negative eigenvalues occurring at many sampling points along many trajectories. It is clearly observed that the trajectories tend to have four negative eigenvalues most frequently, and there is only a very small chance for them to have more than nine negative values. This implies that the maximum dimensionality of the chaotic subspace changing from time to time is not likely to exceed 9.

The order of the magnitude of Δ_a (the barrier height measured from the bottom of a basin, a = PBP, COCT, IST, SKEW) in Table VI is exactly the same as that of ε_a^0 (energy of the bottom of a basin a); $\varepsilon_a^0(\text{PBP}) = -16.505\epsilon$, $\varepsilon_a^0(\text{COCT}) = -15.935\epsilon$, $\varepsilon_a^0(\text{IST}) = -15.593\epsilon$, $\varepsilon_a^0(\text{SKEW}) = -15.533\epsilon$. On the

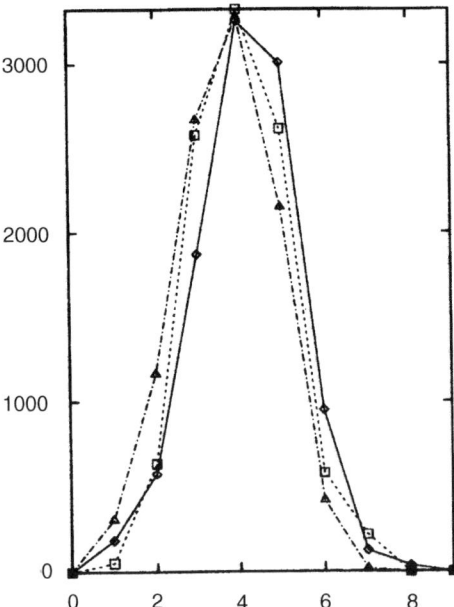

Figure 15. Frequency distribution of the number of negative eigenvalues in the Hessian matrix of the potential function that are counted along running trajectories. The solid, dotted, and dot–dashed lines are, respectively for the energies -11.0ϵ, -11.5ϵ, and -12.0ϵ. (Reproduced from Ref. 11 with permission.)

other hand, the average height $(\overline{V}_a)$ of the critical points is differently ordered. Assuming $\kappa_N = 30$ as above, one can estimate $\overline{V}_a$ as $\overline{V}_a(\text{PBP}) = -14.39\epsilon$, $\overline{V}_a(\text{COCT}) = -14.96\epsilon$, $\overline{V}_a(\text{IST}) = -14.82\epsilon$, and $\overline{V}_a(\text{SKEW}) = -14.86\epsilon$. ($\overline{V}_a$ should not be confused with the relative barrier height $\overline{V}_a - \epsilon_a^0$.) Just for a reference, the energy of the lowest saddle ever found is known to be -15.45ϵ, which bridges between PBP and COCT, while that of the highest saddle is -14.596ϵ locating between PBP and SKEW. Qualitatively we may say that the basin of PBP is surrounded by the critical points which are higher than those of COCT, IST, and SKEW. So, trajectories should tend to wander more easily among the high three basins as far as the barrier height alone is concerned. Remember, however, that the number of the channels M_a is another crucial factor to determine the frequency of isomerization n_a. Further extensive studies are necessary to draw more clear conclusions.

Appendix B: On Ergodicity and Nonergodicity of the Liquid-like Dynamics

Here we make use of the density of states, as numerically obtained through Eqs. (37) and (41), to formulate the ergodicity in the present dynamics.

Ergodicity of isomerization dynamics has been evidenced numerically [19] in the sense that an accumulated residence time $(t^{(a)})$ for trajectories to stay in a basin a is proportional to the volume of phase space $(V^{(a)})$ assigned to the corresponding basin

$$t^{(a)} \propto V^{(a)} \qquad (64)$$

The volume of phase space belonging to each basin can be estimated directly with a random sampling technique.

The role of the relative frequency [26] can be discussed in the context of ergodicity. It has been clarified as shown in Eqs. (23) and (24). On the other hand, the total residence time $t^{(a)}$ does not appear in the rate expressions above. So, we try to connect these two quantities. Let $\rho_a(t)$ be a phase-space population in a basin a at time t. Due to the physical meaning of the relative frequency, they should satisfy a kinetic equation

$$\dot{\rho}_a = \sum_b \rho_b f_{b \to a} - \rho_a \sum_b f_{a \to b} \qquad (65)$$

for each basin. It is then straightforward to show that the stationary solution of these simultaneous differential equations are simply $\rho_a(stationary) = C t^{(a)}$, where C is a constant. Here we have used the microscopic reversibility $n_{a \to b} = n_{b \to a}$ and the relation Eq. (11). Thus the phase-space volume in the stationary state, which can be referred to as an invariant density in phase space [21], is proportional to the accumulated residence time. We thus have reproduced the ergodicity [19].

We then pay an attention to deviation from the ergodicity. For the sake of a short and simple presentation of the statistical behavior of the dynamics, we define the following Shannon entropy:

$$S_{st}(E) = -\sum_a V_{st}^a(E) \log V_{st}^a(E) \qquad (66)$$

where $V_{st}^a(E)$ are renormalized so as to satisfy

$$\sum_a V_{st}^a(E) = 1 \qquad (67)$$

Likewise we consider

$$S_{dy}(E) = -\sum_a t^a(E) \log t^a(E) \qquad (68)$$

where $t^a(E)$ are the normalized accumulated residence time discussed already in Eqs. (12) and (64). In Fig. 16, these entropies are plotted versus the energy E.

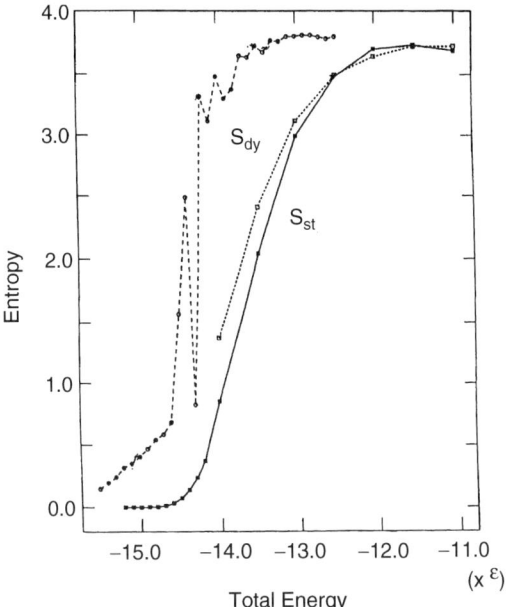

Figure 16. The entropy S_{st} relevant to the size of four potential basins in phase space. $S_{dy}(E)$ represents entropy in terms of the accumulated residence time $t^a(E)$. The dotted curve without a mark symbol represents a Lindeman index.

This figure also shows the Lindemann index as a reference. Since classical trajectories are practically confined in one of the potential basins below $E = -14.0\varepsilon$, $S_{dy}(E)$ is truncated at this energy.

As seen in the graph, $S_{st}(E)$ and $S_{st}(E)$ are similar to each other as is expected. However, a closer look at them shows a considerable difference between them below $E = -12.0\varepsilon$. From this figure and other data not shown here, we can conclude that the system is really ergodic above this energy. But it is not ergodic, even if the Lindemann index suggests that the system is liquid-like above $E = -13.5\varepsilon$. So, being in the liquid-like phase is generally different from being ergodic. This is not surprising if we recall that the Lindemann index merely measures the flexibility of molecular shape by actually detecting the deviation from an averaged structure. We may therefore claim that the liquid-like phase should be defined as the ergodic region that is the energy region higher than -12.0ε. Also, there are two subphases in the coexistence phase (which is in between -14.5ε and -12.0ε): one that is detected by the steep rise of the Lindemann index in the range of $[-14.5\varepsilon, -13.5\varepsilon]$, and the other one in $[-13.5\varepsilon, -12.0\varepsilon]$ that is not sensitive to the Lindemann index but yet nonergodic.

Appendix C: Canonical Temperature

The number of states in Eq. (30) is given as Defining the Boltzmann entropy as

$$v_{st}(E) = V_{st}(E) = \frac{2}{N} C_N \int d\vec{Q}(E - V(\vec{Q}))^{N/2} \quad (69)$$

by integrating $V_{st}(E)$ of Eq. (35). Using the entropy of Eq. (35), one may want to determine a temperature by

$$k_B T_c = \left(\frac{\partial S(E)}{\partial E}\right)^{-1} = \left(\frac{1}{v_{st}(E)} \frac{\partial v_{st}(E)}{\partial E}\right)^{-1} \quad (70)$$

This gives explicitly

$$k_B T_c = \frac{2}{N} \frac{\int d\vec{Q}(E - V(\vec{Q}))^{N/2}}{\int d\vec{Q}(E - V(\vec{Q}))^{(N-2)/2}} \quad (71)$$

which can also be evaluated numerically. This quantity is a canonical temperature by definition. Since we have not yet performed numerical study over this temperature, no clear information can be provided.

This temperature should indicate the way of response of a cluster to a thermal reservoir, when embedded in it. In this regard, it is interesting to recall that this canonical temperature can also be defined for the individual isomers by limiting the $\vec{Q}$-integration to the relevant potential basins. Thus different isomers may have a different canonical temperature. Suppose we have $T_c^{(b)} < T_c^{(a)}$ under a given energy E in the above expression, where $T_c^{(a)}$ is a canonical temperature for an isomer a. Suppose further that this cluster is placed in a reservoir of a temperature T_c satisfying $T_c^{(b)} < T_c < T_c^{(a)}$. Then, this cluster can absorb energy from the reservoir if it is in the form of isomer b, while it will emit in the form of a. If one can assume that the heat reservoir does not affect the isomerization dynamics mechanically, the cluster would show an interesting temperature dependence. This aspect may be worth studying.

VI. LINEAR SURPRISAL REVISITED: A THEORETICAL FOUNDATION

A. Exploring Temperature in Chemical Reaction Dynamics

Having studied the classical dynamics of cluster isomerization in the liquid-like phase, we are now stepping into a new stage of statistical chemical reaction theory. However, before proceeding, we would like to explore "temperature" in general chemical reaction dynamics, since we could define well a temperature to characterize a distribution of canonical ensemble and since this temperature shows an extremely interesting property. So the last section of this review is devoted to an analysis of temperature and the linear surprisal theory from the variational statistical theory.

Elementary processes in chemical dynamics are universally important, besides their own virtues, in that they can link statistical mechanics to deterministic dynamics based on quantum and classical mechanics. The linear surprisal is one of the most outstanding discoveries in this aspect (we only refer to review articles [2–7]), the theoretical foundation of which is not yet well established. In view of our findings in the previous section, it is worth studying a possible origin of the linear surprisal theory in terms of variational statistical theory for microcanonical ensemble.

B. Linear Surprisal

The discovery, the linear surprisal, due to Kinsey, Bernstein, and Levine is about a rule on microcanonical rate constants $(k_j(\varepsilon_j))$ or the associated product distribution $(p_j(\varepsilon_j))$ experimentally observed in a chemical reaction, in which a final state, for instance, in a vibrational level of a given energy ε_j is specified. A statistically estimated product distribution $p_j^0(\varepsilon_j)$ corresponding to $p_j(\varepsilon_j)$ is called the prior distribution, which is usually evaluated in terms of the volume of a relevant classical phase space and is frequently represented in terms of energy parameters. Their remarkable finding [2–5] is an exponential form

$$\frac{p_j}{p_j^0} \propto \exp(-\lambda \varepsilon_j) \qquad (72)$$

in which different λ-parameters can appear depending on physical observables one is looking at. The inverse of λ may be regarded as a temperature in the natural analogy of the Boltzmann distribution. In contrast to the canonical temperature though, there can be many different temperatures depending on what we study. Many examples of the linear surprisal have been shown by Levine, Bernstein, and their groups. For instance, a reaction

$$F + HBr \rightarrow HF(v') + Br \qquad (73)$$

was highlighted by Ben-Shaul [42], where the vibrational quantum number v' of the final product HF serves as the suffix j in Eq. (72). Although $p(v')$ itself significantly deviates from an exponential distribution versus $\varepsilon_{v'}$, the values $-\log(p(v')/p^0(v'))$ come on a straight line with a positive slope. See Fig. 17. This implies that the higher vibrational states deviate more from the simple statistical estimates $p^0(v')$, which is sometimes an indication of population inversion. Thus λ is a negative value (actually $\lambda = -4.0$ was found for the

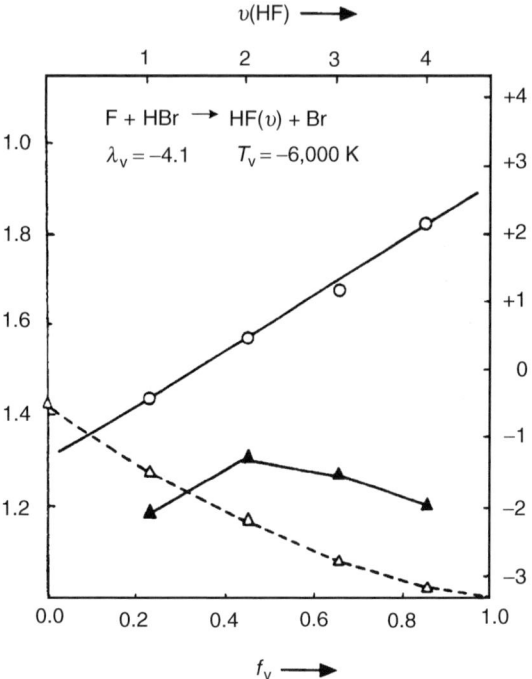

Figure 17. Linear surprisal in the reaction of F + HBr → HF(v') + Br. Black triangles indicate an experimental energy disposal, while the white triangles represent a statistical inference. The linear surprisal is given by the circles connected by a straight line. (Reproduced from Ref. 42 with permission.)

reaction in Eq. (73) [42]), which gives a negative "temperature." This is an empirical fact in the linear surprisal. Many other experimental examples [2–7] have been illustrated to evidence the relationship Eq. (72). On the other hand, counterexamples are also known [43,44]. However, it is exciting to explore why the linear surprisal can appear so universal. Many studies have been devoted to rationalizing or finding the theoretical foundation of Eq. (72). After all of their extensive studies, Levine and Bernstein [2–5] ascribed the above surprising discovery to the maximum entropy principle (MEP) [12].

The MEP is now well known to set an axiom of modern statistical mechanics [12–14]. The Shannon information entropy

$$S = -\sum_i p_i \log p_i \qquad (74)$$

where p_i is a probability to find a state at the ith level, is to be maximized under a condition that a set of physical quantities $\{f_i^{(k)}, i = 1, 2, \ldots, k = 1, 2, \ldots\}$ are experimentally observed as

$$\sum_i p_i f_i^{(k)} = f_k \tag{75}$$

including a trivial observable $f_i^{(k)} \equiv 1$ arising from the normalization condition $\sum_i p_i = 1$. The MEP brings about exponential distribution functions (no mechanism built in to lead to functional forms other than the exponential one!) in the following form:

$$p_i = \exp\left(-\lambda - \sum_k \lambda_k f_i^{(k)}\right) \tag{76}$$

where the exponents $\lambda_k (k = 1, 2, \ldots)$ are associated with the inverse of temperatures for individual observables. Note, however, that the bear populations p_i, rather than a ratio p_j/p_j^0, should satisfy the exponential law. Compare Eqs. (72) and (76).

The necessity of the prior distribution p_j^0 in Eq. (72) has been rationalized by minimizing the so-called entropy deficiency (or cross entropy [45])

$$\Delta S = \sum_i p_i \ln\left(\frac{p_i}{p_i^0}\right) \tag{77}$$

This is based on an anticipation that the deviation of the true entropy, Eq. (74), from an entropy that is estimated beforehand with an unspecified means

$$S_{\text{prior}} = -\sum_i p_i \log p_i^0 \tag{78}$$

should be small and must be minimized under a situation in which we have no way to know how accurate these estimated populations p_i^0 are.

However, paying no attention to physics behind actual events, the MEP simply maximizes information content under the constraints stated above. It is not *physically* clear how the prior distributions should be specified. In fact, several different methods have been proposed to determine a suitable candidate of $p_j^0(\varepsilon_j)$ [46]. It is also not clear why the product distributions themselves do not directly obey the exponential distributions as in Eq. (76) in chemical reactions. It should also be noted that the appearance of exponential distributions is not always the actual case [43,44]. (We are not discussing the notion of surprisal synthesis due to Levine and Bernstein [2–5]).

The information theoretic logic behind the MEP is perfectly self-contained within itself. On the other hand, the linear surprisal (LS) are often valid even in collision processes of energy conserved small systems, which are seemingly far from statistical situation. This suggests that the LS could be more fundamental than the MEP is. Moreover, it is more natural to conceive that the LS may constitute a dynamical foundation of the MEP. Therefore it makes sense to try to reexamine the linear surprisal in terms of more standard physical language. Our study in this line will end up with an understanding that there is no special trick behind the linear surprisal.

Here, we show, in terms of a variational principle of statistical mechanics, that a temperature proportional to λ_k^{-1} can be naturally defined to characterize the first-order feature (fluctuation) around the peak position of a distribution in a state space that is projected onto an appropriate coordinate. It is also shown that the necessity of the so-called prior distribution can naturally result along with its clear physical meaning. Several new features of the linear surprisal theory are uncovered through the analysis.

C. Variational Structure in Microcanonical Ensemble to Determine the Final Energy Disposals of Chemical Reactions and Associated Exponential Distributions

1. Chemical Reaction as a Nonequilibrium Stationary Flow

Let us consider a chemical reaction due to binary collision. To facilitate understanding the situation we suppose, let us begin with a simple reaction

$$X + YZ \to XY + Z \tag{79}$$

where X, Y, and Z denote atoms and/or molecules as in the reaction of Eq. (73). The Lippman–Schwinger scattering equation claims that a time-dependent scattering situation based on a wavepacket dynamics can be transformed into a time-independent state in which stationary waves continue to be incident and get out of the interaction region, thereby making an steady flow. This is more or less similar to the situation of nonequilibrium stationary state in statistical theory.

To formulate this situation in an intuitive perspective, it is convenient to use a one-dimensional curvilinear reaction coordinate (denoted by R) along which the initial channel is smoothly converted to the final one (see Fig. 18) [47,48]. The Hamiltonian defined along R is given as [48]

$$H(p_R, R, \{P_k, Q_k\}) = \sum_{k=1}^{N-1} \left(\tfrac{1}{2} P_k^2 + \tfrac{1}{2} \omega_k(R)^2 Q_k^2 \right) + V_0(R)$$
$$+ \tfrac{1}{2} \frac{[p_R - \sum_{k=1}^{N-1} Q_k P_l B_{k,l}(R)]^2}{[1 + \sum_{k=1}^{N-1} Q_k B_{k,N}(R)]^2} \tag{80}$$

where p_R is a momentum conjugate to R and $\{Q_k | k = 1, 2, \ldots, N-1\}$ is a coordinate system perpendicular to the reaction coordinate with P_k being a

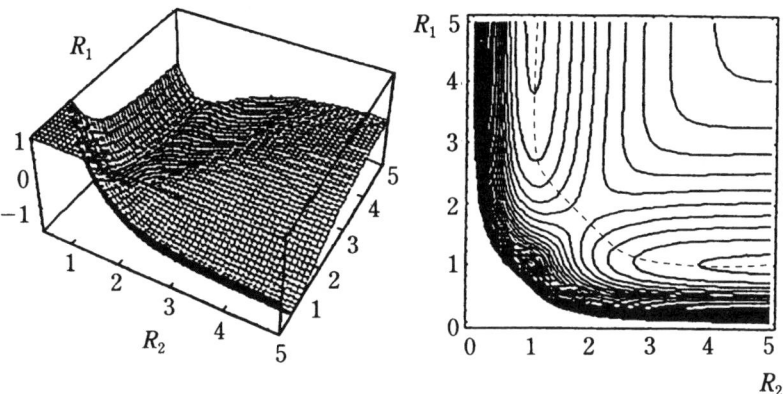

Figure 18. A curvilinear reaction coordinate on a potential surface. Chemical reaction may be viewed as a steady flow that would exhange energy with the local coordinates orthogonal to the reaction coordinate.

momentum conjugate to Q_k. $V_0(R)$ is a potential along R, and $\omega_k(R)$ denotes the frequency of a harmonic potential assumed in the Q_k-coordinate, although the harmonicity is not an essential assumption as far as the present work is concerned. The coordinates transversal to R couple kinematically with each other with an coupling element $B_{k,l}(R)$, which is written as [48]

$$B_{k,l}(R) = \left\langle \frac{\partial \psi_k}{\partial R} \Big| \psi_l \right\rangle \qquad (81)$$

where ψ_k is, for instance, an eigenvector of the force constant matrix from which the component of the reaction coordinate is projected out. As a reaction proceeds tracking the R coordinate, mode mixing among the coordinates takes place depending on the curvature of R. In this way, the vibrational mode YZ in Eq. (79) is smoothly converted to that of XY.

On the basis of the above physical situation presupposed, the entire state space is now represented in three independent modes: (1) a mode of the reaction coordinate R, which is eventually led to translational or dissociative state; (2) an internal state, that is, a vibrational mode under focus (denoted by A), which is to be observed as a linear surprisal; and (3) all other remaining internal modes (collectively denoted as B). A critical region (denoted by ∂R) is supposed to exist somewhere on the R coordinate beyond the transition state. At ∂R the modes A and B are practically separated and uniquely identified. For instance, the path Hamiltonian at a given point $R = R_0$ is

$$H(P_A, Q_A, P_B, Q_B)|_{R=R_0} = \sum_{k=1}^{A,B} \left(\tfrac{1}{2}P_k^2 + \tfrac{1}{2}\omega_k(R_0)^2 Q_k^2\right) + V_0(R_0)$$

$$+ \frac{1}{2} \frac{\sum_{k=1}^{A,B} Q_k P_l B_{k,l}(R_0)^2}{\left[1 + \sum_{k=1}^{A,B} Q_k B_{k,N}(R_0)\right]^2} \tag{82}$$

$$= H_A + H_B + H_{AB}$$

and $H_{AB} \simeq 0$ is assumed at $R_0 = \partial R$. Thus the critical region is not necessarily regarded as the transition state, at which the coupling H_{AB} is usually large.

2. Linear Surprisal as the Most Probable State

We next consider an experimental situation in which the final energy disposals are projected only onto the A mode by integrating information over all the B modes. Since the linear surprisal is usually formulated on the basis of accumulated product distributions rather than reaction rates [2–7], we hence begin with a standard statistical theory for the ratio of a population produced in the product site:

$$P(E) = \frac{1}{\Omega_{\text{total}}(E)} \int_0^E d\varepsilon \int_0^\varepsilon d\varepsilon_A \Omega_A(\varepsilon_A) \Omega_B(\varepsilon - \varepsilon_A) \Omega_{\partial R}(E-\varepsilon) \tag{83}$$

This quantity represents the ratio of a population at the critical region to the density of states before reaction. The variable ε is meant to be the total internal energy that is the sum of ε_A (the energy of the mode A) and ε_B (that of B). $\Omega_A(\varepsilon_A)$, for instance, is the quantum density of states in the A mode at an energy ε_A, which is defined as

$$\Omega_A(\varepsilon_A) = \int \sum_i \delta(\varepsilon_A - e_i^A) \tag{84}$$

where $\{e_i^A | i = 0, 1, 2, \ldots\}$ is the energy level of the A mode at $R = \partial R$. Also, the density of states in the direction of the reaction coordinate R is

$$\Omega_{\partial R}(E-\varepsilon) = \int \delta\left(E - \varepsilon_A - \varepsilon_B - V_0(\partial R) - \tfrac{1}{2}p_R^2\right) dp_R \tag{85}$$

As seen in the reaction path Hamiltonian, Eq. (80), the energies ε_A and ε_B are measured from zero. On the other hand, the linear surprisal Eq. (72) does not care about the origin in the energy coordinate. For the sake of simplicity, therefore, we simply set $V_0(\partial R) = 0$ in what follows. $\Omega_{\text{total}}(E)$ is designated as the total density of states before chemical reaction. Then, the population after reaction to be found in the ith level of the A mode must be proportional to

$$P_i^A(E) = \frac{1}{\Omega_{\text{total}}(E)} \int_0^E d\varepsilon \int_0^\varepsilon d\varepsilon_A \delta(\varepsilon_A - e_i^A) \Omega_A(\varepsilon_A) \Omega_B(\varepsilon - \varepsilon_A) \Omega_{\partial R}(E - \varepsilon) \quad (86)$$

Here we apply the discussion of variational structure to the numerator of Eq. (86) or (83), namely,

$$\hat{\Omega}(E) = \int_0^E d\varepsilon \int_0^\varepsilon d\varepsilon_A \Omega_A(\varepsilon_A) \Omega_B(\varepsilon - \varepsilon_A) \Omega_{\partial R}(E - \varepsilon) \quad (87)$$

As in the preceding section, the microcanonical ensemble is projected onto the internal energies with a form

$$\Xi(E, \varepsilon_A, \varepsilon_B) = \Omega_A(\varepsilon_A) \Omega_B(\varepsilon_B) \Omega_{\partial R}(E - \varepsilon_B - \varepsilon_A) \quad (88)$$

It is natural to assume that $\Omega_A(\varepsilon_A)$ increases as ε_A, while the product $\Omega_B(\varepsilon_B) \Omega_{\partial R}(E - \varepsilon_B - \varepsilon_A)$ decreases. Thus $\Xi(E, \varepsilon_A, \varepsilon_B)$ should be a single-peaked function on the ε_A axis. Let us then consider a statistically mixed state in which energies are exchanged between the A mode (having an energy ε_A) and the $B + R$ modes (an energy $E - \varepsilon_A$). Thus the most probable physical situations (or the most feasible initial conditions) before the system is brought up to ∂R should be prepared at a state in which the *"entropy"* $\ln \Xi(E, \varepsilon_A, \varepsilon_B)$ is maximum. Hence,

$$\frac{\partial}{\partial \varepsilon_A} \ln \Xi(E, \varepsilon_A, \varepsilon_B) = 0 \quad (89)$$

should be satisfied. As usual, one can specify the inverse temperature

$$\frac{\partial}{\partial \varepsilon_A} \ln \Omega_A(\varepsilon_A) = \frac{\partial}{\partial (E - \varepsilon_A)} \ln \Omega_B(\varepsilon_B) \Omega_{\partial R}(E - \varepsilon_B - \varepsilon_A) \equiv \beta_A \quad \text{(at } \varepsilon_A = \varepsilon_A^*)$$

$$(90)$$

The inverse temperature β_A serves as an order parameter to quantify the rate process for the A mode. Note that β_A varies as E changes. Here again [see Eq. (50)], the first-order distribution for $\Omega_A(\varepsilon_A)$ should look

$$\Omega_A(\varepsilon_A) \cong \Omega_A(\varepsilon_A^*) \exp[\beta_A(\varepsilon_A - \varepsilon_A^*)] \quad (91)$$

This first-order population should be brought back into Eq. (86), and all the contributions from different ε_B are to be accumulated. P_i^A is thus rewritten as

$$P_i^A \cong \frac{\exp[-\beta_A \varepsilon_A^*]}{\Omega_{\text{total}}} \Omega_A(\varepsilon_A^*) \int_0^E d\varepsilon \int_0^\varepsilon d\varepsilon_A \delta(\varepsilon_A - e_i^A) \exp[\beta_A \varepsilon_A] \Omega_B(\varepsilon - \varepsilon_A) \Omega_{\partial R}(E - \varepsilon)$$

$$= \frac{\exp[-\beta_A \varepsilon_A^*]}{\Omega_{\text{total}}} \Omega_A(\varepsilon_A^*) \exp[\beta_A e_i^A] \int_0^{E-e_i^A} d\varepsilon_B \Omega_B(\varepsilon_B) \Omega_{\partial R}(E - \varepsilon_B - e_i^A) \quad (92)$$

with a simple transformation $\varepsilon_B = \varepsilon - e_i^A$.

Let us next assume a case in which the quantum energy levels $\{e_i^A | i = 0, 1, 2, \ldots\}$ and $\{e_j^B | j = 0, 1, 2, \ldots\}$ are identified at the critical region ∂R, beyond which these modes are approximately separated from each other. Besides, it is assumed that the levels change only adiabatically beyond ∂R and that the ratios of energy intervals such as $e_i^A - e_j^A$ do not vary largely. For this situation to be valid, the so-called final-state interaction must be small. This is not a pathological assumption in many statistical reactions.

By defining a factor

$$D_i^A = \frac{\exp[-\beta_A \varepsilon_A^*]}{\Omega_{\text{total}}} \Omega_A(\varepsilon_A^*) \int_0^{E-e_i^A} d\varepsilon_B \Omega_B(\varepsilon_B) \Omega_{\partial R}(E - \varepsilon_B - e_i^A) \quad (93)$$

one immediately observes in Eq. (92) an exponential form

$$\frac{P_i^A}{D_i^A} = \exp[\beta_A e_i^A] \quad (94)$$

This is a theoretical prototype of the linear surprisal.

3. The Meaning of the Prior Distribution

There can be different ways to estimate D_i^A. In any case, its physical meaning is very clear; it is proportional to the total number of states producing the state of e_i^A. Hence we call it the degeneracy factor to this state. A typical example to estimate the degeneracy factor is as follows. First, we assume a quantum form for $\Omega_B(\varepsilon_B)$ as

$$\Omega_B(\varepsilon_B) = \sum_j g_j^B \delta(\varepsilon_B - e_j^B) \quad (95)$$

where g_j^B is a degeneracy of the energy level of e_j^B; for instance, $g_j^B = 2j + 1$ for a rotational level. Then we have

$$P_i^A \cong \frac{\exp[-\beta_A \varepsilon_A^*]}{\Omega_{\text{total}}} \Omega_A(\varepsilon_A^*) \exp[\beta_A e_i^A] \sum_j g_j^B \Omega_{\partial R}(E - e_j^B - e_i^A) \quad (96)$$

Furthermore, it is well known that the density of states in a translational mode such as the mode R is given by

$$\Omega_{\partial R}(E - e_j^B - e_i^A) = A_T(E - e_j^B - e_i^A)^{1/2} \tag{97}$$

with a constant factor $A_T = \mu^{3/2}/(2^{1/2}\pi^2\hbar^3)$, where μ is the reduced mass of receding molecules [2–5]. Then we have

$$P_i^A \cong \frac{\exp[-\beta_A \varepsilon_A^*]}{\Omega_{\text{total}}} \Omega_A(\varepsilon_A^*) \exp[\beta_A e_i^A] \sum_j g_j^B A_T (E - e_j^B - e_i^A)^{1/2} \tag{98}$$

Thus one has the original form of the linear surprisal

$$\log \frac{P_i^A}{P_i^{0A}} = \beta_A e_i^A + C \tag{99}$$

with

$$P_i^{0A} = \sum_j g_j^B A_T \left(E - e_j^B - e_i^A\right)^{1/2} \tag{100}$$

and

$$C = \log \frac{\exp[-\beta_A \varepsilon_A^*]}{\Omega_{\text{total}}} \Omega_A(\varepsilon_A^*) \tag{101}$$

P_i^{0A} is exactly one of the standard prior distributions given by Levine and co-workers [2,4,5]. The factor C in Eq. (99) is a constant with respect to the level index i, but it depends on β_A or the total energy. Comparing Eq. (99) with Eq. (72), we have

$$\beta_A = -\lambda \tag{102}$$

which accounts for the "negative" temperature (negative valuedness of λ) in the surprisal plot if β_A is evaluated to be positive.

Finally, we comment on the prior distributions for the corresponding reaction rate k_i^A. To represent the rate, we need the velocity factor along the reaction coordinate in such a way that

$$k_i^A = \frac{1}{\Omega_{\text{total}}(E)} \int_0^E d\varepsilon \int_0^\varepsilon d\varepsilon_A \delta(\varepsilon_A - e_i^A) \Omega_A(\varepsilon_A) \Omega_B(\varepsilon - \varepsilon_A) \Omega_{\partial R}(E - \varepsilon)[E - \varepsilon]^{1/2} \tag{103}$$

After the similar procedure, the result appears as

$$k_i^A \cong \frac{\exp[-\beta_A \varepsilon_A^*]}{\Omega_{\text{total}}} \Omega_A(\varepsilon_A^*) \exp[\beta_A e_i^A] \sum_j g_j^B A_T (E - e_j^B - e_i^A) \quad (104)$$

and thus the prior distribution should have the form

$$k_i^{0A} = \sum_j g_j^B A_T (E - e_j^B - e_i^A) \quad (105)$$

which should be compared with Eq. (100). Then we have

$$\log \frac{k_i^A}{k_i^{0A}} = \beta_A e_i^A + C \quad (106)$$

4. Some New Features of the Surprisal Theory

We thus have arrived at the exponential form of the linear surprisal theory, Eq. (72), without resorting to the information theoretic reasoning [2–5,7]. Since we have tracked a physical pathway rather than an information theoretic reasoning, we have much more to say about the physical insights behind the linear surprisal. It is often stated [7] that the MEP or the linear surprisal theory is not necessarily a predictive theory but useful only to systematize and interpret experimental data. For instance, the temperature parameters show up empirically only after a surprisal plot is made. However, our theoretical development has revealed the following new aspects of the linear surprisal: (1) The product population and the reaction rate can be calculated as in Eq. (92) and Eq. (103), respectively. Furthermore, even the inverse temperature β_A can be predicted theoretically as in Eq. (90). (We do not claim though that the calculations are easy.) (2) The prior distribution appears in an *a priori* manner. As a consequence, it has been explicitly shown that the prior distribution for product populations and that of the corresponding rate should be mutually different. (3) A new feature of the temperature dependence of product populations through the degeneracy factor has been uncovered as in Eq. (93) or Eq. (101). (4) Our approach suggests how to go beyond the simple exponential distributions or the linear surprisal by taking account of the second-order effect or the higher as in Eq. (51).

D. Study of the Surprisal Should be Resumed

We have discussed a variational structure in a microcanonical ensemble and shed a new light on a possible physical origin of the linear surprisal. There can be various classes of variational structures depending on systems under study. In the present review, we have studied a variational distribution of product population among different modes in chemical reactions and have found that the linear surprisal and associated temperatures naturally result. Our derivation to the final result, Eq. (98), consequently shows a series of physical conditions for the linear surprisal to be valid.

In this review we have investigated only one aspect of the gigantic feature of the linear surprisal theory. To our regret, people's interest in the linear surprisal seems to have been almost lost these days, and we are afraid that it is regarded only as an empirically fitting formula. However, as we have discussed above, it is a general theory that should be examined from one's viewpoint to link elementary process and statistical mechanics. We are happy if this review stimulates such studies.

VII. GENERAL CONCLUSIONS

Taking M_7 cluster as an illustrative example, we have reviewed a novel statistical property of isomerization dynamics in the liquid-like phase, which is a prototype of the high-energy multichannel chemical reaction. We project that studies of chemical reaction will be more and more aiming at such multichannel reactions as dissociation dynamics in electron plasma.

The first highlight of the isomerization dynamics has been found in the uniformity of the average passage times (lifetimes), which is a macroscopic manifestation of chaos. One implication of the uniformity is that it would not be very useful to study the precise characteristics of the individual transition states. This is a chemical reaction that is beyond the transition-state concept.

However, we do not know to what extent chaos should be strong to lead to the uniformity. Nor do we have precise information about a relation between the length of the average passage time and a potential function under study. Also, in the very low energy close to the onset of the isomerization, the uniformity tends to deviate to a large extent, which reveals the breakdown of the RRKM premise. We also have found a peculiar short-time behavior that is not assumed by the RRKM theory either. Classical paths are conceived to spend some longer time to find out ways out to the next visiting basins.

To account for the memory-losing stochastic appearance of isomers in the liquid-like phase, we have shown that the successive bifurcation of the reaction tubes is responsible for a geometrical foundation of this phenomenon. With this geometrical view in mind, we have proposed the concept of inter-basin mixing and quantify the rate of bifurcation of the reaction tubes by measuring an extended Liapunov exponent for the rate of relaxing to inter-basin mixing. If a relaxation time to the inter-basin mixing is longer than the corresponding lifetime of molecular-shape conversion, a non-Markovian and less statistical time series of visited basins should be exhibited. For instance, neither protein-folding nor cluster isomerization in the coexistence region is likely to be simply stochastic. From a viewpoint of Hamilton chaos in many-dimensional systems, it is convenient to categorize chaos into two types: chaos inducing the global mixing, which is usually represented in terms of stretching and folding in the Smale horseshoe map, and chaos bringing about inter-basin mixing, which causes the merge and bifurcation of reaction tubes.

Defining the microcanonical temperature as a kinetic energy that maximizes a phase-space distribution when projected onto the potential energy coordinate, we have shown that this temperature can characterize a time scale of structural isomerization dynamics in the liquid-like phase. In particular, it has been found that the local microcanonical temperature bears an Arrhenius-like relation to the inverse of the average lifetime in isomerization of M_7 clusters. Thus, with this temperature one can extract critical information hidden behind the stepwise fluctuation of the kinetic energy of a trajectory in an isomerization process [33]. We have explored a possible origin of the Arrhenius-like relation.

We finally summarize the conditions under which our statistical treatment can be valid. First of all, the dynamics behind is required to be ergodic. Second, in the case where the phase-space distribution function projected onto the potential energy coordinate, $\Xi(\varepsilon)$, is neither a singly nor sharply peaked function, the microcanonical temperature T_m would lose its significance. For instance, a molecule having a very complicated potential landscape, which may have very many potential basins of very different energies as seen in protein-folding dynamics, may have multiple peaks in $\Xi(\varepsilon)$. The local microcanonical temperatures $T_m^{(a)}$ nonetheless may remain meaningful, if $\Xi^{(a)}(\varepsilon)$ can be well-defined in each basin in terms of a single peak and narrow shape. Another difficult problem to treat is a case in which $\Xi(\varepsilon)$ has a large width. This corresponds to large heat capacity and to a large fluctuation in the character of dynamics is expected. Possible examples may be seen in a relatively low energy dynamics on a rough potential surface composed of many very shallow basins, in which identifying each basin does not make sense. Dynamics having a broad and/or anisotropic $\Xi(\varepsilon)$ is nevertheless worth studying extensively, since isomerization in the coexistence phase or dynamics of glassy states of clusters falls in this class.

We finally report some of the major progresses about cluster dynamics in our group. One is on the kinematic effects associated with molecular frame. As is well known, a centrifugal force arises from kinematics in a rotating system, which couples with a potential to give an effective centrifugal barrier (recall an electron motion in hydrogen atom). Similarly, intrinsic kinematic "forces" can be generated in the process of molecular deformation associated with the change of mass configuration of a molecule. This is really the case. For instance, one of such kinematic forces can give a systematic and beautiful account for the so-called trapping phenomenon (or the recrossing problem) at a transition state. To describe these effects, one needs to define the so-called internal space by removing the translational and rotational degrees of freedom from a system, which becomes intrinsically non-Euclidian in general. We have shown that the relevant metric arising from this curved space can indeed bring about a significant effect on the rate process of isomerization dynamics [49,50]. This aspect is reviewed by Yanao and Takatsuka in this volume. Also, such a metric should be taken into account to estimate the volume element in phase space that is originally estimated as Eq. (35) [51].

Another important major progress is semiclassical quantization of the vibrational state of M_7 cluster. As stated in the introduction, the isomerization dynamics can be viewed as a large-amplitude vibrational motion. Even if dynamics in the liquid-like state is extremely chaotic and the isomers appear in a Markovian manner, it can certainly be quantized in principle as a stationary state. Therefore the total vibrational eigenfunction for such a large-amplitude motion should represent a superposition of classical molecular shapes (isomers). It is extremely interesting to investigate how the vibrational spectra and associated eigenfunctions can be built up in this situation. On the other hand, semiclassical quantization of highly chaotic state of more than two degrees of freedom has been virtually impossible [52]. It is now widely recognized that even the periodic orbit theory [53] does not work well to quantize a general two-dimensional system except for specific systems like the stadium billiard. In our own development of semiclassical theory, we recently have devised an amplitude-free quasi-correlation function method that can be applied to a many dimensional system [54–56]. Using the modified Hénon–Heiles potential, we have shown that this method can really quantize the strong chaos [57]. We have applied this method to an M_7 cluster in a preliminary stage [55]. An extensive study in this line and relevant analysis is underway.

Acknowledgments

The original works that have been unified in this review have been done with two of my former students, Dr. Chihiro Seko and Dr. Tomohiro Yanao. I thank Professor Steve Berry for his valuable and encouraging discussions in the early stage of this study. This work has been supported in part by the Grant-in-Aid from the Ministry of Education, Culture, Sports, Science, and Technology of Japan.

References

1. For extensive reviews, see R. S. Berry, *Chem. Rev.* **93**, 2379 (1993); R. S. Berry, *J. Quant. Chem.* **58**, 657 (1996); K. D. Ball, R. S. Berry, R. E. Kunz, F.-Y. Li, A. Proykova and D. J. Wales, *Science* **271**, 963 (1996). See also D. J. Wales, *Science* **271**, 925 (1996).
2. R. B. Bernstein and R. D. Levine, *Adv. At. Mol. Phys.* **11**, 215 (1975).
3. R. D. Levine, *Annu. Rev. Phys. Chem.* **29**, 59 (1978).
4. R. D. Levine and J. L. Kinsey in *Atom-Molecule Collision Theory*, R. B. Bernstein, ed., Plenum, New York, 1979, p. 693.

5. R. D. Levine, *Adv. Chem. Phys.* **47**, 239 (1981).
6. R. K. Nesbet, in *Theoretical Chemistry*, Academic Press, New York, **6B**, 1981, p. 79.
7. J. I. Steinfeld, J. S. Francisco, and W. L. Hase, *Chemical Kinetics and Dynamics*, Prentice-Hall, Englewood Cliffs, NJ, 1989.
8. K. Takatsuka and C. Seko, *J. Chem. Phys.* **105**, 10356 (1996).
9. For progress of the RRKM theory, see the Marcus issue, *J. Chem. Phys.* **101** (1994).
10. K. Takatsuka and C. Seko, *J. Chem. Phys.* **110**, 3263 (1999).
11. K. Takatsuka and T. Yanao, *J. Chem. Phys.* **113**, 2552 (2000).
12. E. T. Jaynes, *Phys. Rev.* **106**, 620 (1957).
13. R. D. Levine and M. Tribus, eds., *The Maximum Entropy Formalism*, MIT Press, Cambridge, MA, 1979.
14. W. T. Grandy, Jr., *Foundations of Statistical Mechanics*, I and II, D. Reidel, Dordrecht, Holland, 1987.
15. K. Takatsuka, *Chem. Phys. Lett.* **345**, 453 (2001).
16. M. R. Hoare and P. Pal, *J. Cryst. Growth.* **17**, 77 (1972).
17. N. Shida, unpublished data.
18. F. H. Stillinger and T. A. Weber, *Phys. Rev. A* **25**, 978 (1982).
19. C. Seko and K. Takatsuka, *J. Chem. Phys.* **104**, 8613 (1996).
20. S. K. Nayak, R. Ramaswany, and C. Chakaravarty, *Phys. Rev.* **E51**, 3376 (1995).
21. A. J. Lichtenberg and M. A. Lieberman, *Regular and Chaotic Dynamics*, Springer, Berlin, 1992.
22. G. Benettin, L. Galgani, and J. M. Strelcyn, *Phys. Rev.* **A14**, 2338 (1976).
23. K. Takatsuka, *Chem. Phys. Lett.* **204**, 491 (1993).
24. K. Takatsuka, *Bull. Chem. Soc. Jpn.* **66**, 3189 (1993).
25. N. Hashimoto and K. Takatsuka, *J. Chem. Phys.* **103**, 6914 (1995).
26. F. G. Amar and R. S. Berry, *J. Chem. Phys.* **85**, 5943 (1986).
27. R. S. Dumont and P. Brumer, *J. Phys. Chem.* **90**, 3509 (1986).
28. M. Berblinger and C. Schlier, *J. Chem. Phys.* **101**, 4750 (1994) and references therein.
29. C. Seko and K. Takatsuka, *J. Chem. Phys.* **108**, 4924 (1998).
30. C. Seko and K. Takatsuka, *J. Chem. Phys.* **109**, 4768 (1998).
31. J. Jellinek, T. L. Beck, and R. S. Berry, *J. Chem. Phys.* **84**, 2783 (1986).
32. R. Kubo, M. Toda, and N. Hashimoto, *Statisitical Physics I and II*, Springer, New York, 1998.
33. (a) N. De Leon, M. A. Mehta, and R. Q. Topper, *J. Chem. Phys.* **94**, 8310 (1991), **94**, 8329 (1992); (b) N. De Leon, *J. Chem. Phys.* **96**, 285 (1992); (c) J. R. Fair, K. R. Wright, and J. S. Hutchinson, *J. Phys. Chem.* **99**, 14707 (1995)
34. T. L. Beck, J. Jellinek, and R. S. Berry, *J. Chem. Phys.* **87**, 545 (1987).
35. E. M. Pearson, T. Halicioglu, and W. A. Tiller, *Phys. Rev. A* **32**, 3030 (1985).
36. M. V. Berry and K. E. Mount, *Rep. Prog. Phys.* **35**, 315 (1972).
37. M. A. Miller and D. J. Wales, *J. Chem. Phys.* **107**, 8568 (1997).
38. J. C. Light, *J. Chem. Phys.* **40**, 3221 (1964): *Discuss. Faraday Soc.* **44**, 14 (1967).
39. T. Yanao and K. Takatsuka, *Chem. Phys. Lett.*, **313**, 633 (1999).
40. R. J. Hinde, R. S. Berry, and D. J. Wales, *J. Chem. Phys.* **96**, 1376 (1992).
41. R. J. Hinde and R. S. Berry, *J. Chem. Phys.* **99**, 2942 (1993).
42. A. Ben-Shaul, *Chem. Phys.* **1**, 244 (1973).

43. H. Horiguchi and S. Tsuchiya, *J. Chem. Phys.* **70**, 762 (1979).
44. S. Kato, R. J. Jaffe, A. Komornicki, and K. Morokuma, *J. Chem. Phys.* **78**, 4567 (1983).
45. J. E. Shore and R. W. Johnson, *IEEE Trans. Inf. Theory.* **IT-26**, 26 (1980).
46. E. Pollak, *J. Chem. Phys.* **68**, 547 (1978).
47. K. Fukui, S. Kato, and H. Fujimoto, *J. Am. Chem. Soc.* **97**, 1 (1975); S. Kato, H. Kato, and K. Fukui, *J. Am. Chem. Soc.* **99**, 684 (1977).
48. W. H. Miller, N. C. Handy, and J. E. Adams, *J. Chem. Phys.* **72**, 99 (1980).
49. T. Yanao and K. Takatsuka, *Phys. Rev. A*, **68**, 032714 (16 pages) (2003); *J. Chem. Phys.* in press (2004).
50. T. Yanao and K. Takatsuka, *Adv. Chem. Phys. Part B* **130**, 87 (2005).
51. H. Teramoto and K. Takatsuka, to be published.
52. (a) M. Tabor, *Chaos and Integrability in Nonlinear Dynamics*, John Wiley & Sons, New York, 1989; (b) L. E. Reichl, *The Transition to Chaos*, Springer, New York, 1992; (c) K. Nakamura, *Quantum Chaos*, Cambridge University Press, Cambridge, 1993; (d) P. Gaspard, *Chaos, Scattering and Statistical Mechanics*, Cambridge University Press, Cambridge, 1993.
53. M. C. Gutzwiller, *J. Math. Phys.* **11**, 1791 (1970); *J. Math. Phys.* **12**, 343 (1971); *Chaos in Classical and Quantum Mechanics*, Springer, Berlin, 1990.
54. K. Takatsuka and A. Inoue, *Phys. Rev. Lett.* **78**, 1404 (1997); A. Inoue-Ushiyama and K. Takatsuka, *Phys. Rev. A* **59**, 3256 (1999).
55. K. Takatsuka and A. Inoue, *Phys. Rev. A* **60**, 112 (1999)
56. K. Takatsuka, *Phys. Rev. E* **64**, 016224 (2001).
57. K. Hotta and K. Takatsuka, *J. Phys. A: Gen. Math.* **36**, 4785–4803 (2003); S. Takahashi and K. Takatsuka, *Phys. Rev. A* **69**, 022110 (2004).

CHAPTER 12

EFFECTS OF AN INTRINSIC METRIC OF MOLECULAR INTERNAL SPACE ON CHEMICAL REACTION DYNAMICS

TOMOHIRO YANAO

Department of Complex Systems Science, Graduate School of Information Science, Nagoya University, 464-8601, Nagoya, Japan

KAZUO TAKATSUKA

Department of Basic Science, Graduate School of Arts and Sciences, University of Tokyo, Komaba, 153-8902, Tokyo, Japan

CONTENTS

I. Introduction
   A. Kinematics in Internal Dynamics of Molecules
   B. Isomerization Dynamics of Atomic Clusters as a Case Study
II. Basic Scheme of the Gauge-Theoretical Formalism for Internal Motions of n-Body Systems
   A. A Familiar Expression for the Rotation–Vibration Kinetic Energy: Gauge-Dependent Expression
   B. Gauge-Invariant Expression for the Kinetic Energy and True Metric of Internal Space
III. Collective Coordinates and Roles of a Metric Force: Structural Isomerization Dynamics of Three-Atom Clusters
   A. The Principal-Axis Hyperspherical Coordinates
   B. Classical Equations of Internal Motion and a Metric Force
   C. Overall Feature of the Isomerization Reaction on a Potential Topography Mapped onto PAHC
   D. Effects of the Democratic Centrifugal Force: Inducing Mass-Balance Asymmetry and Trapping Trajectories Around the Transition State
IV. Quantitative Study on the Effect of the Gauge Field: Suppressing the Isomerization Rate Relative to Dynamics in the Eckart Subspace

---

Geometric Structures of Phase Space in Multidimensional Chaos: A Special Volume of Advances in Chemical Physics, Part B, Volume 130, edited by M. Toda, T Komatsuzaki, T. Konishi, R.S. Berry, and S.A. Rice. Series editor Stuart A. Rice.
ISBN 0-471-71157-8 Copyright © 2005 John Wiley & Sons, Inc.

A. Quest for the Quantitative Role of the Gauge Field
B. Eckart Subspace and Dynamics in It
C. Effects of the Gauge Field on Reaction Rate: A Numerical Experiment
D. Rationale for the Suppressing Effect of the Gauge Field
V. Extension to the Four-Body System
A. PAHC and Classical Equations of Motion
B. Collective Coordinates Dominating the Isomerization Mechanism
C. Trapped Motion Around the Transition State Due to DCF
D. The Suppressing Effect of the Gauge Field Associated with the Eckart Frame
VI. Concluding Remarks
Acknowledgments
References

I. INTRODUCTION

A. Kinematics in Internal Dynamics of Molecules

The internal space of an n-atom $(n \geq 3)$ system in the three-dimensional physical space is generally referred to as a $(3n - 6)$-dimensional space after the elimination of the translational and rotational degrees of freedom. Although the elimination of translation is obvious, the separation of rotational and internal degrees of freedom is highly nontrivial. The so-called "falling cat" phenomenon [1] illustrates this situation strikingly, in which a falling cat can change its orientation even under the conditions of vanishing total angular momentum. Actually, this remarkable phenomenon is quite universal in nature from fundamental three-body systems [2] to various living bodies [3]. Our molecular system is no exception: A molecule inevitably changes its orientation more or less as a result of internal motion even under the conditions of vanishing total angular momentum. Geometric properties of internal space can be deduced through a correct treatment of such a falling cat phenomenon in polyatomic molecules.

A gauge-theoretical formalism for the separation of rotation and internal motion in n-body systems has been developed by Guichardet [4], Tachibana and Iwai [5,6], and Littlejohn and Reinsch [7]. In this formalism, a body-fixed frame (body frame) is employed to specify instantaneous orientation of the system. Even if the total angular momentum of the system is zero, a body frame generally rotates continuously as the system changes its shape. The way of assigning body frame is arbitrary and corresponds to gauge convention in the gauge-theoretical formalism. Referring to such a moving body frame, the $(3n - 6)$-dimensional internal motion for molecular shape can be extracted. A remarkable consequence of the above-mentioned gauge theory is that the gauge fields arise inevitably on internal space in accordance with the choice of body frame. Furthermore, the gauge field couples with the Euclidean metric of configuration space to induce a non-Euclidean metric in internal space. This arises from the fact that the internal space for a polyatomic molecule is intrinsically a "curved" space. Consequently a "metric force" appears to be an important factor in the dynamics of polyatomic molecules.

The aim of this review is twofold. First, we identify the metric force arising from the intrinsic non-Euclidean metric of internal space based on a collective coordinate system called principal-axis hyperspherical coordinates (PAHC) developed by Chapuisat et al. [8–10] and Kuppermann [11]. This coordinate system describes the internal motion in terms of gyration radii and hyper-angles referring to the principal axes of moment of inertia tensor of cluster at each instant. We show that this coordinate system is particularly suitable to elucidate the mechanism of isomerization dynamics of three- and four-atom clusters because only a couple of predominant variables are effectively extracted in the isomerization dynamics. The metric force is identified as "democratic centrifugal force (DCF)" that acts on gyration radii arising from the change of hyperangle called kinematic or democratic rotation [7,12–15]. We find that DCF generally has a tendency to induce asymmetry in mass balance in a molecule along instantaneous principal axes. Therefore, DCF drives a three- or four-atom cluster to change their shape from equilateral (tetrahedral) to collinear (planer rhombus) configurations, thereby helping isomerization reactions to begin. Furthermore, the DCF works to trap the cluster to its transition state. By constructing an effective potential, we present a geometrical view of the origin of such a trapping motion, which should provide a general framework for the recrossing problem in chemical reaction dynamics.

As the second aim, we try to quantify how large these kinematic effects are by concentrating on the kinematic effects associated with the specific choice of the so-called "Eckart frame," which is widely used as a standard body frame for the separation of rotations and internal motions in polyatomic molecules [16–18]. Actually, the monumental work by Eckart [19] dates back to the 1930s, long before the above-mentioned gauge-theoretical formalism. He originally exploited his frame for an approximate separation of rotational and vibrational modes for semirigid molecules around their local equilibrium structures. The Eckart frame is particularly useful for the normal-mode analysis in the vicinity of local equilibrium structure since it (approximately) factors out a $(3n - 6)$-dimensional Euclidean subspace for internal motions. The Eckart idea of constructing the internal subspace is further developed in the theory of "reaction path Hamiltonian" due to Miller et al. [20], where the Eckart condition is assigned along a reaction path. Hinde and Berry [21] had also employed the Eckart procedure for the instantaneous normal mode analysis of internal motions of the Morse and Lennard-Jones clusters.

However, it should be recalled that the separation of internal motions and rotational motions is not globally accomplished by simply invoking the Eckart procedure. When the normal-mode coordinates are employed as internal coordinates referring to the Eckart frame, the correct description of internal motions with these coordinates requires a treatment of gauge fields arising from the specific choice of the Eckart frame. Therefore we elucidate the effects of the gauge field associated with the Eckart frame in the isomerization dynamics of cluster. The motivation of this study lies in the fact that the gauge field arising from the choice of the Eckart frame is often disregarded in the studies of normal-mode analysis for chemical reaction dynamics. In order to quantify a possible error in such theoretical treatments that neglect the gauge field, we compare two dynamics: One is the true dynamics under the gauge field and the other is a constrained dynamics, in which the gauge field is totally eliminated. It turns out that the gauge field generally has an effect of suppressing the rate of isomerization reaction significantly. A theoretical explanation of this suppressing effect is also provided in terms of the PAHC.

B. Isomerization Dynamics of Atomic Clusters as a Case Study

The kinematic effects based on the DCF is most significantly observed in large-amplitude "vibrational" motions and chemical reactions. Hence we select the isomerization dynamics of atomic clusters, which can change molecular shapes from time to time. Therefore this dynamics can be regarded as a large amplitude vibrational motion and a chemical reaction as well. Because of its fundamental importance as a prototype of collective chemical dynamics, it has been widely studied from various aspects. For example, they are studied as a microcanonical analog of solid–liquid phase transitions [22,23], and also as a prototype of multichannel chemical reactions that requires a new treatment in statistical-dynamical theory beyond the transition-state concept [24]. They are also interesting from the perspective of chaos and regularity in Hamiltonian many-body systems [21,25,26].

In the standard picture for chemical reaction dynamics, potential energy hypersurface in multidimensional internal space is usually thought to be the most dominant factor for the reactivities of chemical species [27–30]. This picture is extended to the study of protein dynamics [31,32], where the free energy landscape is also considered to be important. However, it is equally important to investigate the geometrical nature of the internal space itself such as its intrinsic metric. The metric we concentrate on here is the one associated with the expression of the kinetic energy term of polyatomic molecules and is responsible for kinematic or mass effects. In this chapter, it will be clarified that an intrinsic metric associated with a collective coordinate system gives birth to a "metric force" that plays a dominant role in the dynamics of clusters coupling with the usual force arising from potential energy.

The system we study here is a cluster composed of n ($n = 3$ and 4) identical atoms that mutually interact through the pairwise Morse potential, M_n. Total angular momentum of the system is zero throughout. The Hamiltonian $\mathscr{H}/\varepsilon$ of this system has the following dimensionless form:

$$\frac{\mathscr{H}}{\varepsilon} = \frac{1}{2}\sum_{i=1}^{n}(\dot{r}_{si} \cdot \dot{r}_{si}) + \sum_{i<j}[e^{-2(d_{ij}-d_0)} - 2e^{-(d_{ij}-d_0)}] \quad (1)$$

where a three-dimensional vector $r_{si} = (r_{six}, r_{siy}, r_{siz})^T$ is the position vector of the ith particle with respect to a space-fixed frame. ε represents the depth of the Morse potential and d_{ij} is the interparticle distance between the ith and jth atoms. The parameter d_0, which corresponds to the equilibrium distance of the pairwise Morse potential, controls the Hamiltonian and we set this parameter to $d_0 = 6.0$, which provides a potential topography similar to that of the Lennard-Jones potential that is frequently used to model the van der Waals clusters. The masses of all particles can be set to unity. In what follows, our numerical results are presented in the absolute units.

II. BASIC SCHEME OF THE GAUGE-THEORETICAL FORMALISM FOR INTERNAL MOTIONS OF n-BODY SYSTEMS

A. A Familiar Expression for the Rotation–Vibration Kinetic Energy: Gauge-Dependent Expression

We begin with an n-atom molecule in the three-dimensional physical space in this subsection for a rather general argument. To eliminate the translational degrees of freedom, we begin with the $(n-1)$ mass-weighted Jacobi vectors $\{\rho_{s1}, \ldots, \rho_{sn-1}\}$. The subscript s on ρ_{si}, and on other quantities as well, represents a vector relative to a space-fixed frame, whereas a vector ρ_i without subscript s represents a vector relative to a body frame (see below). In the following, we reduce the rotational degrees of freedom for the system of vanishing total angular momentum based on a gauge-theoretical treatment [7]. Let a body frame (body-fixed frame) be represented by a 3 × 3 proper rotation matrix $R \in SO(3)$, whose three-column vectors represent the three orthonormal axes of the frame. R can be parameterized by the Euler angles $\{\theta^\xi\}(\xi = 1, 2, 3)$ and thereby specifies the orientation of the axes with respect to the space-fixed frame. The body frame R is assigned to each configuration in a continuous manner to specify orientation of a system at each instant, and a rule of this assignment corresponds to a gauge convention [7]. The way of assigning a body frame is arbitrary in general, but in this chapter, we are mainly concerned with the so-called principal-axis frame and the Eckart frame. Below, we briefly summarize some basic prerequisite facts that are common to any choice of the body frame.

The vectors $\{\boldsymbol{\rho}_{si}\}$ are related to the mass-weighted Jacobi vectors referred to a body frame R, $\{\boldsymbol{\rho}_i\}$, by

$$\boldsymbol{\rho}_{si} = \mathsf{R}(\{\theta^\xi\})\boldsymbol{\rho}_i(\{q^\mu\}) \qquad (i = 1, \ldots, n-1) \tag{2}$$

where $\{q^\mu\}$ ($\mu = 1, \ldots, 3n-6$) are internal coordinates that specify the molecular shape uniquely and are invariant under translations and rotations of the system. The angular momentum of the system about the center of mass (referred to the space-fixed frame), $\boldsymbol{L}_s = \sum_{i=1}^{n-1} \boldsymbol{\rho}_{si} \times \dot{\boldsymbol{\rho}}_{si}$, and that referred to the body frame $\boldsymbol{L}$ are also related mutually by the relation $\boldsymbol{L}_s = \mathsf{R}\boldsymbol{L}$. Then $\boldsymbol{L}$ can be expressed by use of Eq. (2) and its time derivative as

$$\boldsymbol{L} = \sum_{i=1}^{n-1} \boldsymbol{\rho}_i \times (\boldsymbol{\omega} \times \boldsymbol{\rho}_i) + \sum_{i=1}^{n-1} \boldsymbol{\rho}_i \times \frac{\partial \boldsymbol{\rho}_i}{\partial q^\mu} \dot{q}^\mu \tag{3}$$

where the sum convention is adopted for the index μ from 1 to $3n-6$. Likewise, we always adopt this convention for the indices μ and ν. The three-dimensional vector $\boldsymbol{\omega}$ is the angular velocity of the body frame (referred to the frame itself), whose relationship to the angular velocity matrix $\Omega \equiv \mathsf{R}^T\dot{\mathsf{R}}$ is expressed as

$$\Omega \equiv \begin{pmatrix} 0 & -\omega_3 & \omega_2 \\ \omega_3 & 0 & -\omega_1 \\ -\omega_2 & \omega_1 & 0 \end{pmatrix} \Leftrightarrow \boldsymbol{\omega} \equiv \begin{pmatrix} \omega_1 \\ \omega_2 \\ \omega_3 \end{pmatrix} \tag{4}$$

Equation (3) can be rewritten in a more compact form as

$$\boldsymbol{L} = \mathsf{M}(\boldsymbol{\omega} + \boldsymbol{A}_\mu \dot{q}^\mu) \tag{5}$$

with M being the moment of inertia tensor referred to the body frame

$$M_{\alpha\beta} = \sum_{i=1}^{n-1}[(\boldsymbol{\rho}_i \cdot \boldsymbol{\rho}_i)\delta_{\alpha\beta} - \rho_{i\alpha}\rho_{i\beta}] \tag{6}$$

where $\delta_{\alpha\beta}$ is the Kronecker delta and the indices α and β specify axes of the body frame. $\boldsymbol{A}_\mu$ is a gauge potential defined by

$$\boldsymbol{A}_\mu = \mathsf{M}^{-1}\left(\sum_{i=1}^{n-1} \boldsymbol{\rho}_i \times \frac{\partial \boldsymbol{\rho}_i}{\partial q^\mu}\right) \tag{7}$$

The kinetic energy $K = \sum_{i=1}^{n-1} |\dot{\boldsymbol{\rho}}_{si}|^2/2$ in the Jacobi vectors can also be expressed in terms of the quantities referred to the body frame using the time derivative of Eq. (2) as

$$K = \tfrac{1}{2}(\boldsymbol{\omega}^T\mathsf{M}\boldsymbol{\omega}) + (\boldsymbol{\omega}^T\mathsf{M}\boldsymbol{A}_\mu)\dot{q}^\mu + \tfrac{1}{2}h_{\mu\nu}\dot{q}^\mu\dot{q}^\nu \tag{8}$$

where $h_{\mu\nu}$ is defined as

$$h_{\mu\nu} \equiv \sum_{i=1}^{n-1} \frac{\partial \boldsymbol{\rho}_i}{\partial q^\mu} \cdot \frac{\partial \boldsymbol{\rho}_i}{\partial q^\nu} \qquad (9)$$

The above expressions for the angular momentum and the kinetic energy are familiar in the conventional theory of molecular vibrations [16,19]. The first, second, and third terms on the right-hand side of Eq. (8) are usually referred to as rotational kinetic energy, the Coriolis coupling term, and vibrational kinetic energy, respectively. And $h_{\mu\nu}$ looks like a metric tensor for molecular vibrations. However, it is important to note that the decomposition in Eq. (8), as well as $h_{\mu\nu}$, depends on the choice of body frame—that is, is gauge-dependent. Therefore it is possible to choose a body frame that makes the Coriolis coupling term small for small-amplitude vibration around a local equilibrium structure. This is the spirit of the Eckart frame for the *approximate* separation of rotations and internal motions, in which the metric $h_{\mu\nu}$ happens to be Euclidean for the normal-mode coordinates. This Euclidean metric tensor is often conceived as though it could uniquely characterize the internal motion of a molecule. However, it is crucial to note that the metric tensor $h_{\mu\nu}$ is always dependent on the choice of body frame, and therefore $h_{\mu\nu}$ alone is not appropriate for the precise description of internal motions. For this reason, $h_{\mu\nu}$ is referred to as a pseudometric [7]. Significance of the distinction between the pseudometric $h_{\mu\nu}$ arising from the choice of the Eckart frame and the true metric that is independent of choice of body frame shown in the next subsection will be scrutinized in Section IV.

B. Gauge-Invariant Expression for the Kinetic Energy and True Metric of Internal Space

Littlejohn and Reinsch [7] have reformulated the expression of the kinetic energy as

$$K = \tfrac{1}{2}(\boldsymbol{\omega} + \mathbf{A}_\mu \dot{q}^\mu)^T \mathbf{M}(\boldsymbol{\omega} + \mathbf{A}_\nu \dot{q}^\nu) + \tfrac{1}{2} g_{\mu\nu} \dot{q}^\mu \dot{q}^\nu \qquad (10)$$

where $g_{\mu\nu}$ is defined as

$$g_{\mu\nu} = h_{\mu\nu} - \mathbf{A}_\mu^T \mathbf{M} \mathbf{A}_\nu \qquad (11)$$

This expression of K is gauge-invariant, and both of the first and the second terms on the right-hand side of Eq. (10) are independent of the choice of body frame. The former vanishes if and only if the total angular momentum $\boldsymbol{L}$ is zero [cf. Eq. (5)]. The metric tensor $g_{\mu\nu}$ in Eq. (11) is also gauge-invariant and is the true metric appropriate for the description of internal motions in polyatomic molecules.

Thus, the Lagrangian for an n-atom system of vanishing total angular momentum is reduced to

$$\mathscr{L} = \tfrac{1}{2} g_{\mu\nu} \dot{q}^\mu \dot{q}^\nu - V(\{q^\mu\}) \qquad (12)$$

where we restrict ourselves to a system whose potential term depends only on the internal variables $\{q^\mu\}$ as in our cluster dynamics. The classical equations of motion for the internal coordinates $\{q^\mu\}$ are obtained straightforwardly by applying the Lagrangian Eq. (12) to the Euler–Lagrange equations as will be done in the following sections. Thus, the internal motions of an n-atom system of vanishing total angular momentum are described in terms of only the $3n - 6$ internal variables $\{q^\mu\}$. For more general expression for the equations of motion for a system of nonzero angular momentum, see Ref. 7.

A remarkable fact about the true metric $g_{\mu\nu}$ is that it is essentially non-Euclidean for three- or more-atom systems [7]. This implies that "metric forces" should be arising inevitably in the internal dynamics of polyatomic molecules independently of the forces arising from the potential. Kinematic effects owing to such a metric force and the non-Euclidean nature of internal space are our main concern in this review.

III. COLLECTIVE COORDINATES AND ROLES OF A METRIC FORCE: STRUCTURAL ISOMERIZATION DYNAMICS OF THREE-ATOM CLUSTERS

A. The Principal-Axis Hyperspherical Coordinates

As an explicit materialization of the above gauge-invariant formalism, we adopt here the so-called principal-axis hyperspherical coordinates (PAHC). A three-body atomic cluster M_3 is presumably the best system to begin with to study the kinematic effects inherent to the internal space. The M_3 cluster has two local equilibrium structures on its potential energy surface corresponding to the two permutationally distinct equilateral triangle structures whose potential energy is $V = -3.00\varepsilon$, whereas it bears three permutationally distinct collinear saddle structures constituting a barrier height $V = -2.005\varepsilon$. These are summarized in Fig. 1. The system is laid on the x–y plane with $r_{s1z} = r_{s2z} = r_{s3z} = 0$ without loss of generality.

We start with a 3×2 matrix $\mathbf{W}_s$ composed of the two mass-weighted Jacobi (column) vectors as

$$\mathbf{W}_s \equiv (\boldsymbol{\rho}_{s1}\ \boldsymbol{\rho}_{s2}) \qquad (13)$$

EFFECTS OF AN INTRINSIC METRIC OF MOLECULAR INTERNAL SPACE

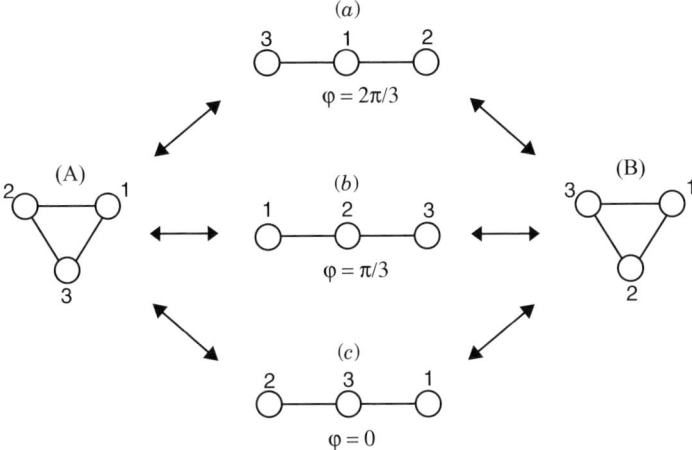

Figure 1. Local equilibrium and saddle structures of a three-atom Morse cluster, M_3. This cluster has two permutationally distinct local equilibrium structures ($V = -3.00\varepsilon$) and three saddle structures ($V = -2.005\varepsilon$). The equilibrium structure is equilateral triangle and the saddle structure is collinear. Values for the hyper-angle φ that specify respective saddle structures are also denoted.

where

$$\boldsymbol{\rho}_{s1} = \sqrt{\mu_1}(\boldsymbol{r}_{s1} - \boldsymbol{r}_{s2}), \qquad \mu_1 = \tfrac{1}{2}$$
$$\boldsymbol{\rho}_{s2} = \sqrt{\mu_2}\left(\frac{\boldsymbol{r}_{s1} + \boldsymbol{r}_{s2}}{2} - \boldsymbol{r}_{s3}\right), \qquad \mu_2 = \tfrac{2}{3} \tag{14}$$

with μ_1 and μ_2 the reduced masses [7,15]. According to the singular value decomposition theorem [33], W_s can be decomposed into a product of three matrices as

$$\mathsf{W}_s = \mathsf{R}\mathsf{N}\mathsf{U}^T \tag{15}$$

$$= (\boldsymbol{e}_1\,\boldsymbol{e}_2\,\boldsymbol{e}_3) \begin{pmatrix} a_1 & 0 \\ 0 & a_2 \\ 0 & 0 \end{pmatrix} \begin{pmatrix} \boldsymbol{u}_1^T \\ \boldsymbol{u}_2^T \end{pmatrix} \tag{16}$$

where R is a 3×3 orthogonal matrix whose orthonormal three-column vectors are $\boldsymbol{e}_1, \boldsymbol{e}_2$, and $\boldsymbol{e}_3$, and U is a 2×2 orthogonal matrix whose orthonormal two-column vectors are $\boldsymbol{u}_1$ and $\boldsymbol{u}_2$. N is a 3×2 diagonal matrix whose diagonal elements a_1 and a_2 are called the singular values of W_s, and off-diagonal elements are all equal to zero. All of the singular values and unit vectors in Eq. (16) must satisfy the following eigenvalue problems [12]:

$$(W_s W_s^T) e_\alpha = a_\alpha^2 e_\alpha \quad (\alpha = 1, 2, 3) \tag{17}$$

$$(W_s^T W_s) u_\beta = a_\beta^2 u_\beta \quad (\beta = 1, 2) \tag{18}$$

respectively. The eigenvalues of $W_s^T W_s$ and two of the eigenvalues of $W_s W_s^T$ are the same and equal to the square of singular values, a_1^2 and a_2^2. We set the order $a_1 \geq a_2$. The eigenvectors e_1, e_2, and e_3, coincide with the principal axes of the instantaneous moment of inertia tensor of the three-atom system because the off-diagonal elements of the moment of inertia tensor [Eq. (6)] coincide with those of the matrix $W_s W_s^T$ except for their sign. For our three-atom system laid on the x–y plane, the third eigenvalue of $W_s W_s^T$, a_3^2, is zero, and the corresponding eigenvector is set to be $e_3 = (0, 0, 1)^T$. At the same time, z-components of e_1 and e_2 are always zero. We restrict both R and U to be a proper rotation matrix. These are parameterized by θ and φ, respectively. φ is referred to as the hyper-angle. To summarize, Eq. (15) is made explicit in terms of these variables as

$$W_s = \begin{pmatrix} \cos\theta & -\sin\theta & 0 \\ \sin\theta & \cos\theta & 0 \\ 0 & 0 & 1 \end{pmatrix} \begin{pmatrix} a_1 & 0 \\ 0 & a_2 \\ 0 & 0 \end{pmatrix} \begin{pmatrix} \cos\varphi & \sin\varphi \\ -\sin\varphi & \cos\varphi \end{pmatrix} \tag{19}$$

In PAHC the leftmost matrix R on the right-hand side of Eq. (15) identifies a body frame (the principal-axis frame). The Jacobi vectors referred to this body frame, ρ_1 and ρ_2, are expressed as [cf. Eq. (2)]

$$(\rho_1 \quad \rho_2) = NU^T = \begin{pmatrix} a_1 \cos\varphi & a_1 \sin\varphi \\ -a_2 \sin\varphi & a_2 \cos\varphi \\ 0 & 0 \end{pmatrix} \tag{20}$$

Thus, the variables a_1, a_2, and φ are coordinates on a (three-dimensional) internal space. The singular values a_1 and a_2 are called "gyration radii" [8], since they represent the mass-weighted length (size) of the system along each principal axis. We let the sign of a_2 classify the permutational isomers of the three-atom cluster [13]. That is, if $\hat{z} \cdot (\rho_{s1} \times \rho_{s2}) > 0$, which is the case for the structure of type (A) in Fig. 1, a_2 is positive. Otherwise (type (B) in Fig. 1), a_2 is negative, where $\hat{z}$ is a unit vector along the positive z-axis. Condition $a_2 = 0$ specifies a collinear molecular shape.

In Eq. (19), the angle θ specifies the orientation of the principal-axis frame of the three-atom system and has nothing to do with the shape of the molecule. The continuous change in θ causes the ordinary rotation of the system without changing the system shape. On the other hand, the continuous change in φ in

Eq. (19), which is called the kinematic or democratic rotation, generally brings about a change in molecular shape exchanging the positions of the constituent atoms in a democratic manner. It is proved [13] that the range of the hyper-angle φ is limited to $0 \leq \varphi < \pi$ for φ to preserve one-to-one correspondence between molecular shape and the internal coordinates.

B. Classical Equations of Internal Motion and a Metric Force

The gauge-invariant metric tensor of internal space and classical equations of internal motion in terms of the PAHC are given explicitly as follows, from which the general properties of a metric force associated with this coordinate system will be deduced. By applying Eq. (20) to Eq. (6), we obtain for the moment of inertia tensor referred to the body frame as

$$\mathbf{M} = \begin{pmatrix} a_2^2 & 0 & 0 \\ 0 & a_1^2 & 0 \\ 0 & 0 & a_1^2 + a_2^2 \end{pmatrix} \quad (21)$$

which is diagonal as is expected. From Eq. (20) and Eq. (21), the gauge potential [Eq. (7)] is calculated for the internal coordinates to be

$$\mathbf{A}_{a_1} = \mathbf{0}, \qquad \mathbf{A}_{a_2} = \mathbf{0}, \qquad \mathbf{A}_{\varphi} = \begin{pmatrix} 0 \\ 0 \\ -\dfrac{2a_1 a_2}{a_1^2 + a_2^2} \end{pmatrix} \quad (22)$$

Since the pseudometric tensor $h_{\mu\nu}$ defined in Eq. (9) for the coordinates (a_1, a_2, φ) becomes

$$(h_{\mu\nu}) = \begin{pmatrix} 1 & 0 & 0 \\ 0 & 1 & 0 \\ 0 & 0 & a_1^2 + a_2^2 \end{pmatrix} \quad (23)$$

in a matrix form, we obtain the true metric tensor $g_{\mu\nu}$ as

$$(g_{\mu\nu}) = \begin{pmatrix} 1 & 0 & 0 \\ 0 & 1 & 0 \\ 0 & 0 & \dfrac{(a_1^2 - a_2^2)^2}{a_1^2 + a_2^2} \end{pmatrix} \quad (24)$$

by applying Eqs. (21), (22), and (23) to Eq. (11). Thus, we see that a subspace composed of the gyration radii a_1 and a_2 is Euclidean, since $g_{11} = g_{22} = 1$ and $g_{12} = g_{21} = 0$. We call this Euclidean space "gyration space," each point on which expresses the mass balance of a system along principal axes.

It is now important to note again that the true metric $g_{\mu\nu}$ in Eq. (24) for the internal coordinates (a_1, a_2, φ) is independent of the choice of body frame, although the PAHC initially refers to the principal-axis frame. Therefore, if another body frame is initially referred to in Eq. (2), then Eqs. (21), (22), and (23) will be altered but Eq. (24) does not change for the internal coordinates (a_1, a_2, φ). With use of this gauge-invariant metric tensor $g_{\mu\nu}$, the Lagrangian for the three-atom system of vanishing total angular momentum is given as

$$\mathscr{L} = \tfrac{1}{2}\left\{\dot{a}_1^2 + \dot{a}_2^2 + \frac{(a_1^2 - a_2^2)^2}{a_1^2 + a_2^2}\dot{\varphi}^2\right\} - V(a_1, a_2, \varphi) \qquad (25)$$

Subsequently, the classical equations of motion are obtained as

$$\ddot{a}_1 = \frac{a_1(a_1^2 + 3a_2^2)(a_1^2 - a_2^2)}{(a_1^2 + a_2^2)^2}\dot{\varphi}^2 - \frac{\partial V}{\partial a_1} \qquad (26)$$

$$\ddot{a}_2 = \frac{a_2(a_2^2 + 3a_1^2)(a_2^2 - a_1^2)}{(a_1^2 + a_2^2)^2}\dot{\varphi}^2 - \frac{\partial V}{\partial a_2} \qquad (27)$$

and

$$\frac{d}{dt}L_D = -\frac{\partial V}{\partial \varphi} \qquad (28)$$

where we have defined the democratic angular momentum L_D as

$$L_D \equiv \frac{\partial \mathscr{L}}{\partial \dot{\varphi}} = \frac{(a_1^2 - a_2^2)^2}{a_1^2 + a_2^2}\dot{\varphi} \qquad (29)$$

L_D is an angular momentum of the democratic rotation and is a constant of geodesic motion in the internal space, since φ becomes cyclic in the Lagrangian of Eq. (25) in the absence of the potential term.

As is evident from the first terms on the right-hand sides of Eqs. (26) and (27), which are proportional to the square of $\dot{\varphi}$, a kind of "centrifugal force" arises in the gyration space every time the democratic rotation occurs. We call this force "democratic centrifugal force (DCF)." Of course, DCF is different from the ordinary centrifugal force, and it arises even in a system of zero angular momentum. Figure 2 shows the field of DCF on gyration space for a selected $\dot{\varphi}$. If $|\dot{\varphi}|$ increases, the strength of the force also increases, keeping the same directionality at each point. It can be seen that the DCF works so as to avoid the degeneracy of the two gyration radii, $a_1 = |a_2|$. The arrows in Fig. 2 tend to align parallel to the positive a_1-axis for large a_1 with small $|a_2|$. These

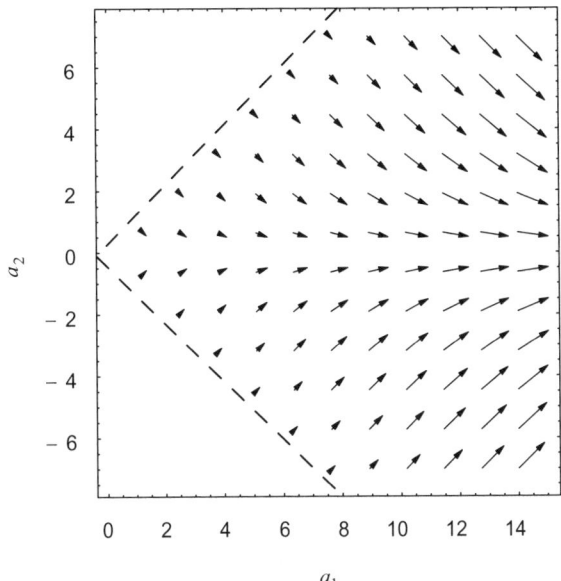

Figure 2. Fields of the democratic centrifugal force (DCF) on the gyration space at a typical value of $\dot{\varphi}$. The broken lines represent degeneracy between the two gyration radii, $a_1 = |a_2|$.

characteristics of the DCF indicate that an isolated three-atom system intrinsically tends to be longer in the longer direction and to be shorter in the shorter direction along the principal axes. (Here, the terms longer and shorter are in the meaning of the gyration radii.) Note that this effect of the DCF is a purely kinematic one and can compete with the potential force as is expected from Eqs. (26) and (27). The roles of the DCF will be studied in Section III.D.

C. Overall Feature of the Isomerization Reaction on a Potential Topography Mapped onto PAHC

Before investigating the roles of the DCF, here we clarify the overall feature of the isomerization dynamics in terms of the PAHC. A typical time evolutions of a_1 and a_2 is shown in Fig. 3a. In this example, the cluster undergoes isomerization three times, passing through the collinear saddle structures, which are characterized by $a_2 = 0$. Here we note that values of the gyration radii at the equilibrium structures (equilateral triangle) are $(a_1, a_2) = (4.24, \pm 4.24)$, where the sign of a_2 characterizes the two permutational isomers as we have defined in Section III.A. The time evolution of hyper-angle φ shown in Fig. 3b is quite characteristic in that φ is almost "locked" to 0 or $\pi/3$ or $2\pi/3$ with small oscillations during such an isomerization motions, whereas φ changes very

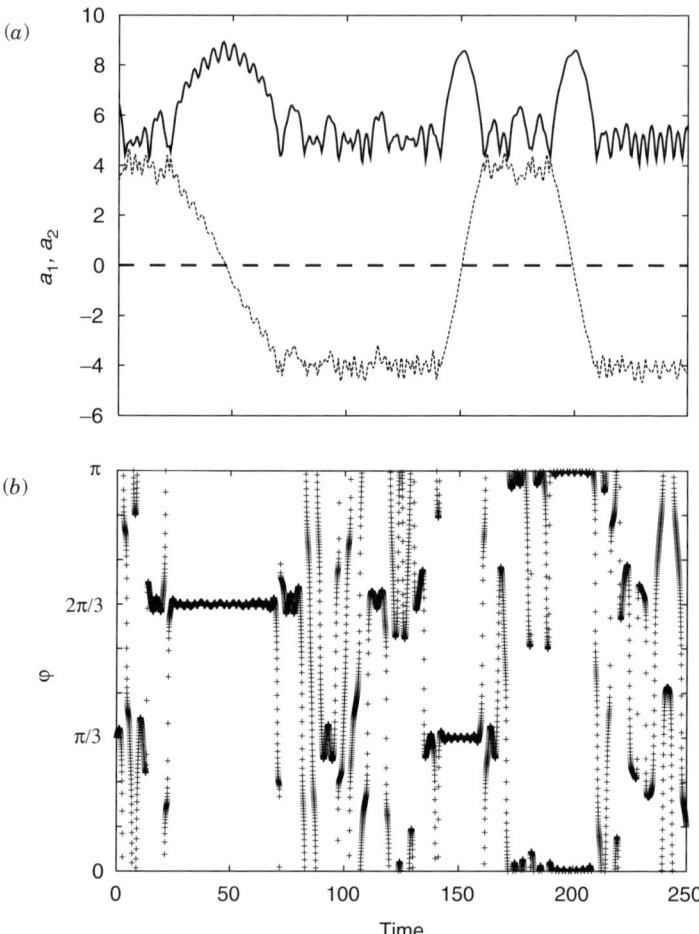

Figure 3. A typical time evolution of the two gyration radii, a_1 and a_2 ($a_1 \geq a_2$), and the hyper-angle φ in the three-atom Morse cluster, M_3, at the total energy $E = -1.80\varepsilon$. The condition $a_2 = 0$ (indicated by the horizontal broken line) corresponds to the collinear configuration. During this period, the saddle crossing has taken place three times. The locked angle $\varphi = 0, \pi/3, 2\pi/3$ specifies the three permutationally distinct saddle structures.

rapidly as the system is in the vicinity of the equilateral-triangle structures. (Note that $\varphi = 0$ and $\varphi = \pi$ are connected since φ is π-periodic.) Substituting these three angles about which φ is locked into Eq. (20), one confirms that they characterize the three permutationally distinct oblate isosceles-triangle structures of the M_3 cluster for arbitrary a_1 and a_2. Therefore the locking phenomenon

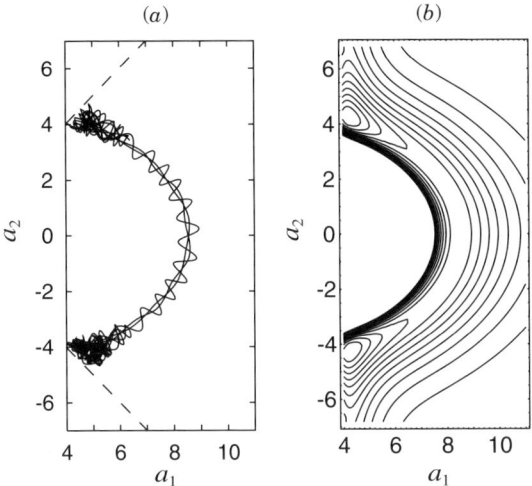

Figure 4. (a) A trajectory on gyration space corresponding to the time evolution in Fig. 3. The broken lines are the degeneracy lines between the two gyration radii, $a_1 = |a_2|$. (b) Potential energy surface on gyration space with hyper-angle φ fixed to 0 or $\pi/3$ or $2\pi/3$. The permutationally distinct local equilibrium structures (equilateral triangles) are located at $(4.24, 4.24)$ and $(4.24, -4.24)$, where the potential energy is $V = -3.00\varepsilon$, and the collinear transition state is located at $(8.48, 0)$, where the potential energy is $V = -2.005\varepsilon$. The energy difference between the neighboring contour lines is 0.231ε.

of the hyper-angle during the isomerizing motions is the reflection of the fact that the cluster undergoes isomerization, mostly keeping the isosceles-triangle symmetry. Thus φ classifies the permutationally distinct reaction channels shown in Fig. 1.

Since the hyper-angle φ is practically locked during the isomerization motion, the reaction path can be extracted on the space of a_1 and a_2, the gyration space. Figure 4a shows the trajectory mapped onto gyration space that corresponds to the dynamics in Fig. 3. In the gyration space, the two local equilibrium structures (equilateral triangle) are at $(a_1, a_2) = (4.24, \pm 4.24)$, and the collinear saddle structures are at $(a_1, a_2) = (8.48, 0)$. We can obtain a rough picture for the reaction path on gyration space from Fig. 4a. In fact, the potential surface on gyration space shown in Fig. 4b, where the hyper-angle φ is fixed to 0 or $\pi/3$ or $2\pi/3$, confirms the existence of the similarly curved reaction path. The topography of the potential energy surface is exactly common to these three angles for φ as a result of permutational symmetry. If the hyper-angle φ is shifted from these three values, the height of potential barrier becomes large and the reaction path is "closed" as shown in Fig. 5, where φ is shifted from these

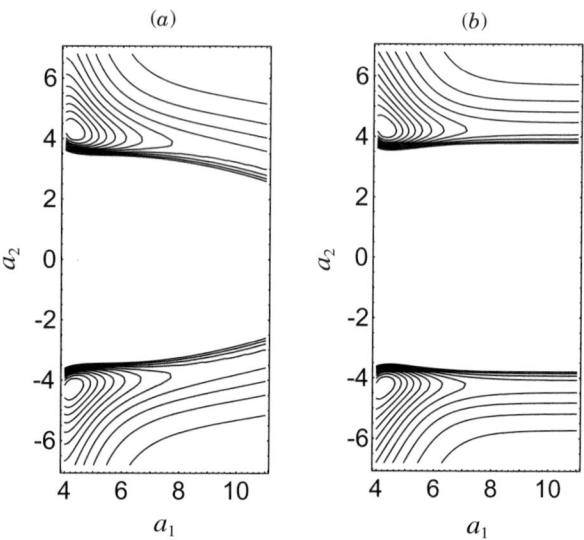

Figure 5. Potential energy surfaces $V(a_1, a_2, \varphi)$ mapped onto the gyration space. The hyper-angle is fixed to (a) $\varphi = (0 \text{ or } \pi/3 \text{ or } 2\pi/3) + \pi/12$, and (b) $\varphi = (0 \text{ or } \pi/3 \text{ or } 2\pi/3) + \pi/6$. The potential energy at $(4.24, \pm 4.24)$ is -3.00ε in both figures. The energy difference between the neighboring contour lines is 0.231ε.

values by $\pi/12$ and $\pi/6$. This is why the isomerization does not take place unless the hyper-angle φ is close to 0 or $\pi/3$ or $2\pi/3$.

On the other hand, the stability of locking of φ can be understood in terms of the potential energy against φ. Figure 6 shows the potential profiles for four given sets of gyration radii, $(a_1, a_2) = (4.24, 4.24), (5.0, 3.9), (6.0, 3.5), (8.48, 0)$, which are picked up along the reaction path in Fig. 4. The potential curve with respect to φ is flat for a trajectory that is in the vicinity of the equilibrium configuration $(a_1, a_2) = (4.24, 4.24)$, where a large and rapid democratic rotation is possible. The potential curve begins to swell as the gyration radii recedes from the equilibrium, and near the collinear transition state $(a_1, a_2) = (8.48, 0)$, only the regions around 0 or $\pi/3$ or $2\pi/3$ for φ are energetically accessible. Therefore, the locking of the hyper-angle tends to become tighter as the system climbs up along the reaction path on the gyration space.

To summarize, reaction mechanism of the M_3 cluster in terms of the PAHC is as follows: In the motions around the local equilibrium structure, a trajectory on the gyration space searches for a chance to get into the reaction path, which is frequently turned on and off due to the rapid democratic rotation. In this regime, dynamics of the cluster is dominated by all of the three internal variables, a_1, a_2,

EFFECTS OF AN INTRINSIC METRIC OF MOLECULAR INTERNAL SPACE 103

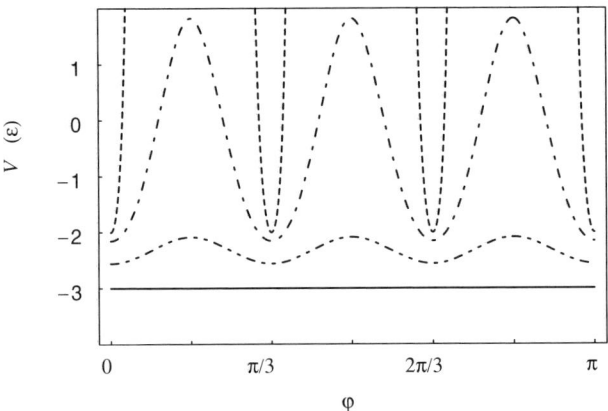

Figure 6. Potential energy curves against the hyper-angle φ with the gyration radii a_1 and a_2 fixed to $(a_1, a_2) = (4.24, 4.24), (5.0, 3.9), (6.0, 3.5), (8.48, 0)$ from the bottom to the top. These values for gyration radii are selected along the reaction coordinate in Fig. 4.

and φ. Once a trajectory on the gyration space successfully gets into the reaction path, the hyper-angle tends to be locked tightly so as to allow the trajectory to pass through the transition state. Thus the two variables, a_1 and a_2, play a role of the collective coordinates.

D. Effects of the Democratic Centrifugal Force: Inducing Mass-Balance Asymmetry and Trapping Trajectories Around the Transition State

We next investigate the roles played by the democratic centrifugal force (DCF) in the above process of structural isomerization. In the vicinity of the local equilibrium points, φ generally tends to vary rapidly as we have seen in Fig. 3. The flatness of potential curve for φ in the vicinity of the equilibrium points as shown in Fig. 6 makes this rapid democratic rotation possible. Another origin of this rapid democratic rotation can be understood by expressing $\dot{\varphi}$ explicitly as

$$\dot{\varphi} = \dot{u}_1 \cdot u_2 = \frac{u_1^T(\dot{W}_s^T W_s + W_s^T \dot{W}_s)u_2}{a_1^2 - a_2^2} \qquad (30)$$

which is obtained by differentiating Eq. (18) for $\beta = 1$ with respect to time and taking scalar product with u_2. Since the numerator of Eq. (30) is generally not zero, $\dot{\varphi}$ tends to be large as a_1 and $|a_2|$ come close to each other and $\dot{\varphi}$ can even diverge. Inserting Eq. (30) into the components of DCF in Eq. (26) and Eq. (27), we see that in such a case, both of these components can also diverge. Therefore the very rapid democratic rotation near the lines of degeneracy, $a_1 = |a_2|$, necessarily generates a very strong DCF on gyration space in the direction shown in Fig. 2 so as to avoid the complete degeneracy. In fact, we find that the trajectory in Fig. 4 is repulsed by the degeneracy lines if we look into this figure carefully. Thus the three-atom cluster has a constant tendency to be in an unsymmetrical mass-balance. Since the directionality of DCF field in Fig. 2 and that of the reaction path in Fig. 4b are nearly parallel to each other except for the saddle region, the DCF helps trajectories to climb up the potential barrier along the reaction path in the gyration space.

In the vicinity of the collinear transition state, the effect of DCF becomes more prominent. Although the hyper-angle φ is locked around 0 or $\pi/3$ or $2\pi/3$ when the system is in this region, φ oscillates within a small range around these angles as can be seen from Fig. 3. This small oscillation can generate a DCF on the gyration space that is as significant as the force arising from the potential, since the DCF in the vicinity of the collinear structure tends to be strong compared to that around the equilateral triangle structures *for a common value of* $\dot{\varphi}$, as is demonstrated by the length of the arrows in Fig. 2. We here note the fact that the arrows in Fig. 2 are directed to the line $a_2 = 0$ from the both sides, resulting in a tendency to be parallel to this line around it. This suggests that the DCF should have an effect of trapping trajectories in the vicinity of the collinear structure. In fact, such tentatively trapped trajectories are frequently observed even at a relatively high internal energy. A typical example of the time evolution of gyration radii for the trapped motion is shown in Fig. 7a. In this figure, the

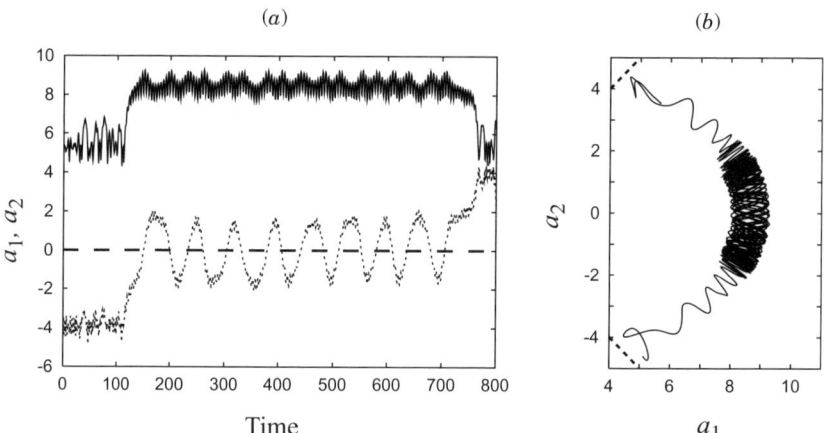

Figure 7. (a) A typical time evolution of gyration radii, a_1 and a_2 ($a_1 \geq a_2$), which shows a "trapped" motion in the vicinity of the collinear saddle structure from $t \approx 150$ to $t \approx 700$ recrossing the so-called "dividing surface" many times. Total internal energy of the trajectory is $E = -1.6\varepsilon$. (b) The corresponding trajectory on the gyration space from $t = 110$ to $t = 770$ in (a).

gyration radii a_1 and a_2 keep close to the values of the collinear saddle structure $(a_1, a_2) = (8.48, 0)$ for a long time before falling to one of the regions of equilibrium structures $((a_1, a_2) = (4.24, \pm 4.24))$. Figure 7b shows the corresponding trapped trajectory on the gyration space in the interval $t = 110 \sim t = 770$ in Fig. 7a. It oscillates in the direction of not only a_1 (corresponding to symmetric and antisymmetric stretching motions) but also a_2 (bending motion).

In order to comprehend the total effects arising from both the bear potential surface and the DCF, we define an effective potential V_{eff} as the sum of the usual potential function V and the potential for the DCF, which is the term proportional to the square of $\dot{\varphi}$ in the Lagrangian Eq. (25) as

$$V_{\text{eff}} \equiv -\frac{1}{2} \frac{(a_1^2 - a_2^2)^2}{a_1^2 + a_2^2} \dot{\varphi}^2 + V(a_1, a_2, \varphi) \qquad (31)$$

The resultant topography of V_{eff} is shown in Fig. 8, where φ is set to 0 or $\pi/3$ or $2\pi/3$ and $|\dot{\varphi}| = 0.05$ for (a) and $|\dot{\varphi}| = 0.15$ for (b), both of which are proper values for $|\dot{\varphi}|$ during the locking of φ. It clearly demonstrates that a new basin does appear around the saddle region of the potential energy surface. The new basin becomes broader and wider as $|\dot{\varphi}|$ becomes larger. Thus the trapped motion is rationalized from the viewpoints of energetics.

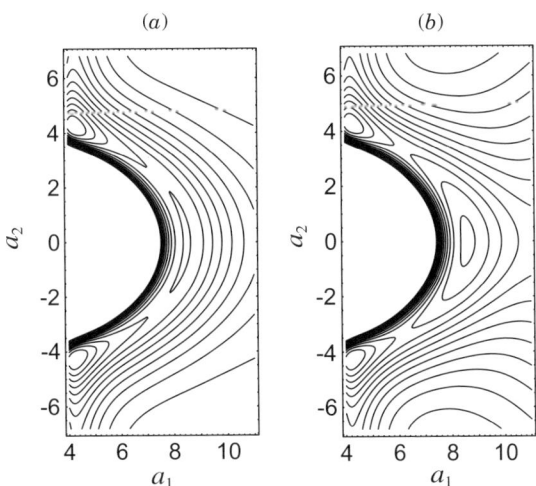

Figure 8. The effective potential energy surface, Eq. (31), mapped onto the gyration space. The hyper-angle φ is fixed to 0 or $\pi/3$ or $2\pi/3$. The democratic angular velocity is set to $|\dot{\varphi}| = 0.05$ in (a) and $|\dot{\varphi}| = 0.15$ in (b). The energy difference between the neighboring contour lines is 0.231ε.

In general, trapped and recrossing motion in the vicinity of a transition state is regarded as one of the main factors that bring about the overestimate of reaction rate in the transition state theory, and is thereby very important. In the past studies of the trapped motions in triatomic chemical reactions such as $H + H_2$, the resonant motion on the "skewed" potential surface has been the main concern within the scheme of the collinear configuration [34,35]. On the other hand, the present approach based on the PAHC provides a more general theoretical basis for the trapped motion not only within the collinear configuration but also in the direction of bending-like motion (direction of the a_2-axis on gyration space) out of the collinear configuration.

Our approach leading to Fig. 8 also provides a new insight into the regularity of the dynamics around the saddle region in van der Waals clusters, which has been discussed for a decade [21,25]. It is widely recognized through numerical investigations that the dynamics in the saddle region are generally more regular than that in the basin region. This can be rationalized naturally from our point of view: First, the effective degrees of freedom that dominate the saddle-crossing motion are reduced from three (a_1, a_2, φ) in the bottom region to two (a_1, a_2) due to the locking of the hyper-angle φ. Second, the basin structure thus found should help to reduce the irregular (chaotic) behavior. In our view, the stability of motion in the vicinity of transition state is a result of the fact that the transition state is coincidentally located on the *focal line* of the DCF—that is, on the line $a_2 = 0$ on the gyration space.

IV. QUANTITATIVE STUDY ON THE EFFECT OF THE GAUGE FIELD: SUPPRESSING THE ISOMERIZATION RATE RELATIVE TO DYNAMICS IN THE ECKART SUBSPACE

A. Quest for the Quantitative Role of the Gauge Field

In this section we demonstrate the quantitative importance of the gauge field in the rate of isomerization reaction by examining the validity of a rather standard picture for internal motion based on the normal mode coordinates that refer to the Eckart frame. In the conventional theories of chemical reaction dynamics, the separation of rotation and internal motions is sometimes assumed by resorting to the so-called Eckart frame [16–19]. It is a key here to construct a $(3n - 6)$-dimensional subspace for molecular vibration in the $(3n - 3)$-dimensional translation-reduced configuration space. We refer to this subspace as the Eckart subspace. (See the next subsection for the explicit definition. Also, distinguish the Eckart frame and the Eckart subspace.) In general, the way of constructing that kind of internal subspace is arbitrary, reflecting the arbitrariness of the choice of body frame. Actually, the pseudometric, $h_{\mu\nu}$, in Eq. (9) is the metric of thus constructed subspace. However, the Eckart subspace together with a potential surface in it does not necessarily describe the internal dynamics well because the role of the gauge field, $\{A_\mu\}$, is not taken into consideration correctly. Thus here we quantify how much the gauge field, $\{A_\mu\}$, arising from the choice of the Eckart frame may contribute to the reaction rate by comparing two systems: One of these is equipped with the full effects of the gauge field (true dynamics) as studied in Section III, and in the other the gauge-field effect is disregarded. The standard classical trajectory simulations of molecular dynamics does not care about the construction of internal space, since they are usually carried out in the $3n$-dimensional original Cartesian space. Nevertheless, such a comparison should be meaningful in that the gauge-field effect is made quantitatively explicit.

B. Eckart Subspace and Dynamics in It

Let the $n-1$ three-dimensional vectors $\{z_{si}\}(i=1,\ldots,n-1)$ be the mass-weighted Jacobi vectors for a reference molecular configuration. The reference configuration is usually set to be a local equilibrium structure of the molecule oriented to a certain orientation. The Eckart subspace is defined as a $(3n-6)$-dimensional subspace in the $(3n-3)$-dimensional translation-reduced configuration space, which is parameterized by Jacobi vectors $\{\rho_i^E\}(i=1,\ldots,n-1)$ with three additional constraint conditions called the Eckart conditions,

$$\sum_{i=1}^{n-1} \rho_i^E \times z_{si} = 0 \qquad (32)$$

The geometrical meaning is that the Eckart subspace $\{\rho_i^E\}$ is perpendicular to the three-dimensional manifold of rigid-body rotation at the reference configuration $\{z_{si}\}$. The Eckart subspace is Euclidean since the conditions in Eq. (32) are linear. Therefore this space can be spanned by vectors $\{n_{i\mu}\}(\mu=1,\ldots,3n-6)$ that specify the $3n-6$ directions of (vibrational) normal modes at the reference configuration $\{z_{si}\}$ in the $(3n-3)$-dimensional configuration space. The vectors $\{n_{i\mu}\}$ are orthonormal as $\sum_{i=1}^{n-1} n_{i\mu} \cdot n_{i\nu} = \delta_{\mu\nu}$. Thus the Eckart subspace can be parameterized by the $3n-6$ normal-mode coordinates $\{q^\mu\}$ as

$$\rho_i^E(\{q^\mu\}) = z_{si} + \sum_{\mu=1}^{3n-6} n_{i\mu} q^\mu \qquad (33)$$

The Eckart frame with respect to the reference configuration $\{z_{si}\}$ for an arbitrary configuration $\{\rho_{si}\}$ is given by a proper rotation matrix $\mathsf{R} \in SO(3)$ that satisfies

$$\sum_{i=1}^{n-1} (\mathsf{R}^T \rho_{si}) \times z_{si} = 0 \qquad (34)$$

The Eckart subspace is determined uniquely if the reference configuration $\{z_{si}\}$ in Eq. (32) is specified. However, the reference configuration does not have to be a local equilibrium structure in general. For instance, it can be set at a point along a trajectory as in the instantaneous normal-mode analysis [21] and can be continuously varied along a reaction coordinate as in the theory of the reaction path Hamiltonian [20]. For our analysis of the M_3 cluster, we simply set the reference configuration to be the equilibrium equilateral triangle structure (type (A) in Fig. 1), in which its two Jacobi vectors defined similarly to Eq. (14), z_{s1} and z_{s2}, are parallel to the x- and y-axis of the space-fixed frame, respectively. Here we consider only an Eckart subspace whose reference configuration is a local equilibrium structure.

The pseudometric of the Eckart subspace is Euclidean: $h_{\mu\nu} = \delta_{\mu\nu}$ for the normal-mode coordinates. Therefore equations of constrained motion to the Eckart subspace under the influence only of the potential topography in it are given by the familiar-looking Newtonian equations,

$$\ddot{q}^\mu = -\frac{\partial V}{\partial q^\mu} \quad (\mu = 1, \ldots, 3n - 6) \tag{35}$$

for the normal-mode coordinates $\{q^\mu\}$. As far as a small-amplitude vibration around an equilibrium structure is concerned, Eq. (35) may be a good approximation for internal motion under vanishing angular momentum. However, even if the total angular momentum is zero, trajectory starting from a point on the Eckart subspace gradually gets away from the subspace and wander widely in the entire $(3n - 3)$-dimensional translation-reduced configuration space. Within the framework of Eq. (35), the "falling cat" phenomenon never takes place. Thus, Eq. (35) is never rigorous and therefore it is important to quantify how large the error can arise in the reaction rate.

The rigorous dynamics in the internal space can be represented even in terms of the normal-mode coordinates by taking an appropriate account of $g_{\mu\nu}$ in Eq. (11). Since $h_{\mu\nu} = \delta_{\mu\nu}$ for the Eckart frame, Eq. (11) is written as

$$g_{\mu\nu} = \delta_{\mu\nu} - A_\mu^T M A_\nu \tag{36}$$

where the moment of inertia tensor M and the gauge potential $\{A_\mu\}$ are those defined in Eqs. (6) and (7), respectively, for the Eckart frame. The true metric tensor $g_{\mu\nu}$ of Eq. (36) is no longer Euclidean due to the second term on the right-hand side of Eq. (36), where the gauge potential $\{A_\mu\}$ for the Eckart frame generally does not vanish except at the reference structure [7]. Thus rigorous equations of motion in internal space (for zero angular momentum) are expressed as

$$g_{\mu\nu}(\ddot{q}^\nu + \Gamma^\nu_{\kappa\lambda}\dot{q}^\kappa\dot{q}^\lambda) = -\frac{\partial V}{\partial q^\mu} \quad (\mu = 1, \ldots, 3n - 6) \tag{37}$$

for the normal-mode coordinates $\{q^\mu\}$, where the Christoffel symbols $\Gamma^\nu_{\kappa\lambda}$ are defined by

$$\Gamma^\nu_{\kappa\lambda} = \tfrac{1}{2} g^{\nu\mu} \left(\frac{\partial g_{\mu\kappa}}{\partial q^\lambda} + \frac{\partial g_{\mu\lambda}}{\partial q^\kappa} - \frac{\partial g_{\kappa\lambda}}{\partial q^\mu} \right) \qquad (38)$$

The dynamics arising from these equations of motion should be compared with the constrained dynamics of Eq. (35).

C. Effects of the Gauge Field on Reaction Rate: A Numerical Experiment

In order to quantify the effects of the gauge field incidental to the choice of Eckart frame, we now compare numerically the reaction rate of isomerization of the M_3 cluster obeying the two different kinds of dynamics; one is the true dynamics obeying Eq. (37) and the other is the constrained dynamics obeying Eq. (35). We run 5000 trajectories of an internal energy $E = -1.6\varepsilon$, which is above the isomerization threshold. Random sampling to prepare the initial conditions in configuration space are made so as to set them to be common to two different dynamics. Here in this particular case study we monitor the time-dependent number of trajectories that remain in the original basin (Fig. 9).

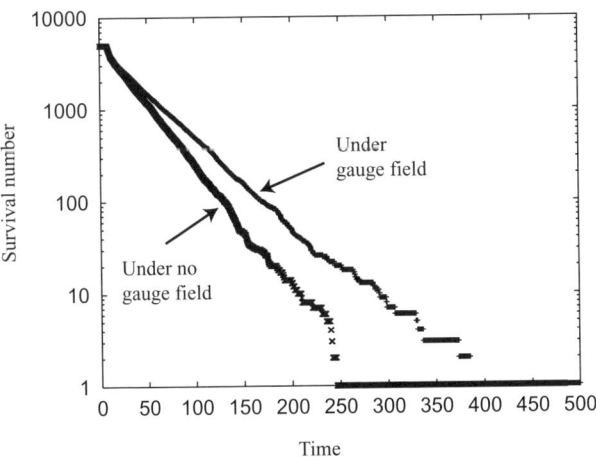

Figure 9. Decay of the number of surviving trajectories of the M_3 against isomerization (see the text for the precise definition of "isomerization"). The lower curve represents the dynamics constrained to the Eckart subspace under no gauge field, Eq. (35), while the upper one indicates the true dynamics under the full gauge field, Eq. (37). All of the initial conditions are randomly sampled in configuration space and are taken to be exactly the same for the two sets of dynamics. Number of the sample trajectories is 5000 and their internal energy is set to $E = -1.6\varepsilon$.

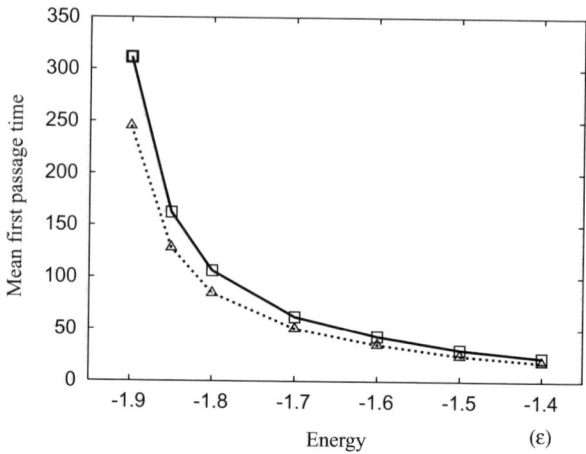

Figure 10. Internal-energy dependence of the mean first passage time (average lifetime) of the M_3 cluster. The lower curve represents the dynamics constrained to the Eckart subspace under no gauge field, Eq. (35), while the upper one shows the data for the true dynamics, Eq. (37).

Practically, a trajectory is discarded from the set of surviving trajectories as soon as it passes across the dividing surface along the collinear configuration, and the recrossing motion is simply disregarded. The mean first passage time—that is, an average time for the trajectories to arrive at the dividing surface—is also measured (Fig. 10).

Figure 9 shows the decay in the number of such surviving trajectories of the M_3 isomers of type (A) in Fig. 1. Since the survival number decays mostly in an exponential manner in the both dynamics, they must be sufficiently stochastic. (The absence of decay in the very short time range is due to the specific initial conditions that all atoms are at rest.) From the figure, it is clear that the decay of the survival number has been significantly suppressed by the gauge field. Figure 10 demonstrates that the suppression can be seen in a wide range of the internal energy. The difference between the mean first passage times in these dynamics amounts to about 20% to 30%, which is far beyond a negligible quantity. These results strongly suggest that one may *overestimate* the reaction rate of the structural isomerization to such a large amount by disregarding the effects of the gauge fields in the internal space.

D. Rationale for the Suppressing Effect of the Gauge Field

We now analyze how the above-mentioned suppressing effect comes about. To do so, we use the parameterization of the Eckart subspace in terms of the PAHC

(a_1, a_2, φ) whose reference configuration is the same as ours defined in Section IV.B, which was given by Littlejohn et al. [13]. The result is

$$
\begin{pmatrix} \rho_1^E & \rho_2^E \end{pmatrix} = \begin{pmatrix} \cos\varphi & -\sin\varphi & 0 \\ \sin\varphi & \cos\varphi & 0 \\ 0 & 0 & 1 \end{pmatrix} \begin{pmatrix} a_1 & 0 \\ 0 & a_2 \\ 0 & 0 \end{pmatrix} \begin{pmatrix} \cos\varphi & \sin\varphi \\ -\sin\varphi & \cos\varphi \end{pmatrix} \quad (39)
$$

$$
= \begin{pmatrix} a_1 \cos^2\varphi + a_2 \sin^2\varphi & (a_1 - a_2)\sin\varphi\cos\varphi \\ (a_1 - a_2)\sin\varphi\cos\varphi & a_1 \sin^2\varphi + a_2 \cos^2\varphi \\ 0 & 0 \end{pmatrix} \quad (40)
$$

With this parameterization and Eq. (9), the pseudometric tensor $h_{\mu\nu}$ of the Eckart subspace is represented as

$$
(h_{\mu\nu}) = \begin{pmatrix} 1 & 0 & 0 \\ 0 & 1 & 0 \\ 0 & 0 & 2(a_1 - a_2)^2 \end{pmatrix} \quad (41)
$$

Then the associated Lagrangian for the dynamics constrained to the Eckart subspace turns out to be

$$
\mathscr{L} = \tfrac{1}{2}\{\dot{a}_1^2 + \dot{a}_2^2 + 2(a_1 - a_2)^2 \dot{\varphi}^2\} - V(a_1, a_2, \varphi) \quad (42)
$$

and the classical equations of motion are

$$
\ddot{a}_1 = 2(a_1 - a_2)\dot{\varphi}^2 - \frac{\partial V}{\partial a_1} \quad (43)
$$

$$
\ddot{a}_2 = 2(a_2 - a_1)\dot{\varphi}^2 - \frac{\partial V}{\partial a_2} \quad (44)
$$

$$
\frac{d}{dt}[2(a_1 - a_2)^2 \dot{\varphi}] = -\frac{\partial V}{\partial \varphi} \quad (45)
$$

These equations should be compared with Eqs. (26)–(28) to see the consequence of the absence of the gauge-field effects. It is noted that the "democratic centrifugal force (DCF)" arises on the gyration space as the first terms of the right-hand sides of Eq. (43) and Eq. (44), but they are different from the original DCF in Eqs. (26) and (27). We now look into the detail of this difference.

The field of the DCF in Eqs. (43) and (44) is shown in Fig. 11 at a selected $\dot{\varphi}$. Comparing Fig. 11 with Fig. 2, we see that these two fields differ significantly in

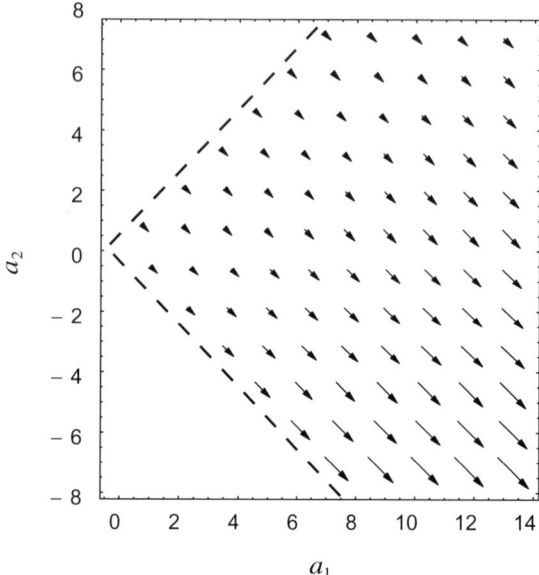

Figure 11. Fields of democratic centrifugal force on the gyration space in the dynamics constrained to the Eckart subspace under no gauge field. The length of each arrow reflects the strength of the force at each point for a given value of democratic angular velocity $\dot{\varphi}$. The broken lines are the degeneracy lines between the two gyration radii, $a_1 = |a_2|$.

their directionality. In the true dynamics under the full gauge-field effect (Fig. 2), the direction of the DCF tends to be more parallel to the positive a_1-axis in the vicinity of the dividing line of the isomerization, the a_1-axis itself, than that in the constrained dynamics without the gauge field (Fig. 11). In the constrained dynamics, the DCF is always parallel to the line $a_1 = -a_2$ and keeps the tendency to lead the molecular geometry to the collinear configuration, provided that a trajectory starts in the upper-half plane. But, at the same time, it keeps to push the trajectory carrying further across the dividing line in clear contrast to the true dynamics. Considering the directionality of the reaction path on gyration space in Fig. 4, we see that the DCF in the dynamics without the gauge field is more favorable for the system to go along the reaction path in the upper-half plane of the gyration space than the DCF in the true dynamics under the gauge field. [Note that the potential topography in terms of the coordinates (a_1, a_2, φ) is exactly common to the two kinds of dynamics because the difference in the two dynamics lies not in their potential but in their metric.]

Thus, the gauge field suppresses the isomerization reaction in such a way that it reduces the effect of DCF to collapse the cluster.

V. EXTENSION TO THE FOUR-BODY SYSTEM

The arguments thus made can be generalized to four-atom clusters. We thereby show the robustness of our findings obtained in the previous sections through the application of PAHC to a four-atom Morse cluster. The local equilibrium structure of this cluster is a regular tetrahedron whose potential energy is $V = -6.0\varepsilon$. There are two permutationally distinct isomers for the equilibrium structure as shown in Fig. 12. The first rank saddle point on the potential energy surface corresponds to the planar rhombus-shaped structure having the potential energy $V = -5.02\varepsilon$. (The square-planar structure is a second-rank saddle.) There are six permutationally distinct saddles on the potential surface which are also depicted in Fig. 12. Structural isomerization proceeds through the vicinities of these saddle structures.

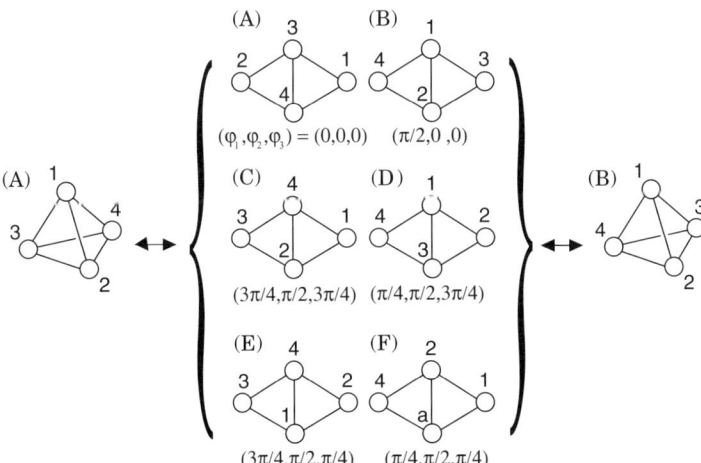

Figure 12. Local equilibrium and saddle structures of a four-atom Morse cluster, M_4. The local equilibrium structures (regular tetrahedron) and the first rank saddle structures (planar rhombus) have the energy -6.00ε and -5.02ε, respectively. φ_1, φ_2, and φ_3 are the hyper-angles that specify each reaction channel. When $\varphi_2 = 0$, φ_1 and φ_3 cannot be determined uniquely and only the summation $\varphi_1 + \varphi_3$ is meaningful due to the convention for Euler angles in Eq. (50). Hence, we set $\varphi_3 = 0$ when $\varphi_2 = 0$.

A. PAHC and Classical Equations of Motion

Similarly to the three-atom system, we resume with the mass-weighted Jacobi vectors defined as

$$\begin{aligned}
\boldsymbol{\rho}_{s1} &= \sqrt{\mu_1}(\boldsymbol{r}_{s1} - \boldsymbol{r}_{s2}), & \mu_1 &= \tfrac{1}{2} \\
\boldsymbol{\rho}_{s2} &= \sqrt{\mu_2}(\boldsymbol{r}_{s3} - \boldsymbol{r}_{s4}), & \mu_2 &= \tfrac{1}{2} \\
\boldsymbol{\rho}_{s3} &= \sqrt{\mu_3}\left(\frac{\boldsymbol{r}_{s1} + \boldsymbol{r}_{s2}}{2} - \frac{\boldsymbol{r}_{s3} + \boldsymbol{r}_{s4}}{2}\right), & \mu_3 &= 1
\end{aligned} \quad (46)$$

This definition of the Jacobi vectors implies that reaction channels are specified by multiples of $\pi/4$ in the hyper-angles as will be shown later.

According to the singular-value decomposition, the 3×3 matrix W_s, whose columns are the three mass-weighted Jacobi vectors, is decomposed as

$$\begin{aligned}
\mathsf{W}_s \equiv (\boldsymbol{\rho}_{s1} \quad \boldsymbol{\rho}_{s2} \quad \boldsymbol{\rho}_{s3}) &= \mathsf{RNU}^T \\
\equiv (\boldsymbol{e}_1 \quad \boldsymbol{e}_2 \quad \boldsymbol{e}_3) &\begin{pmatrix} a_1 & 0 & 0 \\ 0 & a_2 & 0 \\ 0 & 0 & a_3 \end{pmatrix} \begin{pmatrix} \boldsymbol{u}_1^T \\ \boldsymbol{u}_2^T \\ \boldsymbol{u}_3^T \end{pmatrix}
\end{aligned} \quad (47)$$

where both matrices R and U belong to SO(3). The unit vectors $\{\boldsymbol{e}_\alpha\}$ and $\{\boldsymbol{u}_\beta\}(\alpha, \beta = 1, 2, 3)$ and the singular values a_1, a_2, and a_3 are solutions of the eigenvalue problems,

$$(\mathsf{W}_s \mathsf{W}_s^T)\boldsymbol{e}_\alpha = a_\alpha^2 \boldsymbol{e}_\alpha \quad (\alpha = 1, 2, 3) \quad (48)$$

$$(\mathsf{W}_s^T \mathsf{W}_s)\boldsymbol{u}_\beta = a_\beta^2 \boldsymbol{u}_\beta \quad (\beta = 1, 2, 3) \quad (49)$$

We define the order of the singular values as $a_1 \geq a_2 \geq |a_3|$. The planar and collinear configurations give $a_3 = 0$ and $a_2 = a_3 = 0$, respectively. Furthermore, we let the sign of a_3 specify the permutational isomers of the cluster [14]. That is, if $(\det \mathsf{W}_s) = \boldsymbol{\rho}_{s1} \cdot (\boldsymbol{\rho}_{s2} \times \boldsymbol{\rho}_{s3}) \geq 0$, which is the case for isomer (A) in Fig. 12, $a_3 \geq 0$. Otherwise, $a_3 < 0$. Eigenvectors $\boldsymbol{e}_\alpha(\alpha = 1, 2, 3)$ coincide with the principal axes of instantaneous moment of inertia tensor of the four-body system. We thereby refer to the principal-axis frame as a body frame. On the other hand, the triplet of axes $(\boldsymbol{u}_1, \boldsymbol{u}_2, \boldsymbol{u}_3)$ or an SO(3) matrix U constitutes an "internal frame." Rotation of the internal frame in a three-dimensional space, which is the democratic rotation in the four-body system, is parameterized by three

EFFECTS OF AN INTRINSIC METRIC OF MOLECULAR INTERNAL SPACE 115

hyper-angles φ_1, φ_2, and φ_3 in a manner similar to that of the usual Euler angles $(\varphi_1, \varphi_2, \varphi_3)$ as [36]

$$\mathsf{U}^T = \begin{pmatrix} \cos\varphi_1\cos\varphi_3 - \sin\varphi_1\cos\varphi_2\sin\varphi_3 & \cos\varphi_1\sin\varphi_3 - \sin\varphi_1\cos\varphi_2\cos\varphi_3 & \sin\varphi_1\sin\varphi_2 \\ -\sin\varphi_1\cos\varphi_3 - \cos\varphi_1\cos\varphi_2\sin\varphi_3 & -\sin\varphi_1\sin\varphi_3 + \cos\varphi_1\cos\varphi_2\cos\varphi_3 & \cos\varphi_1\sin\varphi_2 \\ \sin\varphi_2\sin\varphi_3 & -\sin\varphi_2\cos\varphi_3 & \cos\varphi_2 \end{pmatrix}$$
(50)

The angular velocity of the internal frame with respect to the frame itself, γ_D, is expressed in the usual manner [36] using the hyper-angles and their time derivatives as

$$\gamma_D \equiv \begin{pmatrix} \gamma_{23} \\ \gamma_{31} \\ \gamma_{12} \end{pmatrix} = \begin{pmatrix} 0 & \cos\varphi_1 & \sin\varphi_1\sin\varphi_2 \\ 0 & -\sin\varphi_1 & \cos\varphi_1\sin\varphi_2 \\ 1 & 0 & \cos\varphi_2 \end{pmatrix} \begin{pmatrix} \dot\varphi_1 \\ \dot\varphi_2 \\ \dot\varphi_3 \end{pmatrix}$$
(51)

where we have defined the components of γ_D as $\gamma_{\alpha\beta} = \dot{\boldsymbol{u}}_\alpha \cdot \boldsymbol{u}_\beta (\alpha, \beta = 1, 2, 3, \alpha \neq \beta)$, which represents the strength of coupling between the αth and βth axes of the internal frame. The components $\gamma_{\alpha\beta}$ are antisymmetric, $\gamma_{\alpha\beta} = -\gamma_{\beta\alpha}$. We also define the 3×3 matrix on the right-hand side of Eq. (51) as G. Thus variables $(a_1, a_2, a_3, \varphi_1, \varphi_2, \varphi_3)$ are the coordinates in the six-dimensional internal space of the four-body system. It has been proved [11,14] that the ranges of the hyper-angles in the internal space are

$$0 \leq \varphi_1, \varphi_3 < \pi, \qquad 0 \leq \varphi_2 \leq \pi$$
(52)

Next, we construct the Lagrangian for the four-body system of vanishing total angular momentum. The moment of inertia tensor M in Eq. (6) is diagonal as

$$\mathsf{M} = \begin{pmatrix} a_2^2 + a_3^2 & 0 & 0 \\ 0 & a_3^2 + a_1^2 & 0 \\ 0 & 0 & a_1^2 + a_2^2 \end{pmatrix}$$
(53)

since we refer to the principal-axis frame R. The gauge potential in our principal-axis gauge is obtained by invoking Eq. (7). The result for the gyration radii a_1, a_2, and a_3 is

$$\mathbf{A}_{a_1} = \mathbf{A}_{a_2} = \mathbf{A}_{a_3} = \mathbf{0}$$
(54)

which is similar to the result of Eq. (22). On the other hand, the components of the gauge potential for the hyper-angles φ_1, φ_2, and φ_3 are expressed collectively in a 3 × 3 matrix form as

$$\begin{pmatrix} A_{\varphi_1} & A_{\varphi_2} & A_{\varphi_3} \end{pmatrix} = \mathsf{M}^{-1}\mathsf{b}\mathsf{G} \tag{55}$$

where a 3 × 3 diagonal matrix b is defined as

$$\mathsf{b} \equiv \begin{pmatrix} -2a_2a_3 & 0 & 0 \\ 0 & -2a_3a_1 & 0 \\ 0 & 0 & -2a_1a_2 \end{pmatrix} \tag{56}$$

and G is the 3 × 3 matrix defined in Eq. (51).

The pseudometric tensor $h_{\mu\nu}$ in Eq. (9) for the internal coordinates $(a_1, a_2, a_3, \varphi_1, \varphi_2, \varphi_3)$ in this order is

$$(h_{\mu\nu}) = \begin{pmatrix} \mathsf{I} & 0 \\ 0 & \mathsf{G}^T\mathsf{M}\mathsf{G} \end{pmatrix} \tag{57}$$

where I is a 3 × 3 unit matrix. Putting Eq. (53), Eq. (54), Eq. (55), and Eq. (57) into Eq. (11), we obtain the true metric of internal space as

$$(g_{\mu\nu}) = \begin{pmatrix} \mathsf{I} & 0 \\ 0 & \mathsf{G}^T(\mathsf{M} - \mathsf{b}\mathsf{M}^{-1}\mathsf{b})\mathsf{G} \end{pmatrix} = \begin{pmatrix} \mathsf{I} & 0 \\ 0 & \mathsf{G}^T\tilde{\mathsf{M}}\mathsf{G} \end{pmatrix} \tag{58}$$

where the diagonal matrix $\tilde{\mathsf{M}}$ is defined as

$$\tilde{\mathsf{M}} \equiv \mathsf{M} - \mathsf{b}\mathsf{M}^{-1}\mathsf{b} = \begin{pmatrix} \dfrac{(a_2^2 - a_3^2)^2}{a_2^2 + a_3^2} & 0 & 0 \\ 0 & \dfrac{(a_3^2 - a_1^2)^2}{a_3^2 + a_1^2} & 0 \\ 0 & 0 & \dfrac{(a_1^2 - a_2^2)^2}{a_1^2 + a_2^2} \end{pmatrix} \tag{59}$$

From Eq. (58), we again confirm that the three-dimensional gyration space for the coordinates (a_1, a_2, a_3) is Euclidean.

In terms of the abbreviated expressions $\boldsymbol{a} = (a_1, a_2, a_3)^T$ and $\boldsymbol{\varphi} = (\varphi_1, \varphi_2, \varphi_3)^T$, the kinetic energy K with zero angular momentum is represented with use of Eq. (58) as

$$2K = \begin{pmatrix} \dot{\boldsymbol{a}}^T & \dot{\boldsymbol{\varphi}}^T \end{pmatrix} \begin{pmatrix} \mathsf{I} & 0 \\ 0 & \mathsf{G}^T\tilde{\mathsf{M}}\mathsf{G} \end{pmatrix} \begin{pmatrix} \dot{\boldsymbol{a}} \\ \dot{\boldsymbol{\varphi}} \end{pmatrix}$$

$$= \begin{pmatrix} \dot{\boldsymbol{a}}^T & \boldsymbol{\gamma}_D^T \end{pmatrix} \begin{pmatrix} \mathsf{I} & 0 \\ 0 & \tilde{\mathsf{M}} \end{pmatrix} \begin{pmatrix} \dot{\boldsymbol{a}} \\ \boldsymbol{\gamma}_D \end{pmatrix} \tag{60}$$

where we have used the democratic angular velocity defined in Eq. (51) in the second equality. Thus, the Lagrangian is obtained as

$$\mathscr{L} = \frac{1}{2}\dot{a}^2 + \frac{1}{2}\gamma_D^T \tilde{M}\gamma_D - V(a,\varphi)$$
$$= \frac{1}{2}(\dot{a}_1^2 + \dot{a}_2^2 + \dot{a}_3^2) + \frac{(a_1^2 - a_2^2)^2}{2(a_1^2 + a_2^2)}\gamma_{12}^2 + \frac{(a_2^2 - a_3^2)^2}{2(a_2^2 + a_3^2)}\gamma_{23}^2 + \frac{(a_3^2 - a_1^2)^2}{2(a_3^2 + a_1^2)}\gamma_{31}^2$$
$$- V(a_1, a_2, a_3, \varphi_1, \varphi_2, \varphi_3) \quad (61)$$

The classical equations of motion are then derived, the gyration components of which are

$$\ddot{a}_1 = \frac{a_1(a_1^2 + 3a_2^2)(a_1^2 - a_2^2)}{(a_1^2 + a_2^2)^2}\gamma_{12}^2 + \frac{a_1(a_1^2 + 3a_3^2)(a_1^2 - a_3^2)}{(a_1^2 + a_3^2)^2}\gamma_{13}^2 - \frac{\partial V}{\partial a_1} \quad (62)$$

$$\ddot{a}_2 = \frac{a_2(a_2^2 + 3a_1^2)(a_2^2 - a_1^2)}{(a_2^2 + a_1^2)^2}\gamma_{21}^2 + \frac{a_2(a_2^2 + 3a_3^2)(a_2^2 - a_3^2)}{(a_2^2 + a_3^2)^2}\gamma_{23}^2 - \frac{\partial V}{\partial a_2} \quad (63)$$

$$\ddot{a}_3 = \frac{a_3(a_3^2 + 3a_2^2)(a_3^2 - a_2^2)}{(a_3^2 + a_2^2)^2}\gamma_{32}^2 + \frac{a_3(a_3^2 + 3a_1^2)(a_3^2 - a_1^2)}{(a_3^2 + a_1^2)^2}\gamma_{31}^2 - \frac{\partial V}{\partial a_3} \quad (64)$$

and for the hyper-angle part, based on the *vielbein* formalism [7], we have

$$\frac{d}{dt}(\tilde{M}\gamma_D) + \gamma_D \times (\tilde{M}\gamma_D) = -(\Lambda^{-1})^T \begin{pmatrix} \partial V/\partial\varphi_1 \\ \partial V/\partial\varphi_2 \\ \partial V/\partial\varphi_3 \end{pmatrix} \quad (65)$$

In a system free of the potential term, the democratic angular momentum vector

$$L_D \equiv U\tilde{M}\gamma_D \quad (66)$$

becomes a constant of motion.

As is evident from the first and the second terms on the right-hand sides of Eqs. (62)–(64), the democratic centrifugal force (DCF) arises to act on the gyration radii, a_1, a_2, and a_3, in response to the democratic rotation, the angular velocity of which is specified by $\{\gamma_{\alpha\beta}\}$. In general, the component $\gamma_{\alpha\beta} = \dot{u}_\alpha \cdot u_\beta (\alpha, \beta = 1,2,3, \alpha \neq \beta)$, which specifies the coupling strength between the α-axis and β-axis of the internal space, yields the field of DCF on all the planes parallel to the a_α–a_β plane in the gyration space. Appearance of this field is exactly the same as that in Fig. 2, where a_1 and a_2 in Fig. 2 are replaced by a_α and a_β ($a_\alpha \geq a_\beta$), respectively. As for the largest gyration radii a_1, DCF works always in the positive direction since $a_1^2 - a_2^2 \geq 0$ and $a_1^2 - a_3^2 \geq 0$ in Eq. (62).

While for a_2, it can work in both positive and negative direction depending on the sum of the first term (always negative) and the second term (always positive) on the right-hand side of Eq. (63). Finally, for the smallest gyration radii a_3, DCF always works in the direction that the absolute value of a_3 is diminished since $a_3^2 - a_2^2 \leq 0$ and $a_3^2 - a_1^2 \leq 0$ in Eq. (64). Thus the four-atom cluster generally tends to be distorted in such a direction that massive directions should become more massive. The role of the DCF in the dynamics of our M_4 cluster is studied in Section V.C.

B. Collective Coordinates Dominating the Isomerization Mechanism

Here we describe the isomerization dynamics of the four-atom Morse cluster, M_4, in terms of the PAHC. At the local equilibrium point of this cluster, which corresponds to a regular tetrahedron structure, the three gyration radii are exactly degenerate, $a_1 = a_2 = |a_3| = 4.24$, reflecting that the system is in an isotropic mass balance. Figure 13a shows a typical time evolution of the three gyration radii. When the system is in the vicinity of the local equilibrium structure, values of a_1, a_2, and $|a_3|$ are mutually close, while a_1 and a_3 deviate largely from their equilibrium values in the isomerizing motions. The smallest gyration radii a_3 reaches zero at an instant that the system crosses the planar structures, which constitute the so-called dividing surface for the isomerization reaction.

Figure 13b shows the time evolutions of the three hyper-angles in the same dynamics as in Fig. 13a. It can be seen that the hyper-angles vary rapidly when the system is in the vicinity of the equilibrium structure, while they are almost locked to certain values during the period of collective isomerization motions. The origin of the rapid democratic rotations in the vicinity of the local equilibrium structure lies in the degeneracy among the three gyration radii there. Similarly to Eq. (30), the components of the democratic angular velocity $\gamma_{\alpha\beta}$ ($\alpha, \beta = 1, 2, 3, \alpha \neq \beta$) are expressed as

$$\gamma_{\alpha\beta} = \dot{\boldsymbol{u}}_\alpha \cdot \boldsymbol{u}_\beta = \frac{\boldsymbol{u}_\alpha^T(\dot{\mathsf{W}}_s^T \mathsf{W}_s + \mathsf{W}_s^T \dot{\mathsf{W}}_s)\boldsymbol{u}_\beta}{a_\alpha^2 - a_\beta^2} \quad (67)$$

which suggests that the coupling between the two axes of the internal frame, $\boldsymbol{u}_\alpha$ and $\boldsymbol{u}_\beta$, should be enhanced in the vicinity of the degeneracy $a_\alpha^2 \approx a_\beta^2$. The flatness of the potential curve with respect to the hyper-angles is, in addition, the necessary condition for the rapid democratic rotation.

It is again remarkable that all of the hyper-angles are locked in the isomerizing intervals similarly to the case of the three-atom cluster. Correspondence between the set of locked angles and the reaction channels is listed in Fig. 12. In our four-atom cluster composed of identical atoms, the angles at which the hyper-angles are locked are associated with the symmetry of

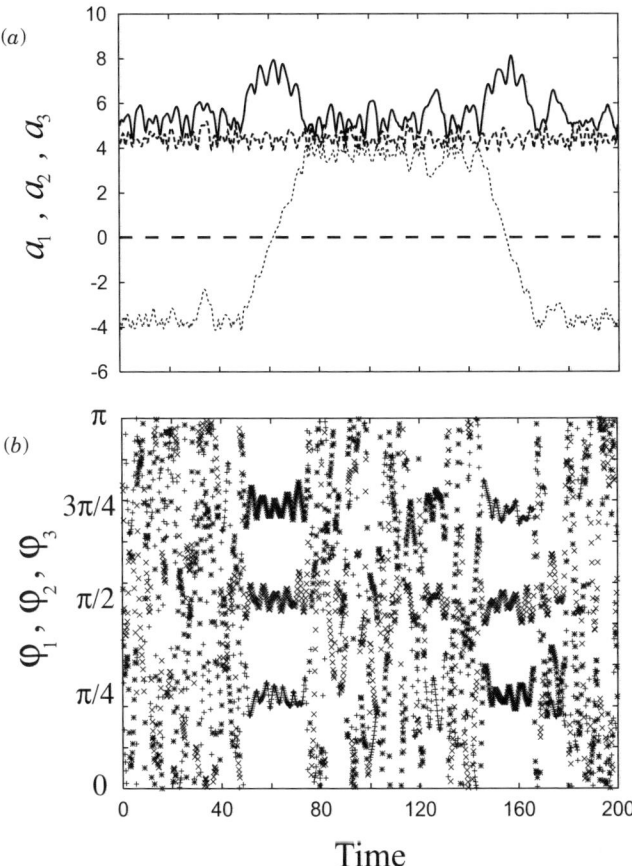

Figure 13. A typical time evolution of the three gyration radii, a_1, a_2, and a_3 ($a_1 \geq a_2 \geq a_3$) (a), and the three hyper-angles, $\varphi_1(+), \varphi_2(\times)$, and $\varphi_3(*)$ (b). Structural isomerization reaction has taken place two times in this example by crossing the planar structures as indicated by the condition $a_3 = 0$. Total internal energy is $E = -4.1\varepsilon$.

puckered rhombus (point group C_{2v}). This fact can be confirmed by Eqs. (47) and (50). The M_4 cluster itself can be continuously deformed to the planar rhombus structure (point group D_{2h}), which is the saddle structure, keeping the puckered-rhombus symmetry. Therefore the observed locking phenomenon of the hyper-angles during the periods of isomerizing motions is the result of the fact that the cluster undergoes the reaction, nearly keeping the puckered rhombus symmetry. Although the locked angles depend on the choice of the Jacobi vectors, the locking phenomenon itself does not.

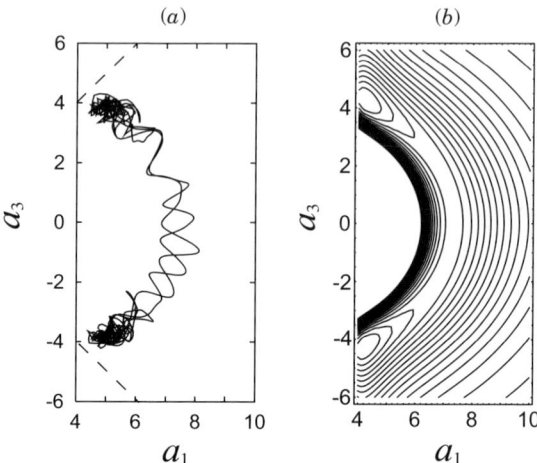

Figure 14. (a) A trajectory on the a_1–a_3 plane corresponding to the time evolution in Fig. 13. The broken lines are the degeneracy lines between a_1 and $|a_3|$. (b) Potential energy surface on the a_1–a_3 plane fixing the remaining gyration radius $a_2 = 4.25$ and all the hyper-angles to one of the sets given in Fig. 12. There exist local potential minima at $(4.24, 4.24)$ and $(4.24, -4.24)$ which correspond to the permutationally distinct tetrahedral structures. The saddle point is located at $(a_1, a_3) = (7.33, 0)$, which corresponds to the planar-rhombus-shaped structure. The energy difference between the neighboring contour lines is 0.278ε.

The locking phenomenon of all the hyper-angles, together with the fact that a_2 does not vary largely during the period of the isomerization motion, indicates that the collective variables dominating the isomerization are the two gyration radii a_1 and a_3. Hence a reaction path can be effectively extracted on the a_1–a_3 plane. Figure 14a shows a trajectory on the a_1–a_3 plane corresponding to the time evolution in Fig. 13, from which we can perceive a rough picture for a reaction pathway. In fact, the reaction path exists as shown in Fig. 14b, which shows the potential energy surface on the a_1–a_3 plane with the remaining internal variables fixed to the values corresponding to the rhombus-shaped saddle structure. That is, $a_2 = 4.25$ and the values for the hyper-angles are given in Fig. 12. On the a_1–a_3 plane in Fig. 14b, two equilibrium points corresponding to the permutationally distinct tetrahedral structures are connected by a curved reaction path that passes through a saddle point corresponding to the planar-rhombus structure. The isomerizing motion of the M_4 cluster proceeds overwhelmingly along this reaction path. The reaction path of Fig. 14b is closed as the hyper-angles deviate from the values listed in Fig. 12. In this way, the hyper-angles switch the reaction gate in the a_1–a_3 space. Thus the mechanism of the reaction is quite similar to the case of the three-atom cluster.

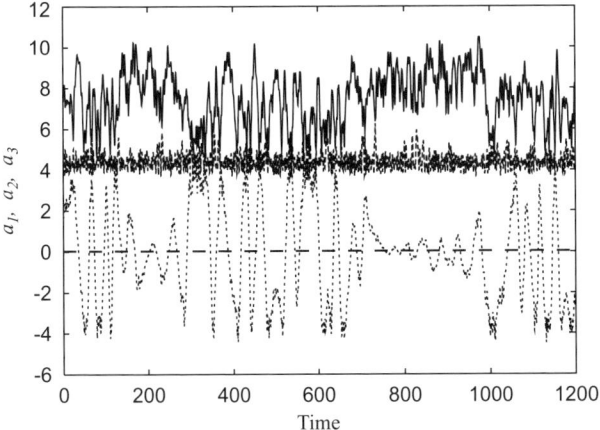

Figure 15. A typical time evolution of the gyration radii, a_1, a_2, and a_3 ($a_1 \geq a_2 \geq a_3$), showing the trapped motion around the planar saddle structure. The four-atom cluster stays in the region around the planar structure from $t \approx 150$ to $t \approx 300$ and from $t \approx 700$ to $t \approx 950$, recrossing the dividing surface again and again. The total internal energy of the trajectory is $E = -3.06\varepsilon$.

C. Trapped Motion Around the Transition State Due to DCF

Similar to the M_3 cluster, M_4 often shows a trapped motion around the planar-rhombus-shaped saddle structure. Figure 15 shows a typical time evolution of the three gyration radii, a_1, a_2, and a_3, where the values of a_1 and $|a_3|$ are kept apart for a while without going to their equilibrium values. This clearly indicates that the M_4 cluster is trapped around the planar-rhombus-shaped saddle structure and recrosses the potential barrier several times. The similar behavior is frequently observed, especially at the high internal energy. Again, this peculiar motion is explained in terms of the DCF. When the system is in the vicinity of the planar-rhombus-shaped saddle structure, the internal frame U in Eq. (47) is usually locked and oscillates slightly around one of the orientations specified by the hyper-angles in Fig. 12. This small oscillation generates a DCF in the gyration space and can affect the behavior of trajectories on the a_1–a_3 plane. As we did in the case of the three-atom cluster, we define an effective potential V_{eff} according to Lagrangian Eq. (61) as

$$V_{\text{eff}} = -\frac{(a_1^2 - a_2^2)^2}{2(a_1^2 + a_2^2)}\gamma_{12}^2 - \frac{(a_2^2 - a_3^2)^2}{2(a_2^2 + a_3^2)}\gamma_{23}^2 - \frac{(a_3^2 - a_1^2)^2}{2(a_3^2 + a_1^2)}\gamma_{31}^2$$
$$+ V(a_1, a_2, a_3, \varphi_1, \varphi_2, \varphi_3) \qquad (68)$$

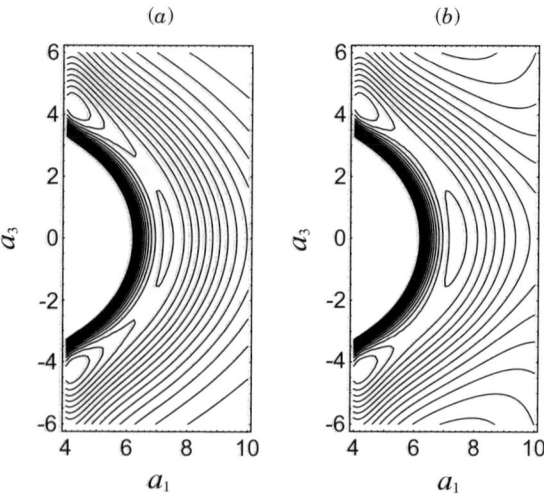

Figure 16. Effective potential energy surfaces, Eq. (68), on the a_1–a_3 plane fixing the remaining four internal variables to the same values as in Fig. 14b. (a) $\gamma_{12} = 0.05, \gamma_{23} = 0.1, \gamma_{31} = 0.05$. (b) $\gamma_{12} = 0.2, \gamma_{23} = 0.1, \gamma_{31} = 0.05$. The energy difference between the neighboring contour lines is 0.278ε.

A typical energy surface of this effective potential on the a_1–a_3 plane is shown in Fig. 16a, where we have set the components of democratic angular velocity as $\gamma_{12} = 0.05$, $\gamma_{23} = 0.1$, $\gamma_{31} = 0.05$, which are typical values for these components during the locking of the hyper-angles. (The remaining internal variables are fixed to the same values as in Fig. 14b.) We clearly see that a basin structure of the effective potential has appeared at the saddle point of the usual potential energy surface, which accounts for the fact that the planar rhombus-shaped saddle structures are stable enough to allow for the trapped motion. Among the six terms of the DCF appearing in Eqs. (62)–(64), those proportional to γ_{32}^2 and γ_{31}^2 on the right-hand side of Eq. (64) are important for the trapped motion because these terms always work so as to diminish the absolute value of a_3 to keep the system planar. The components of the DCF proportional to γ_{12}^2 (or γ_{21}^2), on the other hand, do not act on the smallest gyration radius, a_3, directly but act on a_1 and a_2 to enlarge a_1 and to diminish a_2 [cf. Eqs. (62) and (63)]. Therefore it is expected that the increase of γ_{12}^2 should have an effect of broadening the basin on the effective potential around the saddle point on the a_1–a_3 plane in the direction of positive a_1-axis. This effect can be observed in Fig. 16b, where only the value of γ_{12} is increased to $\gamma_{12} = 0.2$ with other values fixed to the same values as in Fig. 16a. In addition, it is also expected that the components of the

DCF proportional to γ_{12}^2 should have an effect to further distort a trapped planar M_4 cluster into a linear structure.

D. The Suppressing Effect of the Gauge Field Associated with the Eckart Frame

In Section IV, we have demonstrated that the gauge field associated with the Eckart frame has an effect of suppressing the reaction rate of M_3. Here, we confirm that this suppressing effect is general and can be observed in the four-atom cluster M_4. The total angular momentum of the system is again zero.

Our strategy here is similar to that in Section IV.C: We compare the rate of isomerization of the M_4 cluster in the two kinds of dynamics; one is the true dynamics under the influence of the correct gauge field, and the other is the dynamics in which the effect of the gauge field is eliminated. The classical equations of motion for the former dynamics is expressed as in Eq. (37) in the normal-mode coordinates $\{q^\mu\}(\mu = 1, \ldots, 6)$, while that for the latter is in the form of Eq. (35) for the same coordinates. The latter dynamics has a geometrical meaning of the constrained dynamics to the six-dimensional Eckart subspace in the nine-dimensional translation-reduced configuration space of M_4. Here we have set the reference configuration for the Eckart subspace to the local equilibrium configuration of type (A) in Fig. 12.

We have randomly prepared 5000 initial conditions for the dynamics of M_4 in the potential basin of the local equilibrium configuration of type (A) in Fig. 12 with zero initial momenta so as to set them common to the two dynamics. Total internal energy of all the trajectories is set to $E = -3.6\varepsilon$, which is above the isomerization threshold. We have monitored the time-dependent number of the surviving trajectories for the two different dynamics that remain in the initial potential basin. (We have discarded a trajectory from the set of surviving trajectories once it passes across the dividing surface.) The result is shown in Fig. 17. The survival number decays mostly in an exponential manner in the two dynamics, reflecting the fact that they are sufficiently stochastic. It is evident that the existence of the gauge field reduces the reaction rate. The internal-energy dependence of the mean first passage time in the two different dynamics is also measured, which is shown in Fig. 18. Throughout the shown energy range, the isomerization reaction under no gauge field proceeds faster than that under the gauge field effect. Difference in the reaction rates of the two dynamics amounts to about 20% to 40%. These results are in accordance with those in Figs. 9 and 10 for the M_3 cluster. Thus we have confirmed that the gauge field associated with the Eckart frame suppresses the reaction rate significantly not only in the three-atom but also in the four-atom cluster.

The effect of the gauge field associated with the Eckart frame to suppress the reaction rate seems to be more general than is valid for M_3 and M_4. Indeed we have confirmed the similar effect of the gauge field in the isomerization

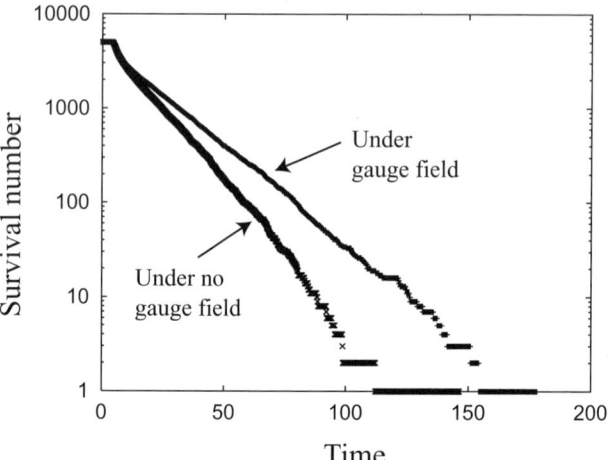

Figure 17. Decay of the number of surviving trajectories of the M_4 cluster staying in the initial potential basin. The lower curve represents the dynamics constrained to the Eckart subspace under no influence of the gauge field, while the upper one indicates the true dynamics under the influence of the gauge field. Number of the sample trajectories is 5000 and their internal energy is set to $E = -3.6\varepsilon$ for the respective dynamics.

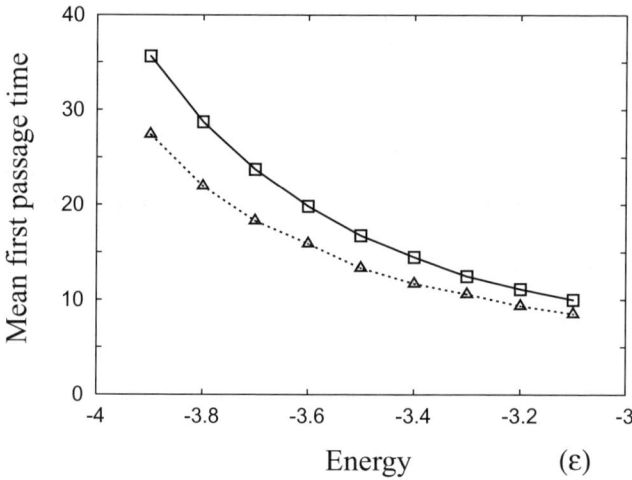

Figure 18. Internal-energy dependence of mean first passage time (average lifetime) of the M_4 cluster. The lower curve (open triangle) represents the dynamics constrained to the Eckart subspace under no gauge field, while the upper curve (open square) corresponds to the result for the true dynamics under the full influence of the gauge field.

dynamics of a seven-atom Morse cluster, M_7. Therefore it is expected that the suppressing effect should be universal, although more general proof is necessary.

VI. CONCLUDING REMARKS

Based on a gauge-theoretical formalism for the separation of rotation and internal motion in many-body systems, we have explored kinematic effects arising from a true metric (Section III) and a gauge field (Section IV) in internal space in the structural isomerization dynamics of three-atom cluster under zero total angular momentum. These results have been extended to the dynamics of four-atom cluster subsequently (Section V). In Section III, a coordinate system called the principal-axis hyperspherical coordinates (PAHC) was employed effectively to elucidate both the mechanism of the isomerization reaction and the kinematic effects in the dynamics. A reaction path has been extracted on a two-dimensional space of gyration radii based on the locking phenomenon of the hyper-angle of the PAHC during the isomerizing collective motions. We have obtained a metric force called "democratic centrifugal force (DCF)," which acts on gyration radii in response to the hyper-angular motion (democratic rotation), as a significant kinematic factor that dominate the collective motion. It has been revealed that the DCF generally has an effect of inducing an asymmetry in mass balance of a molecule along the principal axes; therefore, the triatomic system tends to be collapsed toward the collinear shape, which helps the isomerization reaction to start. A remarkable fact is that the collinear saddle region of the M_3 cluster is kinematically stabilized due to the DCF. In fact, an effective potential surface, which is defined as a sum of an ordinary potential function and the potential of the DCF, holds a new "basin" around the collinear transition state. Owing to this new basin, the M_3 cluster can be trapped in the transition region for a long time and recrosses the dividing surface many times. The present approach would provide a unified view of the recrossing problem and regularity of reactive trajectories in the saddle region.

Following the gauge-invariant study of the isomerization dynamics in terms of the PAHC as above, in Section IV, we have clarified the kinematic effects of the gauge field that arises in internal space based on the Eckart frame. We have found that the gauge field, which is often disregarded in the conventional picture for chemical reaction dynamics, has an effect of suppressing the isomerization reaction of the M_3 cluster significantly. Our numerical results for the M_3 cluster suggest that the reaction rate theories that disregard the effects of the gauge field could *overestimate* the reaction rates up to about 20% to 30%. We have also presented a theoretical explanation for this suppressing effect of the gauge field in terms of the change in the directionality of the DCF in gyration space. Essentially the same effect has been confirmed in the M_4 cluster in Section V.D.

It should also be mentioned that the suppressing effect of the gauge field associated with the Eckart frame on the rate of isomerization reaction has been confirmed universally in larger clusters, such as seven-atom clusters. According to our numerical experiment, the effect of the gauge field tends to be larger for the isomerization reactions during which the system greatly changes its mass balance. Therefore the roles of the gauge field is worth scrutinizing in the dynamics of macromolecules, too.

We believe the present study has effectively demonstrated the significance of kinematic effects in the dynamics of polyatomic molecules. It seems that their significance has been too much underestimated compared to that of the potential energy topography in the conventional picture for chemical reaction dynamics including polymer and protein dynamics. The above results would require a reconsideration of the conventional reaction-rate theories: First, the intrinsic non-Euclidean nature of internal space should be taken into account for the precise estimate of vibrational and rotational partition functions and the density of states that appear in reaction-rate formulae. Second, the trapped and recrossing motions in the vicinity of the transition state supported by the kinematic force, DCF, obviously work against the nonrecrossing hypothesis in the transition-state theory. Furthermore, the newly appeared basin on the effective potential surface of Fig. 8 and Fig. 16 implies the existence of kinematically stable states that are not expected from the bare potential topography alone.

Finally, we would like to point out that the kinematic effects of DCF inducing an asymmetry in mass balance in a many-body system should be of universal significance not only in molecular systems but also in a wide variety of many-body systems. This is because the DCF originates not from the interaction potential but from the intrinsic metric of internal space, which is uniquely determined from the shape and mass balance of a system. It is therefore anticipated that the DCF should be an important factor not only in molecular dynamics but in collective motions in nuclear, celestial, and biological many-body systems.

Acknowledgments

This work has been supported in part by the Grant-in-Aid for the 21st Century COE Program for Frontiers in Fundamental Chemistry from the Ministry of Education, Culture, Sports, Science, and Technology of Japan. One of the authors (T.Y.) has been supported by Research Fellowships of the Japan Society for the Promotion of Science for Young Scientists.

References

1. R. Montgomery, in *The Geometry of Hamiltonian Systems*, T. Ratiu, ed., Springer-Verlag, 1991, p. 403.
2. R. Montgomery, *Nonlinearity* **9**, 1341 (1996).

3. A. Shapere and F. Wilczek, *Phys. Rev. Lett.* **58**, 2051 (1987); *J. Fluid Mech.* **198**, 557 (1989).
4. A. Guichardet, *Ann. Inst. H. Poincaré* **40**, 329 (1984).
5. A. Tachibana and T. Iwai, *Phys. Rev. A* **33**, 2262 (1986).
6. T. Iwai, *Ann. Inst. H. Poincaré* **47**, 199 (1987); T. Iwai, *J. Math. Phys.* **28**, 964 (1987); **28**, 1315 (1987); T. Iwai, *Phys. Lett. A* **162**, 289 (1992).
7. R. G. Littlejohn and M. Reinsch, *Rev. Mod. Phys.* **69**, 213 (1997).
8. X. Chapuisat and A. Nauts, *Phys. Rev. A* **44**, 1328 (1991).
9. X. Chapuisat, *Phys. Rev. A* **45**, 4277 (1992).
10. X. Chapuisat, J. P. Brunet, and A. Nauts, *Chem. Phys. Lett.* **136**, 153 (1987).
11. A. Kuppermann, *J. Phys. Chem.* **100**, 2621 (1996); *J. Phys. Chem. A* **101**, 6368 (1997).
12. R. G. Littlejohn and M. Reinsch, *Phys. Rev. A* **52**, 2035 (1995).
13. R. G. Littlejohn, K. A. Mitchell, V. A. Aquilanti, and S. Cavalli, *Phys. Rev. A* **58**, 3705 (1998).
14. R. G. Littlejohn, K. A. Mitchell, M. Reinsch, V. Aquilanti, and S. Cavalli, *Phys. Rev. A* **58**, 3718 (1998).
15. V. Aquilanti and S. Cavalli, *J. Chem. Phys.* **85**, 1355 (1986).
16. E. B. Wilson, J. C. Decius, and P. C. Cross, *Molecular Vibrations*, McGraw-Hill, New York, 1955.
17. J. D. Louck and H. W. Galbraith, *Rev. Mod. Phys.* **48**, 69 (1976).
18. B. T. Sutcliffe, in *Quantum Dynamics of Molecules*, R. G. Wooley, eds., Plenum, New York, 1980, p. 1.
19. C. Eckart, *Phys. Rev.* **47**, 552 (1935).
20. W. H. Miller, N. C. Handy, and J. E. Adams, *J. Chem. Phys.* **72**, 99 (1980).
21. R. J. Hinde and R. S. Berry, *J. Chem. Phys.* **99**, 2942 (1993).
22. R. S. Berry, T. L. Beck, H. L. Davis, and J. Jellinek, *Adv. Chem. Phys.* **70**, 75 (1988); J. Jellinek, T. L. Beck, and R. S. Berry, *J. Chem. Phys.* **84**, 2783 (1986); F. G. Amar and R. S. Berry, *J. Chem. Phys.* **85**, 5943 (1986).
23. P. Labastie and R. L. Whetten, *Phys. Rev. Lett.* **65**, 1567 (1990).
24. C. Seko and K. Takatsuka, *J. Chem. Phys.* **104**, 8613 (1996); **108**, 4924 (1998); **109**, 4768 (1998); K. Takatsuka and C. Seko, *J. Chem. Phys.* **105**, 10356 (1996); **110**, 3263 (1999); T. Yanao and K. Takatsuka, *Chem Phys. Lett.* **313**, 633 (1999); K. Takatsuka and T. Yanao, *J. Chem. Phys.* **113**, 2552 (2000).
25. C. Amitrano and R. S. Berry, *Phys. Rev. Lett.* **68**, 729 (1992); *Phys. Rev. E* **47**, 3158 (1993); R. J. Hinde and R. S. Berry, and D. J. Wales, *J. Chem. Phys.* **96**, 1376 (1992).
26. T. Komatsuzaki and M. Nagaoka, *J. Chem. Phys.* **105**, 10838 (1996); T. Komatsuzaki and R. S. Berry, *J. Chem. Phys.* **110**, 9160 (1999); **115**, 4105 (2001).
27. R. S. Berry, *Chem. Rev.* **93**, 2379 (1993); P. A. Braier, R. S. Berry, and D. J. Wales, *J. Chem. Phys.* **93**, 8745 (1990); M. A. Miller, J. P. K. Doye, and D. J. Wales, *J. Chem. Phys.* **110**, 328 (1999).
28. F. H. Stillinger and T. A. Weber, *Phys. Rev. A* **25**, 978 (1982); R. A. LaViolette and F. H. Stillinger, *J. Chem. Phys.* **83**, 4079 (1985).
29. J. I. Steinfeld, J. S. Francisco, and W. L. Hase, *Chemical Kinetics and Dynamics*, Prentice-Hall, Englewood Cliffs, NJ, 1989.
30. W. H. Miller, *J. Phys. Chem. A* **102**, 793 (1998); *Faraday Discuss.* **110**, 1 (1998).

31. J. N. Onuchic, Z. A. Luthey-Schulten, and P. G. Wolynes, *Annu. Rev. Phys. Chem.* **48**, 545 (1997).
32. K. A. Dill and H. S. Chan, *Nature Struct. Biol.* **4**, 10 (1997); C. M. Dobson, A Šali, and M. Karplus, *Angew. Chem. Int. Ed.* **37**, 868 (1998).
33. G. Strang, *Linear Algebra and Its Applications*, Academic Press, New York, 1976.
34. J. Costley and P. Pechukas, *J. Chem. Phys.* **77**, 4957 (1982).
35. J. Manz, E. Pollak, and J. Römelt, *Chem. Phys. Lett.* **86**, 26 (1982).
36. H. Goldstein, *Classical Mechanics*, Addison-Wesley, New York, 1980.

CHAPTER 13

ONSET DYNAMICS OF PHASE TRANSITION IN Ar_7

NORIHIRO SHIDA

Omohi College, Graduate School of Engineering, Nagoya Institute of Technology, Gokiso-cho, Showa-ku, Nagoya, 466-8555, Japan

CONTENTS

I. Introduction
II. Method of Analysis
   A. Gradient Extremal
   B. Cell Petition of a Potential Energy Surface
   C. Potential Function, MD Simulation, and Temperature
III. Results and Discussions
   A. Structures
   B. Reaction Paths
   C. Lindemann's Criterion
   D. Configuration Entropy
   E. Extent of the Configuration Space
   F. Liapunov Exponent and KS Entropy
   G. Other Stationary Points
   H. Dynamics on Partitioned Cells
   I. Watershed
   J. Lindemann's Criterion of Ar_5
IV. Summary and Conclusion
Acknowledgment
References

I. INTRODUCTION

Micro-clusters consisting of a few molecules or atoms are in intermediate position between individual molecules and a condensed phase, and their properties can be analyzed as both dynamical and statistical properties. Argon

---

Geometric Structures of Phase Space in Multidimensional Chaos: A Special Volume of Advances in Chemical Physics, Part B, Volume 130, edited by M. Toda, T Komatsuzaki, T. Konishi, R.S. Berry, and S.A. Rice. Series editor Stuart A. Rice.
ISBN 0-471-71157-8   Copyright © 2005 John Wiley & Sons, Inc.

clusters are one of such clusters and have been widely investigated in the last quarter of a century [1–26]. Among various sizes of the argon clusters, Ar_7 is especially interesting since it is the smallest size of the cluster that has a clear nature of phases, such as a solid-like phase or a liquid-like phase [1–4]. The transition between these two phases is thought to be a "melting" and a "freezing" and has been extensively studied by several people [1–20,27], especially by Berry's group.

In the early days of the work, this phase transition was mainly analyzed from the structures of Ar_7 including the conformation change of the isomers, the various energy profiles, and the fluctuation of these properties [1–9,11]. Among these analyses, Lindemann's δ was often used to monitor the phase transition. Most of these calculations carried out in these days are the Molecular Dynamics and the Monte Carlo calculations since the quantum effect was expected to be small for argon atoms [21,22,26]. These types of analyses can be categorized as the analysis of the potential energy features from the various directions.

From the early 1990s, the phase-space dynamics of Ar_7 has been investigated by using the methodologies of the nonlinear dynamics [10,11–20,27]. The primarily interested quantities were the KS entropy and the Liapunov exponents [28]. These quantities are the indexes to express the variety of the phase space and were extensively studied in the phase-space dynamics of Ar_7 [10,12–20,27,29]. Among these, Hinde et al. presented an intuitive picture of the relationship between the potential energy surface (PES) and the dynamics. In spite of this successful work, the KS entropy and the Liapunov exponents themselves were thought to be the unsuitable quantities to express the phase transition since they increase smoothly as the temperature increases and there is no sudden change at the phase transition [10,12,13]. In 1995, Nayak et al. claimed that the largest Liapunov exponent has an analogous behavior with Lindemann's δ and is able to become a good index to analyze the phase transition. The usefulness of the Liapunov exponent was considered again, and an extended concept of the Liapunov exponent was proposed [29].

Generally speaking, the degrees of freedom in many-body systems, such as Ar_7, are too many to analyze the phase-space dynamics, and only limited methods originally developed to investigate chaotic systems with a few degrees of freedom can be applicable for the analysis. Seko et al. calculated the phase volume—that is, the configuration entropy—of Ar_7 and proposed a new concept of the temperature in micro-clusters based on this phase volume [17]. A phase-space analysis seems to be prospective even for many-body systems, such as Ar_7. However, most of the currently available methods concern statistical properties. The methods and quantities that are directly related to the dynamics are expected for a detailed analysis.

A reaction path that was first proposed by Fukui [30] has been successfully used to describe chemical reactions in gas phases. In this method, chemical reactions are described along unique one-dimensional pathways. The reaction path by Fukui's criterion is the steepest descent path in a mass-weighted Cartesian coordinates staring from a saddle point to a local minimum in a PES. In spite of the fact that this reaction path is defined only in a configuration space without referring a momentum space, it works fairly well to describe various types of the chemical reactions. Many researchers suggested that the coexistence and the liquid-like phases of Ar_7 could be characterized by the isomerization dynamics [31]. The quenched structure of each isomer is in the position of the basin of the PES. Therefore, they can be regarded as a reactant or a product in chemical reactions. In this sense, the isomerizationo dynamics are equivalent to a crowd of the elementary processes of the chemical reactions.

In this work, the phase transition of Ar_7, especially the onset dynamics of the melting, is analyzed from various points of view using a new method. This new method is essentially a generalization of the reaction path concept. In this method, the global region of a PES is divided into many cells and each cell contains a unique stationary point as if it were the nameplate of the cell. The PES is, thus, symbolically expressed by the set of stationary points. The real dynamics of Ar_7 are projected onto this symbolic expression of the PES and are analyzed precisely. The gradient extremal criterion is introduced to get the various types of the stationary points. Other aspects, such as the conventional reaction path description for isomerization, Lindemann's δ, Liapunov exponents, KS entropy, configuration entropy, and so on, are also discussed with a new insight. In Section II, the method of the gradient extremal, the way of the cell partition, and other details about the calculations are briefly described. The results and the discussion are in Section III.

II. METHOD OF ANALYSIS

This section describes the methods of calculation used in this work.

A. Gradient Extremal

The gradient extremal is a mathematical nature of a hypersurface [32–34]. The most rigid discussion of the gradient extremal is due to Hoffman et al. [32] in 1986. Assuming an M-dimensional surface,

$$V(\mathbf{x}) = f(x_1, x_2, \ldots, x_3) \tag{1}$$

the term contour subspace is introduced to denote the $(M - 1)$-dimensional subspace defined by the condition $V(x) = $ constant. The gradient extremal is defined as a one-dimensional path where the absolute value of the gradient is extremal on each contour subspace. Mathematically, it is defined as follows:

On the contour subspace $V(x) = V_0$, the gradient extremal path (GEP) passes through the point where the square norm of the gradient is extremal. Thus, introducing the Lagrange multiplier λ, the following equation must be satisfied on the GEP:

$$\frac{\partial}{\partial \mathbf{x}}(\mathbf{g}^T(\mathbf{x})\mathbf{g}(\mathbf{x}) - 2\lambda(V(\mathbf{x}) - V_0)) = 0 \qquad (2)$$

where $\mathbf{g}(x)$ is the gradient vector. Charring out the differentiation, we obtain the following equation;

$$\mathbf{H}(\mathbf{x})\,\mathbf{g}(\mathbf{x}) = \lambda \mathbf{g}(\mathbf{x}) \qquad (3)$$

where $\mathbf{H}(\mathbf{x})$ is the Hessian matrix. Equation (3) is the definition of the GEP. This definition is a local criterion since Eq. (3) is always fulfilled at every point on the GEP. It is contrast to the steepest descent path, which is defined numerically by the steepest descent direction from the top to the bottom. Every stationary point is on the GEPs since $\mathbf{g}(x)$ is zero on this point and Eq. (3) always is satisfied. It can be shown that a GEP starting from a stationary point continues to another stationary point if it exists. Physically, taking the extremal of $\mathbf{g}(x)$ as the minimum, this GEP can be interpreted as the least ascent path and is expected to become the post of the reaction path [34]. Obviously, this path can be calculated from both upward and downward directions, and they give the same result. Jørgensen et al. [33] proposed an iterative algorithm to calculate GEPs. They showed that the direction of the step vector could be approximately obtained by adding some correction to one of the Hessian eigenvectors. However, it should be mentioned that there is always a finite angle between the step vector and the Hessian eigenvector except for the stationary points [32]. In this sense, the eigenvector-following algorithm [35] that is widely used to find saddle points is the approximation of the GEP. In the eigenvector-following algorithm, it is necessary to switch to the Newton–Raphson step near the saddle point since there is no warranty to exist the saddle point on the path. In contrast, no such procedure is required for the GEP. However, the true step vector of the GEP requires the third-order derivatives [32]. In the present work, all the stationary points are needed for the analysis. It is particularly important to use a true step vector to find all the stationary points. Otherwise, some of them may be overlooked. The procedure to find all the stationary points was done by an iterative way. We start from the quenched structures and then climb up for all the possible directions on the various GEPs. If the reached points are unknown stationary points, we restart again from these points and climb up and go down for all the possible directions. This process is repeated until a new point is no longer found.

ONSET DYNAMICS OF PHASE TRANSITION IN Ar_7 133

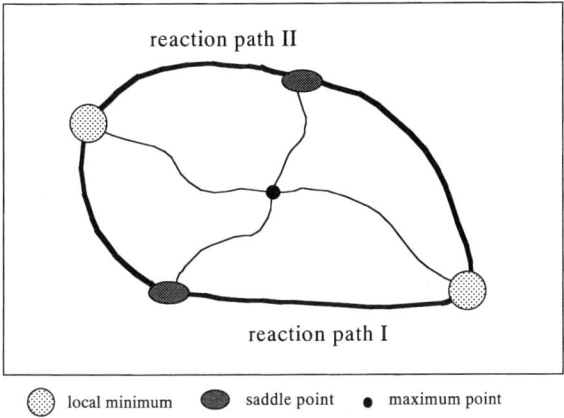

Figure 1. An illustration of a two-dimensional potential energy surface where two local minima and two saddle points exist.

B. Cell Petition of a Potential Energy Surface

A reaction path is characterized by the two local minima and the saddle point of a PES. In other words, a local minimum and the saddle point are directly related to the reaction. For later convenience, the term "class" will be used to express the number of negative Hessian eigenvalues (i.e., a local minimum is a class 0 and a saddle point is a class 1). Mezey [36,37] stated that other stationary points whose class is more than 1 are not directly related to a reaction, but they are characterizing other stationary points with lower classes. His argument can be understood by a typical example in Fig. 1. Figure 1 is a two-dimensional PES where two local minima and two saddle points exist. For this kind of topology, two kinds of different reaction paths can be defined to connect both sides of the local minima. Obviously, these local minima and saddle points are characterizing these reaction paths. However, the remaining maximal point in the middle of the figure also has a special meaning. Namely, the existence of this maximal point guarantees the existence of the two saddle points. Mezey's argument is a generalization of this idea. In addition to the meaning of such stationary points of high classes, there is a practical advantage to use these points for the present analysis. Stationary points with various classes are scattered around the various region of a PES. In this work, a new method to symbolize the global region of a PES was developed by using the nature described above. Assume a one-dimensional potential energy curve like the one in Fig. 2. It is a typical example of a reaction path, where there are the two local minima and the saddle point. Since the second derivative is positive at the local minimum and is negative at the saddle point, an inflection point always exists between the local minimum and

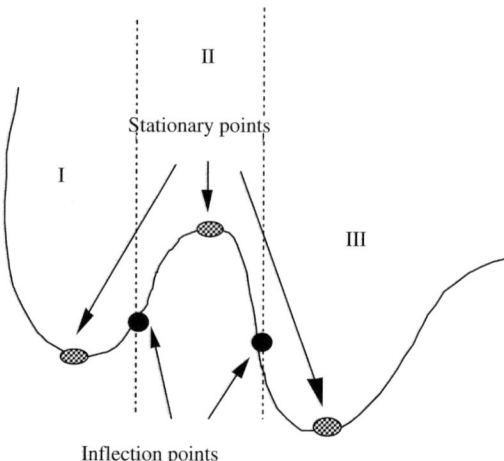

Figure 2. A typical example of a one-dimensional potential energy curve where there are the two local points and the saddle point.

the saddle point. In Fig. 2, there are two inflection points among the three stationary points. We will extend this idea to a multidimensional case. The situation may be illustrated as Fig. 3. In this multidimensional case, the inflection point forms the $(n-1)$-dimensional inflection surface, where n represents the dimension of the PES. And by this inflection surface, the entire region of the PES is divided into many "cells." From the analogy of the one-dimensional case, a unique stationary point always exists in each cell like a cell nucleolus.

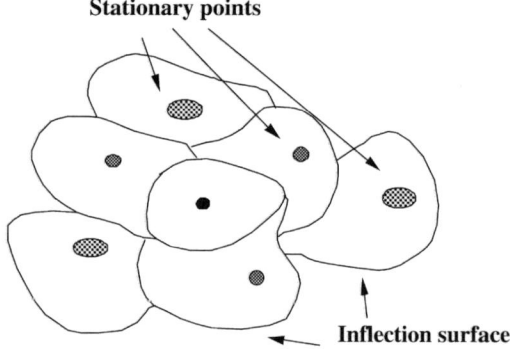

Figure 3. Stationary points and inflection surfaces in a multidimensional potential energy surface.

C. Potential Function, MD Simulation, and Temperature

Throughout the entire analysis in this work, the following pairwise Lennard-Jones potential (6-12) was used as the potential function of Ar_7:

$$V = \sum_{i<j}^{7} 4\varepsilon \left[\left(\frac{\sigma}{r_{ij}}\right)^{12} - \left(\frac{\sigma}{r_{ij}}\right)^{6} \right] \quad (4)$$

where $\sigma = 3.4$ Å and $\varepsilon = 1.67 \times 10^{14}$ erg, which are the common parameters for Ar_7 [38]. For the MD simulations, the usual fourth-order Adams–Moulton predictor–corrector method was used to solve the classical equation of motion numerically. Jellinek et al. [5] defined the "internal temperature" for argon clusters as the long-time average of the kinetic energy as follows:

$$T = \frac{2N}{3N-6} \frac{E_{kin}}{k} \quad (5)$$

A dynamical temperature is an analogy of the temperature in a condensed phase. Seko et al. proposed an alternative definition of "temperature" for clusters to satisfy the statistical mechanics. Since the meaning of the temperature itself is out of the topics in this work, we will just use the dynamical temperature defined in Eq. (5). Figure 4 shows the relationship between the total energy from the

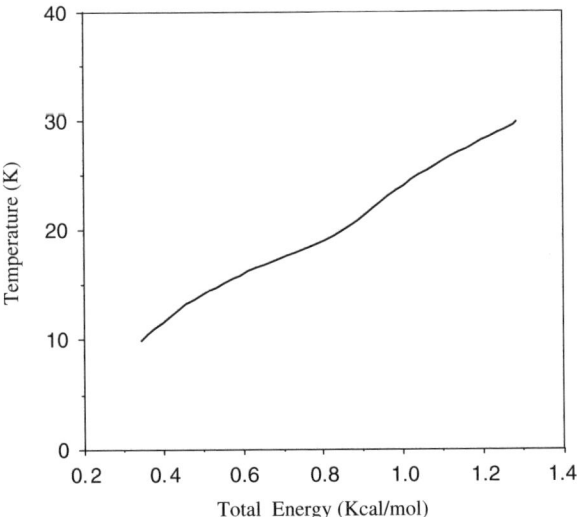

Figure 4. Relationship between dynamical temperature (K) and the total energy from the lowest inherent structure.

lowest inherent structure and the dynamical temperature. According to the convention in condensed phase, we prefer the temperature rather than the total energy in the later discussion. But, it does not have any special meaning.

III. RESULTS AND DISCUSSIONS

A. Structures

According to the previous work, four inherent structures (local minima of the potential energy surface) have been found in the Ar_7 cluster. They were first reported by Hoare and Pal [39]. In this work, these four geometries were reconfirmed again as the starting point for further analyses. These four geometries are shown in Fig. 5. Note that an arbitral permutation among Ar atoms forms the same structure since this cluster is a uni-atom cluster. Thus, each inherent structures actually correspond to the $7! = 5040$ different structures. In Fig. 5, structure (a) has the highest symmetry (Ih) compared with other structures and has the lowest energy. The relative energies of other structures from (a) are 0.14, 0.22, and 0.23 kcal/mol. So, the dynamics of this cluster are expected to be like a lattice vibration trapped into the inherent structures (a) when the energy is very low. Figure 6 shows the 12 geometrical saddle points of the PES—that is, the transition states in the usual sense. These structures and energies have been also reported by Wales et al. [39]. However, some other transition states are newly found in this work by the GEP described in the Section II.A. Each transition state in Fig. 6 has the $7! = 5040$ different structures as well as the inherent structures. The relative energies of the transition states are in 0.26–0.47 kcal/mol.

B. Reaction Paths

For a chemical reaction in a gas phase, the reaction path method described in Section I has been successfully used. Although the dynamics of Ar_7 is not a true chemical reaction, it can be regarded as a crowd of chemical reactions in the coexistence and the liquid-like phases. In Fig. 7, all the reaction paths among the 4 local minima and the 12 saddle points are illustrated, together with the relative energies. In Fig. 7, a tangled connection diagram is found and every local minimum is connecting to the plural transition states. Using Fig. 7, one can pick

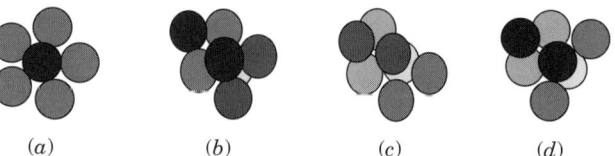

(a)   (b)   (c)   (d)

Figure 5. The quenched structures of Ar_7. (a) Absolute minimum.

ONSET DYNAMICS OF PHASE TRANSITION IN Ar_7 137

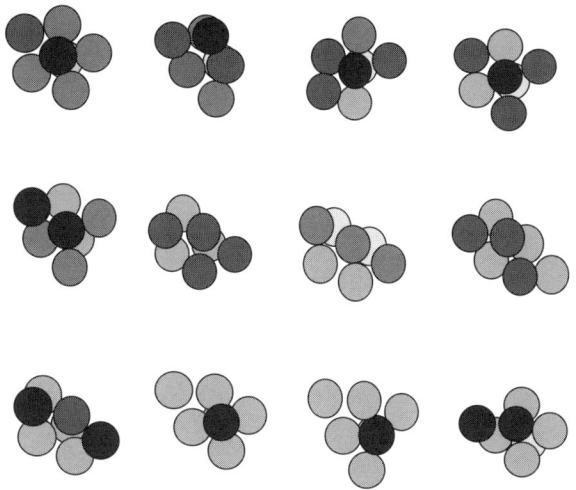

Figure 6. The geometrical saddle points of the potential energy surface of Ar_7.

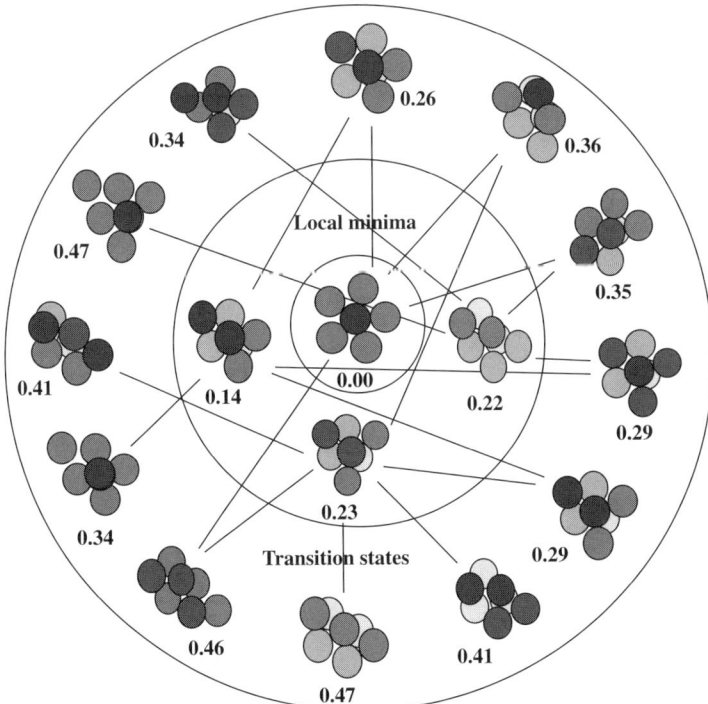

Figure 7. All the reaction paths among the four local minima and the 12 saddle points.

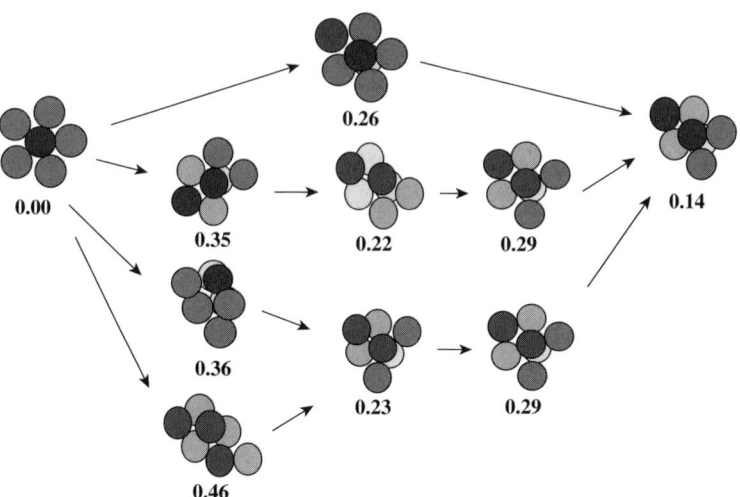

Figure 8. The reaction paths from the lowest inherent structure (*a*) to the second lowest one (*b*).

up the individual reaction paths for the conformation change between the different inherent structures. The example is shown in Fig. 8. In Fig. 8, the reaction paths from the lowest inherent structure (a) to the second lowest one (b) are illustrated. Since the shape of the potential energy surface is so complicated, the multiple pathways exist in this conformation change. Only from the energetic point of view, the uppermost reaction path passing through the transition state that has the energy of 0.23 kcal/mol is expected to become the most dominant one. However, the reaction path analysis only refers the PES and it does not guarantee the true dynamics. This will be discussed later.

C. Lindemann's Criterion

The root mean square fluctuation of the particle distance δ, which is referred to as Lindemann's criterion, has been successfully used to monitor the phase transition of a micro-cluster:

$$\delta = \frac{2}{N(N-1)} \sum_{i<j} \frac{\left(\langle r_{ij}^2 \rangle - \langle r_{ij} \rangle^2\right)^{1/2}}{\langle r_{ij} \rangle} \qquad (6)$$

Equation (6) shows that the dynamics is expected to be a regular motion when δ is small, such as a lattice-like vibration in a solid phase. As the δ value increases,

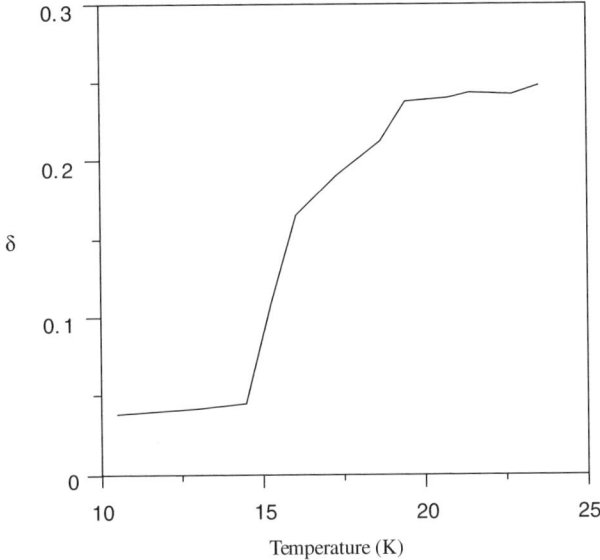

Figure 9. The root mean square fluctuation of the particle distances, δ, which is referred to as Lindemann's criterion against the temperature.

the motion gradually becomes irregular and becomes more like liquid-phase dynamics. For the Ar_7 clusters, several researches have calculated this index to see melting and other phenomena [1–20,27]. The previous work shows that the melting of Ar_7 occurred around ~15 K. In this work, this index was calculated again for the later analysis. These results are shown in Fig. 9. The rapid increment of δ can be found around 15–20 K in Fig. 9. This can be interpreted as the melting of Ar_7, and this energy region corresponds to the coexistence region. We will focus our attention on the onset of the melting. At this onset, the dynamics may suddenly change its dynamics from the lattice vibration trapped into the lowest minimum of the PES to the large-amplitude vibration traveling among various local minima across the transition states. In this sense, the onset dynamics of the melting can be really regarded as a chemical reaction. It may be natural to expect that the reaction paths in Fig. 8 describe the beginning step of this reaction. However, the temperature of 15–20 K corresponds to the total energy of 0.52–0.83 kcal/mol (see Fig. 4). This energy is much higher than the transition energies in Fig. 8. Figure 8 shows that the reaction energy of 0.26 kcal/mol is enough to go across the lowest transition state and is enough to reach the second lowest minimum. This fact implies that the real dynamics is somewhat different from the conventional reaction path description.

D. Configuration Entropy

When the melting occurs, the phase space changes its volume rapidly, and its changes directly reflect on the configuration entropy (S) defined as

$$S = k \ln(W) \tag{7}$$

where W is the classical phase volume or the quantum mechanical density of states. Seko and Takatsuka [17] calculated the configuration entropy based on the Thomas–Fermi statistics and used it to analyze the coexistence region. In this work, the configuration entropy was also calculated again to monitor the melting of Ar_7 in addition to Lindemann's δ by using an alternative method. The method used is the adiabatic switching of the Hamiltonian [40]. The essence of this method is that the density of states always conserves when the total Hamiltonian changes very slowly with respect to time (i.e., adiabatically). So, we start from the exact Hamiltonian of Ar_7; the classical equation of motion is solved numerically by the usual MD-type calculation. During this calculation, the total Hamiltonian of the system $(H(t))$ changes from the exact Hamiltonian (H_0) to the model Hamiltonian (H_1) very slowly:

$$H(t) = (1 - f(t))H_0 + f(t)H_1 \tag{8}$$

In Eq. (8), $f(t)$ is a switching function that satisfies $f(t_0) = 0$ and $f(f_{\text{inal}}) = 1$. In this work,

$$f(t) = t^2/f_{\text{inal}} \tag{9}$$

was used for the $f(t)$. And 15-dimensional harmonic oscillator that has the same force constants as those of exact Hamiltonian at the lowest inherent structure was used for the H_1. After the time evaluation, the total energy of the system no longer conserves. However, the density of states does still conserve. Since the model Hamiltonian is the harmonic oscillator, the density of states can be known from the given total energy, and it also corresponds to the density of states in the original exact Hamiltonian. More detail on this method is in Ref. 40. The results are shown in Fig. 10. In Fig. 10, the configuration entropy increases rapidly around ∼15 K. It is consistent with the results of Lindemann's δ. The fact that the melting never occurred until ∼15 K was reconfirmed.

E. Extent of the Configuration Space

The results from Lindemann's δ and the configuration entropy show that the melting of Ar_7 occurs around ∼15 K. However, the conventional reaction path analysis failed to monitor the correct onset of the melting. From the reaction path

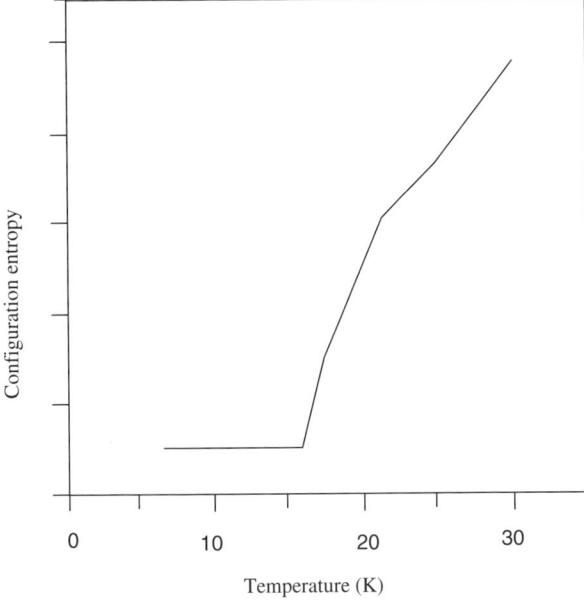

Figure 10. The configuration entropy against the temperature.

analysis, the melting must occur at less than 10 K. This implies that the real pathway of the dynamics is far from the conventional reaction path and it never passes through or around the conventional transition state defined as the geometrical saddle point. Obviously, the real dynamics should be expressed in the phase space, not in the configuration space such as the PES. However, it may be possible to get an intuitive picture about the real dynamics only from the configuration space. Figure 11 illustrates one crude idea to do so. There is a conventional transition state (saddle point) between two local minima of the PES. This transition state is, however, in back of a tall mountain. And from the

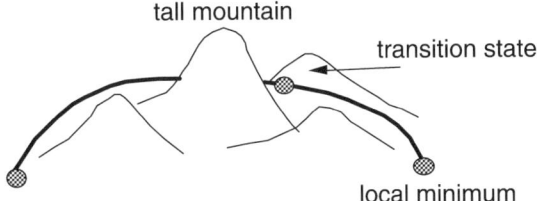

Figure 11. A rough sketch of a conventional transition state (saddle point) between two local minima on the potential energy surface. The transition state is in back of a tall mountain.

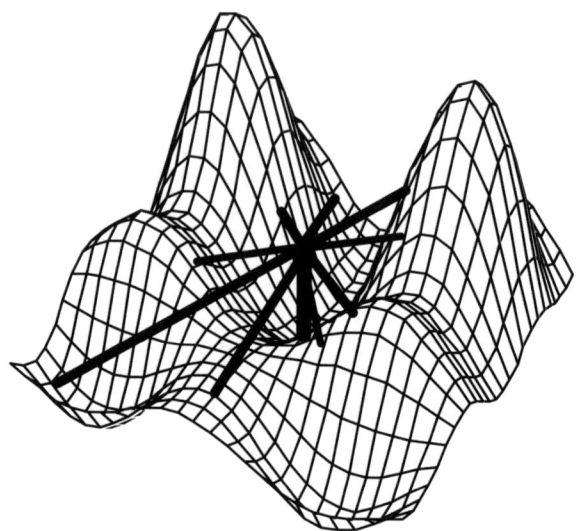

Figure 12. An extent of the configuration space. First, stand at the local minimum. Then, climb vertically, proportional to the given energy, and evaluate the extent of the scenery.

local minima, we cannot look at this transition state. In other words, the reaction path is too curvilinear in the configuration space or the path is too narrow in the phase space. If this is the case, the tall mountain has to become the real transition state instead of the geometrical saddle point. To verify this assumption, the following calculation was carried out. First, we stand at the local minimum. Then, we climb vertically, proportional to the given energy, and evaluate the extent of the scenery. This can be done by shooting a laser gun for various directions and summing up the length of the laser beam (see Fig. 12). The idea of this calculation is that the visible place is the only region we can access. The result of this analysis is shown in Fig. 13. In Fig. 13, a rapid increscent can be found in the extent of the configurational space around ~ 15 K. This behavior is similar to those of Lindemann's δ and the configuration entropy.

F. Liapunov Exponent and KS Entropy

The configuration entropy represents the size of the phase space, and its projection onto the configuration space may correspond to the "extent of the configuration space." However, it does not give any information about the dynamics itself. The Liapunov exponents and the KS entropy that is the positive sum of the Liapunov exponents may give the characteristic of the dynamics. Thus, these properties were calculated in this work to compare the results

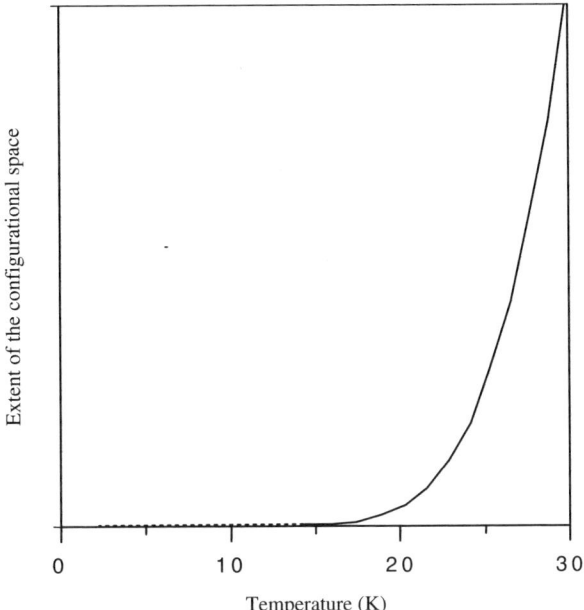

Figure 13. The calculated extent of the configuration space against the temperature.

obtained so far, although they are already reported in several works [10,12–13,17]. The results are shown in Fig. 14a and Fig. 14b. In Fig. 14a, the KS entropy of the Ar_7 gradually increases as the temperature increases and there is no drastic change around ~15 K where the onset of the melting occurs. That is, the dynamics gradually diversifies as the temperature increases. Figure 9b shows the positive Liapunov exponents that are the components of the KS entropy. In Fig. 9b, almost all of the Liapunov exponents shows the similar behavior as the KS entropy. Some of these, such as the second smallest one or the largest one, are appeared to have a minor change of the slope of the graph around ~15 K. Nayak et al. pointed out that the largest Liapunov exponent of Ar_7 has exhibited a variation analogous the Lindemann's δ. This behavior was reconfirmed by the present calculation. In addition, the present results show that the second smallest Liapunov exponent has also a similar behavior with the largest one.

G. Other Stationary Points

The local minima and the saddle points of a PES are directly related to reactions. However, in the PES of Ar_7, there are number of other stationary points with high classes. These stationary points are unstable and have relatively high energies

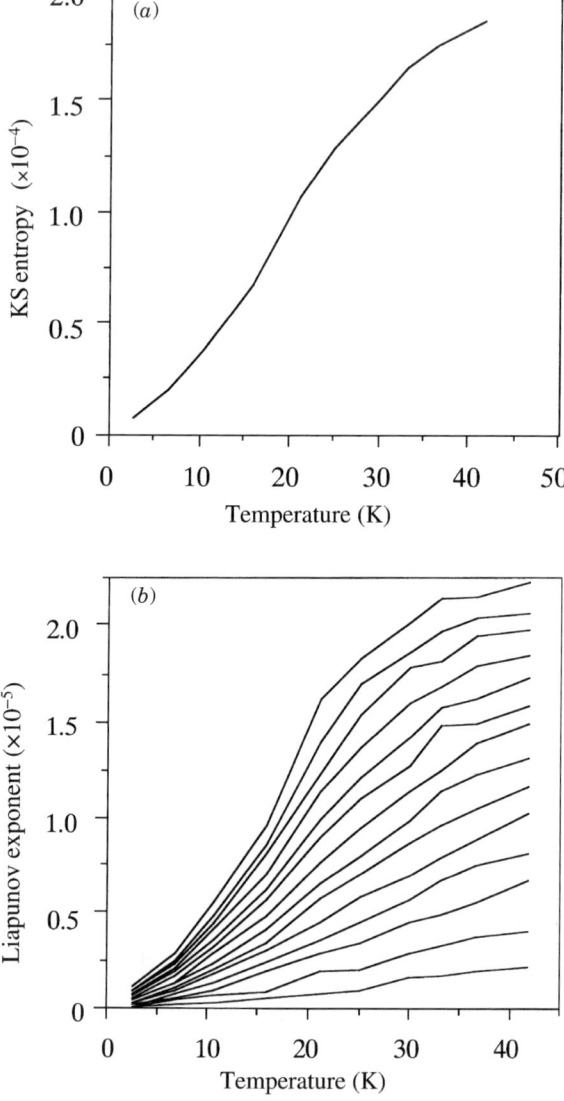

Figure 14. (a) The KS entropy that is the positive sum of the Liapunov exponents against the temperature. (b) The Liapunov exponents against the temperature.

compared with those of low classes. In this work, all of these stationary points were also obtained by the GEP in addition to the local minima and the saddle points. The results are summarized in Tables I and II. Table I shows the total numbers of the stationary points of the same classes. The number of the

TABLE I
The Numbers of the Stationary Points That Have the Same Numbers of the Negative Hessian Eigenvalues

Class	0	1	2	3	4	...	9
No. of stationary points:	4	12	24	48	87	...	8

TABLE II
The Relative Energies from the Lowest Inherent Structure for the Low Classes Together with the Corresponding Temperature

Class:	0	1	2	3
Relative energies (kcal/mol):	0–0.23	0.26–0.47	0.38–0.63	0.52–0.97
Corresponding temperature (K):	0–7.3	8.2–13.5	11.4–20.7	14.6–23.4

stationary points increases as the class increases for the low classes. For example, the numbers of the local minima (class 0), the saddle points (class 1), and the so-called monkey saddle points (class 2) are 4, 12, and 24, respectively. Since the internal degrees of freedom of Ar_7 are 15, the possible highest class is also 15. The stationary points of class 15 are the local maxima. Generally speaking, the numbers of local maxima are very limited since the strict condition is required for the local maxima. Thus, the numbers of the stationary points decrease gradually as the class closes to 15. Table II shows the relative energies from the lowest stationary point (i.e., the lowest inherent structure) for the low classes together with the temperatures converted from the relative energies. As expected before, the relative energies increase as the class increases. Only from the energetic point of view, the melting region of ~15 K corresponds to the energies (temperatures) of the classes 2 and 3.

H. Dynamics on Partitioned Cells

Using the cell partition of the PES described in the Section II.B, the dynamics of Ar_7 were analyzed at the various temperatures. In this calculation, all of the classical trajectories at every time step were classified into either of the cells. Figure 15 shows the classified rate to the cells. At the low temperatures, the cells of class 0 have the highest ratio and monotonously decrease their ratio. The reason is that the energetically accessible region spread out as the temperature increases and, as the result, the relative classification rate decreases. The situation is the same for other cells. For the cells of class 1, the classification rate is small at the low temperatures since the energies of these cells are relatively high and many of the cells are energetically forbidden. As the temperature increases, these cells become accessible and the classification rate increases. However, the temperature increases furthermore, other cells of high classes also biome

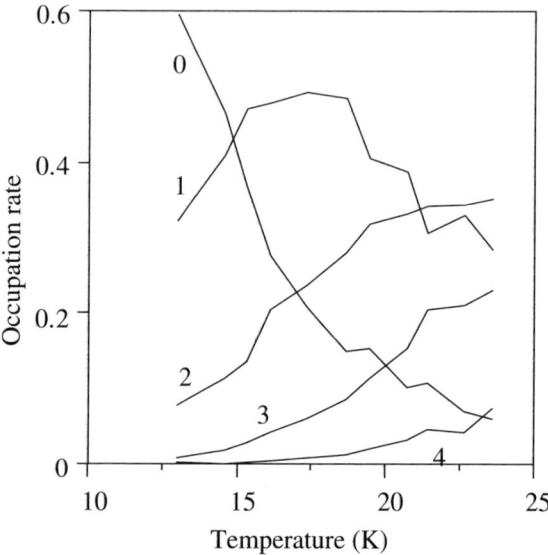

Figure 15. Classified rate of the trajectories to the cells against the temperature.

accessible and the relative rate of class 1 decreases. The classification rate of classes 2, 3, and 4 monotonically increases as the temperature goes up. It should be noted that, for the cells of class 2, there is a minor change of the slope in the graph around 15 K and this behavior is somewhat similar to the situation of Lindemann's δ or the configuration entropy.

I. Watershed

Among all the cells, only a few of them play a role of the transition region between two specified local minima. A rough sketch of an example is illustrated in Fig. 16. In Fig. 16, the closed thick lines are the ridgelines of the watersheds that divide the different inherent structures. It should be noted that the ridgelines across not only the cells of class 1, but also other cells of high classes. In the Section IIIH, the dynamics of Ar_7 were analyzed by the cell partition. However, the results were not so clear to see the onset of the melting. Once the trajectory goes over the transition region, the accessible region increases rapidly since the tall mountain of the watershed is already in the back and the trajectory can go to various places. If this situation occurs, the transition region fades out into the various trajectories and has unclear results. To see the onset of the dynamics more clearly, the following calculations were carried out. In this calculation, the dynamics was calculated until the trajectory escaped from the lowest inherent structure and the cell corresponding to the watershed of this trajectory was

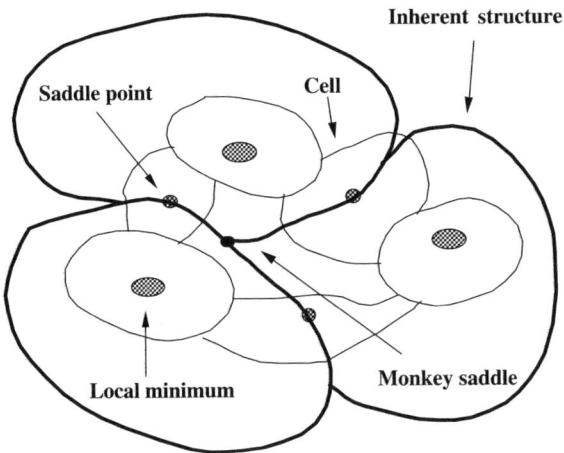

Figure 16. A rough sketch of the relationship among the cells and the watershed.

checked. By changing the initial conditions, the above procedure was repeated until the statistical data were obtained. For this calculation, the temperature of 20 K that was a bit higher than the melting temperature was used to save the computer time. The results are shown in Fig. 17. In Fig. 17, the cells of class 2 have the highest rate for the watershed. It is about 45%. The cells of class 3 have the second highest rate and have about 30%. The cells of the class 1 have only

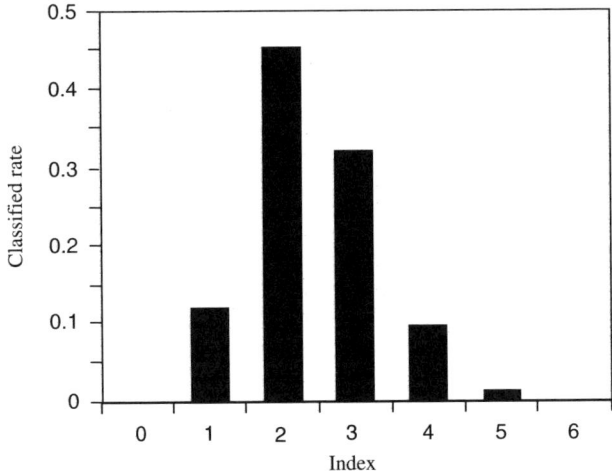

Figure 17. The classified rate of the cells corresponding to the watershed.

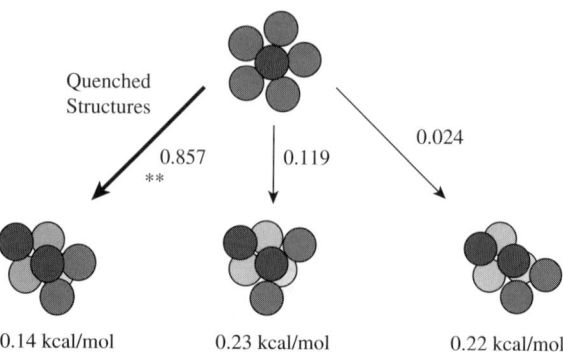

Figure 18. The arrival-quenched structures after the watershed.

about 10%. Figure 18 shows the arrival-quenched structures. About 85% of the trajectories arrived at the second lowest quenched structure, and it is what we expected before.

J. Lindemann's Criterion of Ar_5

In the Section III.I, the watershed of the inherent structures was analyzed so as to find the transition region at the onset of the melting. The results show that the trajectories across the cells of class 1 are only about 10% of the entire trajectories that passed through the watershed. In contrast, the cells of classes 2 and 3 have a relatively high rate. In order to identify the transition region more clearly, Lindemann's δ was calculated again with a modified PES that is the sum of the original PES and an additional potential. The additional potential (U) used here is an exponential function whose center (r_0) is on either of the stationary points as

$$U(\mathbf{r}) = \alpha \, \exp[\beta(\mathbf{r} - \mathbf{r}_0)^2] \tag{10}$$

where $\alpha = 1.60 \times 10^{-13}$ (erg) and $\beta = 0.5$ (Å) were used. This form of the function makes the potential energy around a particular cell go up and shuts out the trajectories. If this potential is put on the transition region, the onset dynamics of the melting is expected to change dramatically. For this calculation, Ar_5 was used instead of Ar_7 to reduce the number of the cells. Figure 19 shows the stationary points of Ar_5. In Ar_5, there is only one global minimum and the arrival minimum is the permutated same inherent structure. The numbers of other stationary points of classes 1, 2, and 3 are 2, 4, and 3, respectively. Figure 20 shows the reaction paths among them. In Fig. 20, the vertical axis represents the

ONSET DYNAMICS OF PHASE TRANSITION IN Ar₇ 149

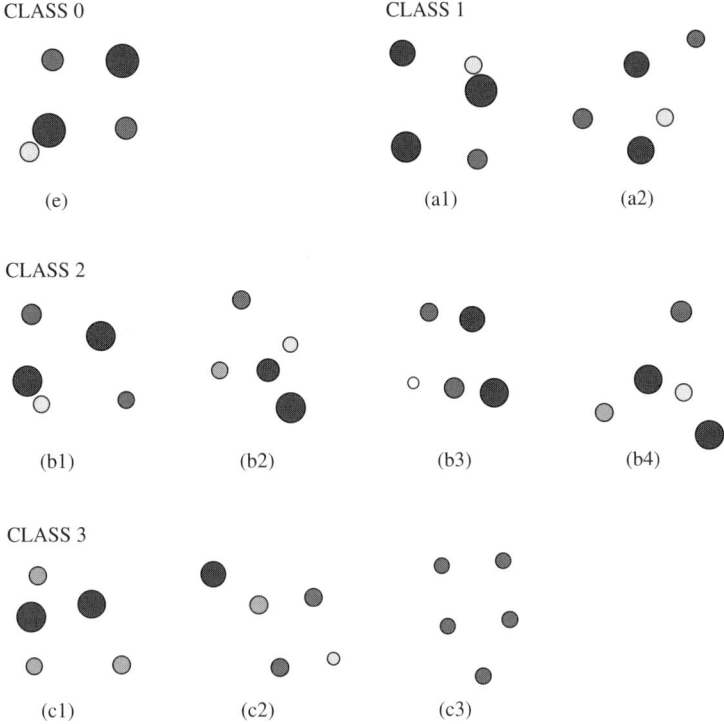

Figure 19. The stationary points of Ar$_5$.

relative energy from the minimum structure (e), and a permutated structure of this minimum (e') is also shown as the arrival structure. The onset dynamics of the melting is expected be the dynamics from the minimum e to the minimum e'. From the reaction path description, there are three different pathways for the low energies. They are $e \to a1 \to e', e \to a2 \to e'$ and $e \to b1 \to e'$. Thus, the cells of $a1$, $a2$, and $b1$ are the candidates of the transition region. Figure 21 shows the Lindemann's δ against the temperature using the original PES. In Fig. 21, the phase transition is already unclear compared with that of Ar$_7$ since the number of the atoms is too small. However, it is not a problem for the present analysis. Figures 22a, 22b, and 22c show Lindemann's δ employing the additional function on $a1$, $a2$, and $b1$, respectively. Comparing these figures with Fig. 21, one can see that only Fig. 20c has a completely different behavior from others. Until 11 K, a large reduction of the Lindemann's δ can be found and this tendency continues until 13 K or so. Figures 21a and 21b do not have such an appearance. From these results, we conclude that the transition region in Ar$_5$ at the onset of the melting is cell $b1$, which is the cell of class 2.

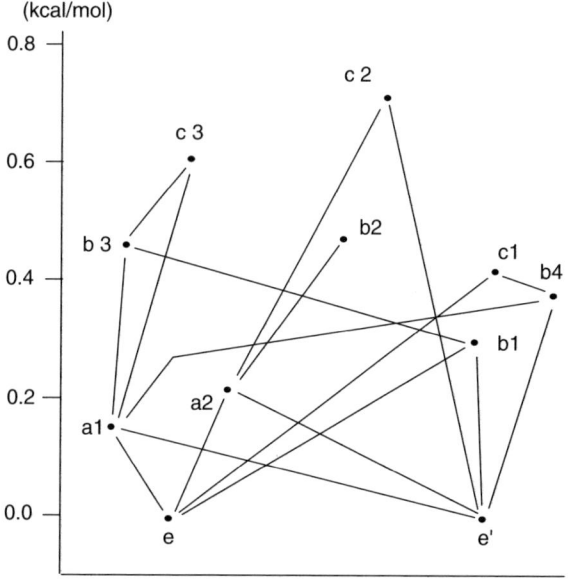

Figure 20. The reaction paths among the stationary points and the relative energies of Ar_5.

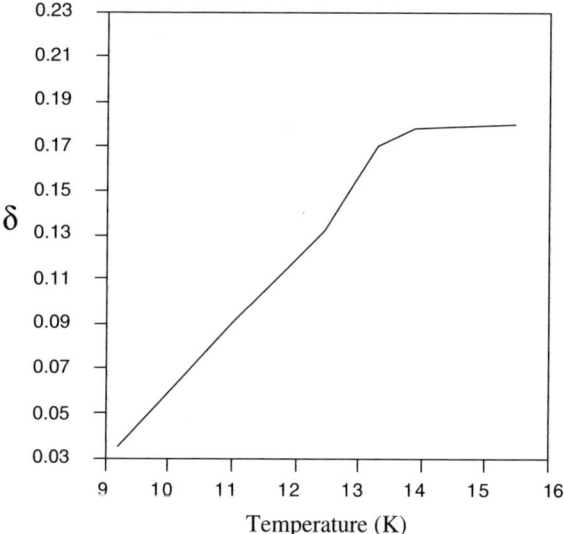

Figure 21. Lindemann's criterion against the temperature using the original potential energy surface.

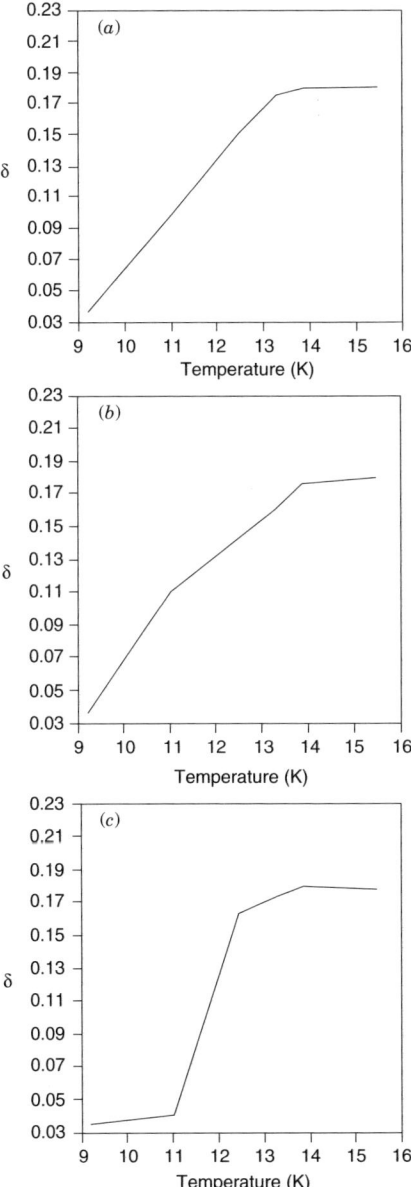

Figure 22. (*a*) Lindemann's criterion against the temperature adding the artificial potential centered on the stationary point $a1$. (*b*) Lindemann's criterion against the temperature adding the artificial potential centered on the stationary point $a2$. (*c*) Lindemann's criterion against the temperature adding the artificial potential centered on the stationary point $b1$.

IV. SUMMARY AND CONCLUSION

The phase transition of Ar_7, especially its onset dynamics, was analyzed from various points of view. The tangled reaction path diagram among all the local minima and the saddle points of the PES was obtained so as to analyze the isomerization dynamics. It was helpful to get an intuitive picture for the isomerization, but gave a poor description for the real dynamics. The configuration entropy and the extent of the configuration space behaved analogous to Lindemann's δ around the onset of the melting. In contrast, the KS entropy gave no sign about the phase transition. However, it was found that the largest and the second smallest Liapunov exponents had small changes of the slope around the melting temperature. Not only the local minima and the saddle point, but also other stationary points that have more than one negative eigenvalue of the Hessians, were obtained by the GEP. These stationary points were used to symbolize the global region of the PES by the cell partition method. The real dynamics was examined by this symbolic expression. The results showed that the region around the stationary points of class 2 was critical at the onset of the melting. About 45% of the trajectories passed through this region to go out of the current inherent structure. These phenomena were reconfirmed by the calculations of the Lindemann's δ of Ar_5. When the trajectories were shut out only from the region around the stationary points of class 2, Lindemann's δ decreased dramatically.

Based on these experiences, it seems to be natural to conclude that the region around the stationary points of class 2—that is, the monkey saddle points— plays a key role of the onset of the melting.

Acknowledgment

This work was supported by the Grant Aid from the Ministry of Education, Science, and Culture of Japan.

References

1. C. L. Briant and J. J. Burton, *J. Chem. Phys.* **63**, 2045 (1975).
2. R. D. Etters and J. Kaelberer, *Phys. Rev. A* **11**, 1068 (1975).
3. J. B. Kaelberer and R. D. Etters, *J. Chem. Phys.* **66**, 3233 (1977).
4. D. Etters and J. Kaelberer, *J. Chem. Phys.* **66**, 5112 (1977).
5. J. Jellinek, T. L. Beck, and R. S. Berry, *J. Chem. Phys.* **84**, 2783 (1986).
6. F. G. Amar and R. S. Berry, *J. Chem. Phys.* **85**, 5943 (1986).
7. H. L. Davis, J. Jellinek, and R. S. Berry, *J. Chem. Phys.* **86**, 6456 (1987).
8. T. L. Beck, J. Jellinek, and R. S. Berry, *J. Chem. Phys.* **87**, 545 (1987).
9. T. L. Beck and R. S. Berry, *J. Chem. Phys.* **88**, 3910 (1988).
10. T. L. Beck, D. M. Leitner, and R. S. Berry, *J. Chem. Phys.* **89**, 1681 (1988).
11. D. J. Wales and R. S. Berry, *J. Chem. Phys.* **92**, 4283 (1990).

12. D. J. Wales and R. S. Berry, *J. Phys. B* **24**, L351 (1991).
13. C. Amitrano and R. S. Berry, *J. Phys. Rev. Lett.* **68**, 729 (1992).
14. R. J. Hinde, R. S. Berry, and D. J. Wales, *J. Chem. Phys.* **96**, 1376 (1992).
15. R. J. Hinde and R. S. Berry, *J. Chem. Phys.* **99**, 2942 (1993).
16. R. S. Berry, *Chem. Rev.* **93**, 2379 (1993).
17. C. Seko and K. Takatsuka, *J. Chem. Phys.* **104**, 8613 (1996).
18. K. Takatsuka and C. Seko, *J. Chem. Phys.* **105**, 10356 (1997).
19. M. A. Miller and D. J. Wales, *J. Chem. Phys.* **107**, 8568 (1997).
20. C. Seko and K. Takatsuka, *J. Chem. Phys.* **109**, 4768 (1998).
21. T. L. Beck, J. D. Doll, and D. L. Freeman, *J. Chem. Phys.* **90**, 5651 (1989).
22. S. W. Rick, D. L. Leitner, J. Doll, D. L. Freeman, and D. D. Frantz, *J. Phys. Phys.* **95**, 6658 (1991).
23. F. H. Stillinger and D. Stillinger, *J. Chem. Phys.* **93**, 6013 (1990).
24. D. Wales, *J. Chem. Phys.* **101**, 3750 (1994).
25. J. P. K. Doye and D. Wales, *J. Chem. Phys.* **102**, 9673 (1995).
26. D. M. Leitner and R. S. Berry, *J. Chem. Phys.* **91**, 3470 (1989).
27. S. K. Nayak and R. Ramaswany, *Phys. Rev. E* **51**, 3376 (1995).
28. W. G. Hoover, H. A. Posch, and S. Bestiale, *J. Chem. Phys.* **87**, 665 (1987).
29. K. Takatsuka and C. Seko, *J. Chem. Phys.* **1110**, 3263 (1999).
30. K. Fukui, *Acc. Chem. Res.* **14**, 368 (1981).
31. In addition to the case of Ar_7, a comprehensible explanations is given in F. H. Stillinger and T. A. Weber, *Science* **225**, 983 (1984).
32. D. K. Hoffman, R. S. Nord, and K. Ruedenberg, *Theor. Chim. Acta* **69**, 265 (1986).
33. P. Jørgensen, H. J. A. Jensen, and T. Helgaker, *Theor. Chim. Acta* **73**, 55 (1988).
34. N, Shida, J. E. Almöf, and P. F. Barbara, *Theor. Chim. Acta* **76**, 7 (1989).
35. C. J. Cerjan and W. H. Miller, *J. Chem. Phys.* **75**, 2800 (1981).
36. P. G. Mezey, *Theor. Chim. Acta* **60**, 97 (1981).
37. P. G. Mezey, *Theor. Chim. Acta* **63**, 9 (1983).
38. A. Rahman, *Phys. Rev. A* **136**, 405 (1964).
39. M. R. Hoare and P. Pal, *Adv. Phys.* **20**, 161 (1971).
40. M. Watanabe and P. Reinhardt, *Phys. Rev. Lett.* **65**, 3301 (1990).

CHAPTER 14

RAPID ALLOYING IN BINARY CLUSTERS: MICROCLUSTER AS A DYNAMIC MATERIAL

YASUSHI SHIMIZU and TAIZO KOBAYASHI

Department of Physical Sciences, Faculty of Science and Engineering, Ritsumeikan University, Kusatsu, 525-8577, Japan

KENSUKE S. IKEDA

Department of Physical Sciences, Faculty of Science and Engineering, Ritsumeikan University, Kusatsu, 525-8577, Japan

SHIN'ICHI SAWADA

School of Science and Technology, Kwansei Gakuin University, Sanda 669-1337, Japan

CONTENTS

I. Introduction
II. Rapid Alloying: A Fast Diffusion Process in Nano-Sized Clusters
   A. Unusual Aspects of RA
   B. A Simple Model for Simulating RA
III. Numerical Results
   A. How Does the RA Proceed in the MD Simulation?
   B. Size Effect
   C. Heat of Solution as a Driving Force for RA
   D. RA Process in Solid Phase
   E. RA as a Diffusion Process: Radial Diffusion and Surface Diffusion
IV. A Microscopic Mechanism of RA
   A. The Floppy Surface Atoms and the Reaction Paths on PES

Geometric Structures of Phase Space in Multidimensional Chaos: A Special Volume of Advances in Chemical Physics, Part B, Volume 130, edited by M. Toda, T Komatsuzaki, T. Konishi, R.S. Berry, and S.A. Rice. Series editor Stuart A. Rice.
ISBN 0-471-71157-8 Copyright © 2005 John Wiley & Sons, Inc.

B. Numerical Enumeration of Reaction Paths
C. A Distribution of the Saddle Point Energy
V. Summary
References

I. INTRODUCTION

One of the most important features of atomic and molecular clusters is that they exhibit neither the properties of bulk nor those of molecules both in their static and dynamical aspects [1]. Such an intermediate property of clusters is often attributed to the facts that the majority of the constituent atoms is located at the surface and that those surface atoms induce large fluctuation in motion. Actually, it was experimentally shown that small Au clusters fluctuate between different multiply twinned and single-crystal structures rather than having fixed structures [2]. Ajayan and Marks [3] pointed out that this property has its origin in a quasi-molten state where the Gibbs free energy surface as a function of the cluster morphology forms quite shallow minima. On the other hand, Sawada and Sugano [4] showed that the structural change of Au clusters observed in experiments is regarded as a floppy motion between local minima of a potential energy surface due to the dynamical nature of clusters. In both viewpoints we may say that small clusters suffer from anomalously large dynamical fluctuations. Because of a large fluctuation, it is hard to give a clear answer for the following naive questions; Are microclusters like solids where atoms are oscillating around their respective equilibrium positions? Are they like liquids where atoms move diffusively? Or, are they fluctuating between different solid phases during the course of their motion [1]? According to the works by Berry and co-workers [5], their microcanonical molecular dynamics (MD) study of a Lennard-Jones cluster reveals that there exists the intermediate phase (called *coexisting phase*) between liquid and solid both in small and relatively large clusters. In addition, by paying an attention to motion of atoms composing a cluster, Cheng and Berry [6] show that clusters tend to have *floaters*, which are freely migrating surface atoms above or outside the outer layer of the cluster. The presence of these *floaters* would be important for the dynamics in an extremely long-time scale beyond microseconds, which is termed a *meso time scale* in this chapter.

In this chapter we discuss a novel fast diffusion process that was experimentally discovered by Yasuda, Mori, and co-workers (YM) [7] in nano-sized binary metal clusters, because it is supposed to be a typical manifestation of an anomalous diffusion process peculiar to microclusters. The authors have reported some numerical results of MD simulation on rapid alloying (RA) [8]. Special attention is paid to the diffusion of atoms in cluster [9]. The aim of the present work is to realize RA by elucidating what kind of diffusive motion is relevant to RA.

The rest of the chapter is organized as follows: In Section II, unusual aspects of RA and the model system for MD simulation are summarized. Section III provides a detailed description of the numerical simulation. A microscopic mechanism of RA is discussed in Section IV, and Section V is a summary.

II. RAPID ALLOYING: A FAST DIFFUSION PROCESS IN NANO-SIZED CLUSTERS

A. Unusual Aspects of RA

Rapid alloying (RA) is a fast diffusion process that was experimentally discovered by Yasuda, Mori, and co-workers (YM) in binary microclusters. By using an evaporator, they deposited individual solute atoms (Cu) on the surface of host nano-sized clusters that are supported by amorphous carbon film at room temperature. YM observed the alloying behavior with a transmission electron microscope in *in situ* condition as schematically described in Fig. 1. In Ref. 7 it is demonstrated that Au clusters promptly changed into highly concentrated, homogeneously mixed (Au–Cu) alloy clusters. RA is similarly observed in various nano-sized binary clusters, such as (Au–Ni), (In–Sb), (Au–Zn), and (Au–Al) [7]. They examined the presence and absence of RA for clusters of different sizes. Consequently, YM summarize the unusual features of RA as follows:

1. *RA Is an Extremely Fast Diffusion Process.* The value of the diffusion coefficient of copper atoms in clusters is approximately nine orders of magnitude larger than that in bulk crystalline alloys. In terms of the relation $x = \sqrt{Dt}$, between the diffusion coefficient D and the time t needed to achieve diffusion of solute atoms across the distance x, YM roughly estimated the value 1.1×10^{-19} m^2/s at least. Note that the value D of copper in the bulk gold is about 2.4×10^{-28} m^2/s at 300 K [10].

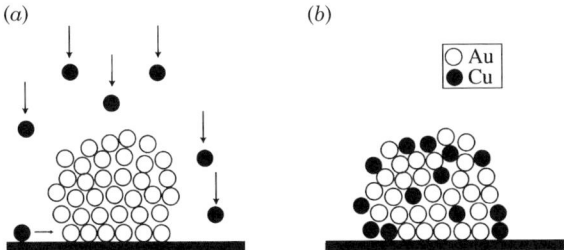

Figure 1. A schematic picture of the *in situ* observation of RA by YM. White and black circles denote Au and Cu atoms, respectively. (*a*) Before the onset of RA, individual Cu atoms are deposited on Au clusters. (*b*) After RA is completed, Cu atoms dissolve into Au clusters to form homogeneously mixed alloy clusters.

2. *Negative Heat of Solution Is a Driving Force for RA.* Negative heat of solution plays a primary role in enhancing and suppressing the RA process. Indeed, RA has never been observed in the combination of the solute and host atoms with *sufficiently large positive* heat of solution. However, it is worth noting that RA occurs even with the atomic species with almost null heat of solution.

3. *Temperature Effect.* Temperature, say T, is also an important parameter controlling RA. At relatively high temperature ($T \sim 245$ K), Cu atoms can dissolve well into the center of a 4-nm-sized Au cluster, while at medium temperatures ($T \sim 215$ K, 165 K), the dissolution of Cu takes place only over a limited, shell-shaped region beneath the surface of a 4-nm-sized cluster, and the thickness of that region where a solid solution is formed decreases with the decrease in temperature.

4. *Size Effect.* With increase in cluster size the occurrence of RA is significantly suppressed. In Au clusters of approximately 10 nm in mean size, the rapid diffusion of Cu atoms takes place only at the shell-shaped region beneath the free surface of an individual cluster, while pure gold is retained at the center of the cluster. In Au clusters of approximately 30 nm in the mean size, the RA does not take place. It should be emphasized that the critical size of the RA increases with the negative heat of solution and temperature.

5. *RA Takes Place in a Solid Phase.* This fact is confirmed by the experimental observation that the location of the twin boundaries in host clusters, which have multiply twinned structure, does not become altered during the RA process. Although the electron beam heating seemingly has considerable influence on RA, the estimated magnitude of the temperature rise in clusters is of the order of 10 K. Thus, it is reasonable to expect that the resulting temperature rise causes no significant effect on RA.

B. A Simple Model for Simulating RA

We numerically investigate the atomistic mechanism behind RA, which could be essentially different from diffusion process in bulk medium. We employ a model that is as simple as possible, since our purpose for simulating RA is not only to reproduce RA in a computer, but also to understand the underlying mechanism for RA. Our model is designed by taking into account the following experimental conditions:

(a) The host clusters are placed on a carbon substrate that plays the role of a *heat reservoir.*

(b) The onset of RA is independent of atomic species, provided that they imply the *negative heat of solution.*

(c) RA is a characteristic common to nano-sized clusters where the number of surface atoms is about 50% of the total number of cluster atoms.

These conditions lead us to the following requirements for the modeling, respectively.

(a') The emitted heat due to the negative heat of solution is immediately released to the substrate on the time scale much shorter than the alloying. In addition, what we would like to clarify is the mechanism of very rapid diffusion of solute atoms into the microclusters, which will not be peculiar to binary clusters but universal in microclusters. Thus, one has to explore the *activation mechanism* behind the diffusion process, as has been done in the investigations of diffusion in bulk media. For these two reasons, it is desirable to perform MD under the temperature-controlled condition. To realize an isothermal condition for a cluster, we modified the velocity scaling method [11]. In short, the method we adapted does not control the instantaneous kinetic temperature, but time-averaged kinetic temperature over 1000 MD time steps. We term the present method the *averaged kinetic temperature controlling method* (AKTCM). (See Ref. 11 for details.)

(b') To pick up the effect of heat of solution, it would be nice to choose a model that has the fewer number of parameters controlling the magnitude of the heat of solution. Then, our model is made to have only one parameter, of which physical meaning is transparent.

(c') The present numerical study aims to elucidate the role of the heat of solution, temperature dependence, and the cluster size upon RA. For this purpose, one needs to explore many trajectories by changing parameters that control these three factors. Recently, it has become possible to compute few trajectories covering a long-time scale for a relatively large system like a nano-sized system. However, carrying out an extensive computation by changing values of parameters and initial conditions is still beyond the recent computational capability. In order to simulate long-time dynamics of a large enough system and obtain many trajectories given by various conditions, we employ the two-dimensional (2D) model even though it is seemingly different from the realistic 3D models.

The potential energy part of the model is given by the Morse potential:

$$V_{kl}(r) = \epsilon_{kl}\{e^{-2\beta_{kl}(r-\sigma_{kl})} - 2e^{-\beta_{kl}(r-\sigma_{kl})}\} \tag{1}$$

where the suffixes k and l specify the two species of atoms, say *host* and *guest* atoms. The host and guest atoms are denoted by A and B, respectively. Because a cluster contains N_A A atoms and N_B B atoms, the total number of atoms is

$N = N_A + N_B$. We choose $\beta_{AA} = \beta_{BB} = \beta_{AB} = 1.3588\,\text{A}^{-1}$, $\epsilon_{AA} = \epsilon_{BB} = \epsilon = 0.3429\,\text{eV} = 3.979 \times 10^3\,\text{K}$ and $\sigma_{AA} = \sigma_{BB} = \sigma_{AB} = \sigma = 2.866\,\text{A}$, which are suitable for Cu atom [13]. The important parameter that controls the heat of solution is ϵ_{AB}, which determines the depth of the pairwise potential function between the A and B atoms. The only free parameter controlling the magnitude of the heat of solution is $\alpha = \epsilon_{AB}/\epsilon_{AA}$. Since the heat of solution, ΔH, is given by $\Delta H = z(1 - \alpha)\epsilon$ where z is a coordination number, our choice for α ($\alpha = 1.2$) provides a negative heat of solution for the binary system. Note that the binary system has a negative heat of solution in the case where the value of α is larger than 1.0.

III. NUMERICAL RESULTS

In this section we demonstrate that the size effect and an important role of heat of solution for RA are reproduced in our concise model. In addition, we examine whether RA is completed in a solid state by MD simulation. On the other hand, the diffusion process of atoms during RA is characterized by evaluating activation energy. In particular, a special attention is put on the direction of diffusing motion. Activation energy of the atomic diffusion along the cluster surface and that along radial direction is estimated respectively, because an important aspect of alloying dynamics is extracted by emphasizing the diffusive motion along these two directions as will be discussed in Section IV.

A. How Does the RA Proceed in the MD Simulation?

Before turning to the size-dependent property of RA, we take a look at a typical alloying behavior observed in isothermal MD simulation. The temperature is set to be 600 K, which is significantly below the melting point. This cluster consists of 47 host atoms and 20 guest atoms, and its composition is denoted by $A_{47}B_{20}$. The melting point of $A_{47}B_{20}$ is evaluated as 700 K in terms of the location of abrupt jump on a caloric curve [11]. The ratio of the number between host and guest atoms is based upon YM's experiments. In Fig. 2, we display the snapshots of atomic configuration in the course of time evolution. The guest atoms begin to penetrate into the cluster at about 57 ns, and a fully alloyed cluster is realized at about 400 ns. The trend of the atomic motion in time evolution is more evident by following the location of the gray circles, which are host atoms initially at the center of $A_{47}B_{20}$. During the alloying process, these gray circles gradually move from the center of the mass to the cluster surface. Moreover, the initial hexagonal shape that consists of these gray circles is unchanged until a part of the hexagon exposes on the active surface. The behavior of gray circles indicates that the core part of the cluster is solid-like, whereas the outer part is in almost liquid state at 600 K. This is a typical alloying behavior numerically observed below the melting point.

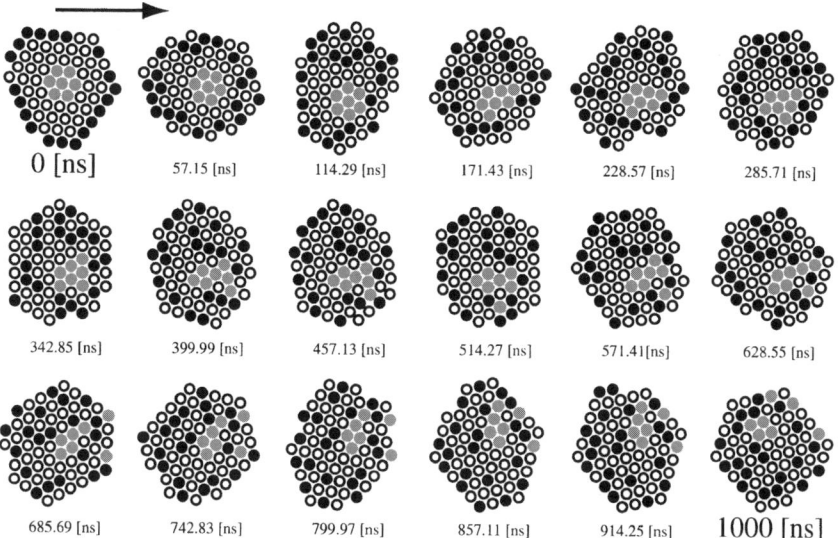

Figure 2. Successive snapshots of a cluster $A_{47}B_{20}$ in the course of time evolution. The upper left is the initial configuration, and the elapsed time are imposed for each snapshot. Guest atoms are shown by black circles, while host atoms are denoted by white circles. Gray circles indicate host atoms that are initially located in the center of the cluster. In 1 μs the cluster is in alloyed state. Gray atoms are gradually moves toward the cluster surface. On the other hand, guest atoms diffuse into the inside.

B. Size Effect

One of the most surprising features of the RA is that the time required for the alloying to be completed is much shorter than that in the bulk medium. This fact implies that the diffusion rate of the atoms into the cluster depends sensitively upon the size of the cluster. The aim of this subsection is to show the qualitative validity of our modeling by reproducing the size dependence of RA. We prepare a medium-sized cluster $A_{95}B_{40}$ and a large one $A_{160}B_{68}$ for comparison.

When we compare the alloying dynamics for different size of clusters, the composition of the two atomic species must be so chosen as to result in the common decrease of potential energy per atom in the ideally alloyed limit, which is given by $\Delta \frac{U}{N} = \epsilon(\alpha - 1)r(1 - r)z$. To satisfy this requirement, the ratio of composition $r = \frac{N_B}{N_A + N_B}$ must be a common value ($r \sim 0.3$ in the present case) in all the different size of clusters to be compared.

In Fig. 3, time evolution of a medium-sized cluster $A_{95}B_{40}$ is displayed. Again, the temperature is controlled to $T = 600$ K by the AKTM. In 1 μs the alloyed state is observed near the cluster surface. However, there is no mixing

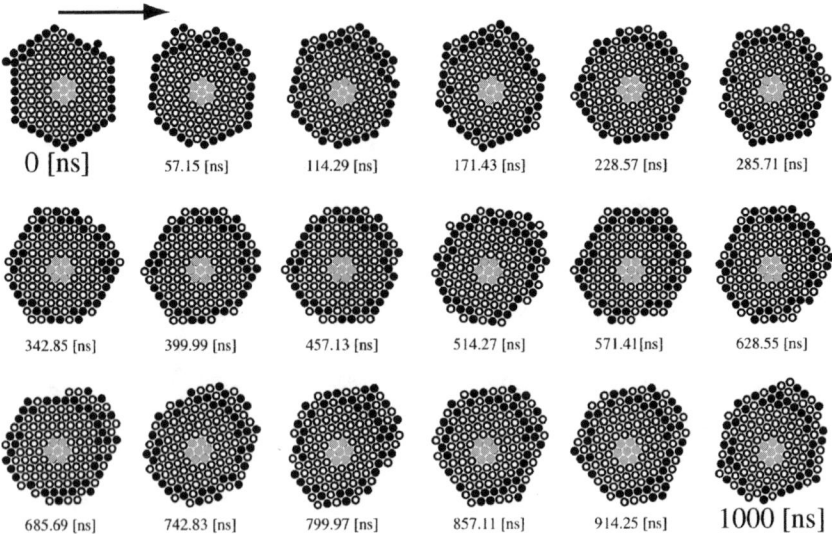

Figure 3. Successive snapshots of a cluster $A_{95}B_{40}$ in the course of time evolution. The upper left is the initial configuration, and the elapsed time is imposed for each snapshot. The guest atoms are shown by black circles, while host atoms are denoted by white circles. Gray circles indicate host atoms that are initially located in the center of the cluster.

between host and guest atoms in the core of the cluster. Finally, unlike $A_{47}B_{20}$, RA is not completed within simulation time. In a large cluster $A_{160}B_{68}$, the alloying is completed in a further limited region near the surface as depicted in Fig. 4. By comparing Fig. 2, Fig. 3, and Fig. 4, it is evident that RA is significantly suppressed as the size of the cluster is increased. To quantify the difference in the alloying process, we introduce the mean distance of the guest atoms, $R(t)$, measured from the center of mass (cm) of the cluster,

$$R(t) = \frac{1}{N_B} \sum_{i \in B-\text{atoms}} |\bar{\mathbf{q}}_i(t)| \tag{2}$$

where $\mathbf{q}_i$ is the position vector of the ith atom. Note that the cm is fixed and is chosen as the origin because the total momentum vanishes. To remove rapid vibrations of atoms around the equilibrium positions (typical period is the inverse of the Debye frequency $\omega_D^{-1} \sim 0.1$ ps), we take a short-time average $\bar{\mathbf{q}}(t) \equiv \frac{1}{\Delta t} \int_{t-\Delta t}^{t} \mathbf{q}(t') dt'$ over a time interval $\Delta t \sim 10\,\text{ps} \gg \omega_D^{-1}$. As is obvious from Fig. 5, the speed of guest atoms placed initially on the surface to diffuse inside the cluster, which is characterized by $\frac{d}{dt}R(t)$, decreases markedly as the

RAPID ALLOYING IN BINARY CLUSTERS 163

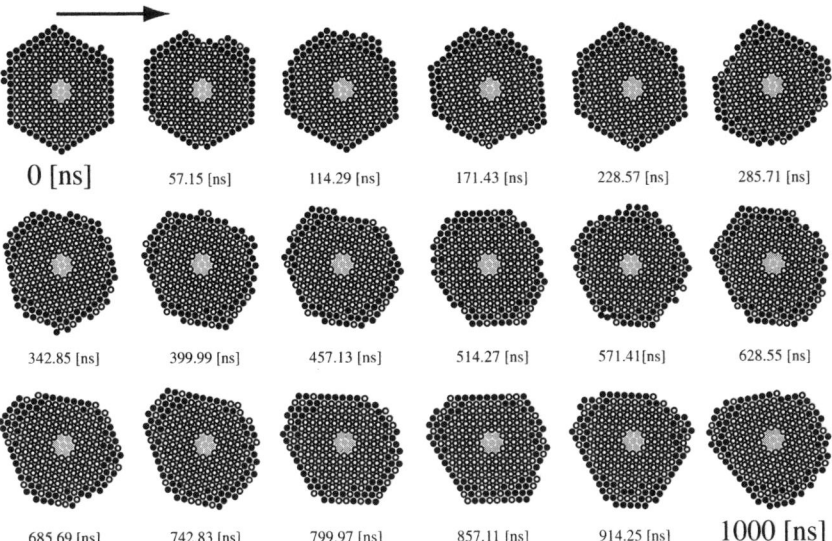

Figure 4. Successive snapshots of a cluster $A_{160}B_{68}$ in the course of time evolution. The upper left is the initial configuration, and the elapsed time is imposed for each snapshot. Guest atoms are shown by black circles, while host atoms are denoted by white circles. The position of gray circles is not altered in 1 µs.

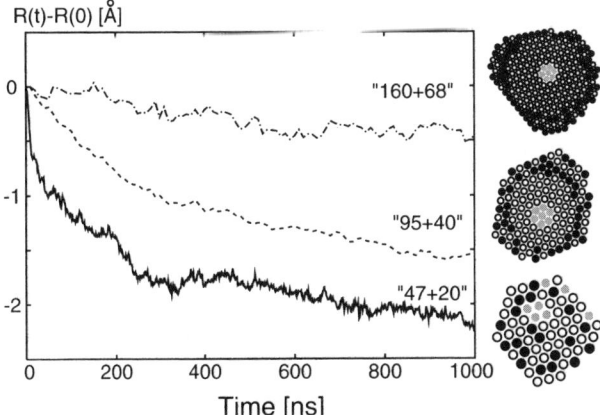

Figure 5. Comparison of time dependence of the mean distance $R(t)$ for the three different sizes of binary clusters $A_{47}B_{20}$, $A_{95}B_{40}$, and $A_{160}B_{68}$ together with their snapshots of the configuration taken at $t = 1000$ ns. The temperature of the systems is maintained at $T = 600$ K.

size of the cluster increases. These observations are the strong evidence that the RA is induced by the smallness of the cluster size.

C. Heat of Solution as a Driving Force for RA

In this subsection we probe the role of the heat of solution in RA by changing the parameter α. In order to evaluate how fast host and guest move to mix, the number of neighboring host atoms per a guest atom, n_B, is defined as

$$n_B(t) = \frac{1}{N_B} \sum_{i \in B-\text{atoms}} N_{A(i)} \qquad (3)$$

where $N_{A(i)}$ is the number of the nearest-neighbor host atoms around the ith B atom at time t. If the host and guest atoms are uniformly mixed, $n_B(t)$ should be $n_B^c = z(1 - r)$, where $r = \frac{N_B}{N_A + N_B}$. The value of $n_B(t)$ generally increases starting from $n_B(0)$ decided by the initial configuration to n_B^c as the alloying proceeds. Note that $n_B(0)$ corresponding to the initial configuration depicted in Fig. 6 is about 2.1, and n_B^c is about 4.2 for the $A_{47}B_{20}$ cluster. In Fig. 6 we show the time evolution of $n_B(t)$ observed in the homogeneous and binary clusters at $T = 600$ K. The homogeneous cluster is represented by $\alpha = 1.0$ and implies null heat of solution. A negative heat of solution is represented by $\alpha = 1.1, 1.2, 1.3,$ and 1.4, while a positive one is given by $\alpha = 0.95$. By comparing the behaviors of $n_B(t)$, one can promptly verify that complete alloying is attained for the cluster

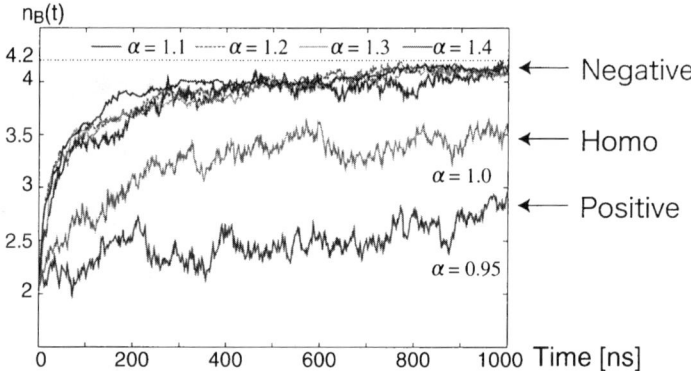

Figure 6. Time dependence of n_B is computed for $\alpha = 0.95$, $\alpha = 1.0$, $\alpha = 1.1$, $\alpha = 1.2$, $\alpha = 1.3$, and $\alpha = 1.4$. The value $n_B(t)$ for $\alpha = 0.95$ (positive heat of solution) does not reach 3.0, while that for $\alpha = 1.1, 1.2, 1.3, 1.4$ (negative heat of solution) almost reaches 4.2, which indicates the achievement of the ideal mixing between host and guest atoms. The value $n_B(t)$ for $\alpha = 1.0$ (null heat of solution) is about 3.5, which is an intermediate value between two cases corresponding to positive and negative heat of solution.

with negative heat of solution. On the contrary, host and guest atoms do not sufficiently mix each other in the cluster with positive heat of solution. The cluster with null heat of solution exhibits the intermediate mixing property. That is, the negative heat of solution indeed plays a key role in the rapid growing process of $n_B(t)$.

D. RA Process in Solid Phase

As pointed out by YM, RA proceeds in solid phase. However, the notion of "solid phase" is ambiguous in a finite system like clusters. Thus, the meaning of "solid phase" should be used in a restrictive sense. In our simulation, we define the melting temperature of a cluster as a point where the all atoms including those located at the center fully fluctuate to fulfill the so-called Lindemann's criterion. In order to verify that $A_{47}B_{20}$ is surely in solid at 600 K, we evaluate how often atoms rearrange in a cluster. We compute the frequency of recombination of the neighboring atoms, which is estimated by the distance index [4]. Distance index is derived from an adjacency matrix $A(t)$, which is the $N \times N$ symmetric matrix whose elements $A_{ij}(t)$ are 1 for $|\bar{r}_{ij}(t)| < \sqrt{2}\sigma$ and zero otherwise, where $\bar{r}_{ij}(t) = |\bar{\mathbf{q}}_i(t) - \bar{\mathbf{q}}_j(t)|$ is the average distance between ith and jth atoms. The distance index $d_i(t)$ of the ith atom is then defined by

$$d_i(t) = \sqrt{\sum_j |A_{ij}(t + \Delta t) - A_{ij}(t)|^2} \qquad (4)$$

This quantity measures the number of the recombining events around the ith atom and characterizes the mobility or the activity of the individual atom, where Δt is taken short enough to resolve successive atomic recombinations whose frequency is $\sim$ 10–100 ps. We usually set $\Delta t \sim$ 10 ps. The accumulated distance index per atom,

$$S(t) = \sum_{n=1}^{\frac{t}{\Delta t}} \left[\sum_{i=1}^{N} \frac{d_i(n\Delta t)}{N} \right] \qquad (5)$$

is used to quantify the mobility of the whole cluster atoms. We assume that a single cluster is divided into shells according to the distance of a target atom from the center atom, which is defined as the atom closest to the cm of the cluster. As presented in Fig. 7, we allocate the shell index to each atom. The degree of activity of atoms in shells is estimated by averaging $S(t)$ over atoms that belong to the same shell. In Fig. 8, time evolution of the accumulated distance index per atom with respect to each shell is depicted. At the center of a cluster, no

166   YASUSHI SHIMIZU ET AL.

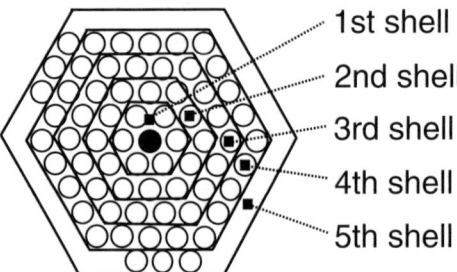

Figure 7. The shell index is allocated to each atom according to the distance from the center atom, which is denoted by the shaded circle. Initial configuration of $A_{47}B_{20}$ consists of five shells.

rearrangement of atoms is observed at 600 K. In contrast to the solid-like behavior of the inner shells, the frequency of the rearranging event at the surface is conspicuously large, because the surface of the cluster (the fifth shell) is almost melting. That is, a cluster $A_{47}B_{20}$ is not completely melting at 600 K, where a typical fast diffusion of guest atoms is numerically observed as shown in Fig. 2 and Fig. 5.

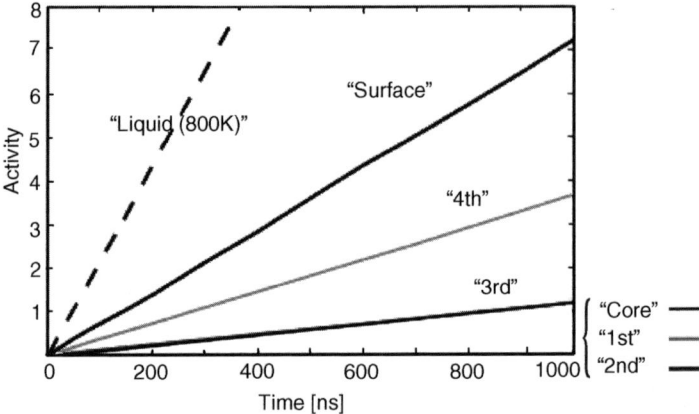

Figure 8. Time evolution of the cumulated number of rearranging events per atom in respective shells at 600 K. The indices specifying the shells are allocated for each shell. The center atoms is denoted by "core." As mentioned in the caption of Fig. 7, the initial configuration of $A_{47}B_{20}$ consists of five shells. The fifth shell is indicated by thick solid line with a caption "surface." The dotted line denotes a typical increasing trend of the cumulated number of rearranging events in a completely melting cluster.

E. RA as a Diffusion Process: Radial Diffusion and Surface Diffusion

In this subsection we examine the mechanism of the very fast diffusion. In the bulk medium the vacancies and interstitial site play a primary role in accelerating the diffusion. However, these diffusion mechanisms are not relevant in microclusters. It is well known that the vacancies created inside the cluster are immediately pushed to the surface. Indeed in our simulation the creation of vacancies inside the cluster is a very rare event even at the temperature close to the melting temperature. Moreover, we cannot find any evidence that the interstitial deformation takes place inside the cluster, and therefore neither of them is responsible for the rapid diffusion into the cluster. The key feature of the cluster that distinguishes the cluster from the bulk medium is that it is surrounded by the surface beyond which no atoms exist. In other words, the outside of the cluster is occupied by vacancies. As a result, the atoms on the surface move very actively along the surface. Such an active movement along the surface will be responsible for the rapid diffusion in the radial direction of the cluster. We focus our attention to the details of the active diffusive motion along the surface of the cluster, and we present a direct evidence that the surface activity controls the radial diffusion. A direct measure of the surface motion is the diffusion rate of the surface atoms

$$D_s(t) = \frac{1}{t}\left\langle \left[\int_0^t r(t)\,d\theta(t)\right]^2 \right\rangle \tag{6}$$

where $\theta(t)$ is the angle of the coordinate vector of a surface atom measured from the center of mass, and $r(t)$ is the instantaneous distance from the center of the cluster. The ensemble average $\langle \cdot \rangle$ is taken over the atoms sampled along the surface. We also computed the accumulated distance index $S(t)$ defined by Eq. (5). Most of the rearranging event of atoms contributing to $S(t)$ occurs on the layer closest to the surface [8]. The distance index per unit time, $\frac{d}{dt}S(t)$, is proportional to the probability of an atom to change its lattice site. As a result, the relation

$$D_s(t) = k\frac{d}{dt}S(t) \tag{7}$$

should hold, and it is numerically confirmed in our simulation [11]. The relation Eq. (7) may be interpreted as the well-known relation between the diffusion coefficient and the mobility [13]. We therefore take $\frac{d}{dt}S(t)$ as the measure of the surface mobility. In the case of $\alpha = 1.0$ the cluster is always in an equilibrium state and the value $\frac{d}{dt}S(t)$ is almost constant. In Fig. 9a we show the time series of $S(t)$, which is Arrhenius scaled by the optimal activation temperature. They

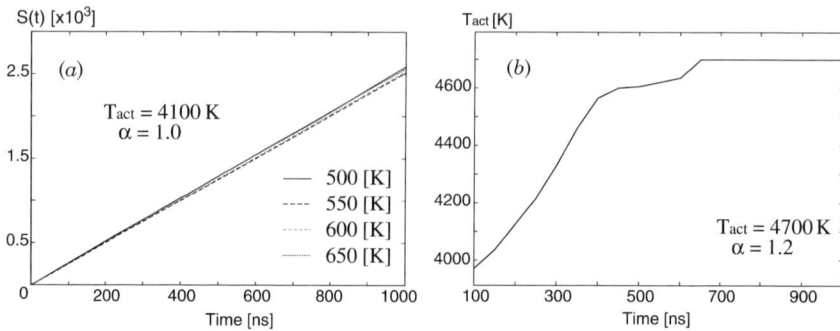

Figure 9. (a) Time evolution of the accumulated distance index $S(t)$. Optimally Arrhenius scaled data $S^{sc}(T)$ for the $\alpha = 1.0$ at 500, 550, and 600 K is plotted. All data are scaled at $T_{scale} = 600$ K to coincide with each other by assuming $T_{act} = 4100$ K. (b) Time dependence of the activation temperature of $S(t)$. The saturating point of T_{act} after increasing trend is bound from the above by the sampling time t_{mx} for $\alpha = 1.2$.

exhibit a nice Arrhenius scaling property, and the optimal activation temperature is decided as $T_{act} = 4100 \pm 100$ K, which is represented by $T_{//}$. (Here we do not explain the way to rescale the numerical data to fit the so-called Arrhenius plot. The detail fitting technique is mentioned in Ref. 11.) The same technique is also applied to the numerical data $R(t)$ for estimating the optimal activation temperature of the radial diffusion. Thus we obtain two sorts of activation temperatures: One characterizes the surface diffusion, that is, $T_{//} \sim 4000$ K; the other one, $T_{\perp} \sim 9000$ K, is associated with the radial diffusion. By comparing the two sorts of the activation temperatures $T_{//}$ and $T_{\perp}$, we extract two activation processes associated with the *parallel* motion along the surface and the *perpendicular* motion in the radial direction. The resulting values of the activation temperature, $T_{//}$ and $T_{\perp}$, are given in Table I. By using ϵ as a unit of energy, activation energy of surface and radial diffusion for $\alpha = 1.0$ are expressed as $E_{//} = 1.03 \pm 0.03\epsilon$ and $E_{\perp} = 2.26 \pm 0.13\epsilon$, respectively. For

TABLE I
Numerically Estimated Values of the Activation Temperature for the Homogeneous ($\alpha = 1.0$) and the Binary ($\alpha = 1.2$) clusters

	$\alpha = 1.0$	$\alpha = 1.2$
$T_{//}$	4100 ± 100 K (1.03ϵ)	$4000 \to 4700 \pm 100$ K
$T_{\perp}$		9000 ± 500 K

$\alpha = 1.2$ the activation energy of surface diffusion is increased by 15%, while that of radial diffusion is not much altered. Note that $T_\perp$ of $\alpha = 1.2$ is slightly higher than $T_\perp$ of $\alpha = 1.0$. A normalization of the activation temperature by the melting temperature implies the following relation:

$$\frac{T_{//}}{T_m} \sim 6, \qquad \frac{T_\perp}{T_m} \sim 13 \tag{8}$$

where the melting temperature is evaluated as $T_m \sim 700$ K.

Precisely speaking, there is a different feature between the activation temperature of homogeneous clusters and that of binary clusters. In the initial stage, T_{act} is 4000 K and rises gradually up to 4700 K for $\alpha = 1.0$ (see Fig. 9b). In case of a binary cluster ($\alpha > 1.0$), however, the activation energy is not constant but depends on time, because the initial state is not in the equilibrium. Indeed, as shown in Fig. 9b the activation temperature, which is decided by the method of optimal Arrhenius scaling, rises significantly as the function of the upper bound of the sampling time t_{mx}, which means that the surface activity reduces as the alloying process proceeds and the system approaches the equilibrium. In the initial stage we have $T_{//} \sim 4000$ K, and it coincides with that of the homogeneous cluster. However, it rises gradually up to $T_{act} = 4700$ K in the final stage of alloying. This is because the constituent atoms of the binary cluster are more tightly bounded in the alloyed binary cluster than those in the homogeneous cluster.

In this way we extract the values of the activation energy with respect to atomic diffusion along the surface and radial direction. The important thing is that these two types of the diffusive motion are closely interacted. To verify the presence of such an interplay, a simulation indicating how the alloying process is affected if the motion along the surface is interrupted by an artificial operation is done in the following way: We start the MD simulation of RA as shown in Fig. 2, and then the mass of some atoms on the surface is switched to a larger mass (typically 300 times larger one) at a certain time $t = t_s$ when about half of the guest atoms initially on the surface are absorbed into inner shells rather than the outermost shell. The atoms whose mass is changed are selected from the host atoms in the outermost shell. Thus, no guest atoms contributing to the alloying undergo the mass-switching. Instantaneous increase of the mass of host atoms decreases their frequencies of vibration, thereby playing the role of the stoppers for the motion of guest atoms along outermost shell on the surface. In this condition we monitor how further penetration of guest atoms into the inner shells is modified afterwards. In Fig. 10 the time evolution of $n_B(t)$ in comparison with that of the free running simulation is exhibited. It is evident that the increase of $n_B(t)$ is notably suppressed and keeps almost the same level at $t = t_s$, although $n_B(t)$ of the free running simulation increases above the value

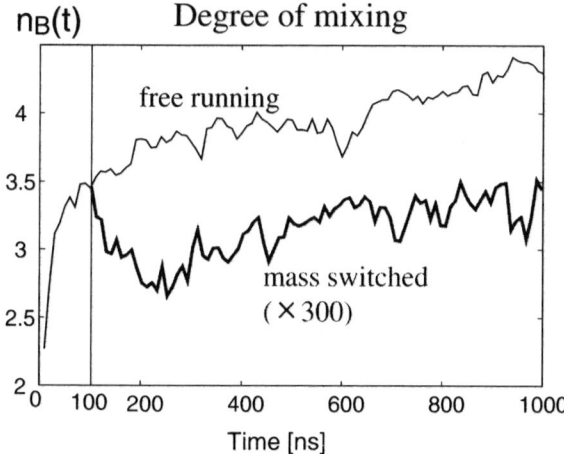

Figure 10. Time dependence of $n_B(t)$ for the free running and the mass switching simulation. In the latter simulation the mass of the surface atoms was switched to a 300 times larger one at $t_s = 100$ ns, which is indicated by a vertical line. A cluster $A_{47}B_{20}$ implying $\alpha = 1.2$ is maintained at the constant temperature $T = 600$ K.

of ideal mixing. The above observation provides direct evidence that any event which occurs inside the cluster is not responsible for the radial motion and the presence of the surface plays the key role in the RA process.

IV. A MICROSCOPIC MECHANISM OF RA

A. The Floppy Surface Atoms and the Reaction Paths on PES

As mentioned in previous section, RA is originated from the frequent rearranging motion of the surface atoms. Their floppy motion is a manifestation of a wandering motion among many local minima that are partitioned by substantially low saddles on the multidimensional potential energy surface (PES) [3,4]. In a cluster containing order of 10^1–10^2 atoms, the floppy atomic motion tends to result from the *floaters* that continue to wander among stable sites on the surface of geometrically packed clusters (see Fig. 11) [6]. Since the majority of constituent atoms of a cluster are located on surface, many *floaters* are expected to be created and annihilated throughout dynamics. Furthermore, these *floaters* may interact with each other and move collectively. Actually, as presented in Figs. 5 and 8, the active surface atoms are able to move almost freely along the cluster surface and diffuse into the inside of a cluster due to an accumulation of rearrangements of surface atoms at substantially below the melting point.

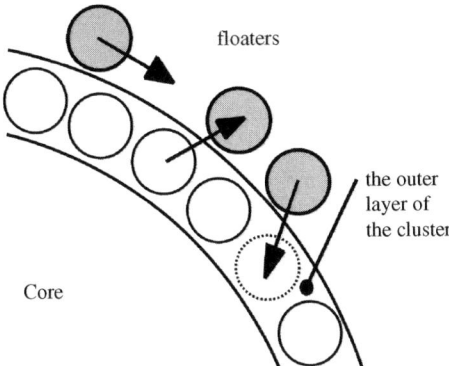

Figure 11. A schematic picture of creation and annihilation of the floaters in the vicinity of the cluster surface. Surface atoms are denoted by white circles, while floaters are indicated by shaded circles. A vacancy is indicated by a dotted circle.

The purpose here is to investigate an elementary process, which dominates RA, by giving a brief sketch of a relationship between motion of surface atoms in a configurational space and that on a multidimensional PES. In particular, special emphasis is placed on an analysis of the pathways connecting a local minimum to the saddles which is climbed over during an isomerization process. In the previous section we obtained the values of the activation energy for the radial and surface diffusion. These two values provide clues to find out a key elementary process for the onset of RA. Alternatively stated, the goal of the present part is to give an interpretation of the values of activation energy by searching for the most relevant reaction paths for causing RA.

B. Numerical Enumeration of Reaction Paths

It is practically difficult and time-consuming to enumerate local minima and saddle points on PES extensively in the case where a system contains at most 10^1–10^2 atoms. Even for a cluster $A_{47}B_{20}$, it is still hard to map each rearranging motion to the corresponding reaction paths. Then we make a restrictive numerical search for the reaction paths. In other words we look for the saddle points that are *reachable* neighbors of the given local minimum in terms of the eigenvector-following method proposed by Cerjan and Miller [14]. In the present work the saddle points are enumerated by starting the searching procedure from a given local minimum on PES. More precisely, a random fluctuation is added to the atomic configurations at a local minimum in order to reach every saddle point that may be randomly distributed in the vicinity of the given local minima. About 10,000 randomly distributed points near a local minimum were used as initial points for the eigenvector-following method. The intensity of random fluctuation is about 7% of the mean separation of the neighboring atoms. Although no one guarantees that the saddle points neighboring to a given local minimum are completely detected by the present method, the authors confirmed that almost all trivial saddles that are noticeable by inspection are successfully found in various sizes and shapes of the 2D Morse clusters. The present method is applied to pick up the saddle points of the Morse cluster $A_{47}B_{20}$. For simplicity we choose the case $\alpha = 1.0$, since the configuration of the saddle for a homogeneous cluster is not much altered from that for a binary cluster.

Two representative clusters that are different in shape are exemplified. One is a *compact* cluster—that is, a geometrically packed structure—which is located

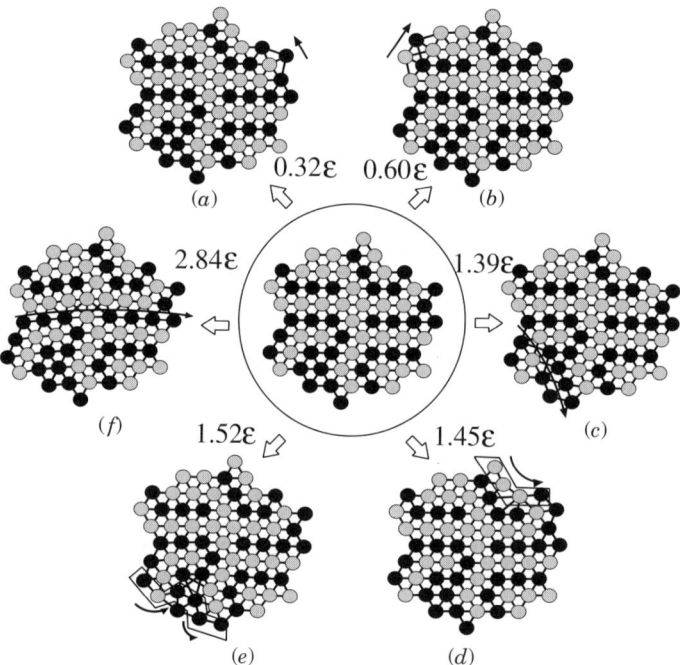

Figure 12. Representative configurations of the cluster atoms corresponding to the saddle points for the noncompact $A_{47}B_{20}$ are exhibited by balls and sticks. Sticks are inserted between pair atoms whose separation is shorter than 1.5σ. In the center circle, the configuration of the noncompact $A_{47}B_{20}$ at a local minima is displayed. Typical atomic configurations of the first-order saddle points are shown in (a)–(f). The direction of the atomic displacement from the initial local minimum to another beyond the saddle point is denoted by arrows. The barrier height is also inserted in the figure. Each atom is colored by black or gray randomly just to identify its location before and after the displacement.

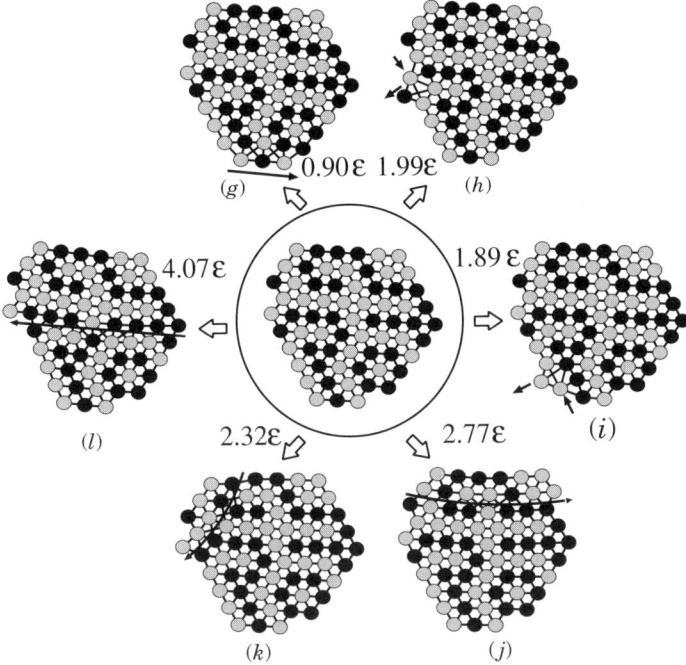

Figure 13. Representative configurations of the cluster atoms corresponding to the first-order saddle points for the compact $A_{47}B_{20}$ are exhibited by balls and sticks in (g)–(l).

at a deep minimum on PES. Another is a *noncompact* cluster that is in a relatively shallow minimum. An important difference between these clusters is presence and absence of *floaters*. The compact cluster has no floater, while the noncompact one is possessed by four *floaters* capable to move almost freely, as shown in Figs. 12 and 13 [15].

C. A Distribution of the Saddle Point Energy

The number of the first- and second-order saddles which are detected for the noncompact $A_{47}B_{20}$ are 89 and 656, respectively. Those for the compact $A_{47}B_{20}$ are 185 and 1020, respectively.

In Fig. 14 the number of the first- and second-order saddles with respect to the value of the barrier height is displayed by the two histograms for the compact and the noncompact $A_{47}B_{20}$. One can immediately notice that the noncompact cluster has more saddles lying in the low-energy region both for the first- and second-order saddles. If one takes into account the presence of many floppy *floaters*, it is not a surprising result to appear more low-lying barriers near the

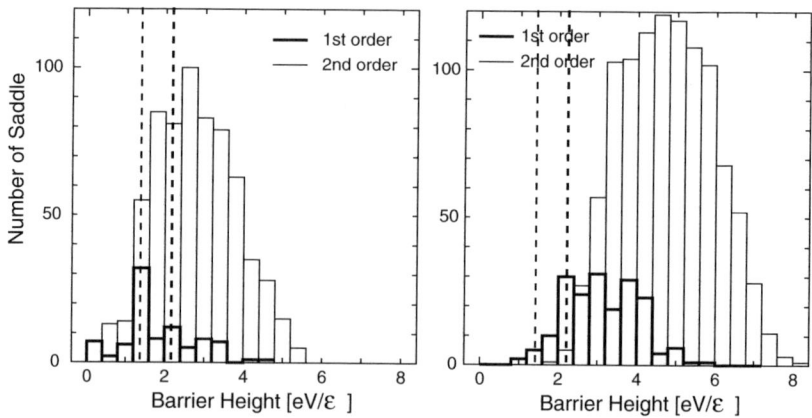

Figure 14. Histograms for the number of the saddle points with respect to the value of the barrier height. The barrier energy is normalized by the depth of the pair potential ϵ. (a) The data of the first- and second-order saddles for the noncompact $A_{47}B_{20}$ are shown by a thick line and a thin line, respectively. (b) The data of the first- and second-order saddles for the compact $A_{47}B_{20}$ are also shown by a thick line and a thin line. The barrier height for the surface diffusion is denoted by a thin dotted line, while that for radial diffusion is shown by a thick dotted line.

local minimum representing the noncompact $A_{47}B_{20}$. In Figs. 12 and 13, the six representative atomic configurations corresponding to the saddles, which are directly connected with the initial local minimum, are also shown. As exhibited in Fig. 12a, the reaction paths crossing the extremely low-lying barrier, whose height is about 0.3ϵ (1200 K), are attributed to the hopping motion of a single *floater*. We call this motion the *floater hopping* (FH). Similarly, the bounded train of a few surface atoms is easy to glide over the cluster surface as shown in Figs. 12b and 13g, and the barrier heights for such motion are considerably low. The barrier height for the gliding motion of two and three atoms are evaluated as about 0.6ϵ (2400 K) and 0.9ϵ (3000 K), respectively. Since the FH costs about 0.3ϵ, the barrier height for the gliding motion in Fig. 12b and 13g are understandable by estimating as $0.6\epsilon (= 0.3\epsilon \times 2)$ and $0.9\epsilon (= 0.3\epsilon \times 3)$, respectively.

Let us compare these data with the values of the activation energy for surface and radial diffusion. In Fig. 14 the barrier height for the surface diffusion is denoted by thin dotted line, while that for radial diffusion is shown by a thick dotted line. For the noncompact cluster, the number of the first-order saddle abruptly increases about 1.0ϵ. On the other hand, for the compact cluster, the number of the first-order saddle jumps at about 2.0ϵ (8000 K). This trend is similarly observed in other compact and non-compact clusters, even though some exceptional cases are also present.

An agreement between the estimated activation energy and the saddle point energy at which there appears the sudden increase of the frequency suggests the followings:

1. The activation energy for surface diffusion is attributed to the barrier implying gliding motion of several surface atoms that form a train as indicated by Figs. 12b, 12d, and 13g. We call such a motion the edge running (ER). In fact, so many events that are classified as ER-type motion are detected by monitoring the animation generated from MD data. Thus, it is plausible that the surface diffusion is contributed mainly by the ER.
2. On the other hand, the activation energy for radial diffusion is attributed to the cost to remove one corner atom and impose another surface atom. A typical example is in Figs. 13h and 13i. These rearrangements result in an *interlayer mixing* (IM) between surface atoms and inner atoms. Such an exchange is the onset of radial diffusion and is often observed in MD simulation. That is, the radial diffusion is efficiently promoted by the IM at the corner.

Let us consider the role of the ER and the IM in attaining RA. As demonstrated in Fig. 9, the radial diffusion is significantly suppressed as the motion of surface atoms is suppressed by an instantaneous increase of their mass. In this sense, the diffusion along the surface direction and that along the radial direction are not independent, but cooperative somehow.

This cooperative relationship is transparent by considering the relation between the ER and the IM in dynamics of a cluster. If a noncompact cluster is prepared as an initial state of time evolution, a noncompact cluster becomes a compact one by rearrangements like the FH and the ER which imply a low-energy barrier. In contrast, a compact cluster spends a long time in that stable state until it gains enough large energy to climb over a barrier corresponding to the IM. The IM is a candidate for the bottleneck of radial diffusion. In other words, a compact cluster configuration that proceeds to radial diffusion is prepared by frequent surface diffusion. That is, the radial diffusion is caused by surface diffusion.

V. SUMMARY

In the present chapter, we demonstrate that our simple Morse model successively reproduces the experimental feature of RA at a qualitative level. Under an isothermal condition, the atomistic process behind RA is characterized by estimating the activation energy. In particular, the surface diffusion and radial diffusion is extracted separately. We also confirm that the active motion of surface atoms is converted into the radial diffusion. Such a numerical observation is reinforced with a reaction path analysis. The ER and the FH contributes to the surface diffusion, while the IM controls the onset of the radial diffusion. In terms of the interplay between two types of diffusion mechanism, RA is completed.

The viewpoint that we would like to stress for the dynamical aspect of a cluster may become clear by comparing with the earth in motion. The earth seems to be solid all the way through. However, the surface of the earth is covered by the lithosphere, which is *floating* on the thick, gooey, molten rock called the asthenosphere. The lithosphere consists of the plates that gradually move around on the top of the asthenosphere. In the extremely long time scale, the solid crust on the plate migrates, and the earth exhibits a liquid-like behavior on its surface. Similarly our findings in this work suggest that, due to the presence of active surface, the microcluster is a *dynamic* material where the fast diffusion like RA comes forward in the *meso-time scale*, even though the microcluster itself exhibits the solid-like behavior. Conversely speaking, another image of the microcluster can be pulled out by emphasizing its *meso-time scale* behaviors.

References

1. S. Sugano and H. Koizumi, *Microcluster Physics*, Springer, Berlin, 1999.
2. J. O. Bovin, R. Wallenberg, and D. J. Smith, *Nature (London)* **317**, 47 (1985); S. Iijima and T. Ichihashi, *Phys. Rev. Lett.* **56**, 616 (1986); M. Mitome, Y. Tanishiro, and K. Takayanagi, *Z. Phys.* **D12**, 45 (1989); P. M. Ajayan and L. D. Marks, *Phys. Rev. Lett.* **63**, 279 (1989).
3. P. M. Ajayan and L. D. Marks, *Phys. Rev. Lett.* **60**, 585 (1988); *Phys. Rev. Lett.* **63**, 279 (1989).
4. S. Sawada and S. Sugano, *Z. Phys.* **D14**, 247 (1984); S. Sawada and S. Sugano, *Z. Phys.* **D20**, 258 (1991); S. Sawada and S. Sugano, *Z. Phys.* **D24**, 377 (1992).
5. J. Jellinek, T. L. Beck, and R. S. Berry, *J. Chem. Phys.* **84**, 2783 (1985); F. G. Amar and R. S. Berry, *J. Chem. Phys.* **85**, 5943 (1986); R. S. Berry, *Chem. Rev.* **93**, 2379 (1993) and references therein.
6. H.-P. Cheng and R. S. Berry, *Phys. Rev.* **A45**, 7969 (1992).
7. H. Yasuda and H. Mori et al., *J. Electron. Microsc.* **41**, 267 (1992); H. Yasuda and H. Mori, *Z. Phys. D* **31**, 131 (1994); *Z. Phys. D* **31**, 209 (1994); H. Mori, H. Yasuda, and T. Kamino, *Philos. Mag. Lett.* **69**, 279 (1994); H. Yasuda, H. Mori, M. Komatsu, and K. Takeda, *J. Appl. Phys.* **73**, 1100 (1993); H. Yasuda and H. Mori, *Phys. Rev. Lett.* **69**, 3747 (1992); H. Yasuda and H. Mori, *Intermetallics* **1**, 35 (1993).
8. Y. Shimizu, K. S. Ikeda, and S. Sawada, *Phys. Rev.* **B64**, 75412 (2001); Y. Shimizu, K. S. Ikeda, and S. Sawada, *Eur. Phys. J.* **D4**, 365 (1998); Erratum *Eur. Phys.* **D6**, 281 (1998); S. Sawada, Y. Shimizu, and K. S. Ikeda, *Phys. Rev.* **B67**, 024204 (2003).
9. T. Kobayashi, K. S. Ikeda, Y. Shimizu, and S. Sawada, *Phys. Rev.* **B66**, 245412 (2002).
10. O. Kubachewski, *Trans. Faraday Soc.* **46**, 713 (1950).
11. T. Kobayashi, K. S. Ikeda, Y. Shimizu, and S. Sawada, *J. Chem. Phys.* **118**, 6552 (2003).
12. L. A. Girifalco and V. G. Weizer, *Phys. Rev.* **114**, 687 (1959).

13. P. G. Shewmon, *Diffusion in Solids*, McGraw-Hill, New York, 1963; L. A. Girifalco, *Atomic Migration in Crystals*, Blaisdell, Waltham, MA, 1964; P. Guiraldenq, *Diffusion dans les Metaux*, Techniques de l'ingenieur, 1978.
14. C. Cerjan and W. H. Miller, *J. Chem. Phys.* **75**, 288 (1981); C. J. Tsai and K. D. Jordan, *J. Phys. Chem.* **97**, 11227 (1993).
15. In the present 2D model, we employ the term "*floater*" in a rather restrictive sense for clarity. That is, a *floater* is defined as a surface atom that has less than 3 neighbors. Thus, the atoms attached two sticks in figures are floaters.

CHAPTER 15

VIBRATIONAL ENERGY RELAXATION (VER) OF A CD STRETCHING MODE IN CYTOCHROME c

HIROSHI FUJISAKI, LINTAO BU, and JOHN E. STRAUB

Department of Chemistry, Boston University, Boston, Massachusetts, 02215, USA

CONTENTS

I. Introduction
II. Vibrational Energy Relaxation (VER)
   A. Perturbation Expansion for the Interaction
   B. General Formula for VER
   C. Use of a Symmetrized Autocorrelation Function
   D. Quantum Correction Factor Method and Other Methods
   E. Approximations for the Force–Force Correlation Function
      1. Taylor Expansion of the Force
      2. Contribution from the First Term
      3. Contribution from the Third Term
III. Application to a CD Stretching Mode in Cytochrome c
   A. Definition of System and Bath
   B. Calculation of the Coupling Constants
   C. Assignment of the "Lifetime" Parameter
   D. Results
      1. Classical Calculation
      2. Quantum Calculation
      3. Discussion
IV. Summary and Further Aspects
Acknowledgments
References

Geometric Structures of Phase Space in Multidimensional Chaos: A Special Volume of Advances in Chemical Physics, Part B, Volume 130, edited by M. Toda, T Komatsuzaki, T. Konishi, R.S. Berry, and S.A. Rice. Series editor Stuart A. Rice.
ISBN 0-471-71157-8 Copyright © 2005 John Wiley & Sons, Inc.

I. INTRODUCTION

Vibrational energy relaxation (VER) is fundamentally important to our understanding of chemical reaction dynamics as it influences reaction rates significantly [1]. In general, estimating VER rates for selected modes in large molecules is a challenging problem because large molecules involve many degrees of freedom and, furthermore, quantum effects cannot be ignored [2]. If we assume a weak interaction between the "system" and the surrounding "bath," however, we can derive an estimate of the VER rate through Fermi's golden rule [3–6]: A VER rate is written as a Fourier transformation of a force–force correlation function. Though it is not trivial to define and justify a separation of a system and a bath, such a formulation has been successfully applied to many VER processes in liquids [7] and in proteins [8].

Here we apply such theories of VER to the problem of estimating the vibrational population relaxation time of a CD stretching mode—in short, a CD mode—in the protein cytochrome c [9]. (We will define the CD mode to be the system and define the remainder of the protein to be the bath.) Recently Romesberg's group succeeded in selectively deuterating a terimnal methyl group of a methionie residue in cytochrome c [10]. The resulting CD mode has a frequency $\omega_S \simeq 2100$ cm^{-1}, which is located in a transparent region of the density of states of the protein. As such, spectroscopic detection of this mode provides clear evidence of the protein dynamics, including the VER of the CD vibrational mode. Note that at room temperature ($T = 300$ K) $\beta\hbar\omega_S \simeq 10$, where $\beta = 1/(k_B T)$; hence quantum effects are not negligible for this mode.

Let us mention a little more about cytochrome c (cyt c). Cyt c is a protein known to exist in mitochondrial inner membranes, chloroplasts of plants, and bacteria [11]. Its functions are related to cell respiration [12]; and cyt c, using its heme molecule, "delivers" an electron from cytochrome bc 1 to cytochrome oxidase—two larger proteins both embedded in a membrane. Recently it was also found that cyt c is released when apoptosis occurs [13]. In this sense, cyt c governs the "life and death" of a cell.

The heme molecule in cyt c has a large oscillator strength, and it serves as a good optical probe. As a result, many spectroscopic experiments have been designed to clarify VER and the (un)folding properties of cyt c [14]. Cyt c is often employed in numerical simulations [15,16] because a high-resolution structure was obtained [17] and its simulation has become feasible. Attempts have also been made to characterize cyt c through ab initio (DFT) calculations [18,19].

VER of the selected CD mode in the terminal methyl group of methionine (Met80) was previously addressed by Bu and Straub [9]: They used equilibrium simulations for cyt c in water with the quantum-correction factor (QCF) method

[20], and they predicted that the VER time for the CD mode is on the order of 0.3 ps. However, their results are approximate: The use of the QCF method is not justified a priori, and their analysis is based on a harmonic model for cyt c. To extend that previous analysis, in this work, we model cyt c in vacuum as a normal mode system and include the lowest anharmonic coupling elements. A similar analysis has been completed for another protein myoglobin by Kidera's group [21] and by Leitner's group [22]. Use of a reduced model Hamiltonian allows us to investigate the VER rate of the CD mode in cyt c more "exactly" and to move beyond the use of quantum correction factors and the harmonic approximation.

This chapter is organized as follows: In Section I.B, we derive the principal VER formula employed in our work, and we mention the related Maradudin–Fein formula. In Section I.C, we apply those theoretical results for the rate of VER to the CD mode in cyt c, and we compare our results with (a) the classical simulation by Bu and Straub and (b) the experiments by Romesberg's group. In Section I.D, we provide a summary of our results, and we discuss further aspects of VER processes in proteins.

II. VIBRATIONAL ENERGY RELAXATION (VER)

A. Perturbation Expansion for the Interaction

We begin with the von Neumann equation for the complete system written as

$$i\hbar \frac{d}{dt}\rho(t) = [\mathcal{H}, \rho(t)] \tag{1}$$

The interaction representation for the von Neumann equation is

$$i\hbar \frac{d}{dt}\tilde{\rho}(t) = [\tilde{\mathcal{V}}(t), \tilde{\rho}(t)] \tag{2}$$

where

$$\mathcal{H} = \mathcal{H}_0 + \mathcal{V} = \mathcal{H}_S + \mathcal{H}_B + \mathcal{V} \tag{3}$$

and

$$\tilde{\rho}(t) \equiv e^{i\mathcal{H}_0 t/\hbar}\rho(t)e^{-i\mathcal{H}_0 t/\hbar}, \quad \tilde{\mathcal{V}}(t) \equiv e^{i\mathcal{H}_0 t/\hbar}\mathcal{V}e^{-i\mathcal{H}_0 t/\hbar} \tag{4}$$

Here $\mathcal{H}_S$ is the system Hamiltonian representing a vibrational mode, $\mathcal{H}_B$ the bath Hamiltonian representing solvent or environmental degrees of freedom, and

$\mathscr{V}$ the interaction Hamiltonian describing the coupling between the system and the bath. An operator with a tilde means the one in the interaction picture. If we *assume* that $\mathscr{V}$ is small in some sense, we can carry out the perturbation expansion for $\mathscr{V}$ as

$$\tilde{\rho}(t) = \rho(0) + \frac{1}{i\hbar}\int_0^t dt'[\tilde{\mathscr{V}}(t'), \tilde{\rho}(t')]$$

$$= \rho(0) + \frac{1}{i\hbar}\int_0^t dt'[\tilde{\mathscr{V}}(t'), \rho(0)]$$

$$+ \frac{1}{(i\hbar)^2}\int_0^t dt' \int_0^{t'} dt''[\tilde{\mathscr{V}}(t'), [\tilde{\mathscr{V}}(t''), \rho(0)]] + \cdots \quad (5)$$

Let us calculate the following probability:

$$P_v(t) \equiv \text{Tr}\{\rho_v \rho(t)\} = \text{Tr}\{\rho_v e^{-i\mathscr{H}_0 t/\hbar} \tilde{\rho}(t) e^{i\mathscr{H}_0 t/\hbar}\} \quad (6)$$

$$\rho_v \equiv |v\rangle\langle v| \otimes 1_B \quad (7)$$

$$\rho(0) = \rho_S \otimes \rho_B = |v_0\rangle\langle v_0| \otimes e^{-\beta \mathscr{H}_B}/Z_B \quad (8)$$

$$Z_B = \text{Tr}_B\{e^{-\beta \mathscr{H}_B}\} \quad (9)$$

where the initial state is assumed to be a direct product state of $\rho_S = |v_0\rangle\langle v_0|$ and $\rho_B = e^{-\beta \mathscr{H}_B}/Z_B$. Here $|v\rangle$ is the vibrational eigenstate for the system Hamiltonian $\mathscr{H}_S$, that is, $\mathscr{H}_S|v\rangle = E_v|v\rangle$. The VER rate $\Gamma_{v_0 \to v}$ may be defined as follows:

$$\Gamma_{v_0 \to v} \equiv \lim_{t \to \infty} \frac{d}{dt} P_v(t) \quad (10)$$

Note that the results derived from this definition are equivalent to those derived from Fermi's golden rule [23]. Hence we refer to them as a Fermi's golden rule formula.

B. General Formula for VER

First we notice that

$$P_v(t) = \text{Tr}\{\rho_v e^{-i\mathscr{H}_0 t/\hbar} \tilde{\rho}(t) e^{i\mathscr{H}_0 t/\hbar}\} = \text{Tr}\{\rho_v \tilde{\rho}(t)\} \quad (11)$$

as ρ_v commutes with $\mathscr{H}_0$. If we assume that $v \neq v_0$, then $\rho_v \rho(0) = 0$. Using this fact, we obtain the lowest (second)-order result

$$P_v(t) \simeq \frac{1}{(i\hbar)^2} \int_0^t dt' \int_0^{t'} dt'' \, \text{Tr}\{\rho_v[\tilde{\mathscr{V}}(t'), [\tilde{\mathscr{V}}(t''), \rho(0)]]\}$$

$$= \frac{1}{\hbar^2} \int_0^t dt' \int_0^{t'} dt'' \, \text{Tr}\{\rho_v \tilde{\mathscr{V}}(t')\rho(0)\tilde{\mathscr{V}}(t'') + \rho_v \tilde{\mathscr{V}}(t'')\rho(0)\tilde{\mathscr{V}}(t')\} \quad (12)$$

$$= \frac{1}{\hbar^2} \int_0^t dt' \int_0^{t'} dt'' [e^{i\omega_{v_0v}(t'-t'')} C(t'-t'')$$

$$+ e^{i\omega_{v_0v}(t''-t')} C(t''-t')]$$

where

$$C(t) \equiv \langle \tilde{\mathscr{V}}_{v_0v}(t) \mathscr{V}_{vv_0}(0) \rangle \equiv \text{Tr}_B\{\rho_B \tilde{\mathscr{V}}_{v_0v}(t) \mathscr{V}_{vv_0}(0)\} \quad (13)$$

$$\tilde{\mathscr{V}}_{vv_0}(t) = \langle v|\tilde{\mathscr{V}}(t)|v_0\rangle \quad (14)$$

$$\omega_{v_0v} = (E_{v_0} - E_v)/\hbar \quad (15)$$

Hence the lowest-order estimate of the VER rate is given by

$$\Gamma_{v_0 \to v} = \lim_{t \to \infty} \frac{1}{\hbar^2} \int_0^t dt'' [e^{i\omega_{v_0v}(t-t'')} C(t-t'') + e^{i\omega_{v_0v}(t''-t)} C(t''-t)]$$

$$= \frac{1}{\hbar^2} \int_{-\infty}^{\infty} dt \, e^{i\omega_{v_0v}t} C(t) \quad (16)$$

If we *assume* that a system variable q is small in some sense, the interaction Hamiltonian is expressed as

$$\mathscr{V} = -q\mathscr{F}(\{q_k\}, \{p_k\}) \quad (17)$$

where $\{q_k\}, \{p_k\}$ are position and momentum variables for the bath. This $\mathscr{F}(\{q_k\}, \{p_k\})$ is a force applied to the system by the bath. Thus we finally obtain the following Fermi's golden rule formula [3–6]:

$$\Gamma_{v_0 \to v} = \frac{|q_{v_0v}|^2}{\hbar^2} \int_{-\infty}^{\infty} dt \, e^{i\omega_{v_0v}t} \langle \tilde{\mathscr{F}}(t)\tilde{\mathscr{F}}(0)\rangle \quad (18)$$

with

$$q_{v_0 v} = \langle v_0 | q | v \rangle \tag{19}$$

$$\tilde{\mathscr{F}}(t) = e^{i\mathscr{H}_B t/\hbar} \tilde{\mathscr{F}} e^{-i\mathscr{H}_B t/\hbar} \tag{20}$$

$$\langle \tilde{\mathscr{F}}(t)\tilde{\mathscr{F}}(0) \rangle = \text{Tr}_B\{\rho_B \tilde{\mathscr{F}}(t)\tilde{\mathscr{F}}(0)\} \tag{21}$$

In most situations, the transition from $v_0 = 1$ to $v = 0$ is considered. In such a case, $q_{10} = \sqrt{\hbar/2m_S\omega_S}$, where m_S is the system mass and $\omega_S = \omega_{10}$ is the system frequency in the harmonic approximation. Hence

$$\Gamma_{1\to 0} = \frac{1}{2m_S\hbar\omega_S} \int_{-\infty}^{\infty} dt\, e^{i\omega_S t} \langle \tilde{\mathscr{F}}(t)\tilde{\mathscr{F}}(0) \rangle \tag{22}$$

C. Use of a Symmetrized Autocorrelation Function

It is useful to define a *symmetrized* force–force correlation function as [3–6]

$$S(t) = \tfrac{1}{2}[\langle \tilde{\mathscr{F}}(t)\tilde{\mathscr{F}}(0) \rangle + \langle \tilde{\mathscr{F}}(0)\tilde{\mathscr{F}}(t) \rangle] \tag{23}$$

Since $S(t)$ is real and symmetric with respect to t, $S(t) = S(-t)$, we consider it to be analogous to $S_{\text{cl}}(t)$, the classical limit of the correlation function. Hereafter we drop the tilde on $\mathscr{F}$ for simplicity. By half-Fourier transforming $S(t)$ with the use of the relation $\langle \mathscr{F}(0)\mathscr{F}(t) \rangle = \langle \mathscr{F}(t - i\beta\hbar)\mathscr{F}(0) \rangle$, we have

$$\int_0^\infty dt\, e^{i\omega t} S(t) = \tfrac{1}{2} \int_0^\infty dt\, e^{i\omega t} \langle \mathscr{F}(t)\mathscr{F}(0) \rangle + \tfrac{1}{2} \int_0^\infty dt\, e^{i\omega t} \langle \mathscr{F}(0)\mathscr{F}(t) \rangle$$

$$= \tfrac{1}{2} \int_0^\infty dt\, e^{i\omega t} \langle \mathscr{F}(t)\mathscr{F}(0) \rangle + \tfrac{1}{2} \int_0^\infty dt\, e^{i\omega t} \langle \mathscr{F}(t - i\beta\hbar)\mathscr{F}(0) \rangle$$

$$= \tfrac{1}{2} \int_0^\infty dt\, e^{i\omega t} \langle \mathscr{F}(t)\mathscr{F}(0) \rangle + \tfrac{1}{2} e^{-\beta\hbar\omega} \int_0^\infty dt\, e^{i\omega t} \langle \mathscr{F}(t)\mathscr{F}(0) \rangle$$

$$= \tfrac{1}{2}(1 + e^{-\beta\hbar\omega}) \int_0^\infty dt\, e^{i\omega t} \langle \mathscr{F}(t)\mathscr{F}(0) \rangle \tag{24}$$

Taking the real parts of both sides, we have

$$\int_0^\infty dt\, \cos(\omega t)\, S(t) = \tfrac{1}{4}(1 + e^{-\beta\hbar\omega}) \int_{-\infty}^\infty dt\, e^{i\omega t} \langle \mathscr{F}(t)\mathscr{F}(0) \rangle \tag{25}$$

where we have used the fact that $S(-t) = S(t)$ is real and $\langle \mathscr{F}(t)\mathscr{F}(0)\rangle^* = \langle \mathscr{F}(0)\mathscr{F}(t)\rangle = \langle \mathscr{F}(-t)\mathscr{F}(0)\rangle$. Hence, Eq. (22) can be rewritten as [6]

$$\Gamma_{1\to 0} = \frac{1}{m_S \hbar \omega_S} \frac{2}{1 + e^{-\beta\hbar\omega_S}} \int_0^\infty dt \, \cos(\omega_S t) S(t) \qquad (26)$$

Note that this expression diverges in the classical limit because $\Gamma_{1\to 0} \propto 1/\hbar$. According to Bader and Berne [4], to make contact with the classical limit, we introduce another VER rate as

$$\begin{aligned}\frac{1}{T_1} &= (1 - e^{-\beta\hbar\omega_S})\Gamma_{1\to 0} \\ &= 2C(\beta, \hbar\omega_S) \int_0^\infty dt \, \cos(\omega_S t) S(t)\end{aligned} \qquad (27)$$

where

$$C(\beta, \hbar\omega_S) = \frac{1}{m_S \hbar \omega_S} \frac{1 - e^{-\beta\hbar\omega_S}}{1 + e^{-\beta\hbar\omega_S}} \qquad (28)$$

This is a final quantum expression, which can be interpreted as an energy relaxation rate and be used to estimate the VER rate.[1]

D. Quantum Correction Factor Method and Other Methods

Though Eq. (27) is exact in a perturbative sense, it is demanding to calculate the *quantum mechanical* force autocorrelation function $S(t)$ even for small molecular systems. Hence, many computational schemes have been developed to approximate the quantum mechanical force autocorrelation function.

Skinner and coworkers advocated to use the quantum correction factor (QCF) method [20], which is the replacement of the above formula Eq. (18) with

$$\Gamma_{v_0 \to v} = Q(\omega_S) \frac{2|q_{v_0 v}|^2}{\hbar^2} \int_0^\infty dt \, \cos(\omega_{v_0 v} t) S_{\rm cl}(t) \qquad (29)$$

where $S_{\rm cl}(t) = \langle \mathscr{F}(t)\mathscr{F}(0)\rangle_{\rm cl}$ and the bracket means a classical ensemble average (not a quantum mechanical average). This approach is very intuitive and easily applicable for large molecular systems because one only needs to calculate the *classical* force autocorrelation function $S_{\rm cl}(t)$ multiplied by an

---

[1] Although the experimental observable is $\Gamma_{1\to 0}$, we note that $1/T_1 \simeq \Gamma_{1\to 0}$ because $\beta\hbar\omega_S \gg 1$ for our case of a CD stretching mode.

appropriate QCF $Q(\omega_S)$. There exist several QCFs corresponding to different VER processes [20].

However, one challenge that arises in the application of the QCF method is that we do not know *a priori* which VER process is dominant for the system considered. Furthermore, it is possible that several VER processes compete [24]. Hence one must be careful in the application of the QCF method, and Skinner and coworkers have provided a number of examples of how this can be accomplished.

In this chapter, we employ the harmonic approximation for the relaxing oscillator, and the vibrational relaxation time T_1^{QCF}. Hence Eq. (29) is transformed to

$$\frac{1}{T_1^{QCF}} = \frac{Q(\omega_S)}{m_S \hbar \omega_S} \int_0^\infty dt \, \cos(\omega_S t) S_{cl}(t) = \frac{Q(\omega_S)}{\beta \hbar \omega_S} \frac{1}{T_1^{cl}} \quad (30)$$

where we have introduced the classical VER rate $1/T_1^{cl}$

$$\frac{1}{T_1^{cl}} = \frac{\beta}{m_S} \int_0^\infty dt \, \cos(\omega_S t) S_{cl}(t) \quad (31)$$

which is the classical limit $\hbar \to 0$ of Eq. (27). This result can also be derived from a classical theory of Brownian motion, and is known as the Landau-Teller-Zwanzig (LTZ) formula.

Alternatively, one may calculates $S(t)$ *itself* systematically using controlled approximations. Calculating a correlation function for large systems has a long history in chemical physics [25], including recent applications to VER processes in liquid [26,27]. The vibrational self-consistent field (VSCF) method [28] will also be useful in this respect.

On the other hand, if we *approximate* $\mathscr{F}$ as a simple function of $\{q_k\}, \{p_k\}$, we can calculate $S(t)$ rather easily and, in a sense, more "exactly." In the next section, we explore such an approach.

E. Approximations for the Force–Force Correlation Function

1. Taylor Expansion of the Force

We can formally Taylor-expand the force as a function of the bath variables $\{q_k\}, \{p_k\}$:

$$\mathscr{F}(\{q_k\}, \{p_k\}) = \sum_k A_k^{(1)} q_k + \sum_k B_k^{(1)} p_k + \sum_{k,k'} A_{k,k'}^{(2)} q_k q_{k'} + \sum_{k,k'} B_{k,k'}^{(2)} p_k p_{k'} + \cdots \quad (32)$$

where the expansion is often truncated in the literature following the first term. Depending on the system–bath interaction considered, higher-order coupling including the third and fourth terms can be relevant. For example, the fourth term appears in benzene to represent the interaction between the CH stretch and CCH wagging motion [29] through the Wilson G matrix [30]. In the case of a CD stretching mode in cyt c, as discussed below, or a CN^- stretching mode in water [24], the third term is relevant for VER.

2. Contribution from the First Term

If the first term $\sum_k A_k^{(1)} q_k$ is dominant in the force, then the force–force correlation function becomes

$$\langle \mathscr{F}(t)\mathscr{F}(0)\rangle = \sum_{k,k'} A_k^{(1)} A_{k'}^{(1)} \langle q_k(t) q_{k'}(0) \rangle = \sum_k \frac{\hbar(A_k^{(1)})^2}{2\omega_k}[(n_k+1)e^{-i\omega_k t} + n_k e^{i\omega_k t}] \tag{33}$$

where we have used

$$q_k(t) = \sqrt{\frac{\hbar}{2\omega_k}}(a_k e^{-i\omega_k t} + a_k^\dagger e^{i\omega_k t}) \tag{34}$$

and $\langle a_k^\dagger a_{k'}\rangle = n_k \delta_{k,k'}$ with $n_k = 1/(e^{\beta\hbar\omega_k} - 1)$ because

$$\rho_B \propto e^{-\beta \mathscr{H}_B} = e^{-\beta \sum_k \hbar\omega_k(a_k^\dagger a_k + 1/2)} \tag{35}$$

Here we have *assumed* that the bath Hamiltonian is an ensemble of harmonic oscillators:

$$\mathscr{H}_B = \sum_k \hbar\omega_k(a_k^\dagger a_k + 1/2) = \sum_k \left(\frac{p_k^2}{2} + \frac{\omega_k^2}{2}q_k^2\right) \tag{36}$$

where ω_k is the kth mode frequency for the bath, and

$$p_k = -i\sqrt{\frac{\hbar\omega_k}{2}}(a_k - a_k^\dagger) \tag{37}$$

Thus we obtain

$$S(t) = \sum_k \frac{\hbar(A_k^{(1)})^2}{2\omega_k}(2n_k+1)\cos\omega_k t \tag{38}$$

and

$$\frac{1}{T_1} = \pi\hbar C(\beta, \hbar\omega_S) \sum_k \frac{(A_k^{(1)})^2}{\omega_k} (2n_k + 1)[\delta(\omega_S - \omega_k) + \delta(\omega_S + \omega_k)] \quad (39)$$

The contribution from the second term $\sum_k B_k^{(1)} p_k$ is calculated in the same way.

3. Contribution from the Third Term

If the third term $\sum_{k,k'} A_{k,k'}^{(2)} q_k q_{k'}$ is dominant in the force, then

$$\langle \mathscr{F}(t)\mathscr{F}(0) \rangle = \sum_{k,k',k'',k'''} A_{k,k'}^{(2)} A_{k'',k'''}^{(2)} \langle q_k(t)q_{k'}(t)q_{k''}(0)q_{k'''}(0) \rangle$$
$$= R_{--}(t) + R_{++}(t) + R_{+-}(t) \quad (40)$$

where

$$R_{--}(t) = \frac{\hbar^2}{4} \sum_{k,k',k'',k'''} D_{k,k',k'',k'''}^{(2)} \langle a_k a_{k'} a_{k''}^\dagger a_{k'''}^\dagger \rangle e^{-i(\omega_k + \omega_{k'})t} \quad (41)$$

$$R_{++}(t) = \frac{\hbar^2}{4} \sum_{k,k',k'',k'''} D_{k,k',k'',k'''}^{(2)} \langle a_k^\dagger a_{k'}^\dagger a_{k''} a_{k'''} \rangle e^{i(\omega_k + \omega_{k'})t} \quad (42)$$

$$R_{+-}(t) = \frac{\hbar^2}{4} \sum_{k,k',k'',k'''} D_{k,k',k'',k'''}^{(2)} \left[\langle a_k a_{k'}^\dagger (a_{k''}^\dagger a_{k'''} + a_{k''} a_{k'''}^\dagger) \rangle e^{-i(\omega_k - \omega_{k'})t} \right.$$
$$\left. + \langle a_k^\dagger a_{k'} (a_{k''}^\dagger a_{k'''} + a_{k''} a_{k'''}^\dagger) \rangle e^{i(\omega_k - \omega_{k'})t} \right] \quad (43)$$

with

$$D_{k,k',k'',k'''}^{(2)} = \frac{A_{k,k'}^{(2)} A_{k'',k'''}^{(2)}}{\sqrt{\omega_k \omega_{k'} \omega_{k''} \omega_{k'''}}} \quad (44)$$

Using the following

$$\langle a_k a_{k'} a_{k''}^\dagger a_{k'''}^\dagger \rangle = (1 + n_k)(1 + n_{k'})(\delta_{kk''}\delta_{k'k'''} + \delta_{kk'''}\delta_{k'k''}) \quad (45)$$

$$\langle a_k^\dagger a_{k'}^\dagger a_{k''} a_{k'''} \rangle = n_k n_{k'}(\delta_{kk''}\delta_{k'k'''} + \delta_{kk'''}\delta_{k'k''}) \quad (46)$$

$$\langle a_k a_{k'}^\dagger (a_{k''}^\dagger a_{k'''} + a_{k''} a_{k'''}^\dagger) \rangle = (1 + n_k)(1 + 2n_{k'})\delta_{kk'}\delta_{k''k'''}$$
$$+ (1 + n_k)n_{k'}(\delta_{kk''}\delta_{k'k'''} + \delta_{kk'''}\delta_{k'k''}) \quad (47)$$

$$\langle a_k^\dagger a_{k'} (a_{k''}^\dagger a_{k'''} + a_{k''} a_{k'''}^\dagger) \rangle = n_k(1 + 2n_{k'})\delta_{kk'}\delta_{k''k'''}$$
$$+ n_k(1 + n_{k'})(\delta_{kk''}\delta_{k'k'''} + \delta_{kk'''}\delta_{k'k''}) \quad (48)$$

we have

$$R_{--}(t) = \frac{\hbar^2}{2} \sum_{k,k'} D^{(2)}_{k,k',k,k'}(1+n_k)(1+n_{k'})e^{-i(\omega_k+\omega_{k'})t} \quad (49)$$

$$R_{++}(t) = \frac{\hbar^2}{2} \sum_{k,k'} D^{(2)}_{k,k',k,k'} n_k n_{k'} e^{i(\omega_k+\omega_{k'})t} \quad (50)$$

$$R_{+-}(t) = \langle \mathscr{F}(0) \rangle^2 + \hbar^2 \sum_{k,k'} D^{(2)}_{k,k',k,k'}(1+n_k)n_{k'} e^{-i(\omega_k-\omega_{k'})t} \quad (51)$$

where we have used $A^{(2)}_{k,k'} = A^{(2)}_{k',k}$, and

$$\langle \mathscr{F}(0) \rangle^2 = \frac{\hbar^2}{4} \sum_{k,k'} D^{(2)}_{k,k,k',k'}(1+2n_k)(1+2n_{k'}) \quad (52)$$

Hence we obtain

$$S(t) = \sum_{k,k'} \left[\zeta^{(+)}_{k,k'} \cos(\omega_k+\omega_{k'})t + \zeta^{(-)}_{k,k'} \cos(\omega_k-\omega_{k'})t \right] + \langle \mathscr{F}(0) \rangle^2 \quad (53)$$

and

$$\frac{1}{T_1} = \pi C(\beta,\hbar\omega_S) \sum_{k,k'} \left\{ \zeta^{(+)}_{k,k'} [\delta(\omega_k+\omega_{k'}-\omega_S) + \delta(\omega_k+\omega_{k'}+\omega_S)] \right. \\ \left. + \zeta^{(-)}_{k,k'} [\delta(\omega_k-\omega_{k'}-\omega_S) + \delta(\omega_k-\omega_{k'}+\omega_S)] \right\} \quad (54)$$

where we have assumed $\omega_S \neq 0$ and defined

$$\zeta^{(+)}_{k,k'} = \frac{\hbar^2}{2} D^{(2)}_{k,k',k,k'}(1+n_k+n_{k'}+2n_k n_{k'}) \quad (55)$$

$$\zeta^{(-)}_{k,k'} = \frac{\hbar^2}{2} D^{(2)}_{k,k',k,k'}(n_k+n_{k'}+2n_k n_{k'}) \quad (56)$$

Though its appearance is rather different, Eq. (54) is equivalent to that derived by Kenkre, Tokmakoff, and Fayer [5] as well as by Shiga and Okazaki [24]. There is also a similar result known as the Maradudin–Fein formula [31]

$$W = W_{\text{decay}} + W_{\text{coll}} \tag{57}$$

$$W_{\text{decay}} = \frac{\pi \hbar}{2 m_S \omega_S} \sum_{k,k'} \frac{(A^{(2)}_{k,k'})^2}{\omega_k \omega_{k'}} (1 + n_k + n_{k'}) \delta(\omega_S - \omega_k - \omega_{k'}) \tag{58}$$

$$W_{\text{coll}} = \frac{\pi \hbar}{m_S \omega_S} \sum_{k,k'} \frac{(A^{(2)}_{k,k'})^2}{\omega_k \omega_{k'}} (n_k - n_{k'}) \delta(\omega_S + \omega_k - \omega_{k'}) \tag{59}$$

which has been utilized to describe VER processes in glasses [32] and in proteins by Leitner's group [22]. As was demonstrated in Ref. 5, this formula is also equivalent to Eq. (54); in the following we make use of Eq. (54). A problem with this formula is that we cannot take its continuum limit in the case of finite systems like proteins. As a remedy, a *width parameter* related to the vibrational lifetime is usually introduced leading to a definite value for the VER rate. We will discuss this problem in Section III.C.

III. APPLICATION TO A CD STRETCHING MODE IN CYTOCHROME c

A. Definition of System and Bath

We take horse heart cytochrome c (cyt c) as an example of how one may estimate the rate of VER for selected modes in proteins. We use the CHARMM program [33] to describe the force field, to minimize the structure, and to calculate the normal modes for the system. Starting from the 1HRC structure for cyt c in Protein Data Bank (PDB) [34], one hydrogen atom of the terminal methyl group of Met80 was deuterated. The energy of the protein structure was minimized in vacuum using the conjugate gradient algorithm. We diagonalized the Hessian matrix (second derivatives of the potential) for that mechanically stable configuration of the protein:

$$K_{ij} = \frac{\partial^2 V_{\text{CHARMM}}}{\partial \bar{x}_i \partial \bar{x}_j} = \frac{1}{\sqrt{m_i m_j}} \frac{\partial^2 V_{\text{CHARMM}}}{\partial x_i \partial x_j} \tag{60}$$

where V_{CHARMM} is the CHARMM potential, and $\bar{x}_i = \sqrt{m_i} x_i$ are mass-weighted Cartesian coordinates. The number of atoms in cyt c is 1745 (myoglobin has 2475 atoms), so the Hessian matrix is 5235 × 5235, and its diagonalization was readily accomplished using the `vibran` facility in CHARMM [33].

The result of this calculation was the density of states (DOS) for the system as shown in Fig. 1. The DOS consists of three regions: (1) below around

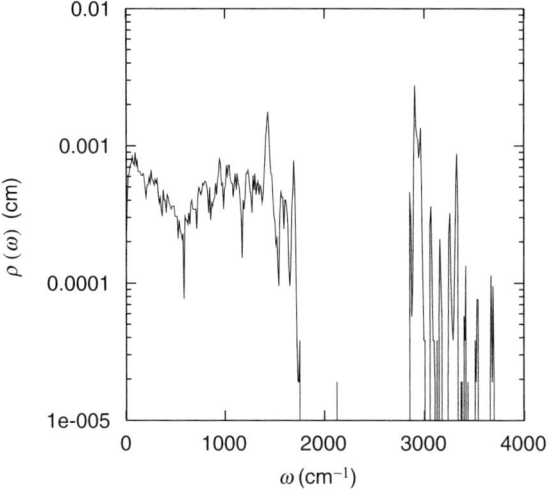

Figure 1. Density of states for cytochrome c in vacuum.

1700 cm^{-1}, (2) from around 1700 cm^{-1} to around 2800 cm^{-1}, and (3) above around 2800 cm^{-1}. The first region corresponds to rotational and torsional motions of the protein, and the third corresponds to bond stretching motions such as CH bonds. The second is rather "transparent," but one can observe one mode localized around the CD bond stretching mode in Met80 with frequency 2129.1 cm^{-1} as shown in Fig. 2a. Hence we refer to this as a CD stretching mode, or CD mode, the dynamics of which is the focus of our study. The other two modes in Fig. 2b and 2c are strongly coupled modes with the CD mode: 3330th mode (1330.9 cm^{-1}), an angle bending mode of Met80, and the 1996th mode (829.9 cm^{-1}), a stretch-bend mode in Met80. In the following, we will discuss the detail of the coupling and how it affects the VER rate.

At this level of description, the system is an ensemble of harmonic oscillators—that is, normal modes. Since we are interested in VER of the CD mode, we represent it as a system

$$\mathcal{H}_S = \frac{p_{CD}^2}{2} + \frac{\omega_{CD}^2}{2} q_{CD}^2 \tag{61}$$

while other degrees of freedom are treated as a bath:

$$\mathcal{H}_B = \sum_k \left(\frac{p_k^2}{2} + \frac{\omega_k^2}{2} q_k^2 \right) \tag{62}$$

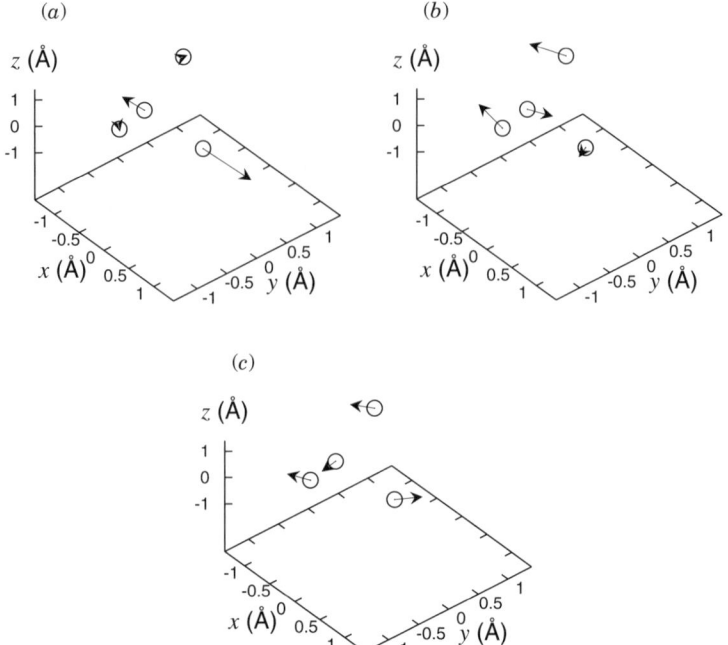

Figure 2. Normal modes of cytochrome c in vacuum. (a) 4357th mode (CD mode) with $\omega = 2129.1$ cm^{-1}, (b) 3330th mode (angle bending mode of Met80) with $\omega = 1330.9$ cm^{-1}, (c) 1996th mode (a stretch-bend mode in Met80) with $\omega = 829.9$ cm^{-1}. Only vectors on the terminal methyl group of Met80 in cyt c are depicted. These modes are strongly coupled with the Fermi resonance.

The interaction between the system and bath is described by the interaction Hamiltonian

$$\mathcal{V} = \mathcal{H}_{\text{cyt c}} - \mathcal{H}_S - \mathcal{H}_B \tag{63}$$

where $\mathcal{H}_{\text{cyt c}}$ is the Hamiltonian for the full cyt c protein. We will discuss the content of $\mathcal{V}$ in the following section.

B. Calculation of the Coupling Constants

As in Eq. (17), we *assume* that the interaction Hamiltonian is of the form

$$\mathcal{V} = -q_{CD}\mathcal{F} \tag{64}$$

and Taylor expand the force as Eq. (32). The first and second terms do not appear because this is a normal mode expansion, and the fourth term does not appear as the original coordinates are Cartesian coordinates. As in the first approximation, we take the force to be

$$\mathscr{F} = \sum_{k,k'} A^{(2)}_{k,k'} q_k q_{k'} \qquad (65)$$

The coupling coefficients $A^{(2)}_{k,l}$ are calculated as

$$A^{(2)}_{k,l} = -\frac{1}{2} \frac{\partial^3 V}{\partial q_{CD} \, \partial q_k \, \partial q_l} \qquad (66)$$

A problem arises: How does one calculate these coupling coefficients? The most direct approach is to use a finite difference method:

$$A^{(2)}_{k,l} \simeq -\frac{1}{2} \frac{V_{+++} - V_{-++} - V_{+-+} - V_{++-} + V_{--+} + V_{-+-} + V_{+--} - V_{---}}{(2\Delta q_{CD})(2\Delta q_k)(2\Delta q_l)} \qquad (67)$$

where $V_{\pm\pm\pm} = V(\pm\Delta q_{CD}, \pm\Delta q_k, \pm\Delta q_l)$. However, this is rather cumbersome. Instead, we use the approximation [22,24]

$$A^{(2)}_{k,l} \simeq -\frac{1}{2} \sum_{ij} U_{ik} U_{jl} \frac{K_{ij}(\Delta q_{CD}) - K_{ij}(-\Delta q_{CD})}{2\Delta q_{CD}} \qquad (68)$$

where U_{ik} is an orthogonal matrix that diagonalizes the Hessian matrix at the mechanically stable structure K_{ij}, and $K_{ij}(\pm\Delta q_{CD})$ is a Hessian matrix calculated at a shifted structure along the direction of the CD mode with a shift $\pm\Delta q_{CD}$. This expression is approximate but readily implemented using the CHARMM facility to compute the Hessian matrix. A comparison between Eqs. (67) and (68) is made in Table I. We also examined the convergence of the results by changing Δq_{CD}, and we found that $\Delta q_{CD} = 0.02$Å is sufficient for the following calculations.

TABLE I
Comparison Between the Finite Difference Method [Eq. (67)] and Eq. (68)[a]

(k, l)	Equation (68)	Finite Difference
(3330, 1996)	22.3	22.4
(3330, 4399)	−29.6	−29.5
(3327, 1996)	−5.7	−5.8
(1996, 678)	0.64	0.63

[a]We have used $\Delta q_{CD} = 0.02$ Å, and $A^{(2)}_{k,l}$ is given in kcal/mol/Å^3.

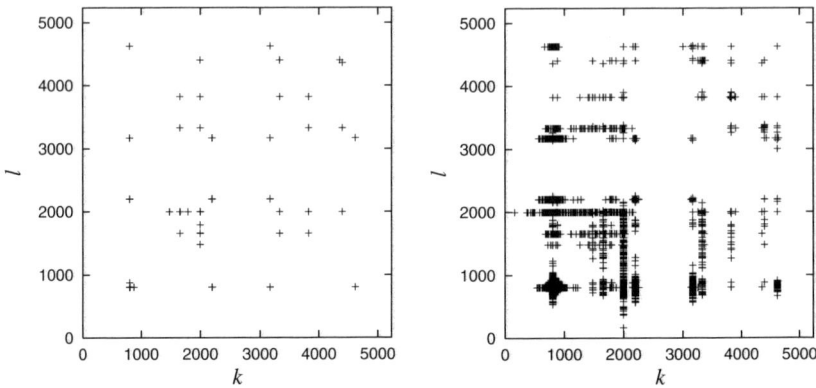

Figure 3. Distribution of the coupling elements. **Left:** $50.0 > |A_{k,l}^{(2)}| > 5.0$. **Right:** $5.0 > |A_{k,l}^{(2)}| > 0.5$. The value of $A_{k,l}^{(2)}$ are given in kcal/mol/Å^3. Note that (k, l) are mode numbers, not wavenumbers.

The numerical results for the coupling elements are shown in Fig. 3. The histogram for the elements is shown in Fig. 4. As one can see from these figures, most of the elements are small, while the largest coupling elements are rather large. See Table II. Note that the combination (3330, 1996) is particularly significant for the CD mode because it approximately satisfies the

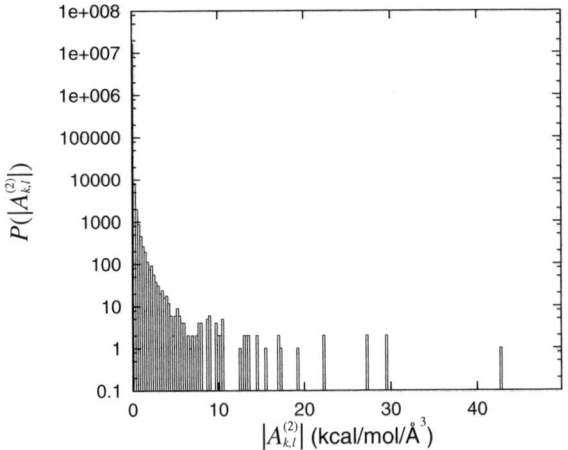

Figure 4. Histogram for the amplitude of the coupling elements.

TABLE II
The Largest Coupling Elements[a]

(k, l)	$|A_{k,l}^{(2)}|$
(1996, 1996)	42.9
(4399, 3330)	29.6
(4622, 3170)	27.3
(3330, 1996)	22.3

[a]The value of $A_{k,l}^{(2)}$ are given in kcal/mol/Å^3.

resonant condition [21]:

$$|\omega_{CD} - \omega_k - \omega_l| \ll \mathcal{O}(|A_{k,l}^{(2)}|) \tag{69}$$

As shown in Figs. 2b and 2c, these modes are localized near the terminal methyl group of Met80 as well as near the CD mode. In such a case, resonant energy transfer (Fermi resonance) is expected as shown by Moritsugu, Miyashita, and Kidera [21]. We have observed similar behavior in cyt c when the CD mode was excited, and the energy immigration to other normal modes facilitated by resonance was followed.

C. Assignment of the "Lifetime" Parameter

We cannot directly evaluate Eq. (54) because it contains delta functions. Evaluation of this expression for a finite system like a protein leads to a null result. To circumvent this problem, we "thaw" the delta function $\delta(x)$ as

$$\delta(x) = \frac{1}{\pi} \frac{\gamma}{\gamma^2 + x^2} \tag{70}$$

using a width parameter γ. Physically this means that each normal mode should have a lifetime $\simeq 1/\gamma$ due to coupling to other degrees of freedom—that is, the surrounding environment including water (or we might be able to interpret $1/\gamma$ as a time resolution). It is difficult to derive γ from first principles, so we take it to be a phenomenological parameter as in the literature [22,32].

As a result, the VER rate, Eq. (54), for the CD mode becomes

$$\frac{1}{T_1} = C(\beta, \hbar\omega_S) \sum_{k,k'} \left[\frac{\gamma \zeta_{k,k'}^{(+)}}{\gamma^2 + (\omega_k + \omega_{k'} - \omega_S)^2} + \frac{\gamma \zeta_{k,k'}^{(+)}}{\gamma^2 + (\omega_k + \omega_{k'} + \omega_S)^2} \right.$$

$$\left. + \frac{\gamma \zeta_{k,k'}^{(-)}}{\gamma^2 + (\omega_k - \omega_{k'} - \omega_S)^2} + \frac{\gamma \zeta_{k,k'}^{(-)}}{\gamma^2 + (\omega_k - \omega_{k'} + \omega_S)^2} \right] \tag{71}$$

We employ this expression in our subsequent calculations.

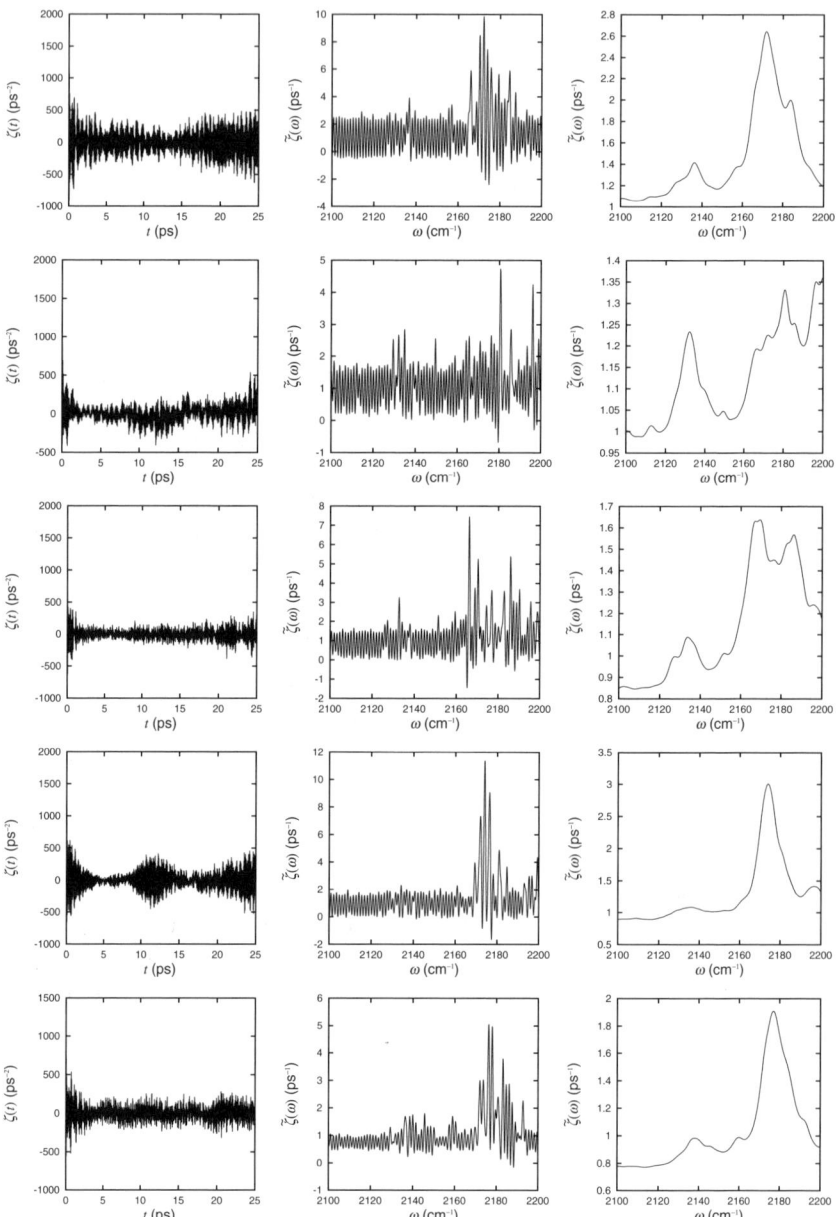

Figure 5. **Left:** Classical data for the force–force correlation function. **Middle:** Fourier spectra for the correlation function. **Right:** The corresponding coarse-grained Fourier spectra. The "lifetime" width parameter $\gamma = 3$ cm^{-1}.

D. Results

1. Classical Calculation

Classical force–force correlation functions and their (cosine) Fourier transformations are shown in the left and middle column of Fig. 5 for five different trajectories. Here we have defined a ζ function as

$$\zeta(t) = \frac{\beta}{m_S} S_{\text{cl}}(t) \tag{72}$$

and have defined its (cosine) Fourier transformation as $\tilde{\zeta}(\omega)$, that is, $1/T_1^{\text{cl}} = \tilde{\zeta}(\omega_S)$. Note that these data are obtained from molecular dynamics simulations of cyt c in water [9]. As can be seen, the correlation functions oscillate wildly, and the (cosine) Fourier transformations are messy. As such, it is difficult to extract a reliable and stable value for the VER rate.

To address this problem, we introduce the window function

$$w(t) = \exp(-\gamma t) \tag{73}$$

The ζ functions are multiplied by this function and are (cosine) Fourier transformed. This corresponds to broadening each peak of a spectrum with a Lorentzian with width γ. The results for five trajectories are shown in the right column of Fig. 5. (The width parameter is taken as $\gamma = 3$ cm^{-1}.) The results in the right column are better behaved than those in the middle column, but there still remain some fluctuations.

According to Bu and Straub simulations of cyt c in water [9], we take $\omega_S = 2135$ cm^{-1} to investigate the γ dependence of the result as shown in the left of Fig. 6. We see that $\tilde{\zeta}(\omega_S) \simeq 1.1 \sim 1.2$ ps^{-1} for $\gamma \simeq 3 \sim 30$ cm^{-1}. Since $Q(\omega_S)/(\beta\hbar\omega_S) \simeq 2.4 \sim 3.0$ for two-phonon processes [9], this corresponds to a VER time of 0.3–0.4 ps according to Eq. (29).[2]

2. Quantum Calculation

We use the formula Eq. (71) as a quantum mechanical estimate of the VER rate. The γ dependence of the result is shown on the right-hand side of Fig. 6. We see that, for $\gamma \simeq 3 \sim 30$ cm^{-1}, the quantum mechanical estimate gives $T_1 \simeq 0.2 \sim 0.3$ ps, which is similar to the classical estimate Eq. (29): $T_1 \simeq 0.3 \sim 0.4$ ps.

In Table III, we list the largest contributions to the VER rate for different width parameters. For the case of $\gamma = 3$ cm^{-1}, the largest contribution is due to modes (3823,1655). This combination of modes is nearly resonant with the CD

---

[2]In the VER calculation of myoglobin [22], Leitner's group took $\gamma = 0.5 \sim 10$ cm^{-1} to be the width, and confirmed that the result is relatively insensitive to the choice of γ in this range.

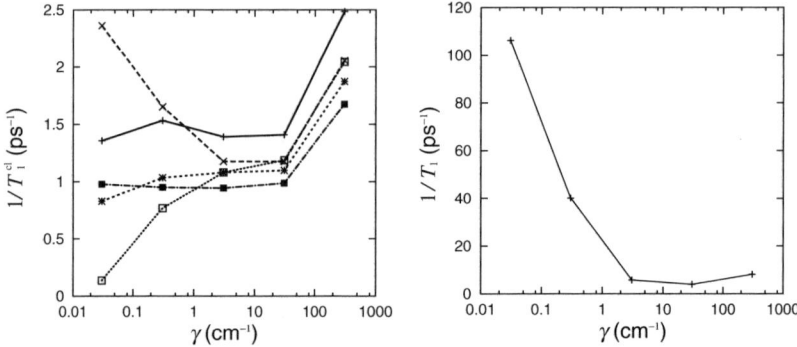

Figure 6. **Left:** Classical VER rate for five trajectories as a function of the "lifetime" width parameter γ. **Right:** VER rate calculated by Eq. (71) as a function of γ.

mode as $|\omega_{3823} + \omega_{1655} - \omega_{CD}| \simeq 0.03$ cm^{-1}. Though the coupling element for the combination is small ($|A^{(2)}_{3823,1655}| = 5.1$ kcal/mol/Å^3), this mode combination contributes significantly to the VER rate. On the other hand, for the case of $\gamma = 30$ cm^{-1}, the largest contribution results from the combination of modes (3330,1996). This combination is somewhat off-resonant—that is, $|\omega_{3330} + \omega_{1996} - \omega_{CD}| = 32$ cm^{-1}—but the coupling element is very large ($|A^{(2)}_{3330,1996}| = 22.3$ kcal/mol/Å^3), and the contribution is significant. In both cases, one combination of modes dominates the VER rate ($\simeq 20\%$), though there are nonnegligible contributions from other combinations of modes.

We have also examined the temperature dependence of the VER rates using Eq. (71). As shown in Fig. 7, for $T < 300$ K there is little temperature dependence as has been addressed in the case of myoglobin [22]. Thus we can say that the relaxation of the CD mode is quantum mechanical rather than thermal because the decay at 300 K is similar to that at 0.3 K.

TABLE III
The Largest Contributions to the VER Rate (in ps^{-1}) for $\gamma = 3$ cm^{-1} (left) and $\gamma = 30$ cm^{-1} (right)

(k,l)	Contribution	(k,l)	Contribution
(3823, 1655)	1.10 (19%)	(3330, 1996)	0.88 (22%)
(3823, 1654)	0.43 (8%)	(3823, 1655)	0.11 (3%)
(3822, 1655)	0.37 (6%)	(3170, 2196)	0.07 (2%)
(3330, 1996)	0.17 (3%)	(1996, 1996)	0.05 (1%)
(3822, 1654)	0.15 (3%)	(3823, 1654)	0.04 (1%)
(3823, 1661)	0.14 (3%)	(3170, 2202)	0.04 (1%)
(3822, 1661)	0.05 (1%)	(3822, 1655)	0.04 (1%)
(3822, 1656)	0.05 (1%)	(3327, 1996)	0.03 (1%)
(3823, 1658)	0.04 (1%)	(3330, 1655)	0.02 (1%)

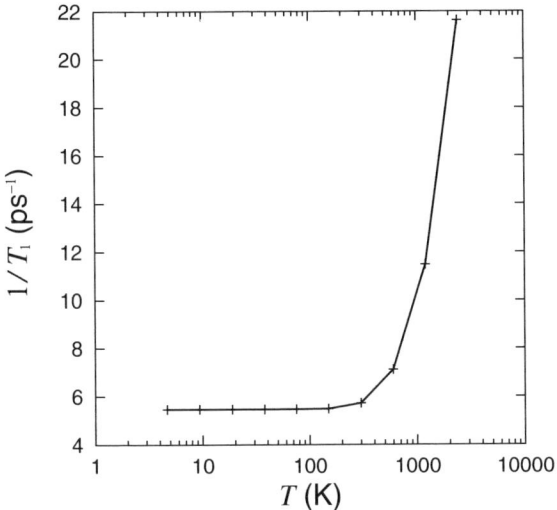

Figure 7. Temperature dependence of the VER rate calculated by Eq. (71). The width parameter is $\gamma = 3$ cm^{-1}.

3. Discussion

We examine the relationship between the theoretical results described above and the corresponding experiments of Romesberg's group, which has studied the spectroscopic properties of the CD mode in cyt c [10]. They measured the shifts and widths of the spectra for different forms of cyt c; the widths of the spectra (FWHM) were found to be $\Delta\omega_{\text{FWHM}} \simeq 6.0$–$13.0$ cm^{-1}. A rough estimate of the VER rate leads to

$$T_1 \sim 5.3/\Delta\omega_{\text{FWHM}} \quad (\text{ps}) \tag{74}$$

which corresponds to $T_1 \simeq 0.4$–0.9 ps. This estimate is similar to the "semi"-classical prediction computed using Eq. (29) and appropriate QCFs (0.3–0.4 ps) and the perturbative quantum mechanical estimate using the reduced model (0.2–0.3 ps). This result appears to justify the use of QCFs and the reduced model in this situation, and it suggests that the effects of the protein solvation (by water) are negligible in describing the VER of the CD mode. Of course, we must be careful in comparing the estimate derived from Eq. (74) because there may be inhomogeneity in the experimental spectra.[3] As such, it is more desirable to

---

[3] We have confirmed that the methyl group does not rotate during the equilibrium simulations. Thus we might exclude the rotation as a possible reason of inhomogeneity.

calculate not only VER rates but spectroscopic observables themselves to compare with experiments.

Finally we discuss the relation between this work and previous work on carbon monoxide myoglobin (MbCO). Though there are many experimental studies on Mb [35], we focus on the experiments of Anfinrud's group [36] and Fayer's group [37] on MbCO. The former group found that the VER time for CO in the heme pocket (photolyzed MbCO) is $\simeq$ 600 ps, whereas the latter group found that the VER time for CO bounded to the heme is $\simeq$ 20 ps. This difference is interpreted as follows: CO is covalently bonded to the heme for the latter case, whereas CO is "floating" in the pocket in the former case; that is, the force applied to CO for the latter case is stronger than that for the former case. This difference in the magnitude of the force causes the slower VER for the "floating" CO. In this respect, the CD bond is expected to be stronger than the CO–heme coupling. This may explain a VER time $\sim$ 0.1 ps, which is similar to the VER times for the CH(CD) stretching modes in benzene (or perdeuterobenzene) [29,38]. It will be interesting to apply a similar reduced model to the analysis of the VER of CO in MbCO.

IV. SUMMARY AND FURTHER ASPECTS

After reviewing the VER rate formula derived from quantum mechanical perturbation calculations, we applied it to the analysis of VER of a CD stretching mode in cyt c. We modeled cyt c in vacuum as a normal mode system with the third-order anharmonic coupling elements, which were calculated from the CHARMM potential. We found that, for the width parameter $\gamma = 3 \sim 30 \text{ cm}^{-1}$, the VER time is 0.2–0.3 ps, which agrees rather well with the previous classical calculation using the quantum correction factor (QCF) method, and is consistent with the experiments by Romesberg's group. This result indicates that the use of QCFs or a reduced model Hamiltonian can be justified a posteriori to describe the VER problem. We decomposed the VER rate into contributions from two modes, and we found that the most significant contribution, which depends on the "lifetime" width parameter, results from modes most resonant with the CD mode.

Finally we note several future directions which should be studied: (a) Our final results for the VER rate depend on a width parameter γ. Unfortunately we do not know which value is the most appropriate for γ. Nonequilibrium simulations (with some quantum corrections [39]) might help this situation, and they are useful to investigate energy pathways or sequential IVR (intramolecular vibrational energy redistribution) [40] in a protein. (b) This work is motivated by pioneering spectroscopic experiments by Romesberg's group. The calculation of the VER rate and the linear or nonlinear response functions, related to absorption or 2D-IR (or 2D-Raman) spectra [41–44], is desirable. (c)

Romesberg's group investigated a spectroscopic change due to the oxidation or reduction of Fe in the heme; such an electron transfer process [45] is fundamental for the functionality of cyt c. To survey this process dynamically, it will be necessary to combine some quantum chemistry (ab initio) calculations with MD simulations [18,46,47].

Acknowledgments

We thank Dr. J. Gong for noting Ref. 5, Prof. A. Kidera, Prof. S. Okazaki, Prof. J. L. Skinner, Prof. D. Coker, Prof. D. M. Leitner, Prof. Y. Mizutani, Dr. T. Takami, Dr. T. Miyadera, Dr. Y. Kawashima, Dr. S. Fuchigami, Mr. H. Teramoto for useful discussions, Prof. A. Stuchebrukhov for sending reprints, and Dr. M. Shigemori for providing perl scripts used in this work. We thank the National Science Foundation (CHE-0316551) and Boston University's Center for Computational Science for generous support of our research.

References

1. B. J. Berne, M. Borkovec, and J. E. Straub, *J. Phys. Chem.* **92**, 3711 (1988); J. I. Steinfeld, J. E. Francisco, and W. L. Hase, *Chemical Kinetics and Dynamics*, Prentice-Hall, Englewood Cliffs, NJ, 1989; A. Stuchebrukhov, S. Ionov, and V. Letokhov, *J. Phys. Chem.* **93**, 5357 (1989); T. Uzer, *Phys. Rep.* **199**, 73 (1991).
2. D. E. Logan and P. G. Wolynes, *J. Chem. Phys.* **93**, 4994 (1990); S. A. Schofield and P. G. Wolynes, *J. Chem. Phys.* **98**, 1123 (1993); S. A. Schofield, P. G. Wolynes, and R. E. Wyatt, *Phys. Rev. Lett.* **74**, 3720 (1995); S. A. Schofield and P. G. Wolynes, *J. Phys. Chem.* **99**, 2753 (1995); D. M. Leitner and P. G. Wolynes, *J. Phys. Chem. A* **101**, 541 (1997).
3. D. W. Oxtoby, *Adv. Chem. Phys.* **47**, 487 (1981).
4. J. S. Bader and B. J. Berne, *J. Chem. Phys.* **100**, 8359 (1994).
5. V. M. Kenkre, A. Tokmakoff, and M. D. Fayer, *J. Chem. Phys.* **101**, 10618 (1994).
6. S. A. Egorov and J. L. Skinner, *J. Chem. Phys.* **105**, 7047 (1996).
7. R. Rey and J. T. Hynes, *J. Chem. Phys.* **104**, 2356 (1996); *J. Chem. Phys.* **108**, 142 (1998).
8. D. E. Sagnella and J. E. Straub, *Biophys. J.* **77**, 70 (1999).
9. L. Bu and J. E. Straub, *Biophys. J.* **85**, 1429 (2003); Erratum, to be published.
10. J. K. Chin, R. Jimenez, and F. Romesberg, *J. Am. Chem. Soc.* **123**, 2426 (2001); *J. Am Chem. Soc.* **124**, 1846 (2002).
11. D. Keilin, *The History of Cell Respirations and Cytochrome*, Cambridge University Press, Cambridge, 1966; R. E. Dickerson, *Sci. Am.* **242**, 136 (1980); G. W. Pettigrew and G. R. Moore, *Cytochromes c: Evolutionary, Structural, and Physiochemical Aspects*, Springer-Verlag, Berlin, 1990.
12. B. Alberts, D. Bray, J. Lewis, M. Raff, and K. Roberts, *Molecular Biology of the Cell*, 3rd ed., Garland, New York, 1994; G. Karp, *Cell and Molecular Biology: Concepts and Experiments*, 3rd ed., John & Sons, New York, Wiley 2002.
13. J. Yang, X. Liu, K. Bhalla, C. N. Kim, A. M. Ibrado, J. Cai, T.-I Peng, D. P. Jones, and X. Wang, *Science* **275**, 1129 (1997); R. M. Kluck, E. Bossy-Wetzel, D. R. Green, and D. D. Newmeyer, *Science* **275**, 1132 (1997).
14. S.-R. Yeh, S. Han, and D. L. Rousseau, *Acc. Chem. Res.* **31**, 727 (1998); S. W. Englander, T. R. Sosnick, L. C. Mayne. M. Shtilerman, P. X. Qi, and Y. Bai, *Acc. Chem. Res.* **31**, 737 (1998);

W. Wang, X. Ye, A. A. Demidov, F. Rosca, T. Sjodin, W. Cao, M. Sheeran, and P. M. Champion, *J. Phys. Chem. B* **104**, 10789 (2000).
15. S. H. Northrup, M. P. Pear, J. A. McCammon, and M. Karplus, *Nature* **286**, 304 (1980); C. F. Wong, C. Zheng, J. Shen, J. A. McCammon, and P. G. Wolynes, *J. Phys. Chem.* **97**, 3100 (1993); A. E. Carcía and G. Hummer, *Proteins: Struct. Funct. Genet.* **36**, 175 (1999); A. E. Cardenas and R. Elber, *Proteins: Struct. Funct. Genet.* **51**, 245 (2003); X. Yu and D. M. Leitner, *J. Chem. Phys.* **119**, 12673 (2003).
16. L. Bu and J. E. Straub, *J. Phys. Chem. B* **107**, 12339 (2003).
17. G. W. Bushnell, G. V. Louie, and G. D. Brayer, *J. Mol. Biol.* **214**, 585 (1990).
18. W. Andreoni, A. Curioni, and T. Mordasini, *IBM J. Res. & Dev.* **45**, 397 (2001).
19. F. Sato, T. Yoshihiro, M. Era, and H. Kashiwagi, *Chem. Phys. Lett.* **341**, 645 (2001).
20. J. L. Skinner and K. Park, *J. Phys. Chem. B* **105**, 6716 (2001).
21. K. Moritsugu, O. Miyashita, and A. Kidera, *Phys. Rev. Lett.* **85**, 3970 (2000); *J. Phys. Chem. B* **107**, 3309 (2003).
22. D. M. Leitner, *Phys. Rev. Lett.* **87**, 188102 (2001); X. Yu and D. M. Leitner, *J. Phys. Chem. B* **107**, 1698 (2003).
23. J. J. Sakurai, *Modern Quantum Mechanics*, 2nd ed., Addison-Wesley, Reading, MA, 1994; G. C. Schatz and M. A. Ratner, *Quantum Mechanics in Chemistry*, Dover, Mineola, NY, 2002.
24. M. Shiga and S. Okazaki, *J. Chem. Phys.* **109**, 3542 (1998); *J. Chem. Phys.* **111**, 5390 (1999); T. Mikami, M. Shiga, and S. Okazaki, *J. Chem. Phys.* **115**, 9797 (2001).
25. W. H. Miller, *J. Phys. Chem. A* **105**, 2942 (2001).
26. Q. Shi and E. Geva, *J. Phys. Chem. A* **107**, 9059 (2003); *J. Phys. Chem. A* **107**, 9070 (2003); *J. Chem. Phys.* **119**, 9030 (2003).
27. H. Kim and P. J. Rossky, *J. Phys. Chem. B* **106**, 8240 (2002); J. A. Poulsen, G. Nyman, and P. J. Rossky, *J. Chem. Phys.* **119**, 12179 (2003).
28. R. B. Gerber, V. Buch, and M. A. Ratner, *J. Chem. Phys.* **77**, 3022 (1982); A. Roitberg, R. B. Gerber, R. Elber, and M. A. Ratner, *Science* **268**, 1319 (1995); S. K. Gregurick, E. Fredj, R. Elber, and R. B. Gerber, *J. Phys. Chem. B* **101**, 8595 (1997); Z. Bihary, R. B. Gerber, and V. A. Apkarian, *J. Chem. Phys.* **115**, 2695 (2001). For a review of the VSCF methods, see P. Jungwirth and R. B. Gerber, *Chem. Rev.* **99**, 1583 (1999).
29. E. L. Sibert III, J. T. Hynes, and W. P. Reinhardt, *J. Chem. Phys.* **81**, 1135 (1984); E. L. Sibert III, W. P. Reinhardt, and J. T. Hynes, *J. Chem. Phys.* **81**, 115 (1984).
30. E. B. Wilson, Jr., J. C. Decius, and P. C. Cross, *Molecular Vibrations: The Theory of Infrared and Raman Vibrational Spectra*, Dover, Mineola, NY, 1980.
31. A. A. Maradudin and A. E. Fein, *Phys. Rev.* **128**, 2589 (1962).
32. J. Fabian and P. B. Allen, *Phys. Rev. Lett.* **77**, 3839 (1996).
33. B. R. Brooks, R. E. Bruccoleri, B. D. Olafson, D. J. States, S. Swaminathan, and M. Karplus, *J. Comp. Chem.* **4**, 187 (1983); A. D. MacKerell, Jr., B. Brooks, C. L. Brooks III, L. Nilsson, B. Roux, Y. Won, and M. Karplus, in *The Encyclopedia of Computational Chemistry*, vol. 1, P. v. R. Schleyer et al., ed., John Wiley & Sons: Chichester, 1998, p. 271.
34. See, for example, http://www.rcsb.org/pdb.
35. Y. Mizutani and T. Kitagawa, *Science* **278**, 443 (1997); *J. Phys. Chem. B* **105**, 10992 (2001); M. D. Fayer, *Annu. Rev. Phys. Chem.* **52**, 315 (2001); X. Ye, A. Demidov, and P. M. Champion, *J. Am. Chem. Soc.* **124**, 5914 (2002); F. Rosca, A. T. N. Kumar, D. Ionascu, X. Ye, A. A. Demidov, T. Sjodin, D. Wharton, D. Barrick, S. G. Sligar, T. Yonetani, and P. M. Champion, *J. Phys. Chem. A* **106**, 3540 (2002).

36. D. E. Sagnella, J. E. Straub, T. A. Jackson, M. Lim, and P. A. Anfinrud, *Proc. Natl. Acad. Sci. USA* **96**, 14324 (1999).
37. J. R. Hill, D. D. Dlott, C. W. Rella, K. A. Peterson, S. M. Decatur, S. G. Boxer, and M. D. Fayer, *J. Phys. Chem.* **100**, 12100 (1996).
38. H. W. Schranz, *J. Mol. Struct. (Theochem.)* **368**, 119 (1996).
39. P. H. Nguyen and G. Stock, *J. Chem. Phys.* **119**, 11350 (2003).
40. K. Someda and S. Fuchigami, *J. Phys. Chem. A* **102**, 9454 (1998); H. Hasegawa and K. Someda, *J. Chem. Phys.* **110**, 11255 (1999).
41. K. Okumura and Y. Tanimura, *Chem. Phys. Lett.* **278**, 175 (1997).
42. T. I. C. Jansen, J. G. Snijder, and K. Duppen, *J. Chem. Phys.* **113**, 307 (2000); *J. Chem. Phys.* **114**, 10910 (2001).
43. K. A. Merchant, W. G. Noid, D. E. Thompson, R. Akiyama, R. F. Loring, and M. D. Fayer, *J. Phys. Chem. B* **107**, 4 (2003).
44. J. Edler and P. Hamm, *J. Chem. Phys.* **119**, 2709 (2003).
45. D. N. Beratan, J. N. Onuchic, J. E. Winkler, and H. B. Gray, *Science* **258**, 1740 (1992); J. J. Regan, S. M. Risser, D. N. Beratan, and J. N. Onuchic, *J. Phys. Chem.* **97**, 13083 (1993).
46. V. Gogonea, D. Suárez, A. van der Vaart, and K. M. Merz, Jr., *Curr. Opin. Struct. Biol.* **11**, 217 (2001).
47. Y. Komeiji, T. Nakano, K. Fukuzawa, Y. Ueno, Y. Inadomi, T. Nemoto, M. Uebayasi, D. G. Fedorov, and K. Kitaura, *Chem. Phys. Lett.* **372**, 342 (2003).

CHAPTER 16

HEAT TRANSPORT IN MOLECULES AND REACTION KINETICS: THE ROLE OF QUANTUM ENERGY FLOW AND LOCALIZATION

DAVID M. LEITNER

Department of Chemistry and Chemical Physics Program, University of Nevada, Reno, Nevada 89557, USA

CONTENTS

I. Introduction
II. Quantum Energy Flow, Localization, and Their Influence on Rates of Unimolecular Reactions
   A. Local Random Matrix Theory (LRMT)
   B. Rice–Ramsperger–Kassel–Marcus (RRKM) Theory
   C. Dynamical Corrections to RRKM Theory from LRMT
   D. Cyclohexane Ring Inversion: A Case Study
III. Thermal Conduction in Clusters and Macromolecules
   A. Energy Diffusion in Clusters and Molecules
   B. Thermal Transport in Water Clusters
   C. Anomalous Subdiffusion of Vibrational Energy in Proteins
   D. Anharmonic Decay of Vibrational States in Proteins
   E. Thermal Transport in Proteins
IV. Summary
Acknowledgments
References

I. INTRODUCTION

The transfer and storage of vibrational energy in large and small molecules mediate a variety of molecular processes. A central motivation for the study of vibrational energy flow in molecules has long been its influence on chemical reaction kinetics in gas and condensed phases [1–11], as well as its role in

---

Geometric Structures of Phase Space in Multidimensional Chaos: A Special Volume of Advances in Chemical Physics, Part B, Volume 130, edited by M. Toda, T Komatsuzaki, T. Konishi, R.S. Berry, and S.A. Rice. Series editor Stuart A. Rice.
ISBN 0-471-71157-8 Copyright © 2005 John Wiley & Sons, Inc.

photochemical reactions [12], allosteric transitions [13], and charge transfer reactions [1,14] in proteins. An understanding of vibrational energy flow in very large molecules can also help us develop a picture of thermal transport in individual macromolecules such as proteins. Proteins function over a narrow temperature range, and thus they require efficient transfer of heat by vibrations. A proper description of vibrational energy flow in molecules, large and small, generally requires a quantum mechanical treatment. Indeed, for most of the vibrational transitions and transport processes that we shall discuss here, the large majority of the vibrational modes of the molecule are barely excited, if at all.

In this chapter we address two processes for which quantum mechanical energy flow in large molecules plays a central role. The first is unimolecular reaction kinetics involving reactants with some tens of vibrational modes. We focus, in particular, on conformational isomerization, for which barriers are typically low, in many cases a few kilocalories per mole. At energies near or even significantly above the barrier to conformational change, only a small number of the few dozen vibrational modes are excited. In this case, conventional rate theories that exploit the simplicity of equilibrium ensembles of reactants, notably Rice–Ramsperger–Kassel–Marcus (RRKM) theory, are by themselves inadequate; they also require dynamical corrections arising from slow vibrational energy flow or localization. The second process that we address in which quantum energy flow plays an important role is thermal transport in large molecules such as proteins. Thermal conduction in proteins is, to a large extent, mediated by anharmonic decay of vibrational states, necessitating a description of pathways in a protein by which energy is transported quantum mechanically from vibrational modes with low excitation to other vibrational modes.

We address first the influence of quantum mechanical vibrational energy flow and localization on chemical reaction kinetics. Although vibrational energy flow in molecules figures prominently in reaction kinetics, many rate theories circumvent addressing energy flow directly. One assumption of RRKM theory, for instance, maintains that energy flows freely among the vibrational degrees of freedom of a molecule, and a more severe assumption holds that vibrational lifetimes are extremely short, shorter than the time for passage from activated complex to product. For reactions involving many modest-sized molecules, such as conformational isomerization of 2-fluoroethanol [4] and photoisomerization of *trans*-stilbene [6], we are forced to entertain the possibility that flow among the vibrations of these molecules may not be ergodic, even at energies surmounting reaction barriers. Locating the quantum mechanical ergodicity transition in a molecule thus becomes a matter of broad relevance in unimolecular reaction rate theory. A statistical description of state energies and coupling among them, called Local Random Matrix Theory (LRMT) [15–21], provides a means to calculate where the ergodicity threshold lies and to characterize the approach to the threshold in many molecules [19–22]. Even if we assume that energy flow is ergodic, the finite rate of energy transfer in the vibrational state space can influence chemical reaction kinetics [3–6,23]. Transition state theories assume that energy redistribution in molecules occurs more rapidly than the intrinsic rate of crossing over from activated complex to product. In many modest-sized molecules such as *trans*-stilbene, quantum mechanical calculations reveal energy transfer rates of about 1 ps^{-1} at energies well above the barrier to photoisomerization [3,6]. This is about an order of magnitude smaller than the barrier-crossing rate. We note that vibrational energy transfer rates of order 1 ps^{-1} or slower are also typically found in proteins [24–28]. For example, pump-probe studies of amide I vibrations, between 1600 and 1700 cm^{-1}, measure lifetimes of 1 ps or longer in a number of proteins at temperatures from 10 K to 310 K [24–26].

To illustrate how LRMT provides dynamical corrections to RRKM estimates for reaction rates, we use LRMT to compute rates of cyclohexane ring inversion. We find that at energies near the barrier to isomerization of about 10 kcal/mol, the energy transfer rate from transition states of the chair conformer to nonreactive states is about an order of magnitude slower than the rate of crossing the barrier from transition states to product. The quantum energy flow threshold is found to lie below the barrier, so that corrections to the RRKM estimate of the isomerization rate arise from finite but sufficiently sluggish energy flow. Dynamical corrections that we calculate reveal interesting qualitative variations of the reaction rate with the frequency of collisions with the solvent environment. As the pressure increases in the gas phase, the rate reaches a plateau [29] that is lower by about an order of magnitude than the RRKM limit. As the pressure is increased further still, in liquids the rate of ring inversion rises [30,31] as collisions enhance the transfer of energy from transition states of cyclohexane. Our results are consistent with results of measurements by NMR in gas and liquid phases [29–31] and results of classical simulations in liquids [7].

In the second part of this chapter, we turn to vibrational energy flow in clusters and very large molecules—in particular, proteins. For such systems we can begin to address vibrational energy transport in macroscopic terms, for instance, by computing thermal transport coefficients for these objects. For many such systems—for example, clusters of water molecules—calculation of, say, the coefficient of thermal conductivity and the thermal diffusivity are relatively straightforward except at very low temperatures, at which finite size effects of these mesoscopic objects become apparent. The coefficient of thermal conductivity can be expressed as the product of the heat capacity per unit volume and energy diffusion coefficient summed over the vibrational modes of the object. The energy diffusion coefficient can be computed at frequency ω with wave packets expressed as a superposition of vibrational modes over a fairly narrow range of frequency, $\omega \pm \delta\omega$. Calculation of the coefficient of thermal conductivity for a protein is less straightforward. Vibrational wave packets propagated in proteins exhibit anomalous subdiffusion as a result of trapping of vibrational energy in side chains. Indeed, the majority of vibrational modes of proteins are localized and thus cannot carry heat. The spread of energy in proteins turns out to resemble energy subdiffusion in fractal objects such as percolation clusters, which have for some time been used to model the vibrations of polymers [32–34]. The theory of vibrations on fractal objects is by now well developed, and it appears to nicely describe vibrational energy transport in proteins, as we shall see below. However, the theory of thermal transport in proteins or other objects whose vibrations resemble those of fractals is less well developed. Building on their pioneering work [35] in describing vibrations on fractals, Alexander, Orbach, and their co-workers proposed a theory for thermal conductivity in fractal objects, which they argued would hold for amorphous systems generally [36–39]. As discussed in this part of the chapter, their theory, while serving as a useful starting point for thermal conduction in proteins (though possibly less so in glasses [40–42]), is nevertheless incomplete. Two essential contributions to thermal conduction in proteins are needed. These are (1) transport of energy by extended vibrational modes whose scaling properties resemble those of fractal objects, which were assumed localized in Refs. 36 and 37; and (2) transport of energy among localized modes of the protein by anharmonic coupling that is not simply phonon-assisted. We thus devote a section of this chapter to describing anharmonic decay of vibrational states in proteins.

Section II provides a summary of Local Random Matrix Theory (LRMT) and its use in locating the quantum ergodicity transition, how this transition is approached, rates of energy transfer above the transition, and how we use this information to estimate rates of unimolecular reactions. As an illustration, we use LRMT to correct RRKM results for the rate of cyclohexane ring inversion in gas and liquid phases. Section III addresses thermal transport in clusters of water molecules and proteins. We present calculations of the coefficient of thermal conductivity and thermal diffusivity as a function of temperature for a cluster of glassy water and for the protein myoglobin. For the calculation of thermal transport coefficients in proteins, we build on and develop further the theory for thermal conduction in fractal objects of Alexander, Orbach, and co-workers [36,37] mentioned above. Part IV presents a summary.

II. QUANTUM ENERGY FLOW, LOCALIZATION, AND THEIR INFLUENCE ON RATES OF UNIMOLECULAR REACTIONS

Spectroscopic studies have identified a range of time scales for intramolecular vibrational redistribution (IVR) in molecules, large and small, all the way from subnanosecond to picoseconds [8,13,24–28,43–45]. Experimental evidence reveals that the IVR rate does not depend on the total density of states of the molecule, but instead on pathways for energy to flow that depend on a local state density. Even in proteins, anharmonic decay of vibrational states occurs on the picosecond time scale or slower [24–28]. Lifetimes of many high-frequency modes appear to be independent of temperature [25,28], indicating that the pathway for energy flow from these modes does not lead directly to the many low-frequency modes of the protein. Vibrational energy may also be localized to a very small subset of states over considerable periods of time in rather sizable molecules [45]. Criteria for ergodicity in quantum mechanical systems thus depend on the connectivity of vibrational states that give rise to a network of resonances [15–23, 46–51]. To address the nature of vibrational energy flow and localization, we summarize a "mesoscopic" theory for vibrational energy flow and localization in molecules, Local Random Matrix Theory (LRMT), which describes quantum energy flow and localization in a vibrational state space that embodies the local nature of coupling and transport.

Our major purpose in outlining LRMT is to highlight results that are useful to carry over into unimolecular reaction rate theories for modest-sized to large molecules. As we do this, it is useful to compare LRMT with other perspectives on IVR [52]. A sizable body of work on the influence of IVR on reaction kinetics has addressed few degree-of-freedom systems by classical mechanics, such as small van der Waals complexes [9,10]. Classical treatments have also been used to account for the role of energy flow in kinetics of unimolecular reactions of modest-sized molecules such as cyclohexane ring inversion [7] in liquids and isomerization of small Ar clusters [53]. There are also studies comparing classical and quantum ergodicity and connections to barrier crossing in Lennard-Jones trimers [54–57]. As dimensionality increases, the complexity of classical phase space transport such as Arnold diffusion poses enormous challenges [11,58], though the scale of complexity may be reduced by accounting for the influence of finite $\hbar$ [59–61]. Other finite $\hbar$ contributions to energy flow arise from dynamical tunneling between regions of phase space separated classically [62]. In many respects, it is useful and perhaps even simpler to begin from a quantum mechanical perspective. Local Random Matrix Theory applies to the quantum mechanical domain of molecules with many vibrations and a modest amount of energy distributed among them—for example, energies corresponding to or significantly exceeding a barrier to conformational change. LRMT is thus well-suited to address the influence of energy flow on unimolecular reaction kinetics of moderate-sized molecules.

A. Local Random Matrix Theory (LRMT)

Our aim is to describe the extent and rate of vibrational energy flow in molecules with perhaps tens of vibrational modes. We do this within the framework of an ensemble theory for molecules, in this case a matrix ensemble where each member is a reasonable representation of the vibrational Hamiltonian of a molecule with many vibrational modes. We thereby address energy flow and localization from a random matrix perspective [63–78], an approach that has been very successful in describing statistical properties of spectra of highly excited molecules. In perhaps the simplest application of random matrix theory, one identifies all conserved quantities and constructs an ensemble of random matrices labeled by good quantum numbers. Statistical properties of eigenstates are then obtained with this ensemble and can be matched to those of a molecule whose dynamics are strongly chaotic [63]. If a quantum number is partially conserved, a matrix ensemble that accounts for this may be constructed and useful statistical information about eigenstates of such systems may be found [63,70–78].

A similar but more complex situation pertains to molecular vibrations in general. At low energy—for instance, energies corresponding to barriers to conformational change of a few kilocalories per mole, only a small number of vibrational modes of the molecule are excited, and anharmonic coupling among vibrational modes is generally small. The largest anharmonic terms are typically low-order terms. These properties suggest a reasonable representation for the Hamiltonian ensemble. A natural zero-order vibrational state space consists of a product space of states labeled by the occupation number of a collection of nonlinear oscillators [15,16] (see Fig. 1). These may, for example, be the normal modes of vibration and the diagonal anharmonic terms of the Hamiltonian. The zero-order states are coupled to each other by matrix elements that arise from low-order anharmonic coupling. In principle, these matrix elements can be calculated for a given molecule and quantum dynamics may be studied [79]. However, it may be that many of the properties that we are interested in can be suitably captured by an ensemble of Hamiltonians, each one a reasonable representation of the one of interest. We thus take the elements of the vibrational Hamiltonian of a molecule as random subject to the constraint that states are coupled locally in the quantum number space. For this reason, we refer to the theory summarized below as Local Random Matrix Theory (LRMT). Several reviews of LRMT have already appeared [21–23]. Here, we merely summarize the important steps toward arriving at predictions relevant to the calculation of rates of unimolecular reactions.

Our N-nonlinear oscillator Hamiltonian is defined by $H = H_0 + V$, where

$$H_0 = \sum_{\alpha=1}^{N} \varepsilon_\alpha(\hat{n}_\alpha) \tag{1a}$$

$$V = \sum_m \Phi_m \prod_\alpha (b_\alpha^\dagger)^{m_\alpha^+} (b_\alpha)^{m_\alpha^-} \tag{1b}$$

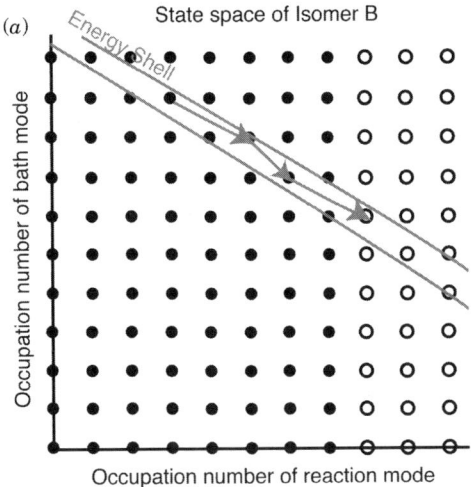

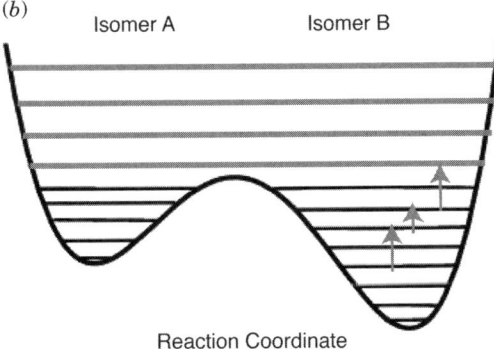

Figure 1. (*a*) A two-dimensional slice through the vibrational quantum number space of a molecule, where coordinates corresponding to the "reaction mode" and one of the bath modes are shown. The energy shell is indicated by two gray lines; states of one of the isomers (isomer B) are indicated by filled circles, and transition states are indicated by open circles. Anharmonicity of a coupled oscillator system favors local transitions in the state space, indicated by the arrows. (*b*) Schematic illustration of the reaction mode of a molecule along which it changes conformation from isomer B to isomer A. Energy for reaction is acquired through coupling to other molecular modes and to the environment. States corresponding to those with sufficient energy for isomerization are referred to as transition states and are indicated by thick, gray levels. Transitions among energy levels of isomer B corresponding to the local transitions depicted in (*a*) are indicated.

where $\mathbf{m} = \{m_1^\pm, m_2^\pm, \ldots\}$. The zero-order Hamiltonian H_0 consists of a sum over the energies of the nonlinear oscillators, where each oscillator has frequency $\omega_\alpha(n_\alpha) = \hbar^{-1} \partial \varepsilon_\alpha / \partial n_\alpha$, and nonlinearity $\omega'_\alpha(n_\alpha) = \hbar^{-1} \partial \varepsilon_\alpha / \partial n_\alpha$, and the number operator is defined by $\hat{n}_\alpha = b_\alpha^+ b_\alpha$. The nonlinearity is assumed to be sufficiently small so that $\hbar |\omega'_\alpha(n_\alpha)| \ll \omega_\alpha(n_\alpha)$, though finite nonlinearity is essential in removing correlations among matrix elements coupling states in the vibrational state space [15]. The set of zero-order energies, $\{\varepsilon_\alpha\}$, and coefficients of V, $\{\Phi_m\}$, are treated as random variables with suitable average and variance. The vibrational Hamiltonian defined by Eq. (1) includes direct resonant coupling terms of arbitrary order. In order for coupling of states in the matrix ensemble to be "local," we assume that the coefficients Φ_m, on average, decay exponentially. A specific form for the off-diagonal matrix elements that embodies this assumption has been proposed for modest-sized organic molecules and supported by computational work by Gruebele and co-workers [20,22] and is given by Eq. (11) below. The larger low-order terms in V couple states close to one another in the vibrational quantum number space. Following Logan and Wolynes [15], we can thus picture the topology of the state space as an N-dimensional lattice where each lattice site is coupled locally to nearby sites by matrix elements arising from low-order anharmonic terms in the potential. The zero-order energy for a site in the vibrational state space is determined by the frequencies and nonlinearities of the vibrational modes, which may in principle be known. However, if one such site is coupled to a fairly large number of sites nearby in quantum number space, it is tempting to assume the zero-order energies of all these sites to be randomly distributed within of order a vibrational frequency. This leads to a tight-binding picture in a many-dimensional vibrational quantum number space, or many-dimensional lattice, with random site energies [15]. The problem of vibrational energy flow in molecules thus resembles the problem of single-particle quantum transport on a many-dimensional disordered lattice, and theoretical approaches to address the condensed phase problem [80] can be brought to bear on describing vibrational energy flow in molecules.

Exploiting this connection, Logan and Wolynes found a transition for energy to flow globally on the energy shell that occurs at a critical value of the product of the anharmonic coupling and local density of states [15]. The problem of energy transport in the vibrational state space of the molecule centers on the self-energy for a particular state, or site on the lattice representing the vibrational state space, specifically the imaginary part of the site self-energy. The imaginary part is finite for extended states and proportional to the energy transfer rate from that state, while it is infinitesimally small for a localized state. To calculate the real and imaginary parts of the self-energy, it is convenient to approximate the topology of the vibrational state space as a Cayley tree. The real and imaginary parts of the site self-energy are expressed by the Feenberg renormalized perturbation series, for which only the lowest-order term is needed for a state space with a Cayley tree topology. They are then obtained self-consistently. Since the matrix elements and zero-order energies are random, the self-consistent procedure is statistical. Two self-consistent approaches have been used to address energy flow and localization in molecules. In one approach, the most probable value for the imaginary part of the self-energy is found self-consistently [15,16]. In a second approach, the average inverse participation ratio, defined by Eq. (3) below, is solved self-consistently [17]. In both cases, the eigenstates of H are assumed localized, so that the most probable value for the imaginary part of the self-energy must be infinitesimally small and the average value for the inverse participation ratio must be finite. A range of molecular parameters are then identified for which the most probable value of the self-energy is in fact infinitesimally small, or, similarly, where the average inverse participation ratio is finite. The result for both self-consistent approaches is the same to within a constant of order 1. Solving for the most probable value of the imaginary part of the self-energy, we find that vibrational energy flow is unrestricted when [15,16]

$$T(E) \equiv \left(\sqrt{\frac{2\pi}{3}} \sum_Q \langle |V_Q| \rangle \rho_Q \right)^2 \geq 1 \qquad (2)$$

while energy is localized in the vibrational state space at energy E when $T(E)$ is less than 1. Here, ρ_Q is the local density of states that lie a distance Q away in quantum number space (see Fig. 1), and $\langle |V_Q| \rangle$ is the average coupling matrix element to such states. The self-consistent analysis reveals that as the size of the molecule increases, the location of the transition becomes more sensitive to higher-order resonances [16,19]. We note that off-resonant transitions may also play a role in locating the IVR transition, in which case we need to include higher-order terms in the Feenberg renormalized perturbation series. Leitner and Wolynes have incorporated off-resonant coupling while maintaining the simplicity of the Cayley tree topology by renormalizing the values of $\langle |V_Q| \rangle$ to account for this contribution to $T(E)$ [16].

LRMT provides useful information on how the transition is approached when $T(E)$ is less than 1 and energy is localized to a finite number of states on the energy shell. The extent of localization of molecular vibrations can be determined spectroscopically by the dilution factor, which is proportional to the inverse participation ratio for state n,

$$\delta_n = \sum_\alpha |c_{n\alpha}|^4 \qquad (3)$$

where $c_{n\alpha}$ are eigenvector components. Equation (3) gives the inverse of the number of vibrational states that overlap a particular zero-order state, n, and is the survival probability of the initially excited state in the infinite time limit.

LRMT provides an analytical form for the dilution factor distribution near a particular energy, which is [17]

$$P(\delta) = \gamma \delta^{-1/2}(1-\delta)^{-3/2} \exp\left(-\frac{\pi\gamma^2\delta}{1-\delta}\right) \tag{4a}$$

$$\gamma = \sqrt{\frac{3T(E)}{2\pi(1-T(E))}}, \quad T(E) < 1 \tag{4b}$$

This apparently broad form for the distribution has been confirmed by numerical calculations by Bigwood and Gruebele on thiophosgene as the transition to extended states is approached [20]. It also characterizes [20] the range of dilution factors observed by Stewart and McDonald [81] for about 20 organic molecules with energy near 3000 cm^{-1}. We note that above the transition, LRMT gives for the probability distribution of $|c_{n\alpha}|^2$ the Porter–Thomas distribution, which is the result expected for quantum ergodic systems [63,82].

Above the transition, $T(E) > 1$, energy flows over all states of the energy shell. Schofield and Wolynes [23,83] have argued that energy flow in the vibrational state space both just above and well beyond the IVR transition can be described by a random walk, a picture that has been supported by numerical calculations over a wide range of time scales [84]. The state-to-state energy transfer rate can be estimated by LRMT. Well above the transition we would expect the rate of quantum energy flow between states of the vibrational state space to be given by $k^q_{IVR}(E) = \frac{2\pi}{\hbar}\sum_Q |V_Q|^2 \rho_Q(E)$, where the superscript q denotes quantum and will be used to distinguish from collisional contributions to energy transfer rates below. More generally, including the region near the transition, we find the energy transfer rate to be [16]

$$k^q_{IVR}(E) = \sqrt{1 - T^{-1}(E)}\, \frac{2\pi}{\hbar} \sum_Q |V_Q|^2 \rho_Q(E) \tag{5}$$

Equation (5) goes over to a Golden Rule-like expression and reveals the locality of energy flow through a crossover region just above the transition, which in practice we find to be quite narrow, particularly when we account for higher-order resonances. While the transition itself is increasingly influenced by higher-order terms the larger the molecule, the influence of high-order anharmonic coupling on vibrational energy transfer rates is less pronounced but can be important in large molecules [16,18].

B. Rice–Ramsperger–Kassel–Marcus (RRKM) Theory

The study of energy flow in molecules has been largely motivated by its role in unimolecular reaction kinetics, such as the kinetics of conformational change of a molecule. The cornerstone of unimolecular rate theory is the Rice–Ramsperger–Kassel–Marcus (RRKM) theory [85]. RRKM theory avoids directly addressing vibrational energy flow in molecules by assuming an equilibrium ensemble at a given energy. An important assumption is that re-equilibration of activated complexes is very rapid, faster than the time the population of activated complexes is depleted by reaction, a vibrational time scale. RRKM theory expresses the microcanonical rate of isomerization as [85]

$$k_{\text{RRKM}}(E) = \frac{N^+(E - E_0)}{h\rho(E)} \quad (6)$$

where $N^+(E - E_0)$ is the number of vibrational states of the transition state with excess energy less than or equal to $E - E_0$, E_0 is the barrier energy, $\rho(E)$ is the density of vibrational states of the reactant at energy E, and h is Planck's constant. As mentioned, an inherent assumption embodied in Eq. (6) is that the reactant remains in microcanonical equilibrium at all times. RRKM theory assumes that reactants that are poised to react will do so on a time scale of the barrier crossing frequency. For this to be true and for the reactant population to remain in equilibrium, the transition states must be repopulated either by IVR or by collisions following reaction on a time scale much faster than the vibrational frequency for crossing the barrier.

The thermal rate is then computed as a function of collision frequency, α. When a strong collision model is assumed [85], the thermal rate is given by

$$k(T) = Q^{-1} \int_{E_0}^{\infty} \frac{dE\, \alpha k(E)\rho(E)e^{-\beta E}}{\alpha + k(E)} \quad (7)$$

where $Q = \int \rho(E)\exp(-\beta E)\, dE$, and $K(E)$ is the microcanonical reaction rate.

C. Dynamical Corrections to RRKM Theory from LRMT

RRKM theory assumes reactants remain in equilibrium at all times. However, a small fraction of vibrational states of the reactant, the transition states, are depopulated by formation of product and, in the absence of collisions, become repopulated only by vibrational energy redistribution.

Consider for simplicity the following two-step mechanism occurring in a molecule with energy, E, which is higher than the reaction barrier [5]. In the first step, energy flows between nonreactive states and the relatively small number of transition states of the reactant, and in the second step the product is formed from transition states. We call the "deactivation" rate (i.e., the rate of energy transfer from a reactive to nonreactive state of the molecule) k_{IVR}, while the formation of product from a transition state occurs with a rate v_R. This two-step mechanism is simply the microcanonical version of the Lindemann mechanism for unimolecular reactions. Such a microcanonical Lindemann mechanism clarifies how the RRKM rate, which we can write as $\nu_R P^*(E)$, where $P^*(E)$ is the fraction of states that are reactive at energy E, should be corrected for finite k_{IVR}. Just as the collision rate appears in the dynamical correction to the transition state theory estimate for the thermal unimolecular reaction rate via the Lindemann mechanism, the microcanonical Lindemann mechanism introduces the dynamical correction $\kappa(E) = (1 + (\nu_R/k_{IVR}(E)))^{-1}$ at energy E. For the more general case in which the product conformer can return to reactant, we must account for the re-equilibration rate of the product conformer as well. The reaction rate is then

$$k(E) = \kappa(E)k_{\text{RRKM}}(E) \qquad (8)$$

where the transmission coefficient providing the dynamical correction to the RRKM rate is [86]

$$\kappa(E) = (1 + (\nu_R/k_{IVR}(E)) + (\nu_P/k_{IVR}(E))^{-1} \qquad (9)$$

We note that this expression assumes that energy transfer is a single exponential process, which, as noted above, is a crude approximation, as simulations reveal that energy relaxation is multiexponential or diffusive [22,84]. The two IVR rates in Eq. (9) are, in general, different—one corresponding to the reactant and the other to the product—but we do not distinguish between them and take for both an average rate of energy transfer from transition states to nonreactive states at energy E for simplicity. The rate of formation of reactant from transition states of the product is ν_P.

The rate of vibrational energy flow in the reactant and product conformer is influenced by the rate of collisions with the solvent environment. Using a strong collision model, the microcanonical rate is still given by $k(E) = \kappa(E)k_{\text{RRKM}}(E)$, where the IVR rate that enters into the transmission coefficient in Eq. (9) incorporates a collision frequency, α:

$$k_{IVR}(E) = k_{IVR}^q(E) + \alpha \qquad (10)$$

D. Cyclohexane Ring Inversion: A Case Study

To illustrate the role of quantum energy flow on rates of conformational isomerization, we calculate the rate of ring inversion of cyclohexane. Measurements of rates of ring inversion by NMR in the vapor phase and in nonpolar liquids reveal an interesting pattern in the variation of the rate over a broad range of collision frequency with solvent. In the vapor phase, the rate increases with increasing pressure and then levels off, in accord with qualitative predictions of RRKM theory. Then, at higher pressures, in nonpolar solvents, the rate of ring inversion rises with increasing pressure. Such a variation can be accommodated by RRKM theory only if the reaction barrier decreases with increasing pressure, but such a possibility has been ruled out by molecular dynamics simulations by Chandler and co-workers [87]. Subsequent classical simulations, freezing higher-frequency vibrational modes of cyclohexane, reveal that the rise in rate with pressure in liquid CS_2 is due to the limited and slow energy flow in cyclohexane [7]. We address here the influence of quantum energy flow on the rate of cyclohexane ring inversion in gas and liquid phases. We use LRMT to compute the energy flow threshold and rate of energy transfer from reactive modes of cyclohexane to calculate dynamical corrections to RRKM theory.

To determine both the location of the IVR transition and the rate of energy transfer, we need to compute the local density of states. Vibrational state densities are computed by normal modes, whose values for cyclohexane we take from Ref. 88. We then obtain the local density of states coupled by all orders of anharmonicity to any given state on the energy shell by direct count. Low-order matrix elements needed to compute the energy flow transition and energy transfer rates could be computed directly using the corresponding potential energy surface. However, since we require only averages of matrix elements, we simply adopt the scaling relations of Gruebele and co-workers [20,89]. They have shown that anharmonic matrix elements, $V_{ii'}$, coupling states $|i\rangle$ and $|i'\rangle$ can be well estimated for organic molecules roughly the size of cyclohexane using some rather simple formulas once the vibrational frequencies are known:

$$V_{ii'} = \prod_\alpha R_\alpha^{d_\alpha} \tag{11a}$$

$$R_\alpha \approx \frac{a^{1/Q}}{b}(\omega_\alpha \bar{n}_\alpha)^{1/2} \tag{11b}$$

$$Q = \sum_\alpha d_\alpha \tag{11c}$$

where d_α is the occupation number difference between two normal modes, α' and α, for the basis states $|i\rangle$ and $|i'\rangle$, $d_\alpha = n_{\alpha'} - n_\alpha$; $\bar{n}_\alpha$ is the geometric mean of the occupation number of mode α in the two states, $|i\rangle$ and $|i'\rangle$; a and b are constants. If $V_{ii'}$ is expressed in cm^{-1}, then $a \approx 3000$ and $b \approx 200$–300. We use $a = 3050$ and $b = 270$ in our calculations. Gruebele's work has shown that Eq. (11) provides good estimates for anharmonic coupling matrix elements for various moderate-sized branched and cyclic organic compounds [20,90].

We first locate the IVR threshold in cyclohexane using Eqs. (2) and (11), which we find to lie near $2200\,cm^{-1}$, well below the isomerization barrier of about $3500\,cm^{-1}$. This means that energy indeed flows freely over the energy

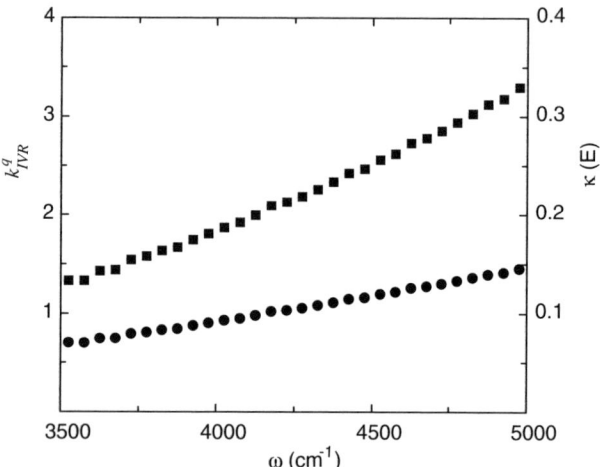

Figure 2. The vibrational energy transfer rate, k_{IVR}^q, from reactive states to nonreactive states at energy E is plotted (squares) above the barrier energy of $3500\,\text{cm}^{-1}$ for cyclohexane chair–boat isomerization. Also plotted (circles) is the corresponding transmission coefficient, $\kappa(E)$, calculated with Eq. (9).

shell at energies above the reaction barrier. To calculate the rate of vibrational energy transfer from transition states to nonreactive states, we identify a reaction mode, which we take to be a low-frequency ring mode at $293\,\text{cm}^{-1}$. Simulations suggest the value of the barrier crossing frequency is likely higher [7], so that the transmission coefficient may well be somewhat smaller than suggested by our calculations. Using Eq. (5), we compute the IVR rate at energy E for transfer of energy from transition states to non-reactive states. We then obtain the transmission coefficient using Eq. (9); the results are plotted in Fig. 2. The thermal rate of ring inversion is calculated with Eq. (7), where the energy-dependent reaction rate in this expression is the product of the RRKM rate, computed with Eq. (6), and the transmission coefficient, computed with Eqs. (9) and (10). We note that Eq. (6) is used to calculate the RRKM rate for chair–boat isomerization; the rate of cyclohexane ring inversion is one-half the rate of chair–boat isomerization, and there are two channels to the boat conformer from the chair conformer.

To compare the computed thermal rate of cyclohexane ring inversion with experimental results, we need to match the collision frequency with pressure. We compare with vapor- and liquid-phase measurements in CS_2 solvents at a temperature of 263 K. A hard-sphere model gives $\approx 10^{-6}\,\text{ps}^{-1}\,\text{torr}^{-1}$ in the gas phase. We find that the experimental data in the gas phase compare well with the calculations if a somewhat lower collision frequency corresponds to the same

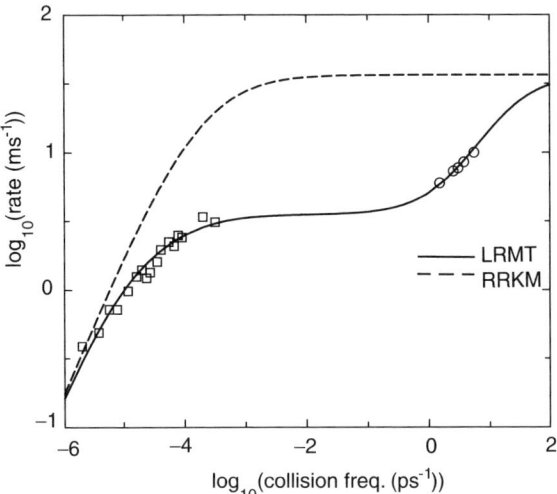

Figure 3. Rate of cyclohexane ring inversion calculated with LRMT at a temperature of 263 K as a function of collision frequency is plotted as the solid curve. The corresponding RRKM calculation gives the dashed curve. The calculations accounting for quantum energy flow in cyclohexane reveal that at low pressures the RRKM rates yield good estimates for the isomerization rate. However, the ring inversion rate eventually appears to saturate at higher pressures in the gas phase to a rate about an order of magnitude below the RRKM estimate, consistent with the measured rates in the gas phase [29], which are plotted in the figure (squares). At higher pressures, corresponding to liquids, the rate rises as the IVR rate increases due to collisions, again consistent with experiment [31], the results of which are also plotted (circles).

pressure, or 1.0×10^{-7} ps^{-1} torr^{-1}. In liquid CS_2, we match collision frequency to viscosity by fitting to experiment, in which case we obtain empirically a value of 2.5 ps^{-1} cP^{-1}. The results are plotted in Fig. 3 along with the gas-phase experimental results of Ross and True [29] and the liquid-phase measurements of Jonas and co-workers [31]. In the gas phase we observe a rise in the rate with pressure at low pressures. The ring inversion rate then levels off at a collision frequency above about 1 ns^{-1} to a value of about 3 ms^{-1}, about an order of magnitude smaller than the RRKM prediction of ≈ 40 ms^{-1}. This is due to a separation of time scales for the fast barrier crossing frequencies, v_R and v_P, the slower quantum energy flow rates from transition to nonreactive states, k_{IVR}^q, and the slower still rates of collision, α. At higher pressures, in the liquid phase, the collision rate exceeds k_{IVR}^q, and k_{IVR} increases with pressure, yielding higher values of the transmission coefficient with pressure and, thus, a faster rate of ring inversion.

Measurements of ring inversion rates in a wide range of environments [29–31] and the consistent picture presented by these calculations lead to a rather compelling case for the influence of quantum energy flow on the isomerization rate. By now we expect this situation to be the norm rather than the exception. Recent measurements by Baer and Potts [91] on rates of axial–equatorial interconversion of methyl- and ethyl-cyclohexanones reveal rates that are orders of magnitude slower than those predicted by RRKM theory. The barriers to conformational change lie around 4 kcal/mol [91]. The IVR threshold would be expected to be fairly near or perhaps somewhat lower than that of cyclohexane, which was found to be just over 6 kcal/mol. The failure of RRKM theory to predict the axial–equatorial interconversion rates is thus due to the relatively high IVR transition energy. In earlier calculations on 2-fluoroethanol and allyl fluoride, we also found the energy flow threshold to lie above barriers to conformational change [4]. Spectroscopic measurements on conformational isomerization of 2-fluoroethanol and allyl fluoride by Pate and co-workers again reveal a rate that is orders of magnitude slower than predicted by RRKM theory [92,93].

Similarly, we have found that quantum energy flow influences significantly the kinetics of *trans*-stilbene photoisomerization [3,6]. Results of jet experiments [94,95] reveal a rate of *trans*-stilbene photoisomerization that is about an order of magnitude slower than RRKM calculations [96], including our own RRKM calculations using a new S_1 surface for *trans*-stilbene [6]. Dynamical corrections calculated by LRMT yield results for the energy-dependent rate well within a factor of two for stilbene and several deuterated isotopomers. The energy flow transition is found by LRMT to lie near $1300\,\text{cm}^{-1}$ [3,6], which happens to be near the reported reaction threshold based on molecular beam studies [94,95]. The calculated IVR threshold is also near the measured IVR threshold, also based on beam studies [43]. The barrier to isomerization on the new S_1 surface lies near $800\,\text{cm}^{-1}$ [6], thus below the energy flow threshold. The reported reaction threshold may in fact then correspond to the IVR threshold. The rise in the rate of photoisomerization by over an order of magnitude with pressures from about 1 to 100 atm in methane buffer gas, observed by Balk and Fleming [97], is to a large extent captured by our LRMT calculations [3,6].

We point out again that qualitative deviations from RRKM theory should not be very surprising when considering reactions of sizable molecules over relatively low barriers, such as conformational isomerization of moderate-sized organic molecules. At energies corresponding to a reaction barrier, only a relatively small number of modes of the molecule are vibrationally excited. Anharmonic coupling may be sufficiently weak so that, even if energy does flow freely over the energy shell, energy transfer into and out of transition states is typically still substantially slower than conversion from activated complex to product. The qualitative variation in the reaction rate with pressure reflects these dynamical processes and in many ways follows that suggested long ago by Rice [98] to be quite general for unimolecular reactions of large molecules. RRKM theory serves as a useful starting point for estimating the rate of conformational change, while LRMT provides dynamical corrections that yield the correct qualitative, and to a large extent quantitative, variations of the rate with pressure in gas and liquid phases.

III. THERMAL CONDUCTION IN CLUSTERS AND MACROMOLECULES

One of our motivations for the study of vibrational energy transport in macromolecules and clusters is its role in thermal transport. By moving to objects that are large on the molecular scale, we can address transport properties in terms of macroscopic concepts such as thermal diffusivity, though the finite size of the object may strongly influence these properties, particularly at low temperature. Thermal transport in proteins is closely linked to function. Since proteins serve as media for chemical reactions in cells, efficient transfer of heat via their vibrational modes is needed to maintain nearly constant temperature. An understanding of thermal transport in macromolecules has also been useful in modeling cooling processes that can influence conformational change, including the reversible denaturation of nucleic acids and proteins by remote heating of attached nanoparticles [160], and modeling structural changes during flash cooling in biomolecular cryocrystallography [161]. There is also considerable interest in thermal transport in finite-sized glasses. For example, the structure of water near the surface of a protein is amorphous and fairly rigid [155]. The problem of thermal transport in materials on the mesoscopic scale has recently risen to the fore due to the acute buildup of heat in very small integrated circuits [99]. This issue and the aim of measuring heat transport at the level of individual phonons [100,101] have motivated a great deal of recent work on the theory of thermal conduction in nanoscale crystalline materials [102–105] and molecules [106,107], as well as the development of computational tools to compute thermal transport coefficients in nanocrystals [108]. Our focus here is on thermal transport in aperiodic systems that are finite but large on the molecular scale. We shall address, in particular, water clusters and proteins.

The coefficient of thermal conductivity, which relates the net energy flux to the thermal gradient, is

$$\kappa = \sum_\alpha C_\alpha D_\alpha \approx \int d\omega\, n(\omega)\, C(\omega)\, D(\omega) \qquad (12)$$

where the sum is over each mode, α, of the object, $n(\omega)$ is the density of states, $C(\omega)$ the heat capacity per unit volume, and $D(\omega)$ the frequency-dependent energy diffusion coefficient. The thermal diffusivity is defined by

$$D_T = \frac{\kappa}{\int d\omega\, n(\omega)\, C(\omega)} \qquad (13)$$

where the denominator is the heat capacity per unit volume at temperature T;

$$n(\omega) = \rho(\omega)N \qquad (14)$$

$$C(\omega) = \frac{\hbar^2\omega^2}{V k_B T^2} \frac{\exp(\hbar\omega/k_B T)}{[\exp(\hbar\omega/k_B T) - 1]^2} \qquad (15)$$

The normalized density of states is $\rho(\omega)$, N is the total number of vibrational modes, T is temperature, k_B is Boltzmann's constant, and V is the volume of the macromolecule or cluster. In what follows we address first thermal transport in a glassy water cluster, then turn to proteins.

A. Energy Diffusion in Clusters and Molecules

Our main focus in computing thermal transport coefficients is calculation of the frequency-dependent energy diffusion coefficient, $D(\omega)$, which appears in Eq. (12). Computation of $D(\omega)$ is relatively straightforward if we express the vibrations of the object in terms of its normal modes. We shall compute $D(\omega)$ with wave packets expressed as superpositions of normal modes, which we then filter to a range of frequencies near ω to determine $D(\omega)$.

Consider the kinetic energy, $E_n(t)$, of atom n at time t. We start with a cold water cluster or protein and introduce a spatially localized excitation in the form of a wave packet placed in the interior of the object. The center of kinetic energy, $\mathbf{R}_0(t)$, of the wave packet is found from

$$\mathbf{R}_0(t) = \frac{\sum_n \mathbf{R}_n E_n(t)}{\sum_n E_n(t)} \qquad (16)$$

The variance is

$$\langle \mathbf{R}^2(t) \rangle = \frac{\sum_n (\mathbf{R}_n - \mathbf{R}_0(t))^2 E_n(t)}{\sum_n E_n(t)} \qquad (17)$$

Our focus is on the propagation of a wave packet by the normal modes of the object. We idealize the initial wave packet as a traveling wave and take the displacement of atom n to initially have the form

$$\mathbf{U}_n(t) = \frac{\mathbf{A}_n}{b^2} \exp\left(-\frac{(\mathbf{R}_n - \mathbf{R}' - \mathbf{v}_0 t)^2}{2b^2}\right) e^{i(\mathbf{Q}_0 \cdot \mathbf{R}_n - \omega_0 t)} \qquad (18)$$

HEAT TRANSPORT IN MOLECULES AND REACTION KINETICS   223

from which displacements and velocities at $t = 0$ are determined. Displacements and velocities for $t > 0$ are then computed in terms of normal modes as [109]

$$\mathbf{U}_n(t) = \sum_\alpha \mathbf{e}_n^\alpha \cos(\omega_\alpha t) \sum_{n'} \mathbf{e}_{n'}^\alpha \cdot \mathbf{U}_{n'}(0) + \sum_\alpha \mathbf{e}_n^\alpha \frac{\sin(\omega_\alpha t)}{\omega_\alpha} \sum_{n'} \mathbf{e}_{n'}^\alpha \cdot \mathbf{V}_{n'}(0) \quad (19a)$$

$$\mathbf{V}_n(t) = \sum_\alpha \mathbf{e}_n^\alpha \cos(\omega_\alpha t) \sum_{n'} \mathbf{e}_{n'}^\alpha \cdot \mathbf{V}_{n'}(0) - \sum_\alpha \mathbf{e}_n^\alpha \omega_\alpha \sin(\omega_\alpha t) \sum_{n'} \mathbf{e}_{n'}^\alpha \cdot \mathbf{U}_{n'}(0)$$

(19b)

We filter the wave packet in frequency around $\bar{\omega} \pm \delta\omega$ with

$$f_\omega = \exp(-(\omega - \bar{\omega})^2/2\delta\omega^2) \quad (20)$$

The positions and velocities are then computed as

$$\mathbf{U}_n(t) = \sum_\alpha \mathbf{e}_n^\alpha \cos(\omega_\alpha t) \sum_{n'} \mathbf{e}_{n'}^\alpha \cdot \mathbf{U}_{n'}(0) f_{\omega_\alpha} + \sum_\alpha \mathbf{e}_n^\alpha \frac{\sin(\omega_\alpha t)}{\omega_\alpha} \sum_{n'} \mathbf{e}_{n'}^\alpha \cdot \mathbf{V}_{n'}(0) f_{\omega_\alpha}$$

(21a)

$$\mathbf{V}_n(t) = \sum_\alpha \mathbf{e}_n^\alpha \cos(\omega_\alpha t) \sum_{n'} \mathbf{e}_{n'}^\alpha \cdot \mathbf{V}_{n'}(0) f_{\omega_\alpha} - \sum_\alpha \mathbf{e}_n^\alpha \omega_\alpha \sin(\omega_\alpha t) \sum_{n'} \mathbf{e}_{n'}^\alpha \cdot \mathbf{U}_{n'}(0) f_{\omega_\alpha}$$

(21b)

B. Thermal Transport in Water Clusters

We calculate thermal transport properties for a water cluster, specifically a cluster of 950 water molecules interacting with the TIP3 potential. The glassy water cluster is formed by removing the central 950 water molecules, which fit roughly in a sphere of radius 18 Å, from a box of water molecules with a density of 1.0 g cm^{-3}. We then quench the structure to its nearest energy minimum and compute the normal modes of the glassy water cluster using the program MOIL [110].

Before carrying out the energy diffusion calculations on the water cluster, it is useful to determine the speed of sound in glassy water, which we compute from the dispersion relation for the water cluster. To compute a dispersion relation, we need to assign a wave number, k, to a normal mode of frequency, ω. We have obtained dispersion relations for proteins [111] via computation of the correlation function for the direction of atomic displacements as a function of distance for individual normal modes of the protein, a function that was studied in earlier work by Nishikawa and Go [112]. Computation of the correlation function allows us to match a wave number of a plane wave to a normal mode

[112]. Consider normal mode α. We can define a direction vector of displacement for an atom, n, whose position is $\mathbf{R}_n$ at the potential minimum as $\mathbf{A}_\alpha(\mathbf{R}_n) = \mathbf{e}_\alpha(\mathbf{R}_n)/|\mathbf{e}_\alpha(\mathbf{R}_n)|$. The correlation function for the direction vector of two atoms spaced a distance $d = |\mathbf{R}_n - \mathbf{R}_{n'}|$ apart is then

$$C_\alpha(d) = \frac{\sum_n \sum_{n'} \mathbf{A}_\alpha(\mathbf{R}_n) \cdot \mathbf{A}_\alpha(\mathbf{R}_{n'}) \delta(d - |\mathbf{R}_n - \mathbf{R}_{n'}|)}{\sum_n \sum_{n'} \delta(d - |\mathbf{R}_n - \mathbf{R}_{n'}|)} \qquad (22)$$

The subscript α simply refers to a particular mode, and we henceforth omit it. The correlation function $C(d)$ is thereby defined as an average over dot products of direction vectors of displacements of pairs of atoms separated in space by distance d. For low-frequency modes of proteins and clusters the correlation function appears to oscillate; the smallest value of d, where $C(d) = 0$, corresponds to half the wavelength of the corresponding plane wave. For example, we plot $C(d)$ for a 20-cm^{-1} normal mode of the water cluster in Fig. 4. The corresponding plane wave has a wavelength of 9.4 Å. We can determine the wave number, k, corresponding to the low-frequency modes of the water cluster by locating the smallest value of d, d_0, where $C(d) = 0$; the wave number of that mode is then $k = 1/2d_0$. We assign wave numbers to the normal modes of the water cluster up to 60 cm^{-1}. A plot of ω versus k for water is shown in Fig. 5.

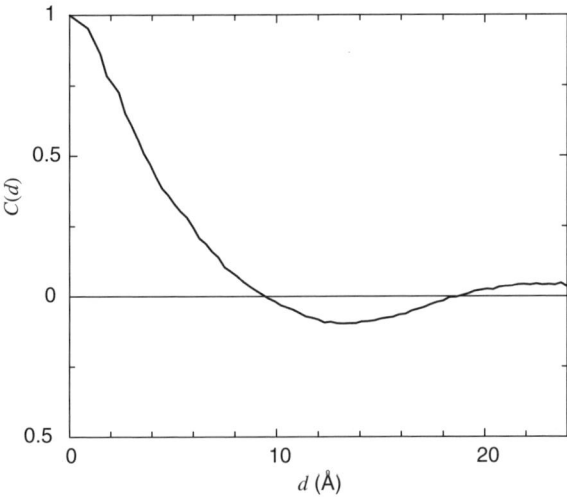

Figure 4. The correlation function for the direction vector of two atoms spaced by d, $C(d)$, defined by Eq. (22), is plotted for a 20 cm^{-1} normal mode of a cluster of 950 water molecules. The smallest value of d where $C(d) = 0$ is $d = 9.4$ Å, which is half the wavelength of this mode.

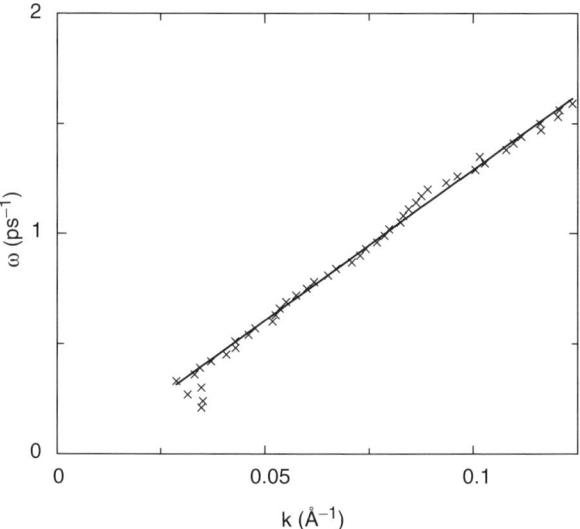

Figure 5. Dispersion relation, ω versus k, is plotted for a cluster of 950 water molecules. A linear fit is plotted with a slope of 14 Å ps^{-1}.

A fit through the points in the plot gives a slope of 14 Å ps^{-1}, which is a reasonable value for the speed of sound in glassy water, consistent with a value of about 15 Å ps^{-1} in water [113].

We turn now to the calculation of $D(\omega)$, which we carry out as follows. The variance of the wave packet as a function of time that we report is an average over results for 10 wave packets, centered initially about a water molecule whose position, $\mathbf{R}'$, lies near the center of mass of the cluster. The width of the initial wave packet is $b = 3$ Å. The magnitude of the wave vector of the initial excitation, Q_0, is 0.63 Å^{-1} and it points $+45°$ from the x-, y- and z-axis; we take $\omega_0 = 9.4$ ps^{-1} for wave packets that include all normal modes, and $v_0 = 15$ Å ps^{-1}, consistent with the speed of sound in glassy water. All components of $\mathbf{A}_n$ for all atoms are taken to be the same, and the magnitude is unimportant because it cancels out when we compute the center of energy and its variance. Displacements and velocities are then propagated with Eq. (21), which give the positions, $\mathbf{R}_n(t)$, and kinetic energies, $E_n(t)$, used in Eqs. (16) and (17) to locate the center of energy and variance in the water cluster.

Using a frequency filter whose width, $\delta\omega$, is 200 cm^{-1}, we compute $D(\omega)$ up to 500 cm^{-1} in intervals of 50 cm^{-1}. The resulting $D(\omega)$ is plotted in Fig. 6. We then sum over the normal modes of the water cluster, whose density, $\rho(\omega)$, is plotted in Fig. 7, to compute the coefficient of thermal conductivity with Eq. (12). The results from temperatures of 20–320 K are plotted in Fig. 8a. The

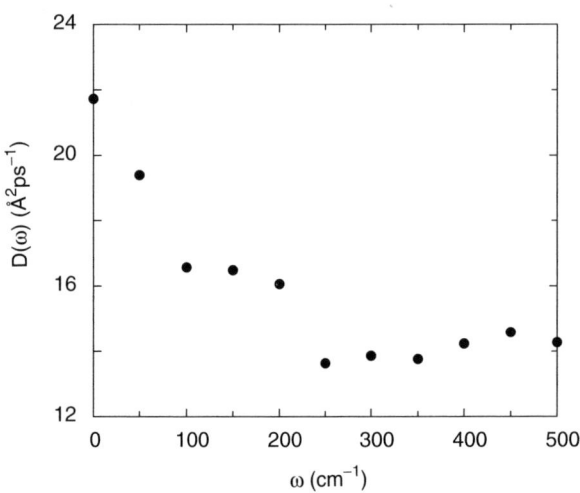

Figure 6. The frequency-dependent diffusion coefficient, $D(\omega)$, computed for a cluster of 950 water molecules.

thermal diffusivity is computed with Eq. (13) over the same temperature range, and the result is plotted in Fig. 8b. We find the coefficient of thermal conductivity for water to be 3.8 mW cm^{-1} K^{-1} at 300 K, which is smaller than the actual value of 6.1 mW cm^{-1} K^{-1} [113]. The difference is due to the difference between the heat capacity computed with normal modes and the heat

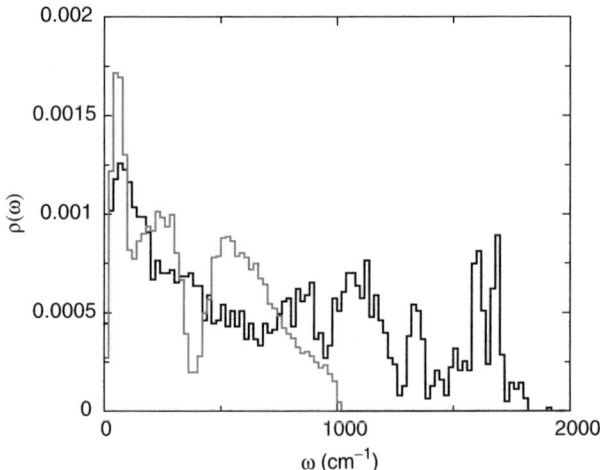

Figure 7. Normal mode density (normalized) computed for cytochrome c (black) up to 2000 cm^{-1} and a cluster of 950 water molecules (gray) up to 1100 cm^{-1}.

HEAT TRANSPORT IN MOLECULES AND REACTION KINETICS   227

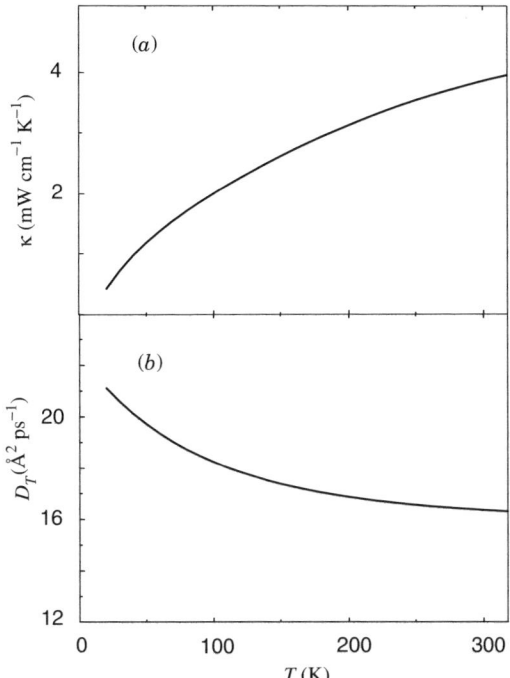

Figure 8. (*a*) Coefficient of thermal conductivity, and (*b*) thermal diffusivity, computed for a cluster of 950 water molecules at temperatures from 20 K to 320 K.

capacity of liquid water, which is larger than the former by about the same factor [113], due to the neglect of both anharmonicity and configurational contributions in our calculation. The computed thermal conductivity should be much more accurate for glassy water, below 200 K, where configurational contributions would be minimal. The thermal diffusivity, given by Eq. (13), should be more reliable in both the glass and liquid regimes, because it is not directly related to the heat capacity. We find good agreement between our calculated value of 16 $\text{Å}^2\,\text{ps}^{-1}$ and the actual value [113] of 15 $\text{Å}^2\,\text{ps}^{-1}$ at 300 K.

C. Anomalous Subdiffusion of Vibrational Energy in Proteins

In the previous section we computed thermal transport coefficients for a water cluster whose size is reasonably similar to that of a typical globular protein. The calculation of thermal transport properties of proteins turns out not to be so simple. For one thing, there is considerable computational and experimental evidence to suggest that energy transport in proteins is non-Brownian.

Stretched exponential relaxation in proteins [114,115], anomalous diffusion of a protein in its configuration space [116], and the fractal nature of protein structural and potential energy fluctuations [117–119] have been characterized as processes resulting from a protein's complex energy landscape. Anomalous subdiffusion is a consequence of distributions of waiting times on a random walk that are especially broad, leading to significant numbers of steps with long trapping times [120–123]. Proteins may encounter many such traps as they explore their energy landscapes, and indeed recent single-molecule experiments by Xie and co-workers [124,125] reveal non-Brownian protein dynamics during chemical reactions at room temperature.

Our focus, however, is on vibrational dynamics rather than the dynamics of configurational change. In our calculation of thermal transport in proteins, we restrict energy transport to the energy carried by normal modes and thereby confine the protein to the bottom of one basin of the energy landscape, just as we did for the water cluster. There is still much to suggest that the flow of energy in the protein may be quite complex, or specifically that the diffusion of energy carried by normal modes might be anomalous. For instance, most normal modes of proteins are spatially localized [112,126–129], leading to trapping of energy in a harmonic protein by localized modes contained in a wave packet. Related to this is the observation that protein structures may be fractal [130,131] (in a sense explained below), and at low frequencies the normal mode density appears to vary with frequency as expected for fractal objects [131,132]. The measured anomalous temperature dependence of electron spin resonance relaxation in proteins is consistent with relaxation in a fractal [133,134]. Molecular dynamics simulations on myoglobin and cytochrome c reveal highly anisotropic flow of energy [135–137], which could indicate anomalous diffusion.

We first investigate the diffusion of energy in proteins as we did for the water cluster, by expanding a vibrational excitation in the form of a wave packet as a superposition of normal modes of the protein. For the time being, we include the contributions of all normal modes rather than filtering a particular band of modes. In the following, we consider energy diffusion in three proteins, myoglobin, cytochrome c, and green fluorescent protein (GFP), whose structures range from helical (myoglobin) to β-barrel (GFP).

Normal modes for each of the proteins were obtained with the program MOIL [110]. Figure 7 shows the normalized density of normal modes of cytochrome c together with those for a cluster of 950 water molecules. The densities for myoglobin and GFP are similar to that for cytochrome c. Vibrational frequencies of each protein range from about 5 to about $1850\,\text{cm}^{-1}$. There are also higher-frequency modes above $3000\,\text{cm}^{-1}$ corresponding to CH, NH, and OH stretches, but only modes lying within the band of frequencies up to $\approx 2000\,\text{cm}^{-1}$ will be considered.

HEAT TRANSPORT IN MOLECULES AND REACTION KINETICS   229

In describing the normal modes of a protein, it is instructive to compare them conceptually with those of a simple model of a polymer, such as a chain of atoms, both periodic and aperiodic. In a harmonic periodic chain, the normal modes carry energy without resistance from one end of the 1D crystal to the other. On the other hand, the vast majority of normal modes of an aperiodic chain are spatially localized [138]. Protein molecules, which are of course not periodic, can be better characterized as an aperiodic chain of atoms, and most normal modes of proteins are likewise localized in space [111,112,126–128]. If a normal mode α is exponentially localized, then the vibrational amplitude of atoms in mode α decays from the center of excitation, $\mathbf{R}_0$, as

$$|\mathbf{e}_\alpha(\mathbf{R}_n)| \propto \exp(-|\mathbf{R}_n - \mathbf{R}_0|/\xi) \qquad (23)$$

where $|\mathbf{e}_\alpha(\mathbf{R}_n)|$ is the magnitude of the displacement of atom n, located at $\mathbf{R}_n$; ξ is the localization length; and $\mathbf{R}_0$ is the position of the atom overlapping the largest component of the normal mode vector. To determine ξ, we calculate $\ln |\mathbf{e}_\alpha(\mathbf{R}_n)|$ for all atoms and plot it against $|\mathbf{R}_n - \mathbf{R}_0|$. A linear fit gives ξ.

In this way, we have calculated ξ for all the normal modes of cytochrome c in 20-cm^{-1} intervals, and we plotted the results in Fig. 9. We have done the same for a water cluster, in this case a cluster of 735 molecules whose radius is about 15 Å, and also plotted the localization length as a function of mode frequency in Fig. 9. Turning first to cytochrome c, we observe that at sufficiently low frequency, below about 150 cm^{-1}, the normal modes appear to be essentially

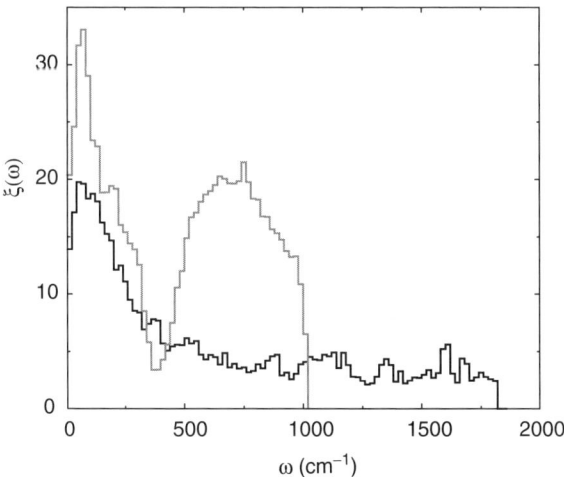

Figure 9. Localization length as a function of mode frequency for cytochrome c (black) and a cluster of 735 water molecules (gray).

delocalized over the protein, with $\xi \approx 14\text{–}20$ Å, comparable to the protein radius. Above $150\,\text{cm}^{-1}$, ξ decreases to values between 2.5 Å and 5 Å. As for the water cluster, we find the translational modes below $300\,\text{cm}^{-1}$ and the librational modes from $500\,\text{cm}^{-1}$ to $1000\,\text{cm}^{-1}$ to be extended over the cluster, consistent with other numerical results on water [139]. Only in the crossover region from about $300\,\text{cm}^{-1}$ to $500\,\text{cm}^{-1}$, where the mode density is quite small, are the vibrational modes of water localized. We note that these localized modes are barely accessible at 300 K and thus would contribute little to thermal conduction. However, in cytochrome c and other globular proteins, the thermally accessible localized modes appear approximately at $150\,\text{cm}^{-1}$ and cannot transport heat through the protein.

We now address the diffusion of energy in proteins, which we shall see below exhibit anomalous subdiffusion. Anomalous subdiffusion of vibrational energy in polymers such as proteins has been anticipated for some time by theories for transport on percolation clusters [33,140]. Because of dangling ends and side chains of polymers, some energy may be trapped for periods that are long compared to the transit time of the rest of the energy through the object. A random walker whose waiting time distribution has a sufficiently long tail executes anomalous subdiffusion. We thus examine how the mean square displacement of kinetic energy, $\langle R^2 \rangle$, in a protein varies with time for a general diffusive process,

$$\langle R^2 \rangle \sim t^{2\nu} \tag{24}$$

where $\nu = 1/2$ corresponds to normal diffusion while for proteins we find that $\nu < 1/2$.

Anomalous subdiffusion occurs on percolation clusters or on objects that in a statistical sense can be described as fractal, by which we mean that self-similarity describes simply the scaling of mass with length. Connections between ν, the fractal dimension of the cluster, D, and the spectral dimension, $\bar{d}$, have been established, relations that were originally derived by Alexander and Orbach [35], who developed a theory of vibrational excitations on fractal objects which they called fractons. An elegant scaling argument by Rammal and Toulouse [140] also leads to these relations, and we summarize their results.

Consider a fractal object, whose parts scale with length in a way that is different from the d-dimensional Euclidean space in which it is embedded. For instance, the mass may scale with length, L, as

$$M \sim L^D \tag{25}$$

where D is the fractal dimension, which is not necessarily d. Let the object have a definite size or length scale, L. Let the density of normal modes vary with mode frequency, ω, as a power law,

$$\rho_L(\omega) \propto \omega^{\bar{d}-1} \qquad (26)$$

where $\bar{d}$ is called the spectral dimension. Changing the length scale of the object from $L \to L/b$, the mode density scales as

$$\rho_{L/b}(\omega) = b^{-D}\rho_L(\omega) \qquad (27)$$

For a lattice of atoms in d-dimensional space connected to nearest neighbors by bonds of similar strength, $\bar{d} = D = d$. In this case, we have the simple dispersion relation, $\omega \sim k$, where k is the wave number.

Rammal and Toulouse [140] introduce a scaling assumption for mode frequency on a fractal. Changing the length scale of the object from $L \to L/b$, the frequency changes as

$$\omega(L/b) = b^a \omega(L) \qquad (28)$$

Then

$$\rho_{L/b}(\omega) = b^{-a}\rho_L(\omega b^{-a}) \qquad (29)$$

Equating Eqs. (27) and (29), along with using Eq. (26) for ρ_L, leads to

$$a = D/\bar{d} \qquad (30)$$

Taking the length scales L and b to be a wavelength, the dispersion relation given by Eqs. (28) and (30) is

$$\omega \sim k^{D/\bar{d}} \qquad (31)$$

The scaling of length with time is obtained by noting that on a given length scale, L, we have $t \approx D_L^{-1} k_L^{-2}$, so that in Eq. (24) we can substitute $t \to \omega^{-2}$. Since $\omega \sim L^{-D/\bar{d}}$, substitution of $L^{-D/\bar{d}}$ for ω into Eq. (24) leads to

$$\nu = \frac{\bar{d}}{2D} \qquad (32)$$

The ratio $D/\bar{d}$ for a protein can be found by computing both the dispersion relation, $\omega(k)$, and by propagating a vibrational excitation in the protein. The spectral dimension, $\bar{d}$, can also be found by computing the density of states and fitting to Eq. (26). The fractal dimension, D, as defined by Eq. (25), can be found by examining how the mass contained inside a sphere varies with radius, L.

Alexander and Orbach [35] conjectured based on numerical evidence at hand that for fractal objects embedded in a two- or three-dimensional Euclidean space, $\bar{d} \approx 4/3$. We find similar values for the three proteins examined below. In fact the value of $\bar{d}$ depends on the size of the protein and approaches 2 for large proteins with thousands of residues [162].

We turn now to the calculation of the energy diffusion coefficient in cytochrome c, myoglobin and GFP, which we carry out just as for the water cluster, only for now without filtering specific bands of frequencies. The variance of the wave packet as a function of time that we report is again an average over results for 10 wave packets, centered initially about a backbone atom whose position, $\mathbf{R}'$, lies near the center of mass of the protein. The width of the initial wave packet is $b = 3\,\text{Å}$. The magnitude of the wave vector of the initial excitation, Q_0, is $0.63\,\text{Å}^{-1}$, and it points $+45°$ from the x-, y-, and z-axis; we take $\omega_0 = 9.4\,\text{ps}^{-1}$ for wave packets that include all normal modes, and $\mathbf{v}_0 = 0\,\text{Å}\,\text{ps}^{-1}$ for simplicity. All components of $\mathbf{A}_n$ for all atoms are taken to be the same, and the magnitude is unimportant because it cancels out when we compute the center of energy and its variance. Displacements and velocities are then propagated with Eq. (19), which give the positions, $\mathbf{R}_n(t)$, and kinetic energies, $E_n(t)$, used in Eqs. (16) and (17) to locate the center of energy and variance in the protein.

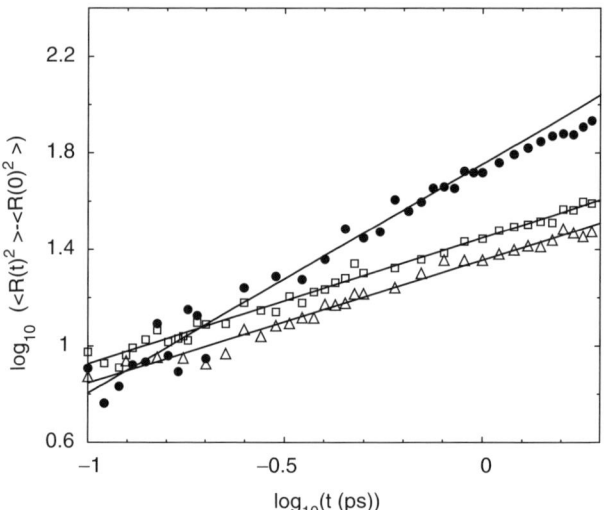

Figure 10. Plot of $\log(\langle R^2(t)\rangle - \langle R^2(0)\rangle)$ versus $\log(t)$ for a cluster of 735 water molecules (circles), cytochrome c (triangles), and cytochrome c hydrated by 400 water molecules (squares). Plotted fit to water data from 0.1 ps to 0.9 ps has a slope of 0.95 ± 0.05, indicating normal diffusion. The slope of the plotted fit to both cytochrome c results from 0.1 ps to 3.0 ps is 0.52 ± 0.01, indicating anomalous subdiffusion with exponent $\nu = 0.26$.

In Fig. 10, we plot the variance of the kinetic energy, $\log(\langle R^2(t) \rangle - \langle R^2(0) \rangle)$ versus $\log(t)$, for cluster of 735 water molecules and cytochrome c, which in one case is dehydrated and in the other is hydrated with 400 water molecules. Results for two other proteins, green fluorescent protein (GFP) and myoglobin, will be discussed below. Consider first the spreading of the wave packet in the water cluster. The slope of the line fit to data from 0.1 to 0.9 ps is 0.95 ± 0.05, close to 1 so that ν is 1/2. Energy diffusion in glassy water appears to be normal. At times beyond about 0.9 ps the variance begins to saturate as the energy spreads out to all atoms of the cluster.

Energy diffusion in proteins is strikingly different than in glassy water. In Fig. 10, we find a slope of 0.52 ± 0.01 for times from 0.1 ps to 2 ps for dehydrated cytochrome c, after which the spreading of energy appears to saturate. We observe the same slope for cytochrome c hydrated by 400 water for times from 0.1 ps to 3.0 ps. The presence of hydration water does not appear to affect ν, although the spreading appears to be slightly faster in the hydrated case. We note that for GFP and myoglobin we find from times of 0.1 ps to 3.0 ps slopes of 0.63 ± 0.01 and 0.58 ± 0.01, respectively [141]. Overall we find anomalous subdiffusion of vibrational energy in each protein, with fairly similar values of ν, ranging from about 0.26 for cytochrome c to about 0.32 for GFP.

We can also examine how the center of energy of the wave packet relaxes. As we find anomalous diffusion of energy in proteins, we expect relaxation processes to follow stretched exponential kinetics [142]. Consider a property, $\Phi(t)$, which is 1 at $t = 0$ and approaches 0 as $t \to \infty$. Stretched exponential kinetics implies

$$\Phi(t) = \exp(-(t/\tau)^\beta) \qquad (33)$$

where $\beta = 2\nu$ [142]. Let $\Phi(t) = d(t)/d(0)$, where $d(t)$ is the distance at time t of the center of energy from its value at long times. In Fig. 11 we plot the relaxation of the center of energy computed for both myoglobin and cytochrome c, the latter hydrated by 400 water molecules. We find that the center of energy hardly moves after ≈25 ps, so we average its position at 25 ps up to 100 ps to obtain its long-time position. Using $\beta = 0.52$, the value of 2ν found by plotting the energy variance versus time for cytochrome c in Fig. 10 (the value for myoglobin is actually found to be 0.58), we find that Eq. (33) fits the data well with a relaxation time, τ, of 11 ps for both molecules.

As already pointed out, the exponent ν, calculated for three proteins above, is related to the fractal dimension of the object, D, and the spectral dimension, $\bar{d}$, by $\nu = \bar{d}/2D$ [35,140]. The latter is defined from the form in which the density of states varies with frequency as $\rho(\omega) \sim \omega^{\bar{d}-1}$. We thus seek a value for the spectral dimension, $\bar{d}$, which for many fractal objects is $\bar{d} \approx 4/3$ [35]. In Fig. 12, we plot $\log \rho(\omega)$ versus $\log \omega$ for the normal modes of cytochrome c, myoglobin, and GFP; the normal mode density is computed in 2-cm^{-1} intervals

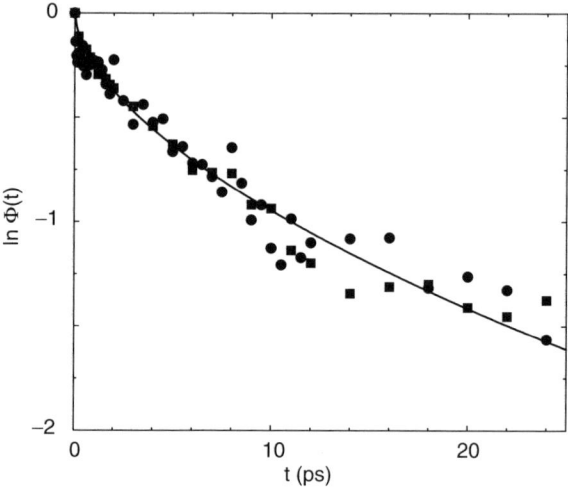

Figure 11. Relaxation, $\Phi(t)$, of the center of energy is plotted for wave packets propagated by the normal modes of cytochrome c hydrated by 400 water molecules (circles) and myoglobin (squares). Curve is a stretched exponential, Eq. (33), with $\beta = 2\nu = 0.52$, the value fit to the computed energy diffusion data for cytochrome c plotted in Fig. 10, and time constant, $\tau = 11$ ps.

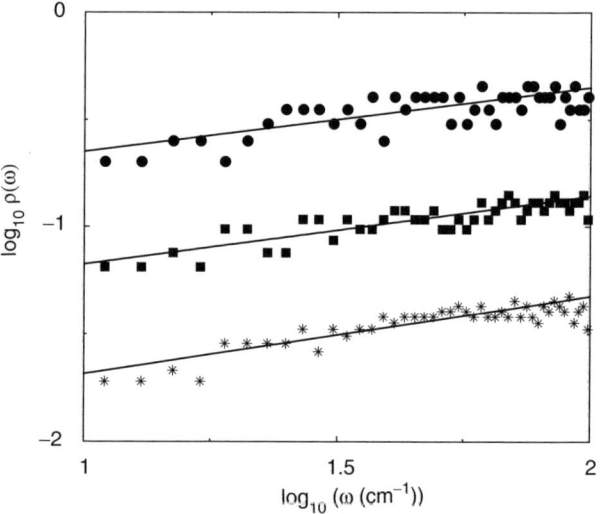

Figure 12. $\log \rho(\omega)$ versus $\log \omega$ for the normal modes of cytochrome c (circles), myoglobin (squares), and GFP (stars). Units for $\log \rho(\omega)$ are arbitrary. Values of the spectral dimension, $\bar{d}$, obtained from linear fits to the data for ω from $13\,\text{cm}^{-1}$ to $80\,\text{cm}^{-1}$ are 1.30, 1.30, and 1.36, respectively.

for each. From $13\,\text{cm}^{-1}$ up to $80\,\text{cm}^{-1}$, the scaling of the density with mode frequency appears to follow a power law with an exponent of $\bar{d} - 1$. We find $\bar{d} = 1.30 \pm 0.05$, 1.30 ± 0.04, and 1.36 ± 0.04, for cytochrome c, myoglobin and GFP, respectively. Similar values for $\bar{d}$ were obtained earlier from theoretical work on models for protein molecules [131,132].

Using $\nu = \bar{d}/2D$, we estimate the fractal dimension, D, to be 2.50 ± 0.15, 2.24 ± 0.10, and 2.16 ± 0.10 for cytochrome c, myoglobin, and GFP, respectively.

The same value of D should also be obtained from the dispersion relation for each of these proteins. To compute a dispersion relation for a protein, we need to assign a wave number, k, to a normal mode of frequency, ω. Just as we did for the water cluster, we compute the correlation function for the direction of atomic displacements as a function of distance for individual normal modes of each protein, which allows us to match a wave number of a plane wave to a normal mode. We thereby assign wave numbers to the normal modes of cytochrome c, myoglobin, and GFP up to $80\,\text{cm}^{-1}$. A plot of $\log(\omega)$ versus $\log(k)$ for myoglobin is shown in Fig. 13a.

Other than at very low frequency, below about $13\,\text{cm}^{-1}$, the points plotted in Fig. 13a fall reasonably close to the line fit to them. The slope, $a = D/\bar{d}$, of the line is the exponent of the dispersion relation given by Eq. (31). The value of exponent a for myoglobin is 1.69 ± 0.03. Similarly, we find that $a = 1.75$ and 1.56 for cytochrome c and GFP, respectively, with the same uncertainty as for myoglobin. Using 1.69 for the exponent, we plot the dispersion relation, ω versus k^a, for myoglobin in Fig 13b, where we observe the points to fit well to a straight line, particularly those with frequency greater than $13\,\text{cm}^{-1}$. The line extrapolates close to the origin, as it should. The slopes of the lines that we get in this way are 110 ± 2 for cytochrome c, 71 ± 1 for GFP, and 91 ± 2 for myoglobin.

The dimension D can again be computed, this time with the exponent $a = D/\bar{d}$. Using the values for the spectral dimension, $\bar{d}$, obtained from the density of states, we find D for cytochrome c, myoglobin, and GFP to be 2.28 ± 0.12, 2.20 ± 0.10, and 2.12 ± 0.09, respectively. Each of these values matches, within the stated errors, the values of D obtained from the computed scaling of energy diffusion.

To assign physical meaning to D, we associate D with the mass fractal dimension of the protein, defined by $M \sim L^D$. Then D is the slope of $\log M$ versus $\log L$, where M is the mass of all protein atoms enclosed by concentric spheres of radius L. For a completely collapsed, space-filling polymer, D is 3, but D may be generally less than 3 in proteins due to a combination of compact structures and spacing between them. Figure 14 shows $\log M$ versus $\log L$ for cytochrome c. The concentric spheres of radius L are centered on one of the C_α's of the protein backbone. We compute M for values of L from 2 Å to 16 Å. The plot appears linear with slope 2.15. Continuing in this way, we compute

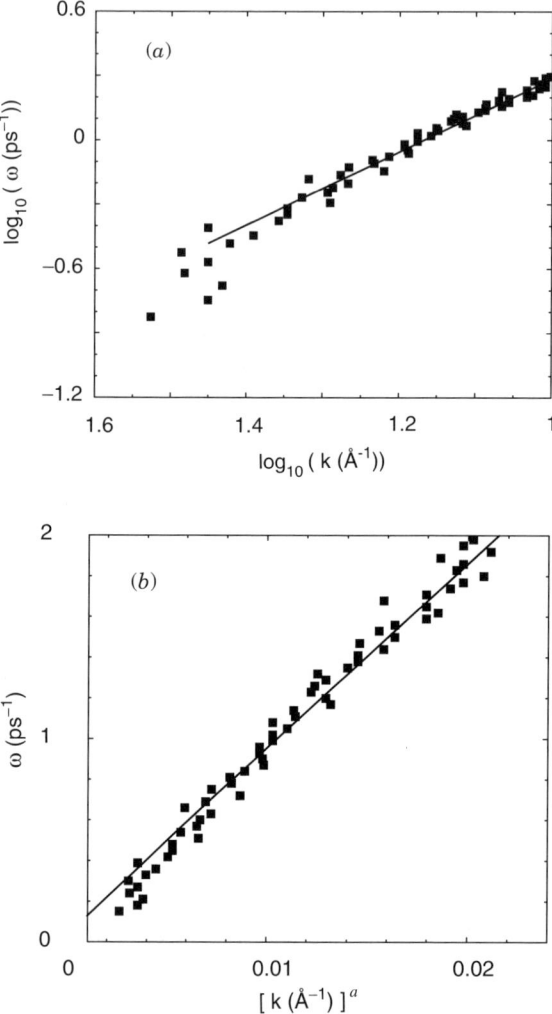

Figure 13. (a) $\log(\omega)$ versus $\log(k)$ computed for myoglobin. A linear fit in the range $13\,\text{cm}^{-1} < \omega < 80\,\text{cm}^{-1}$ is plotted through the data, with a slope of $a = 1.69$. (b) ω versus k^a is plotted for myoglobin using value of $a = 1.69$, obtained in (a).

$\log M$ versus $\log L$ by taking as centers all the C_α's. The average value of the slopes of these plots, $D \pm$ a standard deviation, is 2.31 ± 0.16. Similarly, we find 2.34 ± 0.18 for myoglobin and 2.42 ± 0.20 for GFP. These values for D, in the neighborhood of 2.3, are smaller than the values around 3 previously estimated with less extensive averaging over the whole protein for a number of

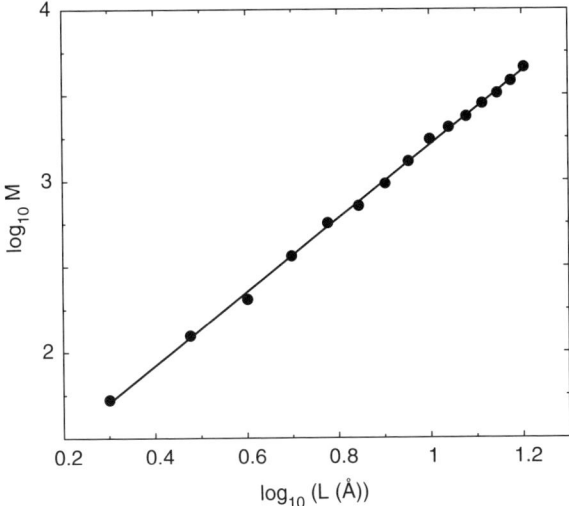

Figure 14. Plot of log M versus log L for cytochrome c, where values of M are the masses of all protein atoms enclosed by concentric spheres of radius L centered at a backbone atom. The slope of the line fit to the data for L from 2 Å to 16 Å gives the mass fractal dimension, in this case 2.15.

globular proteins including myoglobin [130,131]. The values for the mass fractal dimension that we compute for each protein are in fair agreement with values of D determined above by energy diffusion and dispersion, which gave results around 2.3 ± 0.2, so that it may be possible to associate D determined by energy diffusion with mass fractal dimension.
The results of this section are summarized in Table I.

TABLE I
Computed Values for the Spectral Dimension, $\bar{d}$; Exponent, ν, Characterizing Anomalous Subdiffusion in Eq. (24); Exponent, a, Given by Eq. (30), Characterizing the Dispersion Relation for $\omega(k)$; the Fractal Dimensions, D, Obtained from These Values (in Parentheses); and the Mass Fractal Dimension, D, Listed in the Rightmost Column, Computed with Eq. (25) for Each Protein

Protein	$\bar{d}$	ν	a	D
Cytochrome c	1.30 ± 0.05	0.26 ± 0.01	1.75 ± 0.03	2.31 ± 0.16
		($D = 2.50 \pm 0.15$)	($D = 2.28 \pm 0.12$)	
Myoglobin	1.30 ± 0.04	0.29 ± 0.01	1.69 ± 0.03	2.34 ± 0.18
		($D = 2.24 \pm 0.10$)	($D = 2.20 \pm 0.10$)	
GFP	1.36 ± 0.04	0.32 ± 0.01	1.56 ± 0.02	2.42 ± 0.20
		($D = 2.16 \pm 0.10$)	($D = 2.12 \pm 0.09$)	

D. Anharmonic Decay of Vibrational States in Proteins

We turn now to calculation of the lifetimes of the vibrational modes. We consider only the contribution of cubic anharmonic terms in the potential energy to the lifetime, an assumption that is valid at sufficiently low temperatures. The energy transfer rate from mode α, W_α, can be calculated with the Golden Rule, written as the sum of terms that can be described as decay and collision [143], the former typically very much larger except at low frequency where both terms are comparable. The anharmonic decay rate of vibrational mode α is then the sum of these two terms:

$$W_\alpha^{\text{decay}} = \frac{\hbar\pi}{8\omega_\alpha} \sum_{\beta,\gamma} \frac{|\Phi_{\alpha\beta\gamma}|^2}{\omega_\beta\omega_\gamma}(1 + n_\beta + n_\gamma)\delta(\omega_\alpha - \omega_\beta - \omega_\gamma) \quad (34)$$

$$W_\alpha^{\text{coll}} = \frac{\hbar\pi}{4\omega_\alpha} \sum_{\beta,\gamma} \frac{|\Phi_{\alpha\beta\gamma}|^2}{\omega_\beta\omega_\gamma}(n_\beta - n_\gamma)\delta(\omega_\alpha + \omega_\beta - \omega_\gamma) \quad (35)$$

where n_α is the occupation number of mode α, which at temperature T we take to be $n_\alpha = (\exp(\hbar\omega_\alpha/k_B T) - 1)^{-1}$. The matrix elements $\Phi_{\alpha\beta\gamma}$ appear as the coefficients of the cubic terms in the expansion of the interatomic potential in normal coordinates, computed numerically as

$$\Phi_{\alpha\beta\gamma} = (\partial^2 V/\partial Q_\alpha \partial Q_\beta|_{Q_0+\delta Q_\gamma} - \partial^2 V/\partial Q_\alpha \partial Q_\beta|_{Q_0-\delta Q_\gamma})/2\delta Q_\gamma \quad (36)$$

where Q_α is a mass-weighted normal coordinate, and Q_0 is the equilibrium position of the protein in normal coordinates. Low-order anharmonic terms have also been computed for small proteins to calculate their contribution to spectral shifts at low temperature [144].

In Fig. 15 we plot the vibrational energy transfer rate for myoglobin and cytochrome c as a function of frequency at 15 K and 135 K. Rates in both proteins at the same temperature are similar, though there are interesting differences over this range of frequency that will be discussed elsewhere. Particularly noteworthy is the general insensitivity of the decay rates to temperature above about 500 cm^{-1}. Significant temperature dependence of the anharmonic decay rate implies that energy is flowing directly into low-frequency modes. The apparently small temperature dependence suggests that, given the high density of low-frequency modes, matrix elements coupling a given high-frequency mode to a pair of other modes are small if the frequency of one of the pair is small. For the matrix element coupling a triple of modes to be appreciable, the three modes must overlap in space [145–151]. As mode frequency increases, the normal mode vibrations generally become more localized, as we have already seen. If energy in a high-frequency localized

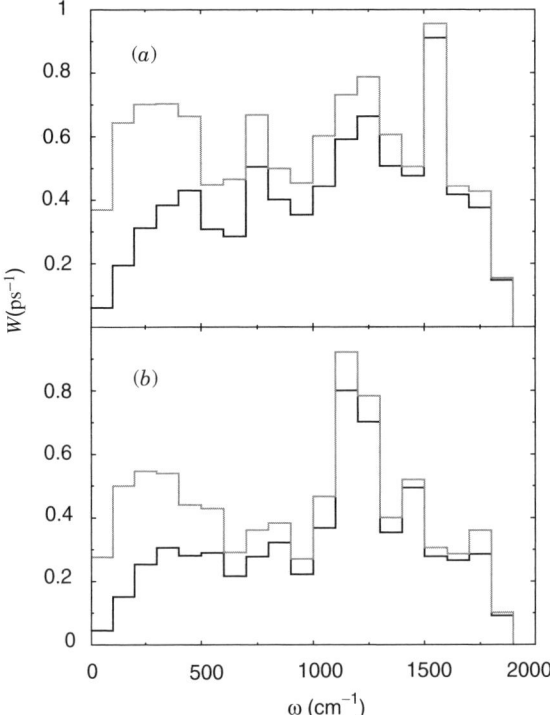

Figure 15. Vibrational energy transfer rate (ps^{-1}) of normal modes of (a) myoglobin and (b) cytochrome c at 15 K (black) and 135 K (gray).

mode, α, decays into a low-frequency mode of the protein, the rest of the energy must decay into a localized mode whose frequency is similar to ω_α. However, as discussed below, localized modes with similar frequencies rarely overlap in space. As a consequence, energy transfer to a localized mode with similar frequency and the remainder to a low-frequency mode occurs slowly, and the vibrational energy transfer rate from high-frequency modes is only weakly dependent on temperature.

Temperature-independent anharmonic decay rates of high-frequency modes of myoglobin have been observed in time-resolved spectroscopic studies. Pump-probe vibrational spectra of the amide I band, between 1600 and 1700 cm^{-1}, measured at temperatures from 6 K to 310 K reveal decay rates ranging only from $\approx$0.5 to 1 ps^{-1} [25], similar to the values we calculate. Similarly, pump-probe studies on myoglobin-CO reveal that the decay of the CO stretch, about 1950 cm^{-1}, is also essentially independent of temperature over the same temperature range [28].

We now address the "repulsion" of localized normal modes of proteins. As illustrated by cytochrome c above, the large majority of normal modes of globular proteins are spatially localized. As such, the vibrations of proteins have much in common with those of 1D disordered systems. One important consequence of strong localization of normal modes in 1D disordered systems is that frequencies of normal modes whose localization centers overlap in space are generally very different [145–147]. This trend gives the appearance of "repulsion" of mode frequencies between pairs of nearby localized modes. Such mode repulsion has important consequences on the temperature dependence of the anharmonic decay rates of normal modes of proteins [111,148,149], as seen above.

The influence of spatial localization of two normal modes computed for cytochrome c on the frequency difference of these vibrational modes is shown in Fig. 16. We locate the largest component of each normal mode, α, and calculate the probability, $P(\Delta\omega)$, that for another mode, β, whose largest component lies a certain distance away from the largest component of mode α, the difference in frequency between them is $\Delta\omega = |\omega_\alpha - \omega_\beta|$. We consider only localized modes, α, whose frequency, ω_α, falls between $1000\,\text{cm}^{-1}$ and $2000\,\text{cm}^{-1}$. Probabilities are calculated for $\Delta\omega$ in intervals of $20\,\text{cm}^{-1}$ to $\Delta\omega = 600\,\text{cm}^{-1}$. The solid-line histogram is an average over all pairs of modes of cytochrome c, regardless of the distance between the largest components of modes α and β. We

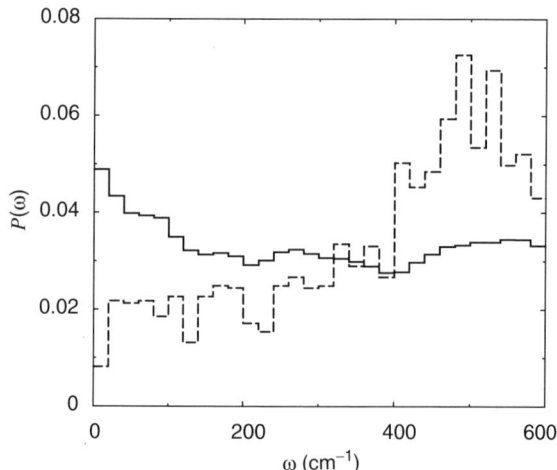

Figure 16. Probability, $P(\Delta\omega)$, of finding a pair of vibrational modes of cytochrome c with frequency difference $\Delta\omega$, when at least one of the modes is localized with frequency ω in the range $1000\,\text{cm}^{-1} < \omega < 2000\,\text{cm}^{-1}$. Solid line corresponds to any pair of modes with frequency difference $\Delta\omega$, regardless of their distance from one another in space. Dashed line corresponds to pairs of modes whose largest components are restricted to lie within 2 Å of each other.

observe that for any pair of modes α and β there is a slightly greater chance that the difference between their vibrational frequencies is small, say, less than $100 \, \text{cm}^{-1}$, than large, say, $500-600 \, \text{cm}^{-1}$. However, if we only consider pairs of modes whose largest components overlap atoms that lie less than $2 \, \text{Å}$ from each other, we obtain the dashed-line histogram. There we see that if pairs of localized modes lie close in space, there is a propensity for their frequency differences to be large, in this case around $500 \, \text{cm}^{-1}$, rather than small, say, below $200 \, \text{cm}^{-1}$. Such a propensity diminishes as we consider modes whose largest components lie farther away from each other. For distances between $4 \, \text{Å}$ and $5 \, \text{Å}$, for instance, we find $P(\Delta\omega)$ to be nearly the same as the solid histogram. We note that similar distributions were found for myoglobin [111].

E. Thermal Transport in Proteins

Both the propagation of vibrational wave packets in proteins and the dispersion relations can be understood and described in terms of scaling relations originally derived for fractal objects. The theory of vibrations on fractal objects, pioneered by Alexander and Orbach [35] to model vibrations in polymers and percolation clusters, was developed further by them and their co-workers [36–39] to address thermal conductivity in these and other amorphous materials. We begin this section by very briefly summarizing the main idea behind their theory. We find that it is in fact a good starting point for calculating thermal transport coefficients of proteins with the addition of two contributions, which we address here.

The main idea behind the theory of Alexander, Orbach, and co-workers [36,37] for thermal conductivity of fractal objects is the separation of vibrational modes into propagating phonon modes and localized "fracton" modes. The long-wavelength phonon modes carry heat while the localized fracton modes cannot other than by anharmonic coupling to phonon modes. The coefficient of thermal conductivity thus contains two terms, one that is expressed in terms of thermal conduction via phonon modes and a second that incorporates the phonon-assisted transport from fracton to other fracton modes. Separation of the phonon and fracton modes occurs for a fractal object at a crossover wavelength on the order of the length scale of the object itself. The crossover wave number scales with the size of the object according to the dispersion relation given by Eq. (31).

The coefficient of thermal conductivity obtained by Alexander, Orbach, and co-workers is then

$$\kappa(T) = \kappa_{\text{phonon}}(T) + \kappa_{\text{hop}}(T) \tag{37}$$

and their microscopic theory provides a means to calculate $\kappa_{\text{hop}}(T)$. In their theory, the hopping term arises from phonon-assistant transport of energy among the localized "fracton" modes, where

$$\kappa_{\text{hop}} = \sum_{\alpha=\text{fracton}} C_\alpha D_\alpha \qquad (38)$$

The diffusion coefficient is $D_\alpha = (1/\tau_\alpha)R^2$, where $1/\tau_\alpha$ is the phonon-assisted fracton-hopping time and R is the mean hopping distance for fracton modes with frequency near ω_α. The rate of phonon-assisted fracton hopping is calculated with Eqs. (34) and (35) specifically for the cases where one fracton decays into another fracton and a phonon, as well as for the reverse process.

Separation of the thermal conductivity into two contributions, one from the phonon modes and the other from the phonon-assisted hopping of energy among localized fracton modes, does not seem to be sufficient for proteins. We have seen that the fracton modes appear at very low frequencies, around 13 cm^{-1} for myoglobin, cytochrome c, and GFP (cf Fig. 12). The crossover frequency for proteins of this size separating phonon and fracton modes lies near 13 cm^{-1}. However, most fracton modes between 13 and 150 cm^{-1} are extended over these proteins, and they become increasingly localized only at higher frequencies. The phonon–fracton crossover, ω_c, thus lies at very much lower frequency than the crossover frequency, ω_l, at which the localization length of the fracton becomes comparable to the size of the protein. Since both ω_c and ω_l scale with the size of the system in the same way, there will always be a range of mode frequencies over which the fracton modes are essentially delocalized over the protein, no matter how large the protein. A similar situation is encountered with one-dimensional disordered objects of N atoms, for which all eigenmodes are localized, but the localization length of the lowest $\sim N^{1/2}$ modes is longer than the object itself [138], contributing to the anomalous length dependence of the thermal conductivity of 1D glasses [138,147]. We must thus address directly the role of extended fracton modes in thermal conduction in proteins. The thermal conductivity then has three contributions,

$$\kappa(T) = \kappa_{\text{phonon}}(T) + \kappa_{\text{deloc.fractons}}(T) + \kappa_{\text{hop}}(T) \qquad (39)$$

The first two terms arise from modes that are extended over the protein. In proteins the size of cytochrome c, myoglobin, and GFP, which range, respectively, from 103 to 228 amino acids, almost all extended modes are fracton modes. The number of "phonon" modes in proteins of this size is very small, only 24 modes from 4 to 13 cm^{-1} for GFP, the largest of the three proteins. We cannot isolate the dynamics of a wave packet in terms of just these modes for an object this small. Thus, the contribution of extended modes to thermal conduction in proteins arises almost entirely from delocalized fractons. We give an expression for this contribution below.

A second problem with the theory of Alexander, Orbach, and co-workers is that κ_{hop}, the contribution of energy transport among localized modes via anharmonic coupling to thermal conduction, is incomplete. In their theory, anharmonicity contributes to thermal conduction via coupling only between localized (fracton) and propagating (phonon) modes. However, Fabian and Allen [145,146] have shown that anharmonic decay from localized to other localized modes occurs at a rate that is of the same order as phonon-assisted transport between localized modes. In fact, Leitner [111,147–149] has shown that the former processes are typically very much faster due to weak anharmonic coupling among a triple of modes of which two are high and one is low in frequency. Energy diffusion by anharmonic decay of a localized mode to nearby localized modes must therefore also contribute to thermal conduction. We have generalized the energy transfer picture of Alexander, Orbach, and co-workers to include "hopping" of energy over the protein by energy transfer from localized to other localized modes by anharmonic coupling [111].

We shall calculate here the coefficient of thermal conductivity and thermal diffusivity for myoglobin. We then need to calculate for each mode of myoglobin the heat capacity with Eq. (15) and to estimate the frequency-dependent energy diffusion coefficient, $D(\omega)$. We are assuming that the heat capacity, calculated per unit volume of protein, is a function only of its vibrational energy. To estimate the volume of myoglobin, we assume for simplicity that the protein is a sphere with radius 17 Å [111].

To calculate the frequency-dependent energy diffusion coefficient, we begin with the normal modes of the protein. In this case, $D(\omega)$ can only be defined for the extended modes, where $\omega < 150 \, \text{cm}^{-1}$, in which case

$$D_h(\omega) = \tfrac{1}{3} v(\omega) l(\omega) \tag{40}$$

where v is the group velocity of a wave packet built up of modes with frequencies near ω, l represents the mean free path, and the subscript h refers to the harmonic, normal mode limit. The mean free path, $l(\omega)$, can extend upwards to a characteristic length of the protein, say, 20 Å, down to a typical interatomic spacing of roughly 1.5 Å. The velocity, v, is obtained from the dispersion relation given by Eq. (31),

$$v(\omega) = d\omega/dk = v_0 \omega^{1-\bar{d}/D} \tag{41}$$

The mean free path scales with frequency as

$$l(\omega) = l_0 \omega^{-\bar{d}/D} \tag{42}$$

We thus need the constants v_0 and l_0. The former is found for myoglobin from the dispersion relation computed above (Fig. 13), where we found $\omega \approx 91 k^{\bar{d}/d}$ for ω in units of ps^{-1} and k in units of Å^{-1}. This gives $v_0 = 24.15$ Å ps$^{-\bar{d}/D}$. To find l_0, we have calculated the energy diffusion coefficients for myoglobin using the wave packet propagation method described in Sec. IIIA. The wave packets were propagated at $\omega = 50\,\text{cm}^{-1}$ and $\omega = 100\,\text{cm}^{-1}$ with a filter of $\delta \omega = 100\,\text{cm}^{-1}$. Averaging over 10 wave packets, we find for myoglobin values of $D\,(50\,\text{cm}^{-1}) = 24.2$ Å^2 ps^{-1} and $D\,((100\,\text{cm}^{-1}) = 21.7$ Å^2 ps^{-1}, which has the expected scaling with mode frequency, $D \sim \omega^{1-2\bar{d}/D}$. These values yield, using Eqs. (40)–(42), a constant $l_0 = 3.23$ Å ps$^{-\bar{d}/D}$. We plot $v(\omega)$, calculated with Eq. (41), and $D(\omega)$, calculated with Eqs. (40)–(42), in Fig. 17.

We now calculate the coefficient of thermal conductivity in the harmonic limit as

$$\kappa_h(T) \approx \kappa_{\text{deloc.fracton}} = \sum_{\alpha = \text{deloc.fracton}} C_\alpha D_\alpha \approx \int_0^{\omega_l} d\omega\, n(\omega) C(\omega) D_h(\omega) \qquad (43)$$

where we use $\omega_l = 150\,\text{cm}^{-1}$ as the upper limit, above which the normal modes are localized. We shall carry out the sum explicitly for myoglobin. Our result for

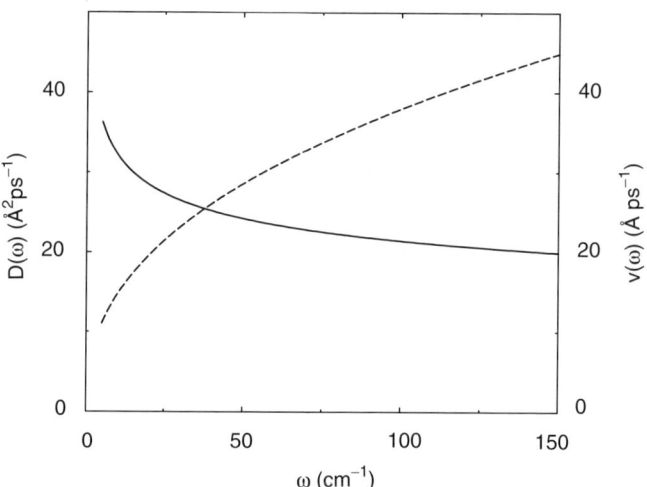

Figure 17. The group velocity (dashed curve) and frequency-dependent diffusion coefficient, calculated with Eqs. (40)–(42) using parameters obtained for myoglobin, are plotted for frequencies up to $150\,\text{cm}^{-1}$.

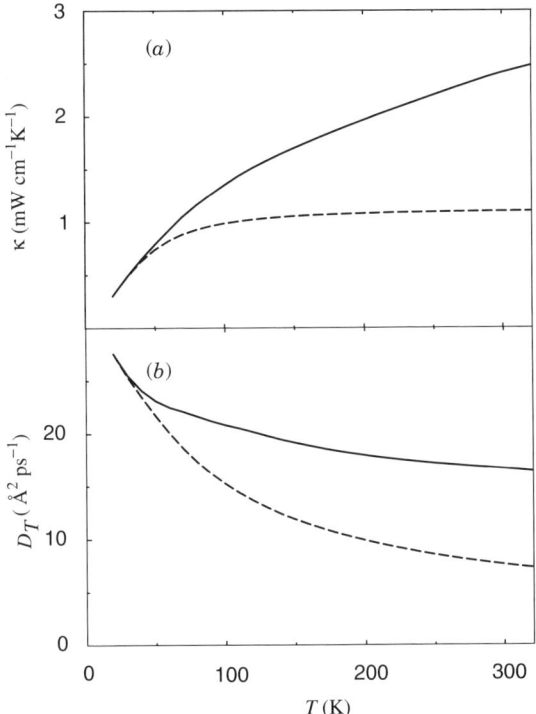

Figure 18. The coefficient of thermal conductivity calculated for myoglobin from $T = 20$ K to 320 K is plotted as a solid curve in (a). The dashed curve gives the thermal conductivity if no anharmonic coupling of normal modes is included in the calculation. (b) Thermal diffusivity calculated for myoglobin is plotted from 20 K to 320 K with (solid) and without (dashed) contributions of anharmonicity.

κ_h is plotted in Fig. 18a as a dashed curve for temperatures from 20 K to 320 K. This is the thermal conductivity of myoglobin in the harmonic limit, where only extended modes contribute to heat flow. We observe that κ_h increases with T until $T \approx 100$ K, at which point it begins to saturate, increasing with T only slowly in approaching its limiting value of about 1.1 mW cm^{-1} K^{-1}. The thermal diffusivity in the harmonic limit is plotted in Fig. 18b as the dashed curve. Unlike the thermal conductivity, the thermal diffusivity does not appear to approach a limiting value in the harmonic limit, decreasing to about 7 Å ps^{-1} by 300 K.

We address now the influence of the localized normal modes, with frequencies greater than about 150 cm^{-1}, on heat transport in the protein [111]. If the normal modes of myoglobin were not localized above $150\,\text{cm}^{-1}$, κ_h could be calculated by integrating Eq. (43) over the entire band of frequencies. In this case, the thermal conductivity would continue to increase with temperature above 100 K. The normal modes above $150\,\text{cm}^{-1}$ are in fact localized, but anharmonicity gives rise to finite lifetimes, which we have seen to be on the order of 1–10 ps for these modes. Energy localized to a normal mode is transferred to other modes on this time scale. To estimate the role of anharmonicity in thermal diffusion, we have built on a hopping model used by Alexander, Orbach, and co-workers [36,37] to describe energy transfer among localized modes of fractal objects and amorphous systems. Their model only accounts for phonon-assisted transfer of energy between localized modes, whereas we account in an approximate fashion for all processes that transfer energy to and from localized modes.

If, near a given frequency, ω, the localization length has a typical value $\xi(\omega)$, then energy can diffuse this distance before a transition to other modes takes place. If the anharmonic transition rate, $W(\omega)$, is sufficiently slow, then $\xi(\omega)$ is effectively the distance over which energy spreads in a time $W^{-1}(\omega)$. Energy that flows into a mode of similar frequency can travel about the same length and spends about the same amount of time, $W^{-1}(\omega)$, in that mode. Since the modes that couple most strongly by cubic anharmonicity to the one at frequency ω have frequencies typically in the range $\approx \omega/3$ to $2\omega/3$ [147,148], the localization lengths of such coupled modes are of the same order. We take as an approximation to the mean free path the localization length, $\xi(\omega)$. The diffusion coefficient due to anharmonic transitions is then

$$D_a(\omega) \approx \tfrac{1}{3}\xi^2(\omega)W(\omega) \qquad (44)$$

where the subscript a indicates energy transfer due to anharmonic coupling of normal modes. If, however, $W^{-1}(\omega)$ is sufficiently short, then a vibrational excitation will not have spread as far as $\xi(\omega)$ before a transition to other modes takes place. When anharmonic decay is rapid, transitions occur before the effects of localization influence diffusion of energy. In this case, $D(\omega)$ is given by $D_h(\omega)$.

We can estimate which diffusion coefficient to use in the regime of localized normal modes, $D_a(\omega)$ or $D_h(\omega)$, for $\omega > \omega_l \approx 150\,\text{cm}^{-1}$, by calculating the time for a vibrational excitation to diffuse a distance $\xi(\omega)$. This time, t^*, can be estimated as

$$t^*(\omega) = \tfrac{1}{3}\xi^2(\omega)\,D_h^{-1}(\omega) \qquad (45)$$

where for $D_h(\omega)$ we use Eq. (40). Then for $\omega > 150\,\text{cm}^{-1}$, where normal modes of myoglobin are localized, we take the energy diffusion coefficient to be

$$D(\omega) = D_h(\omega), \quad \text{if } W^{-1}(\omega) \leq t^*(\omega) \qquad (46a)$$

$$D(\omega) = D_a(\omega), \quad \text{if } W^{-1}(\omega) > t^*(\omega) \qquad (46b)$$

where $t^*(\omega)$ is calculated using Eq. (45). Equation (46a) represents the case where rapid anharmonic decay allows energy diffusion to occur without the restriction of localization, since mode lifetimes are shorter than the time it would take for energy to diffuse the length of the normal mode. We note that in principle, W could be so large as to establish the mean free path, which would then be about $v(\omega)W^{-1}(\omega)$. However, for $v > 20\,\text{Å ps}^{-1}$ and $W \approx 1\,\text{ps}^{-1}$ for thermally accessible modes, the mean free path due to anharmonic decay is greater than the localization length of the modes and thus does not influence energy diffusion. Equation (46b) represents the case of anharmonic energy transfer processes sufficiently slow to influence diffusion. In the limit $W = 0$, there is no anharmonic decay of localized modes, and only the extended modes contribute to thermal transport.

In Fig. 18a we plot κ, which has been calculated by summing Eq. (12) over the full band of frequencies for myoglobin. In the extended regime—that is, vibrational frequencies of $\omega < 150\,\text{cm}^{-1}$—$D(\omega)$ appearing in Eq. (12) is just $D_h(\omega)$, given by Eq. (40). For $\omega > 150\,\text{cm}^{-1}$, we calculate $D(\omega)$ using the criterion of Eq. (46). The localization lengths that enter into the calculation of t^* in Eq (45) are those computed for myoglobin using Eq. (23) [111], and they are similar to those plotted for cytochrome c in Fig. 9. The energy transfer rates, $W(\omega)$, are those plotted in Fig. 15 at $T = 135\,\text{K}$, apart from a correction that we make for the temperature. To account for the temperature variation of $W(\omega)$, we assume that $W(\omega)$ is independent of temperature for $\omega \geq 500\,\text{cm}^{-1}$, whereas we assume it is proportional to temperature for $\omega < 500\,\text{cm}^{-1}$, fairly reasonable given the temperature variations we observed in our calculation of $W(\omega)$ at $T = 15\,\text{K}$ and $135\,\text{K}$, shown in Fig. 15. The value of the thermal conductivity at $300\,\text{K}$, $2.4\,\text{mW cm}^{-1}\,\text{K}^{-1}$, is more than double the value in the harmonic limit, highlighting the importance of anharmonicity toward enhancement of heat flow in proteins. The thermal conductivity that we have calculated for myoglobin is less than half the thermal conductivity of H_2O at $300\,\text{K}$, which is $6.1\,\text{mW cm}^{-1}\,\text{K}^{-1}$ [113].

The thermal diffusivity, D_T, of myoglobin can be calculated in terms of κ with Eq. (13). Results are plotted in Fig. 18b from $20\,\text{K}$ to $320\,\text{K}$. At $300\,\text{K}$ we find the thermal diffusivity of myoglobin to be $15\,\text{Å}^2\,\text{ps}^{-1}$, which can be compared with the value of $15\,\text{Å}^2\,\text{ps}^{-1}$ for water [113]. We note also that the anharmonic, "hopping" contribution makes up more than half of the thermal diffusivity at $300\,\text{K}$. We see also that with inclusion of anharmonicity in the calculation of D_T, its value drops relatively slowly with increasing temperature, compared with the faster drop in D_T with increasing T seen for the harmonic calculation. The value of D_T that we obtain is consistent with the 20 to 30-ps time for heat from a photoexcited heme of myoglobin to reach the surrounding solvent, measured by Hochstrasser and co-workers [152], a time scale corroborated by simulations [135,136]. Our result is also comparable to the value of the thermal diffusivity, 7 $\text{Å}^2\,\text{ps}^{-1}$, estimated from a simulated cooling of the photosynthetic reaction center of *Rhodospseudomonas viridis* between 200 K and 300 K by Tesch and Schulten [153].

Since the values of the thermal diffusivity for myoglobin and water are close, the discrepancy between their thermal conductivities is largely due to differences in their respective heat capacities. At 300 K the specific heat of water is 4.2 J $\text{K}^{-1}\,\text{g}^{-1}$, whereas we calculate the specific heat of myoglobin to be 1.0 J $\text{K}^{-1}\,\text{g}^{-1}$ at 300 K. This latter value is reasonably close to measured specific heats for proteins. For example, the specific heat of lysozyme in dilute aqueous solution is 1.5 J $\text{K}^{-1}\,\text{g}^{-1}$, and it is 1.3 J $\text{K}^{-1}\,\text{g}^{-1}$ for the dry protein [154,155]. Still, we should bear in mind that at 300 K the partial specific heat of a hydrated protein may be some 50% larger than the specific heat of a dry protein [156], so that the coefficient of thermal conductivity may be correspondingly larger.

While we have only carried out calculations for myoglobin, we note that for the two other proteins studied in this chapter, cytochrome c and GFP, the latter structurally very different than myoglobin, the diffusion of energy and sound velocities scale with frequency in similar (though not identical) ways. Since rates of anharmonic decay have a similar frequency dependence, and since these proteins are not all that different in size, we expect the thermal transport coefficients for these proteins to be similar.

IV. SUMMARY

We have explored in this chapter how quantum mechanical energy flow in moderate-sized to large molecules influences kinetics of unimolecular reactions and thermal conduction. In the first part of this chapter we addressed vibrational energy flow in moderate-sized molecules, and we also discussed its influence on kinetics of conformational isomerization. In the second part we examined the dynamics of vibrational energy flow through clusters of water molecules and through proteins, and we computed thermal transport coefficients for these objects.

For moderate-sized molecules with tens of vibrational modes, vibrational energy flow is conveniently described in a vibrational quantum number space. A statistical theory for the vibrational Hamiltonian, called Local Random Matrix Theory (LRMT), exploits the local coupling in the state space. LRMT predicts and locates the IVR transition and the range of energies over which the vibrational states of a molecule are localized. We have seen that the localization threshold often lies higher than barriers to conformation isomerization, so that RRKM theory could not be expected to reliably predict microcanonical reaction rates. Even when the IVR threshold lies at energies significantly lower than a barrier to isomerization, we may still find substantial dynamical corrections to RRKM theory due to sufficiently slow intramolecular vibrational redistribution. In particular, so long as vibrational energy transfer rates from reactive states of the vibrational state space to nonreactive states are slower than rates to cross from reactive states to product, dynamical corrections to RRKM theory are significant. In this respect, IVR rates need not be all that slow to influence isomerization kinetics. We have illustrated this effect on the rate of cyclohexane ring inversion, where we use LRMT to compute corrections to RRKM theory. Dynamical corrections due to sluggish IVR influence thermal rates of isomerization over a wide range of frequency of collisions with the solvent environment, spanning gas, and condensed phases. Dynamical corrections lead to rates of cyclohexane ring inversion in, say, CS_2 solvent that are about an order of magnitude smaller than RRKM rates in the gas-phase high pressure limit. The rate of ring inversion then rises at higher collision frequencies, corresponding to liquid CS_2, where the IVR rate, and thus the transmission coefficient, becomes proportional to the collision frequency. These features, captured by LRTM, have been observed in NMR measurements [29–31]. We recall that LRMT aims to describe energy flow and localization in molecules whose total energy corresponds to a small number of quanta per mode. In this domain, oscillators representing the vibrations of a molecule are only weakly coupled to one another. LRMT is thus well-suited to describe the nature and extent of energy flow in the energy regime where reactions over relatively low barriers take place—hence the success of LRMT in correcting equilibrium rate theories for energy flow, as here for cyclohexane isomerization and for a number of other reactions [3,4,6], including the photoisomerization of *trans*-stilbene [3,6], discussed above. In this respect, LRMT may also help guide us in controlling chemical reactions with light fields by identifying conditions where mode-selective chemistry might most feasibly occur. Hints of mode-selective chemistry in dipeptides [157] suggest the significance of unraveling energy flow pathways toward controlling reactions in biomolecules.

In the second part of the chapter, we have examined the spread of vibrational energy through coordinate space in systems that are large on the molecular scale—in particular, clusters of hundreds of water molecules and proteins—and computed thermal transport coefficients for these systems. The coefficient of thermal conductivity is given by the product of the heat capacity per unit volume and the energy diffusion coefficient summed over all vibrational modes. For the water clusters, the frequency-dependent energy diffusion coefficient was obtained by propagating wave packets that were expanded in terms of the normal modes of vibration of the cluster filtered over a select frequency range. In this way we obtained the thermal transport coefficients for a cluster of nearly 1000 water molecules. The computed value of the thermal conductivity for the water cluster near 300 K appears to be low, which can be attributed to the small value of the heat capacity for liquid water obtained with a harmonic model. Computing equilibrium and nonequilibrium thermodynamic properties in terms of normal modes is attractive [158,159], though most broadly useful for these water clusters in the glassy state. Still, the computed thermal diffusivity for liquid water at 300 K is close to its actual value of 15 Å^2 ps^{-1}, indicating that this property arises from the network vibrations of water.

Thermal transport in proteins is more complex than in water. Energy diffusion in proteins is non-Brownian on the subpicosecond time scale due to the wide range of trapping times as vibrational energy flows through a protein. The diffusion of vibrational energy in proteins, the scaling of mode density with frequency, and the scaling of mode frequency with wave number are all well characterized by the corresponding scaling relations for fractal objects. The theory of vibrations on fractals, pioneered by Alexander and Orbach [35], who coined the term *fractons*, provides a foundation for us to describe thermal transport in proteins. Alexander, Orbach, and their co-workers [36,37] put forth a theory to calculate the thermal conductivity of a fractal, which we find provides a useful starting point for describing thermal conductivity in proteins, once two important additional contributions are included.

The first contribution arises from vibrational modes characteristic of fractals (fracton modes) that are delocalized over the whole protein. Such modes make up nearly all extended modes of the protein. We have included the contribution of these modes to the thermal conductivity, and we have carried out here a specific calculation for the protein myoglobin. The second important contribution arises from anharmonic coupling among localized vibrational modes of the protein. Most vibrational modes of proteins are localized and contribute to thermal conduction through transport of energy to other localized modes to which they are anharmonically coupled. Alexander, Orbach, and coworkers accounted for such an anharmonically induced "hopping" mechanism only through phonon-assisted hopping [36,37]. However, we have seen here that coupling between pairs of high-frequency localized modes and low-frequency modes, phonon or otherwise, is typically weak. Indeed, anharmonic decay of a localized mode to another pair of localized modes is typically much faster than decay to a localized mode and a low frequency phonon-like mode. Thus, a hopping model addressing contributions of cubic anharmonicity to thermal conduction must also include such processes, which we incorporate into our calculation.

In myoglobin, we find that the anharmonic contribution significantly enhances thermal conduction over that in the harmonic limit, by more than a factor of 2 at 300 K. Moreover, the thermal conductivity rises with temperature for temperatures beyond 300 K as a result of anharmonicity, whereas it appears to saturate around 100 K if we neglect the contribution of anharmonic coupling of vibrational modes. The value for the thermal conductivity of myoglobin at 300 K is about half the value for water. The value for the thermal diffusivity that we calculate for myoglobin is the same as the value for water. Thermal transport coefficients for other proteins will be presented elsewhere.

Acknowledgments

Since we submitted this chapter we have slightly refined our calculation of thermal transport coefficients for myoglobin and green fluorescent protein [163], as well as for water [164].

This work was supported by the National Science Foundation (NSF CHE-0112631), a New Faculty Award from the Camille and Henry Dreyfus Foundation, and a Research Innovation Award from the Research Corporation. During the course of writing this chapter, the author benefited from stimulating discussions with the participants of two workshops on chemical reaction dynamics in complex systems, at Kobe University and at Kyoto University in October 2003, which were sponsored by the Japan–U.S. Cooperative Science Program of the Japan Society for the Promotion of Science, the Yukawa Institute at Kyoto University, and the Inoue Foundation for Science.

References

1. J. N. Onuchic and P. G. Wolynes, *J. Phys. Chem.* **92**, 6495 (1988).
2. B. J. Berne, M. Borcovec, and J. E. Straub, *J. Phys. Chem.* **92**, 3711 (1988).
3. D. M. Leitner and P. G. Wolynes, *Chem. Phys. Lett.* **280**, 411 (1997).
4. D. M. Leitner, *Int. J. Quant. Chem.* **75**, 523 (1999).
5. S. Nordholm, *Chem. Phys.* **137**, 109 (1989).
6. D. M. Leitner, B. Levine, J. Quenneville, T. J. Martínez, and P. G. Wolynes, *J. Phys. Chem. A* **107**, 10706 (2003).
7. R. A. Kuharski, D. Chandler, J. A. Montgomery, F. Rabii, and S. J. Singer, *J. Phys. Chem.* **92**, 3261 (1988).
8. J. C. Keske and B. H. Pate, *Annu. Rev. Phys. Chem.* **51**, 323 (2000).
9. M. J. Davis and S. K. Gray, *J. Chem. Phys.* **84**, 5389 (1986); S. K. Gray, S. A. Rice, and M. J. Davis, *J. Phys. Chem.* **90**, 3470 (1986).
10. R. E. Gillilan and G. S. Ezra, *J. Chem. Phys.* **94**, 2648 (1991).
11. M. Toda, *Adv. Chem. Phys.* **123**, 153 (2002).
12. D. Xu, C. H. Martin, and K. Schulten, *Biophys. J.* **70**, 453 (1996).
13. R. J. D. Miller, *Annu. Rev. Phys. Chem.* **42**, 581 (1991).
14. D. Antoniou and S. D. Schwartz, *Proc. Natl. Acad. Sci. USA* **94**, 12360 (1997).
15. D. E. Logan and P. G. Wolynes, *J. Chem. Phys.* **93**, 4994 (1990).
16. D. M. Leitner and P. G. Wolynes, *J. Chem. Phys.* **105**, 11226 (1996).
17. D. M. Leitner and P. G. Wolynes, *Chem. Phys. Lett.* **258**, 18 (1996).
18. D. M. Leitner and P. G. Wolynes, *J. Phys. Chem. A* **101**, 541 (1997).

19. D. M. Leitner and P. G. Wolynes, *Phys. Rev. Lett.* **76**, 216 (1996).
20. R. Bigwood, M. Gruebele, D. M. Leitner, and P. G. Wolynes, *Proc. Natl. Acad. Sci. USA* **95**, 5960 (1998).
21. D. M. Leitner and P. G. Wolynes, *ACH-Models in Chemistry*, **134**, 663 (1997).
22. M. Gruebele, *Adv. Chem. Phys.* **114**, 193 (2000).
23. S. A. Schofield and P. G. Wolynes, in *Dynamics of Molecules and Chemical Reactions*, R. E. Wyatt and J. Z. H. Zhang, eds., Marcel Dekker, New York, 1996, p. 123.
24. P. Hamm, M. Lim, and R. M. Hochstrasser, *J. Phys. Chem. B* **102**, 6123 (1998).
25. K. A. Peterson, C. W. Rella, J. R. Engholm, and H. A. Schwettman, *J. Phys. Chem. B* **103**, 557 (1999).
26. A. Xie, L. van der Meer, W. Hoff, and R. H. Austin, *Phys. Rev. Lett.* **84**, 5435 (2000).
27. A. Xie, A. G. F. van der Meer, R. H. Austin, *Phys. Rev. Lett.* **88**, 018102 (2002).
28. M. D. Fayer, *Annu. Rev. Phys. Chem.* **52**, 315 (2001); C. W. Rella, K. D. Rector, A. Kwok, A.; J. R. Hill, H. A. Schwettman, D. D. Dlott, and M. D. Fayer, *J. Phys. Chem.* **100**, 15620 (1996).
29. B. D. Ross and N. S. True, *J. Am. Chem. Soc.* **105**, 1382, 4871 (1983).
30. D. L. Hasha, J. Eguchi, and J. Jonas, *J. Am. Chem. Soc.* **104**, 2290 (1982).
31. D. M. Campbell, M. Mackowiak, and J. Jonas, *J. Chem. Phys.* **96**, 2717 (1992).
32. T. Nakayama, K. Yakubo, and R. L. Orbach, *Rev. Mod. Phys.* **66**, 381 (1994).
33. S. Havlin and D. Ben-Avraham, *Adv. Phys.* **36**, 695 (1987).
34. T. S. Chow, *Mesoscopic Physics of Complex Materials*, Springer-Verlag, New York, 2000.
35. S. Alexander and R. Orbach, *J. Phys. Lett.* **43**, L-625 (1982).
36. S. Alexander, O. Entin-Wohlman, and R. Orbach, *Phys. Rev. B* **34**, 2726 (1986).
37. A. Jagannathan, R. Orbach, and O. Entin-Wohlman, *Phys. Rev. B* **39**, 13465 (1989).
38. R. Orbach, *Science* **231**, 814 (1986).
39. R. Orbach, *Philos. Mag. B* **65**, 289 (1992).
40. P. B. Allen, J. L. Feldman, J. Fabian, F. Wooten, *Philos. Mag. B* **79**, 1715 (1999).
41. P. Sheng and M. Zhou, *Science* **253**, 539 (1991); P. Sheng, M. Zhou, and Z. Q. Zhang, *Phys. Rev. Lett.* **72**, 234 (1994).
42. S. N. Taraskin, S. R. Elliott, *Phys. Rev. B* **61**, 12017, 12031 (2000).
43. P. M. Felker, W. R. Lambert, and A. H. Zewail, *J. Chem. Phys.* **82**, 3003 (1985).
44. S. L. Schultz, J. Qian, and J. M. Jean, *J. Phys. Chem. A* **101**, 1000 (1997).
45. K. K. Lehmann, G. Scoles and B. H. Pate, *Annu. Rev. Phys. Chem.* **45**, 241 (1994).
46. K. S. J. Nordholm and S. A. Rice, *J. Chem. Phys.* **61**, 203 (1974); **61**, 768 (1974); **62**, 157 (1975).
47. K. G. Kay, *J. Chem. Phys.* **72**, 5955 (1980).
48. M. Kuzmin and A. A. Stuchebruckhov, in *Laser Spectroscopic of Highly Vibrationally Excited Molecules*, V. S. Letokhov, ed., Hilger, New York, 1989.
49. A. A. Stuchebrukhov and R. A. Marcus, *J. Chem. Phys.* **85**, 307 (1993)
50. R. Bigwood and M. Gruebele, *Chem. Phys. Lett.* **235**, 604 (1995).
51. M. Gruebele, *J. Phys. Chem.* **100**, 12183 (1996).
52. T. Uzer, *Phys. Rep.* **199**, 73 (1991).
53. T. Komatsuzaki and R. S. Berry, *Adv. Chem. Phys.* **123**, 79 (2002).
54. T. L. Beck, D. M. Leitner, and R. S. Berry, *J. Chem. Phys.* **89**, 1681 (1988).
55. D. M. Leitner, R. S. Berry, and R. M. Whitnell, *J. Chem. Phys.* **91**, 3470 (1989).

56. D. M. Leitner, J. D. Doll, and R. M. Whitnell, *J. Chem. Phys.* **94**, 6644 (1991); *ibid.* **96**, 9239 (1992).
57. C. Chakravarty, R. J. Hinde, D. M. Leitner, and D. J. Wales, *Phys. Rev. E* **56**, 363 (1997).
58. C. C. Martens, M. J. Davis, and G. S. Ezra, *Chem. Phys. Lett.* **142**, 519 (1987).
59. V. Ya. Demikhovskii, F. M. Izrailev, and A. I. Malyshev, *Phys. Rev. Lett.* **88**, 154101 (2002); *Phys. Rev. E* **66**, 036211 (2002).
60. D. M. Leitner and P. G. Wolynes, *Phys. Rev. Lett* **79**, 55 (1997); D. M. Leitner and P. G. Wolynes, *Chem. Phys. Lett.* **276**, 289 (1997).
61. D. M. Leitner and P. Schmelcher, *Phys. Rev. A* **58**, R3383 (1998).
62. M. J. Davis and E. J. Heller, *J. Phys. Chem.* **85**, 307 (1981); E. J. Heller, *J. Chem. Phys.* **99**, 2625 (1995).
63. T. A. Brody, J. Flores, J. B. French, P. A. Mello, A. Pandey, and S. S. M. Wong, *Rev. Mod. Phys.* **53**, 385 (1981); C. E. Porter, *Statistical Theories of Spectra: Fluctuations*, Academic Press, New York, 1965; M. L. Mehta, *Random Matrices*, Academic Press, New York, 1991.
64. T. Zimmermann, H. Köppel, L. S. Cederbaum, G. Persch, and W. Demtröder, *Phys. Rev. Lett.* **61**, 3 (1988).
65. D. M. Leitner, H. Köppel, and L. S. Cederbaum, *J. Chem. Phys.* **104**, 434 (1996).
66. R. W. Field, S. L. Coy, and S. A. B. Solina, *Prog. Theor. Phys. Suppl.* **116**, 143 (1994).
67. R. Hernandez, W. H. Miller, C. B. Moore, and W. F. Polik, *J. Chem. Phys.* **99**, 950 (1993).
68. E. J. Heller and S. A. Rice, *J. Chem. Phys.* **61**, 936 (1974)
69. W. M. Gelbart, S. A. Rice, and K. F. Freed, *J. Chem. Phys.* **57**, 4699 (1972).
70. D. M. Leitner, *Phys. Rev. E* **48**, 2536 (1993).
71. D. M. Leitner, H. Köppel, and L. S. Cederbaum, *Phys. Rev. Lett.* **73**, 2970 (1994).
72. O. Bohigas, S. Tomsovic, and D. Ullmo, *Phys. Rep.* **223**, 43 (1993)
73. T. Guhr and H. A. Weidenmüller, *Ann. Phys. (N.Y.)* **199**, 412 (1990).
74. T. Guhr, A. Muller-Groeling, and H. A. Weidenmuller, *Phys. Rep.* **299**, 190 (1998).
75. C. Ellegaard et al., *Phys. Rev. Lett.* **77**, 4918 (1996).
76. D. M. Leitner, *Phys. Rev. E* **56**, 4890 (1997).
77. J. Go and D. S. Perry, *J. Chem. Phys.* **103**, 5194 (1995); D. S. Perry, G. A. Bethardy, M. J. Davis, and J. Go, *Faraday Disc. Chem. Soc.* **102**, 215 (1995).
78. D. Burleigh and E. L. Sibert, *J. Chem. Phys.* **98**, 8419 (1993).
79. R. E. Wyatt and C. Iung, in *Dynamics of Molecules and Chemical Reactions*, R. E. Wyatt and J. Z. H. Zhang, eds., Marcel Dekker, New York, 1996, p. 59.
80. R. Abou-Chacra, P. W. Anderson, and D. J. Thouless, *Phys. Rev. B* **36**, 4135 (1987).
81. G. M. Stewart and J. D. McDonald, *J. Chem. Phys.* **78**, 3907 (1983).
82. R. D. Levine, *Adv. Chem. Phys.* **70**, 53 (1987).
83. S. A. Schofield and P. G. Wolynes, *J. Chem. Phys.* **98**, 1123 (1993); *J. Phys. Chem.* **99**, 2753 (1995).
84. S. A. Schofield, P. G. Wolynes, and R. E. Wyatt, *Phys. Rev. Lett.* **74**, 3720 (1995).
85. P. J. Robinson and Holbrook, *Unimolecular Reactions*, John Wiley & Sons, London, 1972; W. Forst, *Theory of Unimolecular Reactions*, Academic, New York, 1972; T. Baer and W. L. Hase, *Unimolecular Reaction Dynamics: Theory and Experiments*, Oxford, New York, 1996.
86. S. H. Northrup and J. T. Hynes, *J. Chem. Phys.* **73**, 2700 (1980).

87. S. J. Singer, R. A. Kuharski, and D. Chandler, *J. Phys. Chem.* **90**, 6015 (1986).
88. K. B. Wiberg and A. Shrake, *Spectrochim. Acta* **29A**, 584 (1973).
89. D. Madsen, R. Pearman, and M. Gruebele, *J. Chem. Phys.* **106**, 5874 (1997).
90. R. Pearman and M. Gruebele, *J. Chem. Phys.* **108**, 6561 (1998).
91. T. Baer and A. R. Potts, *J. Phys. Chem A* **104**, 9397 (2000).
92. D. A. McWhorter, E. Hudspeth, and B. H. Pate, *J. Chem. Phys.* **110**, 2000 (1999).
93. D. A. McWhorter and B. H. Pate, *J. Phys. Chem. A* **102**, 8786, 8795 (1998).
94. J. A. Syage, W. R. Lambert, P. M. Felker, A. H. Zewail, and R. M. Hochstrasser, *Chem. Phys. Lett.* **88**, 266 (1982).
95. P. M. Felker and A. H. Zewail, *J. Phys. Chem.* **89**, 5402 (1985).
96. L. R. Kundkar, R. A. Marcus, and A. H. Zewail, *J. Phys. Chem.* **87**, 2473 (1983).
97. M. W. Balk and G. R. Fleming, *J. Phys. Chem.* **90**, 3975 (1986).
98. O. K. Rice, *Z. Phys. Chem.* **7**, 226 (1930).
99. D. G. Cahill et al., *J. Appl. Phys.* **93**, 793 (2003).
100. T. S. Tinghe, J. M. Worlock, and M. L.. Roukes, *Appl. Phys. Lett.* **70**, 2687 (1997).
101. K. Schwab, E. A. Henriksen, J. M. Worlock, and M. L. Roukes, *Nature (London)* **404**, 974 (2000).
102. D. E. Angelescu, M. C. Cross, and M. L. Roukes, *Superlattices Microstruct.* 23, 673 (1998).
103. M. P. Blencowe, *Phys. Rev. B* **59**, 4992 (1999).
104. L. G. C. Rego and G. Kirczenow, *Phys. Rev. Lett.* **81**, 232 (1998).
105. D. M. Leitner and P. G. Wolynes, *Phys. Rev. E* **61**, 2902 (2000).
106. D. Segal, A. Nitzan, and P. Hänggi, *J. Chem. Phys.* **119**, 6840 (2003).
107. A. Buldum, D. M. Leitner, and S. Ciraci, *Europhys. Lett.* **47**, 208 (1999).
108. P. K. Schelling, S. R. Phillpot, and P. Keplinski, *Phys. Rev. B* **65**, 144306 (2002).
109. P. B. Allen and J. Kelner, *Am. J. Phys.* **66**, 497 (1998).
110. R. Elber et al., *Comput. Phys. Commun.* **91**, 159 (1995).
111. X. Yu and D. M. Leitner, *J. Phys. Chem. B* **107**, 1698 (2003).
112. T. Nishikawa and N. Go, *Proteins: Struct. Func. Gene.* **2**, 308 (1987).
113. N. E. Dorsey, *Properties of Ordinary Water Substance*, Reinhold, New York; 1940; D. R. Lide, ed. *CRC Handbook of Chemistry and Physics*, 71st ed., CRC Press, Boston, 1992.
114. H. Frauenfelder, S. G. Sligar, and P. G. Wolynes, *Science*, **254**, 1598 (1991).
115. I. E. T. Iben, D. Braunstein, W. Doster, H. Frauenfelder, M. K. Hong, J. B. Johnson, S. Luck, P. Ormos, A. Schulte, P. J. Steinbach, A. H. Xie, and R. D. Young, *Phys. Rev. Lett.* **62**, 1916 (1989).
116. A. E. García, R. Blumenfeld, G. Hummer, and J. A. Krumhansl, *Physica D* **107**, 225 (1997).
117. D. A. Lidar, D. Thirumalai, R. Elber, and R. B. Gerber, *Phys. Rev. E* **59**, 2231 (1999).
118. T. Y. Shen, K. Tai, and J. A. McCammon, *Phys. Rev. E* **63**, 041902 (2001).
119. P. Carlini, A. R. Bizzarri, and S. Cannistraro, *Physica D* **165**, 242 (2002).
120. J.-P. Bouchaud and A. Georges, *Phys. Rep.* **195**, 127 (1990).
121. H. Scher and E. W. Montroll, *Phys. Rev. B* **12**, 2455 (1975).
122. R. Metzler and J. Klafter, *Phys. Rep.* **339**, 1 (2000)
123. R. Metzler, E. Barkai, and J. Klafter, *Phys. Rev. Lett.* **82**, 3563 (1999).
124. H. Yang et al., *Science* **302**, 262 (2003).

125. X. S. Xie, *J. Chem. Phys.* **117**, 11024 (2002); H. Yang and X. S. Xie, *J. Chem. Phys.* **117**, 10965 (2002).
126. B. R. Brooks and M. Karplus, *Proc. Natl. Acad. Sci. USA* **80**, 6571 (1983).
127. M. Levitt, C. Sander, P. S. Stern, *J. Mol. Biol.* **181**, 423 (1985).
128. C. L. Brooks, M. Karplus, B. M. Pettitt, *Adv. Chem. Phys.* **71**, 1 (1988).
129. J. A. McCammon, and S. C. Harvey, *Dynamics of Proteins and Nucleic Acids*, Cambridge University Press, Cambridge, 1987.
130. T. G. Dewey, *Fractals in Molecular Biophysics*, Oxford University Press, New York, 1997.
131. R. Elber, in *The Fractal Approach to Heterogeneous Chemistry*, D. Avnir, ed., John Wiley & Sons, New York, 1989, p. 345.
132. R. Elber and M. Karplus, *Phys. Rev. Lett.* **56**, 394 (1986).
133. H. J. Stapleton, J. P. Allen, C. P. Flynn, D. G. Stinson, and S. R. Kurtz, *Phys. Rev. Lett.* **45**, 1456 (1980).
134. A. R. Drews, B. D. Thayer, H. J. Stapleton, G. C. Wagner, G. Giugliarelli, and S. Cannistraro, *Biophys. J.* **57**, 157 (1990).
135. D. E. Sagnella, J. E. Straub, and D. Thirumalai, *J. Chem. Phys.* **113**, 7702 (2000).
136. D. E. Sagnella and J. E. Straub, *J. Phys. Chem. B* **105**, 7057 (2001).
137. L. Bu and J. E. Straub, *J. Phys. Chem. B* **107**, 12339 (2003); Q. Wang, C. F. Wong and H. Rabitz, *Biophys. J.* **75**, 60 (1998).
138. H. Matsuda and K. Ishii, *Suppl. Prog. Theor. Phys.* **45**, 56 (1970).
139. M. Cho, G. R. Fleming, S. Saito, I. Ohmine, and R. M. Stratt, *J. Chem. Phys.* **100**, 6672 (1994).
140. R. Rammal and G. Toulouse, *J. Phys. Lett.* **44**, L-13 (1983).
141. X. Yu and D. M. Leitner, *J. Chem. Phys.* **119**, 12673 (2003).
142. T. S. Chow, *Mesoscopic Physics of Complex Materials*, Springer-Verlag, New York, 2000, p. 90.
143. A. A. Maradudin and A. E. Fein, *Phys. Rev.* **128**, 2589 (1962).
144. A. Roitberg, R. B. Gerber, R. Elber, and M. A. Ratner, *Science* **268**, 1319 (1995); A. Roitberg, R. B. Gerber, and M. A. Ratner, *J. Phys. Chem. B* **101**, 1700 (1997).
145. J. Fabian, *Phys. Rev. B* **55**, R3328 (1997).
146. J. Fabian and P. B. Allen, *Phys. Rev. Lett.* **79**, 1885 (1997).
147. D. M. Leitner, *Phys. Rev. B* **64**, 094201 (2001).
148. D. M. Leitner, *J. Phys. Chem. A* **106**, 10870 (2002).
149. D. M. Leitner, *Phys. Rev. Lett* **87**, 188102 (2001).
150. K. Moritsugu, O. Miyashita, and A. Kidera, *Phys. Rev. Lett.* **85**, 3970 (2000).
151. K. Moritsugu, O. Miyashita, and A. Kidera, *J. Phys. Chem. B* **107**, 3309 (2003).
152. T. Q. Lian, B. Locke, Y. Kholodenko, and R. M. Hochstrasser, *J. Phys. Chem.* **98**, 11648 (1994).
153. M. Tesch and K. Schulten, *Chem. Phys. Lett.* **169**, 97 (1990)
154. P.-H. Yang and J. A. Rupley, *Biochemistry* **18**, 2654 (1979).
155. J. A. Rupley and G. Careri, *Adv. Prot. Chem.* **41**, 37 (1991).
156. X. Yu, J. Park, and D. M. Leitner, *J. Phys. Chem. B* **107**, 12820 (2003).
157. B. C. Dian, A. Longarte, and T. S. Zwier, *Science* **296**, 2369 (2002).
158. D. J. Wales and I. Ohmine, *J. Chem. Phys.* **98**, 7245 (1993).

159. A. Pohorille, L. R. Pratt, R. A. LaViolette, M. A. Wilson, and R. D. MacElroy, *J. Chem. Phys.* **87**, 6070 (1987).
160. K. Hamad-Schifferli, et al., *Nature* **415**, 152 (2002).
161. B. Halle, *Proc. Natl. Acad. Sci. (USA)* **101**, 4793 (2004).
162. R. Burioni, et al., *Proteins: Struct, Funct and Bioinformatics* **55**, 529 (2004).
163. X. Yu and D. M. Leitner, *J. Chem. Phys.* (in press).
164. X. Yu and D. M. Leitner, *Chem. Phys. Lett.* **398**, 480 (2004).

CHAPTER 17

REGULARITY IN CHAOTIC TRANSITIONS ON MULTIBASIN LANDSCAPES

TAMIKI KOMATSUZAKI, KYOKO HOSHINO, and YASUHIRO MATSUNAGA

Nonlinear Science Laboratory, Department of Earth and Planetary Sciences, Faculty of Science, Kobe University, Nada, Kobe, 657-8501, Japan

CONTENTS

I. Introduction
II. Minimalistic 46-Bead Protein Models
III. Nonstationarity of Energy Fluctuations on Protein Landscapes
IV. State-Space Structure of Multibasin Dynamics
   A. Embedding: Phase-Space Reconstruction
   B. False Nearest Neighbors
   C. Average Mutual Information
   D. Temperature Dependency in Dimensionality of Folding Dynamics
V. Concluding Remarks and Future Prospects
VI. Appendix
   A. Embedology
   B. Berendsen Algorithm for Constant-Temperature MD Simulation
Acknowledgments
References

I. INTRODUCTION

Any complexity observed for dynamics and kinetics of clusters, liquids, glasses, and biomolecules results from the time evolution of systems involving multiple transitions through saddles (not necessarily solely first-rank saddles but also higher-rank saddles) on rugged multidimensional multibasin potential energy landscapes. The "energy landscape" perspective holds great promise for

---

Geometric Structures of Phase Space in Multidimensional Chaos: A Special Volume of Advances in Chemical Physics, Part B, Volume 130, edited by M. Toda, T Komatsuzaki, T. Konishi, R.S. Berry, and S.A. Rice. Series editor Stuart A. Rice.
ISBN 0-471-71157-8 Copyright © 2005 John Wiley & Sons, Inc.

resolving the most important contemporary problems on kinetics and thermodynamics of clusters, liquids, glasses, and biomolecules [1]. In this chapter, we instead focus on establishing the description of the dynamical events that occur on the multidimensional rugged, multibasin energy landscapes. The elementary components of transitions on (rugged) multibasin potential energy surfaces (PESs) are passages from one local minimum to one another across a saddle— that is, (conventional) chemical reactions. The introduction of the concept of transition state (TS) in the 1930s [2–4] had great successes in enhancing the understanding of the kinetics of chemical reactions. The transition state is defined as a *hypersurface* decomposing the space into two distinct regions, reactant and product. A central assumption in a common class of chemical reaction theories [2–6] is that there exists a quasi-equilibrium between the reactant and a system crossing the TS from the reactant to the product. In the other words, the reacting system is assumed to attain a local equilibration within the reactant well more quickly than it finds a passage to the TS. Moreover, in all forms of the TS theories, it is assumed that any reactive trajectories should pass only once through the TS—that is, dividing hypersurface (generally in phase space) on the way from reactants to products before being "captured" in the products. This is known as the "no return" assumption. The concept of TS gave us a magnifying glass to decompose the evolution of the reactions into, first, how a reacting system reaches there from the reactant state by getting a certain amount of thermal energy and, second, how the system leaves there after its arrival—for example, its passage velocity and pattern of crossings. People have interpreted the physical (classical) origin of the deviation of the experimentally observed kinetics from the theoretical estimations (as represented by the so-called transmission coefficient $\kappa(<1)$) in terms of the violation(s) of either/both the assumptions of the local equilibrium (in the region of reactants) and the no-return (in the region of saddles).

Recall that typical reactions involving chemical bond formation and/or cleavage may have, as a typical activation energy, a few tens of kilocalories per mole, while an average energy associated with a single degree of freedom (DOF) is about 0.6 kcal/mol at room temperature. We may anticipate that such chemical reactions are regarded as very rare events, and most (strongly chaotic) reacting systems likely move through all the accessible phase space in the reactant basin before finding the elusive transition state that is localized somewhere in the vicinity of a first-rank saddle. However, if the energy barrier of the reacting system becomes comparable with the average energy of a single DOF, such a common scenario implicitly assumed so far should no longer be valid. There is no guarantee that any "component" of global reaction dynamics on multibasin potential landscapes (i.e., two-basin chemical reactions) persists in such conventional situations to enable us to begin with the local equilibrium assumption.

Chemists have long envisioned global reactions on multibasin energy landscapes to be decomposable into a sequence of two-basin chemical reactions across (first-rank) saddles linking the successive pairs of potential basins so that each reaction takes place in a dynamically independent fashion; that is, local equilibrium is postulated to be attained quickly in each basin before the system goes to the next so the system loses all dynamical memories. However, it is natural to expect that multibasin energy landscapes associated with a large exothermicity tend to bring about nonstatistical behavior even in a deep potential basin. Hase and co-workers [7] explored this nonstatistical feature of a gas-phase reaction $OH^- + CH_3F \rightarrow CH_3OH + F^-$ by *ab initio* direct dynamics. The potential energy surface is composed of three basins associated with a large exothermicity, and one of the basins is a deep minimum in the product exit channel arising from the $CH_3OH \cdots F^-$ hydrogen-bonding complex. They found that the majority of the trajectories (90%) avoid being trapped in this potential energy minimum and dissociate directly to products. Such dynamical, nonstatistical behavior is beyond the availability of conventional, statistical reaction rate theories. Moreover, at high energies above the lowest, presumably (but not necessarily) first-rank saddle, the system may pass over higher-rank saddles on the PES. This provides us with a new, untouched, fundamental problem that should be inherent to all the relaxation dynamics on a multibasin PES, especially with a large exothermicity: a situation in which the system finds higher-rank saddles densely distributed in the regions of high potential energies and would pass through such regions at least as frequently as through the lowest, first-rank saddles [8,9]. This will require going back to the fundamental question of what the transition state is—that is, whether a dividing hypersurface could still exist or be definable as separating the space of the system into the distinct regions identifiable with the individual stable states. Let us first remind what most chemical reaction rate theories implicitly assume:

1. A reacting system moves on a single PES or a single "free-energy" surface from the reactant state, passing across a single *first-rank* saddle, to the product state. In other terms, the system never passes through any higher-rank saddle along the dynamical evolution of reactions (or it's simply negligible even if they may exist).
2. The validity of the transition state concept that separates the phase space into the two distinct regions has not been clarified in cases in which the system passes through higher-rank saddles at least as frequently as through the lowest, first-rank, saddle.
3. The sequence of passages across the (first-rank) saddles linking the successive distinct basins on multibasin energy surfaces is dynamically independent; that is, local equilibrium is attained quickly in each basin before the system goes to the next so the system loses all dynamical memories.

4. The reaction bottleneck or the transition state is located somewhere in the vicinity of the first-rank saddle, and is not delocalized through the whole region of configurational (or phase) space. [Note, however, that a delocalized, rigorous transition state certainly exists for some situations (e.g., Refs. 10–12).]

5. The reaction coordinate for describing chemical reactions has been defined globally, and the reactive or nonreactive nature associated with each DOF is invariant in time along the evolution of those reactions. For reaction or relaxation dynamics on multibasin landscapes composed of a huge number of sequential two-basin chemical reactions, a reactive DOF defined locally for describing the passage through one basin to another should not persist its nature as reaction coordinate and change to become one of the bath DOFs for another transition across another saddle. In other words, "What is the physical condition that enables us to describe the dynamical evolution in terms of a single, global reaction coordinate?" is one of the nontrivial untouched questions to be resolved.

How can we extract a set of best reaction coordinate(s) from the configuration space (or in general the state space) to reveal the complexity of multibasin transitions, involving the possibility to persist *dynamical memory* among some sequential saddle crossings? Although there is no general means to overcome this problem yet, one may be able to address the following two strategies.

1. Geometrical Structure of the Phase Space Transport for Multibasin Transitions. If nonstatistical, dynamical correlation exists through sequential transitions across the distinct saddles, it must be associated with some approximate, invariant structure buried in the state space. The well-known, invariant sets that persist robustly under any perturbation in chaotic dynamical systems are normally hyperbolic invariant manifolds (NHIM), regarded as the generalization of "saddle" in the state space, and the stable and unstable invariant "cylindrical" manifolds associated with the "saddle." Recent developments in chemical reaction dynamics enable us to calculate such (approximate) invariant sets in the state space for many DOFs systems [13–20]. One can identify each invariant cylinder associated with the first-rank (and higher-rank if necessary) saddle linking the distinct energy minima (or lower-rank saddle). The intersections between the stable cylinder approaching one NHIM (associated with one configurational saddle) and the unstable cylinder departing from another NHIM (with another configurational saddle) can tell how a bundle of trajectories passing through each distinct saddle is dynamically connected. In the case of gas-phase multibasin reactions, to elucidate the mechanism of the intersection provides us with a clue to establish the control of sequential chemical reactions. It is expected that there exists the strong dependency of the initial condition "within the NHIM" to bring the system either to another NHIM or back to the same NHIM. That is, the *skeleton* composed of the invariant sets sheds light on not only the origin of dynamical correlations but also the multi-transition pathways on the state space [21].

2. The Postulation of a Certain Global or Local Collective Coordinate(s). For proteins with rugged, multibasin PES, the so-called principal component (PC) analysis [22–25], which determines a set of linear, collective coordinates to best represent most fluctuations of the system, has often been studied in understanding cooperative behavior of protein systems.

Suppose we have n data points associated with k variables—in our case, n instantaneous structures of $3N$ Cartesian coordinates of N particles along a molecular dynamics (MD) trajectory. Let **D** be a $3N \times n$ matrix, whose elements D_{im} are defined as the deviation of the (mass-weighted) ith Cartesian coordinate $q_i(t_m)$ at a time $t_m (1 \leq m \leq n)$ from the time average $\langle q_i \rangle$, that is,

$$D_{im} = q_i(t_m) - \langle q_i \rangle \tag{1}$$

where

$$\langle q_i \rangle = \frac{1}{n} \sum_{m=1}^{n} q_i(t_m) \tag{2}$$

Then, using a $3N \times 3N$ variance–covariance matrix **R**,

$$\mathbf{R} = \frac{1}{n} \mathbf{D} \mathbf{D}^\mathrm{T} \tag{3}$$

(where $\mathbf{D}^\mathrm{T}$ is the transpose matrix of **D**), whose second moment element R_{ij} is

$$R_{ij} = \frac{1}{n} \sum_{m=1}^{n} (q_i(t_m) - \langle q_i \rangle)(q_j(t_m) - \langle q_j \rangle) \tag{4}$$

we can define a set of principal components **Q** by using the eigenvectors **U** that diagonalize **R**:

$$\mathbf{R}\mathbf{U} = \mathbf{U}\mathbf{r} \, (\mathbf{U}^\mathrm{T}\mathbf{U} = \mathbf{I}) \tag{5}$$

The eigenvalue r_i, the ith element of the diagonal matrix **r**, represents the variance of the ith collective coordinate Q_i,

$$Q_i = \sum_{j=1}^{3N} U_{ji} q_j \tag{6}$$

The larger the r_i, the more Q_i represents the configurational protein fluctuations. The principal components **Q** are sorted in order of decreasing variance, $r_1 \geq r_2, \ldots, \geq r_{3N}$.

The PC analysis (PCA) originally developed in the statistical problem can provide us with not only collective coordinates but also the conjugate momenta for any microcanonical system: Let a set of the original coordinates and the conjugate momenta be $(\mathbf{p}, \mathbf{q})$ and the new collective coordinates $\mathbf{Q}(Q_i = \sum_j U_{ji}q_j)$ and the conjugate momenta **P**. By assuming a generating function $F(\mathbf{q}, \mathbf{P})$, we obtain

$$F = \sum_{i,j} U_{ji} q_j P_i \quad (7)$$

$$\frac{\partial F}{\partial P_i} = Q_i = \sum_j U_{ji} q_j \quad (8)$$

$$\frac{\partial F}{\partial q_i} = p_i = \sum_j U_{ij} P_j \quad (9)$$

$$\mathbf{p} = \mathbf{UP} \quad (10)$$

Therefore, the new momenta **P** can be written as

$$\mathbf{P} = \mathbf{U}^{-1}\mathbf{p} = \mathbf{U}^{-1}\mathbf{p} \quad (11)$$

$$P_i = \sum_{j=1}^{3N} U_{ji} p_j \quad (12)$$

A complementary approach [26,27] to replace Eq. (3) by an $n \times n$ matrix, $\mathbf{R} = \frac{1}{3N}\mathbf{D}^T\mathbf{D}$, has also been studied by Becker and Karplus [28], and Elmaci and Berry [29] attempted to visualize the complexity of the multidimensional energy landscapes of proteins.

Here we overview some theoretical studies using an MD simulation of complex dynamics occurring on a multibasin rugged PES;

Liquid Water. Ohmine and his coworkers [30,31] have extensively explored the hydrogen-bond rearrangement dynamics in liquid water. The PES for a system with many particles consists of an enormous number of potential wells. They surveyed a set of minimum-energy configurations in the individual potential wells, referred to as inherent structures [32,33], that can be obtained by the sequential quenching of the total system along MD trajectories. The configurational change of a system can be separated into the transition among the inherent structures and the vibrational motions around the inherent structures. Figure 1 depicts water molecule's displacements in a transition

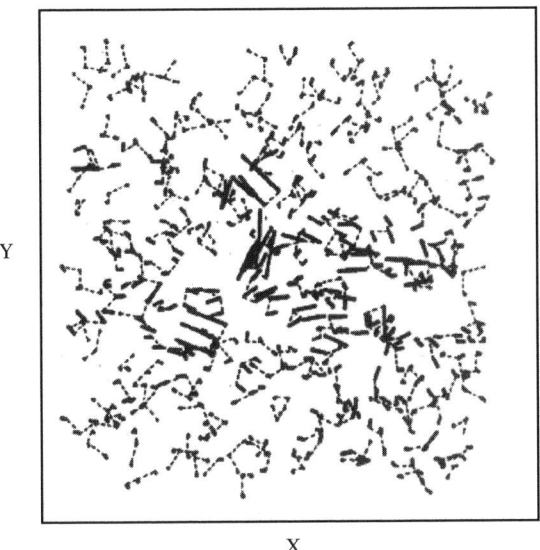

Y

X

Figure 1. Collective motions from a "basin" to "basin" passage in liquid water. [Reprinted with permission from I. Ohmine and H. Tanaka, *Chemical Reviews* **93**, 2545 (1993). Copyright © 1993, American Chemical Society.]

from an inherent structure to the next one along an MD trajectory for 216 water molecules at temperature 298 K. The heavy solid lines indicate the displacements of individual atoms of H_2O, and the dotted lines indicate the OH bonds before the transition. They found that the transitions between each inherent structure occur intermittently, associated with $1/f$-type potential energy fluctuation [34]. It was also shown that by removing the vibrational components of water molecules by averaging the system for a short period, of a time scale of librational motion, $\sim 10^{-12}$ s, the transitions among the potential basins are regarded not as occurring diffusively but as in a cooperative dynamical fashion, associated with collective molecular motions. We believe this indicates that, as shown in Ar_6 isomerization dynamics [13–20], passages of the system from one basin to another tend to be regularized to yield cooperative motions after washing out (coarse graining) the vibrational components even for the rugged PES.

Proteins. The questions, "What kinds of mechanisms carry a protein into an unique native state?" and "What is the best reaction coordinate to describe the dynamics of protein folding?" have been among the most intriguing subjects over the past decades in biological physics [35,36]. The process of protein folding may be interpreted as a normal Brownian process of a few collective coordinates on a thermodynamic energy landscape like a funnel [37–39]. The diffusive nature of dynamics, however, depends on the choice of the viewpoint from which one sees the events (e.g., Refs. 40 and 41). The fraction of native contacts Q is often taken as a reaction coordinate or global order parameter. However, it is not self-evident, as discussed by Karplus [36], that Q is always appropriate to represent the progress of folding. That is, many different sets of contacts may yield the same value of Q and there exists the nontrivial question, "Are the motions along that coordinate sufficiently slow enough to average out all the dynamical contributions of all the other DOFs, resulting in an effective single dominant free energy barrier of folding?"

The multiplicity of basins and ruggedness on the free-energy profile of proteins has simply been regarded as retarding a system to fold to the native state, in which the kinetics is controlled by escape rates from different low-lying energy traps, yielding nonexponential folding kinetics (e.g., Ref. 42). However, it has been recently pointed out both theoretically [43–45] and experimentally [46] that the existence of ruggedness or an intermediate state induced from nonnative contacts can instead accelerate the folding kinetics, but the sufficient interpretation has not been presented for this kinetic acceleration effect of the ruggedness [47].

Note that it has been assumed implicitly for most kinds of chemical reactions involving multibasin transitions that, just after passing through a saddle, the system's energy will dissipate quickly to the surrounding thermal bath, compared with the time scale of the individual passage across the saddle, and local equilibrium can be attained in each basin *irrespective of the height of barrier and the contributions from higher-rank saddles*. There exists no general answer to which circumstances the non-statistical behavior of dynamics is suppressed or becomes negligible.

García and Hummer [40] showed how non-Brownian, strange kinetics [48–51] emerges in multibasin trajectories generated by all-atom MD simulations of cytochrome c in aqueous solution at 300–550 K for at least 1.5 ns. They analyzed the mean square displacement (MSD) autocorrelation function in terms of molecule optimal dynamic coordinates (MODCs) that are essentially the same as PC. They found that the MSD along some slow MODCs manifestly exhibits non-Brownian dynamics between a temperature at which the protein is in the native state and a temperature above melting where the hydrophobic effect is large and mostly enthalpic, having a power law time dependence with an exponent of nonunity: about 0.5 for times shorter than 100 ps and 1.75 for longer times (see Fig. 2 at 300 K). They elucidated that the transitions from one basin to another (I–VII in the figure) occur on a time scale of 100 ps on average. This implies that protein motions are more suppressed and cover less configurational space than a normal Brownian process on a short time scale while the proteins move about inside a basin, but they become more enhanced

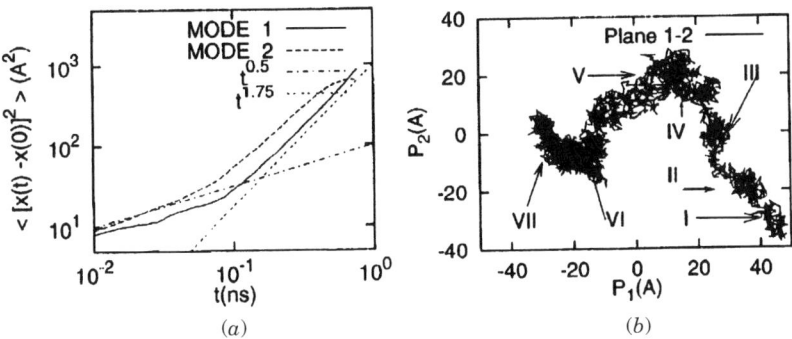

Figure 2. Strange kinetics for cyt c at $T = 300K$: (a) a log–log plot of the MSD autocorrelation function for the best two MODCs and two simple power law curves with exponents 0.5 and 1.75. (b) the projection on the principal place of the best two MODCs. The transitions from one (large) basin to another (I–VII) occur on a time scale of 0.1 ns on average. [From A. E. García and G. Hummer, *Proteins* **36**, 175 (1999). Copyright © 1999, Wiley-Liss.] This material is used by permission of Wiley-Liss, a subsidiary of John Wiley & Sons.

as a ballistic concerted motion on a longer time scale, associated with transitions among basins. It was also found that such strange, dynamical behavior also exists at 360 K and 430 K where transitions occur among (super) basins, a family of small basins (e.g., basins I–VII at 300 K). At a temperature where unfolding occurs in ∼ns, normal Brownian dynamics emerges along the MODCs. Carlini et al. [52] also found such suppressed diffusions along some slow MODCs at 300 K in plastocyanin from 1 ps to 200–400 ps for which the probability of the potential energy around its average value is very satisfactorily described by a Gaussian, indicating that the system resides within a single basin during such durations.

One of the plausible interpretations we could deduce from their findings is that, as shown in Ar_6 isomerization dynamics [13–20] and intermittent, collective motions of water molecules among the inherent structures in liquid water [30,31], transitions from a basin to one another occur cooperatively as dynamical processes if they occur through a "bottleneck" among "(super) basins," a family of an enormous number of small basins through which the system can frequently go without any significant barriers.

Most observed kinetics of multibasin protein dynamics is a consequence of averaging over an ensemble of many activated barrier crossings with multiple time scales. The direct observation [53–55] of dynamical behavior of a single molecule, so far buried in an ensemble average, should provide us with a new magnifying glass, which enables us to explicitly "see" the dynamical behavior inherent to the composite molecules in complex systems.

In the following sections, we present our recent studies on multibasin protein folding dynamics on minimalistic protein landscapes. Here, we focus on the question of how the time series of *scalar* quantities can shed light on the underlying configurational energy landscapes and geometry of the state space [41].

II. MINIMALISTIC 46-BEAD PROTEIN MODELS

We use a coarse-grained, off-lattice "three-color, 46-bead" protein model [29,56,57] as an illustrative vehicle of multibasin transitions [41]. This model is composed of hydrophilic (L), hydrophobic (B), and neutral (N) beads, interacting with the following potential:

$$V = V_r + V_\theta + V_\Phi + V_R \tag{13}$$

$$= \sum_i^{\text{bonds}} K_r (r_i - r_0^i)^2 \tag{14}$$

$$+ \sum_i^{\text{angles}} K_\theta (\theta_i - \theta_0^i)^2 \tag{15}$$

$$+ \sum_i^{\text{torsional}} [A(1 + \cos \Phi_i) + B(1 + \cos 3\Phi_i)] \tag{16}$$

$$+ \sum_{i<j-3}^{\text{nonbonding pairs}} 4\epsilon S_1 \left[\left(\frac{\sigma}{R_{ij}}\right)^{12} - S_2 \left(\frac{\sigma}{R_{ij}}\right)^6 \right] \tag{17}$$

where the van der Waals (vdW) interactions are used to mimic the hydrophilic, hydrophobic, and neutral characters of the beads: $S_1 = S_2 = 1$ for BB (attractive) interactions, $S_1 = 2/3$ and $S_2 = -1$ for LL and LB (repulsive) interactions, and $S_1 = 1$ and $S_2 = 0$ for all the other pairs involving N, expressing only excluded volume interactions. For the bond stretching and angle bending force constants we use $K_r = 231, 2\epsilon\sigma^{-2}$ and $K_\theta = 20\epsilon(\text{rad})^{-2}$, with the equilibrium bond length $r_0^i = \sigma$ and the equilibrium bond angle $\theta_0^i = 1.8326\,\text{rad}$. The units of energy, temperature, the bead mass, time, and frequency are ϵ, ϵ/k_B, M, $t^* = \sigma\sqrt{M/\epsilon}$, and $1/t^*$, unless otherwise noted. The sequence $B_9N_3(LB)_4N_3B_9N_3(LB)_5L$ folds into a lowest energy β-barrel structure with four strands, as shown in Fig. 3 [58,59]. Folding into this structure is ensured by setting up the torsional potential so that there are stiff *trans* preferences in the four strands, while at the loop regions consisting of neutral beads the torsional potential has a small barrier with no preference for any of the three rotameric states. In particular, $A = B = 1.2\epsilon$, except for torsional angles involving two or more neutral residues where $A = 0$

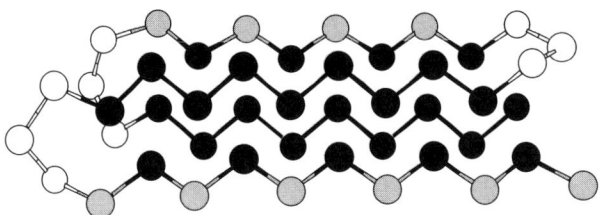

Figure 3. The global minimum structure of a 46-bead model: black, gray, and white circle means hydrophobic, hydrophilic, and neutral bead, respectively.

and $B = 0.2\epsilon$. The rigid bonds of the original model by Honeycutt and Thirumalai [58,59] are replaced with stiff but harmonic, spring-like bonds [56]. Here we employed a constant-temperature MD simulation algorithm developed by Berendsen et al. [60] with a time step $0.0025t^*$ and a coupling time of $5t^*$. This method does not involve any explicit stochastic variable in the equation of motion, and it can control the temperature well with minimal local disturbance to the system (see also Section VI.B, Appendix). The simulation at each temperature is preceded by a 10^5-step equilibration starting from the final conformation at the higher temperature. At each temperature, the trajectory data are collected at every 100 steps for up to 10^7 steps ($= 25000t^*$). A wide range of temperatures was studied, ranging from 0.2 to 5.0, where $T = 0.72$ is regarded as the "collapse" temperature of this model and $T = 0.6$ as the "folding" temperature of the Gō-like model.

The topography of the energy landscape of this 46-bead model, referred as to the BLN model hereinafter, has been surveyed in terms of the so-called disconnectivity graph [1,57]. The disconnectivity graph analysis focuses on the sequence structures of local minima and the (first-rank) saddle point linking them. As shown in Fig. 4, at a given discrete series of total energy $E_1 < E_2 < E_3 < \cdots$, the minima can be classified into disjoint sets, termed *superbasin* [28], whose members are mutually accessible, connected by pathways where the energy never exceeds E_i. Each superbasin is represented by a node, and lines are depicted between the nodes at energies E_i and E_{i+1} if they are the same superbasin or superbasins that become accessible to each other at the higher energy E_{i+1}. As shown in the disconnectivity graph of the low-energy regions of the BLN model in Fig. 5a, the PES exhibits not an ideal "funnel-like," but a rather frustrated, potential energy topography [42,57]. Figure 5b shows the disconnectivity graph of the low-energy regions of a less-frustrated Gō-like BLN model [42,57], in which only native contact pairs of hydrophobic beads separated by less than 1.167σ in the global minimum possess attractive vdW interactions, while the interactions between all the other pairs are repulsive, responsible for excluded volume. One can see that the Gō-like BLN

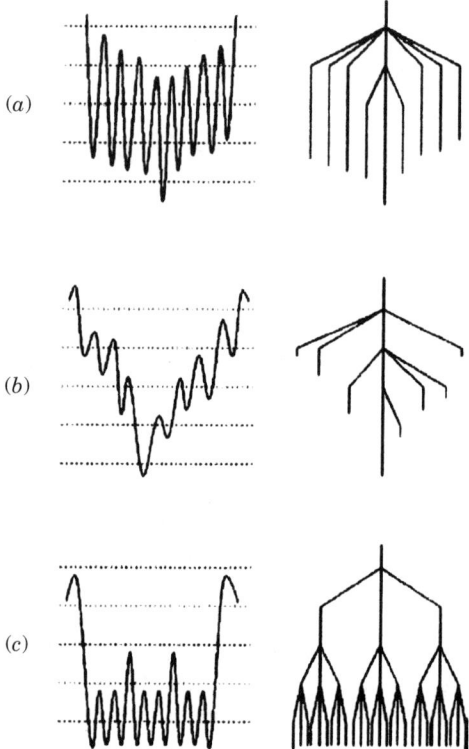

Figure 4. Schematic examples of PESs and the corresponding disconnectivity graphs. The dotted lines indicate the energies at which the superbasin analysis was performed. (a) A gently sloping funnel with high barriers ("willow tree"), (b) a steeper funnel with lower barriers ("palm tree"), and (c) a "rough" landscape where the barrier heights are larger than the typical energy difference between successive minima ("banyan tree"). [Reprinted with permission from M. Miller and D. J. Wales, *J. Chem. Phys.* **111**, 6610 (1999). Copyright © 1999, American Institute of Physics.]

model exhibits an ideal funnel-like topography in which low-lying energy minima are not separated from the global minimum by large barriers. Here, the density of minima per unit energy is larger for the BLN model than for the Gō-like BLN model, resulting in the fact that the highest-energy minima among the sampled configurations is still compact for the BLN model while the β-barrel strands at the highest-energy minima is considerably unfolded for the Gō-like BLN model.

Note that information about density of states associated with either the local minima or the saddles are not included in the disconnectivity graph, and the graph generally depends on the choice of the spacing between the energies at which the superbasin analysis is performed. There does not exist any meaning about "one dimensional" horizontal axis.

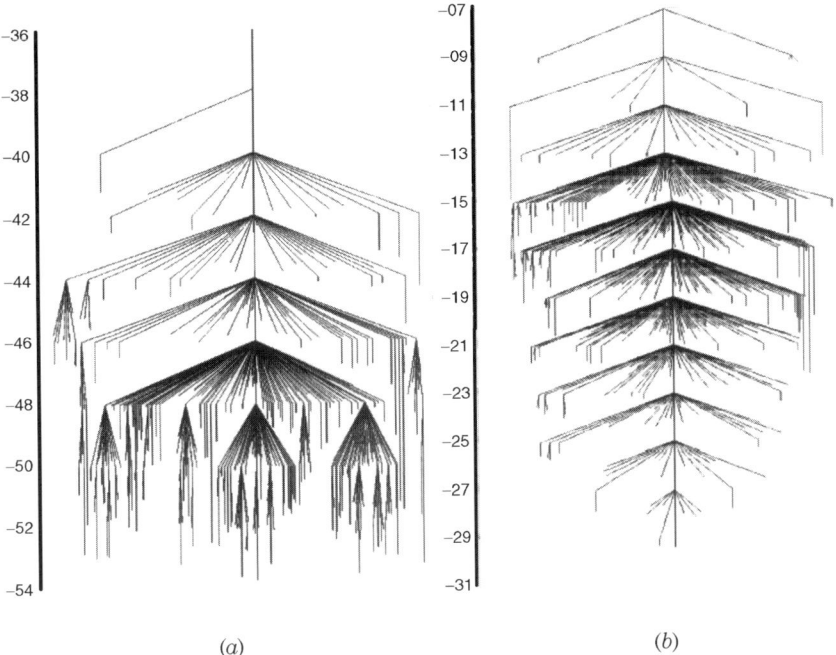

Figure 5. Disconnectivity graphs for (a) BLN model in terms of a sample of 500 minima and 636 saddles, and for (b) Gō-like BLN model a sample of 500 minima and 805 saddles. The energy is in the units of ϵ. [Reprinted with permission from M. Miller and D. J. Wales, *J. Chem. Phys.* **111**, 6610 (1999). Copyright © 1999, American Institute of Physics.]

Nymeyer, García, and Onuchic [42] investigated how the energy landscape of these protein models affects their folding kinetics. They found that the folding rate on the minimally frustrated "funnel" landscape (Gō-like BLN model) exhibits single exponential behavior at the folding temperature T_f and nonexponential kinetics does not emerge until the temperature is much lower than T_f, while that of a highly frustrated BLN landscape starts to deviate from exponential behavior at just below the collapse temperature T_c, at which the kinetics is controlled by escape rates from different low-lying energy traps. It was also found that the Gō-like model manifestly exhibits a two-state-like transition about the folding temperature, because of its less-frustrated energy landscape. The observed distinction in the folding kinetics is consistent with the different topographical feature of these energy landscapes observed in the disconnectivity graphs. Thus, one can discuss how the degrees of funnel affect the global multibasin dynamics and how the distinctive feature of multibasin dynamics can be detected from their single-molecule time series.

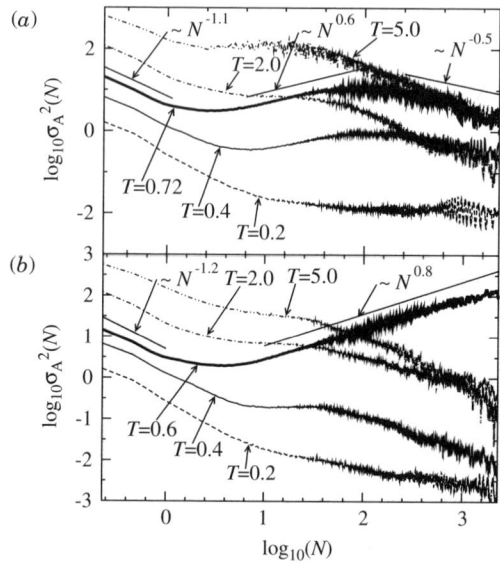

Figure 6. The Allan variance of potential energy fluctuations. (*a*) The original BLN model and (*b*) the Gō-like BLN model.

III. NONSTATIONARITY OF ENERGY FLUCTUATIONS ON PROTEIN LANDSCAPES

Figure 6 shows the so-called Allan variance [61–63] of time series of the potential energies $V(t)$ for these models at several temperatures. The Allan variance $\sigma_A^2(N)$ is defined by

$$\sigma_A^2(N) = \frac{1}{2}\left\langle \left(\frac{1}{N}\sum_{i=1}^{N} s(i) - \frac{1}{N}\sum_{i=1}^{N} s(i+N)\right)^2 \right\rangle \qquad (18)$$

and measures the degree of nonstationarity of a given time series $s(t)$: If $s(t)$ is stationary, the following scaling relation should be satisfied obeying the law of large numbers,

$$\sigma_A^2(N) \sim N^{-\gamma}(\gamma : \text{a positive constant}) \qquad (19)$$

which guarantees the validity of the Wiener–Khinchin theorem. If $s(t)$ is nonstationary, then $\gamma < 0$. These figures demonstrate that, for both the BLN and Gō-like models, significant nonstationary features emerge about the collapse and

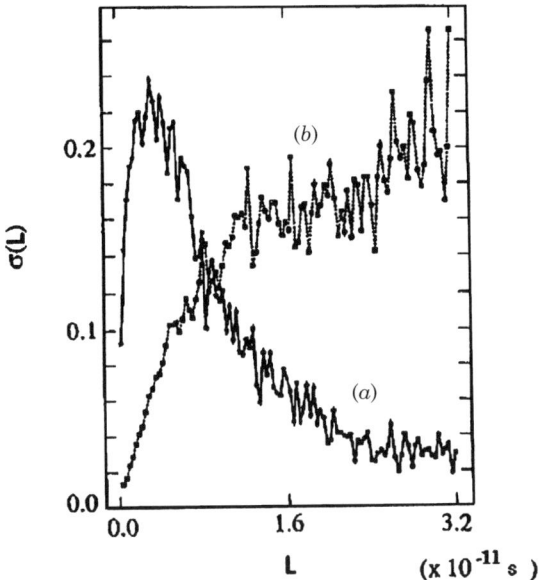

Figure 7. The Allan variances for time series of a geometrical parameter of Ar_7 for the cases of (*a*) the liquid-like state and (*b*) coexistence region where the frequent transitions between solid-like and liquid-like states are observed. The Allan variance for the solid-like state is qualitatively the same aside from the magnitude. [Reprinted with permission from C. Seko and K. Takatsuka, *J. Chem. Phys.* **104**, 8613 (1996). Copyright © 1996, American Institute of Physics.]

folding temperatures in $V(t)$, which diminish departing from these transition temperatures. The emergence of the nonstationary feature is more pronounced in the Gō-like model than in the BLN model. In both models at these transition temperatures, $\sigma_A^2(N)$ obeys the simple scaling relation $\sim N^{-\gamma}$ for $10^{-0.64}t^* \leq t \leq 10^0 t^* (= 1.6t^* \simeq 10^3$ simulation steps) after which the index γ changes from positive [stationary regime: $\gamma \simeq 1.1$(BLN), $\gamma \simeq 1.2$(Gō-like)] to negative [nonstationary regime: $\gamma \simeq -0.6$(BLN), $\gamma \simeq -0.8$(Gō-like)]. The nonstationary regime turns to another stationary regime around $100t^* (\simeq 10^5$ steps) for the BLN model, while it even continues for more than 10^5 steps for the Gō-like model. By using the Allan variance, Seko and Takatsuka [62] found in Ar_7 isomerization (see Fig. 7) that a similar nonstationarity (i.e., $\gamma < 0$) also emerges at the transition temperature from solid-like to liquid-like phases, while the simple scaling relation $\sim N^{-\gamma}$ ($\gamma > 0$) holds at the other temperatures. However, at the Ar_7 transition temperature, there do not exist any transitions between stationary and nonstationary regimes equivalent to those observed in both protein models. This apparent distinction may arise from the generic feature inherent to the hierarchical energy landscapes in proteins: For a short time period at the

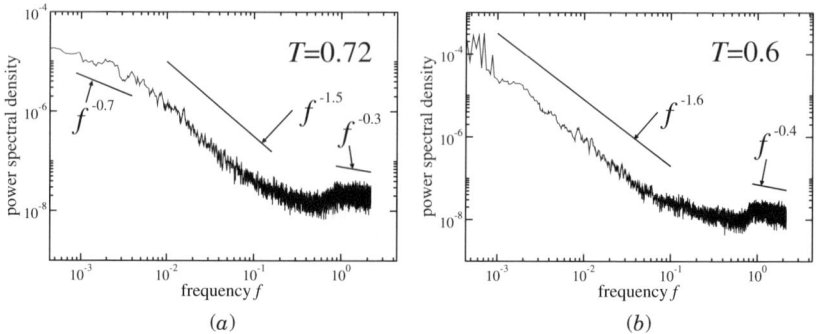

Figure 8. The power spectra of potential energy fluctuations at transition temperatures. (*a*) The original BLN model. (*b*) The Gō-like BLN model.

transition temperature, proteins move about only within an individual rugged basin, and they exhibit *local* stationary behavior due to the fast (chaotic) fluctuations. However, as the time goes by, the protein begins to cross among basins more frequently, bringing about slow large fluctuations.

Figure 8 depicts the power spectra $S(f)$ of the potential energy fluctuations of the original BLN and the Gō-like model at the collapse and folding temperatures.

$$S(f) = \frac{1}{T} \left| \int_0^T dt e^{2\pi i f t} V(t) \right|^2 \qquad (20)$$

Both models exhibit a $1/f$-noise like spectral density that can be expressed as $1/f^\alpha$, where $\alpha = 1.5$ for the BLN model and 1.6 for the Gō-like model. It is noteworthy that in the case of the Gō-like model this behavior persists over three decades of frequency extending into the low-frequency regions, while for the BLN model the $1/f$-like spectrum extends for only two decades in frequency with the crossover to white noise occurring at higher frequencies. This supports the conclusions from the analysis of the Allan variance that the nonstationarity of the Gō-like model is enhanced at the transition temperature.

Figure 9 presents exponents α for both models as a function of temperature. The Gō-like model exhibits a much sharper transition to a white noise spectrum than the BLN model as the temperature departs from the transition temperature. This finding also mirrors the results obtained from the Allan variance where the Gō-model exhibits a sharp transition to nonstationarity around the transition temperature, while this transition is diffuse in the case of the BLN model. It was also shown [64,65] in a two-state-like helix-coil transition of a helical polypeptide that a $1/f$-noise structure of the potential energy fluctuations

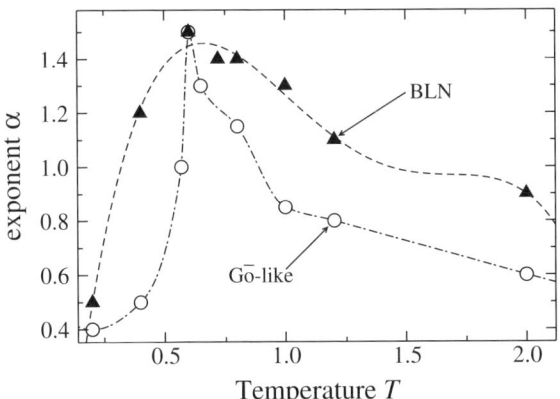

Figure 9. The power α of $1/f^\alpha$ as a function of temperature for the original BLN and the Gō-like BLN models.

emerges at the transition temperature while away from that temperature the power α of $1/f^\alpha$ goes to zero, indicating a transition to a Markovian process. The appearance of $1/f$ noise has been linked by many authors to the existence of multiple-relaxation processes or intermittency [30,64,66–68]. Marinari et al. [69] show that a $1/f$ noise also arises from a random walk on a random self-similar landscape. On the other hand, Brownian motion on a simple double-well surface would give $1/f^2$ noise. In order to clarify the origin of the observed $1/f$ structure, we computed the Allan variance and the power spectra by overdamped Langevin dynamics simulations [70] using the same length of the simulation time and the same potentials. For those Brownian dynamics, the exponents γ for $\sigma_A^2(N) \sim N^{-\gamma}$ and α for $S(f) \sim 1/f^\alpha$ of the potential energy fluctuations showed less nonstationarity than for our MD simulation involving the effect of inertia—that is, at the transition temperatures $\gamma \simeq -0.4$ (BLN) ~ -0.2(Gō-like) for the intermediate nonstationary regimes. (Here, for the first stationary regime at the short time scale $\gamma \sim 0.2$ for both the models.) α are insensitive to the temperature for both the models, $\alpha \sim 0.5 - 0.7$ at $T = 0.2 - 5.0$. This indicates that the emergence of a $1/f$-type spectrum is not due to Brownian motion on a self-similar, fractal landscape but arises from a different origin. It has been interpreted for the origin of $1/f$ that it arises from intermittency, that is, intermittent dynamics yields $1/f^2$ spectra when the jump time is close to zero, but when the jumps have a finite duration a smaller exponent $\alpha < 2$ is expected [30,67,68].

The duality between the power spectrum and the Allan variance has been intriguing in the context of nonstationary chaos by Aizawa and co-workers [63,67,68,71]. Nonstationary chaos reveals a strong long-range order, so that the

TABLE I
Aizawa's Relation Between the Index α in the Power Spectrum and the Index γ in the Allan Variance

Regime	Time $N[t^*]$	γ	α	$1 - \gamma$
Original BLN				
First stationary regime	$10^{-0.64}$–10^0	1.1	0.3	−0.1(+0.4)
Nonstationary regime	$10^{0.8}$–10^2	−0.6	1.5	1.6(−0.1)
Second stationary regime	$10^{2.4}$–$10^{3.4}$	0.5	0.7	0.5(+0.2)
Gō-like BLN				
First stationary regime	$10^{-0.64}$–10^0	1.2	0.4	−0.2(+0.6)
Nonstationary regime	10^1–10^3	−0.8	1.6	1.8(−0.2)

The numbers in the parentheses are $\alpha - (1 - \gamma)$. They should go to zero if the Aizawa's rule $\alpha = (1 - \gamma)$ holds exactly.

Liapunov exponent or Kolmogorov–Sinai entropy cannot be available to elucidate its mechanism. From a practical viewpoint, the Allan variance is one of the versatile means to characterize the feature of nonstationary chaos. Aizawa derived for the modified Bernoulli system the existence of a relation $\alpha = 1 - \gamma$ between the index α in the power spectrum $(S(f) \sim f^{-\alpha})$ and the index γ in the Allan variance $(\sigma_A^2(N) \sim N^{-\gamma})$ [71]. Although no mathematical foundation has not been established yet for a general class of dynamical systems, this relation holds approximately in many other systems—for example, a one-dimensional Lennard-Jones chain [72] and a helix-coil transition of a polypeptide [65]. In Table I we can also see that a similar trend exists approximately between γ and α for *each* stationary and nonstationary regime of potential fluctuations in the different time scales at the transition temperature in Figs. 6 and 8. Note, however, that at the short time scales for which the system is considered to move about inside the basins with a stationary fluctuation of the potential energy, the relation of $\alpha = 1 - \gamma$ holds only fairly well when compared with the system at the longer time scales. This might signal that the system's dynamics is not "uniformly chaotic," as seen in the modified Bernoulli system, within the basins on the multidimensional energy landscape. It was found recently [73] that, although the protein dynamics is uniformly chaotic in the microscopic level, there exists hierarchical structure in dynamics in which the system exhibits a regulatory behavior at the coarse-grained scale.

Nonstationarity of time series $s(t)$ implies that even if the "time average" of $s(t)$ would look converged for a certain time regime, it should become a totally different value by extending the time window for which the "time average" of $s(t)$ will be estimated. That is, rigorously, the concept of *time average* never

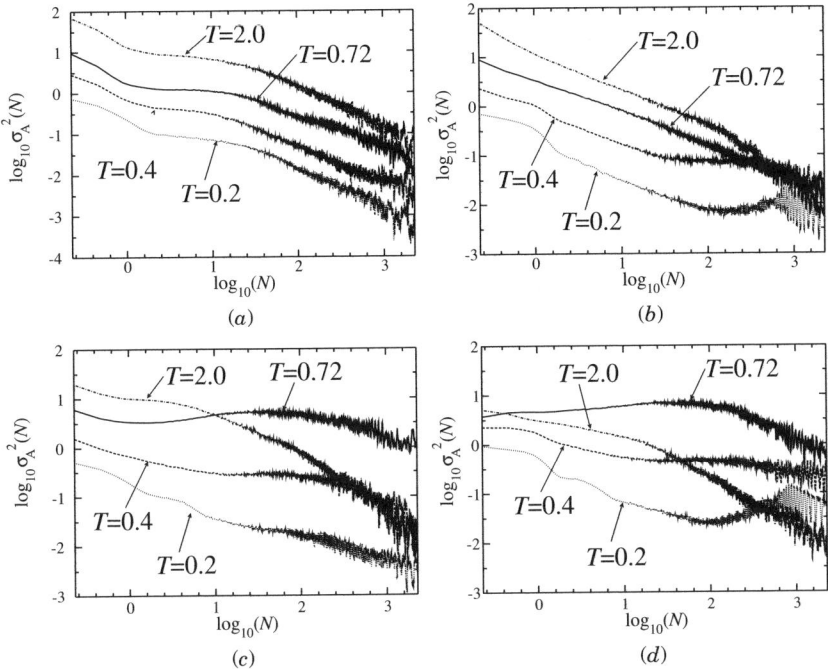

Figure 10. The Allan variances of the individual energy fluctuations of (*a*) bond energy V_r, (*b*) bending energy V_θ, (*c*) torsional angle energy V_Φ, and (*d*) nonbonded vdW interaction energy V_R at $T = 0.2, 0.4, 0.72,$ and 2.0 for the original BLN model.

holds and it can not be replaced by an ensemble average over the *whole* phase space. However, if one can identify stationary regime(s) in the time series $s(t)$, it is expected that the concept of "time average" locally holds in such time domain(s) where the concept of ergodicity can also exist *locally*.

In the following, we will use solely the Allan variance to further inquire into the nonstationarity of the multibasin transition processes on many-dimensional protein landscapes.

The nonstationarity observed in the potential energy fluctuations in Fig. 6 arises from the total contribution of all energy components and all residue beads. To shed light on the role of each energy component and DOFs on the nonstationary behavior in the multibasin transition processes, the Allan variances of the individual energy components for different DOFs are employed in the following. Figures 10 and 11 show the Allan variances of the individual fluctuations of each energy component—that is, bond energy V_r [Eq. (14)], bending energy V_θ [Eq. (15)], torsional angle energy V_Φ [Eq. (16)], and

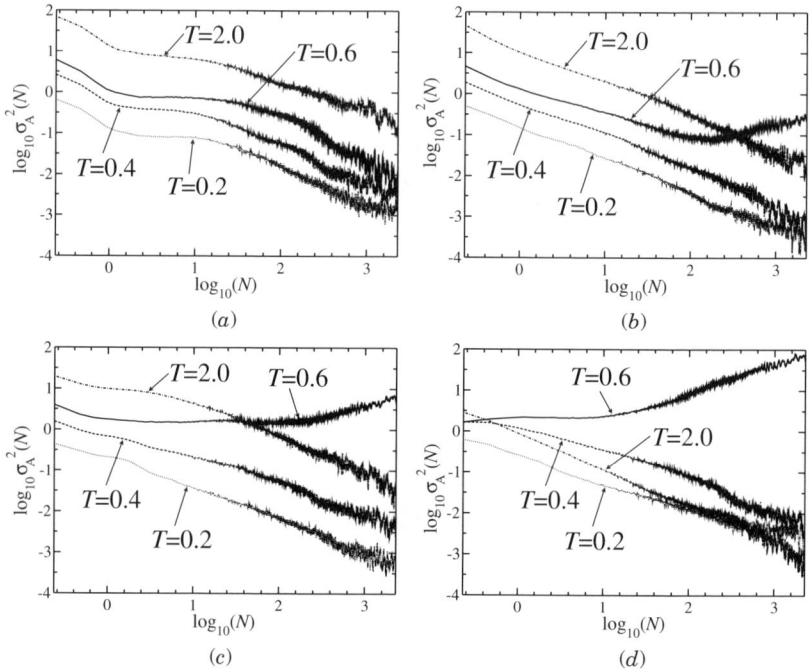

Figure 11. The Allan variances of the individual fluctuations of (a) bond energy V_r, (b) bending energy V_θ, (c) torsional angle energy V_Φ, and (d) nonbonded vdW interaction energy V_R at $T = 0.2, 0.4, 0.6$, and 2.0 for the Gō-like BLN model.

nonbonded vdW interaction energy V_R [Eq. (17)] at $T = 0.2 - 2.0$ for the original BLN and the Gō-like BLN models, respectively.

First, one can see for both the models that the nonstationary behavior observed in the total potential energy fluctuation at the transition temperature, $T = 0.72$ (BLN) and $T = 0.6$ (Gō-like BLN), mainly arises from the contributions of the torsional angle and nonbonded vdW interaction terms. There are no significant differences between the Allan variances of the bond energy fluctuations of these two distinct models. They can be regarded as almost stationary up to all simulated time scales except a "plateau" region for $10^{0.2}t^* \leq t \leq 10^{1.2}t^*$ enhanced at the transition temperature. Here a corresponding stationary phase observed in a short time regime, $10^{-0.6}t^* \leq t \leq 10^{0.2}t^*$, for the total potential energy fluctuations is not found in the nonbonded vdW interaction energy fluctuation at the transition temperature, while all the other energy components are associated with the stationary phase in that short time regime. At temperatures below the transition temperature, $T = 0.2$ and 0.4, as

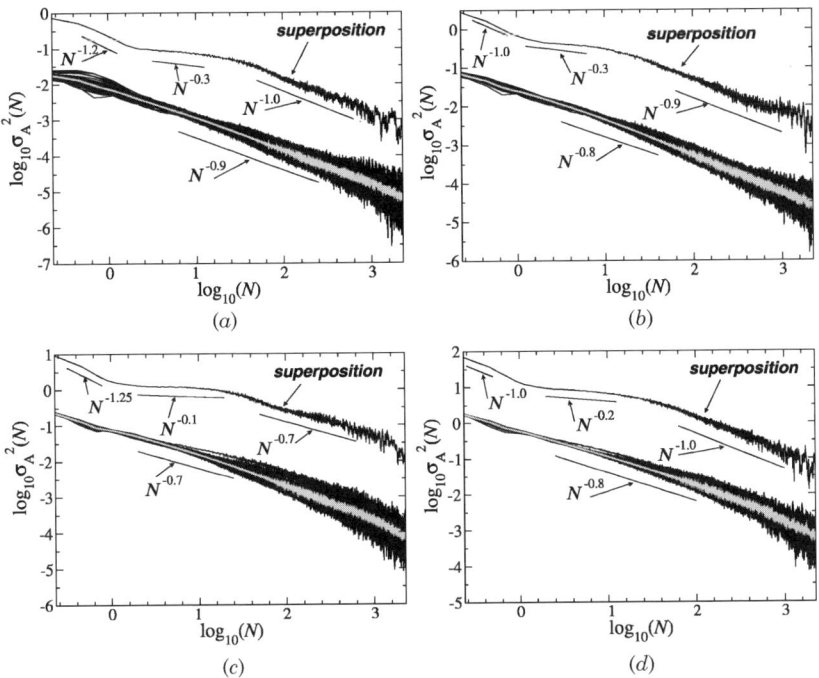

Figure 12. The Allan variances for $V_{r_i}(t)$ and the superposition $V_\mathbf{r}(t)(=\sum_{i=1}^{45} V_{r_i}(t))$ at (a) $T = 0.2$, (b) $T = 0.4$, (c) $T = 0.72$, and (d) $T = 2.0$ for the original BLN model. The gray line indicates the average of the Allan variances for $V_{r_i}(t)$.

expected from their disconnectivity graphs, the fluctuations of $V_\theta(t)$, $V_\Phi(t)$, and $V_\mathbf{R}(t)$ obey the law of large numbers *quite well* for the Gō-like BLN model, but those for the original BLN model don't because of the highly frustrated "glassy" nature of the underlying energy landscape. Note that for the original BLN model the fluctuation in the bending energy $V_\theta(t)$ tends to better obey the simple law of large numbers when temperature increases, although the $V_\theta(t)$ fluctuation for the Gō-like BLN model obeys the law of large numbers quite well at both above and below the folding temperature while the stationary regime turns to a nonstationary regime in the $V_\theta(t)$ fluctuation at $t \simeq 10^2 t^*$ at the folding temperature.

Bond Energy Fluctuation. Figures 12 and 13 show the Allan variances of each bond energy fluctuation $V_{r_i}(t)$ and their superpositions $V_\mathbf{r}(t)(=\sum V_{r_i}(t))$ at $T = 0.2, 0.4, 0.72$, and 2.0 for the original BLN and at $T = 0.2, 0.4, 0.6$, and 2.0 for the Gō-like BLN models, respectively. In these figures (and also Figs. 14–18),

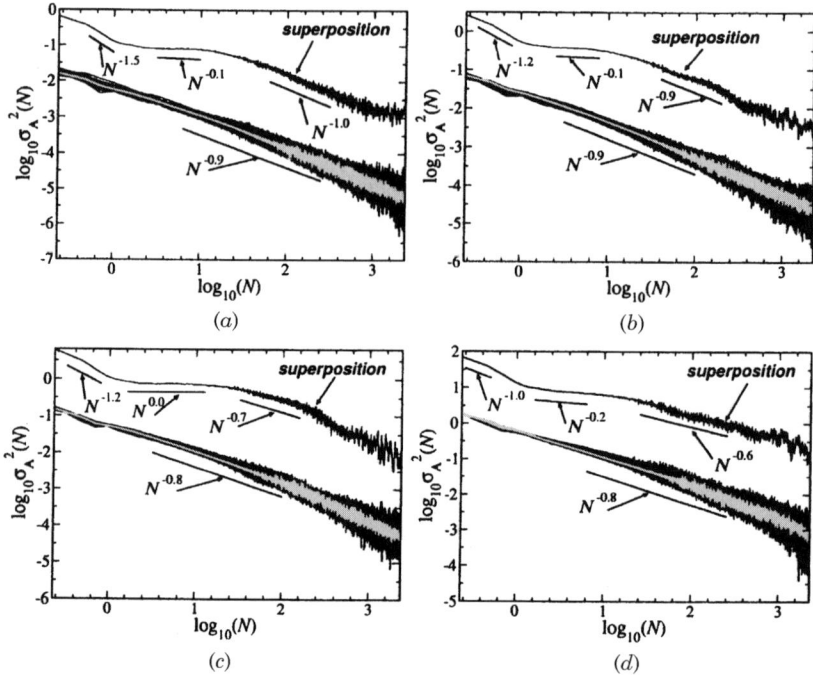

Figure 13. The Allan variances for $V_{r_i}(t)$ and the superposition $V_\mathbf{r}(t)(=\sum_{i=1}^{45} V_{r_i}(t))$ at (a) $T = 0.2$, (b) $T = 0.4$, (c) $T = 0.6$, and (d) $T = 2.0$ for the Gō-like BLN model. The gray line indicates the average of the Allan variances for $V_{r_i}(t)$.

exponents γ in $\sigma_A^2(N) \sim N^{-\gamma}$ are also presented for the superposition and the average of the Allan variances of all individual energy component fluctuations (represented as bold gray lines in the figures). While the individual fluctuation of $V_{r_i}(t)$ exhibits a single strong stationarity $0.7 \le \gamma \le 0.9$ spanning the whole time scale of the simulation for both the models at both the temperatures, the Allan variances of the superpositions do not necessarily follow a similar strong stationarity, but rather introduce "plateau" regimes for an intermediate time domain around $10^{0.2}t^* \le t \le 10^{1.2}t^*$ with $\gamma \simeq 0.1$ at $T = 0.72$ for the BLN model and $\gamma \simeq 0.0$ at $T = 0.6$ for the Gō-like BLN model, with a longer plateau compared with the original BLN model. These plateaus should imply that there exists cooperativity among the *stationary* stretching motions, which are "enhanced" especially when the protein collapses (BLN) or folds (Gō-like BLN). Note also that, after such plateau regimes, the Allan variances of the superposition $V_\mathbf{r}(t)$ just mimic those of the individual bond energy fluctuation $V_{r_i}(t)$, indicating the loss of cooperativity for such a longer time scale.

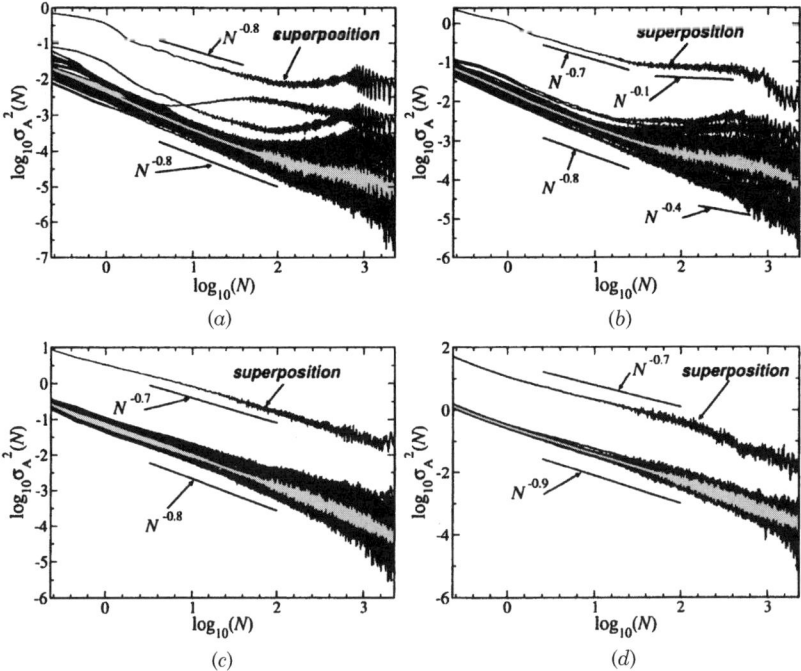

Figure 14. The Allan variances for $V_{\theta_i}(t)$ and the superposition $V_\theta(t) (= \sum_{i=1}^{44} V_{\theta_i}(t))$ at (a) $T = 0.2$, (b) $T = 0.4$, (c) $T = 0.72$, and (d) $T = 2.0$ for the original BLN model. The gray line indicates the average of the Allan variances for $V_{\theta_i}(t)$.

Bending Energy Fluctuation. Figures 14 and 15 show the Allan variances of each bending energy fluctuation $V_{\theta_i}(t)$ and their superpositions $V_\theta(t)$ $(= \sum V_{\theta_i}(t))$ at $T = 0.2$–2.0 for the original BLN and the Gō-like BLN models, respectively. These plots reflect the topographical nature of the underlying multidimensional energy landscapes *along* each bending mode composed of neighboring three beads.

The existence of strong stationarity obeying the law of large numbers well for $10^{-0.6}t^* \leq t \leq 10^{3.4}t^*$ observed in some bending energy fluctuations $V_{\theta_i}(t)$ at a given temperature suggests that the system feels no significant energy barriers on that thermodynamic potential, resulting in a single basin. On the other hand, turning from stationary to nonstationary behavior seen in some $V_{\theta_i}(t)$ fluctuations at a certain time scale (for example, $t \sim 10^2 t^*$ for Gō-like BLN model at $T = 0.2$) should indicate the existence of energy barriers along the corresponding bending DOF so as to yield a transition from one basin to another at the corresponding time scale.

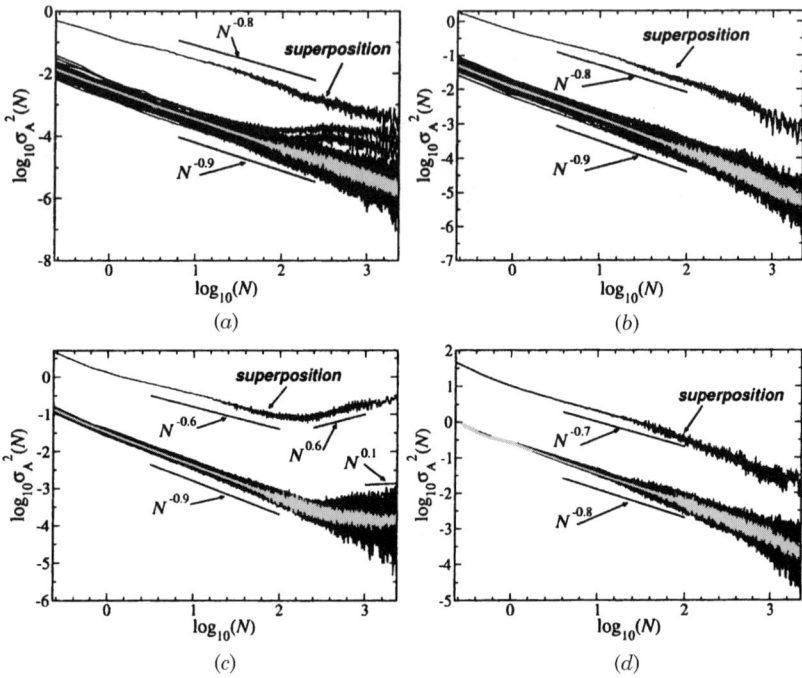

Figure 15. The Allan variances for $V_{\theta_i}(t)$ and the superposition $V_\theta(t) (= \sum_{i=1}^{44} V_{\theta_i}(t))$ at (a) $T = 0.2$, (b) $T = 0.4$, (c) $T = 0.6$, and (d) $T = 2.0$ for the Gō-like BLN model. The gray line indicates the average of the Allan variances for $V_{\theta_i}(t)$.

Below the transition temperature, the energy landscapes in the bending DOFs space look more rugged in the original BLN model than in the Gō-like BLN model (see also gray lines, the average of the Allan variance of $V_{\theta_i}(t)$, in the figures). In the Gō-like BLN model, although some $V_{\theta_i}(t)$ exhibit a weak nonstationarity where the stationary fluctuations in the short time period turn to nonstationary fluctuations around $t \sim 10^2 t^*$ at $T = 0.2$, all $V_{\theta_i}(t)$ fluctuations turn to a strong stationarity at $T = 0.4$. This indicates that most bending DOFs θ_i feel effective *single* basins on the thermodynamic potential at $T = 0.4$ for the Gō-like BLN model while there still exist significant energy barriers separating several small basins for the original BLN model at that temperature. Figure 16a shows the moieties of the protein that exhibit strong stationary behavior in fluctuation of the original BLN model at $T \leq 0.4$. One can see for the BLN model that some bending movements θ_i along which the system feels no significant barriers are well classified as the bead sequence L – B – L-type (L:hydrophilic, B:hydrophobic) despite the uniform bending force constant K_θ independent of the kinds of beads.

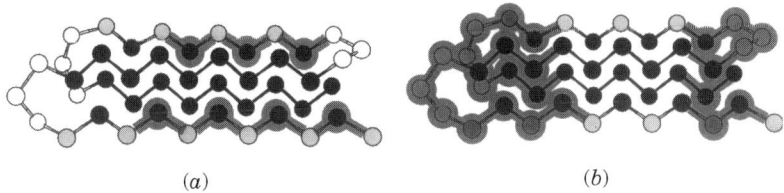

Figure 16. The moieties in the bead sequences exhibiting a strong stationarity in bending energy fluctuation. The center bead B and the two bonds A–B and B–C in a three-bead sequence A–B–C showing a strong stationarity through all the simulated time scale are shaded. (*a*) BLN at $T = 0.4$: The bead sequences exhibiting strong stationary behavior through the whole simulated time scale are $\theta_{L_{15}B_{16}L_{17}}$, $\theta_{L_{17}B_{18}L_{19}}$, $\theta_{L_{19}B_{20}N_{21}}$, $\theta_{L_{38}B_{39}L_{40}}$, $\theta_{L_{40}B_{41}L_{42}}$, $\theta_{L_{42}B_{43}L_{44}}$, $\theta_{L_{44}B_{45}L_{46}}$. (*b*) Gō-like BLN at $T = 0.6$: the bead sequences exhibiting such strong stationary behavior are $\theta_{B_1B_2B_3}$, $\theta_{B_7B_8B_9}$, $\theta_{B_8B_9N_{10}}$, $\theta_{B_9N_{10}N_{11}}$, $\theta_{N_{10}N_{11}N_{12}}$, $\theta_{N_{11}N_{12}L_{13}}$, $\theta_{N_{12}L_{13}B_{14}}$, $\theta_{L_{13}B_{14}L_{15}}$, $\theta_{L_{19}B_{20}N_{21}}$, $\theta_{B_{20}N_{21}N_{22}}$, $\theta_{N_{21}N_{22}N_{23}}$, $\theta_{N_{22}N_{23}B_{24}}$, $\theta_{N_{23}B_{24}B_{25}}$, $\theta_{B_{29}B_{30}B_{31}}$, $\theta_{B_{30}B_{31}B_{32}}$, $\theta_{B_{31}B_{32}N_{33}}$, $\theta_{B_{32}N_{33}N_{34}}$, $\theta_{N_{33}N_{34}N_{35}}$, $\theta_{N_{34}N_{35}L_{36}}$, $\theta_{N_{35}L_{36}B_{37}}$, $\theta_{L_{36}B_{37}L_{38}}$, $\theta_{B_{37}L_{38}B_{39}}$, $\theta_{L_{38}B_{39}L_{40}}$, $\theta_{B_{43}L_{44}B_{45}}$, $\theta_{L_{44}B_{45}L_{46}}$.

At the transition temperature, all bending energy fluctuations $V_{\theta_i}(t)$ get together into single, stationary behavior for the BLN model. This indicates that most bending DOFs θ_i feel effective *single* basins on the thermodynamic potential at $T = 0.72$, which may be separated by high energy barriers much larger than the order of $k_B T$. In the Gō-like BLN model at the folding temperature $T = 0.6$, most bending energy fluctuations $V_{\theta_i}(t)$ behave as stationary, obeying the law of large numbers. The observation of nonstationarity at $t > 10^{2.4} t^*$ in some $V_{\theta_i}(t)$ at the folding temperature implies that, the highly frustrated BLN energy landscape, those bending motions in contrast to, belonging to the "core" of protein excluding the loop regions and the edge regions, make the system cross over the energy barriers on a time scale of $\sim 10^{2.4} t^*$.

At $T = 2.0$ above the transition temperatures for both the models, all bending fluctuations exhibit a strong stationarity. This implies that all bending DOFs feel single basins on those thermodynamic potential and that the energy barriers over which the system requires to cross when $t \sim 10^{2.4} T^*$ at the folding temperature for Gō-like BLN model become negligibly small, compared with $k_B T$, resulting in the collapse of the small basins to single (super)basins along all the bending DOFs.

Note here that the stationarity observed for the fluctuation of the superposed $V_\theta(t)$ and the individual $V_{\theta_i}(t)$ possess almost the same exponent γ for the BLN model at $T = 0.72$—that is, 0.7 (V_θ) and 0.8 (V_{θ_i}). At the folding temperature $T = 0.6$ for the Gō-like BLN model, the apparent distinction of exponent γ between the individual $V_{\theta_i}(t)$ (e.g., $\gamma \simeq 0.9$ at $t < 10^2 t^*$ for Gō-like BLN) and the superposed $V_\theta(t)$ (e.g., $\gamma \simeq 0.6$ at $t < 10^2 t^*$ for Gō-like BLN) should

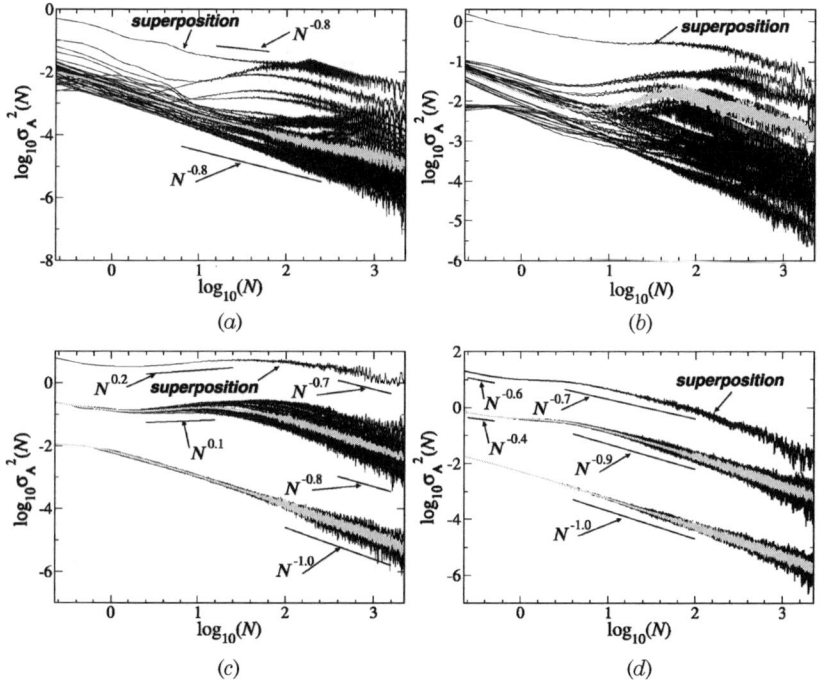

Figure 17. The Allan variances for $V_{\Phi_i}(t)$ and the superposition $V_\Phi(t)(=\sum_{i=1}^{43} V_{\Phi_i}(t))$ at (a) $T = 0.2$, (b) $T = 0.4$, (c) $T = 0.72$, and (d) $T = 2.0$ for the original BLN model. The gray line indicates the average of the Allan variances for $V_{\Phi_i}(t)$.

suggest the existence of cooperativity among the bending motions, which looks more pronounced for the Gō-like BLN model where the protein folds than for the BLN model.

Torsional Angle Energy Fluctuation. The other remarkable consequence can be observed in the Allan variances of the torsional angle energy fluctuation $V_{\Phi_i}(t)$ at the transition temperature. Figures 17 and 18 show, respectively, the Allan variances of the individual torsional angle energy fluctuation $V_{\Phi_i}(t)$ and their superposition $V_\Phi(t)(=\sum V_{\Phi_i}(t))$ at $T = 0.2 \sim 2.0$ for the original BLN, and the Gō-like BLN models.

Below the transition temperature, the individual $V_{\Phi_i}(t)$ exhibit somewhat different behavior in the Allan variances, implying that the energy landscapes in the torsional angle DOFs space look much more rugged compared with stretching and bending DOFs spaces for both the models [see also the average of the Allan variance of $V_{\theta_i}(t)$ and $V_{\Phi_i}(t)$ at $T = 0.4$].

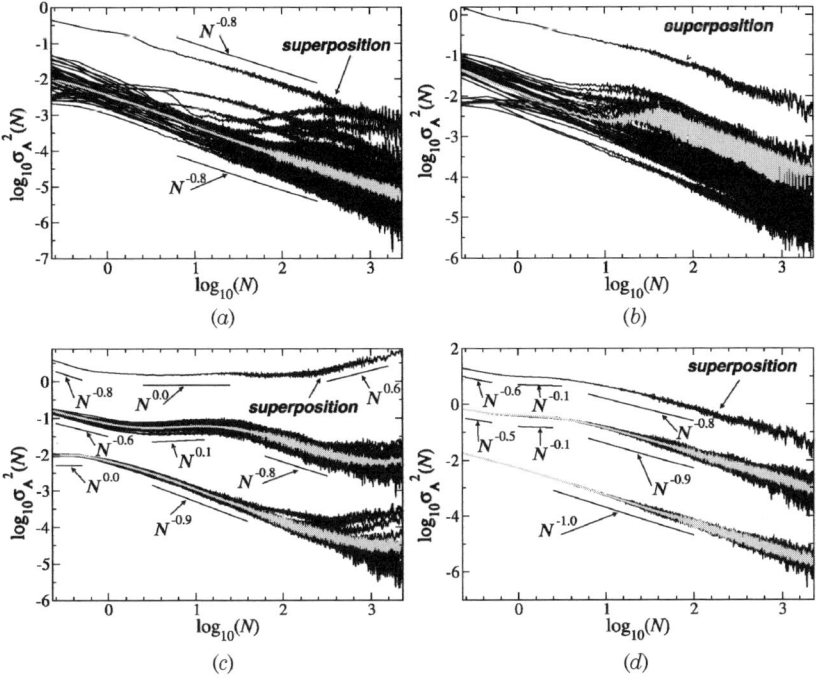

Figure 18. The Allan variances for $V_{\Phi_i}(t)$ and the superposition $V_{\Phi}(t) (= \sum_{i=1}^{43} V_{\Phi_i}(t))$ at (a) $T = 0.2$, (b) $T = 0.4$, (c) $T = 0.6$, and (d) $T = 2.0$ for the Gō-like BLN model. The gray line indicates the average of the Allan variances for $V_{\Phi_i}(t)$.

At the transition temperature, the Allan variances of the individual $V_{\Phi_i}(t)$ apparently collapse into two families: The first family exhibits an almost stationary behavior (i.e., $\sigma_A^2(N) \sim N^{-1}$), for a wide range of time scales, and the second family exhibits a rather complex pattern. In the latter, a first weak stationary phase is observed in the region $0 \leq t \leq 10^{0.2} t^*$, and this turns into an almost-flat nonstationary phase $\sigma_A^2(N) \sim N^{0.1}$ in an intermediate time region of $10^{0.2} t^* \leq t \leq 10^{1.1} t^*$ and eventually falls into a strong stationary phase $\sigma_A^2(N) \sim N^{-0.8}$. The moieties exhibiting strong stationary behavior in torsional angle energy fluctuation are well localized in the loop region at the transition temperature for both the models (see Fig. 19). This "localization" of a strong stationarity onto that *loop* region could arise from small barriers along the torsional movements of bead sequences involving two or more neutral residues.

The collapsing of the torsional angle energy fluctuations into these two families implies that, irrespective of the kinds of beads comprising the

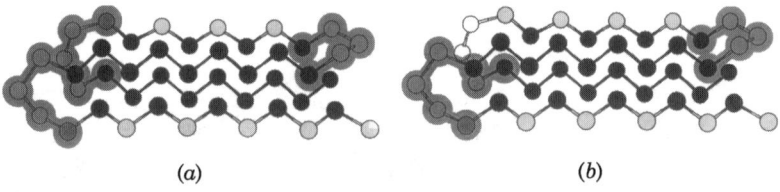

(a)   (b)

Figure 19. The moieties in the bead sequences exhibiting a strong stationarity in torsional angle energy fluctuation at the transition temperature. The middle bond B–C, and the second and third beads B and C in a four-bead sequence A–B–C–D showing a strong stationarity through all the simulated time scale are shaded. (a) BLN at $T = 0.72$: All the bead sequences showing a single, strong stationarity $\gamma \simeq 1.0$ constitute the loop region, $\Phi_{B_8 B_9 N_{10} N_{11}}$, $\Phi_{B_9 N_{10} N_{11} N_{12}}$, $\Phi_{N_{10} N_{11} N_{12} L_{13}}$, $\Phi_{N_{11} N_{12} L_{13} B_{14}}$, $\Phi_{L_{19} B_{20} N_{21} N_{22}}$, $\Phi_{B_{20} N_{21} N_{22} N_{23}}$, $\Phi_{N_{21} N_{22} N_{23} B_{24}}$, $\Phi_{N_{22} N_{23} B_{24} B_{25}}$, $\Phi_{B_{31} B_{32} N_{33} N_{34}}$, $\Phi_{B_{32} N_{33} N_{34} N_{35}}$, $\Phi_{N_{33} N_{34} N_{35} L_{36}}$, $\Phi_{N_{34} N_{35} L_{36} B_{37}}$. (b) Gō-like BLN at $T = 0.6$: Similarly, strong, uniform stationary behavior is localized in the loop region, although some bead sequences turn to a weak nonstationarity after $t \sim 10^{2.2} t^*$. Here, the bead sequences in the loop regions exhibiting strong stationary behavior through the whole simulated time scale are $\Phi_{B_8 B_9 N_{10} N_{11}}$, $\Phi_{L_{19} B_{20} N_{21} N_{22}}$, $\Phi_{N_{21} N_{22} N_{23} B_{24}}$, $\Phi_{N_{22} N_{23} B_{24} B_{25}}$, $\Phi_{B_{31} B_{32} N_{33} N_{34}}$, $\Phi_{B_{32} N_{33} N_{34} N_{35}}$, $\Phi_{N_{34} N_{35} L_{36} B_{37}}$.

corresponding four-bead sequences and the different locations along the sequence space, the $V_{\Phi_i}(t)$ fluctuations at the loop and nonloop regions yield similar stationary and nonstationary fluctuations, respectively. The nonstationarity $\sigma_A^2(N) \sim N^{0.6}$ observed after $t \sim 10^{2.4} t^*$ for the superposed $V_\Phi(t)$ fluctuation is totally different from the individual $V_{\Phi_i}(t)$ fluctuation at the folding temperature $T = 0.6$ for the Gō-like BLN model. This indicates that there exists a strong cooperativity among the torsional movements to establish the strong nonstationary multibasin transition, especially on the less-frustrated funnel energy landscape of the Gō-like BLN model.

The significant nonstationarity in the non loop region arises from the nature of the underlying energy landscapes [57]. Many significantly different structures, though maintaining the energy as almost the same, can be constructed from common motifs such as the four strands in the global minimum. In fact, some low-energy minima differ only by the relative positions of the two purely hydrophobic strands. These strands exhibit a number of stable configurations related by *parallel sliding* [56]. However, the sliding requires disrupting all the nonbonded interactions at once. Instead, the shortest path between such structures typically proceeds through at least 10 separate rearrangements. A reorientation of the hydrophobic strands also yields many significant different low-energy structures among which a transition process involves many steps and a high barrier. The topographical nature of the energy landscapes in the core and non loop regime gives rise to a strong nonstationary behavior up to the transition temperature.

Above the transition temperature, all the torsional movements tend to exhibit a strong stationarity including the non loop region.

Although the Allan variance of the individual nonbonded vdW interaction energy $V_{R_i}(t)$ fluctuations are not shown, the collapsing of $V_{R_i}(t)$ into some small set of families at the transition temperature, as in the analyses of torsional angle energy fluctuation, was not found for either the BLN and Gō-like BLN models.

While the disconnectivity graph apparently grasps a certain topographical feature of the underlying energy hypersurfaces, all dynamical events in proteins are essentially multidimensional on the *multidimensional* energy landscapes. As shown above, the Allan variance analyses can shed light on the multi-dimensional nature of the underlying complex energy landscapes. The Allan variance can also address if the difference of the secondary structure, (i.e., α-helix or β-sheet) provides any distinction of nonstationarity of the folding processes. The Allan variance analysis was achieved for the folding simulation of a simple Gō-like model of three small proteins, λ Repressor (**1lmb**), SH3 protein (**1srl**), Protein G (**2gb1**). It was found that as the protein contains more β-sheet moiety (i.e., **1lmb** → **2gb1** → **1srl**), the folding dynamics brings about more slow large nonstationary fluctuations at the transition temperature (it must arise from the *nonlocal* character of β sheet stabilization, compared with α-helix) and that a significant nonstationarity of the potential energy fluctuation emerges at the transition temperature while away from that temperature it vanishes, just as for the folding transition of the BLN and Gō-like BLN models and the helix-coil transition of a helical polypeptide [64], [74]. The significant distinctions in exponents γ ($\sigma_A^2(N) \sim N^{-\gamma}$) between the Allan variances of the superposition $V_\mathbf{r}(t)$, $V_\theta(t)$, and $V_\Phi(t)$ and the average of the Allan variances of the individual energy components of $V_{r_i}(t)$, $V_{\theta_i}(t)$, and $V_{\Phi_i}(t)$ tell us that there is cooperativity among individual DOFs, especially enhanced for an intermediate time scale at the transition temperature. This strongly shows us the limitation of description in terms of those internal DOFs to look deeply into the nature of multidimensional protein dynamics.

IV. STATE-SPACE STRUCTURE OF MULTIBASIN DYNAMICS

The Allan variance analyses of energy fluctuations can tell us about the nature of the configurational energy landscapes and the existence of (multidimensional) cooperativity among individual DOFs, enhanced for an intermediate time scale at the transition temperature. However, what can we learn or deduce from an (observed) scalar time series about the geometrical aspects of the underlying multi-dimensional *state (or phase) space* buried in the observations? The so-called embedding theorems attributed to Whitney [75] and Takens [76] provide us with a clue to the answer of such a question (see also Section VI.A and Refs. 77–80).

A. Embedding: Phase-Space Reconstruction

Suppose that we have a nonlinear dynamical system, that is, first-order ordinary differential equations,

$$\frac{d\mathbf{x}(t)}{dt} = \mathbf{F}(\mathbf{x}(t)) \qquad (22)$$

where $\mathbf{x}(t)$ and $\mathbf{F}$ are the vector representations of the multidimensional state space variables $(x_1(t), x_2(t), \ldots)$ and the one-to-one map $(F_1, F_2, \ldots)$, respectively. It is supposed, that all DOFs more or less influence one another through explicit or implicit couplings, and all the trajectories are confined to a compact and smooth manifold of dimension D_0 in the state space $\mathbf{x}$. Suppose a scalar quantity $s(t)$ with an infinitesimal precision and $s(t)$ is derived by a smooth transformation h from $\mathbf{x}(t)$, i.e., $s(t) = h(\mathbf{x}(t))$. The embedding theorem states that, in principle, from the knowledge of the time series $s(t)$ of an infinite length an equivalent state space can be reconstructed preserving the differential properties of the state space of the original multivariate variables $\mathbf{x}(t)$: The equivalent new space $\mathbf{y}_d(t)$ composed of the univariate $s(t)$, the so-called time delay coordinate system, is given by

$$\mathbf{y}_d(t) = (s(t), s(t+\tau), \ldots, s(t+(d-1)\tau)) \qquad (23)$$

where the d-dimensional state-space variables $\mathbf{y}_d(t)$ are represented in a discrete form with a time interval τ without loss of generality. How can one find the *unknown* dimension D_0 from the time series of $s(t)$? If there *actually* exists a dynamical system behind the observation of $s(t)$, any orbit $\mathbf{y}$ in the reconstructed state space should never cross with itself because of the uniqueness of the solution (remember that there exists no self-intersection in the phase space of Hamiltonian systems). If a smaller dimension d is used to reconstruct the state space, the orbit $\mathbf{y}$ will have self-intersections and cannot be "untangled" or "unfolded" due to the insufficient size of the chosen dimensionality. In other words, if a minimum dimension to unfold the orbit in the time delay coordinate system is reached, this implies that a state space $\mathbf{y}$ equivalent to the original $\mathbf{x}$ can be reconstructed in such a sense that it has the same differential properties of the original manifolds.

The embedding theorem holds irrespective of the choice of the delay time τ, but, in practice, the observed time series are always contaminated by noise, computer round-off errors, or a finite resolution of observations, and they are sampled up to a certain *finite* time. This requires us to choose an "optimal" time delay τ for an observable s.

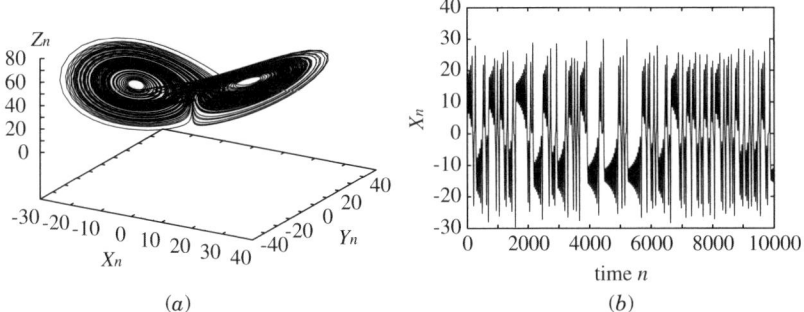

Figure 20. (a) Lorenz attractor and (b) time series of x_n.

Casdagli et al. discussed some guides for choosing the optimal delay time τ in terms of two concepts, redundancy and irrelevance [81]. Here let us explain their statements by using the Lorenz equation:

$$\frac{dx}{dt} = -\sigma x + \sigma y$$
$$\frac{dy}{dt} = -xz + rx - y \qquad (24)$$
$$\frac{dz}{dt} = xy - bz$$

where $\sigma = 10, b = 8/3$ and $r = 28$. This gives us a well-known strange attractor where $D_0 \simeq 2.06$ in terms of the box counting dimension (Fig. 20a). Suppose we were able to access only one variable [e.g., $x(t)$] or a certain scalar quantity $s(t)$ derived by a smooth transformation h from x, y, z, $s(t) = h(x(t), y(t), z(t))$ (and we know *neither* the form of Eq. (24) *nor* if such a dynamical system exists). For instance, let us consider the time series of $x(t)$, $x_n = x(n\Delta t)$ (Δt is the time step) (Fig. 20b). If τ is much shorter than a characteristic time scale inherent to the system in question, the successive points $\mathbf{y}(n\Delta t), \mathbf{y}(n\Delta t + \tau), \mathbf{y}(n\Delta t + 2\tau), \ldots$ in the time delay space will "collapse" onto a diagonal "region," $\mathbf{y}(n\Delta t) \simeq \mathbf{y}(n\Delta t + \tau) \simeq \mathbf{y}(n\Delta t + 2\tau) \simeq \ldots$, inside which it might be impossible or very hard to distinguish the points from each other because of the limited, finite resolution of $x(t)$. That is, too short a τ implies that the dynamical information the time series $s(t)$ would originally possess tends to be *redundant* and one cannot retrieve any information about the geometric feature of the state space because of the unavoidable finite resolution (Fig. 21a).

On the other hand, if τ is too large, the successive points may become numerically tantamount to being "random," even if $s(t)$ originally arises from a

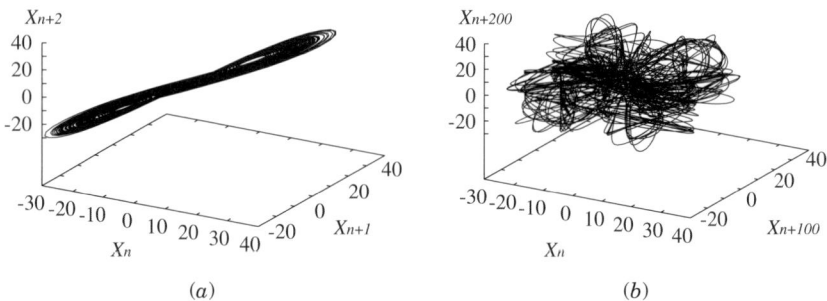

Figure 21. The three-dimensional time delay coordinate systems reconstructed by x_n with two different delay times τ ((a) $\tau = 1$, (b) $\tau = 100$).

deterministic system with a finite dimensionality. That is, due to the orbital instability inherent to chaos, when τ becomes very large, one cannot obtain the precision required to extract the deterministic structure from the time series and to prevent an exponential divergence. That is, practically, any information retrieved from the time series becomes *irrelevant* to the geometrical feature of the state space (Fig. 21b). From this example, the task of choosing an optimal delay time τ should balance these two extremes such that the resulting delay coordinates are maximally independent (*nonredundant*), yet the dynamical relationship between the coordinates is preserved (*relevant*).

What is the implication of a dimension reached for the reconstructed time delay space $\mathbf{y}$ from an observable $s(t)$? The embedding dimension, denoted as d_E hereinafter, is a *global* dimension equal to a sufficient number of the (delay time) coordinates to unfold any arbitrary orbit $\mathbf{y}(t)$ on the smooth and compact D_0-dimensional manifold from self-intersections due to a "wrong" projection to a lower-dimensional space. Takens' theorem ensures $d_E > 2D_0$ is a sufficient condition (not a necessary condition) of the embedding dimension by which any arbitrary orbit $\mathbf{y}(t)$ can be unfolded on the delay time coordinate system. In practice, it is often possible to reconstruct the state space with a smaller dimension (e.g., a Lorenz attractor with $D_0 \simeq 2.06$), which can be reconstructed in terms of a three-dimensional delay coordinate system [77]. If one measures several observables or quantities from the *same* system to find the embedding dimension d_E, each measurement along with its timelag provides a different nonlinear mapping of the original state-space variables $\mathbf{x}(t)$ into a d_E-dimensional reconstructed space. Each should, in principle, reach the same global dimension inherent in the underlying dynamical structure under a certain condition—for example, an infinite sampling length and an infinitesimal resolution. For a finite time evolution, however, each measurement may see different local topography of the reconstructed state space and result in a different *local* dimensionality, referred as to the minimum embedding dimension d_L ($d_L \leq d_E$).

B. False Nearest Neighbors

Here we show the algorithm to determine the minimum embedding dimension d_L for a given *finite* scalar time series $s(t)$ with a given "optimal" delay time τ. If the correct minimum embedding dimension d_L is chosen, the reconstructed state space retains the topological properties of the manifolds, where no intersection exists between any orbit in the delay coordinate system because of the uniqueness of solution in the original state space. When d is smaller than d_L, there should exist intersections between some orbits in the delay coordinate system. Most algorithms to determine d_L are in principle based on this property of dynamical system: As the *tentative* dimension d increases, one must reach at a convergence of the dimension above which all the intersections are removed, and it is expected that this converged dimension could be identified as the minimum embedding dimension d_L.

However, in practice, the intersections quite likely disappear before reaching the desired dimension d_L because of the finite length of the time series for high-dimensional dynamical systems. To overcome this, Abarbanel and co-workers [82] proposed a new concept, the so-called false nearest neighbors (FNN) as follows: One can interpret that the nearest-neighbor point of any point in the reconstructed state space **y** exists *only* due to the topological properties of the underlying dynamical structure. In the other words, if the chosen dimension is smaller than d_L, the nearest-neighbor points should occur due to the *false* projections onto a space of the wrong dimension.

Let $\mathbf{y}_d(t_{n(i,d)})$ be the nearest-neighbor point of a point $\mathbf{y}_d(t_i)$ in a d-dimensional delay coordinate system (hereinafter, d is designated as a tentative dimensions):

$$\mathbf{y}_d(t_i) = (s(t_i), s(t_i + \tau), \ldots, s(t_i + \tau(d-1))) \qquad (25)$$

$$\mathbf{y}_d(t_{n(i,d)}) = \left(s(t_{n(i,d)}), s(t_{n(i,d)} + \tau), \ldots, s(t_{n(i,d)} + \tau(d-1))\right) \qquad (26)$$

where $n(i,d)$ denotes the index of the nearest neighbor of the ith point in the d-dimensional space. The square of the Euclidean distance between the nearest-neighbor points is given by, in d-dimensions,

$$R_d(t_i)^2 = \sum_{k=1}^{d} \left[s(t_i + \tau(k-1)) - s(t_{n(i,d)} + \tau(k-1))\right]^2 \qquad (27)$$

and, in $(d+1)$-dimensions,

$$R_{d+1}(t_i)^2 = \sum_{k=1}^{d+1} \left[s(t_i + \tau(k-1)) - s(t_{n(i,d)} + \tau(k-1))\right]^2 \quad (28)$$

$$= R_d(t_i)^2 + \left[s(t_i + \tau d) - s(t_{n(i,d)} + \tau d)\right]^2 \quad (29)$$

The displacement of the Euclidean distance between the nearest-neighbor points from d-dimensions to $(d+1)$-dimensions relative to the distance in d-dimensions,

$$\sqrt{\frac{R_{d+1}(t_i)^2 - R_d(t_i)^2}{R_d(t_i)^2}} = \frac{|s(t_i + \tau d) - s(t_{n(i,d)} + \tau d)|}{R_d(t_i)} \quad (30)$$

can classify which nearest neighbor (NN) is false or true, that is,

$$\begin{aligned} \text{If} \quad & \frac{|s(t_i + \tau d) - s(t_{n(i,d)} + \tau d)|}{R_d(t_i)} > R_T, \quad \text{then, } \textit{false } \text{NN} \\ \text{If} \quad & \frac{|s(t_i + \tau d) - s(t_{n(i,d)} + \tau d)|}{R_d(t_i)} < R_T, \quad \text{then, } \textit{true } \text{NN} \end{aligned} \quad (31)$$

Here R_T is an empirical threshold value, which should, in principle, vary with the number of data points for the finite data sets of interest. The minimum embedding dimension d_L can be estimated by surveying the convergent behavior of the ratio of number of FNN with respect to the total number of NN as a function of d. However, in practice, all time series of interest are *finite* and are always contaminated by noise, round-off errors, and so forth. Both stochastic and high-dimensional chaotic time series may result in convergence of the percentage of the FNNs at a certain *high* dimension, and it is very difficult to identify the origin of this convergence—that is, whether the convergence arises from a high-dimensional chaos or from a stochastic, but *finite*, time series. An additional diagnosis was introduced to discriminate a stochastic time series from a high-dimensional chaos as follows:

$$\begin{aligned} \text{If} \quad & \frac{|s(t_i + \tau d) - s(t_{n(i,d)} + \tau d)|}{R_A} > A_{tol}, \quad \text{then, } \textit{false } \text{NN} \\ \text{If} \quad & \frac{|s(t_i + \tau d) - s(t_{n(i,d)} + \tau d)|}{R_A} < A_{tol}, \quad \text{then, } \textit{true } \text{NN} \end{aligned} \quad (32)$$

where

$$R_A^2 = \langle (s(t_i) - \langle s \rangle)^2 \rangle, \qquad \langle s \rangle \equiv \frac{1}{N} \sum_{i=1}^{N} s(t_i) \quad (33)$$

Here, N is the total number of time steps; R_A, the root mean square value of the data about its mean, is aimed at reflecting the "size" of the manifold, and A_{tol} is set to be about 2. If $\sqrt{R_{d+1}(t_i)^2 - R_d(t_i)^2}(= |s(t_i + \tau d) - s(t_{n(i,d)} + \tau d)|)$ is greater than twice as large as the "size" of the manifold, it means that the nearest-neighbor point traverses from one "edge" to another "edge" by $d \to d+1$, and can be regarded as *false* NN arisen from some stochastic processes. The minimum embedding dimension d_L can be evaluated in terms of both criteria for a given time series. However, it was found for many cases [83,84] that the choices of R_T and A_{tol} are quite sensitive to yield the desired d_L.

Cao's Algorithm. Cao proposed [83,84] a different scheme on behalf of introducing arbitrary empirical criteria: He introduced a new quantity $a(i,d)$ that estimates how the ith point along the orbit in the d-dimensional delay coordinate space moves away from its nearest neighbor when the tentative dimension increases by 1, that is, $d \to d+1$:

$$a(i,d) = \frac{||\mathbf{y}_{d+1}(t_i) - \mathbf{y}_{d+1}(t_{n(i,d)})||}{||\mathbf{y}_d(t_i) - \mathbf{y}_d(t_{n(i,d)})||} \quad (i = 1, 2, \ldots, N - \tau d) \quad (34)$$

$$||\mathbf{y}_d(t_k) - \mathbf{y}_d(t_l)|| = \left(\sum_{j=0}^{d-1}[s(t_k + j\tau) - s(t_l + j\tau)]^2\right)^{1/2} \quad (35)$$

To diagnose the convergence of the dimensionality required to "unfold" the whole orbit $\mathbf{y}(t)$, the following new indicator $E1$ was introduced—that is, the ratio of the mean value of all $a(i,d)$ on the d-dimensional and the $(d+1)$-dimensional spaces.

$$E1(d) = \frac{E(d+1)}{E(d)} \quad (36)$$

where

$$E(d) = \frac{1}{N - \tau d}\sum_{i=1}^{N-\tau d} a(i,d) \quad (37)$$

When $E1(d)$ converges to unity for d larger than a certain value d_0, one can choose $d_0 + 1$ as the minimum embedding dimension d_L. In practice, an apparent convergence of $E1(d)$ cannot be avoided for stochastic, but finite, time series. To identify the origin of the convergence in $E1(d)$ and discriminate a stochastic time series from a high-dimensional chaos, a supplemental diagnosis $E2$ was also

Figure 22. The growth of the two nearby orbits in a *tentative* d-dimensional state space.

proposed: The growth of displacement between $\mathbf{y}_d(t_i)$ and $\mathbf{y}_d(t_{n(i,d)})$, $|\mathbf{y}_d(t_i) - \mathbf{y}_d(t_{n(i,d)})|$, along the dynamical evolution in time τ (Fig. 22) is written by

$$|\mathbf{y}_d(t_i + \tau) - \mathbf{y}_d(t_{n(i,d)} + \tau)| - |\mathbf{y}_d(t_i) - \mathbf{y}_d(t_{n(i,d)})|$$
$$= \sum_{k=1}^{d} |s(t_i + \tau k) - s(t_{n(i,d)} + \tau k)| - \sum_{k=0}^{d-1} |s(t_i + \tau k) - s(t_{n(i,d)} + \tau k)|$$
$$= |s(t_i + \tau d) - s(t_{n(i,d)} + \tau d)| - |s(t_i) - s(t_{n(i,d)})|$$
$$\cong |s(t_i + \tau d) - s(t_{n(i,d)} + \tau d)|$$

A quantity $E^*(d)$ defined by

$$E^*(d) = \frac{1}{N - \tau d} \sum_{i=1}^{N - \tau d} |s(t_i + \tau d) - s(t_{n(i,d)} + \tau d)| \qquad (38)$$

is, hence, expected to elucidate how the nearest-neighbor points will, on average, move away from one another in a finite time τ in the delay coordinate space of a chosen dimension d (just as in the spirit of the local Liapunov exponent). The new supplemental diagnosis $E2$ is defined by

$$E2(d) = \frac{E^*(d+1)}{E^*(d)} \qquad (39)$$

If the time series $s(t)$ is driven by a stochastic system, the time evolution of the variable should be almost independent of the past value, and one can expect $E^*(d)$ to be almost constant irrespective of the choice of d. Thus, the diagnosis $E2$ will be around unity, independently of d. However, if $s(t)$ arises from a dynamical system, $s(t)$ is expected to give rise to a large $E^*(d)$ at small values of $d (\ll d_L)$ due to the existence of false nearest neighbors, resulting in a significant dependence of $E2$ on d for certain low values of d where the FNNs have not been eliminated. As d increases, $E^*(d)$ will also increase due to a larger separation τd, but, after reaching the minimum embedding dimension—that is, preserving the topological features of its original manifold—E^* is expected to be unaffected by

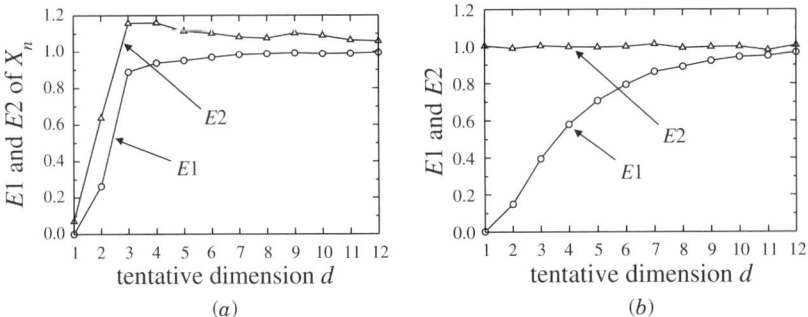

Figure 23. $E1$ and $E2$ plots (a) Lorenz attractor. (b) Time series produced by a (pseudo) random number generator.

the change $d \to d+1$, resulting in the convergence of $E2$ to approximately unity (see Fig. 23).

C. Average Mutual Information

To choose the delay time τ, we used a prescription [85] that is based on the concept of average mutual information in information theory. The average mutual information I_{AB} between measurements A and B is defined by

$$I_{AB} \equiv \sum_{a,b} P_{AB}(a,b) I_{AB}(a,b) \qquad (40)$$

$$= \sum_{a,b} P_{AB}(a,b) \log_2 \left[\frac{P_{AB}(a,b)}{P_A(a) P_B(b)} \right] \qquad (41)$$

where $P_{AB}(a,b)$ is the joint probability density that a and b will be observed in measurements A and B, respectively, and $P_A(a)$ and $P_B(b)$ are the corresponding individual probability densities. $I_{AB}(a,b)$ is the amount learned by the observation of a in measurement A about b in measurement B (and vice versa) in bits. If these two measurements are completely independent—that is, if $P_{AB}(a,b) = P_A(a) P_B(b)$—then $I_{AB}(a,b) = 0$. The average mutual information I_{AB}, an average of $I_{AB}(a,b)$ over all measurements $\{a\}$ and $\{b\}$, thus implies the degree of mutual correlation of the two measurements A and B. As a tool to determine an optimal delay time τ, Fraser and Swinney [85] suggested the *time-delayed mutual information*; that is, it is the information we already possess about the value of $s(t + \tau)$ if we know $s(t)$. Practically, for the computation of the mutual information, we create a histogram for the probability distribution on the interval explored by the data $s(t)$. Let P_i be the probability that the signal assumes a value inside the ith bin of the histogram, and let P_{ij} be the probability

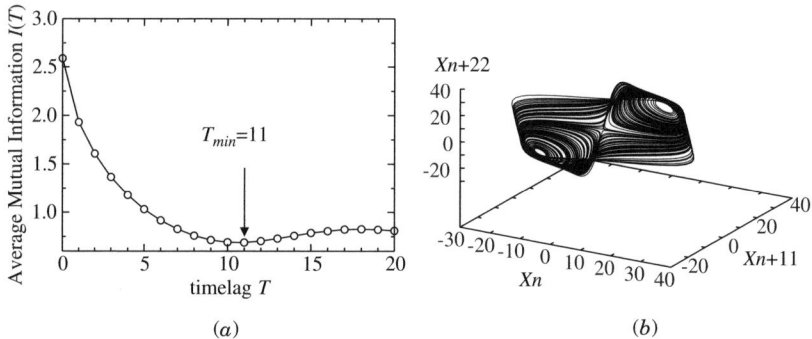

Figure 24. (*a*) The average mutual information of x_n and (*b*) the three-dimensional delay coordinate system with $\tau = T_{\min}$ for Lorenz attractor.

that $s(t)$ is in the ith bin and $s(t+T)$ is in the jth bin. Then the mutual information for time delay T is defined as follows:

$$I(T) = \sum_{i,j} P_{ij}(T) \log_2 \left[\frac{P_{ij}(T)}{P_i P_j} \right] \quad (42)$$

They proposed to take the time $T_{\min}$ of the first minimum of $I(T)$ as an optimal delay time τ. This choice of the delay time is expected to balance the previously mentioned constraints, relevance, and nonredundancy by maximizing vector independence between $s(t)$ and $s(t+T)$ over relatively small lag times. Figure 24 shows the first minimum of $I(T)$ and the reconstructed delay coordinate system of the Lorenz attractor. One can see that this reconstructed system using $T_{\min}$ as τ can better retrieve the topological feature of the original state space (Fig. 20a) than those using the other choice of τ (Fig. 21a,b). It has been known empirically [77,78] that this simple prescription using $I(T)$ provides a good delay time τ for practical purposes and works better than using the linear autocorrelation function $\langle (s(t) - \langle s \rangle)(s(t+T) - \langle s \rangle) \rangle$ where $\langle \cdot \rangle$ is the time average.

D. Temperature Dependency in Dimensionality of Folding Dynamics

The Allan variance analyses of energy fluctuations strongly suggests that we utilize some collective coordinates on behalf of the internal DOFs to elucidate the nature of protein folding dynamics. Here, let us examine the versatility of principal component (PC) analysis in scrutinizing the heterogeneous cooperativity and dimensionality of multibasin dynamics.

It was found [41] that 90% of the total configurational fluctuations can well be represented in terms of about 10 principal components at each temperature.

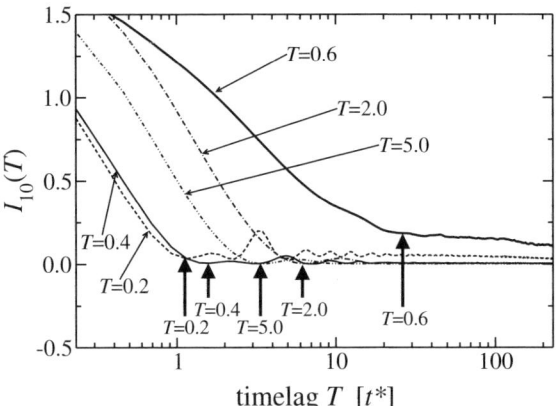

Figure 25. The average mutual information of the principal components $Q_{10}(t)$ and $I_{10}(t)$ for the Gō-like BLN model at $T = 0.2$–5.0. The horizontal axis represents the logarithm of time t, along which each arrow indicates the time of the first minimum of $I_{10}(t)$ at each temperature.

At $T = 0.2$–0.4, most of the system fluctuations are approximately localized on a few PCs for both the models. As the temperature increases, the system fluctuations become more delocalized, spreading over a wider range of the PCs. Note, however, that the number of the PCs required to represent the total configurational fluctuation of the systems are largest about their transition temperatures, $T = 0.72$ (BLN) and $T = 0.6$ (Gō-like).

Then how long does memory persist in the principal components $\mathbf{Q}(t)$ and what are the characteristic time scales inherent to them? The average mutual information $I_i(t)$ of the individual principal component $Q_i(t)$ is one of the

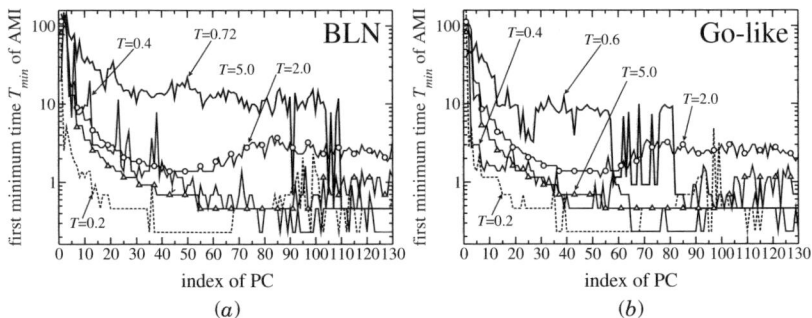

Figure 26. The first minimum time τ of the average mutual information of all principal components. (*a*) The original BLN model. (*b*) The Gō-like BLN model.

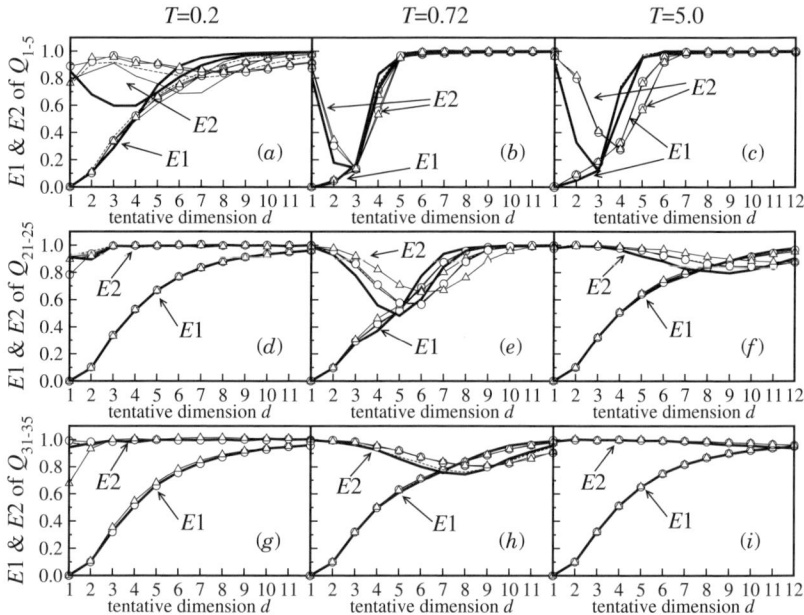

Figure 27. $E1(d)$ and $E2(d)$ of several principal components as a function of the tentative dimension d at $T = 0.2, 0.72, 5.0$ for the original BLN model: (a) Q_1–Q_5 at $T = 0.2$; (b) Q_1–Q_5 at $T = 0.72$; (c) Q_1–Q_5 at $T = 5.0$; (d) Q_{21}–Q_{25} at $T = 0.2$; (e) Q_{21}–Q_{25} at $T = 0.72$; (f) Q_{21}–Q_{25} at $T = 5.0$; (g) Q_{31}–Q_{35} at $T = 0.2$; (h) Q_{31}–Q_{35} at $T = 0.72$; (i) Q_{31}–Q_{35} at $T = 5.0$.

versatile means to measure the degree of mutual correlation of $Q_i(t_0)$ and $Q_i(t_0 + t)$. Figure 25 depicts, as a representative example, $I_{10}(t)$ as a (logarithmic) function of time at $T = 0.2$–5.0 for the Gō-like BLN model. The times $t_{\min}$ of the first minima of $I_{10}(t)$ are indicated by arrows. $I_{10}(t)$ has a longer tail in t at $T = 0.6$ than at the other temperatures, resulting in a longer $t_{\min}$. Figure 26 shows the first minimum time $t_{\min}$ of each PC for both the BLN and Gō-like models at $T = 0.2$–5.0. In general, one can expect that the longer $t_{\min}$, the longer the memory persistence in the signal $Q_i(t)$. Except for the PCs with very small variances, whose indexes are greater than ~ 100 for the BLN and ~ 80 for the Gō-like models, $t_{\min}$ of almost of all the other PCs exhibit a longer increase at the transition temperatures, $T = 0.72$ (BLN) and $T = 0.6$ (Gō-like), than those at the other temperatures.

Now, let us look deeply into the question of what the dimensionality of the state space is, buried in the complexity of the time series of the protein dynamics. Here we apply Cao's embedding technique to every principal component time series for the original BLN model at a wide range of

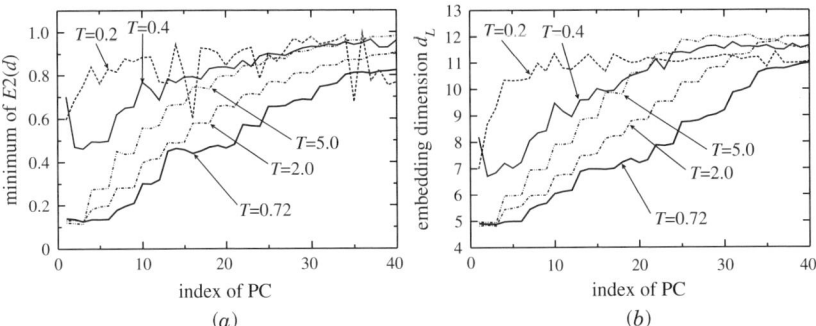

Figure 28. (a) The minimum value of $E2(d)$ and (b) the minimum embedding dimension d_L of the principal components at $T = 0.2-5.0$ for the original BLN model.

temperatures. Figure 27 shows $E1(d)$ and $E2(d)$ of several principal components as a function of the tentative dimension d. In this figure, (a), (b) and (c) display the principal components Q_1-Q_5, (d), (e), and (f) $Q_{21} \sim Q_{25}$, and (g), (h), and (i) $Q_{31}-Q_{35}$ at $T = 0.2, 0.72$, and 5.0, respectively. The more the variance (eigenvalue of the variance–covariance matrix **R**) of the PC decreases—that is, the more the index of the PC increases—the more $E2(d)$ flattens to unity, resulting in a "random" time series with an "infinite" dimensionality. This implies that even if one might reach a finite dimensionality in terms of the $E1(d)$, such dimensionality should artificially arise from the finiteness of the sampling length. An observation from the embedding analysis, perhaps even more striking, appears in the behavior at the collapse temperature: Compared with the other temperatures, the model protein more likely preserves nonrandom characteristics for a wider range of principal components; that is, the flattening of $E2(d)$ is more suppressed. Figure 28 shows the minimum of $E2(d)$, $E2_{min}$, and the minimum embedding dimension d_L of each principal component at several temperatures. Here, d_L is determined by the convergence of $E1(d)$ to within 95% of unity. $E2_{min}$ (=$\min\{E2(1), E(2), \ldots, E2(d)\}$) may indicate the degree of "nonrandomness" of the time series; as $E2_{min}$ goes to one, $E2(d)$ becomes flatter independently of d. Figure 28a shows that as the temperature departs from the collapse temperature, $E2_{min}$ of all the PCs approach unity. The smaller the variances of the PC, the closer $E2_{min}$ to unity. Figure 28b confirms the conclusions from Figure 28a: The larger the variance of the PC, the lower dimensionality is necessary to reconstruct the state space, and away from the collapse temperature the dimensionality of the state space increases.

Let us go back to the fundamental question, What is the implication of the dimension d_L evaluated by scalar *finite* time series of each principal component of the *same* system? If one measures several observables from the *same* system to find the embedding dimension, each measurement along with its timelags provides a different nonlinear mapping of the original state space variables $\mathbf{x}(t)$ into a d-dimensional reconstructed space. Takens' theorem tells us that irrespective of the choice of the observables, each should reach a same *global* dimension d_E if one were able to deal with time series with an infinite sampling length and an infinitesimal resolution. However, all real-time series data do not satisfy these mathematical requirements and each measurement that may see different local topographies of the state space results in a different *local* dimensionality: For example, if the variance of an observable $s(t)$ is almost the same size as or lesser than the (finite) resolution of measurements, one has no way to know the geometric information of the state space.

In principle, the time series of the principal components analyzed in this chapter should not be random, because all the time series originally arise from the deterministic equation of motion. In this report, at each temperature for both the models, we analyzed the time series collected at every 100 simulation steps for up to 10^7 steps. That is, we could access the time series $\mathbf{Q}(t)$ not with an infinitesimal resolution but with only a finite resolution. The flattening observed in $E2(d)$ for some principal components with small variances (enhanced when departing from the transition temperature) indicates that the small and fast chaotic fluctuations of the principal components make the $\mathbf{Q}(t)$ "lose" memory of the process in such a short duration, resulting in a "random" stochastic time series. One can not extract the geometric information of the underlying dynamical structure from any observable composed of such "lost" DOFs: Note that the corresponding time series $\mathbf{Q}(t)$ collected at every step exhibits a strong d-dependency of $E2(d)$ and yields a finite dimension at these temperatures. At much lower temperatures than $T = 0.2$ where the system can only move about a single local minimum—say, $T = 0.01$—the system's motion is well represented by normal modes (i.e., fully regular dynamics) and all the $E2(d)$ exhibit strong d-dependencies, and both the $E1(d)$ and $E2(d)$ quickly converge to unity at a small dimension, irrespective of the time period to record the time series.

All real observables are always contaminated by noise or by computer round-off errors with a finite resolution of observations, and they are sampled up to a certain *finite* time. It would be meaningless to inquire into the absolute figure of the embedding dimension d_L from a given finite time series, especially for high-dimensional systems. However, an analysis of the *dependence of d_L* on the chosen observables (e.g., coordinates, projections, etc.) can shed light on the dynamical structure of multidimensional complex systems. Figure 28 tells us that dynamics of the larger-amplitude principal components tend to be represented by dynamical systems with fewer DOFs. That is, roughly speaking, the larger the amplitude of the principal component, the smaller the (relative) number of "effectively coupled modes." At the collapse temperature the lowest few PCs can be considered to represent transitions between the collapsed state and the denatured state, associated with the large amplitudes. This implies that the dynamics of transition between two "superbasins" is more regularized than those in the "superbasin" (where those "superbasins" are not necessarily defined in the configurational space but generally in the state space).

At the end, we will make a comment on the essential difficulty in elucidating the minimum embedding dimension for high-dimensional systems: Except for small dynamical systems like the Lorenz attractor, it becomes impossible for many-dimensional systems to yield sufficient sampling to find the nearest neighbors of a point $\mathbf{y}_d(t_i)$ besides the adjacent $\mathbf{y}_d(t_i + 1)$ or $\mathbf{y}_d(t_i - 1)$, when d becomes very large, resulting in an apparent convergence at a certain high dimension, but still smaller than the desired high dimension. One plausible prescription might be to replace the searching of the nearest neighbors at every single (recorded) step by that at selected steps—for example, every 10 (recorded) steps [86]. Similarly, such a sampling problem arising from an insufficient search for the nearest neighbors that converges to an adjacent point may also make it difficult to interpret the difference of d_L of several measurements from the *same* system. That is, different delay times τ of different measurements from the *same* system effectively yield different sampling lengths, which may result in an apparent difference of d_L arising not from the different *local* dimensionality of the dynamical structure but rather from the difference of insufficient samples. We examined the minimum embedding dimension d_L of each principal component with a delay time τ at each temperature, whose sampling length N_{samp} is rescaled from the total sampled step ($= 10^5$) so as to be approximately equal to the minimum sampling length of $Q_1(t)$ at the transition temperature T_c with the maximum delay time τ_{max}:

$$N_{\text{samp}} \simeq \frac{\tau}{\tau_{\text{max}}} 10^5 \qquad (43)$$

The results using the rescaled sampling length are qualitatively similar to the original ones; that is, the larger the variance of the PC, the lower the dimensionality, and away from the collapse temperature, the dimensionality of the PC increases. Here, the average number of the nearest neighbors (besides adjacent points) of each principal component measurement are more likely to be equal to each other at all temperatures except that at T_c they are relatively lower than those at the other temperatures. This might indicate that the observed distinctions in d_L at different temperatures reflect not only the difference of local dimensionality of the state space structures but perhaps also the difference in some nonstationary features of the finite time series especially at the transition temperature; that is, the nonstationarity enhanced at the transition temperature might require a longer time, in comparison with the other temperatures, to obtain reliable sampling of the nearest neighbors.

V. CONCLUDING REMARKS AND FUTURE PROSPECTS

Energy landscape theories have succeeded in quantifying and rationalizing several observed complex kinetics [39]. The recent experimental developments in single molecule spectroscopy [53–55] hold great promise for revealing the complexity of multibasin dynamics inherent in a single molecule. All the clues for complexity of dynamics whenever they arise from nonstatistical behavior should be buried in the underlying geometrical structure of the state space. Tsallis nonadditive statistical mechanics [87] may be able to take into account a certain "heterogeneity" of the (global) phase-space structure. In the context of recrossing behavior of transitions in chemical reactions, variational transition state theory [5,6,88] to optimize a given configurational dividing surface so as to minimize the reactive flux through it might also be able to reflect the nonstatistical, dynamical behavior in the region of saddles. However, neither of these methodologies could ever provide us with any deeper understanding about the state space or, in the other terms, the physical foundation of their consequences. The embedding time-series analysis should be one of the most versatile means to address the geometrical aspects of the state space for such manybody systems. By using Cao's embedding algorithm combined with principal component analysis, it was revealed that the fast fluctuations with small amplitudes make the time series of 70–80% of the principal components almost "random" only in a hundred simulation steps, and that the closer the temperature to the collapse temperature, the more the stochastic feature of the principal components is suppressed through a wider range of DOFs. This may indicate that even on a rugged multibasin energy landscape, when the "basin-to-basin" transitions occur through a "small bottleneck" as rare events, regulatory behavior is brought about during the course of transitions, as seen in the other systems—for example, argon cluster [13–20], liquid water [30,31], and cytochrome c [40]. Some of the unresolved issues to be addressed concerning embedding time-series analysis and principal component analysis are summarized as follows:

1. Although many methods have been developed so far to choose the time delay τ and to determine the minimum embedding dimension, there exists no general method, especially for many-body systems where "modes" associated with hierarchical time and space scales are not necessarily decoupled from one another. The question of which geometrical information at a certain hierarchy on the state space can be "reconstructed" is not trivial at all for real finite time-series data with finite resolution.

2. The application of embedding time-series analysis to multivariate observables [75,84,86] is desired in order to extract a good projection, revealing the dynamical structure from a limited set of observables. It is known [89,90] that an application of the embedding analysis to time series involving intermittency, like those of the Gō-like model, is not straightforward and involves many problems that need to be overcome.
3. There does not exist delay embedding theorem for Hamiltonian systems.
4. Principal component analysis (PCA), in addition to similar techniques (e.g., MODC), is one of the most versatile means to address the question of which coordinate(s) or projection(s) is(are) best to trace the underlying mechanism of dynamics in multidimensional complex systems. To what extent can the PCs actually mimic dynamical behavior of a set of the original variables in terms of a reduced set of the new variables? The answer, in principle, depends on the similarity of the distribution of the original data set to a "Gaussian." If the protein moves about only within a single basin, PCA is expected to work quite well [24,25] to describe multidimensional dynamical behavior by a reduced *small* set of the PCs. As the temperature increases, the protein may find reaction paths from one large (super) basin to another, and the transitions may take place through a curved or winding path. It is difficult for the standard PCA algorithm to represent a curved or winding distribution in a reduced small set of the PCs, even while the PCA could still draw a rough sketch of the basin-to-basin transitions.
5. As far as using the (conventional) microscopic Liapunov exponent analysis, one can never quantify any regularized, cooperative behavior of complex dynamics that should emerge at coarse-grained scales observed on rugged, multibasin energy landscapes. Recently, to analyze the dynamical structure at a coarse-grained level for protein folding dynamics, we introduced the concept of the so-called collective chaos [91,92] or finite-size Liapunov exponent [93–96], combined with the principal components (PCs). It was found [73] that although protein dynamics is regarded as "uniformly" strong chaos and the microscopic Liapunov exponents do not depend on the kinds of PCs, regularized, cooperative behavior is confined to a low-dimensional cylindrical manifold associated with small, finite-size Liapunov exponents.
6. There is no versatile technique to transform a full system to a coarse-grained system with a reduced set of DOFs, which still maintains the topographical feature of the energy landscapes of the original full system. There is one possible means to address this coarse-graining problem interms of the collective coordinates and the conjugate momenta in terms of the principal components, Eqs. (6) and (12). It was reported [24,25] that

the largest $\sim 20\%$ of the total principal components are enough to reproduce the variance of the data for typical protein molecules. However, the question "How can the number of DOFs be reduced as little as possible, while maintaining the topographical feature of the energy landscapes of proteins?" has not been discussed so far. Our recent investigation [97] revealed a funnel energy landscape can be well reproduced in terms of a reduced set of PC compared with a more frustrated energy landscape.

7. It was revealed recently by an MD simulation of human lysozyme in solution [98] that, while the rotational relaxation time of the "randomly" moving individual water molecules is about 3.8 ps on average, the rotational diffusion of local dipole field at a "solvent site" around human lysozyme was strongly restrained up to 256 ps, to yield a long-term dynamical memory resulting from the first-layer solvents. Here, the local dipole field is defined as a short time average of the dipole moment vectors of many individual water molecules at a solvent site through which they transiently pass or visit. This indicates that water molecules do not necessarily retard the protein motions as "friction" but can also mediate them. A principal component analysis of protein motions in solution has not taken into account directly such a "positive" contribution of solvent DOFs. Plotkin and Wolynes [99] addressed how the inclusion of the solvent DOFs into the reaction coordinate is essential to describe the passage from the denatured state to the native state. To reveal the regularity of protein motions mediated by surrounding solvents, a new principal component analysis incorporating the solvent DOFs is required to address this subtle problem.

These are some of several challenging, but very fascinating, subjects to still be addressed and overcome for the forthcoming future in multibasin dynamics problems.

VI. APPENDIX

A. Embedology

What can a scalar time series tell us about the multivariate state (or phase) space buried in the observations? The so-called embedology attributed to Whitney [75] and Takens [76] provides us with an essential clue to the answer of such a question. A detailed description of the mathematical proof is beyond the scope of this review, and here we focus on describing the brief concept and methodology.

Embedding. Assume that any trajectories are confined to an attractor A in the k-dimensional Euclidean state space $\mathbb{R}^k$, where trajectories are described by a

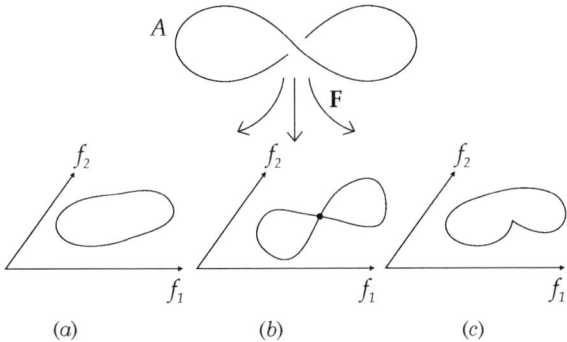

Figure 29. Several mappings **F** of one-dimensional attractor **A** onto two-dimensional spaces (f_1, f_2): **F** is called (*a*) an embedding if and only if **F** is a one-to-one (i.e., if $\mathbf{x} \neq \mathbf{y}$ are points on **A**, and then $\mathbf{F}(\mathbf{x}) \neq \mathbf{F}(\mathbf{y})$ in $\mathbb{R}^m$) and preserves the differential property of the original **A** (the latter is referred to as *immersion*). Namely, (*b*) and (*c*) correspond to an immersion that fails to be a one-to-one, and a one-to-one that fails to be an immersion, respectively.

deterministic dynamical system—that is, k-differential equations. For example, as shown in Fig. 29, suppose a one-dimensional attractor, a twisted circle, in $\mathbb{R}^k$ and project it onto a certain two-dimensional space $\mathbb{R}^2$. Which is the best map or projection, denoted by **F** (**F** : $\mathbb{R}^k \to \mathbb{R}^2$) in that figure, to persist the *topological properties of the original attractor?* One can see that only the map **F** in Fig. 29a preserves the topological properties in the two-dimensional space so that **F** is *one-to-one* (any distinct points on the original attractor **A** in $\mathbb{R}^k$ are mapped to some distinct points on the *mapped* **A** in $\mathbb{R}^2$) and preserves its differential property.

As shown in Fig. 29a, the mapping of an original attractor in $\mathbb{R}^k$ onto a (reduced) m-dimensional space $\mathbb{R}^m$, while preserving the topological properties of the original attractor, is called *embedding*, and the original attractor is regarded as *reconstructed* or *embedded* in the m-dimensional state space. The one-to-one property reflects the uniqueness of solution of a deterministic dynamical system which enables us to predict the dynamical evolution of the system of interest in terms of the embedded state space.

Embedding Theorem. Now suppose there exists an arbitrary d-dimensional smooth and compact manifold **A** in $\mathbb{R}^k$.

> Let us articulate some technical terminologies here. A d-dimensional *manifold* is a set that locally resembles Euclidean space $\mathbb{R}^d$, and each point on a manifold must have a neighborhood around itself that looks like a small piece of $\mathbb{R}^d$. The letters **D** and **O** are 1-(dimensional) manifolds, but the letters **A** and **X** are not. The surfaces of balls and doughnuts are 2-(dimensional) manifolds. In the strict definition of

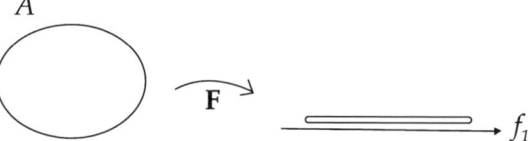

Figure 30. An observation **F** of a 1-manifold **A** in $\mathbb{R}^1$, (f_1).

manifold, although the letters **L** and **U** fail to be 1-manifolds because their end points do not have a small neighborhood around them that looks like $\mathbb{R}^d$, this type of set is called a *manifold with boundary* (A Möbius band is a 2-manifold with boundary because the edges look locally like a line, not a plane). As just a twisted circle shown in Fig. 29, the term *smooth* here stands for the existence of derivatives of, at least, first order of the manifold. By stating that a manifold is *compact*, one means that a subset of points in the Euclidean space $\mathbb{R}^k$ that lie on that manifold is both closed and bounded; for example, while the letter **D** is a compact but not smooth 1-manifold, **O** is a smooth and compact 1-manifold.

Then *what will be the minimum number of dimensions* $m(< k)$ *required to embed a d-dimensional manifold* **A** *lying in* $\mathbb{R}^k$? Whitney's embedding theorem [75] states a condition that ensures producing the embedding of the d-dimensional manifold **A** in $\mathbb{R}^k$ onto a reduced state space $\mathbb{R}^m$. Here, in order to capture its essence, let us consider an example of a one-dimensional manifold **A**, a twisted circle, that will be observed in $\mathbb{R}^{m'}(m' = 1, 2, 3)$. As shown in Fig. 30, when the 1-manifold **A** is projected on to a one-dimensional Euclidean space $\mathbb{R}^1$, self-intersections (i.e., not one-to-one) occur at almost every point inevitably. For the

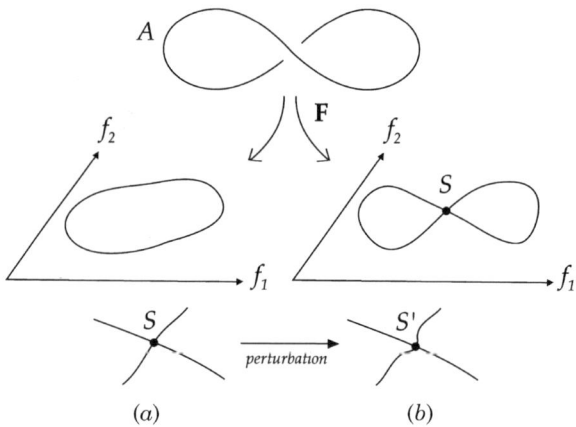

Figure 31. An observation **F** of a 1-manifold **A** in $\mathbb{R}^2$, (f_1, f_2).

REGULARITY IN CHAOTIC TRANSITIONS ON MULTIBASIN LANDSCAPES   305

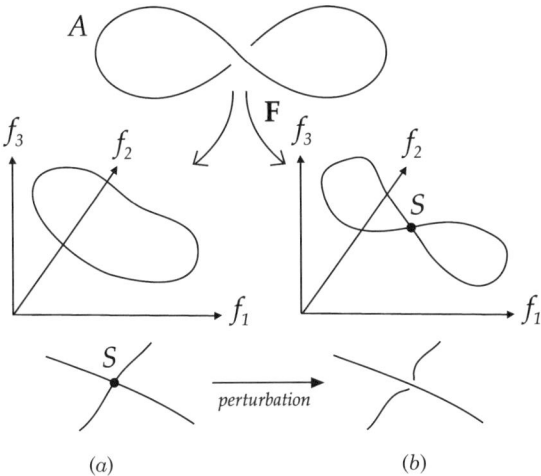

Figure 32. An observation **F** of a 1-manifold **A** in $\mathbb{R}^3$, (f_1, f_2, f_3).

projections of **A** onto the two-dimensional Euclidean space $\mathbb{R}^2$, one might be able to embed the manifold **A** as in Fig. 31a, due to some *a priori* knowledge with respect to the topological character of the state space or by luck. However, it is naturally expected that one often has a self-intersection in the $\mathbb{R}^2$ like in Fig. 31b, and, if it does exist, the intersection cannot be unfolded by any perturbation. [Just imagine that one takes a (2-dimensional) picture of a twisted rubber band. Here one observes a self-intersection in the figure. Then, any small change of the direction of the viewpoint in the measurement cannot untangle the intersection.] Figure 32 shows the projections of **A** onto the three-dimensional Euclidean space $\mathbb{R}^3$. Although it is also possible for "a twisted rubber band" to have a self-intersection in an $\mathbb{R}^3$ as in Fig. 32b, it is easily expected that a perturbation in the measurement will, *in generic*, untangle the intersection in $\mathbb{R}^3$. This shows that three-dimensional state space $\mathbb{R}^3$ is enough to embed the 1-manifold, with preservation of the topological character of the original manifold.

This illustrates Whitney's embedding theorem [75] to provide us with a *sufficient* condition to yield the minimum number of dimension $m(> 2d)$, the so-called *embedding dimension*, required to embed a d-dimensional manifold. The mathematical description of the embedding theorem is as follows:

Assume that there exists a d-dimensional smooth and compact manifold **A** in $\mathbb{R}^k$. If $m > 2d$, a set of embedding maps **F** from $\mathbb{R}^k$ into $\mathbb{R}^m$ consists of an *open* and *dense* set. Here a condition that a set of the maps is open and dense (sometimes the properties of an open and dense set is referred to as *generic*) results in that, given an embedding map **F**, **F** will persist as an embedding under any arbitrarily small

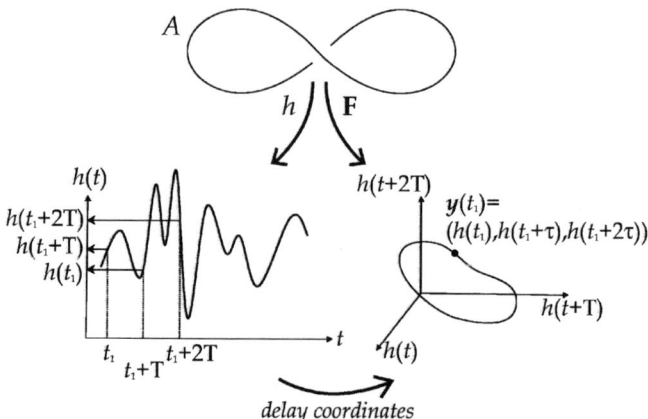

Figure 33. A schematic diagram of the delay coordinate reconstruction.

perturbation, and every smooth map (whether it is an embedding or not) can arbitrarily be close to an embedding.

Delay Coordinate. The Whitney embedding theorem tells us how many simultaneous measurements, $(f_1, f_2, f_3, \ldots)$ we need to embed an underlying manifold of a dynamical system. However, in practice, it is usual to observe only one or a few of the dynamical quantities. The theorem also requires that the different simultaneous measurements are independent. It might be rather difficult even for bright experimentalists to choose observables which are indeed independent mutually. In such cases, Takens' *delay coordinates* approach is one of the most profitable means to overcome the difficulty of simultaneous independent measurements. Figure 33 shows a schematic diagram of reconstructing the state space in terms of the delay coordinate system. Here, the measurement $h : \mathbb{R}^k \to \mathbb{R}^1$ gives us a single time series $h(t)$ in the time domain. The delay coordinate system **y** is defined as

$$\mathbf{y} = [h(\mathbf{x}), h(\mathbf{g}^T(\mathbf{x})), \ldots, h(\mathbf{g}^{(m-1)T}(\mathbf{x}))] = [h(t), h(t+T), \ldots, h(t+(m-1)T)] \tag{44}$$

where **g** denotes the dynamical system for which **A** is the manifold and $\mathbf{g}^T(\mathbf{x}(t))$ means an evolution operation to bring $\mathbf{x}(t)$ to the state T time units later, $\mathbf{x}(t+T)$. As shown in Fig. 33, given T, one can prescribe, for example, three distinct figures $h(t_1)$, $h(t_1 + T)$, and $h(t_1 + 2T)$ at a time t_1, resulting in a single point in the three-dimensional delay coordinate system. Takens' embedding theorem [76] ensures the following:

Assume that **A** is a d-dimensional smooth and compact manifold in $\mathbb{R}^k$. If $m > 2d$ and the delay coordinate map $\mathbf{F} : \mathbb{R}^k \to \mathbb{R}^m$ is made from a generic measurement h and generic time delay T, then $\mathbf{F}$ is an embedding. Here the genericity required on T is necessary because, for instance, if **A** is a periodic orbit with a period equal to the time delay T, then any delay coordinate map cannot be one-to-one.

This embedding theorem holds irrespective of the choice of the delay time T if one could observe an infinitely long time series at infinitesimally small resolution.

General Versions of the Embedding Theorems. The embedding theorems attributed to Whitney and Takens showed that for a d-dimensional manifold **A** we can *generically* produce the embedding map from $\mathbb{R}^k$ into $\mathbb{R}^m$ ($m > 2d$). Most systems we may observe in nature are more or less chaotic, where trajectories are asymptotic to fractal sets rather than manifolds of an integer dimension (e.g., the box counting dimension of Lorenz strange attractor $\simeq 2.06$). Sauer and co-workers [79,80] showed that their embedding theorem also holds for "attractors" **A** which are fractal sets rather than smooth manifolds by interpreting a noninteger fractional dimension d in terms of the so-called box-counting dimension. Suppose an attractor **A** lies in the k-dimensional Euclidean space $\mathbb{R}^k$ which is divided by k-dimensional boxes placed at k-dimensional grid points whose coordinates are ϵ-multiples of integers. Then, the box-counting dimension $boxdim(\mathbf{A})$ is defined by

$$boxdim(\mathbf{A}) = \lim_{\epsilon \to 0} \frac{\log N(\epsilon)}{-\log \epsilon} \qquad (45)$$

where $N(\epsilon)$ is the number of boxes that intersect **A**. If the limit does not exist, the upper (lower) box-counting dimension can be defined by replacing the "lim" by the "lim inf" ("lim sup"). When the box-counting dimension exists, an approximate scaling law

$$N(\epsilon) \sim \epsilon^{-d} \qquad (46)$$

holds where $d = boxdim(\mathbf{A})$. Figure 34 illustrates how to calculate $boxdim\mathbf{A}$ by using the well-known Hénon attractor in $\mathbb{R}^2$. Readers can also imagine that the box-counting dimension will be two irrespective of the size of ϵ (scale-invariant) if points on **A** ergodically distribute through the whole region of $\mathbb{R}^2$. The proof of the theorem could be sketched as follows: Let B_1 and B_2 be arbitrarily chosen, k-dimensional boxes (ϵ each side) that intersect the set **A**, as shown in Fig. 35. When B_1 and B_2 are mapped from $\mathbb{R}^k$ to $\mathbb{R}^m$ by a smooth map $\mathbf{F}$, the probability that the images of B_1 and B_2 in $\mathbb{R}^m$ overlap each other is proportional to $\sim \epsilon^m$ (i.e., the inverse of total number of the "boxes" in $\mathbb{R}^m$) where the boxes in $\mathbb{R}^k$

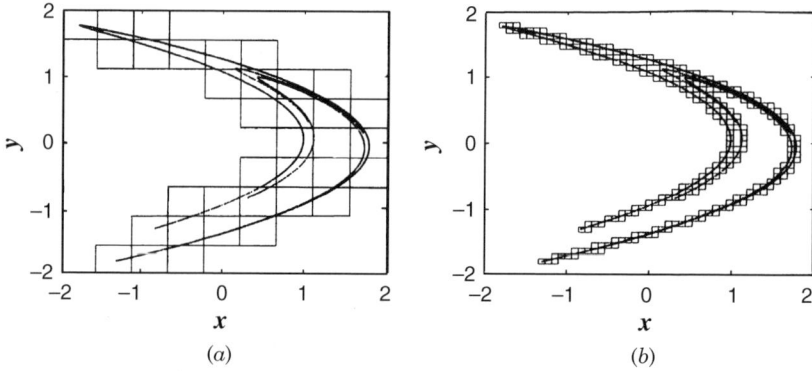

Figure 34. An example of calculating the box-counting dimension for Hénon attractor. (a) $\epsilon = 4/10$, $N(\epsilon) = 35$. (b) $\epsilon = 4/50$, $N(\epsilon) = 270$.

mapped by **F** will be deformed on $\mathbb{R}^m$). The number of boxes that cover **A** of box-counting dimension d is proportional to $\sim \epsilon^{-d}$. The number of pairs of boxes covering **A**, randomly chosen from a set of $N(\epsilon) \sim \epsilon^{-d}$, will be proportional to $\sim (\epsilon^{-d})^2$. Hence, the probability to find an intersection in an image of **A** in $\mathbb{R}^m$ is proportional to $\sim (\epsilon^{-d})^2 \epsilon^m = \epsilon^{m-2d}$. This implies that if $m > 2d$, the probability to choose a **F** that fails to be one-to-one is negligible for small ϵ.

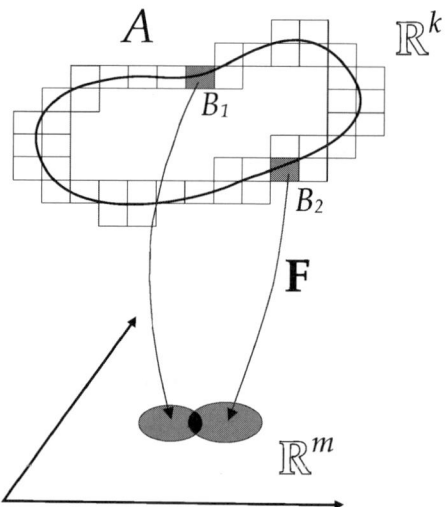

Figure 35. A schematic picture of B_1 and B_2 boxes (ϵ on a side) mapped from $\mathbb{R}^k$ into $\mathbb{R}^m$ by $\mathbf{F} : \mathbb{R}^k \to \mathbb{R}^m$. The probability that the images overlap is proportional to $\sim \epsilon^m$.

One must have noticed that the above discussions are based not on the concept of topology of the set used for the original theorem by Whitney and Takens but on the concept of probability. Although the properties of *open* and *dense* subsets are referred to as *generic*, some *open* and *dense* sets have arbitrarily small Lebesgue measure *even* in Euclidean spaces, resulting in probability-zero, in practice! This implies that the topological arguments are not necessarily appropriate to produce a versatile result on the likelihood of embedding.

Sauer et al. [79] developed a new concept, the so-called *prevalence*, on infinite-dimensional spaces, with a particular emphasis on function spaces, which corresponds to the concept of *almost every* (i.e., with the set of exceptions of Lebesgue measure zero) on finite-dimensional spaces. They generalized Whitney and Takens embedding theorems to ensure that they still hold for *non*-manifold A with noninteger fractal dimension with minor modifications. For a further detailed discussion, one should see Ref. 79.

B. Berendsen Algorithm for Constant-Temperature MD Simulation

Here we derive the Berendsen algorithm to control the temperature T with minimal local disturbance to the system without any explicit stochastic variable, in which the system is weakly coupled to an external heat bath [60]. Suppose a Langevin equation at a desired temperature T_0

$$m_i \dot{v}_i = F_i - m_i \gamma_i v_i + R_i(t) \tag{47}$$

where F_i and m_i are, respectively, the system force and the mass for *i*th DOF and R_i is a Gaussian stochastic force with zero mean and with intensity

$$\langle R_i(t) R_j(t+\tau) \rangle = 2 m_i \gamma_i k_B T_0 \delta(\tau) \delta_{ij} \tag{48}$$

The damping constants γ_i to determine the strength of coupling to the heat bath is set to be γ for all the DOFs of the system, and k_B is the Boltzmann constant.

The time dependence of T can be derived from the derivative of the total kinetic energy K:

$$\frac{dK}{dt} = \lim_{\Delta t \to 0} \frac{\sum_{i=1}^{3N} \frac{1}{2} m_i v_i^2(t+\Delta t) - \sum_{i=1}^{3N} \frac{1}{2} m_i v_i^2(t)}{\Delta t} \tag{49}$$

where N is the number of particles, and

$$\Delta v_i = v_i(t+\Delta t) - v_i(t) = \frac{1}{m_i} \int_t^{t+\Delta t} [F_i(t') - m_i \gamma v_i(t') + R_i(t')] \, dt' \tag{50}$$

By assuming that the system obeys a Langevin equation at a shorter time scale than Δt, that is, $R_i(t')$ is uncorrelated with $v_i(t)$ and $R_i(t)$ for $t' > t$ and

$$\int_t^{t+\Delta t} dt' \int_t^{t+\Delta t} dt'' \left(\frac{1}{3N} \sum_{i=1}^{3N} R_i(t') R_i(t'') \right) = 2 m_i \gamma k_B T_0 \Delta t \tag{51}$$

we obtain

$$\frac{dK}{dt} = \sum_{i=1}^{3N} v_i F_i + 2\gamma \left(\frac{3N}{2} k_B T_0 - K \right) \tag{52}$$

$$= \sum_{i=1}^{3N} v_i F_i + 3N \gamma k_B (T_0 - T) = \sum m_i v_i \dot{v}_i \tag{53}$$

where (kinetic) temperature is $T = 2K/3Nk_B$. Hence, we obtain the following equation of motion to control temperature T to a desired T_0 without adding local stochastic components:

$$m_i \dot{v}_i = F_i + m_i \gamma \left(\frac{T_0}{T} - 1 \right) v_i \tag{54}$$

Acknowledgments

We wish to thank Professors R. Stephen Berry, Mikito Toda, M. Sasai, A. Kidera, and A. Kitao for their continuous, insightful discussions and criticisms. We acknowledge Dr. Konstantin S. Kostov for his valuable contribution, especially in the $1/f$ power spectra analysis of potential energy fluctuations of Brownian Dynamics. We also thank Mr. Gareth Rylance for his critical reading of this manuscript.

Parts of this work were supported by the Japan Society for the Promotion of Science, Grant-in-Aid for Research on Priority Areas "Genome Information Science," and "Control of Molecules in Intense Laser Fields" of the Ministry of Education, Science, Sports, and Culture of Japan, Sumitomo Foundation of Science, Inoue Foundation of Science, Hyogo Science Foundation and 21st century COE (Center of Excellence) of "Origin and Evolution of Planetary Systems (Kobe University)," MEXT.

References

1. D. J. Wales, *Energy Landscapes: With Applications to Clusters, Biomolecules and Glasses*, Cambridge University Press, Cambridge, 2003.
2. H. Eyring, *J. Chem. Phys.* **3**, 107 (1935).
3. M. G. Evans and M. Polanyi, *Trans. Faraday Soc.* **31**, 875 (1935).
4. E. Wigner, *J. Chem. Phys.* **5**, 720 (1938).
5. J. C. Keck, *Adv. Chem. Phys.* **13**, 85 (1967).
6. D. G. Truhlar and B. C. Garrett, *Acc. Chem. Res.* **13**, 440 (1980).

7. L. Sun, K. Song, and W. L. Hase, *Science* **296**, 875 (2002).
8. N. Shida, *Adv. Chem. Phys. Part B* **130**, 129 (2005).
9. K. Takatsuka, *Adv. Chem. Phys. Part B* **130**, 25 (2005).
10. M. J. Davis and S. K. Gray, *J. Chem. Phys.* **84**, 5389 (1986).
11. S. K. Gray and S. A. Rice, *J. Chem. Phys.* **87**, 2020 (1987).
12. M. Zhao and S. A. Rice, *J. Chem. Phys.* **96**, 6654 (1992).
13. T. Komatsuzaki and R. S. Berry, *J. Chem. Phys.* **110**, 9160 (1999).
14. T. Komatsuzaki and R. S. Berry, *J. Chem. Phys.* **116**, 862 (2002).
15. T. Komatsuzaki and R. S. Berry, *Phys. Chem. Chem. Phys.* **1**, 1387 (1999).
16. T. Komatsuzaki and R. S. Berry, *J. Mol. Struct. (Theochem.)* **506**, 55 (2000).
17. T. Komatsuzaki and R. S. Berry, *Proc. Natl. Acad. Sci. USA* **78**, 7666 (2001).
18. T. Komatsuzaki and R. S. Berry, *J. Chem. Phys.* **115**, 4105 (2001).
19. T. Komatsuzaki and R. S. Berry, *Adv. Chem. Phys.* **123**, 79 (2002).
20. T. Komatsuzaki and R. S. Berry, *J. Phys. Chem. A* **45**, 10945 (2002).
21. T. Komatsuzaki and R. S. Berry, *Adv. Chem. Phys. Part A* **130**, 143 (2005).
22. O. M. Becker, Jr., A. D. MacKerell, B. Roux, and M. Watanabe, ed., *Computational Biochemistry and Biophysics*, Marcel Dekker, New York, 2001.
23. R. M. Levy, A. R. Srinivasan, W. K. Olson, and J. A. McCammon, *Biopolymers* **23**, 1099 (1984).
24. A. Kitao, F. Hirata, and N. Gō, *Chem. Phys.* **158**, 447 (1991).
25. A. Kitao and N. Gō, *Curr. Opin. Struct. Biol.* **9**, 164 (1999).
26. J. C. Gower, *Biometrika* **53**, 325 (1966).
27. J. C. Gower, *Biometrika* **55**, 582 (1968).
28. O. M. Becker and M. Karplus, *J. Chem. Phys.* **106**, 1495 (1997).
29. N. Elmaci and R. S. Berry, *J. Chem. Phys.* **110**, 10606 (1999).
30. I. Ohmine, *J. Phys. Chem.* **99**, 6767 (1995).
31. I. Ohmine and S. Saito, *Acc. Chem. Res.* **32**, 741 (1999).
32. F. H. Stillinger and T. A. Weber, *Phys. Rev. A* **25**, 97 (1982).
33. F. H. Stillinger and T. A. Weber, *Phys. Rev. A* **28**, 2408 (1983).
34. A. Shudo and S. Saito, *Adv. Chem. Phys. Part B* **130**, 375 (2005).
35. A. Fersht, *Structure and Mechanism in Protein Science*, W. H. Freeman, San Francisco, 1998.
36. M. Karplus, *J. Phys. Chem. B* **104**, 11 (2000).
37. N. Gō, *Annu. Rev. Biophys. Bioeng.* **12**, 183 (1983).
38. J. D. Bryngelson and P. G. Wolynes, *J. Phys. Chem.* **93**, 6902 (1989).
39. S. Takada, *Proc. Natl. Acad. Sci. USA* **96**, 11698 (1999).
40. A. E. García and G. Hummer, *Proteins* **36**, 175 (1999).
41. Y. Matsunaga, K. S. Kostov, and T. Komatsuzaki, *J. Phys. Chem. A* **106**, 10898 (2002).
42. H. Nymeyer, A. E. García, and J. N. Onuchic, *Proc. Natl. Acad. Sci. USA* **95**, 5921 (1998).
43. C. Wagner and T. Kiefhaber, *Proc. Natl. Acad. Sci. USA* **96**, 6716 (1999).
44. S. S. Plotkin, *Proteins* **45**, 337 (2001).
45. B. Jun and D. L. Weaver, *J. Chem. Phys.* **116**, 418 (2002).
46. D. Hamada, F. Chiti, J. I. Guijarro, M. Kataoka, N. Taddei, and C. M. Dobson, *Nature Struct. Biol.* **7**, 58 (2000).

47. M. Shimono and T. Komatsuzaki, to be submitted for publication.
48. M. F. Shlesinger, G. M. Zaslavsky, and J. Klafter, *Nature* **363**, 31 (1993).
49. M. F. Shlesinger, in *Encyclopedia of Applied Physics*, Vol. 16, VHC Publishers, 1996, p. 45.
50. G. Zumofen and J. Klafter, *Physica D* **69**, 436 (1993).
51. I. M. Sokolov, J. Klafter, and A. Blumen, *Physics Today* **55**, 48 (2002).
52. P. Carlini, A. R. Bizzarri, and S. Cannistraro, *Physica D* **165**, 242 (2002).
53. T. Yanagida, K. Kitamura, H. Tanaka, A. H. Iwane, and S. Esaki, *Curr. Opin. Cell Biol.* **12**, 20 (2000).
54. S. Weiss, *Nature Struct. Biol.* **7**, 724 (2000).
55. Y. Ishii, T. Yoshida, T. Funatsu, T. Wazawa, and T. Yanagida, *Chem. Phys.* **247**, 163 (1999).
56. R. S. Berry, N. Elmaci, J. P. Rose, and B. Vekhter, *Proc. Natl. Acad. Sci. USA* **94**, 9520 (1997).
57. D. J. Wales, J. P. K. Doye, M. A. Miller, P. N. Mortenson, and T. R. Walsh, *Adv. Chem. Phys.* **115**, 1 (2000).
58. J. D. Honeycutt and D. Thirumalai, *Biopolymers* **32**, 695 (1992).
59. Z. Guo, C. L. Brooks III, and E. M. Boczko, *Proc. Natl. Acad. Sci. USA* **94**, 10161 (1997).
60. H. J. C. Berendsen, J. P. M. Postma, W. F. van Gunsteren, A. DiNola, and J. R. Haak, *J. Chem. Phys.* **81**, 3684 (1984).
61. D. W. Allan, *Proc. IEEE* **54**, 221 (1966).
62. C. Seko and K. Takatsuka, *J. Chem. Phys.* **104**, 8613 (1996).
63. K. Tanaka and Y. Aizawa, *Prog. Theor. Phys.* **90**, 547 (1993).
64. M. Takano, T. Takahashi, and K. Nagayama, *Phys. Rev. Lett.* **80**, 5691 (1998).
65. M. Takano, H. K. Nakamura, K. Nagayama, and A. Suyama, *J. Chem. Phys.* **118**, 10312 (2003).
66. M. B. Weissman, *Rev. Mod. Phys.* **60**, 573 (1988).
67. Y. Aizawa, Y. Kikuchi, T. Harayama, K. Yamamoto, M. Ota, and T. Tanaka, *Prog. Theor. Phys. Suppl.* **36**, 985 (1989).
68. Y. Aizawa, *Chaos, Solitons, and Fractal* **11**, 263 (2000).
69. E. G. Marinari, G. Parisi, D. Ruelle, and P. Windey, *Commun. Math. Phys.* **89**, 1 (1982).
70. W. F. van Gunsteren and H. J. C. Berendsen, *Mol. Phys.* **45**, 637 (1982).
71. Y. Aizawa, *Prog. Theor. Phys. Suppl.* **99**, 149 (1998).
72. T. Okabe and H. Yamada, *Chaos, Solitons and Fractals* **9**, 1755 (1998).
73. T. Komatsuzaki, Y. Matsunaga and M. Toda, to be submitted for publication.
74. K. Hoshino, Y. Matsunaga, N. Koga, S. Takada, and T. Komatsuzaki, unpublished.
75. H. Whitney, *Ann. Math.* **37**, 645 (1936).
76. F. Takens, in *Lecture Notes in Mathematics*, 898. Springer, New York, 1981.
77. H. D-I. Abarbanel, *Analysis of Observed Chaotic Data*, Springer-Verlag, New York, 1995.
78. H. Kantz and T. Schreiber, *Nonlinear Time Series Analysis*, Cambridge, 1997.
79. T. Sauer, J. A. Yorke, and M. Casdagli, *J. Stat. Phys.* **65**, 579 (1991).
80. K. T. Alligood, T. D. Sauer, and J. A. Yorke, *Chaos: An Introduction to Dynamical Systems*, Springer-Verlag, New York 1996.
81. M. Casdagli, S. Eubank, J. D. Farmer, and J. Gibson, *Physica D* **51**, 1134 (1991).
82. M. B. Kennel, R. Brown, and H. D-I. Abarbanel, *Phys. Rev. A* **45**, 3403 (1992).

83. L. Cao, *Physica D* **110**, 43 (1997).
84. L. Cao, A. Mees, K. Judd, and G. Froyland, *Int. J. Bifurcat. Chaos* **8**, 1491 (1998).
85. A. M. Fraser and H. L. Swinney, *Phys. Rev. A* **33**, 1134 (1986).
86. D. M. Walker and N. B. Tufillaro, *Phys. Rev. E* **60j**, 4008 (1999).
87. C. Tsallis, *J. Stat. Phys.* **52**, 479 (1988).
88. J. Villa and D. G. Truhlar, *Theor. Chem. Acc.* **97**, 317 (1997).
89. N. Platt, E. A. Spiegel, and C. Tresser, *Phys. Rev. Lett.* **70**, 279 (1993).
90. J. G. von Hardenberg, F. Paparella, N. Platt, A. Provenzale, E. A. Spiegel, and C. Tresser, *Phys. Rev. E* **55**, 58 (1997).
91. T. Shibata and K. Kaneko, *Phys. Rev. Lett.* **81**, 4116 (1998).
92. T. Shibata, T. Chawanya, and K. Kaneko, *Phys. Rev. Lett.* **82**, 4424 (1999).
93. G. Paladin, M. Serva, and A. Vulpiani, *Phys. Rev. Lett.* **74**, 66 (1995).
94. E. Aurell, G. Boffetta, A. Crisanti, G. Paladin, and A. Vulpiani, *Phys. Rev. Lett.* **77**, 1262 (1996).
95. E. Aurell, G. Boffetta, A. Crisanti, G. Paladin, and A. Vulpiani, *J. Phys. A: Math. Gen.* **30**, 1 (1997).
96. G. Boffetta, A. Crisanti, F. Paparella, A. Provenzale, and A. Vulpiani, *Physica D* **116**, 301 (1998).
97. T. Komatsuzaki, K. Hoshino, Y. Matsunaga, G. J. Rylance, R. L. Johnston, and D. J. Wales, *J. Chem. Phys.* in press.
98. T. Yokomizo, S. Yagihara, and J. Higo, *Chem. Phys. Lett.* **374**, 453 (2003).
99. S. S. Plotkin and P. G. Wolynes, *Phys. Rev. Lett.* **80**, 5015 (1998).

CHAPTER 18

NONMETRIC MULTIDIMENSIONAL SCALING AS A DATA-MINING TOOL: NEW ALGORITHM AND NEW TARGETS

Y-H. TAGUCHI

Department of Physics, Faculty of Science and Technology, Chuo University, Bunkyo-ku, Tokyo, 112-8551, Japan; and Institute for Science and Technology, Chuo University, Bunkyo-ku, Tokyo, 112-8551, Japan

YOSHITSUGU OONO

Department of Physics, University of Illinois at Urbana-Champaign, Urbana, Illinois 61801-3080, USA

CONTENTS

- I. Introduction
- II. nMDS: The Concept
- III. Algorithm
- IV. Evaluating Obtained Configurations
  - A. Pointwise Criterion
  - B. Estimation of the Number of Effectively Embedded Points
- V. An Illustrative Example
- VI. Molecular Taxonomy Example
- VII. Comparison of Molecular and Morphological Classifications
  - A. Ferns
  - B. Green Autotrophs
- VIII. Biodiversity of Soil Bacteria
- IX. Brain Wave
- X. Protein Family

Geometric Structures of Phase Space in Multidimensional Chaos: A Special Volume of Advances in Chemical Physics, Part B, Volume 130, edited by M. Toda, T Komatsuzaki, T. Konishi, R.S. Berry, and S.A. Rice. Series editor Stuart A. Rice.
ISBN 0-471-71157-8 Copyright © 2005 John Wiley & Sons, Inc.

XI. Microarray Data
   A. How Much Information Does the Set of All Available Microarray Data Have?
   B. Temporal Patterns of Gene Expression
XII. Concluding Remarks
References

I. INTRODUCTION

Rapid progress in information technology and nanotechnology has made it routine work to obtain a huge dataset of GB order about biological or economical systems. Even inexpensive personal computers have at least 10-GB hard disk and a few-GHz clock speed. This capability is tantamount to collecting more than 10^9 data points every nanosecond; that is, we may have a $10^9 \times 10^9$ data matrix every second. For example, a WWW search engine Google can list as many as a few gigapages for every query. Or, it is practical to measure the expression levels of 10^4 genes under a few hundred different experimental conditions. Thus, getting huge empirical datasets is no longer difficult. The real problem lies in how we cope with the flood of data.

For a huge dataset to make sense to us humans, it is mandatory that the information in the original dataset must be distilled to a "humane size" so that we can recognize patterns. This is the essence of *data mining*. A collection of multivariate analysis (MVA) methods have been developed to this end (see, for example, Ref. 1). A typical and the most classic example is the linear regression analysis that tries to find a linear relationship or linear pattern

$$y_i = ax_i + b$$

between x_i and y_i characterizing the structure of the dataset $\mathscr{D} = \{(x_i, y_i)\}$. That is, a large set $\mathscr{D}$ is represented by only two numbers a and b; we would succeed in a significant data reduction, if the fit is reasonable. A typical class of MVA methods reduces a huge dataset into a set of small number of quantities, or, equivalently, extract simple patterns underlying the dataset. However, even with the Bayesian approach, the class of patterns or functions that are used to reduce the dataset must be given beforehand. In this sense this class of MVA is not truly data-driven.

Another important class of MVA is represented by cluster analysis methods and principal component analysis (PCA). The latter is a representative of data reduction methods that exploit linear algebra. We do not, however, believe all the important patterns can be captured by linear algebraic methods. Linear mathematical methods are ideal for data compression, because to recover the original data distortion is undesirable. Thus, data compression is essentially applied Fourier analysis [2]. In contrast, data mining is a kind of pattern recognition, and so the methods for it should not be confined to linear mathematical methods. We will comment on the cluster analysis later.

In this chapter, we would like to demonstrate that one of the very old MVA tools, nonmetric multidimensional scaling (nMDS) [3], can work well as an unsupervised truly data-driven method for data reduction. We first explain an efficient maximally nonmetric algorithm [4] and then demonstrate its superiority to linear MVA methods. We also demonstrate that the subsequent application of linear MVA after data reduction by nMDS can often be a powerful data mining technique.

In Section II the concept of nMDS is outlined, and our efficient algorithm is explained in Section III. Some statistical evaluation methods of the nMDS analysis results are proposed in Section IV. After these preparations, Section V illustrates that nMDS can reproduce metric results without any metric input. This fact is well known, but our demonstration uses a very large dataset. Section VI illustrates an nMDS applications to taxonomy. It is shown that quite a mechanical treatment of the base sequence data can produce relations of cichlids in Lake Tanganyika and Lake Malawi that are consistent with a detailed molecular phylogenetic analysis result. Section VII is also about taxonomic applications. A possibility of correlating genetic and phenotypic characteristics through common objects is suggested. Section VIII illustrates an analysis of electroencephalogram that demonstrates the capability of nMDS to detect subtle patterns. Section IX is similar in this respect, but the analysis of soil bacterium diversity results in a practical diagnosis of the biological stability of soil. Up to this point (except in Section V), the examples do not exploit one of the merits of our efficient algorithm, namely, capability of treating numerous (more than a few hundred) objects with inexpensive laptop computers. Thus, in Section X the analysis of about 400 AAA proteins demonstrates that nMDS could reveal protein functionalities, and in Section XI microarray data (with as many as 3000 genes) are analyzed for gene expression patterns. The final section is a summary.

II. nMDS: THE CONCEPT

We are interested in a data-mining or data reduction method that is as unsupervised as possible. In other words, we are interested in the data analysis method that is maximally data-driven. We do not mean that we insist that data-mining methods must be unsupervised, but we believe that supervising or guidance could do best with methods that can already allow sufficiently powerful data reduction in the unsupervised mode.

The aim of data reduction is to extract salient patterns underlying the given dataset. For humans to recognize a structure, we should take advantage of our own superb geometric pattern recognition capability. Then, we must explore methods for geometrical representations of data in an appropriate space to aid our vision. PCA is of course such a method, but as already discussed it is a linear method. Since the data need not have any natural or intrinsic vector space structure, we do not wish to be confined within the linear algebraic methodology.

Let $\mathscr{D} = \{\alpha_i\}$ be a set of objects. We assume that there is a way to (at least qualitatively) compare the objects (e.g., for their similarities) and that the comparison result (e.g., the collection of similarities or dissimilarities among the objects) $\mathscr{C}$ is obtainable. Our aim is to find a geometric configuration (constellation) of $\mathscr{D}$ in a certain space S that is in a certain sense compatible with $\mathscr{C}$, so it is hoped that our powerful vision could be used to discern significant patterns in the resultant constellation. It is, however, difficult to visualize nonmetric spaces. Therefore, it is sensible to assume that visualization requires S to be a metric space. This is not a limitation in practice, because we may assume in many applications that the topological space is "regular enough" (e.g., regular, with a countable basis) to be metrizable. We prepare a metric space S and a map $f : \mathscr{D} \to S$ such that the metric relations among the image points and $\mathscr{C}$ are, in a certain sense, compatible. If $f(\mathscr{D})$ has a simple structure (e.g., with a small dimensionality), then the outcome greatly facilitates our pattern recognition. There have already been such methods for a long time: multidimensional scaling (MDS) methods [5].

Cluster analysis methods may also be considered as geometrical representation methods of the underlying structure in $\mathscr{C}$. In this case, we may interpret that S is a Cayley tree with the metric measured in the number of segments in branches. Cluster analyses are often computationally efficient, but unless the data are strictly genealogical, the methods tend to discard too much information. Such a problem may be illustrated by a trivial example: If a cluster analysis method is applied to analyzing a time-series data, obtained clusters may reflect a certain aspect of the time evolution of the system. However, needless to say, structures representing the time dependence cannot automatically be obtained by the cluster analysis, because the relations (e.g., time-ordering) among clusters that are not directly connected cannot be specified by the branching structure.

There are two kinds of MDS: metric and nonmetric [5]. The metric MDS requires that $\mathscr{C}$ is a set of metric relations (that is, $\mathscr{D}$ is assumed to be already in a metric space), and then the method becomes essentially linear algebraic.

Since cluster analysis methods are of limited nature, and MVA methods are not really data-driven, we must conclude that these methods as well as the metric MDS methods are not qualified as maximally data-driven data-mining aids. Therefore, our only hope is to promote the methods broadly called nonmetric MDS (nMDS) as a versatile general reduction methodology for relational data.

Let us assume that we have a set $\mathscr{D}$ consisting of N objects and certain dissimilarity δ_{ij} between objects i and j in $\mathscr{D}$. The totality of the available δ_{ij} is $\mathscr{C}$. nMDS places N points corresponding to the objects in some metric space S such that the metric relation among these points in S is in a certain sense compatible with $\mathscr{C}$. Depending on (I) the choice of the metric space S and (II) the interpretation of "compatibility" with $\mathscr{C}$, there are many different nMDS methods [5].

We never assume *a priori* that the objects are in a metric space at all. Therefore, we do not assume that $\{\delta_{ij}\}$ are directly interpretable as the collection of metric data. However, we assume that the inequality relations such as $\delta_{ij} > \delta_{nm}$ are meaningful; otherwise, we must conclude that $\mathscr{C}$ does not contain any useful information.[1] It should be noted that the inequality does not imply that δ_{ij}'s must be numbers; here we use δ_{ij} to denote a kind of relation. For example, in the application of nMDS in psychology, the original application field of the method, the inequality implies that a subject feels that the relation between pair ij is more remote than that between pair mn, even if she cannot assign any numerical values to δ_{ij}'s. Thus, we assume that all the information in $\mathscr{C}$ is exhausted by the ranking order of δ_{ij} among $\mathscr{C}$.

The reader might question the generality of this assumption; crucial information may have been discarded by only using the ranking order of $\{\delta_{ij}\}$ as the information contained in $\mathscr{C}$ instead of the raw data. As can be seen in Section V, if $\{\delta_{ij}\}$ is the actual metric distance set in a Euclidean space, our nMDS recovers the original configuration of the objects in this space generically very accurately. Therefore, we may safely assume that even the metric information, if contained in the data, is well-preserved by this choice of "reduced" use of the actual $\mathscr{C}$.

The traditional interpretation in nMDS of the compatibility of $\mathscr{C}$ and the geometrical structure $f(\mathscr{D})$ in S is that the distances among the points in $f(\mathscr{D})$ have the same ordering as the dissimilarities in $\mathscr{C}$. More precisely, let d_{ij} be the distance between $f(i)$ and $f(j)$ in S, and let F be the map induced by f: $\delta_{ij} \to F(\delta_{ij}) = d_{ij}$. Then, the compatibility is that F is order-preserving: If $\delta_{ij} > \delta_{mn}$, then $F(\delta_{ij}) > F(\delta_{mn})$. There are many ways to quantify the deviation from this ideal compatibility condition. Thus, there are many versions of nMDS.

In our nMDS realization we use the ordinary Euclidean space for choice (I). As to enforcing compatibility (II), we choose the minimization of rank discrepancies. Let $r(a)$ denote the rank of $a \in A$ in the ascending ordering in A. Then, we try to minimize $\sum [r(\delta_{ij}) - r(F(\delta_{ij}))]^2$.

nMDS methods were invented originally to analyze psychological data [3], and they have been used mainly in social sciences and humanities. However, most MVA textbooks and references lack detailed discussions of nMDS. For example, one of the latest MVA textbooks [1] contains only one superficial section devoted to the topic. Except for a few cases, nMDS has never been used in physical sciences; outside the humanities, perhaps only in ecology has nMDS

---
[1]We need not assume that inequality relations hold between every pair of object pairs: that is, we allow many missing data.

been used; in bioinformatics there are a few examples, but the use has always been confined to examples with not very many objects (less than 50). We wish to treat at least a few thousand objects with an inexpensive laptop computer. Our algorithm for nMDS seems to be suitable to this end. The Fortran source code is down-loadable from http://www.granular.com/MDS/.

III. ALGORITHM

As explained in the preceding section, we try to embed N objects in d-Euclidean space E_d so that the ranking order of δ_{ij} among $\{\delta_{ij}\}$ can be as identical as possible to that of d_{ij} among $\{d_{ij}\}$, where d_{ij} is the distance of the two points corresponding to objects i and j in E_d. The choice of d will be discussed later (e.g., Section XI).

The algorithm we have been using for almost 10 years to realize the above embedding is as follows:

Suppose δ_{ij} is the kth largest among $\{\delta_{ij}\}$ and d_{ij} is the T_kth largest among $\{d_{ij}\}$. Our algorithm is based on the fundamental idea of finding the positions of N points corresponding to the objects in E_d that minimize

$$\Delta = \sum_k (T_k - k)^2$$

Here, the summation is over all the object pairs (or more generally all the available object pairs). The sum is from $k = 1$ to the cardinality of $\mathscr{C}$. A straightforward idea may be to use Δ as a *potential energy* and to invent some overdamped dynamics driven by the *forces* due to the potential.

nmds ($\{\vec{x}_i\}$, $\{\delta_{ij}\}$):
  $\{\vec{x}_i\} \leftarrow$ random numbers;
  repeat
    $\{d_{ij}\} \leftarrow |\{\vec{x}_i\} - \{\vec{x}_j\}|$ /*compute distance*/
    $\{\vec{e}_{ij}\} \leftarrow \frac{\{\vec{x}_i\} - \{\vec{x}_j\}}{\{d_{ij}\}}$
    for $i = 1$ to N
      for $j = 1$ to N
        $\Delta R \leftarrow$ [rank of d_{ij}] - [rank of δ_{ij}];
        $\vec{x}_i \leftarrow \vec{x}_i - ds \cdot \Delta R \cdot \vec{e}_{ij}$;
  until (convergence)

Here, $\vec{x}_i$ (point i) is the image of embedding of object i in E_d, and $\vec{e}_{ij}$ is the unit vector pointing point i from point j in E_d. $ds \ll 1$ is the time increment. When the rank of d_{ij} is greater than the rank of δ_{ij} (i.e., $\Delta R > 0$), d_{ij} should be shortened. In order to accomplish this, we move point i toward point j by the amount proportional to the rank mismatch. When $\Delta R < 0$, we move point i in the opposite direction. We apply this procedure to all points and repeat the whole procedure until the configuration converges with a suitable convergence criterion.

The CPU time necessary for nMDS with our algorithm is typically a few minutes for a few hundred objects, when we use a laptop PC with a Celeron chip (646-MHz, 256-MB RAM, 20-GB HD). Since CPU time is proportional to the square of the total number of objects, a few thousand objects can be treated within a few hours even if we use this inexpensive PC. If we use desktop PCs with a few-gigahertz clocks, the CPU time would be reduced to tens of minutes even for a few thousand objects.

The source code is available on the web (Fortran source codes): http://www.granular.com/MDS/. Also our algorithm is employed as an nMDS module by PAST (http://folk.uio.no/ohammer/past).

IV. EVALUATING OBTAINED CONFIGURATIONS

We propose two types of evaluation of the "goodness" of the configurations that we obtain by the above-mentioned nMDS algorithm. The first one is a pointwise criterion to evaluate individual objects, and the second one is based on the estimation of the number of correctly embedded objects. In the traditional nMDS approaches, there was a quantity—called the stress—that is the target of minimization. We have, however, eliminated such an artificial quantity introduced only for the computation sake from our scheme (see also Ref. 6). This is the reason why we need new criteria.

A. Pointwise Criterion

As a pointwise criterion that judges how well a particular object is embedded, we compute the square sum of the rank discrepancies for pairs between the point corresponding to the object and other points. Thus, for object i we compute

$$\Delta(i) = \sum_{k=1}^{N-1}(k - T_k)^2$$

where k is the rank of the dissimilarity δ_{ij} among the dissimilarities for the $N-1$ pairs (i,s) with $s \in \{1, 2, \ldots, N\}\backslash\{i\}$, and T_k is the rank of d_{ij} among $\{d_{is}\}$ with the same choice of s. We can also define a "dual" local discrepancy for object i as

$$\Delta'(i) = \sum_{k=1}^{N-1}(k - T_k)^2$$

where k is the rank of the dissimilarity δ_{ij} among the dissimilarities for the $N-1$ pairs (s,j) with $s \in \{1, 2, \ldots, N\}\backslash\{j\}$, and T_k is the rank of d_{sj} among $\{d_{sj}\}$ with the same choice of s. Here, $j \in \{1, 2, \ldots, N\}\backslash\{i\}$. We could use the collection of

$\{\Delta'(i)\}$ for all the objects as a minimization target instead of the global Δ. This local scheme is advantageous if the distribution of the points in S is not expected to be very uniform [7].

When d_{ij} is independent of δ_{ij}, it is known [8] that $\Delta(i)$ obeys a normal distribution with mean $[(N-1)^3 - (N-1)]/6$ and variance $(N-1)N^2(N-2)^2/36$. Therefore, if the probability of $\Delta(i)$ under this null hypothesis is small enough (e.g., less than 0.5%), we can reject this hypothesis and confirm that $\{d_{ij}\}$ is related to $\{\delta_{ij}\}$. This means that object i is placed at a suitable location in the embedded space with a certain confidence level.

B. Estimation of the Number of Effectively Embedded Points

It is possible to estimate the number of effectively embedded points. We use the following quantity for the estimation of the number of effectively embedded points:

$$K \equiv \sum_{\langle \alpha, \beta \rangle} \text{sign}[(d_\alpha - d_\beta)(\delta_\alpha - \delta_\beta)]$$

Here, the summation is over all the pairs of the object pairs $\alpha, \beta \in \{(i,j)\}$ ($i < j$). There are $M \equiv N(N-1)/2$ object pairs. Thus, the number of terms in the summation is $M(M-1)/2$. When there is no rank mismatch, K is equal to $M(M-1)/2$. However, generally it is less than $M(M-1)/2$.

If only N' ($< N$) points are embedded correctly, the above sum is divided into two parts. The first part is

$$K' = \sum_{\langle \alpha', \beta' \rangle} \text{sign}[(d_{\alpha'} - d_{\beta'})(\delta_{\alpha'} - \delta_{\beta'})]$$

where both α' and β' are pairs between correctly embedded points. In this case, $\text{sign}(d_{\alpha'} - d_{\beta'})(\delta_{\alpha'} - \delta_{\beta'})$ is always equal to 1, so the sum is $M'(M'-1)/2$ ($M' = N'(N'-1)/2$). The remaining part consists of the sum of $M(M-1)/2 - M'(M'-1)/2$ terms. However, each term is expected to be ± 1 randomly, so, due to the central limit theorem, the remaining part is of order M at most. Thus,

$$K = K' + O[M]$$

For good embedding, N' should be of the same order of N, so M' must be the same order of M. Consequently, since K is of order M'^2, if N is large enough, we can ignore the second term. Thus, we can estimate the number of correctly embedded objects as

$$N' \simeq \sqrt{2\sqrt{2K}}$$

We use $\sqrt{2\sqrt{2K}}/N$ as the ratio of the correctly embedded objects to the total number of objects.

In the above we have assumed that all the dissimilarity pairs are available for simplicity, and it is easy to modify the estimate accordingly when not all the dissimilarities are available.

V. AN ILLUSTRATIVE EXAMPLE

To demonstrate that our nMDS can handle numerous objects and, at the same time, that the rank information of the metric dissimilarities (= actual distances) can correctly reconstruct the original metric relations among the objects (modulo scaling, rotation, and orientation, needless to say), we choose the configuration of 1000 cities around the world.

For these cities, we define the dissimilarities between cities i and j as

$$\delta_{ij} = -\cos\theta_{ij}$$

where θ_{ij} is the angle between position vectors of these two cities when the origin is taken to be the center of the globe. We must emphasize that as the input information we never use the actual values of δ_{ij}, but use only their ranks in $\{\delta_{ij}\}$. The 3D embedding result is shown in (Fig. 1). The output configuration of nMDS

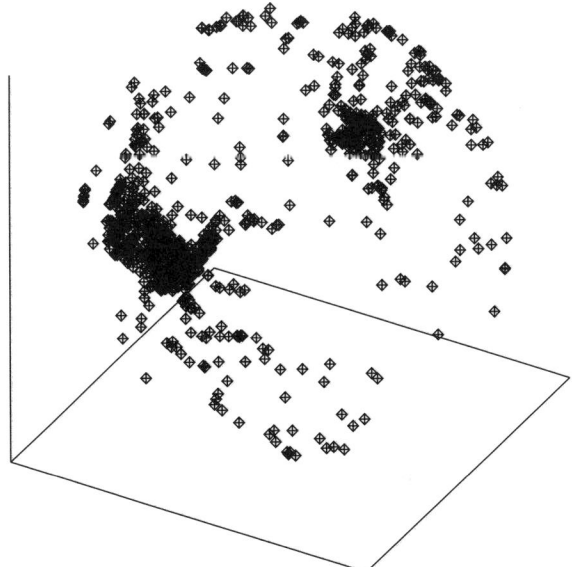

Figure 1. One thousand cities on the globe. ◇: original positions, +: embedded positions. The United States and Europe can be recognized.

cannot generally have correct scale, absolute position, or orientation. Therefore, to compare the output with the original input in this example, we must appropriately scale, rotate, and choose the orientation of the figure. The output of our nMDS scheme correctly recovers the sphericity of the globe and the accurate locations of the cities on it (Fig. 1).

Since the positions of the cities in the embedding space E_3 can be changed continuously and since the input data are only a countable rank information, it is in principle impossible to recover the correct city configurations uniquely. However, for all practical purposes, the remaining latitudes (leeways) are negligible, if we have sufficiently many objects (say, more than 50 in E_3).

VI. MOLECULAR TAXONOMY EXAMPLE

To illustrate the application of our maximally nonmetric MDS to nonmetric data, we analyze the DNA sequence for the NADH subunit 2 of cichlids in Lake Tanganyika and Lake Malawi [9]. The original dataset consists of nucleotide sequences of length $L = 1044$ for 31 species (see the caption of Fig. 2 for the list).

The dissimilarities δ_{ij} are defined as the Hamming distance between the base sequences of species i and j:

$$\delta_{ij} = L - \sum_g \delta[s_{ig}, s_{jg}]$$

where s_{ig} denotes the base (A, T, G, or C) at the gth position of the DNA sequence of species i and $\delta[a, b] = \delta_{ab}$ (Kronecker's δ; not the dissimilarity) (after a suitable alignment). One might criticize that this dissimilarity is not a reasonable choice, because usually the number of uncommon bases is not proportional to the time since the speciation occurred. However, since what we need is just the ranks of δ_{ij}'s among the totality of dissimilarity data $\{\delta_{ij}\}$, any definition that does not alter the ranking gives the same result. In the current example the rate of base substitution is not expected to alter the sequence distances significantly, because we are not studying the so-called deep homologies. Thus, we believe that our results obtained by nMDS is robust and is not affected by a particular definition of distance between the sequences. In this example, $N = 31, ds = 0.1 \times N^{-3}$, and the numbers of iterations are 100 (E_2) and 1000 (E_3, E_4). Note that we have never optimized the algorithm or the parameters in it.

Embedding into E_2 was attempted five times. The results are summarized in Table I. Each run started with a distinct random initial configuration.

Clearly, all solutions have a few worse-than-1%-level species according to the pointwise embedding criterion. Thus, as a whole, $D = 2$ cannot be regarded as a good embedding dimension. One should note that the sum of the total

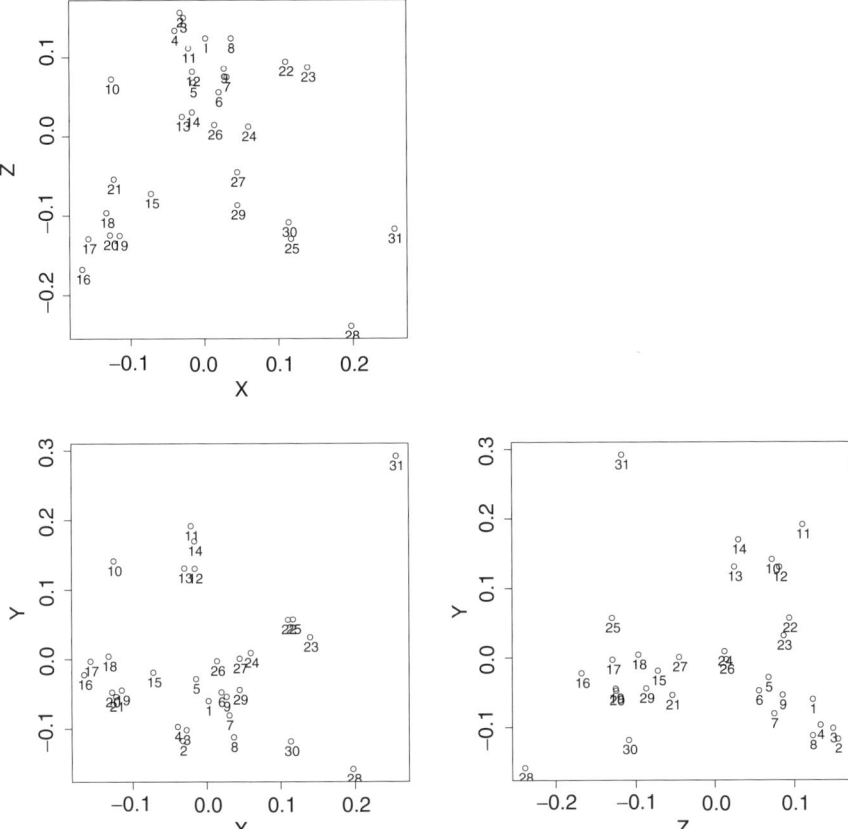

Figure 2. Two-dimensional projections of the obtained 3D configuration—that is, projection to the XY plane, YZ plane, and ZX plane (the directions of the axes are arbitrary)—for cichlid DNA data. Major clusters correspond to the phylogenetic clades proposed by Hasegawa and Kishino (see text). Numbers denote species (names in parentheses other than Malawi represents tribes): 1, *Pseudotropheus zebra* (Malawi); 2, *Buccochromis lepturus* (Malawi); 3, *Champsochromis spilorhynchus* (Malawi); 4, *Lethrinops auritus* (Malawi); 5, *Rhamphochromis sp.* (Malawi); 6, *Lobochilotes labiatus* (Tropheini); 7, *Petrochromis orthognathus* (Tropheini); 8, *Gnathochromis pfefferi* (Limnochromini); 9, *Tropheus moorii* (Tropheini); 10, *Callochromis macrops* (Ectodini); 11, *Cardiopharynx schoutedeni* (Ectodini); 12, *Opthalmotilapia ventralis* (Ectodini); 13, *Xenotilapia flavipinnus* (Ectodini); 14, *Xenotilapia sima* (Ectodini); 15, *Chalinochromis popeleni* (Lamprologini); 16, *Julidochromis marlieri* (Lamprologini); 17, *Telmatochromis temporalis* (Lamprologini); 18, *Neolamprologus brichardi* (Lamprologini); 19, *Neolamprologus tetracanthus* (Lamprologini); 20, *Lamprologus callipterus* (Lamprologini); 21, *Lepidiolamprologus elongatus* (Lamprologini); 22, *Perissodus microlepis 1* (Perissodini); 23, *Perissodus microlepis 2* (Perissodini); 24, *Cyphotilapia frontosa* (Tropheini); 25, *Tanganicodus irsacae* (Eretmodini); 26, *Limnochromis auritus* (Limnochromini); 27, *Paracyprichromis brieni* (Cyprichromini); 28, *Oreochromis niloticus* (Tilapiini); 29, *Tylochromis polylepis* (Tylochromini); 30, *Boulengerochromis microlepis* (Tilapiini); 31, *Bathybates sp.* (Bathybatini).

TABLE I
Results of nMDS Embedding into E_2 of the DNA Sequence Data for Cichlids in both Lake Tanganyika and Lake Malawi[a]

Runs	>0.5%	>1%	>5%	Total	N'	Total + N'
1	—	30	10, 28, 31	4	25.6	29.6
2	—	30	10, 28, 31	4	27.0	31.0
3	30, 21	17, 19, 20, 29	16, 18, 28, 31	10	24.0	34.0
4	—	16, 31	18, 28	4	25.4	29.4
5	10	25, 29	11, 28, 30, 31	7	25.7	33.4

[a]See figure caption of Fig. 2 for the species list. Confidence levels of the embedded objects (species in this case) explained in Section IV.A are shown. N' is the estimated number of correctly embedded species introduced in Section IV.B. Note that not correctly embeddable species are largely the same from run to run.

number of worse-than-0.5%-level species and the estimated number of correctly embedded species is approximately equal to the total number of species, 31, although this asymptotic equality is not expected to hold for the total number of objects as small as in this example. Thus, our two evaluation criteria of goodness proposed in Section IV seem to be consistent. Although the embedding into E_2 is not quite satisfactory, it should be noted that the not successfully embedded objects in Table I are largely the same from run to run. That is, the failure of embedding is due to a few particular species that refuse to be in E_2.

For embedding into E_3, species 28 was with the 0.5–5% confidence level in four out of five runs and species 31 was with the 5% confidence level in three out of five runs. Although a few additional species may be regarded as worse-than-0.5% level species, in two runs only one species (28 or 31) was not embedded correctly. The estimated number of correctly embedded species N' is always about 28.5 independent of the initial conditions.

Even for E_4, we could get only one 0.5% confidence level solution for all species out of five runs; for the rest, species 31 was with 0.5–1% confidence level, and in three runs species 28 was with 0.5–1% confidence level. Note that these species are clear outliers.

Figure 2 illustrates one outcome of the 3D embedding attempts. All the clades recognized by the phylogenetic analysis due to Hasegawa and Kishino [10] ($\{1, 2, \ldots, 9\}$, $\{10, 11, \ldots, 14\}$, $\{15, 16, \ldots, 21\}$ and others) can be recognized as well-defined clusters in Fig. 2. For the reference sake, we give the Shepard plot for this particular 3D embedding (Fig. 3). The correspondence between d_{ij} and δ_{ij} is not as bad as expected from the criteria of the goodness of embedding.

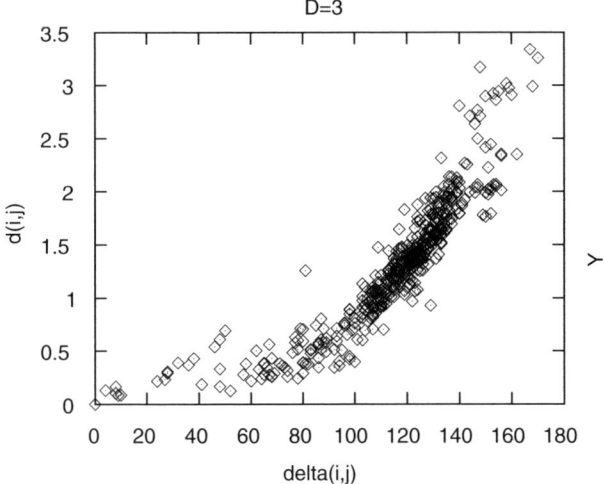

Figure 3. Shepard plot for 3D embedding of the DNA sequence data for cichlids shown in Fig. 3.

The main purpose of Hasegawa and Kishino [10] is to show that the clade $\{1, 2, \ldots, 5\}$ is monophyletic and is separated from the clade $\{6, \ldots, 9\}$. This separation can be recognized as the existing separation plane between these two groups. Thus, in the current example, our analysis simply based on the genetic distance and the result based on detailed phylogenetic analysis agree. However, generally speaking, these two approaches need not agree. Even if the genetic distance is small, species cannot be regarded as phylogenetically close, when there are side branches between them (phylogenetic convention; all the paraphyletic groups are regarded as not proper groups). On the other hand, in our analysis, short genetic distance means closeness. Although it is obvious that closeness in genetic distance alone does not necessarily define a good taxonomic group, we must not forget that the phylogenetic convention need not be biologically sensible for taxonomic studies. Banning all the paraphyletic groups is not necessarily rational.

Although the Shepard plot is reasonable and the outcome may have biologically clear implications, 3D embedding results seem to depend somewhat on initial conditions. Still, the groups recognized in the run illustrated here (or the ones recognized by a much more standard molecular biological analysis) are always separated from run to run. This may be the rule, if the dataset consists of tightly related subgroups; the effective number of data to fix the global geometry could be severely reduced. The following example, although preliminary and the dataset is not updated for the group, may illustrate

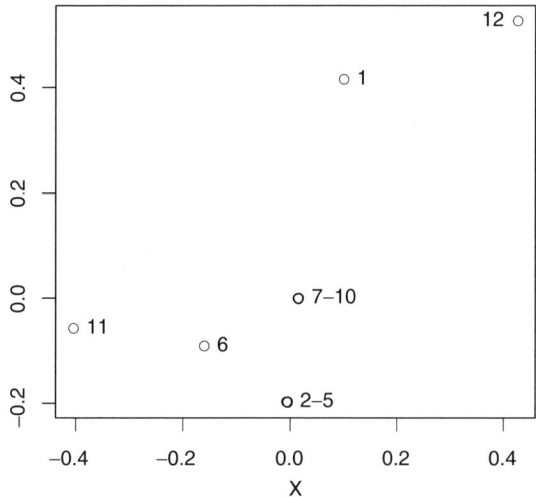

Figure 4. Two-dimensional embeding of some primates. 1, *Lemur catta*; 2, *Homo*; 3, *Pan*; 4, *Gorilla*; 5, *Pongo*; 6, *Hylobates*; 7, *Macaca fuscata*; 8, *M. mulatta*; 9, *M. fascicularis*; 10, *M. sylvanus*; 11, *Saimiri sciureus*; 12, *Tarsius syrichta*.

this point well. We applied nMDS to a small dataset of primates (12 species) [11] (see the caption in Fig. 4).

The dataset consists of mitochondrial DNA sequences with 898 bases. This embedding is not very good because three out of eleven species have failed the pointwise statistical test. However, the estimated number of correctly embedded species is 10.8 and 3D embedding is essentially similar to this, so we show here the 2D result. As can be seen easily, *Homo*, *Pan*, *Gorilla*, and *Pongo* make a single point. Also genus *Macaca* makes a single point. The result suggests that from a purely genetic point of view, it is unreasonable to classify Hominoidea into this many genera. The point of this embedding is that the Hominoidea and *Macaca* clusters are tight and well-defined, but the overall embedding result is not stable as discussed above.

The cichlid example contains only about 30 objects, and we have not investigated the biological significance of the outcome. Since about 90% of the objects are correctly embedded, we may conclude that nMDS has succeeded in visually capturing the information in the data. However, the real power of our method would be shown by the cases with more than 1000 species.

The data discussed in this section have been analyzed traditionally in terms of phylogenetic trees. As we have seen in the cichlid data, the phylogenetic analysis and our nMDS analysis results are compatible. We may expect complementary relations: nMDS cannot directly infer phylogeny, but seems to capture extra relations that cannot be captured by tree diagrams. We must have more experiences with large-scale datasets.

VII. COMPARISON OF MOLECULAR AND MORPHOLOGICAL CLASSIFICATIONS

The relation between genotypes and phenotypes is an issue of central importance in biology (e.g., Ref. 12). It is difficult to compare them in an abstract fashion. One way may be to choose a group of organisms and use it as a digital–analogue converter. Here, we suggest a method to reduce the raw datasets by nMDS before comparison. Preliminary examples follow.

A. Ferns

The first example is the comparison of phenotypes and genotypes of the seedless vascular plants listed below (*Cycas* is listed as an outgroup).[2] It contains 51 species mostly in Pterophyta. For these species we have both DNA sequence and phenotypic character tables [13,14]. The phenotypic character table contains 77 characters that take integer values ranging from 0 to 5. DNA sequences (rbcL) have 1206 bases. The dissimilarity based on phenotypic characters is defined as

$$\delta_{ij} = L - \sum_p \delta[s_{ip}, s_{jp}]$$

where

$$\delta[s_{ip}, s_{jp}] = \begin{cases} 1 & s_{ip} = s_{jp}, \\ 0 & s_{ip} \neq s_{jp}, \\ 0.5 & \text{if } s_{ip} \text{ and/or } s_{jp} \text{ takes more than} \\ & \text{one values and they partially match} \end{cases}$$

---

[2]*Anemia mexicana, Cephalomanes thysanostomum, Asplenium filipes, Lygodium japonicum, Azolla caroliniana, Angiopteris evecta, Blechnum occidentale, Marsilea quadrifolia, Cheiropleuria bicuspis, Matonia pectinata, Cyathea lepifera, Metaxya rostrata, Blotiella pubescens, Botrychium strictum, Dennstaedtia punctilobula, Osmunda cinnamomea, Histiopteris incisa, Ceratopteris thalictroides, Lindsaea odorata, Plagiogyria japonica, Lonchitis hirsuta, Loxogramme grammitoides, Microlepia strigosa, Polypodium australe, Monachosorum henryi, Psilotum nudum, Pteridium aquilinum, Acrostichum aureum, Calochlaena dubia, Adiantum raddianum, Dicksonia antarctica, Coniogramme japonica, Dipteris conjugata, Platyzoma microphyllum, Davallia mariesii, Pteris fauriei, Elaphoglossum hybridum, Taenitis blechnoides, Nephrolepis cordifolia, Salvinia cucullata, Onoclea sensibilis, Actinostachys digitata, Rumohra adiantiformis, Thelypteris beddomei, Diplopterygium glaucum, Vittaria flexuosa, Stromatopteris moniliformi, Lycopodium digitatum, Micropolypodium okuboi, Equisetum arvense, Cycas circinalis.*

and $L = 77$ is the total number of phenotypic characters. The dissimilarity based on the DNA sequences is defined essentially as the Hamming distance:

$$\delta_{ij} = L - \sum_g \delta[s_{ig}, s_{jg}]$$

where

$$\delta[s_{ig}, s_{jg}] = \begin{cases} 1 & s_{ig} = s_{jg} \\ 0 & s_{ig} \neq s_{jg} \end{cases}$$

The nMDS embeddings of the species in E_2 with both dissimilarities are successful (except for *Cycas*, an outgroup representative in this set, failing to pass the pointwise criterion) (Figs. 5 and 6). The estimated number of correctly embedded plants is 46.7 when based on phenotypic characters and 46.4 when based on DNA. They correspond to more than 90% successful embedding. Applying PCA to these results, we have found that the contributions of the first principal component are 69% for phenotypes and 72% for DNA sequences. Thus, these configurations are dominated by single factors.

In order to measure the commonness between the two configurations, we have applied the canonical correlation analysis (CCA) [1]. The CCA applies linear transformations to the original two sets of coordinate systems to produce new coordinate systems such that the correlation coefficients between jth coordinates in both sets is the largest. CCA applied to the nMDS-generated 2D coordinate systems produced the first coordinates with the correlation coefficient 0.79 and the second with 0.32. These values may be understood as quantitative measures of the commonness between the two different types of

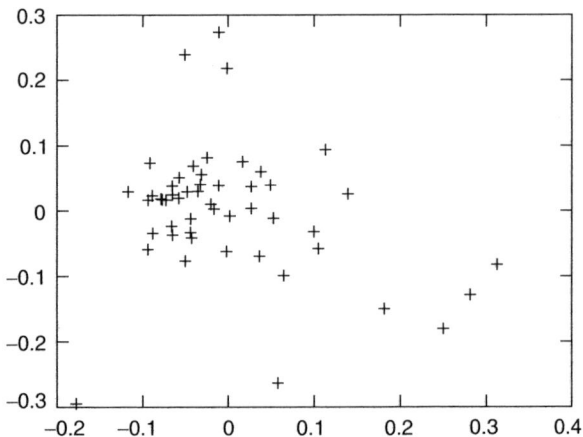

Figure 5. 2D embedding of some seedless vascular plants based on phenotypic characters.

NONMETRIC MULTIDIMENSIONAL SCALING AS A DATA-MINING TOOL 331

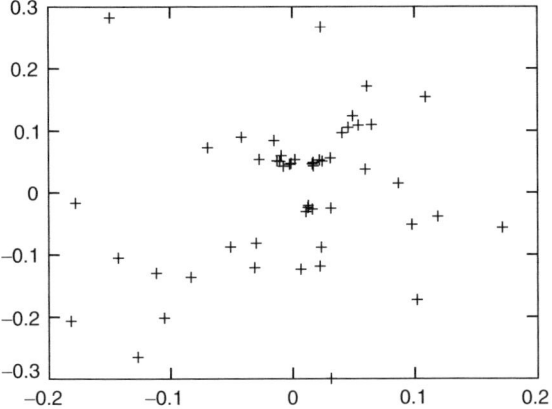

Figure 6. Two-dimensional embedding of some seedless vascular plants based on DNA sequence.

datasets. Without using nMDS, such an analysis is difficult because DNA sequences and phenotypes are hard to compare directly. nMDS enables us to use a group of organisms as an analogue–digital converter, allowing us to compare genotypes and phenotypes as a whole.

It is also possible to merge both embedding coordinates. We denote 2D coordinates for each species as (x_i^P, y_i^P) due to phenotypes and (x_i^D, y_i^D) due to DNA sequences. Applying PCA to $(x_i^P, y_i^P, x_i^D, y_i^D)$, we have found that the cumulative contribution up to the second principal component is 81%. Thus this 2D configuration can express merged reduced embedding of two different types of datasets (see Fig. 7). Here, we could use nMDS as well. However, if PCA

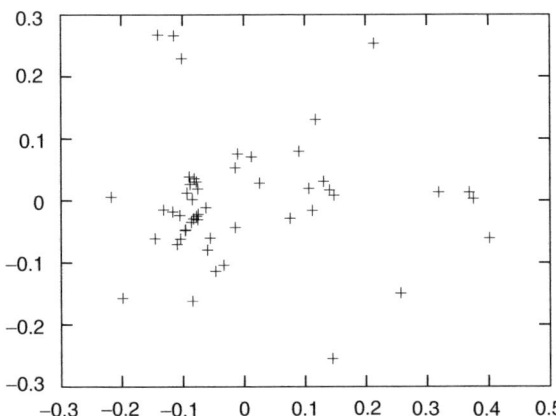

Figure 7. Two-dimensional embedding of some seedless vascular plants with the merged coordinates.

works well, the resulting configuration can be the solution of Nmds with the distance as the dissimilarity.

B. Green Autotrophs

The second example is the analysis of a group including Bryophyta and Chlorophyta.[3] For these species, we have both DNA sequence and phenotype character tables [15,16]. The number of species included is 59, for which 110 phenotypic characters are listed. Each phenotypic character has five evaluation letters from a to e. Some species take more than one value for some characters. Although the definition of δ_{ij} is similar to that used in the preceding subsection, it is modified as

$$\delta_{ij} = \frac{L' - \sum_c \delta[s_{ic}, s_{jc}]}{L'}$$

where L' is the number of phenotypic characters for which both species i and j have some values assigned. This modification is necessary because there are quite a few missing values in the phenotype character table. DNA sequences have 2179 bases (nuclear SSU and LSU rDNA) [17]. The dissimilarity δ_{ij} for the DNA sequences is the same as in the preceding subsection.

The nMDS embeddings of the species in E_2 with both dissimilarities are successful with only one species failing to pass the pointwise criterion. The estimated numbers of correctly embedded species is 55.5 due to phenotypes and 53.63 due to DNA. These numbers imply more than 90% successful embedding. Figures 8 and 9 show 2D embedding results. A PCA analysis of them tells us that the first principal component contributes 92% for phenotypes and 82% for DNA sequences. Thus, these configurations are one-dimensional. A CCA indicates that the first correlation coefficient is 0.49 and the second is 0.26. This implies that for species discussed in this subsection there is less commonness

---

[3]*Glycine max, Cephaleuros parasiticus, Tetraselmis carteriifor, Oryza sativa, Characium vacuolatum, Enteromorpha intestinal, Zamia pumila, Dunaliella parva, Ulva fasciata, Psilotum, Chlamydomonas reinhardt, Ulothrix zonata, Equisetum arvense, Volvox carteri, Cymopolia barbata, Atrichum, Chlorococcopsis min, Batophora erstedtii, Notothylas breutellii, Draparnaldia plumosa, Codium decorticatum, Phaeoceros laevis, Uronema belkae, Cladophoropsis membrano, Porella pinnata, Chlamydomonas moewusii, Blastophysa rhizopus, Conocephalum conicum, Stephanosphaera pluvial, Trentepohlia sp., Asterella tenella, Carteria radiosa, Scenedesmus obliquus, Riccia, Gonium pectorale, Characium hindakii, Klebsormidium flaccidum, Chlorella kessleri, Chlorella fusca, Coleochaete nitellarum, Chlorella vulgaris, Ankistrodesmus falcatus, Fissidens taxifolius, Prototheca wickerhamii, Pseudotrebouxia gigante, Plagiomnium cuspidatum, Chlorella protothecoide, Pleurastrum terrestre, Micromonas pusila, Chlorella minutissima, Characium perforatum, Mantoniella squamata, Neochloris aquaticus, Parietochloris pseudoal, Nephroselmis pyriformis, Neochloris vigenis, Friedmannia israelensis, Pedinomonas minutissima, Pediastrum duplex.*

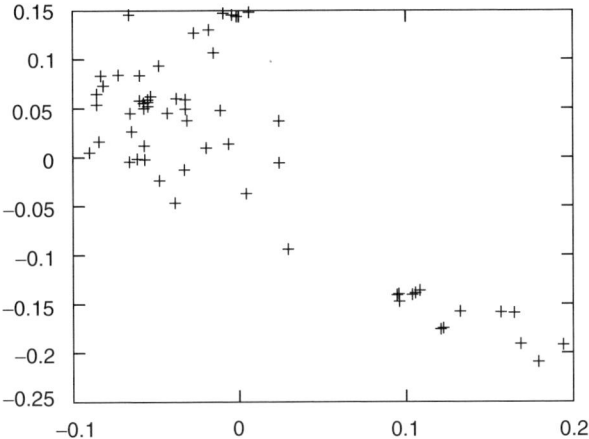

Figure 8. Two-dimensional embedding of the green autotrophs by phenotype.

between phenotype results and DNA sequence results than discussed in the preceding subsection.

It is of course too early to conclude anything definite from these preliminary attempts, but they may suggest that similar phenotypic characters are associated with similar genetic structures only among phylogenetically closely related organisms.

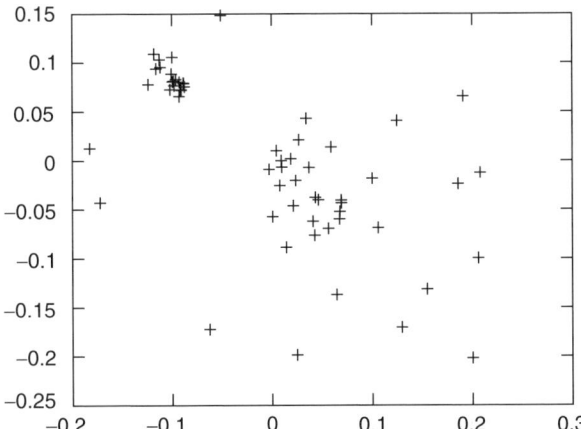

Figure 9. Two-dimensional embedding of the green autotrophs by DNA sequence.

VIII. BIODIVERSITY OF SOIL BACTERIA[4]

In this section we show an example that nMDS can reveal a hidden structure in the data that may be of practical significance. In the soil is a community of bacteria that seems to collectively characterize the agricultural quality of the soil. However, there are very few scientifically recognized species of bacteria. Yokoyama [18] tried to characterize the soil quantitatively through classifying soil bacteria according to their utilization patterns of carbon resources.[5] In the present example, the raw data are given as s_{ir}, where i denotes the bacterial "species" and r denotes the carbon resource; it takes 1 if bacterium i can consume resource r; otherwise, it is 0. The dissimilarity δ_{ij} is defined as the Hamming distance between $\vec{s}_i$ and $\vec{s}_j$, that is,

$$\delta_{ij} = \sum_r |s_{ir} - s_{jr}|$$

The number of carbon resources are 96, which is given by the manufacturer of the test kit.

Here we analyze a particular sample containing 47 cultivable "species." It is not guaranteed that all of them are distinct. The embedding result into E_2 is in Fig. 10, where no bacterium fails to pass the pointwise criterion (all better than 0.5%). The number of correctly embedded bacteria is estimated to be 44.4, corresponding to 94%; 2D embedding is successful. A PCA analysis of the embedded configuration gives the first principal component that explains 85% of the total variance. That is, the nMDS result in E_2 is largely one-dimensional. This result is consistent with the impression given by Fig. 10. This principal component has turned out to be proportional to the number C of the carbon resources that a particular bacterium can consume. More precisely, if we introduce

$$C_i \equiv \sum_r s_{ir}$$

the number of carbon resources that bacterium i can utilize, along with a variable X_i defined by the regression formula

$$X_i = a(\text{first principal component}) + b(\text{second principal component})$$

---

[4]In collaboration with K. Yokoyama of National Agricultural Research Center Hokkaido, National Agriculture and Bio-oriented Research Organization, Japan.
[5]Used plates are produced by BIOLOG Inc.

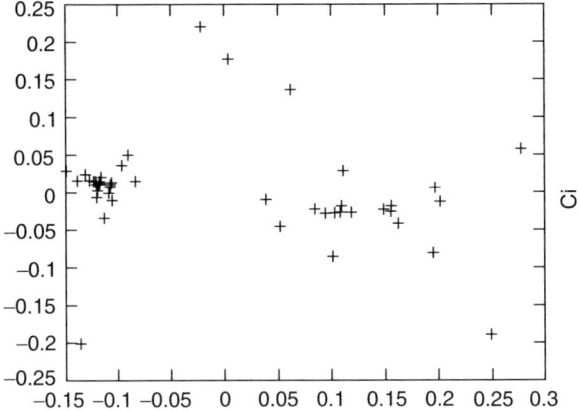

Figure 10. Two-dimensional embedding of 47 bacteria due to the carbon resource consuming capability.

with $a : b \simeq 1 : 0.12$, we see $C_i \propto X_i$ (Fig. 11). The result might suggest that the combination of usable carbon resources defines a niche for a bacterium.

The reader must have wondered what if we apply PCA directly to the original dissimilarity dataset $\{s_{ir}\}$. In the current case, the direct application of PCA gives the same conclusion that the first principal component is a linear

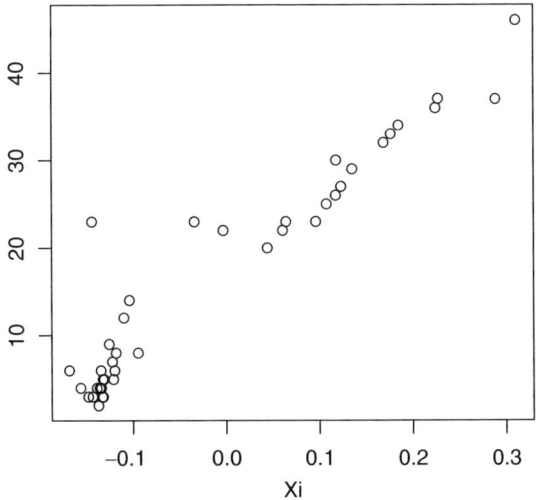

Figure 11. The relationship between the bacterium coordinate X_i and the number of carbon resources C_i that bacterium i can consume.

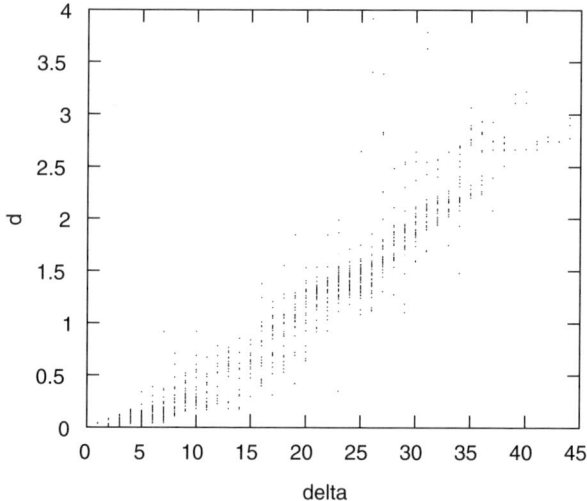

Figure 12. The Shepard plot $\delta_{ii'}$ versus $d_{ii'}$.

function of C. However, the first principal component explains only a half of the total variance, far below the result obtained after nMDS above. That is, the effect of C is not dominant enough in the direct PCA result. In Fig. 12, we plot δ_{ij} versus d_{ij} (Shepard plot). d_{ij} is clearly a linear function of δ_{ij}—that is, the Hamming distance. The Hamming distance is not proportional to the Euclidean distance,

$$\sqrt{\sum_j (s_{ij} - s_{i'j})^2} \not\propto \sum_j |s_{ij} - s_{i'j}|$$

although there is a general monotone relation between them. Since PCA is an orthogonal coordinate transformation, if the Hamming distance is a linear function of C, the first principal component cannot be a linear function of C. Therefore, the direct PCA gives a *weaker* result than the PCA after nMDS. We believe PCA analysis *after* nMDS is a powerful data reduction method. More generally put, linear MVA methods applied after nMDS are more powerful than their direct use.

From this example, we understand why nMDS can be a powerful reduction tool; in contrast to the conventional MVA methods, nMDS assumes less and is more data-driven.

The embedding of bacteria in E_2 for different soils has suggested a practical method to diagnose the quality of the soil (for the agricultural purpose). It has

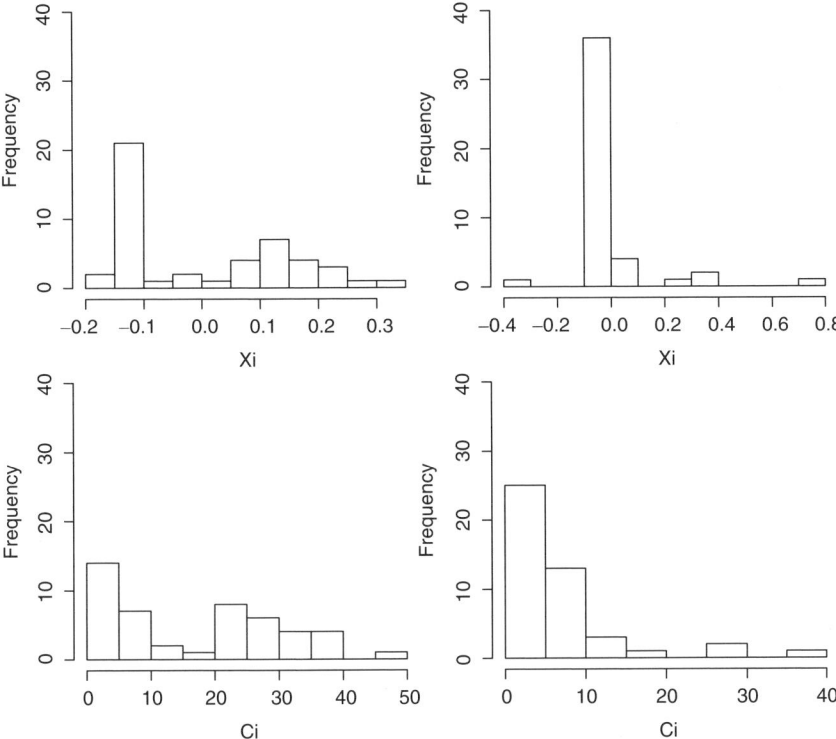

Figure 13. Left column: An example of sickness resistant soil even without applying chemicals **Right column:** An example of soil whose sickness is suppressed by application of chemicals. **Upper row:** X_i. **Lower row:** C_i

been realized that the so-called sick soils invariably give very uneven embedding results (e.g., tightly coupled cluster with several extreme outliers) in contradistinction to the good soils. Figure 13 shows the distributions of bacteria along the X-axis and the C-axis for two distinct soil samples. One of them is from a sickness-suppressed (by chemicals) field—that is, a field whose natural ecological condition is totally destroyed. Such fields become sick as soon as application of chemicals is stopped. The other is from a field without sickness under the natural ecological condition. Clearly, the latter has more biodiversity along these axes (more even). Such a feature is being applied to diagnosis of the agricultural quality of the field soil in practice. Apart from practical usefulness of such studies, we may conclude that certain biodiversity can suppress sickness without application of chemicals. However, we do not know yet how to create appropriate biodiversity in the field soil.

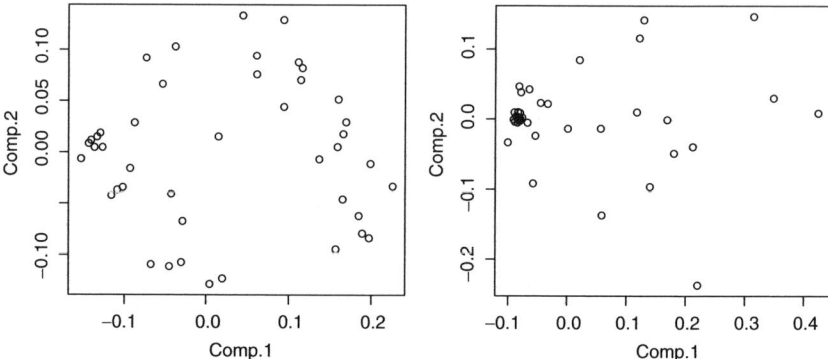

Figure 14. Comparison of 2D embeddings (162 hours later than that of Fig. 13). **Left:** The soil that can suppress sickness without chemicals. **Right:** The soil whose sickness is suppressed by chemicals. Coordinates are rotated by PCA.

In order to see the difference between distributions of these two cases more clearly, in Fig. 14 we exhibit 2D embedding results after a further 162-hour incubation since the situation exhibited in Fig. 13 to enhance the distinction. Clearly, the soil example without chemical has much more even distribution.

IX. BRAIN WAVE

nMDS may be used to analyze dynamical data as well. Here, the feature is illustrated with the analysis of brain wave, or electroencephalogram (EEG). In EEG, 10–100 probes are attached to the scalp to detect voltage changes as a function of time. Here, we have analyzed the result with 64 probes. Each probe (channel) gives a set of data consisting of 256 voltages measured every 1 msec. Let s_{it} be the voltage output at time point t for channel i. As the dissimilarity δ_{ij}, we employ the Euclidean distance between the vectors $(s_{i0}, \ldots, s_{it}, \ldots, s_{i256})$ and $(s_{j0}, \ldots, s_{jt}, \ldots, s_{j256})$ averaged over 10 independent experiments. In these experiments, the subjects were shown two figures with a time interval of 0.3 sec and were asked whether these two were identical or not [19].

We have applied our nMDS to both the cases with identical figures and with distinct figures (Figs. 15–18). For 2D embedding, the estimated number of correctly embedded channels is 60, which is more than 90% of the total number of the channels. We may conclude that these embeddings are very good. Applying PCA to these obtained configurations, we have found that the contribution of the first principal component is 89% for identical figures and 85% for distinct figures. In order to understand the implications of these embedding results, we have plotted the first and second principal components on the channel position map of the scalp.

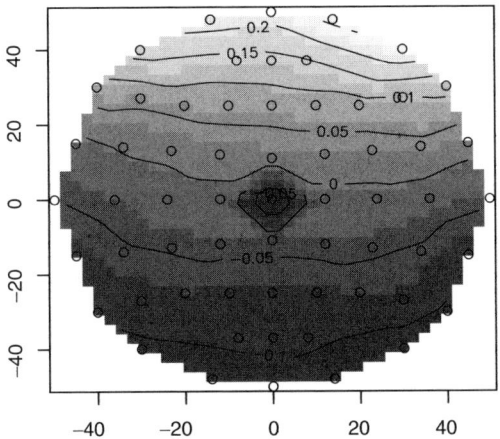

Figure 15. Contour plot of the first principal component of the embedding configuration of brain wave (identical figure cases). Open circles indicate the channel positions.

The first principal component is thought to be due to eye muscle movements: that is, the component has nothing to do with the brain activity itself. On the other hand, the second principal component reflects some activity of visual cortex located on the dorsal side of brain, although not all the peaks are interpretable. Thus, nMDS can separate signals of interest even though they are masked by a one-order-larger component.

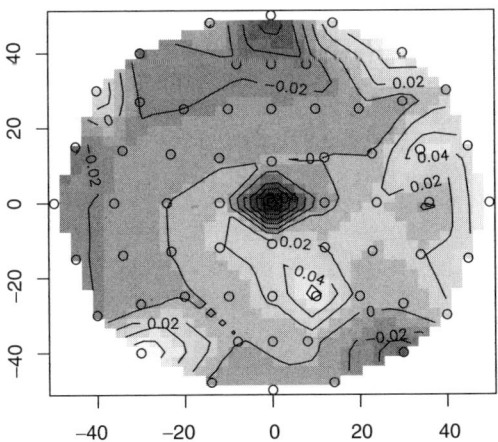

Figure 16. Contour plot of the second principal component of the nMDS embedding result for the brain wave (identical figure cases). Open circles indicate the channel positions. The front is on the top.

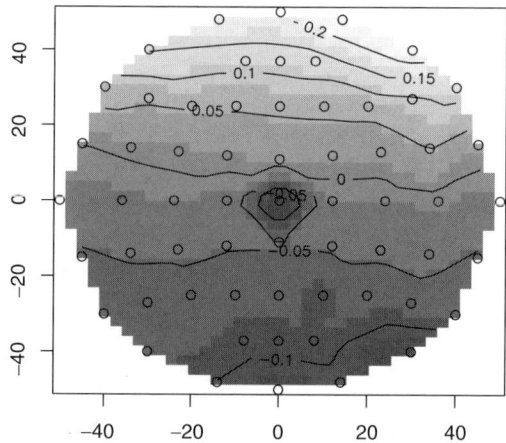

Figure 17. Contour plot of the first principal component of the nMDS embedding result for the brain wave (distinct figure cases). Open circles indicate the channel positions.

The reader might wonder what computationally much faster PCA alone could achieve. The first principal component obtained by a direct application of PCA has the contribution of 77% and the second 7%. In Figs. 19 and 20, we have shown the results of the direct PCA results for identical figure case. There is no clear separation between the signals from eye movements and those from

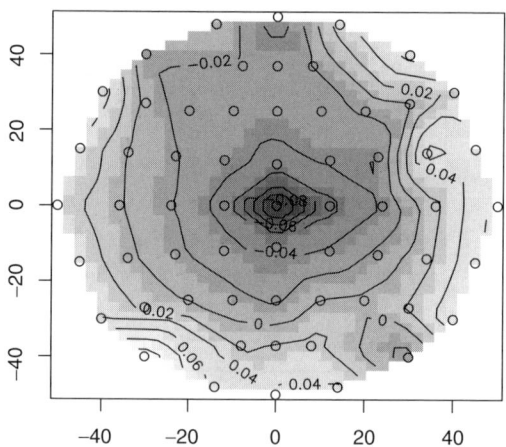

Figure 18. Contour plot of the second principal component of the embedding configuration of brain wave (distinct figure case). Open circles indicate the channel positions.

NONMETRIC MULTIDIMENSIONAL SCALING AS A DATA-MINING TOOL 341

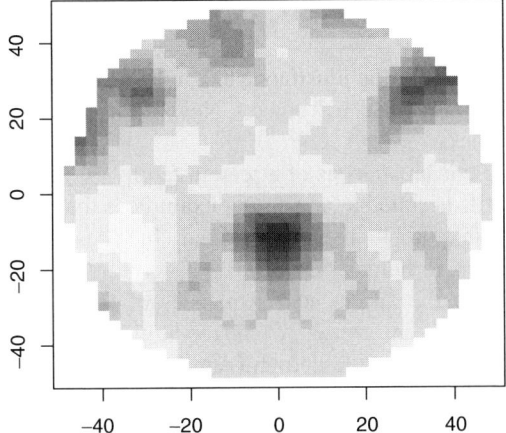

Figure 19. Contour plot of the first principal component by the direct PCA (same figures case).

visual cortex. Thus, this example demonstrates the superiority of the nMDS augmented with linear MVA methods as suggested before.

In the original work [19], such a clear separation between signal and noise was not achieved.

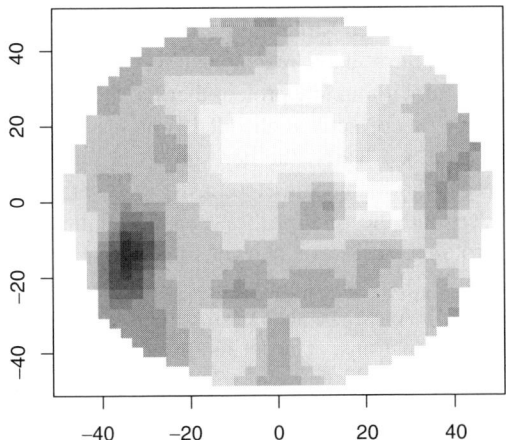

Figure 20. Contour plot of the second principal component by the direct PCA (same figures case).

X. PROTEIN FAMILY

Up to this point, except for the illustrative example of the cities on the globe, all the examples contain less than 100 objects. Now, we wish to move on to a much larger sized examples.

The first example is an analysis of a protein family. A set of proteins sharing representative motifs is called a family. Our example is the AAA protein family, which includes ubiquitous members found in organisms ranging from unicellular organisms to human beings.

We have obtained the amino acid sequences of the AAA family proteins from http://aaa-proteins.uni-graz.at/AAA/Tree.html. It contains 444 proteins belonging presumably to the AAA protein family. After alignment, 667 residues including gaps can be compared. The dissimilarity δ_{ij} is defined to be 667 (the number of matched amino acid between protein i and protein j); essentially, we use the Hamming distance.

When we embed them into E_2 (respectively, E_3), 23 (respectively, 12) proteins failed to pass the pointwise statistical test. The estimated number of correctly embedded proteins is 415 (respectively, 420), which is more than 90% of all proteins. Thus, we conclude that nMDS can reasonably embed hundreds of objects into a few dimensional space. In Fig. 21, we show a 3D embedding result. Although their axes are selected by PCA, there is no appreciable anisotropy; the distribution is spherical, but is quite nonuniform.

The central tight cluster turns out to represent the 'true' AAA proteins and surrounding cloud includes their distant relatives. Although it is sometimes difficult to set the boundary between the "true" ones and distant relatives, nMDS seems to separate them almost automatically.

In order to see the detailed structure of the central tight cluster, the 340 proteins in the cluster have been reanalyzed by nMDS. This time, the 2D embedding is successful without any protein that fails to pass the pointwise test; the estimated number of correctly embedded proteins is about 326, which is more than 90% (Fig. 22). Clearly, the core group is further divided into two main subgroups, each of which seems to consist of several further subgroups.

In order to understand the relation between the embedding result and the functionality of proteins, we have shown some proteins with their known functionalities (Fig. 23, the functionalities are due to http://aaa-proteins.uni-graz.at/AAA/Tree.html). As can be seen easily, the proteins cluster in the nMDS result share similar functions. Clustering can be done by the popular hierarchical clustering analysis, but the relations among obtained clusters is often hard to recognize. This is especially true when there are several uppermost clusters. However, the nMDS result clearly shows us the relations among clusters. There are two big branches. One consists of "meiosis/mitochondria" and "cell division cycle/centrosome/ER homotypic fusion." The other consists

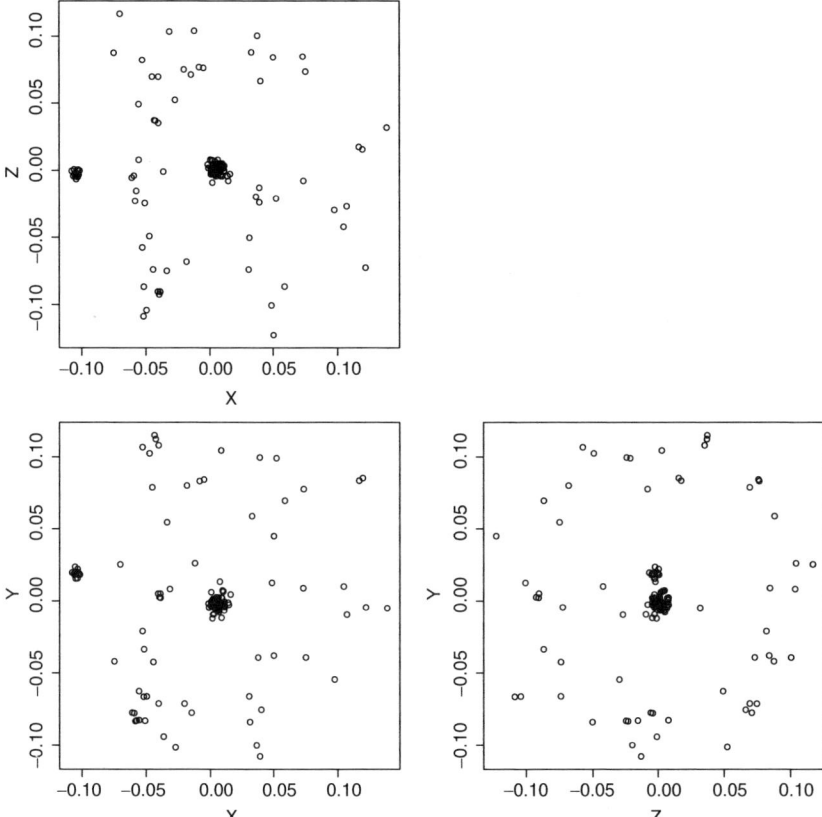

Figure 21. Three-dimensional embedding of the AAA protein family; Orthogonal axes are arbitrary.

of "subunits of the 26S proteasome," "metalloproteases," and "secretion/neurotransmission." These substructures seem to suggest some sort of directional modifications/evolutions from the common ancestor protein. This is hard to see from the tree diagrams. Furthermore, 3D analysis would give us the clusters more clearly. Detailed studies will be published elsewhere.

XI. MICROARRAY DATA

Each cDNA microarray experiment can give us information about the relative populations of mRNAs for thousands of genes. This implies that without extensive data mining it is often hard to recognize any useful information from

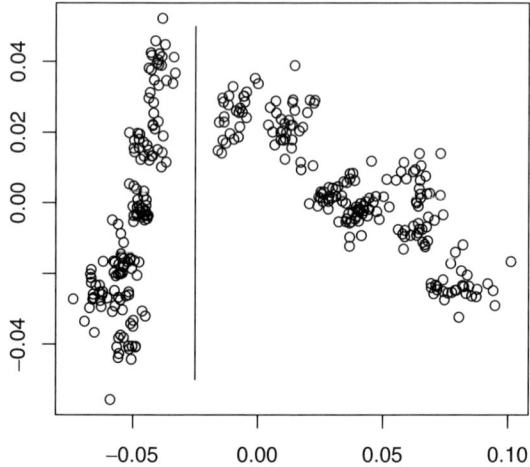

Figure 22. Two-dimensional embedding of the proteins in the "true" AAA family These are located in the central tight cluster in Fig. 21. The vertical line is only for the guide of eyes.

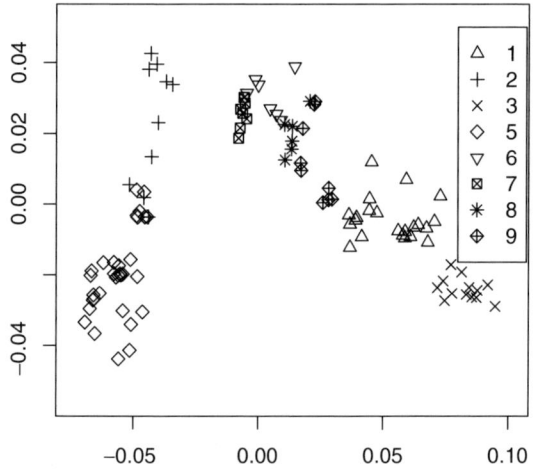

Figure 23. Proteins with known functionalities. 1, metalloproteases; 2, meiosis/mitochondria; 3, secretion/neurotransmission; 5, cell division cycle/centrosome/ER homotypic fusion; 6–9, subunits of the 26S proteasome (s4:6, s6:7, s7:8, s8:9) Configurations are the same as those shown in Fig. 22, but proteins without known functionalities are omitted.

the experimental results. As a maximally unsupervised data-mining method, we have attempted to use the nMDS method to microarray data. Since the application of MDS is a very natural idea, everyone believes that this has been done extensively, but actually, there are only a few examples; the examples are all about classifying types of cancers and other "phenotypes," and the mutual relations of genes have not been explored. This is perhaps because the number of target objects analyzable with readily available nMDS softwares is severely limited.

A. How Much Information Does the Set of All Available Microarray Data Have?

As an example, let us study the collection of 553 microarray experiments under various conditions for 19,738 genes of *Caenorhabditis elegans* available from http://www.sciencemag.org/feature/data/kim1061603/kimbig.zip as data file kimbig.txt.[6] Each datum consists of the logarithm of ratios of two fluorescence signal intensities I_1 and I_2:

$$s_{ge} = \log(I_1/I_2)$$

where g indicates genes and e microarray experiments. Some genes have no data at all, which we have omitted from the dataset. Recall that nMDS is quite robust against missing data. We have 18,479 genes in the dataset. Following Ref. 20, which advocates VixInsight to analyze the microarray data, we use the correlation coefficients between $\{s_{ge}\}_e$ and $\{s_{g'e}\}_e$ as the dissimilarity.

We embed (a) randomly chosen 3000 genes and (b) top 3000 genes in the original file (this may be regarded as another random choice).

We conclude that we can embed the genes in E_3, because:

(i) Most genes pass the pointwise significance test with the confidence level less than 0.5%.
(ii) The estimated number of correctly embedded genes are about 80%.
(iii) The E_4 embedding does not have more information than the E_3 result (see the next paragraph).

The obtained configuration is a thick spherical shell (see Fig 24), and the convergence to this geometry is quick. Also dependence on initial configuration must be weak, because the final result does not have strong dependence on gene samples (a) or (b) as explained later. Furthermore, the embedded results are significantly different from the purely random data.

---

[6]However, this data file does not mention the nature of each experiment.

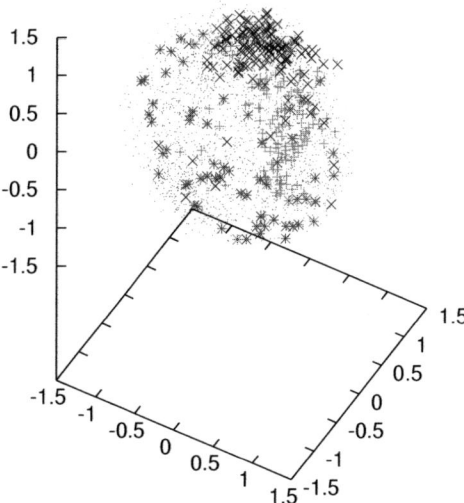

Figure 24. Three-dimensional embedding of microarray data (randomly selected 3000 genes).

The combination of nMDS and PCA reveals important information in this case as well. In E_D the principal component vectors are named as $\mathbf{x}_i^D$, ($i = 1, \ldots, D$), where i denotes the ith principal component. Let r_D be the maximum value among the correlation coefficients between $\mathbf{x}_{D+1}^{D+1}$ and $\mathbf{x}_i^D$s. If r_D is small, then the first E_D may be understood as a subspace of E_{D+1} that is uncorrelated with the rest. In this sense, the information captured in E_D is "closed" and stable. In our case, $r_2 = 0.38$, $r_3 = 0.13$, and $r_4 = 0.017$. Since r_D^2 is a measure of information that has not been captured in the $(D-1)$-dimensional subspace of E_D. Thus, we regard embedding to E_3 as the stable meaningful embedding.

To explore the biological significance of the obtained configuration, we study the distribution of functionally related gene groups: (1) sperm-enriched, (2) G-protein, and (3) biosynthesis. We have chosen these large groups, because they can be sampled well with 1000 or 3000 gene samples. The distribution of 3000 genes (a) embedded in E_3 may be seen in Fig. 24. The other 3000 genes (b) also gave very similar distribution. Thus, we may conclude that the distribution is insensitive to the samples. This suggests that the microarray dataset we are analyzing does not contain so much information as expected.

This is corroborated by the fact that the results using only 1000 randomly chosen genes are indistinguishable from the 3000 gene results.

The result of nMDS should capture more information than can be captured by cluster analyses, if we can find stable spatial patterns. Therefore, an important question is, What is the meaning of the embedding coordinates? This is generally a difficult and delicate question, because the answer is not given by the embedded result itself, but solely through its interpretation. Furthermore, to ask the meaning of a gene is almost absurd because the role of the genes is strongly context-dependent (as is well known, even for the so-called key genes such as *Pax 6*).

Here, to demonstrate that the spatial configuration is not arbitrary, let us study the relation between $\vec{s}_g = \{s_{ge}\}$ (the original expression dataset for g) and the PCA coordinates X_g for gene g determined from the E_3 embedding result (another use of linear MVA subsequent to nMDS). Other genes should also have some effects on this coordinate, but to the crudest level we may regard them as annihilating themselves due to mutual statistical interferences. Actually, we can choose a small subset ε of microarray experiments such that

$$\tilde{X}_g \equiv c_0 + \sum_{e \in \varepsilon} c_e s_{ge}$$

approximates X_g sufficiently accurately. We have found that with an appropriate set ε of size 10, the correlation coefficient between X_g and $\tilde{X}_g$ can be 0.9. As shown in Fig. 25, $\tilde{X}$ can distinguish sperm-enriched genes from G-protein genes.

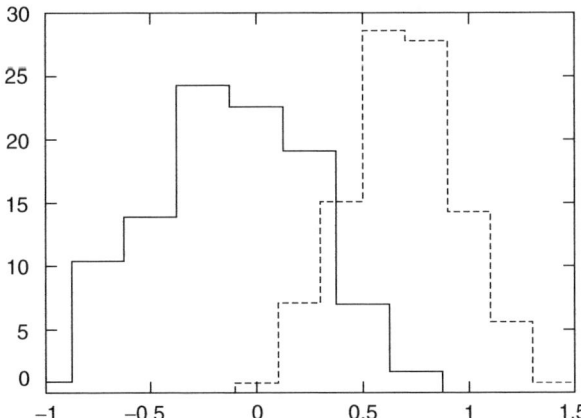

Figure 25. Spatial distribution of $\tilde{X}_g$ for sperm-enriched (solid line) and G-protein (broken line) genes in the random 3000 list. Their relationship in E_3 is correctly captured by the $\tilde{X}$-coordinate. The sperm-enriched genes are not shown here, but their relative locations in E_3 are also correctly reproduced.

Thus, we may conclude that $\tilde{X}$ represents gene relationships correctly. Actually, the same choice of 10 experiments ε for the 3000 gene set (a) can be used to represent the embedding result of the 3000 gene set (b) correctly by adjusting the coefficients c_k. To compare the two embedded results, appropriate scaling and rotation, and so on, are required, so the coefficients for (a) and (b) cannot agree in general. However, they are approximately linearly related, indicating that a common embedding rule in terms of ε may be used for both sets.

However, we should not forget that there are other small subsets of experiments as good as ε used above to define $\tilde{X}$. This implies that the dataset is very redundant. An inevitable conclusion seems to be that the information we can extract from the totality of the microarray experiments can actually be obtained from a much fewer number of experiments. Furthermore, there is no need of analyzing all the genes; it has been shown that 3000 genes are more than enough; we suspect that actually much fewer genes suffice.

However, the authors suspect that the aforementioned pessimistic conclusion is simply because numerous experiments with a wide range of experimental conditions are analyzed blindly together. The reason for our doing this is simply because there is a claim that useful information can still be extracted from the totality of experiments with a clever data-mining technique [20]. Just as can be seen from Fig. 24, we can find clustering of genes with related functions. Therefore, as is claimed in Ref 20, perhaps it is possible to identify some clusters of genes. However, such interpretable clusters seem to be exceptional. A much better defined set of experiments should be analyzed to avoid destructive interference among expression data. The next subsection illustrates such a case.

B. Temporal Patterns of Gene Expression

This subsection outlines the nMDS analysis of the microarray data on the gene transcriptional response of cell cycle-synchronized human fibroblasts to serum [21]. The dissimilarities are calculated as in the previous subsection. The data have been extensively analyzed with the aid of cluster analyses [21]. In contrast to the cluster analysis, for this example, nMDS clearly captures the time-dependence of the gene expression levels as if time correlation functions are explicitly analyzed (Fig. 26).

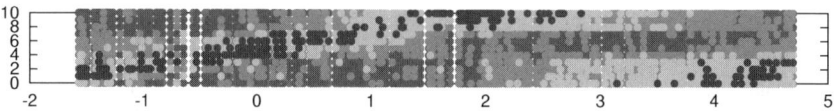

Figure 26. Microarray data visualized by nMDS. Vertical axis: observation times (from bottom to top). Horizontal axis: angle along ring-like configuration (see Fig. 27).

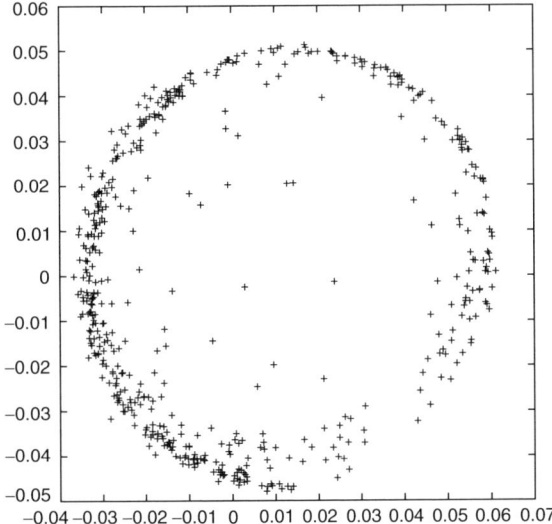

Figure 27. Two-dimensional embedding of the cell cycle data.

This dataset contains 517 cell-cycle-related genes. The microarray data are taken at the following time points: 15 min, 30 min, 1 hr, 2 hr, 4 hr, 6 hr, 8 hr, 12 hr, 16 hr, 20 hr, 24 hr. Applying nMDS to this dataset, we have obtained the ring-like configuration as shown in Fig. 27. The estimated number of correctly embedded genes is 481 corresponding to 93% of the total number of the genes. The number of the genes that fail to pass the pointwise statistical test is only one. Thus, we can conclude that this 2D embedding is excellent. The genes with their expression peaks occurring simultaneously are clustered at a particular location along the ring. Thus, the expression peak position moves in an anti-clockwise manner along the ring as time goes by (Fig. 28).

The direct application of PCA to the present data cannot produce any ring-like configuration. Further details of this research can be found at Ref. 22.

XII. CONCLUDING REMARKS

The main purpose of this chapter is to advertise an efficient nonmetric multidimensional scaling scheme with possible new applications. In particular,

1. We have explained an efficient algorithm that is maximally nonmetric in the sense that it relies only on the rank order of dissimilarities. A few

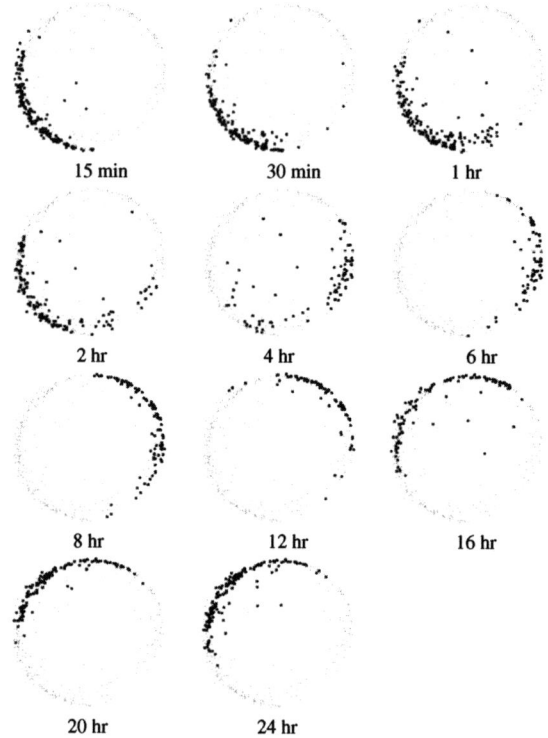

Figure 28. Temporal patterns of gene expression levels visualized with the aid of nMDS. Larger black dots indicate expression peaks.

thousand objects could be practically treated with an inexpensive laptop computer.
2. We have introduced some statistical criteria to evaluate the reliability of the results.
3. The usefulness of linear multivariate analysis methods such as PCA applied subsequently to the reduced and distilled results due to nMDS is advocated.
4. Various applications have illustrated that nMDS is a powerful means for detecting hidden patterns and relations in large-scale datasets without supervising.

We expect that nMDS will become one of the standard data-mining techniques for very large datasets. To this end, further improvements and broadening the scope of MDS are still needed.

References

1. A. C. Rencher, *Methods of Multivariate Analysis*, 2nd ed,, John Wiley & Sons, New York.
2. D. L. Donoho, M. Vetterli, R. A. DeVore, and I. Daubechies, *IEEE Trans. Inf. Theory* **44**, 2435 (1998).
3. J. B. Kruskal and R. N. Shepard, *Am. Psychol.* **19**, 557 (1964).
4. Y-h. Taguchi, Y. Oono, and K. Yokoyama, *Proc. Inst. Stat. Math.* **49**(1), 133 (2001) (in Japanese).
5. I. Borg and P. Groenen, *Modern Multidimensional Scaling*, Springer, New York, 1997.
6. It should be noted that there is a different stress-free proposal: M. W. Trosset, *J. Classif.* **15**, 15 (1998).
7. S. Rajaram, unpublished (2004).
8. E. L. Lehman, *Nonparametrics*, Holden-Day, San Francisco, 1975.
9. T. D. Kocher, J. A. Conroy, K. R. McKaye, J. R. Stauffer, and S. F. Lockwood, *Mol. Phyl. Evol.* **4**, 420 (1995).
10. M. Hasegawa and H. Kishino *Molecular Phylogenetics*, Iwanami, Tokyo 1996 (in Japanese).
11. K. Hayasaka, T. Gojobori, and S. Horai, *Mol. Biol. Evol.* **5**, 626 (1988).
12. M. Barbieri, *The Organic Codes — An Introduction to Semantic Biology*, Cambridge University Press, New York, 2002.
13. Both DNA sequence and phenotypes tables are taken from http://ucjeps.berkeley.edu/ bryolab/ GPphylo/ferndata/. DNA sequence: P95rbcl.nex, phenotypes: P95morph.nex.
14. K. M. Pryer, A. R. Smith, and J. E. Skog. *Am. Fern J.* **85**(4), 205 (1995).
15. Both DNA sequence and phenotypes tables are taken from morph: http://ucjeps.berkeley. edu/ bryolab/GPphylo/Nexus/GPMorph.txt. DNA sequence: http://ucjeps.berkeley. edu/bryolab/ GPphylo/Nexus/GPMolecular.txt.
16. B. D. Mishler, L. A. Lewis, M. A. Buchheim, K. S. Renzaglia, D. J. Garbary, C. F. Delwiche, F. W. Zechman,T. S. Kantz, and R. L. Chapman. *Ann. Missouri Botanical Garden* **81**, 451 (1994).
17. http://www.science.siu.edu/landplants/Alignments/Alignments.html.
18. K. Yokoyama, *Soil Microorganisms* **47**, 1 (1996) (in Japanese).
19. X. L. Zhang, H. Begleiter, B. Porjesz, W. Wang, and A. Litke, *Brain Res. Bull.* **38**, 531 (1995).
20. S. K. Kim, J. Lund, M. Kiraly, K. Duke, M. Jiang, J. M. Stuart, A. Eizinger, B. N. Wylie, and G. S. Davidson, *Science* **293** 2087 (2001).
21. V. R. Iyer, M. B. Eisen, D. T. Ross, G. Schuler, T. Moore, J. C. F. Lee, J. M.Trent, L. M. Staudt, J. S. Hudson Jr., M. S. Boguski, D. Lashkari, D. Shalon, D. Botstein, and P. O. Brown, *Science* **283**, 83 (1999).
22. Y-h. Taguchi and Y. Oono, *Bioinformatics*, in press.

CHAPTER 19

GENERALIZATION OF THE FLUCTUATION–DISSIPATION THEOREM FOR EXCESS HEAT PRODUCTION

HIROSHI H. HASEGAWA

Department of Mathematical Sciences, Ibaraki University, Mito, 310-8512, Japan; and Center for Studies in Statistical Mechanics and Complex Systems, The University of Texas at Austin, Austin, Texas 78712, USA

YOSHIKAZU OHTAKI

Department of Mathematical Sciences, Ibaraki University, Mito, 310-8512, Japan

CONTENTS

I. Introduction
II. Thermodynamics for the Boltzmann Equilibrium Distribution
   A. Thermodynamics
   B. The Fluctuation–Dissipation Theorem
   C. The Area of Hysteresis Loop
      1. Limit of Long Period
      2. Limit of Short Period
III. The Fluctuation–Dissipation Theorem for the Superstatistical Equilibrium Distributions
IV. Anomalous Behavior of Variance of Nonergodic Adiabatic Invariant
   A. The Fluctuation–Dissipation Theorem for the Microcanonical Distribution
   B. Adiabatic Invariant for a Simple Hamiltonian Chaotic System
V. Conclusions and Remarks
Appendix A: Jarzynski's Nonequilibrium Work Relation
Appendix B: The Second Law from the Fluctuation–Dissipation Theorem
Acknowledgments
References

Geometric Structures of Phase Space in Multidimensional Chaos: A Special Volume of Advances in Chemical Physics, Part B, Volume 130, edited by M. Toda, T Komatsuzaki, T. Konishi, R.S. Berry, and S.A. Rice. Series editor Stuart A. Rice.
ISBN 0-471-71157-8 Copyright © 2005 John Wiley & Sons, Inc.

I. INTRODUCTION

Over the past several years, there has been a renewed interest in thermodynamics and many scientists have considered it from new points of view [1–8]. Thermodynamics is a universal effective theory [9]. It does not depend on the details of underlying dynamics. The first law is the conservation of energy. The second law is the nonnegativeness of excess heat production. It is valid for wide classes of Markov processes in which systems approach to the Boltzmann equilibrium distribution.

When we consider the dependence of excess heat production on an external transformation, we can connect thermodynamic quantities and underlying dynamics. We have derived the theorem similar to the fluctuation–dissipation theorem [10]. The theorem shows that thermodynamic entropy production such as excess heat can be written as a correlation function between Einstein–Shanon entropy functions. Through the correlation function the thermodynamic entropy production is related to the underlying dynamics.

We have obtained several interesting results from the theorem: If the period of the external transformation is much longer than the relaxation time, then thermodynamic entropy production is proportional to the ratio of the period and relaxation time. The relaxation time is proportional to the inverse of the Kolmogorov–Sinai entropy for small strongly chaotic systems. Thermodynamic entropy production is proportional to the inverse of the dynamical entropy [11]. On the other hand, thermodynamic entropy production is proportional to the dynamical entropy when the period of the external transformation is much shorter than the relaxation time. Furthermore, we found fractional scaling of the excess heat for long-period external transformations, when the system has long-time correlation such as $1/f^\alpha$ noise. Since excess heat is measured as the area of a hysteresis loop [12], these properties can be confirmed in experiments.

In this chapter we will extend the fluctuation–dissipation theorem for general equilibrium distributions. We will consider two typical equilibrium distributions. One is the superstatistical equilibrium distribution [13]. The other is the microcanonical equilibrium distribution.

When a system has long-time correlation, for which we expect fractional power scaling of excess heat, our assumption of the Boltzmann equilibrium distribution may always not be valid. Actually some power distributions such as the Tsallis distribution [14] have been reported at the edge of chaos [15]. A superstatistical equilibrium distribution is written as a superposition of Boltzmann distributions with different temperatures. Beck and Cohen [13] considered many types of distributions for the inverse of temperature. For example, they chose Gaussian, uniform, gamma, log-normal, and others. In particular, the Tsallis distribution is realized for gamma distribution. We will show that excess heat can be written as a superposition of correlation functions with the different temperatures in the generalized fluctuation–dissipation theorem.

The other distribution is the microcanonical equilibrium distribution. More than 15 years ago, Ott–Brown–Grebogi pointed out fractional scaling of deviation from ergodic adiabatic invariants in Hamiltonian chaotic systems [16, 17]. We will reconsider not only ergodic adiabatic invariants but also nonergodic adiabatic invariants, which are important in the mixed phase space. We will show results of our numerical simulation in which a nonergodic adiabatic invariant corresponding to uniform distribution is broken in the mixed phase space.

In Section II we will review thermodynamics and the fluctuation–dissipation theorem for excess heat production based on the Boltzmann equilibrium distribution. We will also mention the nonequilibrium work relation by Jarzynski. In Section III, we will extend the fluctuation–dissipation theorem for the superstatisitcal equilibrium distribution. The fluctuation–dissipation theorem can be written as a superposition of correlation functions with different temperatures. When the decay constant of a correlation function depends on temperature, we can expect various behaviors in the excess heat. In Section IV, we will consider the case of the microcanonical equilibrium distribution. We will numerically show the breaking of nonergodic adiabatic invariant in the mixed phase space. In the last section, we will conclude and comment.

II. THERMODYNAMICS FOR THE BOLTZMANN EQUILIBRIUM DISTRIBUTION

A. Thermodynamics

In this subsection we will review the thermodynamics for the Boltzmann equilibrium distribution [7].

We consider a system governed by the time evolution operator, $L(x, c(t))$, acting on a probability density, $\rho(x, t)$, in a Markov process:

$$i\frac{\partial \rho(x,t)}{\partial t} = L(x, c(t))\rho(x, t) \qquad (1)$$

For a Hamiltonian dynamical system, $L(x, c(t))$ is the Liouvillian. We control the parameter $c(t)$ as an external transformation.

We assume that the system approaches the Boltzmann equilibrium distribution. We define the Hamiltonian $H(x, c)$, free energy $F_\beta(c)$, and Einstein–Shanon entropy $S_\beta(x, c)$ as follows:

$$\rho_\beta(x, c) = \exp[-S_\beta(x, c)] = \exp[\beta\{F_\beta(c) - H(x, c)\}] \qquad (2)$$

where $\int dx \rho_\beta(x, c) = 1$.

We change the parameter $c(t)$ from c_0 at $t = 0$ to c_T at $t = T$. The first law of thermodynamics is given as

$$\langle \Delta H \rangle = \langle W \rangle + \langle D \rangle \tag{3}$$

where change of energy $\langle \Delta H \rangle$, work $\langle W \rangle$, and dissipation $\langle D \rangle$ are, respectively, defined as follows:

$$\langle \Delta H \rangle \equiv \langle H(x, c_T) \rangle_{\rho(x,T)} - \langle H(x, c_0) \rangle_{\rho(x,0)} \tag{4}$$

$$\langle W \rangle \equiv \int_0^T dt \left\langle \frac{\partial H(x, c(t))}{\partial t} \right\rangle_{\rho(x,t)} \tag{5}$$

$$\langle D \rangle \equiv \int_0^T dt \int dx \langle H(x, c(t)) \rangle_{\partial_t \rho(x,t)} \tag{6}$$

Note that $\langle \cdot \rangle_{\rho(x,t)} \equiv \int dx \cdot \rho(x,t)$.

We consider the work in a quasi-static (QS) process. In this kind of process, the probability density is given as $\rho(x,t) = \rho_\beta(x, c(t))$. The work becomes the difference between the initial and final free potential:

$$\langle W_{QS} \rangle = F_\beta(c(T)) - F_\beta(c(0)) \tag{7}$$

The excess heat ΔW is the additional work in a non-quasi-static process. It is given as the difference between work in a non-quasi-static process and work in the quasi-static process:

$$\langle \Delta W \rangle = \langle W \rangle - \langle W_{QS} \rangle \tag{8}$$

$$= \int_0^T dt \left[\left\langle \frac{\partial V(x, c(t))}{\partial t} \right\rangle_{\rho(x,t)} - \frac{dF_\beta(c(t))}{dt} \right]$$

$$= \int_0^T dt \left\langle \frac{\partial S_\beta(x, c(t))}{\partial t} \right\rangle_{\rho(x,t)} \tag{9}$$

The excess heat production becomes the average change of the entropy.
The second law of thermodynamics is

$$\langle \Delta W \rangle \geq 0 \tag{10}$$

Thanks to work by Jarzynski [3], excess heat production is known to be positive in a wide class of Markov processes. In Appendix A, we will demonstrate Jarzynski's nonequilibrium work relation.

B. The Fluctuation–Dissipation Theorem

In this subsection we will demonstrate the fluctuation–dissipation theorem for excess heat [10].

We rewrite the excess heat production using the deviation from the Boltzmann equilibrium distribution, $\delta\rho(x,t) \equiv \rho(x,t) - \rho_\beta(x,c(t))$:

$$\langle \delta W(t) \rangle = \int dx \frac{\partial H(x,c(t))}{\partial t} \delta\rho(x,t)$$

$$= \int dx \frac{1}{\beta} \frac{\partial S_\beta(x,c(t))}{\partial t} \delta\rho(x,t) \quad (11)$$

where we used $\int dx \delta\rho(x,t) = 0$.

From the equation for the probability density, the deviation is governed by the following equation:

$$i\frac{\partial \delta\rho(x,t)}{\partial t} = L(x,c(t))\delta\rho(x,t) + i\frac{\partial S_\beta(x,c(t))}{\partial t}\rho_\beta(x,c(t)) \quad (12)$$

where we used the property of the equilibrium distribution, $L(x,c(t))\rho_\beta(x,c(t)) = 0$.

The formal solution is given exactly as

$$\delta\rho(x,t) = \int_0^t dt' U(t,t') \frac{\partial S_\beta(x,c(t'))}{\partial t'} \rho_\beta(x,c(t')) + U(t,0)\delta\rho(x,0) \quad (13)$$

where the transfer operator is defined as

$$U(t,t') \equiv T \exp\left[-i \int_{t'}^t d\tilde{t} L(x,c(\tilde{t}))\right] \quad (14)$$

where T is an operator which indicates to take time-ordered products.

By substituting Eq. (13) into Eq. (11),

$$\langle \delta W(t) \rangle = \int_0^t dt' C_\beta(t,t') \quad (15)$$

where the autocorrelation function is defined as

$$C_\beta(t,t') \equiv \left\langle \frac{1}{\beta} \frac{\partial S_\beta(c(t))}{\partial t} U(t,t') \frac{\partial S_\beta(c(t'))}{\partial t'} \right\rangle_{\rho_\beta(x,c(t'))} \quad (16)$$

where we chose $\delta\rho(x,0) = 0$.

Finally we obtain the fluctuation–dissipation theorem for excess heat,

$$\langle \Delta W \rangle = \int_0^T dt \int_0^t dt' C_\beta(t,t') \tag{17}$$

C. The Area of Hysteresis Loop

Excess heat ΔW for a periodic external transformation is measured as the area of a hysteresis loop A. We will consider both the long-period and short-period limit. For simplicity, we consider one period with a constant change of the parameter—that is, $|dc(t)/dt| = 2\Delta c/T$ where $\Delta c \equiv |c_{T/2} - c_0|$ and the sign of the parameter is changed at $t = T/2$.

1. Limit of Long Period

We consider the following four typical cases of the strength of the auto-correlation $C(t,0)$ in the limit of long period:

(i) Exponential decay ($C(t,0) \sim \exp[-\gamma t]$, $T \gg 1/\gamma$):

$$A \sim \frac{\Delta c^2}{\gamma T} \tag{18}$$

The area of the hysteresis loop decays as $1/\gamma T$.

It is interesting how the thermodynamic entropy production is related to the dynamical entropy production. The dynamical entropy production such as the Kolmogorov–Sinai entropy is almost the same as the decay constant γ in a small chaotic system. Since a smaller Kolmogorov–Sinai entropy produces a longer decay time, we can expect greater thermodynamic entropy [11]. This means that the thermodynamic entropy production is proportional to the inverse of the dynamical entropy production in this case. In a large chaotic system, the decay constant γ is $\sim D/L^2$, where D is a diffusion coefficient and L is the size of system. We can expect less thermodynamic entropy production for stronger diffusion.

(ii) Power decay ($C(t,0) \sim 1/t^{1-\alpha}$, $\alpha < 0$):

$$A \sim \frac{\Delta c^2}{T} \tag{19}$$

The area of the hysteresis loop decays as $1/T$.

(iii) Power decay ($C(t,0) \sim 1/t$ $\alpha = 0$):

$$A \sim \Delta c^2 \frac{\log T}{T} \tag{20}$$

There is a logarithmic correction for $\alpha = 0$.

(iv) Power decay ($C(t,0) \sim 1/t^{1-\alpha}$, $0 < \alpha < 1$):

$$\Delta W \sim \frac{\Delta c^2}{T^{1-\alpha}} \qquad (21)$$

The area of the hysteresis loop decays with fractional power; that is, it is proportional to $1/T^{1-\alpha}$. The power α is the same as that of the $1/f^\alpha$ noise.

In a separate article [10] we considered a modified version of a one-dimensional intermittent chaotic system, the Manneville–Pomeau map [18]. We numerically confirmed the fractional power scaling of excess heat production.

2. Limit of Short Period

We consider the following two typical cases of the strength of the autocorrelation $C(t,0)$ in the limit of short period:

(i) Exponential decay ($C(t,0) \sim \exp[-\gamma t]$, $T \gg 1/\gamma$):

$$A \sim \Delta c^2 \gamma T \qquad (22)$$

The area of the hysteresis loop is proportional to γT, so it increases for greater T. The thermodynamic entropy production is proportional to the dynamical entropy production in this case [22, 23]. In a large chaotic system, we can expect less thermodynamic entropy production for weak diffusion.

(ii) Constant minus power increase ($C(t,0) \sim 1 - T_c/t^{1-\alpha}$, $\alpha > 1$, $t < T_c$), where T_c is a cutoff time:

$$A \sim T^{\alpha-1} \quad \text{for } T < T_c \qquad (23)$$

The area of the hysteresis loop increases as $T^{\alpha-1}$ [19]. The power α is the same as that of the $1/f^\alpha$ noise. This is the case of nonstationary Hamiltonian chaos in the mixed phase space [20]. The power of the $1/f^\alpha$ is greater than 1, $\alpha > 1$. We will return this point in Section IV.

III. THE FLUCTUATION–DISSIPATION THEOREM FOR THE SUPERSTATISTICAL EQUILIBRIUM DISTRIBUTIONS

In this subsection we will consider generalized equilibrium distributions in superstatistics,

$$\rho_{gs}(x,c) = \int d\beta \pi(\beta) \rho_\beta(x,c) \qquad (24)$$

where $\rho_\beta(x,c) = \exp[\beta\{F_\beta(c) - H(x,c)\}]$ is the normalized Boltzmann distribution with the inverse of temperature β. The probability distribution of the inverse of temperature $\pi(\beta)$ is positive $\pi(\beta) \geq 0$ and integrable $\int d\beta \pi(\beta) = 1$.

The time evolution operator satisfies $L(x,c)\rho_{ge}(x,c) = 0$. In general, $L(x,c)\rho_\beta(x,c) \neq 0$ for the Boltzmann distribution.

We consider the work in a quasi-static (QS) process. The probability density is given as $\rho(x,t) = \rho_{ge}(x,c(t))$.

$$\langle W_{QS}\rangle = \int_0^T dt \left\langle \frac{\partial H(x,c(t))}{\partial t}\right\rangle_{\rho_{ge}(x,c(t))}$$

$$= \left[\int_0^T dt \left\langle \frac{\partial H(x,c(t))}{\partial t}\right\rangle_{\rho_\beta(x,c(t))}\right]_\beta \quad (25)$$

where $[\cdot]_\beta \equiv \int d\beta \cdot \pi(\beta)$.

Using the equation

$$0 = \frac{1}{\beta}\frac{\partial}{\partial t}\int dx \rho_\beta(x,c(t)) = \left\langle \frac{\partial H(x,c(t))}{\partial t}\right\rangle_{\rho_\beta(x,c(t))} - \frac{dF_\beta(c(t))}{dt}$$

the work becomes the difference between the initial and final free energy averaged over the inverse of temperature:

$$\langle W_{QS}\rangle = [F_\beta(c(T))]_\beta - [F_\beta(c(0))]_\beta \quad (26)$$

The excess heat is given as

$$\langle \Delta W\rangle = \langle W\rangle - \langle W_{QS}\rangle \quad (27)$$

We rewrite the excess heat production using the deviation from the generalized equilibrium distribution, $\delta\rho(x,t) \equiv \rho(x,t) - \rho_{ge}(x,c(t))$:

$$\langle \delta W(t)\rangle = \int dx \frac{\partial H(x,c(t))}{\partial t}\delta\rho(x,t) \quad (28)$$

For $\rho(x,0) = \rho_{ge}(x,c(0))$,

$$\delta\rho(x,t) = -\int_{t'}^t dt' U(t,t')\frac{\partial}{\partial t'}\rho_{ge}(x,c(t'))$$

$$= \left[\int_{t'}^t dt' U(t,t')\frac{\partial S_\beta(x,c(t'))}{\partial t'}\rho_\beta(x,c(t'))\right]_\beta \quad (29)$$

We rewrite the excess heat production using Eq. (29),

$$\langle \delta W(t) \rangle = \left[\int dx \int_{t'}^{t} dt \frac{\partial H(x, c(t))}{\partial t} U(t, t') \frac{\partial S_\beta(x, c(t'))}{\partial t'} \rho_\beta(x, c(t')) \right]_\beta$$

$$= \left[\int_{t'}^{t} dt \left\langle \frac{1}{\beta} \frac{\partial S_\beta(x, c(t))}{\partial t} U(t, t') \frac{\partial S_\beta(x, c(t'))}{\partial t'} \right\rangle_{\rho_\beta(x, c(t'))} \right]_\beta \quad (30)$$

where we used $\int dx U(t, t') \frac{\partial S(x, c(t'))}{\partial t'} \rho_\beta(x, c(t')) = 0$.

Finally we obtain the fluctuation–dissipation theorem,

$$\langle \Delta W \rangle = \int_0^T dt \int_0^t dt' [C_\beta(t, t')]_\beta \quad (31)$$

The fluctuation–dissipation theorem is just given as an average of correlation functions over β. It is not trivial, since the time evolution of $\rho_\beta(x, c(0))$ cannot be separated from other β in general.

Before we close this section, we will comment on the area of the hysteresis loop. When the decay constant of a correlation functions depends on the inverse of temperature, we can expect various behavior for the area of the hysteresis loop. In the case of the Tsallis distribution, the inverse of temperature is distributed as a gamma distribution. If the decay constant is proportional to the temperature, the area of the histeresis loop decays as a modified Bessel function for the large period of external transformation. On the other hand, if the decay constant is proportional to the inverse of temperature, we can expect the fractional power scaling.

IV. ANOMALOUS BEHAVIOR OF VARIANCE OF NONERGODIC ADIABATIC INVARIANT

In this section we will consider an nonergodic adiabatic invariant for a chaotic Hamiltonian system. Specifically, we are interested in the mixed phase space, in which tori and chaotic seas coexist.

A. The Fluctuation–Dissipation Theorem for the Microcanonical Distribution

We will first consider the fluctuation–dissipation theorem for the microcanonical distribution. The microcanonical equilibrium distribution is given as

$$\rho_{mc}(x, c) = \kappa \delta(H(x, c) - E_0) \quad (32)$$

where $\kappa^{-1} = \int dx \delta(H(x, c) - E_0)$.

The microcanonical distribution is formally written in the superstatistical form

$$\rho_{mc}(x,c) = \frac{\kappa}{2\pi i}\int_{-i\infty}^{i\infty} d\beta \exp[\beta\{E_0 - H(x,c)\}] \tag{33}$$

However, β is a complex variable so that we cannot treat the microcanonical distribution as the superstatistical distribution.

The time evolution operator $L(x,c(t))$ is the Liouvillian that satisfies $L(x,c)\rho_{mc}(x,c) = 0$.

We consider the work in a quasi-static (QS) process and define an adiabatic invariant. In the quasi-static process, we assume that the probability density is given as $\rho(x,t) = \rho_{mc}(x,c(t))$. Then,

$$\langle W_{QS}\rangle = E_0(T) - E_0(0) \tag{34}$$

where $E_0(t)$ satisfies the equation

$$\frac{\partial E_0(t)}{\partial t} = \int dx \frac{\partial H(x,c(t))}{\partial t}\rho_{mc}(x,c(t)) \tag{35}$$

Since we can rewrite Eq. (35) as

$$\int dx \left\{\frac{\partial H(x,c(t))}{\partial t} - \frac{\partial E_0(t)}{\partial t}\right\}\delta(H(x,c) - E_0) = 0 \tag{36}$$

we define an adiabatic invariant

$$\mu(E,t) \equiv \int dx\, \Theta(E - H(x,c(t))) \tag{37}$$

where $\Theta(x)$ is the step function. The adiabatic invariant satisfies

$$\mu(E_0(t),t) = \mu(E_0(0),0) \tag{38}$$

in a quasi-static process.

We rewrite the excess heat production using the deviation from the microcanonical distribution, $\delta\rho(x,t) \equiv \rho(x,t) - \rho_{mc}(x,c(t))$:

$$\langle \delta W(t)\rangle = \int dx \frac{\partial H(x,c(t))}{\partial t}\delta\rho(x,t)$$
$$= \int dx\left\{\frac{\partial H(x,c(t))}{\partial t} - \frac{\partial E_0(t)}{\partial t}\right\}\delta\rho(x,t) \tag{39}$$

For $\rho(x,0) = \rho_{mc}(x,c(0)) = \kappa(0)\delta(H(x,c(0)) - E_0(0))$,

$$\rho(x,t) = \kappa(0)\delta(H(x,c(t)) - E_0(t))$$
$$- \kappa(0)\int_0^t dt' U(t,t')\frac{\partial}{\partial t'}\delta(H(x,c(t')) - E_0(t')) \quad (40)$$

By substituting Eq. (40) into Eq. (39), we obtain

$$\langle \delta W(t) \rangle = \left\{\frac{\kappa(0)}{\kappa(t)} - 1\right\}\int dx \left\{\frac{\partial H(x,c(t))}{\partial t} - \frac{\partial E_0(t)}{\partial t}\right\}\rho_{mc}(x,c(t))$$
$$+ \kappa(0)\int_0^t dt' \int dx \left\{\frac{\partial H(x,c(t))}{\partial t} - \frac{\partial E_0(t)}{\partial t}\right\}U(t,t')$$
$$\times \left\{\frac{\partial H(x,c(t'))}{\partial t'} - \frac{\partial E_0(t')}{\partial t'}\right\}\frac{\partial}{\partial E_0(t')}\delta(H(x,c(t')) - E_0(t')) \quad (41)$$

Since $\int dx \frac{\partial}{\partial t}\rho_{mc}(x,c(t)) = 0$, the first term in Eq. (41) vanishes.
Finally we obtain the fluctuation–dissipation theorem for the microcanonical distribution,

$$\langle \Delta W \rangle = \int_0^T dt \int_0^t dt' \frac{\kappa(0)}{\kappa(t')}\left\langle \left\{\frac{\partial H(c(t))}{\partial t} - \frac{\partial E_0(t)}{\partial t}\right\}\right.$$
$$\left. \times U(t,t')\left\{\frac{\partial H(c(t'))}{\partial t'} - \frac{\partial E_0(t')}{\partial t'}\right\}\right\rangle_{\partial_{E_0}\rho_{mc}(c(t'))} \quad (42)$$

B. Adiabatic Invariant for a Simple Hamiltonian Chaotic System

Now we consider the adiabatic invariant for a simple Hamiltonian chaotic system. The Hamiltonian is defined as

$$H(x,p,J,\alpha) = \frac{p^2}{2}(1 - \Theta_\epsilon(\alpha)) + \frac{K}{2\pi\epsilon}\cos(2\pi x)\Theta_\epsilon(\alpha) + J \quad (43)$$

where $p,x,\alpha \in [-1/2, 1/2]$ with periodic boundary conditions, $J \in [-J_0, J_0)$ where we choose J_0 to be sufficiently large compared to K/ϵ, and

$$\Theta_\epsilon(\alpha) = \begin{cases} 1 & \text{if } -\epsilon/2 < \alpha < \epsilon/2 \\ 0 & \text{if } \epsilon/2 \leq \alpha \leq 1 - \epsilon/2 \end{cases}$$

Later, we will take the limit of $\epsilon \to 0$.

By solving the equations of motion, we obtain the well-known standard map,

$$p_{n+1} = p_n + K\sin(2\pi x_n) \quad (44)$$

$$x_{n+1} = x_n + p_{n+1} \quad (45)$$

$$J_{n+1} = J_n + \frac{p_n^2}{2} - \frac{p_{n+1}^2}{2} \quad (46)$$

where $n = 0, 1, 2, \ldots, p_n \equiv p(n - \epsilon), x_n \equiv x(n)$ and $J_n \equiv J(n - \epsilon)$. Since $\dot{\alpha} = 1$, we choose $\alpha(t) = t \bmod 1$.

We will calculate the adiabatic invariant,

$$\mu(E) = \iiint_{-1/2}^{1/2} dx\,dp\,d\alpha \int_{-J_0}^{J_0} dJ\Theta(H(x,p,J,\alpha) - E) \quad (47)$$

By substituting Eq. (43) into Eq. (47),

$$\mu(E) = \left\{ \int_{-1/2}^{-\epsilon/2} d\alpha + \int_{\epsilon/2}^{1/2} d\alpha \right\} \iint_{-1/2}^{1/2} dx\,dp \int_{-J_0}^{E-p^2/2} dJ$$

$$+ \int_{-\epsilon/2}^{\epsilon/2} d\alpha \iint_{-1/2}^{1/2} dx\,dp \int_{-J_0}^{E-K\cos(2\pi x)/(2\pi\epsilon)} dJ$$

$$\to E + J_0 - \tfrac{1}{24} \quad \text{for } \epsilon \to 0 \quad (48)$$

We consider a periodic change of the stochastic parameter $K(t)$ as an external transformation to know how good the adiabatic invariant is. Since $E_0(t)$ is the value of the energy at time t determined by the conservation of $\mu(E_0(t), t) = \mu(E_0(0), 0)$, the deviation and the variance of the adiabatic invariance are simply related to those of the Hamiltonian as

$$\langle \Delta\mu(E) \rangle = \langle \Delta H \rangle = \langle \Delta W \rangle \quad (49)$$

$$\langle (\Delta\mu(E))^2 \rangle = \langle (\Delta H)^2 \rangle \quad (50)$$

where

$$\langle (\Delta H)^2 \rangle$$
$$\equiv \int_0^T dt \int_0^t dt' \left\langle \left\{ \frac{\partial H(c(t))}{\partial t} - \frac{\partial E_0(t)}{\partial t} \right\} U(t,t') \left\{ \frac{\partial H(c(t'))}{\partial t'} - \frac{\partial E_0(t')}{\partial t'} \right\} \right\rangle_{\rho(t')} \quad (51)$$

We have numerically estimated the goodness of an adiabatic invariant.

The choice of an initial distribution is a delicate problem. For large K ($K > 4$), the phase space is mainly chaotic, so the adiabatic invariant can be considered as ergodic. In this case we may choose the initial distribution as the invariant measure given by the long-time average of a trajectory. But for small K ($K < 1$), the phase space is divided by many tori, so it is not reasonable to choose the initial distribution as the invariant measure given by the long-time average of a trajectory. To avoid a mathematically delicate problem, we choose a simple uniform density as the initial distribution. The uniform density is manifestly invariant for all K. Therefore the deviation of the adiabatic invariant vanishes. However, the variance does not vanish in general.

We have controlled the stochastic parameter $K(t)$ as

$$K(t) = K_0 + \Delta K \left(1 - \left|2\frac{t}{T} - 1\right|\right), \qquad 0 \leq t \leq T \tag{52}$$

where we choose $\Delta K = 0.05$ and we change the period T from 2^3 to 2^{15}.

The number of particles is 10^5. All particles have a common initial energy. The initial positions are distributed uniformly and randomly in the x–p space. We have simulated variances $\langle(\Delta H)^2\rangle$ for $K_0 = 0.1$, 0.8, 0.95, 2.0, and 4.0 as illustrated in Figs. 1 to 5. We have confirmed that the change of particle number did not affect our numerical results.

We find that the variance of the adiabatic invariant is a nice quantity to measure the complexity in phase space. In a strong chaotic case such as $K \sim 4$, we can expect goodness of the ergodic adiabatic invariant. The correlation decays in a relatively short time. The variance decays as shown in Fig. 5. In the

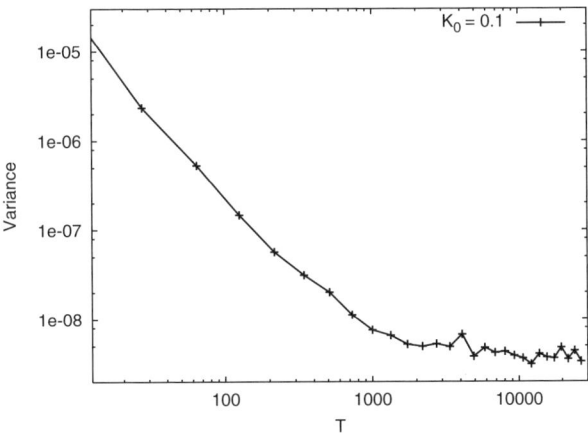

Figure 1. The variance of the adiabatic invariant for $K_0 = 0.1$.

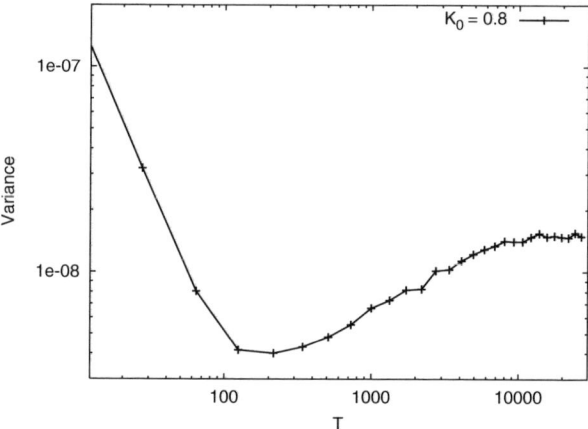

Figure 2. The variance of the adiabatic invariant for $K_0 = 0.8$.

near-regular region $K \sim 0.1$, the motion of particles is almost quasi-periodic in phase space. The variance may be written as a sum of many quasi-periodic modes. The cancellation of these modes makes the decay of variance as is shown in Fig. 1. In the case of a mixed phase space, we expect extremely long-time correlation. It corresponds to the case in the limit of short period discussed in Section II. We can expect an anomalous increase up to a cutoff time T_c as similar as Eq. (23). For $K \sim 0.8$, the cutoff time is estimated as $T_c \sim 2^{14}$, since the anomalous increase stops at this time as shown in Fig. 2. In the near-critical

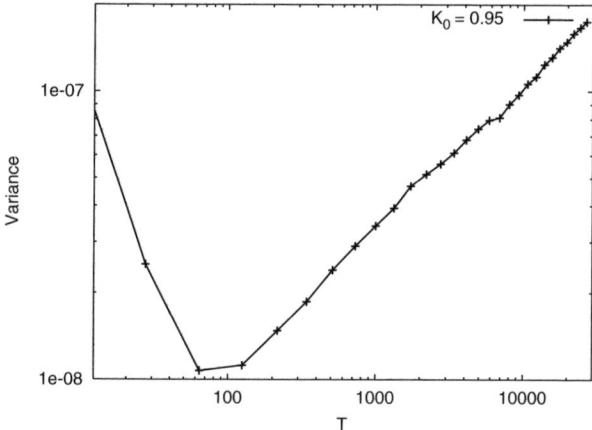

Figure 3. The variance of the adiabatic invariant for $K_0 = 0.95$.

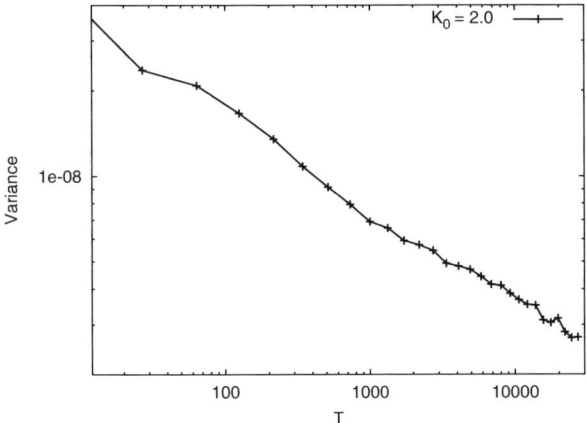

Figure 4. The variance of the adiabatic invariant for $K_0 = 2.0$.

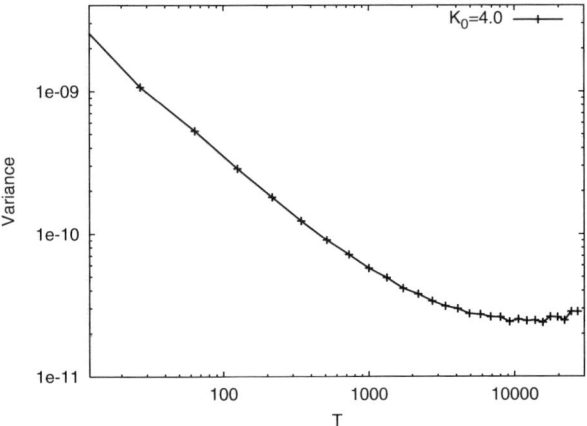

Figure 5. The variance of the adiabatic invariant for $K_0 = 4.0$.

stochastic parameter $K \sim 0.97$, the cutoff time becomes extremely long, since the anomalous increase does not saturate up to $T = 2^{15}$ as is shown in Fig. 3.

V. CONCLUSIONS AND REMARKS

In this chapter we have extended the fluctuation–dissipation theorem with respect to excess heat for general equilibrium distributions. We considered the two typical equilibrium distributions. One was the superstatistical equilibrium distribution. The other was the microcanonical equilibrium distribution.

A superstatistical equilibrium distribution is written as a superposition of Boltzmann distributions with different temperatures. We showed that the excess heat could be written as a superposition of correlation functions with different temperatures using the generalized fluctuation–dissipation theorem. When a relaxation time depends on a temperature, we can expect various behaviors for the area of the hysteresis loop from the fluctuation–dissipation theorem.

With regard to the microcanonical equilibrium distribution and the extension of the fluctuation–dissipation theorem, we considered a nonergodic adiabatic invariant in a simple Hamiltonian chaotic system. We numerically demonstrated the breaking of the nonergodic adiabatic invariant in the mixed phase space. The variance of the nonergodic adiabatic invariant can be considered as a measure for complexity of the mixed phase space.

In this chapter we considered only a small Hamiltonian system whose Poincaré map is the standard map defined on the unit square. It is interesting to consider Hamiltonian systems in a large phase space in which diffusion appears. Specifically, we are interested how the accelerator mode, which causes the anomalous diffusion in the standard map, affects the breaking of the adiabatic invariant. We will continue this study in a forthcoming article [21].

APPENDIX A: JARZYNSKI'S NONEQUILIBRIUM WORK RELATION

In Appendix A, we will demonstrate Jarzynski's nonequilibrium work relation. Jarzynski's nonequilibrium work relation is based on the following equality:

$$\int dx \rho_\beta(x, c(T)) = \int dx \exp\left[\int_0^T dt \frac{\partial S_\beta(x, c(t))}{\partial t}\right] \rho_\beta(x, c(0)) = 1 \quad (53)$$

From $L(x, c(t))\rho_\beta(x, c(t)) = 0$,

$$\int dx \text{T} \exp\left[\int_0^T dt \left\{\frac{\partial S_\beta(x, c(t))}{\partial t} - iL(x, c(t))\right\}\right] \rho_\beta(x, c(0)) = 1 \quad (54)$$

Here we write a path or a trajectory as $\{x\} = (x_N, x_{N-1}, \ldots, x_1, x_0)$, where $x_n = x(nT/N)$. By analogy with the path integral formalism, we can rewrite Eq. (54) for large N,

$$\int \mathscr{D}x P(\{x\}) \exp\left[\int_0^T dt \left\{\frac{\partial F_\beta(c(t))}{\partial t} - \frac{\partial H(x(t), c(t))}{\partial t}\right\}\right] = 1 \quad (55)$$

where

$$\mathscr{D}x \equiv dx_0 dx_1 \cdots dx_N$$
$$P(\{x\}) \equiv P(x_N|x_{N-1})P(x_{N-1}|x_{N-2}) \cdots P(x_1|x_0)\rho_\beta(x_0, c_0)$$
$$P(x_{n+1}|x_n) \equiv \exp[-iL(x_{n+1}, c_{n+1})T/N]\delta(x_{n+1} - x_n)$$
$$c_n \equiv c\left(n\frac{T}{N}\right)$$

Finally we obtain Jarznski's nonequilibrium work relation,

$$\langle \exp[-\beta W]\rangle_{\{x\}} = \exp[-\beta\{F_\beta(c(T)) - F_\beta(c(0))\}] \quad (56)$$

From the Jensen inequality, $1 = \langle \exp[-\beta W]\rangle_{\{x\}} \geq \exp[-\beta\langle W\rangle_{\{x\}}]$, the second law is valid.

APPENDIX B: THE SECOND LAW FROM THE FLUCTUATION–DISSIPATION THEOREM

In Appendix B, we will demonstrate the second law from the fluctuation–dissipation theorem.

We consider a trajectory $\{x\}$ by analogy with Eq. (55) in Appendix A. Then we can rewrite Eq. (17) as

$$\langle \Delta W \rangle = \frac{1}{\beta} \int_0^T dt \int_0^t dt' \\ \times \left\langle \frac{\partial S_\beta(x(t), c(t))}{\partial t} \exp\left[-\int_{t'}^t d\tau \frac{\partial S_\beta(x(\tau), c(\tau))}{\partial \tau}\right] \frac{\partial S_\beta(x(t'), c(t'))}{\partial t'} \right\rangle_{\{\rho_\beta\}} \quad (57)$$

where

$$\langle \cdot \rangle_{\{\rho_\beta\}} \equiv \int \mathscr{D}x \cdot P_\beta(\{x\})$$
$$P_\beta(\{x\}) \equiv P_\beta(x_N|x_{N-1})P_\beta(x_{N-1}|x_{N-2}) \cdots P_\beta(x_1|x_0)\rho_\beta(x_0, c_0)$$
$$P_\beta(x_{n+1}|x_n) \equiv \exp[\{-iL(x_{n+1}, c_{n+1}) + \partial_t S_\beta(x_{n+1}, c_{n+1})\}T/N]\delta(x_{n+1} - x_n)$$

Note that the probability distribution to find x_n is always the Boltzmann equilibrium distribution,

$$\int dx_0 \cdots dx_{n-1} P_\beta(x_n|x_{n-1}) P_\beta(x_{n-1}|x_{n-2}) \cdots P_\beta(x_1|x_0)\rho_\beta(x_0, c_0) = \rho_\beta(x_n, c_n)$$

Here,

$$\int_0^T dt \int_0^t dt' \frac{\partial S_\beta(x(t),c(t))}{\partial t} \exp\left[-\int_{t'}^t d\tau \frac{\partial S_\beta(x(\tau),c(\tau))}{\partial \tau}\right] \frac{\partial S_\beta(x(t'),c(t'))}{\partial t'}$$
$$= \exp[-\Delta S_\beta(T)] - 1 + \Delta S_\beta(T)$$

where $\Delta S_\beta(T) \equiv \int_0^T d\tau \partial_\tau S_\beta(x(\tau),c(\tau))$.
From the inequality, $\exp[x] \geq 1 + x$ for $x \in \mathbf{R}$, the second law is valid:

$$\langle \Delta W \rangle = \langle \exp[-\Delta S_\beta(T)] - 1 + \Delta S_\beta(T) \rangle_{\{\rho_\beta\}} \geq 0 \tag{58}$$

In a near-quasi-static process,

$$\langle \Delta W \rangle = \int_0^T dt \int_0^t dt' C_\beta(t,t') \sim \tfrac{1}{2} \langle (\Delta S_\beta(T))^2 \rangle_{\{\rho_\beta\}} \tag{59}$$

This is the reason why we call the theorem the fluctuation–dissipation theorem for excess heat production.

Acknowledgments

One of authors (HHH) thanks Prof. Mikito Toda and Prof. Tamiki Komatsuzaki for their hospitality and support in the Kyoto conference. He deeply appreciates Prof. Yoji Aizawa and Dr. Chun-Biu Li for their useful suggestions about the adiabatic invariant. He also deeply appreciates Dr. Hiroshi Ando for his suggestion about the proof in Appendix B. He thanks K. Nelson for correcting our draft. This work was supported by Engineering Research Program of the Office of Basic Energy Sciences at the US Department of Energy Grant No. DE-FG03-94ER14465.

References

1. D. Evans, E. Cohen, and G. Morriss, *Phys. Rev. Lett.* **71**, 2401 (1993).
2. G. Gallavotti and E. Cohen, *Phys. Rev. Lett.* **74**, 2694 (1995).
3. C. Jarzynski, *Phys. Rev. Lett.* **78**, 2690 (1997); *Phys. Rev.* **E56**, 5018 (1997).
4. G. E. Crooks, *Phys. Rev.* **E60**, 2721 (1999).
5. J. L. Lebowitz and H. Spohn, *J. Stat. Phys.* **95**, 333 (1999).
6. Y. Oono and M. Paniconi, *Prog. Theor. Phys. Suppl.* **S130**, 29 (1998).
7. K. Sekimoto, *J. Phys. Soc. Japan* **66**, 1234 (1997); K. Sekimoto and S. Sasa, *J. Phys. Soc. Japan* **66**, 3326.
8. T. Hatano and S. Sasa, *Phys. Rev. Lett.* **86**, 3463 (2001).
9. P. Glansdorff and I. Prigogine, *Thermodynamics of Structure, Stability and Fluctuations*, Wiley-Interscience, New York, 1971; D. Kondepudi and I. Prigogine, *Modern Thermodynamics*, John Wiley & Sons, Chichester, 1998.
10. H. H. Hasegawa, C.-B. Li and Y. Ohtaki, *Phys. Lett.* **A307**, 222 (2003).
11. H. H. Hasegawa, *Adv. Chem. Phys.* **122**, 21 (2002).

12. J. Suen and J. Erskin, *Phys. Rev. Lett.* **78**, 3567 (1997); *Phys. Rev.* **B59**, 4249 (1999) and references therein.
13. C. Beck and E. D. D. Cohen, cond-mat/0205097(2003) and references therein.
14. C. Tsallis, *J. Stat. Phys.* **52**, 479 (1988).
15. E. Mayoral and A. Robledo, cond-mat/0401128(2004) and references therein.
16. E. Ott, *Phys. Rev. Lett.* **42**, 1628 (1979); R. Brown, E. Ott, and C. Grebogi, *J. Stat. Phys.* **49**, 511 (1987).
17. C. Jarzynski, *Phys. Rev.* **A46**, 7498 (1992).
18. Y. Pomeau and P. Manneville, *C. Math. Phys.* **74**, 189 (1980).
19. Y. Ohtaki and H. H. Hasegawa, *Bussei Kenkyu* **77**, 926 (2002).
20. Y. Aizawa et al., *Prog. Theor. Phys. Suppl.* **79**, 96 (1984); *Prog. Theor. Phys. Suppl.* **98**, 36 (1989).
21. H. H. Hasegawa and Y. Ohtaki, in preparation.
22. M. Dzugutov, E. Aurell, and A. Vulpiani, *Phys. Rev. Lett.* **81**, 1762 (1998).
23. V. Latora and M. Baranger, *Phys. Rev. Lett.* **82**, 520 (1999).

PART III

NEW DIRECTIONS IN MULTIDIMENSIONAL CHAOS AND EVOLUTIONARY REACTIONS

CHAPTER 20

SLOW RELAXATION IN HAMILTONIAN SYSTEMS WITH INTERNAL DEGREES OF FREEDOM

AKIRA SHUDO

Department of Physics, Tokyo Metropolitan University, Minami-Ohsawa, Hachioji, Tokyo, 192-0397, Japan

SHINJI SAITO

Department of Chemistry, Nagoya University, Furo-cho, Chikusa-ku, Nagoya, 464-8602, Japan

CONTENTS

I. Introduction
II. Anomalous Transport in Hamiltonian Systems with Mixed Phase Space
III. Slow Relaxation in Molecular Systems
IV. Nearly Integrable Pictures
V. Alternative Scenario: Hamiltonian Systems with Internal Degrees of Freedom
VI. Validity of Hypothesized Scenario
VII. Conclusions and Discussions
Acknowledgments
References

I. INTRODUCTION

Since a celebrated work of Fermi–Pasta–Ulam (FPU), computer experiments have provided powerful tools to attack the ergodic problem of dynamical

---

Geometric Structures of Phase Space in Multidimensional Chaos: A Special Volume of Advances in Chemical Physics, Part B, Volume 130, edited by M. Toda, T Komatsuzaki, T. Konishi, R.S. Berry, and S.A. Rice. Series editor Stuart A. Rice.
ISBN 0-471-71157-8 Copyright © 2005 John Wiley & Sons, Inc.

systems [1]. The study of ergodicity in Hamiltonian systems is particularly important since it is tightly connected with the foundation of statistical mechanics. The original motivation of FPU was to verify, employing nonlinear lattice vibration models, that equilibrium states or equipartition of every modes may immediately be achieved only if small nonlinear couplings are present [2]. However, as is now well known, what they observed was the breakdown of ergodicity. More precisely, they have discovered that the energy initially assigned to a certain normal mode of the linear oscillator does not necessarily transfer to every other modes, but the energy transfer occurs only within limited numbers of modes. Even the recurrence of energy to an initially excited mode was found.

The result was accepted as surprising because the prediction of statistical mechanics, which assumes the so-called ergodic hypothesis as its basis, is believed to be satisfactory enough not only up to qualitative but also quantitative levels. However, from different viewpoints, the result might be quite reasonable and plausible; if all the nonlinear systems tend to equipartitioned states, they eventually approach the thermal equilibrium as the time proceeds, which means that no structures and patterns are created and no dynamics emerges in macroscopic levels. But, this is not the case at all. We know plenty of systems that are actually in nonequilibrium and far from thermal equilibrium states on earth. One can say that the breakdown of ergodicity is a necessary condition to create all the structures and pattern formations and changes of the states, including biological systems.

Computer experiments after FPU have indeed been made to examine under what conditions, how the system breaks the ergodicity, and how long nonequilibrium states persist. In particular, nonlinear lattice systems, first investigated by FPU to model the vibrational oscillations around an equilibrium point of solids, have been used as a canonical model to explore such issues [2].

In addition to computer simulations, what drives the research in this direction is elaborated perturbation theories developed almost simultaneously. In particular, the Kolmogorov–Arnold–Moser (KAM) theorem, which has shown the existence of invariant tori under a small perturbation to completely integrable systems, and the Nekhoroshev theorem, which has proved exponentially long-time stability of trajectories close to completely integrable ones, are landmarks in this field. Although a lot of works have been done, there still remain unsolved important questions, and the Hamiltonian system is being studied as one of important branches in the theory of dynamical systems [3–5].

On the other hand, thanks to rapid progress of the computer power, numerical simulations now become applicable to much wider classes and much complicated systems, ranging from the dynamics of atoms and molecules in various phases to that of biological systems. As a matter of fact, such studies yield a lot of examples demonstrating the breakdown of ergodicity in their dynamics.

However, since motivations of molecular dynamics (MD) simulations are slightly different from those in the theory of dynamical systems, many issues, which look common in both fields, are discussed separately and in different communities. There would be several reasons. MD simulations are often performed to pursue system specific natures, whereas the study of the dynamical system mainly focuses on universality of systems. As a result, in MD studies, it is not a primary task to find the simplest possible canonical model, on the other hand, setting and classifying standard forms is the first relevant task to be done in the theory of dynamical systems. The tools or languages used to analyze the data also differ from each other. As discussed in this chapter, the potential-based argument is developed and the potential energy landscape picture is used as a common language in MD simulations. On the contrary, the phase space geometry and trajectories in phase space are of main interest in the study of dynamical systems. Furthermore, in MD simulations, systems far from equilibrium states are often studied, while, as the FPU model does so, the motions near equilibrium points are most closely examined in the latter.

Under these circumstances, the present chapter will be devoted to (a) reconsidering the subjects mostly studied in MD simulations in the light of the theory of dynamical systems and (b) discussing several possible pathways to connect these seemingly different viewpoints. In particular, our attempt will be done by focusing on the slow dynamics first observed and extensively studied in MD simulations, and we will present an alternative picture to explain its origin within a general framework in the sense of dynamical systems. The slow dynamics is an important issue both in ergodic problems of Hamiltonian dynamics and in MD simulations of condensed phases; particularly the dynamics in supercooled or glassy states is currently one of active areas. A concrete example we will specifically examine is liquid water dynamics in room temperatures, but our motivation is not limited to the understanding of a specific system.

The organization of this chapter is as follows. As a rigid basis for later discussions, we first present a brief survey of known results of Hamiltonian systems, especially related with subjects discussed in this chapter. We introduce several scenarios under which the anomalous transport or slow relaxation occurs mainly in few-degrees-of-freedom Hamiltonian systems. There are several theories on the geometry of phase space, and the origin of slow dynamics can be attributed to them. As another point of view, we next present the results of perturbation theories in Hamiltonian systems, which ensure the breakdown of ergodicity or long time adiabaticity. In the final part of this section, a simple two-dimensional model describing nonequilibrium nature of transport is introduced in order to discuss a possible connection to potential based arguments mainly done in the interpretation of MD data.

Section III presents several concrete examples of molecular systems in which slow relaxation has actually been observed. A subject we will be particularly interested in is the slow dynamics of liquid water. Water is one of the most fundamental molecules not only because it broadly exists on earth but because most chemical reactions in condensed phase proceed in water [6, 7]. It was discovered that the total potential energy fluctuation shows intermittent behavior, yielding the $1/f$-type power spectrum [8,9]. We discuss similarity of intermittent motions in liquid water with slow motions often found in supercooled and glassy liquids. The raggedness of the potential energy landscape will be mentioned as a plausible picture to understand the slow dynamics in those systems.

The fact that slow relaxation is observed in liquid water surely provides us with interesting issues worth examining as a problem of Hamiltonian systems. Section IV is devoted to explaining why we believe so. The most extensively studied and well-established situation in the theory of Hamiltonian systems is the system close to completely integrable systems. In fact, the arguments presented in Section II assume nearly integrable forms, and we understand, though not completely, the origin for intermittent motions in such situations. However, the intermittent dynamics found in liquid water is clearly beyond the scope of the nearly integrable Hamiltonian picture. We will present several reasons why we cannot regard the nearly integrable Hamiltonian as a model of those slow motions.

In Section V, we will give an alternative picture, instead of the nearly integrable picture, which may explain and solve our issue. The idea is again based on perturbation theories, whose technical parts are essentially the same as those in nearly integrable Hamiltonian systems, but the small parameter introduced in the perturbation expansion is different. The scenario is based on an idea originally proposed by Boltzmann and Jeans to explain freezing of high-frequency motions in gases [10,11]. It was recently reinterpreted as exponentially long time stability of Nekhoroshev type [12,13].

In Section VI, along with the scenario introduced in Section V, we reexamine slow motions in liquid water, together with the check of predictability of our hypothesized interpretation. It will be shown that almost all the data examined there are compatible with our scenario, and some simulations using artificial molecules, which are obtained by introducing and changing several parameters in real molecules, are done to test the proposed hypothesis [14].

In Section VII we conclude our results and discuss several issues arising from our proposals. We revisit our original motivation—that is, to find a simple model, in the sense of dynamical systems, that captures several common aspects of slow dynamics in liquid water, or more generally supercooled liquids or glasses. Our attempt is to make clear the relation and compatibility between the potential energy landscape picture and phase space theories in the Hamiltonian dynamics. Importance of heterogeneity of the system is discussed in several respects. Unclarified and unsolved points that still remain but should be considered as crucial issues in slow dynamics in molecular systems are listed.

II. ANOMALOUS TRANSPORT IN HAMILTONIAN SYSTEMS WITH MIXED PHASE SPACE

MD simulations with a constant energy is nothing but Hamiltonian dynamics. Recent accumulation of MD simulations will certainly contribute to our further understanding of Hamiltonian systems, especially in higher dimensions. The purpose of this section is to sketch briefly how the slow relaxation process emerges in the Hamiltonian dynamics, and especially to show that transport properties of phase-space trajectories reflect various underlying invariant structures.

If a system is uniformly hyperbolic, every point in phase space has both stable and unstable directions, and the maximum Lyapunov exponent with respect the maximum entropy measure is positive. The system has the mixing property and is therefore ergodic. The correlation function of observables also shows exponential decay. Uniformly hyperbolicity, which is sometimes rephrased as *strong chaos* in physical literature, is a well-established class of systems and is controllable by means of many mathematical tools [15]. In hyperbolic systems, there are no sources to make the relaxation process slow.

It is true that the hyperbolic system is an ideal dynamical system to understand from where randomness comes into the completely deterministic law and why the loss of memory is inevitable in the chaotic system, but *generic* physical and chemical systems do not belong strictly to such ideal systems. They are not uniformly hyperbolic, meaning that invariant structures are *heterogeneously* distributed in phase space, and there may not exist a lower bound of instability. It is believed that dynamical systems of such classes are certainly to be explored for our understanding of dynamical aspects of all relevant physical and chemical phenomena.

However, as compared to hyperbolic systems, it is rather difficult to specify more explicitly than just to say they are *nonhyperbolic systems*. After developing the theory of hyperbolic systems, the studies of nonhyperbolic systems are now the next targets in mathematical studies, either of which will be hard problems however.

As for Hamiltonian systems, the most well understood situation is either completely integrable or fully chaotic systems—more specifically stated, an anti-integrable limit [16]. One reasonable approach to nonhyperbolic systems would be to investigate the system that is sufficiently close to the well-understood extremes by applying various perturbation schemes. If one takes a fully chaotic system as a limiting case, one can prove the existence of chaotic orbits by continuation of these orbits from the limiting case [16]. Techniques used there might be simple as compared to perturbation techniques used in nearly intergrable systems, but it cannot be applied to the system far from the limiting case. As the system approaches truly nonhyperbolic regimes, a series of bifurcations happens and one is necessarily involved in combinatorics problems, which are quite hard to be solved in general. The idea of *pruning front*, which is a two-dimensional analog of the kneading theory of Milnor and Thurston [17], has been proposed [18], but to apply it to nonhyperbolic situations is far from accomplished [19–21] and is even hopeless in some sense [22].

On the other hand, if the perturbation theory is applied to the completely integrable system, the existence of invariant tori can be discussed. More precisely, consider N-degrees-of-freedom Hamiltonians under a small perturbation in the standard action–angle form,

$$H(\vec{I}) = H_0(\vec{I}) + \varepsilon H_1(\vec{I}, \vec{\theta}) \qquad (1)$$

where $\vec{I} = (I_1, \ldots, I_N)$, and $\vec{\theta} = (\theta_1, \ldots, \theta_N)$ are action and angle coordinates, respectively. Reflecting the fact that the perturbation strength ε is small, the Hamiltonian of the above form is referred as the *nearly integrable system*. The KAM (Kolmogorov–Arnold–Moser) theorem claims that invariant tori of unperturbed Hamiltonian $H_0(\vec{I})$ survive under sufficiently small ε, if unperturbed Hamiltonian $H_0(I)$ satisfies the nondegenerate condition,

$$\left| \frac{\partial^2 H_0}{\partial I_i \, \partial I_j} \right| \neq 0 \qquad (1 \leq i,j \leq N) \qquad (2)$$

and rotation numbers $\vec{\omega}$ of quasi-periodic motions on tori are sufficiently irrational in the sense that it satisfies the Diophantine condition,

$$|\vec{k} \cdot \vec{\omega}| > \gamma |\vec{k}|^\tau \qquad (3)$$

for some positive constants γ and τ.

It is important to note that a set of rotation numbers $\vec{\omega}$ satisfying the Diophantine condition has positive Lebesgue measure in $\mathbf{R}^N$. This means that, in principle, KAM tori are observable invariant objects in N-dimensional phase space. But, the original KAM theorem has not made further arguments concerning how much KAM tori remain in phase space as a function of the perturbation strength ε or the degrees of freedom N. This topic will be one of our interest throughout the present chapter.

In addition, we should remark that invariant tori we often found in numerical simulations are not truly KAM tori guaranteed rigorously in the mathematical theorem. The perturbation strength ε is so small and any chaotic orbits cannot be detected in phase space if we perform numerical simulations under the original condition of the KAM theorem as for the smallness of ε.

For these reasons, one may suspect how the KAM theorem is useful to our understanding of MD simulations. But this is not the case because, even though the theory does not cover the situation we just want to know, typical or limiting situations that are fully understandable and completely analyzable contribute to solid building blocks from where we can start to develop our theory.

One more strong basis in nearly integrable Hamiltonian systems is the Nekhoroshev's theorem, which gives, rather than stability of orbits in the infinitely long time interval as the KAM theorem asserts, exponentially long time stability of orbits [23]. The Hamiltonian to which the theorem of Nekhoroshev is applicable is again the nearly integrable form, Eq. (1). Under the steepness condition, which was replaced by a stronger and natural condition—the quasi-convex condition—the variation of the action variables in the unperturbed Hamiltonian $H_0(\vec{I})$ is bounded as

$$|\vec{I}_k(t) - \vec{I}_k(0)| < A\varepsilon^\alpha \quad (1 \leq k \leq N) \qquad (4)$$

for

$$|t| < B \exp\left[\frac{\varepsilon^*}{\varepsilon}\right]^\beta \qquad (5)$$

Here α and β are some constants depending on the system size N, as discussed in the following section. The proof of the theorem is reduced to evaluate the remainder terms after successive canonical perturbations, and it is possible to make the remainder exponentially small.

The perturbation strength ε for which the Nekhoroshev's theorem holds is also so small that it cannot be applied to realistic physical and chemical situations. Indeed it was shown that the range of perturbation strength is much smaller than the situation where the power spectrum density of observables exhibits a continuous one [24]. This means that, in its rigorous sense, the Nekhoroshev's theorem can only be applied to sufficiently weak perturbed systems. For the same reason as mentioned above, Nekhoroshev's theorem is nevertheless a key guiding principle to sticky or stagnant motions in nearly integrable Hamiltonian systems.

For instance, as for the recurrence phenomenon discovered in the FPU nonlinear lattices [25], whose explicit Hamiltonian will be presented in Section IV, there may exist a gap between the results of numerical simulations and the mathematical theorem, but a rigorous result certainly plays a significant role and theoretical arguments based on it can be deduced [26].

In two-dimensional Hamiltonian systems, the trajectories can be visualized by means of the Poincarè surface of section plot. It is also possible to study two-dimensional Hamiltonian systems using the two-dimensional symplectic mapping. A typical phase space portrait of generic nonhyperbolic phase space is

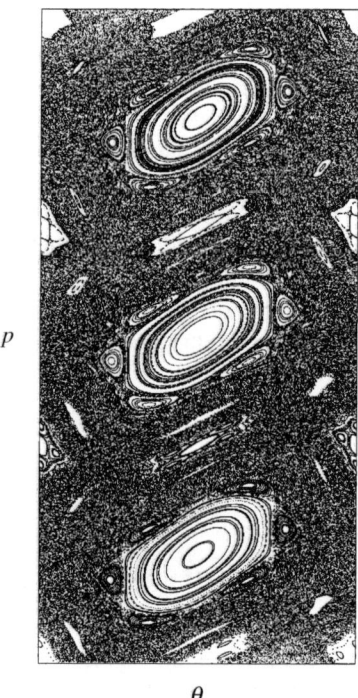

Figure 1. Typical mixed phase space for the two-dimensional mapping. Quasi-periodic and chaotic regions coexist. The motion around tori is sticky and the orbit sometimes takes a long time to get through the sticky region.

displayed in Fig. 1, in which one can see the coexistence of quasi-periodic and chaotic orbits. Such phase space is often called *mixed phase space*.

In contrast to hyperbolic systems, the phase space structure in the mixed system is quite intricate and inhomogeneous, which brings about transport phenomena and relaxation processes essentially different from uniformly hyperbolic cases [3]. A remarkable fact is that qualitatively different classes of motions such as quasi-periodic motions on invariant tori and stochastic motions in chaotic seas coexist in a single phase space. The ordered motions associated with invariant tori are embedded in disordered motions in a self-similar way. The geometry of phase space then reflects the dynamics.

If the resonant tori, which are the invariant tori whose rotational numbers are rational, are broken under perturbations, the pairs of elliptic and hyperbolic cycles are created in the resonance zone. This fact is known as a result of the Poincarè–Birkhoff theorem [4], which holds only if the twist condition, Eq. (2), is satisfied. Around elliptic cycles thus created, new types of tori, which are resonant tori or island of tori, appear. Here we mean by "new" the ones that are not guaranteed to exist by the KAM theorem. Resonant tori accompany smaller-scale resonant tori around them, and the hierarchy of island structures continue to infinitesimal fine scales.

Anomalies found in relaxation processes in the presence of such small islands were indicated and discussed in many publications [3,27,28]. *Stickiness* around small islands was measured by trapping time statistics [29]. The author of Ref. 29 found, by studying two-dimensional polynomial mappings, that the decay of the correlation function is algebraic for small perturbation regimes and suggested that the diffusion constant is infinite, meaning that the distribution function of orbits is no longer governed by a standard diffusion equation.

One more important ingredient in discussing the transport process in mixed phase space is so-called *cantori*. They are invariant sets in which the motion has an irrational frequency. They resemble invariant circles, but they have an infinite number of gaps in them. The existence was proposed by Percival [30] and Aubry [31]. They gave an explicit example, and a proof of their existence has been given afterwards [32–34].

Cantori have an infinitely many gaps and they do not play the role of complete barriers as the KAM tori do, but when most of gaps in a cantorus are very small, orbits take a long time to get through. So, cantori present partial barriers in phase space, especially just after critical parameter values where invariant KAM tori break up the gaps are so small that cantori give strong partial barriers. The flux through the barriers of the cantori can be evaluated by Mather's differences in action [35], which gives bounds on transport between regions separated by a cantorus. Combining such locally estimated flux with assumptions on a global way of arrangement of cantori, the origin of slow decay of correlations has been well explained [35]. The assumption they made there is that cantori divide irregular components of phase space into subregions, within each of which the motion is strongly mixing, and that the noble cantori form the most dominant bottleneck and thus govern the transport.

Meiss and Ott [36] have developed these arguments further and proposed a model based on the Markov tree structure. They have also assumed that partial barriers are formed and cantori with small gaps divide the region around the outermost KAM torus into infinitely many states. An essential point in their argument is to give a labeling scheme to these states and to specify transition probabilities between adjacent states. In particular, each state is viewed as the node of tree, and the transition between these states is assumed to obey the Markov process.

As mentioned in the introduction, the power spectrum density is used to probe the long-time correlation decay. Appearance of $1/f^v$-type spectra is an indication that there are, in principle, infinitely many time scales in the relaxation process. Geisel et al. [37] gave an example of mixed Hamiltonian systems displaying the $1/f$-type spectrum and proposed a mechanism to generate such a type of spectrum. They have employed a two-dimensional Hamiltonian with periodic potential on a square lattice,

$$H = \tfrac{1}{2}(p_1^2 + p_2^2) + A + B(\cos q_1 + \cos q_2) + C \cos q_1 \cos q_2 \qquad (6)$$

where A, B and C are adjustable parameters. As shown in Fig. 2, $1/f$ spectrum is clearly seen for a certain set of parameters [37]. Note that a two-dimensional periodic potential is reminiscent of Lorentz gas model, which will be discussed as a model of nonequilibrium transport.

An idea to explain the appearance of $1/f$ spectrum is based again on a hierarchy of nested cantori [37]. Geisel et al. assumed that a sequence of cantori accumulates to a critical KAM torus and introduced the renewal Markov process as the transition dynamics over the partial barriers. It was also pointed out that the Markov model together with underlying hierarchical tree structures is not sufficient to generate genuine $1/f$ spectrum especially in the case of systems with the compact phase space. But it is true that the existence of partial barriers in phase space is commonly used as an essential ingredient for these modelings.

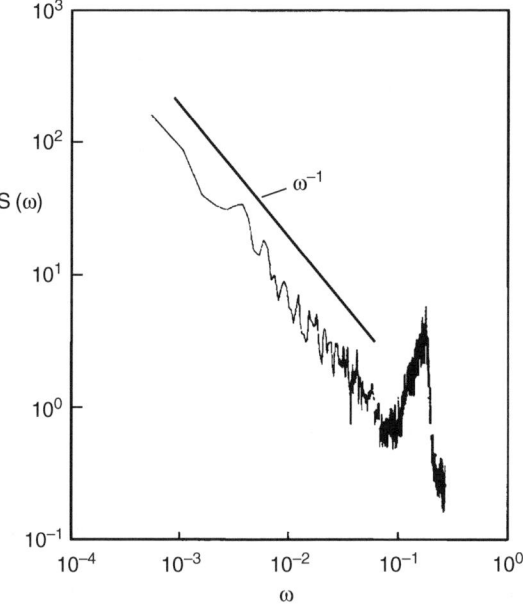

Figure 2. The velocity power spectral density for the system (b). [Reprinted with permission from T. Geisel, A. Zacharl, and G. Radons, *Phys. Rev. Lett.* **59**, 2503–2506 (1987). Copyright 1987 by American Physical Society.]

Along a similar line, self-similarity of phase space is used to explain the origin of power low decay [38,39].

On the other hand, it is possible to explain stickiness of orbits in a different way [26]. As mentioned above, in nearly integrable systems, the Nekhoroshev theory gives exponentially long time stability. If we may apply it to an individual invariant torus in phase space, it is possible to evaluate how long the orbits are trapped around a torus. Assuming the stagnant layer, which is defined as an annulus outside the final KAM torus, one can estimate the first passage time distribution in the stagnant layer as [26]

$$P(T) \sim \frac{1}{T \log T} \qquad (7)$$

where $P(T)\,dT = \#\{$orbits that escape from the stagnant layer within T and $T+dT\}$. Here the final KAM torus is defined as an outermost KAM torus surrounding each elliptic point and borders torus and chaotic regions. In this argument, the smoothness of the final KAM torus is implicitly assumed. An estimation, Eq. (7), indicates that the mean value of the first passage time diverges. The orbits around the KAM torus are so strongly localized around the final KAM, then it turns out that the invariant measure that governs such a strong sticky motion is not a usual Lebesgue measure but a singular one. Furthermore, if one can approximate the motion inside the stagnant layer by a white Gaussian process, the power spectrum density of the dynamical variable is predicted as

$$S(f) \sim f^{-2} \qquad (8)$$

in the low-frequency limit ($f \ll 1$), which means that the stagnant motion obeys the *nonstationary process* [26].

It is not so clear how the scenario based on cantori or the self-similar island chain is compatible with long-time stability derived by the Nekhoroshev theory. We can say that the former is concerned with the geometrical structure of phase space, while the latter is concerned with adiabaticity of the orbits. The existence of cantori can be proved in terms of the variational principle for the action functional, and exponentially long-time stability is achieved by controlling the remainder term of the perturbative expansion. However, it is at least true that the region where all these scenarios took place when the system is close to the integrable limit. Cantori work as partial barriers only when the gaps in each cantorus are sufficiently small; otherwise they allow free passage of orbits, and the states separated by partial barriers cannot be well-defined. As the perturbation strength is increased, small elliptic islands eventually bifurcate and turn to hyperbolic saddles. In a similar way, exponential stability of Nekhoroshev is also given in weakly perturbed regimes. One cannot expect that sticky or stagnant motions occur in strongly perturbed regimes. In the case of the standard mapping, for example, the normal diffusion is recovered if the perturbation strength is large enough.

Here we should mention the importance of dimensionality of phase space. In two-dimensional phase space, KAM curves can encircle the two-dimensional regions and confine the orbits surrounded by them. However, in the case of the system with more than two dimensions, KAM curves do not serve as the barrier of phase space. Likewise, the partial barriers do not form bottlenecks. The possibility of the Arnold diffusion may be taken into account in more than two dimensions, but the Arnold diffusion is usually discussed instead in relation with the Nekhoroshev-type argument, not considered as a consequence of partial barriers discussed here.

It should be noted that the Nekhoroshev's theory is not limited to two-dimensional systems, but rather it holds in general dimensions. Therefore, in the system with many degrees of freedom, the Nekhoroshev's theorem may explain sticky motions. This is also the case with the KAM theory. As given before, the statement of the KAM theorem is not limited to the Hamiltonian with few degrees of freedom. In Section IV, we will discuss to what extend these perturbation theories have capability to predict the slow motions in many-dimensional systems.

In the final part of this section, we present another simple toy model that describes the nonequilibrium nature of transport [40]. The Lorentz gas model, as shown in Fig. 3, is a two-dimensional billiard where a point particle

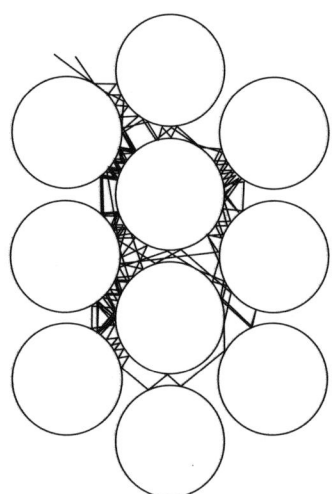

Figure 3. Trajectory of a particle in the hexagonal array of disks.

undergoes elastic collisions on hard disks which are put as a two-dimensional array. The erratic scattering gives rise to diffusive motion in extended two-dimensional configurational space, and it provides one of the simplest possible models for the *deterministic diffusion* [40].

There are two important classes depending on configuration of periodic disks. The first one is the case where the distance of each disk is sufficiently large and the channels between disks allow the ballistic motion, meaning that a particle can move to infinity without colliding any disk. Such a situation is referred to as the *infinite-horizon* configuration. In this configuration, because the collisional horizon of these trajectories diverges, the diffusion coefficient is infinite [41]. On the other hand, the second class is called *finite-horizon* configuration, where the distance between disks is so small that ballistic motions are forbidden. In the finite-horizon case, it was proved that the diffusion coefficient is positive and finite [41].

The most remarkable fact in the Lorentz gas model would be that one can obtain detailed information on the transport analytically. As is found in Fig. 3, the dynamics inside the array of disks is always hyperbolic due to the defocusing character of the collisions. Using such properties, it has been proved that the Lorentz gas has the mixing property [41]. Furthermore, in the case of the Lorentz gas model with a finite-horizon configuration, in addition to the finiteness of the diffusion coefficient, the trajectories can be regarded as a standard Brownian process after an appropriate scaling [41,42]. Concerning the decay of correlation, a rigorous estimate tells us that it is fast and of stretched exponential type at least with respect to the mapping, not the flow on the real configurational plane [43].

The periodic Lorentz gas model with finite-horizon configuration is of particular interest since it resembles the situation of diffusive motions in multi-valley potential surfaces of molecular systems. The local dynamics inside the array of disks is strongly hyperbolic as mentioned above, which reminds us of the intrabasin mixing within a potential well. If the distance between disks is sufficiently small, the channel between arrays might play the role of the bottleneck on the configuration space. The interbasin diffusion process may be modeled by the large-scale diffusion represented by the Lorentz gas model. The similarity will be discussed more closely in the final section.

III. SLOW RELAXATION IN MOLECULAR SYSTEMS

As mentioned in the introduction, water is a quite ubiquitous substance on earth, and most chemical reactions proceed in water. As a problem in chemistry, water is especially relevant and thus should be investigated in great detail. However, if one views water as an object studied within the theory of dynamical systems, it is rather specific and not generic. The first task we should do would be to extract the most intrinsic characters, which may be taken as an issue of the dynamical system, from the results obtained mostly in MD simulations [8,9].

We may list differences between the liquid water system and the FPU model; the latter will be examined in the next section as a representative system in the study of many-dimensional Hamiltonian systems. The most important difference would be that the FPU model describes a lattice vibration around an equilibrium point and the potential energy function possesses a single minimum, whereas there are infinitely many local potential minima and the potential energy landscape generally becomes ragged in the case of the liquid water system. The reason why the character of the potential landscape could be so important is that the raggedness is considered as an origin of slow motions in liquid water or supercooled liquids.

In a series of works by Stillinger and co-workers [44,45], it has explicitly been indicated that the ragged landscape of the potential function controls the real dynamics, specifically the structural changes of liquid water. They proposed a novel method to separate structural changes in dynamics from the vibrational motions by quenching the instantaneous trajectories. It has been shown that the time scale of structural changes is slower than vibrational motions, which may be regarded as noises. They call *inherent structure* (IS for abbreviation) such ragged potential hypersurfaces.

IS is not owned only by liquid water, but it is common to many other molecular systems. In particular, the ragged landscape picture is useful and now even popular to understand cluster dynamics. Extensive works by Berry clarified how the structural transition or isomerization process of Ar cluster takes place reflecting the raggedness of potential functions [46].

The study of liquid water has further been pushed by Ohmine and Tanaka and a lot of pieces of evidence with respect to specificity of liquid water have been pointed out [8,9]. They also performed extensive numerical simulations and found that collective motions, the existence of which was suggested in the preceding work [44], occur in an intermittent way. They also identified the rearrangement of *hydrogen-bond network* and found that it is closely related with collective motions.

Moreover, in the work of Sasai et al. [46], long-time correlation decay in liquid water dynamics, which is our main concern in this chapter, was discovered. They attributed it to the intermittent dynamics, and thereby to collective motions in liquid water [8]. They have measured the power spectrum density of the total energy fluctuation and found that it has an algebraic decaying part as given in Fig. 4a. This reflects intermittent temporal energy fluctuation shown in Fig. 4b. The hump observed around 20–$1000\,\text{cm}^{-1}$ comes from translational and vibrational modes, and around $1.5\,\text{cm}^{-1}$ the spectrum is of a white-noise type. The latter frequency corresponds to the motion of 20 ps, which is roughly interpreted as the time scale of the rearrangement dynamics. Since the lifetime

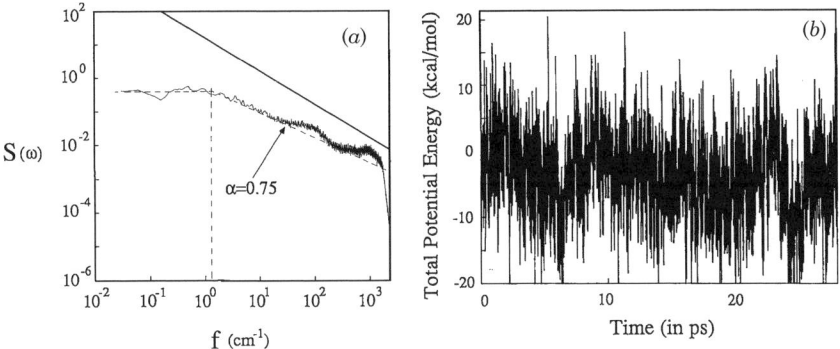

Figure 4. (a) The power spectrum density of total system potential in liquid water. The simulation was performed for 64 water molecules under periodic boundary conditions. The temperature is 297 K. The slope of the solid line and the dashed line fitted to experimental data is −1 and 0.75, respectively. (b) The corresponding total energy fluctuation. [Reprinted with permission from *J. Chem. Phys.* **96**, 3045–3053 (1992). Copyright 1992 by American Institute of Physics.]

of individual hydrogen bonds is about a few picoseconds, the algebraic decaying part in the spectrum shows that the slow relaxation process in liquid water occurs at longer time scales than that of translational and librational motions of individual molecules. In fact, the algebraic decaying part disappears when one sees the spectrum for potential energy fluctuation of individual molecules [47].

In contrast to liquid water, liquid argon does not exhibit such a long-time correlation decay. Figure 5a shows that fluctuation of total of potential

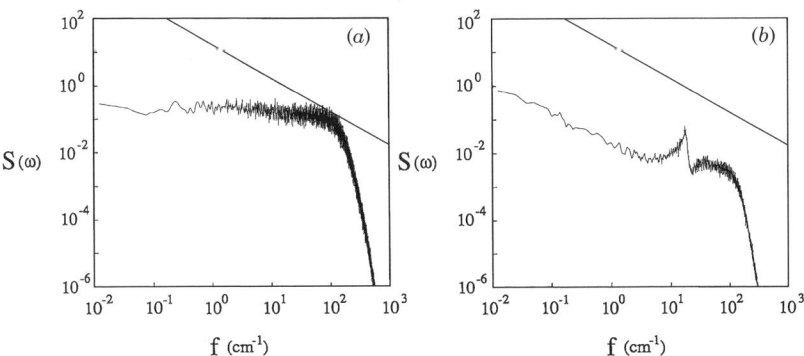

Figure 5. (a) The power spectrum density of total system potential in liquid argon. The simulation was performed for 108 argon molecules under periodic boundary conditions. The temperature is 95 K. The slope of the solid line is 1.0. (b) The power spectrum density of total system potential in an argon cluster. The system size is 108 and the temperature is 35 K. [Reprinted with permission from *J. Chem. Phys.* **96**, 3045–3053 (1992). Copyright 1992 by American Institute of Physics.]

energies of argon atoms is of a white noise type [47], so we can expect that fully chaotic and mixing dynamics is achieved in multidimensional phase space.

Much more instructive in our following discussion is to see the argon cluster case. As presented in Fig. 5b, an algebraic decaying regime appears in the spectrum of the total system potential energy fluctuation. Clusters differ from liquid systems in that clusters have both core and surface parts, which cause *inhomogeneity* in the system. The potential energy height distribution for surface displacements is very different from that for the core displacements [48], which must give rise to different characters in their motions. Structural isomerization of argon and other clusters proceeds in ragged potential surfaces, and a ragged landscape is created because of inhomogeneity of the system.

In the study of Hamiltonian systems, systems with finitely many degrees of freedom such as clusters have not been seriously investigated so far. Two-dimensional systems, or two-dimensional symplectic mappings discussed in the previous section, are the simplest situations where chaos appears in phase space, and they could be analyzed in great detail. On the other hand, a primary focus of FPU problems is put on the studies of ergodicity in the thermodynamical limit. In actual numerical studies, one can only deal with finite systems, but we usually view various finite size effects as secondary in comparison to the properties appearing as the system size goes to infinity. Alternatively stated, we want to avoid a finite size effect since it might have complicated origins. In this sense, one may say that clusters are most poorly understood systems. In cluster systems, the effect of the boundary cannot be neglected, or rather the presence of finite size effects is an essential ingredient. Here we are using the term *inhomogeneity* in a vague sense, and later we will discuss its origin more closely.

An explanation for slow relaxation in liquid water, not from the theory of dynamical systems but from specific natures of water, has been presented in the same series of articles [8,9]. Their interpretation is based on (a) the formation of three-dimensional hydrogen-bond networks and (b) collective motions or large-scale structural changes in the network systems. As is well known, the hydrogen bond is specific in that it is strongly directional and relatively tight. In addition, the hexagonal structure of ice partially survives even in the liquid phase, which forms the network of hydrogen bonding. They demonstrated in what spatial scale and how frequently such rearrangement dynamics occurs by employing several methods and measures; the distance matrix for IS, which probes the correlation between successive transitions among IS [49], direct monitoring of the displacement of individual molecules on the configurational space [49], and network analysis which examines topological connectivity of hydrogen bonding [50]. They present a very sharp contrast to the case of liquid argon in which normal diffusive motions are observed.

Slow relaxation of water was further discussed in more broad contexts. In particular, since the IS picture is expected to become appropriate as the temperature is decreased, special attention has been paid to water in supercooled states, and dynamics in supercooled states has been investigated in relation to applicability of the mode-coupling theory [51]. It was found that the bond lifetime of individual molecules obeys the thermal process, whereas the bond correlation function shows power-law behavior [52,53]. The behavior below or above the temperature at which the mode-coupling theory can be applied was also studied and the transition between IS structures, which is just the network rearrangement dynamics just mentioned above, has clearly been identified in supercooled regions.

The IS concerns information on global aspects of the potential energy landscape, but for the complemental purpose, the instantaneous normal mode (INM) analysis has also been used to extract local information on the curvature of the potential energy surface along a representative trajectory. The INM is defined in analogy with the normal mode analysis in solids; that is, they are the eigenvectors obtained by the diagonalization of the Hessian matrix, which is given as second derivative of the potential energy. Since typical liquid configurations are not local minima of the potential energy function, all the eigenvalues of the Hessian matrix are not necessarily positive; instead they contain information about unstable motions in the dynamics [54]. The cases with negative eigenvalues are further classified into either shoulder modes or double-well modes [55]. It was found there that the diffusion constant depends on the fraction of directions connecting neighboring local minima.

Close studies on water have been and are being done due to its special relevance in chemical reactions, as emphasized in the first part of this section. However, the origin of slow diffusional transitions between interbasin dynamics has not been completely understood. Specificities of water—that is, a three-dimensional hydrogen bond network—certainly play a crucial role in generating intermittent dynamics, and therefore such a behavior seems to be attributed to the specific nature of water. However, if we recall that the separation of the time scale between intrabasin fast librational motions and interbasin slow diffusion would be a manifestation of ragged or multivalley structures of the potential energy landscape, we are tempted to conclude that such a behavior may not be limited merely to the dynamics of liquid or supercooled water, because, as mentioned, the existence of multi-valley minima is not owned only by water. Therefore, sufficient conditions, or more essential ingredients, for slow relaxation in liquids or especially in supercooled liquids should be explored in more generic contexts. The simplest possible or minimal models to realize such behaviors must be found.

Recent studies on slow dynamics in glassy states, and also supercooled states, can be viewed on this line of research. The most important issue in the dynamics in supercooled liquids is to understand why they exhibit nonexponential relaxation. Nonexponential relaxation implies the existence of multi-time scales, and two different explanations are possible in principle: The first one is the *homogeneous* scenario in which all the molecules have an identical relaxation time that is not intrinsically nonexponential, and in the second one the nonexponential relaxation process is due to a superposition of individual exponential relaxations with different characteristic time scales. The latter is thus called the *heterogeneous scenario.*

MD simulations of supercooled liquids are revealing that the heterogeneous scenario is more likely to explain the nonexponential relaxation, and a bunch of articles on this subject were published even within the past few years. *Dynamical heterogeneity* is one of the central issues in the problem of glass-forming liquids.

One important point we should stress, in conjunction with our current interest, is that similar slow relaxation as liquid water is observed in much simpler model systems: The binary mixture of Lennard-Jones liquids, which consist of two species of particles, is now studied extensively as a toy model of glass-forming liquids. It is simulated after careful preparation of simulation conditions to avoid crystallization. Also, the modified Lennard-Jones model glass, in which a many-body interaction potential is added to the standard pairwise Lennard-Jones potential, is also studied as a model system satisfying desired features.

Particularly interesting in simulations of these simple systems is emergence of stringlike rearrangements of groups of particles that move distances much smaller than the average interparticle distance [56]. This reminds us of the rearrangement dynamics of liquid water mentioned above. Both are kinds of collective motions, though the stringlike movements are one-directional motions but the rearrangement dynamics of liquid water occurs in the three-dimensional network. In the vicinity of the mode coupling temperature, it was also found that the dynamics can be separated into vibrational motions around inherent structures and the transition between inherent structures. This picture is entirely consistent with that discussed in liquid water.

These results commonly assert that slow relaxation behaviors observed in supercooled liquids are understood by the inherent dynamics picture; these are regarded as potentially driven processes, and they are likely to occur as the temperature is decreased and the system approaches the glassy states.

IV. NEARLY INTEGRABLE PICTURES

A main message in the previous section is that MD simulations of liquids provide a bunch of uncooked materials to be analyzed as ergodic problems of dynamical systems. In particular, anomalous diffusion or slow relaxation discussed so far should be reexamined and reinterpreted within more generic frameworks. However, there still seems to exist an unsurmountable gap between what the MD simulation studies provide and what is needed in the theory of dynamical systems. One of main sources of such a gap is that MD simulation studies mainly discuss the dynamics on configurational spaces, while the theory of dynamical systems is concerned with phase space of the system. Several mechanisms explained in Section II are described in terms of the language of phase-space trajectories, but the potential landscape picture or the IS tells us how particles behave on potential surfaces. It may be true that a weak correspondence is present between them, but the relation is not straightforward, as mentioned in the final section. We should recall that the progress of the study of dynamical systems was motivated by the fact that the picture based on configurational space does not necessarily reflect the phase-space dynamics in a precise way and sometimes overlooks an essential point.

We therefore begin our discussion by examining the validity of the nearly integrable picture since, as mentioned, it would be the most legitimate approach to treat a class of systems that can generate slow dynamics. Specifically, we take the FPU model as a representative model of nearly integrable Hamiltonian systems. Its explicit form is given as

$$H(\vec{Q},\vec{P}) = \sum_{i=1}^{N} \frac{P_i^2}{2m} + m\Omega^2 \sum_{i=0}^{N} \left[\frac{1}{2}(Q_{i+1} - Q_i)^2 + \frac{\delta}{\gamma}(Q_{i+1} - Q_i)^r \right] \quad (9)$$

Here the $r = 3$ case corresponds to the so-called FPU α model and the $r = 4$ case to the FPU β model. The degree of nonlinearity is adjusted either by the coupling constant δ or by the total energy E of the system, but a convenient way of parameterization is to take the dimensionless coupling constant,

$$\varepsilon = \delta \left(\frac{E}{m\omega^2 N} \right)^{r/2-1} \quad (10)$$

Introducing the normal coordinates, the Hamiltonian is reduced into the form

$$H(\vec{p},\vec{q}) = \frac{1}{2} \sum_{i=1}^{N} (p_i^2 + \omega_i^2 q_i^2) + \varepsilon V(\vec{q},\vec{p}) \quad (11)$$

where $V(\vec{q},\vec{p})$ stands for a suitable polynomial function. Note that the above normal form takes a typical nearly integrable form Eq. (1).

The most surprising discovery of FPU's experiments was the breakdown of ergodicity in spite of the presence of nonlinearity [1,2]. What they found was, rather than an equilibrium state being achieved, recurrence to initially excited modes or oscillatory exchange of the energies between certain limited modes.

The explanation of recurrence so observed was given in several ways: One is by taking a continuous limit to reduce the original system to completely integrable partial differential equations, and the other is to interpret it as a manifestation of the KAM theorem [21]. In either case, a basic idea is to regard that the system is sufficiently close to a completely integrable system. In fact, as introduced in Section II, the KAM theorem is the statement for the nearly integrable Hamiltonian (2.1), meaning that the breakdown of ergodicity can be explained within perturbation schemes. The Nekhoroshev theory is constructed essentially on the same idea. It also carries out the perturbation expansion with respect to the same small parameter, and exponentially long-time stability thus deduced concerns the deviation of action variables $I(t)$, which are the action variables in the integrable limit. Also several mechanisms to explain anomalous transport in phase space, though they might rely somewhat on specificities of dimensionality of phase space, are based on self-similar structures such as cantori. They play the role of strong partial barriers especially in the vicinity of KAM tori or in the nearly integrable regime.

In view of the nearly integrable picture, the observations of liquid water dynamics presented in the previous section, however, invoke several questions listed below:

(i) In contrast to the FPU model, the intermittent behavior of water dynamics was found in MD simulations in liquid phase, and even in room temperatures. In order to regard a given system as a nearly integrable system, one should find a completely integrable limit somewhere close to the system under consideration. More generally, identification of the small parameter is needed if one develops the perturbative argument. In molecular systems, the most natural parameter controlling the integrability or measuring the distance from the completely integrable limit would be the total energy, or the temperature of the system. That is, *the closest* completely integrable system is just the crystal state or the glassy states. However, evidently, liquids at room temperatures are very far from such states. What one naturally expects is very strong mixing and global chaos in phase space.

In fact, extensive numerical and analytical works done on FPU models revealed that the phase space volume of chaotic regions decreases and also the Lyapunov exponents of the typical trajectories increase, implying that strongly mixing is achieved as the increase of the perturbation strength or the total energy of the system. In particular, it was pointed out that there seems to exist a stochasticity threshold, or more definitely the strong stochasticity threshold (SST)[57]. Below the SST, recurrence or induction phenomena can still be observed and anomalous transport is expected to take place, but above the SST, strong chaos with exponential correlation decays simply dominates in phase space.

Here there might be an objection when one recalls some specific features of water: Even in the liquid phase, the ratio of molecules connected via hydrogen bonding amounts to almost 90%, and the three-dimensional network structure is formed as a backbone. This might bring us an impression that liquid water is in a *weakly chaotic regime*. It should be noted, however, that in liquid phase, as a consequence of frustration of hydrogen bondings, bonding partners exchange each other so frequently that an aspect of dynamics is entirely different from that of solid phase, where only small vibrational motions are excited around its equilibrium configuration. The latter dynamics can be modeled by FPU-type systems, but it would not be the case in liquid phase. Therefore what we should clarify is whether or not *weakly chaotic* is equivalent to *nearly integrable* and whether or not the theory of nearly integrable Hamiltonian systems is applicable even in liquid water.

(ii) The system size of MD simulations of liquid water is several tens or several hundreds, and they were performed under the periodic boundary condition. We intend to perform these simulations in order to know the dynamics with infinitely many degrees of freedom or in the thermodynamical limit. The motivation is exactly the same as that of the FPU problem, so we can refer to many works to concerning the system size dependence of the relaxation property [58–62].

As the increase of the computer power, both size and accuracy of simulations have incredibly grown, and also many indicators to probe ergodicity and mixing have been proposed. One should first note, nevertheless, that the situation is still far from being clear and appropriately understood, even after 50 years have passed since the first attempt by FPU.

The recurrence phenomenon found by FPU implies the breakdown of equipartition, which is vital to the foundation of statistical mechanics. There are several works reporting a failure of equipartition even at large N [58]. In order to measure the degree of equipartition, the spectral entropy,

$$S \equiv -\sum_k \rho_k \log \rho_k, \quad \rho_k \equiv \frac{E_k}{\sum_i E_i} \quad (12)$$

where $E_k = (1/2)(p_k^2 + \omega_k^2 q_k^2)(k = 1, 2, \ldots, N)$, has often been employed. The spectral entropy ranges from zero, corresponding to only single-mode excitation, to $\log n$ where complete equipartition is achieved. Several authors asserted that for sufficiently low energy regions the breakdown of equipartition occurs by demonstrating that the spectral entropy does not depend on the system size N [59]. Other articles, however, reported quite different aspects, that is, for a relevant class of initial conditions, the total energy, not the energy per particle, determines the ergodic behavior of the system. There is also a discussion on SST in the large system size limit [63]: One inquires whether the border between

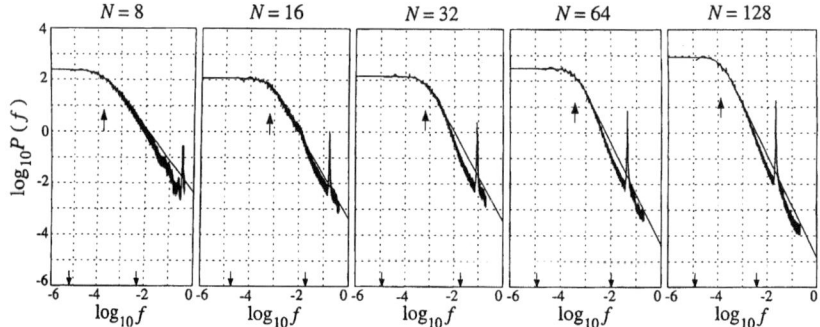

Figure 6. The power spectrum density of the lowest mode energy for the FPU β model. The initial value of each mode energy is fixed at $E = 0.5$. The system size is given as $N = 8, 16, 32, 64, 128$ from left to right, respectively. The line $F(f) = A/\{1 + (f/f_s)^\alpha\}$ is obtained by fitting the power spectrum density in the frequency between two ↓ marks. The shoulder at $f = f_s$ is indicated by a ↑ mark. [Reprinted with permission from *Jpn. J. Appl. Phys.* **35** 2387–2393 (1996). Copyright 1996 by The Institute of Pure and Applied Physics.]

strong stochasticity and quasi-integrable regimes is just the crossover or a strict threshold in the thermodynamical limit.

As mentioned, the long-time correlation decay in liquid water was explicitly identified in measuring the power spectrum. Similarly, the system size dependence of the power spectrum was calculated in the FPU β model [64]. The simulation was carried out from generic initial conditions. As shown in Fig. 6, in either case, if the degree of coupling strength is not small, the power spectrum becomes Lorentzian type, suggesting that exponentially fast relaxation governs the dynamics, which is consistent with the analyses employing different measures. But, it addition to such a general aspect, the authors gave an interesting observation that the characteristic time scale of exponential relaxation becomes large as the system size is increased. This may be related to an indication that energy exchanges depend only on time scales that grow proportionally to the system size N [64].

In any case, since the large system size limit seems to be a delicate issue, in addition to numerical experiments, analytical studies are strongly desired to get more convincing conclusion. Here we see what the perturbation theory tells us on this issue.

KAM tori are definite obstacles for ergodicity. As the volume of KAM tori gets larger, the ratio of initial conditions whose trajectories do not show mixing behavior becomes larger. Thus, we first ask how the total volume of KAM tori occupying the phase space depends on the system size. Several estimations have indeed been given with regard to this matter [65]: Some estimates yield $\varepsilon_c(N) \sim \exp(-BN\ln N), B > 0$ [66], and another ones yield

$\varepsilon_c(N) \sim \exp[-A(\ln N)^{2+\xi}], A, \xi > 0$ [67], where $\varepsilon_c(N)$ denote the stochasticity threshold for the system size N; that is, a set of invariant tori of positive measure exists for $0 \le \varepsilon < \varepsilon_c(N)$. An older result also theoretically analyzed the threshold at which the equipartition is achieved, and it asserted that the KAM theorem is irrelevant for the thermodynamical limit [69]. Either result, except for some special systems [68], suggests that the regions of KAM tori shrink quickly and do not work as real obstacles for ergodicity.

Concerning the Nekhoroshev theorem, the relevant constants that appear in the Nekhoroshev estimate are α and β. To the authors' knowledge, the best estimates are given, for example, in Ref. 70; specifically, the most critical parameter β is given as $1/N$. As indicated, there are enough reasons to believe that, for generic systems, such an estimate cannot be further improved [71]. This N dependence has actually been observed numerically in the case of an FPU model [72]. Therefore, adiabaticity in the sense of Nekhoroshev, though the estimate only gives its lower bounds, will not survive in the thermodynamical limit, but, as discussed in the next section, this does not necessarily mean that Nekhoroshev-type arguments are of no use in the description of systems with macroscopic degrees of freedoms; instead the basic idea is quite broadly applicable.

Considering the points (i) and (ii), we must conclude that the slow relaxation observed liquid water is clearly outside the scope of nearly integrable Hamiltonians, since Hamiltonians of the form given in Eq. (1) inevitably yield strongly chaotic and mixing behavior if the system size is sufficiently large and in strongly perturbed regimes.

(iii) An additional point in which we cannot regard liquid water as being within a nearly integrable regime is the fact that relaxation properties depend sharply on what kinds of observables one monitors. As shown in Fig. 4, the total potential energy exhibits a $1/f$-type spectrum whereas potential energy of individual molecules obeys a simple exponential decay (Debye-type relaxation) [47]. Also dielectric relaxation, which reflects the orientational relaxation of molecules, shows a Lorentzian-type spectrum density. As a probe more suitable for laboratory experiments, Raman scattering and neutron scattering have been evaluated in numerical simulations, and the results differ from each other.

Analogous aspects can be seen in supercooled liquids, in which the separation of the dynamics into intrabasin fast vibrations and interbasin slow diffusion typically takes place. The separation of the time scale means that if one measures a certain observable related to interbasin transitions, one usually observe slow relaxation, but if one sees the dynamics of individual molecules, different correlation properties may be detected.

Bounded motions on KAM tori, or Nekhoroshev-type long-time stability could, however, hardly explain such variety of time scales, because the trajectory on a KAM torus is confined on N-dimensional subspaces in

$2N$-dimensional phase space and shows quasi-periodic motions in all directions. Similarly, long-time freezing predicted by the Nekhoroshev theorem occurs for all the action variables as in Eq. (4). If we are interested only in intrinsic relaxation properties and exclude the effect of transformation of the spectrum due to averaging procedures or the change of variables, any kinds of observables should have the same relaxation properties as the microscopic variables. In the case of liquid water and several supercooled liquids, the most microscopic variables such as vibrational motions of individual molecules show fast relaxation, and the slow mode is found in more macroscopic variables. Therefore, assuming the nearly integrable form as a model Hamiltonian of those molecular systems could hardly be accepted in this sense.

V. ALTERNATIVE SCENARIO: HAMILTONIAN SYSTEMS WITH INTERNAL DEGREES OF FREEDOM

The potential functions of FPU models well approximate even Lennard-Jones or Morse potential, which is often used as interaction potentials of real molecules, at least in the lower-energy region. Therefore, even if the potential function is replaced by such realistic functions, what we can expect would be almost the same as that observed in FPU models as long as the total energy is not so large. Speculations given in the previous section strongly suggest that an entirely different mechanism works as an origin of slow relaxation of liquid water.

An important difference between nonlinear lattice vibration models and molecular systems is that the latter have *heterogeneity* in their models, while the former do not. It would be difficult to specify the term heterogeneity definitely, but one can say that the FPU model is a homogeneous extreme and contains any seeds of heterogeneity in the model, because it is composed of identical particles that interact with each other via a common potential function. On the contrary, molecular systems of our interest sometimes have internal degrees of freedom, or are composed of several species of molecules or potentials. The purpose of this section is to show that introducing heterogeneity into the system leads to an alternative view that possibly solves several controversial points addressed in the previous section.

We especially focus on the presence of internal degrees of freedom, since, except for monatomic molecules such as argon, almost all molecules including water molecules have internal structures, which distinguish specificities of molecules. As a purely mathematical setting of ergodic problems, the presence of internal structures has been regarded as a secondary factor, but our claim presented here is that the presence of internal structures is a crucial ingredient. The idea originates from a series of works [12,13,73], in which the importance of internal degrees of freedom was pointed out.

The authors of Ref. 13 introduced the following Hamiltonian:

$$H(p,x,\pi,\xi) = h_\omega(\pi,\xi) + \hat{h}(p,x) + f(p,x,\pi,\xi) \tag{13}$$

where $h_\omega(\pi,\xi)$ denotes a set of harmonic oscillators with a set of frequencies $\omega = (\omega_1, \omega_2, \ldots, \omega_\nu)$,

$$h_\omega(\pi,\xi) = \frac{1}{2}\sum_{l=1}^{\nu}(\pi_l^2 + \omega_l^2 \xi_l^2) \tag{14}$$

and $\hat{h}(p,x)$ represents a generic N-dimensional system. $f(p,x,\pi,\xi)$ stands for a coupling term that is assumed to be of order ξ and, thus, to vanish for $\xi = 0$.

The Hamiltonian $H(p,x,\pi,\xi)$ is a model of diatomic molecules. $\hat{h}(p,x)$ represents the translational degrees of freedom and $h_\omega(\pi,\xi)$ represents the internal vibrations of the molecules. If all the molecules are identical, we can assume that all frequencies are set to be equal. The internal part $h_\omega(\pi,\xi)$ takes the form of uncoupled harmonic oscillators, so it looks specific. But this is not the case because all the nonlinear terms can be absorbed into the coupling term $f(p,x,\pi,\xi)$.

What we should stress here is that the Hamiltonian $H(p,x,\pi,\xi)$ does not take a nearly integrable form, since $\hat{h}(p,x)$ admits an arbitrary nonlinear system. Thus, it can generate strong chaos as a whole. Even in such a case the authors showed that the perturbation theory with respect to an appropriate small parameter can be applied, and they proved under certain conditions that exponentially long-time stability can be seen in the dynamics if suitably chosen variables are monitored.

The small parameter in the nearly integrable Hamiltonian was just the ratio between the perturbative part $H_1(I,\theta)$ and the completely integrable part $H_0(I)$. In the present case, introducing the ratio of the time scale between two sub-systems $\hat{h}(p,x)$ and $h_\omega(\pi,\xi)$ as

$$\lambda \equiv \frac{\omega}{\Omega} \tag{15}$$

where $\Omega = (\Omega_1, \Omega_2, \ldots, \Omega_\nu)$ denotes the inverse of a typical time scale of the constraint system $\hat{h}(p,x)$, we can identify $1/\lambda$ as a small parameter if the time scale of internal degrees of freedom $h_\omega(\pi,\xi)$ is much faster than that of $\hat{h}(p,x)$.

Under these settings, the following estimate holds [13]. That is, there exist positive constants A and λ_* such that for every $\lambda > \lambda_*$, one has

$$|h_\omega(\pi'(t),\xi'(t)) - h_\omega(\pi'(0),\xi'(0))| < O(\lambda^{-2}) \tag{16}$$

$$|\hat{h}(p'(t),x'(t)) - \hat{h}(p'(0),x'(0))| < O(\lambda^{-1}) \tag{17}$$

for

$$|t| < A \exp\left(\frac{\lambda}{\lambda^*}\right)^\alpha \tag{18}$$

Here the variables (π', ξ') and (p', x') represent new variables obtained from appropriate canonical transformations from (π, ξ) and (p, x).

A brief sketch of how the Nekhoroshev-type perturbation technique is available in their proof is as follows. First, by the scale change of canonical variables,

$$\xi_l = (\lambda \Omega_l)^{-1/2} \tilde{\xi}_l, \quad \pi_l = (\lambda \Omega_l)^{1/2} \tilde{\pi}_l, \quad (1 \leq l \leq \nu) \tag{19}$$

the Hamiltonian is transformed into

$$H(p, x, \tilde{\pi}, \tilde{\xi}, \lambda) = \lambda h_\Omega(\tilde{\pi}, \tilde{\xi}) + \hat{h}(p, x) + \frac{1}{\lambda} f_\lambda(p, x, \tilde{\pi}, \tilde{\xi}) \tag{20}$$

where

$$\lambda h_\Omega(\tilde{\pi}, \tilde{\xi}) = \frac{1}{2} \sum_{l=1}^{\nu} \Omega_l (\tilde{\pi}_l^2 + \tilde{\xi}_l^2) \tag{21}$$

Here, the scale transformation, Eq. (7), gives rise to the factor $\lambda^{-1/2}$, and furthermore the $\lambda^{-1/2}$ factor comes from the fact that f itself is of order $\lambda^{-1/2}$ since f is assumed to have the form $\xi \tilde{f}$.

Following the same prescription as the one used by Nekhoroshev—that is, the reduction of the Hamiltonian into a normal form via a near to identity canonical transformation—one obtains the form

$$H'(p', x', \pi,' \xi') = \lambda h_\Omega(\pi' \xi') + \hat{h}(p', x') + Z(p', x', \pi', \xi', \lambda) \\ + R(p', x', \pi,' \xi', \lambda) \tag{22}$$

where Z satisfies $\{h_\Omega, Z\} = 0$, so Z is in normal form. The normalization process is performed until the remainder term becomes exponentially small.

The small parameter λ^{-1} in this perturbation scheme represents the ratio between the time scales of two subsystems $h_\omega(\pi, \xi)$ and $\hat{h}(p, x)$, so an intuitive interpretation for the estimate given in Eq. (4) is rather easy. If the time scale of vibrational degrees of freedom is much faster than that of translational degrees of freedom, then the energy transfer between two subsystems hardly occurs. As $\lambda \to \infty$, the freezing of energies of subsystems is achieved, which is a sort of the adiabatic limit.

The separation of the time scale and the decoupling of different modes often occur in molecular systems. Eliminating the fast degrees of freedom is therefore a commonly used procedure in constructing the model of molecular systems. If the time scale of internal vibrations between atoms is so fast, then we usually neglect such a degree of freedom. In fact, water molecules employed in MD simulations do not have internal vibrational motions, and they are approximated just by rigid rotors.

In this way, the above statement may be acceptable not only in a qualitative level, but the most important implication of the mathematical statement is that the character of freezing has been shown to be quantitatively the same as adiabaticity predicted by the Nekhoroshev estimate in the nearly integrable system. This solves, at least partially, question (iii) posed in the previous section, because the model Hamiltonian [Eq. (1)] just describes the situation where not all the degrees of freedom do not necessarily show adiabaticity; but only a limited number of variables, just the energies of two subsystems in this case, are almost frozen. It is true that there may be, in principle, many other possibilities and the proposed one is not a unique way as for the division of phase space into lower-dimensional subspaces, but the separation induced by the internal structure of molecules is the most natural and plausible candidate.

It should also be pointed out that system size dependence of adiabaticity, question (ii) in the previous section, can also be within our scope. More precisely, it has been shown that if all the particles are identical, that is, $\Omega_1 = \Omega_2 = \cdots = \Omega_v$, the exponent α that appears in the estimate in Eq. (4) does not depend on the system size N [13]. This implies that the freezing of energies between subsystems persists in the large system size limit and thus can be observed macroscopically. Note, however, that one does not seem to avoid N-dependence of another constant A as discussed below [13,71].

This sort of freezing mechanism and the resulting long-time stability, not infinite time, were first discussed by Boltzmann and Jeans [10,11]. They intended to interpret anomalies of specific heats in polyatomic gases. An interpretation of such anomalies based on classical mechanics is now being hidden behind the foundation of quantum mechanics, but the mechanism of freezing of high frequency motions could widely be observed in classical dynamics of atoms and molecules. One can say that the perturbation argument introduced here is a modern interpretation for the *Boltzmann–Jeans* conjecture [12,13,73].

The prediction based on the perturbation theory should open to numerical verifications. The modified FPU model, in which identical particles in the original FPU model are replaced by those with alternating masses, has been studied to check the validity of perturbative estimate, although detailed conditions of the modified FPU model are not exactly the same as those given as Eq. (1). The Hamiltonian of the modified FPU model is given as

$$H(\vec{Q},\vec{P}) = \sum_{i=0}^{N} \frac{P_i^2}{2m_i} + \sum_{i=0}^{N} \left[\tfrac{1}{2}(Q_{i+1} - Q_i)^2 + \frac{\beta}{4}(Q_{i+1} - Q_i)^4 \right] \quad (23)$$

with $m_i = 1$ for i odd and $m_i = m < 1$ for i even. As is well known, such alternating masses generate a frequency spectrum with disjoint branches, which are called the acoustic and optical branch, respectively. As m gets smaller, the gap between these two modes becomes larger, so that the separation of two subsystems is expected to be evident.

Careful numerical simulations reported in Ref. 72 revealed that nonequipartition of energy is indeed realized, though the theoretical analysis predicts that a stability estimate of exponential type would hold at most in an interval of specific energy shrinking to zero in the thermodynamical limit. Therefore, this numerical experiment suggests that theoretical estimates for the system size dependence of several constants are not optimal ones. Precisely, it was confirmed that the exponent α does not depend on N, but other constants seem to have much weaker size dependence than theoretically derived. This result contains a remarkable message that the freezing of this kind is observable macroscopically. Recall that the freezing in the large system size limit has not been achieved in the ordinary FPU model ($m = 1$).

As in the study of water dynamics, the power spectrum density is useful to detect the long-time correlation or to detect decay slow energy transfers between optical and acoustic modes in the model given in Eq. (11). The relaxation inside optical modes is much faster than that in acoustic modes, since the frequency spectrum in optical modes is sharply localized and almost resonant while the spectrum is broadly spread in acoustic modes.

We can also expect the separation of the time scales in a model for diatomic molecules. Instead of alternating masses, we put alternating potential functions as

$$H(\vec{Q}, \vec{P}) = \sum_{i=0}^{N} \frac{P_i^2}{2m} + \sum_{i=0}^{N} \left[\frac{\kappa_i}{2} (Q_{i+1} - Q_i)^2 + \frac{\beta}{4} (Q_{i+1} - Q_i)^4 \right] \quad (24)$$

with $\kappa_i = 1$ for i odd and $\kappa_i = \kappa < 1$ for i even. Figure 7a gives the power spectrum density for the energy fluctuation of the acoustic subsystem, while in Fig. 7b we present those for individual motions in each subsystem [74]. As clearly seen in Fig. 7a, under a proper choice of the system parameter, the $1/f$-type spectrum clearly appears while the Lorentzian form is observed when one probes individual motions. The result clearly demonstrates that characters of relaxation depend on the observable one measures. If one monitors the relaxation related to individual degrees of freedom, exponentially fast relaxation will be observed, but if one probes the energy exchange between optical and acoustic subsystems, the relaxation will be found to occur in an intermittent manner.

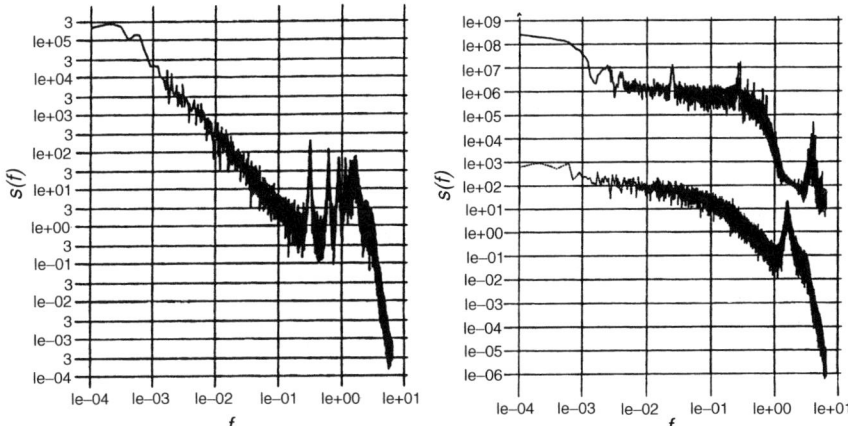

Figure 7. The power spectrum density of energy fluctuation of (*a*) the acoustic subsystem $\hat{h}$ and (*b*) an individual mode in optical and acoustic subsystem. In the simulation, the ratio of perturbation strength is taken as $\lambda = \sqrt{1/\kappa} = 5.0$.

VI. VALIDITY OF HYPOTHESIZED SCENARIO

In the previous section, we have shown that switching the picture from the nearly integrable Hamiltonian to the Hamiltonian with internal structures may make it possible to solve several controversial issues listed in Section IV. In this section we shall examine the validity of an alternative scenario by reconsidering the analyses done in MD simulations of liquid water. As mentioned in Section III, since a water molecule is modeled by a rigid rotor, and has both translational and rotational degrees of freedom. So, the equation of motion involves the Euler equation for the rigid body, coupled with ordinary Hamiltonian equations describing the translational motions. The precise Hamiltonian is therefore different from that of the Hamiltonian in Eq. (1), but they are common in that the systems have internal structures, and the separation of the time scale between subsystems appears if system parameters are appropriately set.

The potential function used in MD calculations of liquid water is the so-called SPC potential—that is, a semiempirical potential yielding the almost correct radial distribution function for oxygen atoms at room temperatures. However, we emphasize that what we want to see is characters that are independent of detailed tunings of the system such as potential functions, bond lengths, bond angles, and even masses of atoms. The properties depending sensitively on them are attributed to system specificities and are outside our present concern.

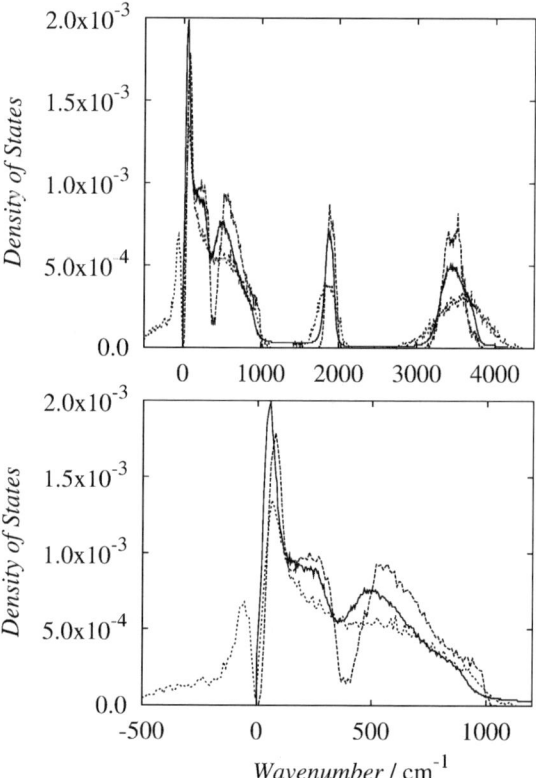

Figure 8. Instantaneous normal mode spectrum of liquid water. Solid, dashed, and dashed–dotted lines are calculated from the velocity correlation function, INM, and QNM, respectively. The system size in numerical simulations is 216.

The first task we have to verify is to see whether the separation of the time scale is noticeably large or not in liquid water. The INM analysis, the utility of which was mentioned in Section III, is one of useful tools to probe the frequency distribution along a trajectory. Figure 8 shows that the density of states for instantaneous and quenched normal mode spectra for liquid water at room temperature. The latter is the normal mode spectrum for IS. It is calculated by quenching the trajectory from each instantaneous configuration. The quenched normal modes cannot be negative because they are the normal modes for local minima, whereas, as mentioned, the normal modes for instantaneous configuration may have imaginary frequency that corresponds to the negative part in Fig. 8. Of much interest for the present concern is that the density profile is composed of three peaks: ~ 60, ~ 200, and ~ 500–600 cm^{-1}. The first two peaks come from O–O–O bending and O–O stretching, respectively, and the third one corresponds to librational motions. The assignment of these modes is done using the data for the far-infrared spectrum, Raman scattering, and incoherent neutron scattering spectra [6,7]. In the overall spectrum profile, the normal modes below 300 cm^{-1} are identified as translational motions, whereas those above 300 cm^{-1} correspond to librational motions. A remarkable fact is that the separation between translational and librational modes is not so conspicuous in other liquids—for example, CS_2, CH_3CN, and CH_3Cl, and so on. Therefore one can say that liquid water is specific in that the moment of interia is small as compared with other molecules and the librational motions are easily to be exited. As is shown in Fig. 9, if we replace hydrogen by deuterium in water molecules, the distance between translational and rotational peaks is reduced as a result of the increase of the moment of interia [75].

The normal mode analysis thus shows that the time scale separation between translational and rotational motions is sensible, so that liquid water meets a necessary condition for the Boltzmann–Jeans-type scenario. We should remark that rotational motions referred to here are not real rotation but just librational or flipping motions of molecules. This is in a contrast to the situation the Boltzmann–Jeans-type scenario assumes. The original Boltzmann-Jeans conjecture was proposed for the interpretation for freezing of high-frequency motions in *gases*, in which real rotational motions take place. However, since a crucial point of the argument is just the presence of different time scales, a type

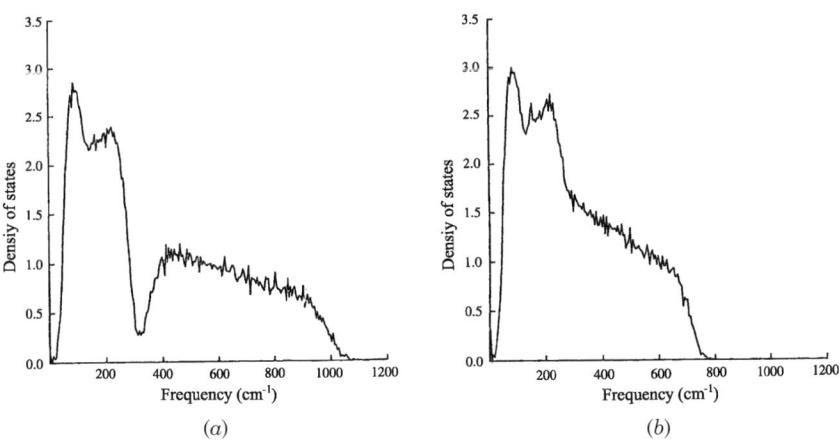

Figure 9. Instantaneous normal mode spectrum of (*a*) liquid water H_2O and (*b*) liquid deuterium D_2O. The system size in numerical simulations was 64 and density of state was obtained over 79 sample averages. [Reprinted with permission from *J. Chem. Phys.* **87**, 6070–6077 (1987). Copyright 1987 by American Institute of Physics.]

of motions does not matter. In fact, the modified FPU system given in Eq. (11) models the lattice vibration of solids in the same manner as does the original FPU system. Similarly, an array of rigid rotors densely distributed on two- or three-dimensional space could be its higher-dimensional version. If the density of rotors is high enough, they are too packed to make free rotations and they can at most librate with small amplitudes. However, if the moment of interia is sufficiently small with the total mass being fixed, we can expect that the time scale of librational motions differs from that of translational motions.

With local information given by INM analysis in mind, we next see the character of rotational relaxation in liquid water. The most familiar way to see this, not only for numerical simulations [76–78] but for laboratory experiments, is to measure dielectric relaxation, by means of which total or individual dipole moments can be probed [79,80]. Figure 10 gives power spectra of the total dipole moment fluctuation of liquid water, together with the case of water cluster, $(H_2O)_{108}$. The spectral profile for liquid water is nearly fitted to the Lorentzian, which is consistent with a direct calculation of the correlation function of rotational motions. The exponential decaying behavior of dielectric relaxation was actually verified in laboratory experiments [79,80]. On the other hand, the profile for water cluster deviates from the Lorentzian function. As stated in Section III, the dynamics of finite systems may be more difficult to be understood.

The result of dielectric relaxation is also consistent with our hypothesized scenario. If the hypothesis works in an ideal sense, rotational relaxation should

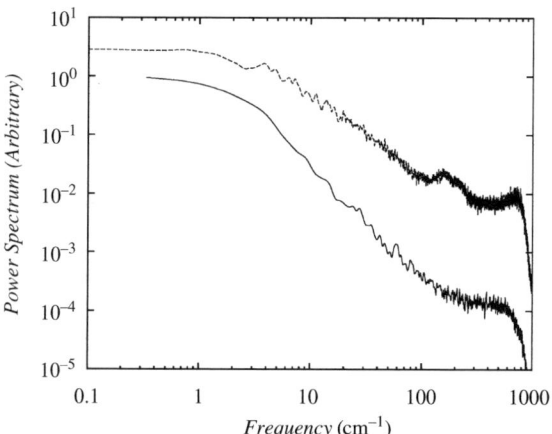

Figure 10. The power spectrum density of the total dipole moment fluctuation of liquid water (solid line) and water cluster $(H_2O)_{108}$ (dashed line). The simulation of liquid water was performed for 216 water molecules under the periodic boundary condition.

obey exponential decaying law, since there is no *a priori* reason for the breakdown of strong mixing in rotational subspaces. Total dipole moment is obtained just by summing up, in the sense of vectors, that of individual dipole moments. In the absence of particular correlation among individual molecules, the character of correlation decay of individual dipole moments should be carried over that of total dipoles.

Other quantities such as depolarized Raman scattering, which probes individual polarization and induced polarization effects, and neutron scattering data, which are related to dynamical structure factors, are also calculated in MD simulations [81]. They are quite useful to detect the presence of hydrogen bond network dynamics and are suitable for the comparison with experiments. In the present context, the result of depolarized Raman scattering is somewhat puzzling because the power spectrum of the total traceless anisotropy of liquid water again shows $1/f$-type spectrum, although the former is expressed by the second-order Legendre polynomial and the latter by the first order, which means that both capture the fluctuation of rotational (librational) motions as dielectric relaxation. The origin of difference is unclear, but, as discussed below, it seems that analysis of rotational motions needs careful considerations.

In order to see the validity of our working hypothesis more explicitly, it is better to observe temporal variation of the energy directly. As demonstrated in Fig. 4b, the total potential energy fluctuates intermittently, which causes the long-time tail in the corresponding power spectrum (see also Fig. 4a). If bottlenecks or partial barriers exist between translational and rotational subspaces, behaviors similar to the total potential energy must be observed also for total rotational and translational energies, and as a result the power spectra should have algebraic parts. This is indeed the case. As is shown in Fig. 11, the total translational and rotational energies fluctuate in an intermittent manner and exhibit long-time fluctuation similar to the total potential energy (see Fig. 4b).

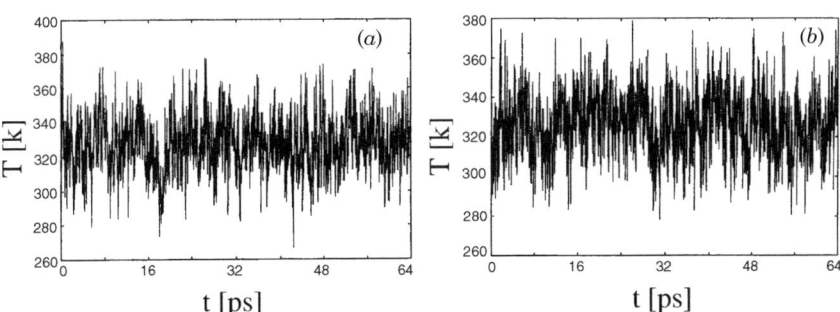

Figure 11. Temporal fluctuation of (*a*) the total translational and (*b*) total rotational energy in liquid water. The simulation was performed for 216 water molecules with $T = 305$ K.

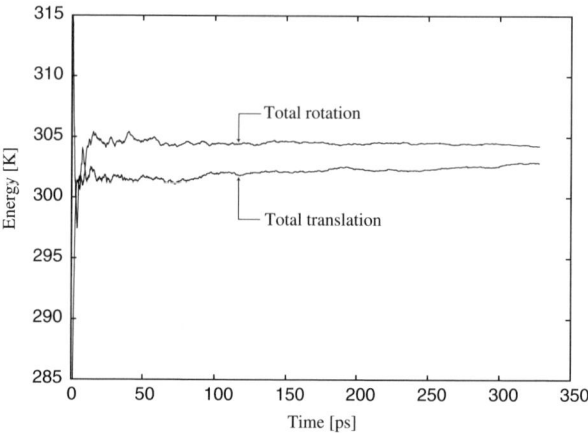

Figure 12. Temporal behavior of averaged energies for liquid water. The number of molecules is 216 and the SPC potential was used for simulations. The averaged energy is defined by $E(t) = (1/t) \int_0^t \sum_{i=1}^n E^{(i)}(t') \, dt'$, where $E^{(i)}(t)$ represents energy for individual molecules. Upper and lower curves, respectively, represent rotational and translational energies.

We can also see the slow energy transfer between rotational and translational subsystems by tracing how an equipartition of both modes is achieved. In Fig. 12, we present the temporal behavior of averaged rotational and translational energies. In this sample, after very large fluctuations in the initial stage, it takes quite a long time to reach an equipartitioned state. Similarly, it sometimes happens that equipartition is not realized even within several nanoseconds.

The power spectrum shows that long-time correlation component is actually contained in the energy fluctuation. As shown in Fig. 13, the power spectra of the temporal behavior of total translational and rotational energies clearly have power-law decaying parts, while those of individual molecules are close to simple Debye-type spectra, implying that strong chaos dominates in the subspace of rotational motions. This result is consistent with that of the diatomic lattice model shown in Fig. 7. Close observation of Fig. 13 reveals that, as compared to individual relaxation in the translational subspace, the power spectrum profile of individual rotational motions is not exactly Lorentzian-shape but some correlations among individual rotational motions are implied. This is possibly due to the fact the hydrogen-bond network is formed in the case of liquid water, while no such correlations are found in optical subspaces in Fig. 7. The origin of such correlations and the relation to hydrogen-bond network dynamics discussed in Refs. 8 and 9 should be clarified in the future study.

As discussed in the previous section, the system size dependence is quite important to judge whether or not the long-time correlation decay persists even

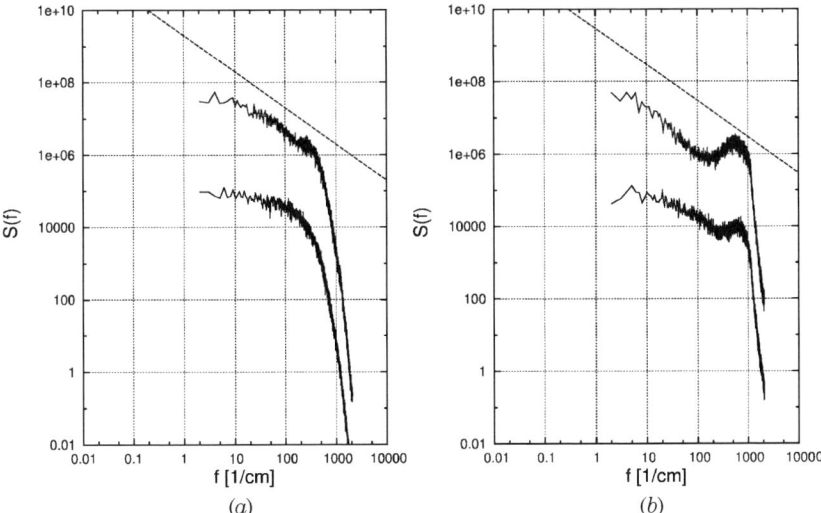

Figure 13. (*a*) The power spectrum density for total (upper) and individual (lower) translational energies. The number of molecules is 216 and SPC potential is used for simulations. The temperature is 305 K. (*b*) The power spectrum density for total (*upper*) and individual (*lower*) rotational energies.

in the thermodynamical limit. In the case of liquid water, systematic calculations show that the power-law decaying regime in each power spectrum remains almost the same as those shown in Fig. 13, though equipartitioned states are more likely to be realized as the system size increases [14]. Figure 14 depicts the difference between rotational and translational energies decreases almost exponentially as the system size increases. This result suggests that even if equipartition of rotational and translational subspaces is achieved in the large system size limit, very slow energy transfers between these subsystems robustly remain.

In any case, an interpretation based on our working hypothesis is consistent with the fact that the power-law decay is particularly observed in limited observables such as the energy transfer between two subsystems. Slow relaxation is observed if one monitors quantities associated with such a pathway.

An advantage of numerical experiments is that one can design artificial models and test them together with introducing some system parameters. This strategy would be particularly suitable for our purpose. We hypothesize that the bottleneck between translational and rotational modes is a source of slow motions, and intermittent fluctuations should disappear if each of them is suppressed. The simulation performed in Ref. 9 exactly presents side evidence

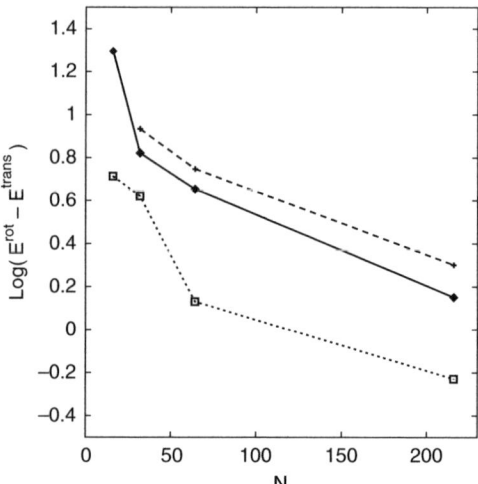

Figure 14. Difference of averaged energies $E(t)$ between total translational and total rotational motions plotted as a function of the system size. The difference was measured at $t = 320$ ps. Filled diamonds, squares, and crosses represent water, oxygen and alcohol molecules, respectively. The number of molecules is 216 and the temperature is 305 K.

of this prediction. In observing Fig. 15, they have concluded that total potential energy fluctuation loses an intermittent character either when rotational motions are suppressed or when translational motions are suppressed.

As additional pieces of evidence, we replace rigid rotators by point particles, keeping the radial distribution function unchanged by adjusting the interacting

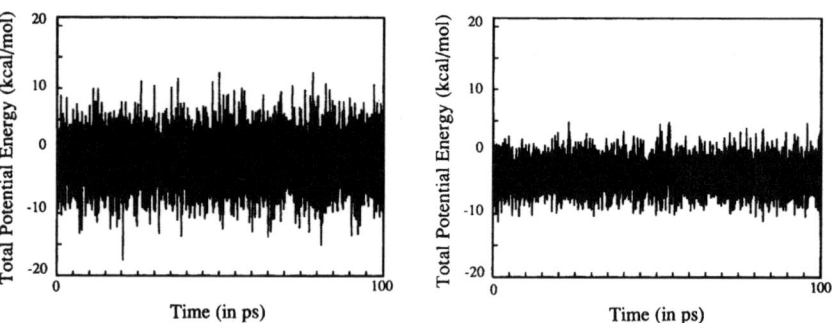

Figure 15. The total potential energy fluctuation for the case (**left**) where rotational motions are suppressed, and the case (**right**) where the translational motions are suppressed. The simulation was performed for 64 water molecules with TIPS potential. The temperature is 298 K. [Reprinted with permission from *Chem. Rev.* **93**, 2545–2566 (1993). Copyright 1993 by American Chemical Society.]

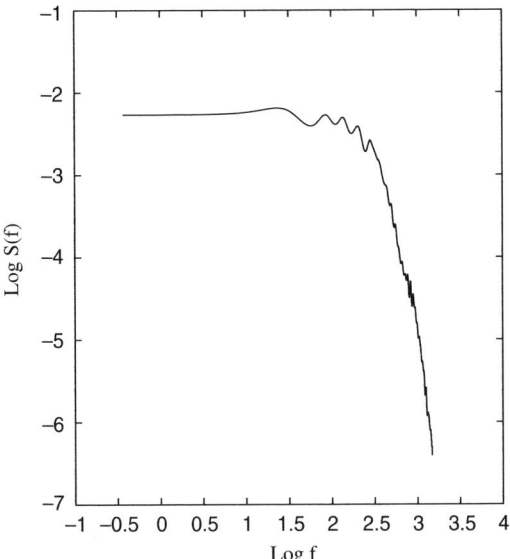

Figure 16. The power spectrum density for molecules for which rigid rotators are replaced by point particles, keeping the radial distribution function of liquid water unchanged.

potential between molecules appropriately [82]. The result is presented in Fig. 16. We can see that power-law decay of the total potential energy completely disappears if we remove the internal degrees of freedom. Also, if one changes the mass ratio γ of hydrogen and oxygen molecules, keeping other parameters fixed, the correlation length of total translational and rotational energies becomes larger with a decrease of the internal moment [14]. This is also an expected result since excitation of fast rotational motions accelerates large separation of the time scale.

In this way, the proposed hypothesis explains the results of MD simulations and is indeed compatible with analyses made for liquid water. One can conversely say that liquid water dynamics may provide a good example to realize an alternative origin of slow relaxation, that is, the Boltzmann–Jeans scenario. A more strict test the hypothesis should pass is *predictability* of the hypothesis to other molecules. As mentioned already, the hypothesis proposed here does not assume any specificities of water molecules, but instead a simple dynamical mechanism that might be more universally existing. Therefore, if our hypothesis is so generic or robust, meaning that subtle tunings of interaction potentials, shape of molecules and even spatial dimensions on which molecules are confined are not needed, we may expect that similar slow relaxation is observed in other molecules that have internal degrees of freedom with separation of time scales.

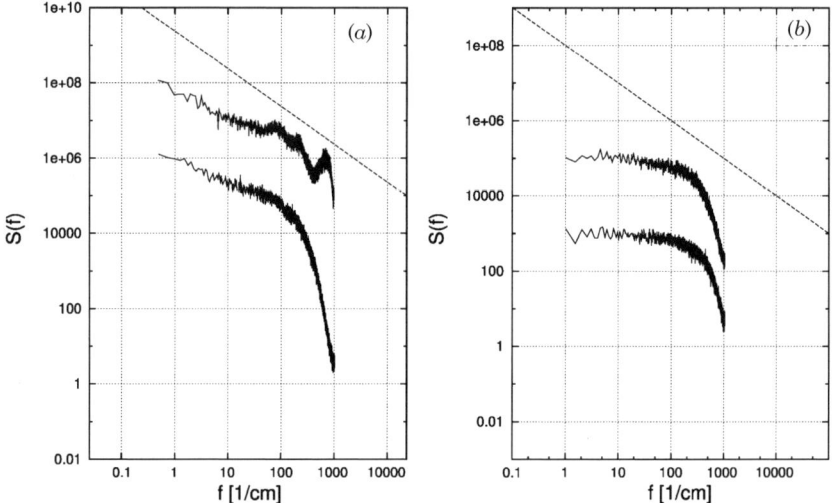

Figure 17. The power spectrum density for total rotational (*upper*) and total translational (*lower*) energies in the case of (*a*) alcohol and (*b*) oxygen molecules. In both figures, the spectra for translational energies are shifted below for clarity. The number of molecules is 216 in each case.

In Fig. 17, we give the power spectra of total rotational and translational energies for liquid methyl alcohol and oxygen. Methyl alcohol is an example of molecules that show relatively large separation of the time scale between rotational and translational modes, and oxygen molecules are the ones not showing so sharp separation. Both are directly checked by examining instantaneous normal mode analyses [14]. Note that in both cases, though not presented here, power spectra of individual molecules are close to Debye-type ones, similar to those in liquid water [14]. In contrast, the power spectrum of the total rotational and translational energies for methyl alcohol has a power-law decaying part, while those for oxygen molecules do not have power-law regime. This again implies that the degree of time-scale separation actually controls the relaxation properties. The present result provides concrete evidence showing that power-law decaying behavior is not limited to liquid water. So, it is natural to expect that the hypothesized mechanism exists much more universally.

VII. CONCLUSIONS AND DISCUSSIONS

In this chapter we have discussed the origin of slow relaxation observed MD simulations for liquids or supercooled liquids in the light of the theory of dynamical systems. We have introduced several established scenarios that explain anomalous transports and sticky motions in phase space, and then we examined the validity of these scenarios as a possible explanation particularly for the $1/f$-type long-time relaxation and intermittent motions in room temperature discovered in liquid water dynamics.

One explanation for anomalous diffusion in Hamiltonian dynamics is the presence of self-similar invariant sets or hierarchical structures formed in phase space that play the role of partial barriers. They slow down the normal diffusion. A different explanation for intermittent behavior is given by the existence of deformed and approximate adiabatic invariants in phase space. They are shown in terms of elaborated perturbation theories such as the KAM and Nekhoroshev theorems.

It has not been established, however, that the former scenario based on the geometry of phase space holds in many-dimensional systems because the effect of dimensionality of partial barriers is not clear yet. It might be limited to a few-degrees-of-freedom system, though the partial barriers possibly survive even in many-dimensional phase space. It would be difficult to relate the first scenario directly with what is observed in the system in liquid or supercooled states since they have large-dimensional phase space, for which similar analyses as done in few-degree-of-freedom systems are almost impossible.

The perturbative arguments could be, on the other hand, applied more generally, but the problem there is to make clear how the critical perturbation strength, below which generic trajectories have infinitely long or exponentially long-time stability, behaves as a function of the system size. If stable or sticky regions shrink to zero in the large system size limit, these theories predict nothing about the slow dynamics in the large system. This is actually a long-standing issue in the ergodic problem in Hamiltonian systems, and a lot of numerical as well as analytical works have been devoted to solving this issue. The current status of the perturbation theory tells us that stability of such a type does not survive in strongly perturbed regimes, and the region of sticky motions does not persist in the thermodynamical limit.

A problem we have posed here was to extract an essential ingredient to produce slow motions often observed in molecular systems, and at the same time to design the simplest possible model producing intermittent motions of liquids or supercooled liquids. Especially in supercooled liquids, a lot of works show that the separation of the time scale takes place between intrabasin fast vibrations and interbasin slow motions. The strong mixing is realized in each potential basin, while bottlenecks exist in the regions connecting basins and they prevent interbasin transitions. The latter can be understood as α-relaxation, and the former β-relaxation, in the theory of glasses. It is commonly believed that the degree of such a separation is enhanced as the decrease of temperature. Therefore, focusing such a typical situation would be a promising strategy to our purpose.

A plausible and the most widely accepted explanation for such a dynamics is based on raggedness of the potential landscape. The "basin" referred to in the description is the basin formed by the potential well, and "bottleneck" stands for the channel connecting two potential wells. These are intuitively clear, and the potential landscape picture may successfully capture qualitative aspects of the problem. But this is not automatically linked to our understanding for what is going on in the phase space. In the following, we give several nontrivial and unclarified aspects of the potential landscape picture, and we show that constructing a simple modeling that can purely represent the potential landscape picture is not a easy task at all and can possibly involve many nontrivial and controversial problems.

The first point is concerned with the dynamics in an individual potential well. As is well known, the dynamics sometimes depends on the shape of the potential in a sensitive way. As an example, we compare the following two-dimensional models, both of which have the double-well potential; nevertheless, the characters of dynamics are entirely different [83]. The Hamiltonian of the first example is given as

$$H(q_1, q_1, p_1, p_2) = \tfrac{1}{2}(p_1^2 + p_2^2) + V(q_1, q_2) \qquad (25)$$

with the potential,

$$V(q_1, q_2) = (q_1^2 - 1)^2 + q_2^2 + \lambda q_1^2 q_2^2 \qquad (26)$$

The contours of potential function, together with typical trajectories, are displayed in Fig. 18a. A saddle point of the potential function is located between two

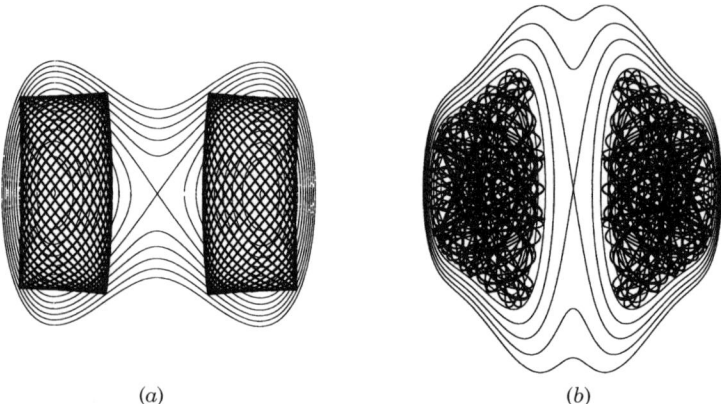

(a) (b)

Figure 18. Contours of the potential function (a) Eq. (26) and (b) Eq. (27), together with typical trajectories in each potential well. The energy of a trajectory in (a) is above the barrier energy. [Reprinted with permission from *Tunneling in complex systems* ed. by S. Tomsovic **96**, 35–100 (1998). Copyright 1998 by World Scientific Publishing Co.]

potential wells. Since the dynamics in the vicinity of the bottom of each potential well is approximated by simple oscillations around a normal equilibrium point, the motions become quasi-periodic. Even when the total energy of a trajectory is above that of the centered saddle point, which allows the transition between two potential wells at least energetically, there is a set of trajectories confined on the KAM tori, so KAM tori work as dynamical barriers. Typically, stochastic regions are developed around a saddle point of the potential function.

In contrast, if we replace the potential function by

$$V(q_1, q_2) = (q_1^2 - 1)^4 + q_1^2 q_2^2 + \mu q_2^2 \qquad (27)$$

then the character so drastically changes that, rather than regular trajectories, chaotic trajectories dominate, even though the potential also has a single saddle in the middle of double wells. The emergence of chaotic orbits is roughly understood because, as shown in Fig. 18b, the potential contours have negative curvature parts.

From these observations, we learn that even simple double-well potential systems demonstrate a variety of dynamics, depending on the shape of potential functions. Note that this sort of question—that is, what information in the potential function is necessary and sufficient conditions which make the system chaotic—dates back to just the beginning of the study of chaos in few-degrees-of-freedom systems.

The second nontrivial point is to design or control raggedness of potential functions. One might have an idea that one of the simplest modelings to produce intra- and interbasin like motions are the Lorentz gas model introduced in Section II. Although it is a billiard model in two-dimensional configuration space, it certainly captures an aspect of intra- and interbasin motions. In fact, if the particle moves a unit cell encircled by disks, strong mixing occurs due to the dispersing nature of the boundary, but the channels between disks work as bottlenecks. The time scale of intrabasin mixing is roughly estimated by the Kolmogorov–Sinai entropy in a unit cell, and the diffusion over much global scale is controlled by the width of channels. As mentioned in Section II. the total decay rate of the correlation function obeys a stretched exponential function.

However, there are essential differences between the Lorentz gas model and IS structures in many-dimensional molecular systems. The most obvious difference is the size distribution of basins. In the Lorentz gas model, the size of unit cells is identical but the basin size, as well as the depth of potentials, of molecular systems is believed to range quite broadly, and possibly distributed in a self-similar way, reflecting that local potential minima increase exponentially as a function of the system size. As a result, one expects that the diffusion among multibasin structures bears different characters. The situation of the latter is somewhat similar to the diffusion among self-similarly formed bottlenecks in phase space. Algebraic dependence of the correlation decay was attributed to self-similarity of invariant structures. However, such an argument is still at a level of analogy, and the prescription of controlling self-similar structures in phase space cannot be applied to the control of many-dimensional potential energy surface.

The third point is related to the first point; a difficulty in developing an appropriate theory that describes the separation of motions into fast and slow motions. As shown in the above simple two-dimensional model, the dynamics in each potential well depends on the potential shape, but in the very vicinity of potential bottoms, the potential function can be approximated by quadratic plus some nonlinear terms. This means that small oscillations around equilibrium points are almost modeled by nonlinear lattice vibrations or their variants at most. As discussed in Section IV, if the total energy is small enough, which is equivalent to a small coupling strength limit, nonlinear lattice vibrations around equilibrium points show quasi-periodic behavior. More precisely, phase-space trajectories are confined on KAM tori or bounded in sticky layers. Otherwise stated, if we assume that the low-temperature limit is an ideal situation of the potential energy landscape picture, the interbasin transition cannot occur, instead only slow dynamics in the sense of nearly integrable pictures will be realized. It is needless to say that slow motions of such a type are not like the ones observed in supercooled liquids. Therefore, the analogy with the Lorentz gas model only holds in a certain intermediate temperature regime where the dynamics shows strongly chaotic behavior at least in each potential well. This might be realized either because the coupling strength is above the strong stochasticity threshold or because the dimensionality of the system makes the volume of stable regions shrink to zero, but at the same time the total energy is not so high that the trajectories feel bottlenecks formed in the potential energy function.

The potential energy landscape picture may certainly be a plausible scenario to explain the separation of the time scale and intermittent behaviors in molecular systems. However, as discussed here, this is still a picture or an analogy, not a theory, so it admits a wide range of interpretations because a concrete minimal model or a limiting case realizing such situations is still missing. In relation to this fact, this picture may be hard to be modeled since it seems to be difficult to develop the perturbation theory, which requires a certain small parameter by which the perturbation expansion is developed. In the Lorentz gas model, the distance between periodically arranged disks will be such a small parameter. But even if we can neglect the discrepancy indicated before and the analogy with the Lorentz gas model holds, it is not clear at all how one can introduce or find such an appropriate small parameter in Hamiltonians of molecular systems, or by which parameter one can control width, depth, or distribution of bottlenecks. The total energy cannot be such a small parameter for the reason stressed above.

In contrast, an alternative scenario we have hypothesized in this chapter is based on the perturbation theory, that is analogous to that applied in nearly integrable systems. Scenarios presented in Section II more or less assume the closeness from the regular invariant components in phase space, or the closeness from completely integrable systems. On the other hand, if the system has internal degrees of freedom, several different time scales induced by each internal degree are naturally introduced, and ratios between them can be small parameters in the perturbation expansion. Recent works introduced in Section V have been done to give an interpretation of freezing of high-frequency vibrations of gases and anomalies of the specific heat thus induced. Such an issue was raised and intuitive discussions were made by Boltzmann and Jeans almost 100 years ago. A relevant point of the discussion is to regard the system as that composed of several subsystems, each of which corresponds to an internal degree of freedom, and carry out the perturbation expansion by taking the ratio of the time scale of each subsystem as a small parameter. The original system assumed by Boltzmann and Jeans was gases. But an essential point is the presence of separation of time scales, and therefore the argument is applicable even in liquid or solid phase. The total energy of the system is irrelevant in this scheme.

As demonstrated in Section VI, various numerical data performed in MD simulations of liquid water were reexamined in the light of the hypothesized scenario and were shown to be consistent with it. Furthermore, it was predicted that another molecules with internal structures must show similar intermittent behaviors only if separation of time scales is large enough. Such a prediction has indeed been confirmed by performing MD simulations of oxygen and alcohol. What was measured is not exactly the same quantities assumed in the mathematical statement; nevertheless, the present results strongly suggest that a necessary condition for the existence of slow relaxation is generic, and intermittent dynamics can be detected if one chooses the observable appropriately.

Finally, we discuss the relation between the proposed hypothesis and the potential landscape picture. Raggedness of the potential energy surface originates from heterogeneity of the system. Heterogeneity may sometimes have its origin in the finiteness of systems. Argon clusters are typical examples showing the ragged potential landscape, in spite of simple monatomic molecules with no internal structures. The raggedness in this case is a result of the boundary effect, so heterogeneity comes in the potential landscape through the finiteness of the system. In fact, in bulk systems, it is believed that raggedness is blurred and a single global minimum dominates the potential surface. Even if the total energy is decreased, glassy states will not be realized in infinite simple liquids.

The presence of internal structures obviously brings about heterogeneity into the system, and for this reason one can say that heterogeneity is an origin of slow motions. However, the concept of dynamical heterogeneity itself is still quite vaguely specified: Is it heterogeneity of what? Potential surfaces, invariant components in phase space, or other aspects of dynamics? Also, how is heterogeneity measured and how is it quantified? Diatomic lattices have optical and acoustic modes, and relaxation nature within each subsystem differs from that of the energy transfer between them. Binary mixture of Lennard-Jones supercooled liquids attains the glassy state since it is composed of two species of atoms. It may be true that both certainly originate from a certain kind of heterogeneity, and so we might expect that the ragged potential landscape picture should have connections with the Boltzmann–Jeans-type scenario. But, it is not clear at all to what extent and how one can generalize the present argument based on the presence of internal degrees of freedom to different types of heterogeneity. It is also uncertain whether any sort of perturbation theories is compatible with dynamical heterogeneity recently discussed. All these questions are far from completely understood.

Acknowledgments

We thank R. Yasuda and K. Ichiki for their collaborations. One of the author (AS) thanks H. Ito for giving us intense lectures on the perturbation theory of Hamiltonian systems.

References

1. E. Fermi, J. Pasta, and S. Ulam, Los Alamos Report LA-1940, 1955, unpublished; in *Collected Papers of Enrico Fermi*, by E. Sergé, University of Chicago Press, Chicago, 1965, p. 978.
2. J. Ford, *Phys. Rep.* **213**, 271 (1992).
3. *Hamiltonian dynamical system*, a reprint selection compiled and introduced by R. S. MacKay and J. D. Meiss, Adam Hilger, Bristol, 1987.
4. A. J. Lichtenberg and M. A. Lieberman, *Regular and Chaotic Dynamics*, Springer, New York, 1994.
5. *Hamiltonian Systems with Three or More Degrees of Freedom*, C. Simó, ed., Kluwer Academic Publishers, Dordrecht, 1999.
6. D. Eisenberg and W. Kauzmann, *The Structure and Properties of Water*, Oxford University Press, London, 1969.
7. *Water, a Comprehensive Treatise*, Vols. 1–17, F. Franks, ed., Plenum, New York, 1972–1982.
8. I. Ohmine, H. Tanaka, and P. G. Wolynes, *J. Chem. Phys.* **89**, 5852 (1988); H. Tanaka and I. Ohmine, *J. Chem. Phys.* **91**, 6318 (1989); I. Ohmine and H. Tanaka, *J. Chem. Phys.* **93**, 8138 (1990).
9. I. Ohmine and H. Tanaka, *Chem. Rev.* **93**, 2545 (1993).
10. L. Boltzmann, *Nature* **51**, 413 (1895).
11. J. H. Jeans, *Philos. Mag.* **6**, 279 (1903); *Philos. Mag.* **10**, 91(1905).
12. G. Benettin, L. Galgani, and A. Giorgilli, *Commun. Math. Phys.* **113**, 87 (1987); O. Baldan and G. Benettin, *J. Stat. Phys.* **62**, 201 (1991).

13. G. Benettin, L. Galgani, and A. Giorgilli, *Commun. Math. Phys.* **121**, 557 (1989).
14. A. Shudo, K. Ichiki, and S. Saito, to be submitted.
15. A. Katok and B. Hasselblatt, *Introduction to the Modern Theory of Dynamical Systems*, Cambridge University Press, Cambridge, 1999.
16. S. Aubry, in R. McGehee and K. R. Meyer, eds., *Twist Mappings and Their Applications*, Vol. 44 of The IMA Volumes of Mathematics and Its Applications, Springer, Berlin, 1992, p. 7; *Physica D* **86**, 284 (1995).
17. J. Milnor and W. Thurston, *On Iterated Maps of the Interval, Dynamical Systems*, College Park, MD, 1986–87, J. C. Alexander ed. Lecture Notes in Math. Vol. 1342 (Springer-Verlag, Berlin, 1988) p. 465.
18. P. Cvitanović, *Physica D* **51**, 138 (1991).
19. Y. Ishii, *Nonlinearity* **10**, 731 (1997).
20. A. de Carvalho and T. Hall, *Nonlinearity* **15**, R19–R68 (2002).
21. Q. Wang and L.-S. Young, *Commun. Math. Phys.* **218**, 1 (2001).
22. A. Sannami, *Nonexistence of Symbolic Representation for Discrete Dynamical Systems*, preprint, 2003.
23. N. N. Nekhoroshev, *Russ. Math. Surv.* **32**, 1 (1977).
24. M. Guzzo and G. Benettin, *A Spectral Formulation of the Nekhoroshev Theorem and Its Relevance for Numerical and Experimental Analysis*, preprint.
25. N. Saitô, N. Ooyama, Y. Aizawa, and H. Hirooka, *Prog. Theor. Phys. Suppl.* **45**, 209 (1970).
26. Y. Aizawa et al., *Prog. Theor. Phys. Suppl.* **98**, 36 (1989).
27. B. V. Chirikov and D. L. Shepelyanski, *Physica D* **13**, 394 (1984).
28. B. V. Chirikov, *Phys. Rep.* **52**, 263 (1979).
29. C. F. F. Karney, *Physica D* **8**, 360 (1983).
30. I. C. Percival, in *Nonlinear Dynamics and Beam-Beam Interaction*, M. Month and J. C. Herrera, eds., *Am. Institute of Physics Conf . Proc.* **57**, 302 (1979).
31. S. Aubry, in *Soliton and Condensed Matter Physics*, A. R. Bishop and T. Schneider, eds., Springer, Berlin, 1987, p. 264.
32. S. Aubry and P. Y. Le Daeron, *Physica D* **8** , 381 (1983).
33. J. N. Mather, *Topology* **21**, 457 (1982).
34. A. Katok, *Ergodic Theory & Dyn. Sys.* **2**, 185 (1982).
35. R. S. MacKay, J. D. Meiss, and I. C. Percival, *Physica D* **13**, 55 (1984).
36. J. D. Meiss and E. Ott, *Physica D* **20**, 387 (1986).
37. T. Geisel, A. Zacharl and G. Radons, *Phys. Rev. Lett.* **59**, 2503 (1987).
38. Y. Aizawa, *Prog. Theor. Phys.* **71**, 1419 (1984).
39. Y. Y. Yamaguchi, and T. Konishi, *Prog. Theor. Phys.* **99**, 139 (1998).
40. P. Gaspard, *Chaos, Scattering and Statistical Mechanics*, Cambridge University Press, Cambridge, 1998.
41. L. Bunimovich and Ya. G. Sinai, *Commun. Math. Phys.* **78**, 247 (1980) ; *Commun. Math. Phys.* **78**, 479 (1980); *Commun. Math. Phys.* **107**, 357 (1986).
42. L. Bunimovich, *Sov. Phys. JETP* **62**, 842 (1985).
43. N. I. Chernov, *J. Stat. Phys.* **74**, 11 (1994).
44. A. Rahman and F. H. Stillinger, *J. Chem. Phys.* **55**, 3336 (1971).
45. F. H. Stillinger, T. A. Weber, *J. Phys. Chem.* **87**, 2833 (1983); *Science* **225**, 983 (1984).

46. R. S. Berry, *Chem. Rev.* **93**, 2379 (1993) and references therein.
47. M. Sasai, I. Ohmine, and R. Ramaswarmy, *J. Chem. Phys.* **96**, 3045 (1992).
48. O. Kitao, I. Ohmine, and K. Nakanishi, unpublished work.
49. S. Saito, M. Matsumoto, and I. Ohmine, *Adv. Classical Trajectory Methods*, **4**, 105 (1999).
50. M. Matsumoto and I. Ohmine, *J. Chem. Phys.* **104**, 2705 (1996).
51. W. Gotze, in *Liquids, Freezing and Glass Transition*, edited by J. P. Hansen, D. Levesque and J. Zinn-Justin, eds., North-Holland, Amsterdam, 1991.
52. S. D. Bembenek and B. B. Laird, *Phys. Rev. Lett.* **74**, 936 (1995).
53. F. W. Starr, J. K. Nielse, and H. E. Stanley, *Phys. Rev. Lett.* **82**, 2294 (1999); *Phys. Rev. E* **62**, 579 (2000).
54. M. Cho, G. R. Fleming, S. Saito, I. Ohmine, and R. M. Stratt, *J. Chem. Phys.* **100**, 6672 (1994).
55. E. La Nave, A. Scala, F. W Starr, H. E. Stanley, and F. Sciortino, *Phys. Rev. Lett.* **84**, 4605 (2000); *Phys. Rev. E* **64**, 036102-1 (2001).
56. T. B. Schroder, S. Sastry, J. C. Dyre, and S. C. Glotzer, *J. Chem. Phys.* **112**, 9834 (2000).
57. M. Pettini and M. Landolfi, *Phys. Rev. A* **41**, 768 (1990); M. Pettini and Monica Cerruti-Sola, *Phys. Rev. A* **44**, 975 (1991).
58. R. Livi, M. Pettini, S. Ruffo, M. Sparpaglione, and A. Vulpiani, *Phys. Rev. A* **31**, 1039 (1985); R. Livi, M. Pettini, S. Ruffo, and A. Vulpiani, *Phys. Rev. A* **31**, 2740 (1985).
59. M. Pettini, *Phys. Rev. A* **47**, 828 (1993).
60. S. Flach and G. Mutschke, *Phys. Rev. E* **49**, 5018 (1994).
61. M. D'Alessandro, A. D'Aquino, and A. Tenenbaum, *Phys. Rev. E* **62**, 4809 (2000).
62. J. De Luca, A. Lichtenberg, and S. Ruffo, *Phys. Rev. E* **51**, 2877 (1995); *Phys. Rev. E* **54**, 2329 (1996); *Phys. Rev. E* **60**, 3781 (1999).
63. H. Kantz, *Physica D* **39**, 322 (1989); H. Kantz, R. Livi, and S. Ruffo, *J. Stat. Phys.* **76**, 627 (1994).
64. H. Kawamura, N. Fuchikami, D. Choi, and S. Ishioka, *Jpn. J. Appl. Phys.* **35**, 2387 (1996).
65. H. Rüssmann, in *Dynamical Systems, Theory and Applications*, J. Moser, ed., Lecture Notes in Physics, Vol. 38, Springer, Berlin, 1975.
66. L. Chierchia and G. Gallavotti, *Nuovo Cimento* B**67**, 277 (1982); G. Benettin, L. Galgani, A. Giorgilli, and J. J. Strelcyn, *Nuovo Cimento* B**79**, 201 (1984).
67. M. Vittot, unpublished.
68. E. Wayne, *Commun. Math. Phys.* **96**, 311 (1984); *Commun. Math. Phys.* **96**, 331 (1984).
69. F. M. Izrailev and B. V. Chirikov, *Dokl. Akad. Nauk SSSR* **166**, 57 (1966) [*Sov. Phys. Dokl.* **11**, 30 (1966)].
70. J. Pöschel, *Math. Z.* **213**, 187 (1993); P. Lochak, *Russian Math. Surv.* **47**, 57 (1992); G. Benettin, L. Galgani, and A. Giorgilli, *Celest. Mech.* **37**, 1 (1985); G. Benettin and G. Gallavotti, *J. Stat. Phys.* **44**, 293 (1986).
71. A. Giorgilli, *Proceedings for the International Congress of Mathematicians*, Vol. III, Berlin, 1998.
72. L. Galgani, A. Giorgilli, A. Martinoli, and S. Vanzini, *Physica D* **59**, 334 (1992).
73. G. Benettin, *Prog. Theor. Phys. Suppl.* **116**, 207 (1994).
74. S. Sekine, master's thesis, Tokyo Metropolitan University, 1999.
75. A. Pohorille, L. R. Pratt, R. A. LaViolette, M. A. Wilson, and R. D. MacElroy, *J. Chem. Phys.* **87**, 6070 (1987).

76. M. Neumann, *J. Chem. Phys.* **82**, 5663 (1985); *J. Chem. Phys.*, **85**, 1567 (1986).
77. J. Anderson, J. J. Ullo, and S. Yip, *J. Chem. Phys.* **87**, 1726 (1987).
78. S. Saito and I. Ohmine, *J. Chem. Phys.* **101**, 6063 (1994).
79. J. Barthel, K. Bachuber, R. Buchner, and H. Hetzenauer, *Chem. Phys. Lett.* **165**, 369 (1990).
80. J. T. Kindt and C. A. Schmuttenmaer, *J. Phys. Chem.* **100**, 10393 (1996).
81. S. Saito and I. Ohmine, *J. Chem. Phys.* **102**, 3566 (1995).
82. T. Head-Gordon and F. H. Stillinger, *J. Chem. Phys.* **98**, 3313 (1993).
83. S. C. Creagh, in *Tunneling in Complex Systems*, S. Tomsovic, ed., World Scientific, Singapore, 1998, p. 35.

CHAPTER 21

SLOW DYNAMICS IN MULTIDIMENSIONAL PHASE SPACE: ARNOLD MODEL REVISITED

TETSURO KONISHI

Department of Physics, Nagoya University, Nagoya, 464-8602, Japan

CONTENTS

I. Introduction
II. Global Motion in Systems with Many Degrees of Freedom
III. Arnold Model and Its Dynamics
IV. Numerical Method
V. Numerical Results
VI. Summary
Acknowledgment
References

I. INTRODUCTION

The dynamics of systems with many degrees of freedom is an interesting and important subject in various fields of research and ordinary life, ranging from the dynamics of molecules to satellite orbiting. Progress in nonlinear science and dynamical systems has revealed the intricate dynamics occurring in nonintegrable systems.

Studies in nonlinear dynamics developed rapidly in the last quarter of the twentieth century, and we now have quite a lot of knowledge about the dynamics of systems with a few degrees of freedom.

What we are really interested in now is the dynamics of systems with many degrees of freedom, which has many aspects that are different from

---

Geometric Structures of Phase Space in Multidimensional Chaos: A Special Volume of Advances in Chemical Physics, Part B, Volume 130, edited by M. Toda, T Komatsuzaki, T. Konishi, R.S. Berry, and S.A. Rice. Series editor Stuart A. Rice.
ISBN 0-471-71157-8 Copyright © 2005 John Wiley & Sons, Inc.

those of small systems, and these differences create unique and interesting behavior.

One of the phenomena that is unique in systems with many degrees of freedom is that the state of the system can undergo large changes regardless of the existence of invariant sets that are remnants of periodic/quasi-periodic orbits of the unperturbed system [1]. Arnold showed this phenomenon for a specific model (shown in Section III) and calculated the upper bound of the rate of change in the action variable by linear perturbation theory. This phenomenon is sometimes called "Arnold diffusion" [2].

Arnold diffusion is important not just in abstract dynamical system theory, but also in realistic systems such as astrodynamics and chemical physics [3,4].

One of the important features of Arnold diffusion is that it is supposed to be quite slow. This is shown in the original estimate of Arnold [1]. Since Arnold diffusion is an essential process for relaxation in nearly integrable (i.e., weakly coupled) systems with many degrees of freedom, most relaxations can proceed in this very slow time scale. The slow relaxation is acceptable when the system we are concerned is the solar system or some other celestial object. But, does Arnold diffusion really occur in molecular systems in the course of relaxation or reaction? Do the systems really show slow relaxation or not? What is the phase-space structure and how does the system move in there? Answers to all these questions are still unknown.

Although the term "Arnold diffusion" is quite popular, it is unfortunately not well understood, although it is about 40 years since its discovery. For example, the original estimate is given as an inequality, and the rate of the actual motion is not known. Second, the range of validity (or the observability) is not known. Moreover, although it is sometimes called "Arnold diffusion" whether it is diffusive or not is not understood.

Thus we really need to clarify what "Arnold diffusion" really is.

Although these questions appear simple and easy to confirm, they are in fact numerically tough problems because of the smallness of the rate of change of the variable.

In this chapter we would like to solve these problems by reexamining the original model proposed by Arnold by making use of a powerful numerical tool called "multi-precision library."

This chapter is organized as follows. In Section II we briefly review general background of the problem. In Section III we introduce the model and review its dynamics. In Section IV we look at the numerical tools needed to deal with extremely small quantities. Numerical results obtained with these tools are given in Section V, and Section VI summarizes and discusses the results.

II. GLOBAL MOTION IN SYSTEMS WITH MANY DEGREES OF FREEDOM

Before going into details of the model, we first briefly explain the background. We write the system we are interested in as a perturbed system to an ideal Hamiltonian H_0:

$$H(I, \varphi) = H_0(I) + \varepsilon H_1(I, \varphi) \tag{1}$$

$$I = (I_1, \ldots, I_N), \quad \varphi = (\varphi_1, \ldots, \varphi_N), \quad |\varepsilon| \ll 1 \tag{2}$$

We are interested in the case $|\varepsilon| \ll 1$ because if $|\varepsilon|$ is large, then the dynamics tends to become more and more irregular and eventually well-approximated by stochastic dynamics (such as the Langevin equation) and a statistical description will be valid.

When $\varepsilon = 0$ the equation of motion reads

$$\dot{I}_i = -\frac{\partial H_0}{\partial \varphi_i} = 0 \tag{3}$$

$$\dot{\varphi}_i = \frac{\partial H_0}{\partial I_i} \tag{4}$$

In this case the system is called "integrable." From the first equation we have $I_i = $ const. for all i; that is, I_i are constants of motion. Thus $\omega_i(I) \equiv \partial H_0/\partial I_i$ are also constants, and we have $\varphi_i = \varphi_{i0} + \omega_i(I)t$. In this way the dynamics of the original N-degrees-of-freedom system H_0 is decomposed into those of N-uncoupled rotators. Every point in the phase space of the system is on a regular orbit (torus).

When $\varepsilon \neq 0$ but $|\varepsilon| \ll 1$ the system is called "nearly integrable." For a generic perturbation H_1 we no longer have global constants of motion ("global" means "does not depend on initial conditions"). For small ε, one may think that the dynamics will be quite similar to that of $\varepsilon = 0$. Indeed, one can have stronger results. For sufficiently small ε (with several mathematical restrictions), the KAM theorem [5–7] guarantees that most of the regular orbits in H_0 are preserved in H.

Since most of the phase space is filled by regular orbits, irregular orbits (chaotic orbits) are allowed to exist in only limited regions. (Note that we are using the words "regular orbits" and "chaotic orbits" in the mathematical sense. That is, "regular orbits" are regular in $-\infty < t < \infty$. Hence the set of regular orbits is invariant in time, and the set of chaotic orbits is also invariant in time.)

For chaotic orbits some of the conserved quantities in H_0 are no longer conserved, and the state of the system tends to move in the phase space, rather than be bound to tori. But, as we have seen, there still exist regular orbits in the phase space as is shown by the KAM theorem. Sometimes the regular orbits (also called "KAM tori") can be topological obstacles for the drift motion of chaotic orbits.

The dimension of the phase space is $2N$, as we have N I's and N φ's. If we still have global conservation laws (such as conservation of energy), the net dimension of the phase space is $d_{\text{phasespace}} = 2N -$ (number of conservation laws). The problem is whether a KAM torus can divide the phase space (or energy surface) into two disjoint parts or not. The necessary condition for global drift of chaotic motion is

$$d_{\text{phasespace}} - d_{\text{KAM}} > 1 \qquad (5)$$

where d_{KAM} is the dimension of a KAM torus, and $d_{\text{KAM}} = N$ since we have $N\varphi$'s that can move freely on a KAM torus. The inequality (5) can be understood as follows. If $d_{\text{phasespace}} - d_{\text{KAM}} = 1$ we can take a coordinate (say x) in the direction normal to the KAM torus. Take $x = 0$ on the KAM torus. Then $x > 0$ or $x < 0$ represents regions inside or outside of the KAM torus.

$d_{\text{phasespace}} = 2N$ for symplectic map, and $d_{\text{phasespace}} = 2N - 1$ for systems with energy conservation. In this way we have

$$N > 1 \,(\text{for symplectic maps}), \quad N > 2 \,(\text{for systems with energy conservation}) \qquad (6)$$

for a necessary condition for global drift. When the above inequality does not hold, chaotic orbits are bound to KAM tori and no global drift (and no microcanonical measure) is realized. This situation is illustrated in Fig. 1.

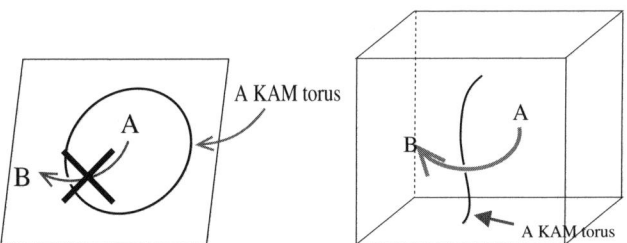

Figure 1. Schematic description showing that a KAM torus can or cannot block the global motion of a chaotic orbit in low- (**left**) or high-dimensional (**right**) phase space.

Up to now we just looked at the topological aspect of the problem and have understood that in systems with many degrees of freedom, KAM tori cannot block the global motion of the system in the entire state space (energy surface). Then how can the system move "beyond" the KAM tori?

The first answer was given by Arnold [1] by using a method now called the "Melnikov–Arnold integral" for a specific model, which we will describe in the next section. The mechanism is now called "Arnold diffusion" and has since been a subject of great interest [2–4, 8–19].

The perturbation H_1 causes resonant interaction between variables:

$$\sum_{i=1}^{N} m_i \omega_i = 0 \qquad (m_i \text{ is an integer}) \tag{7}$$

Arnold diffusion is a motion along a resonance. One can observe motion across the resonance by using a "frequency map" [20]. Also, motion at the crossing points of resonances are interesting [21–23].

III. ARNOLD MODEL AND ITS DYNAMICS

In this chapter we call the following model the "Arnold model":

$$H = \tfrac{1}{2}(I_1^2 + I_2^2) + \varepsilon(\cos\varphi_1 - 1)[1 + \mu(\sin\varphi_2 + \cos t)] \tag{8}$$
$$\varepsilon, \mu \in \mathbf{R} \tag{9}$$

Here (I_1, φ_1) represents a pendulum that has $(0,0)$ as an unstable fixed point. (Note that this definition of the angle is different from that for an ordinary pendulum.) (I_2, φ_2) represents a rotator.

Now let us see the dynamics of this model.

Suppose we have $\mu = 0$. Then the model is decomposed into a pendulum and a free rotator, and the action variable of the rotator I_2 is a constant of motion:

$$H = H_{\text{pendulum}}(I_1, \varphi_1) + H_{\text{rotator}}(I_2)$$
$$= \tfrac{1}{2}I_1^2 + \varepsilon(\cos\varphi_1 - 1) + \tfrac{1}{2}I_2^2 \tag{10}$$
$$I_2 = \text{const.}$$

Now suppose we have a small coupling between the pendulum and the rotator (i.e., $0 \le \mu \ll 1$). Suppose we take the initial condition so that we set the pendulum at a state near its unstable fixed point $(0,0)$, and the action variable of the rotator is $I_2 = I_2^{(0)} \equiv \omega$. Then, as time proceeds, the pendulum repeats swinging. At the same time, because of the coupling μ, the action variable I_2 changes its value. When the pendulum comes back to the neighborhood of

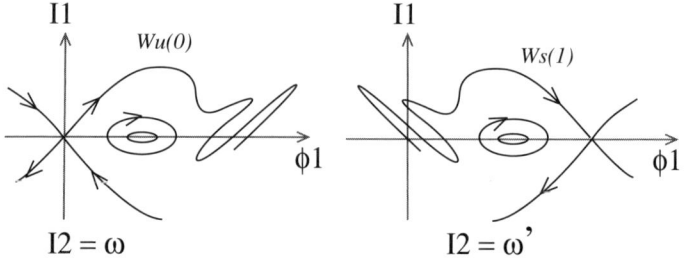

Figure 2. Schematic figure of unstable and stable manifolds in the Arnold model, Eq. (8).

its initial condition, I_2 may no longer be its initial value $I_2^{(0)}$ but may take different value. Let us call this value $I_2^{(1)}$.

In other words, an unstable manifold $W_u^{(0)}$ at $I_2 = I_2^{(0)}$ and a stable manifold $W_s^{(1)}$ at $I_2 = I_2^{(1)}$ cross. This situation is roughly illustrated in Figs. 2 and 3.

We have a sequence of similar crossings

$$W_u^{(0)} \to W_s^{(1)} \to W_u^{(1)} \to W_s^{(2)} \to W_u^{(2)} \to W_s^{(3)} \to \cdots \quad (11)$$

which is called a "transition chain."

Using the fact that μ is small, Arnold estimated the difference

$$\Delta I_2 \equiv I_2^{(1)} - I_2^{(0)} \quad (12)$$

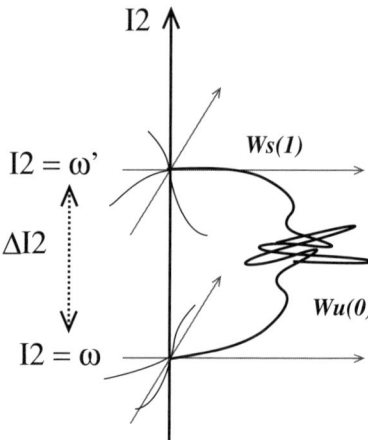

Figure 3. Schematic figure of unstable and stable manifolds in the Arnold model, Eq. (8) (*continued*).

by linear perturbation theory with respect to μ [1]. By calculating what is called the Melnikov (or Melikov–Arnold) integral, Arnold obtained

$$\left| I_2'^2 - I_2^2 \right| \equiv \left| \omega'^2 - \omega^2 \right| < \mu \cdot 4\pi\omega^2 \left(\sinh \frac{\omega\pi}{2\sqrt{\varepsilon}} \right)^{-1} \quad (13)$$

$$\sim 4\pi\mu \exp\left(-\pi\omega/2\sqrt{\varepsilon}\right) \quad (14)$$

This estimation implies two important phenomena.

- First, in the course of temporal evolution, the coupling μ gives I_2 large differences from the initial value regardless of the remaining KAM tori. This dynamics occurs only in systems with many degrees of freedom, and this change of I_2 is the original realization of "Arnold diffusion" where the state in the phase space can travel "beyond" KAM tori.
- Second, in the estimate (14), ΔI_2 depends on ε (hyperbolicity of the pendulum) in a very singular way. (Note that ε is NOT the coupling between the pendulum and the rotator.) As we decrease ε, the rate of the change of I_2 decreases rapidly. From (14), we see that the time it takes for I_2 to change $\mathcal{O}(1)$ is about

$$T \propto \exp\left(\pi\omega/2\sqrt{\varepsilon}\right) \quad (15)$$

which is extremely long for small ε. This is why "Arnold diffusion is slow."

(Incidentally the singular factor in (15) reminds one of the "Nekhoroshev inequality" [24–28]

$$\|I(t) - I(0)\| \leq c\epsilon^\alpha \quad \text{for} \quad |t| \leq T_N \equiv c'\epsilon^{-1} \exp(\xi(1/\epsilon)^\beta) \quad (16)$$

where $c, c', \xi, \alpha, \beta$ are positive constants. A concrete and qualitative relation between Arnold diffusion and the Nekhoroshev inequality (16) is not well understood yet.)

These phenomena lead us to a rather complicated situation. The first phenomenon reminds us of ergodicity, the realization of microcanonical distribution in systems with many degrees of freedom and the validity of statistical mechanics. We know that KAM tori cannot divide the phase space (or energy surface) for systems with many degrees of freedom, and the first phenomenon tells us that two neighborhoods in different parts of the phase space are connected not only topologically but also dynamically. In this sense the phenomenon can be considered as an elementary process of relaxation in systems with many degrees of freedom.

On the other hand, the slowness implies that the relaxation (or the time for statistical mechanics to be observed) may be extremely long, as long as the lifetime of the universe.

In addition, we often hear the term "Arnold diffusion" used to represent the phenomena described above. But it is not clear whether the dynamics is truly "diffusive":

$$|I_2(t) - I_2(0)|^2 \propto t \qquad (17)$$

In fact, Chirikov calculated the diffusion coefficient for a similar model [2], but whether or not the dynamics is diffusive in the general situation or in other models is, as far as the author knows, not clear.

These questions come from the insufficiency of knowledge about the dynamics of I_2, and sometimes exaggerated use of terminology. In this chapter we try to closely examine the dynamics by going back to the original model of Arnold, Eq. (8).

IV. NUMERICAL METHOD

In this section we describe the numerical method we employed to calculate the temporal evolution: multi-precision.

Usually, analytical estimation is supported by numerical computation. However, ordinary numerical computation is useless in this case to check the singular behavior because of limited precision. We would like to numerically calculate the difference of action I_2 and examine that it obeys the estimate (15): $\Delta I_2 \propto \exp(-\pi\omega/2\sqrt{\varepsilon})$ for $\omega = 1$. If $\varepsilon = 0.01$, the factor is $\sim 1.5 \times 10^{-7}$; and if $\varepsilon = 0.001$, the factor is as small as 10^{-22}. This goes beyond the usual level of round-off error of double precision numerical computation.

There are two methods that we may use for problems that require extremely high precision. One is "validated computation" and the other is "multi-precision." "Validated computation" is a method to obtain *rigorous* results with numerical computation. It is based on operation for intervals rather than for numbers. Usual numerical computation on real numbers says, for example, $x = 0.123$, but this just means that the value of x will be near 0.123 and it does not mean that the value of x is exactly 0.123. On the other hand, with the use of validated computation we can rigorously say that the value of x satisfies , for example, $x \in (0.122, 0.124)$.

Although validated computation is a powerful tool, it seems not to be all-purpose, or it may take time to reformulate the problem one want to solve to fit validated computation. Hence we adopt the second tool, a multi-precision library.

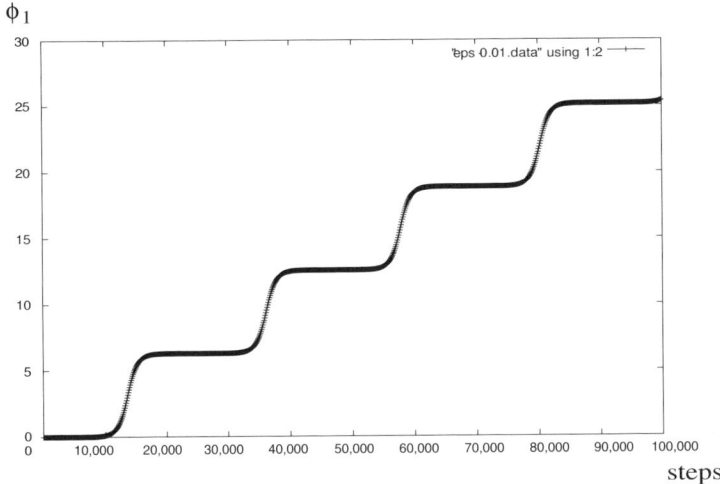

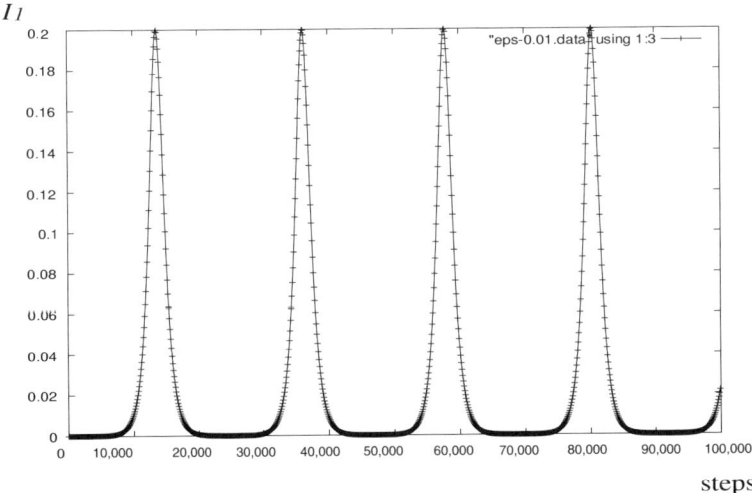

Figure 4. Temporal evolution of φ_1, I_1 in Eq. (8). $\varepsilon = 0.01$, $\mu = 4 \times 10^{-5}$. (I_2, φ_2) for the same time sequence is shown in Fig. 5.

In usual numerical computation the precision of variables is uniquely defined in the realization of the language (`C`, `FORTRAN`, etc.). On the other hand, by using a multi-precision library, users can define the precision of variables arbitrarily (as far as the computer resources permit). Here we adopt a multi-precision library called `cln` [29], which can be used in `C++`.

ϕ_2

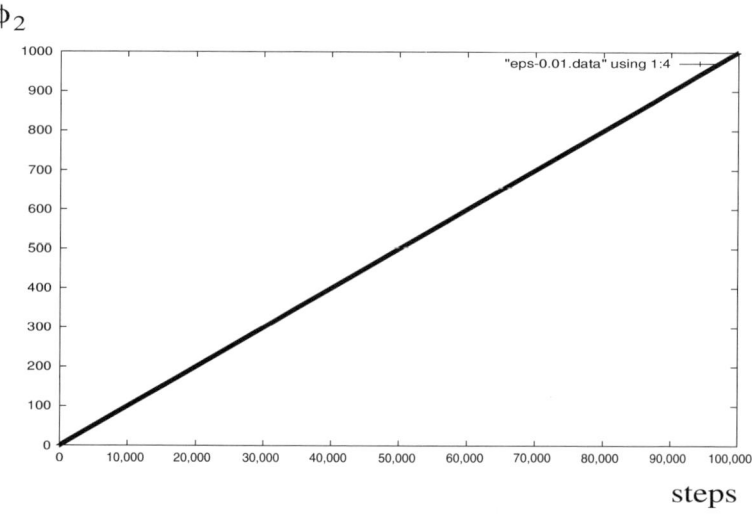

steps

$I_2 - 1$

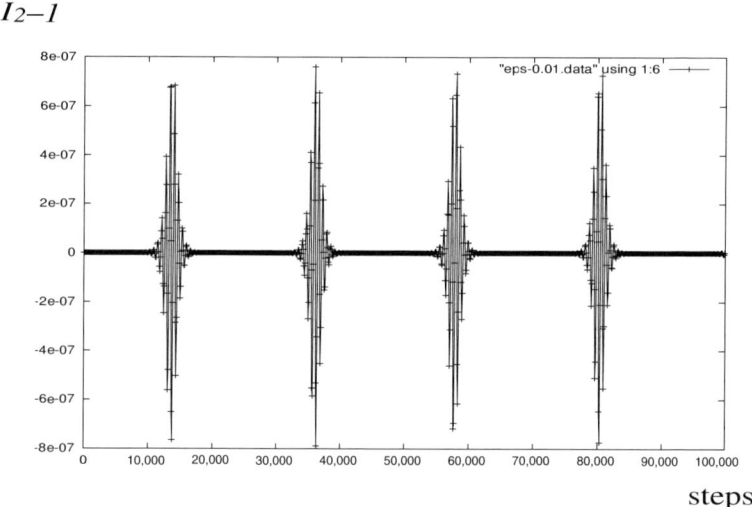

steps

Figure 5. Temporal evolution of φ_1, I_1, φ_2, I_2 in Eq. (8). $\varepsilon = 0.01$, $\mu = 4 \times 10^{-5}$. (I_1, φ_1) for the same time sequence is shown in Fig. 4.

V. NUMERICAL RESULTS

Solving the equation of motion for the model (8) with typically 1000 decimal digits using a multi-precision library cln eighth-order symplectic integrator, we obtain a time slice $\Delta t = 10^{-2}$ or 10th order with $\Delta t = 5 \times 10^{-3}$ for $\varepsilon = 0.004$. A typical evolution of $(\varphi_i(t), I_i(t))$ is shown in Figs. 4, 5, and 6.

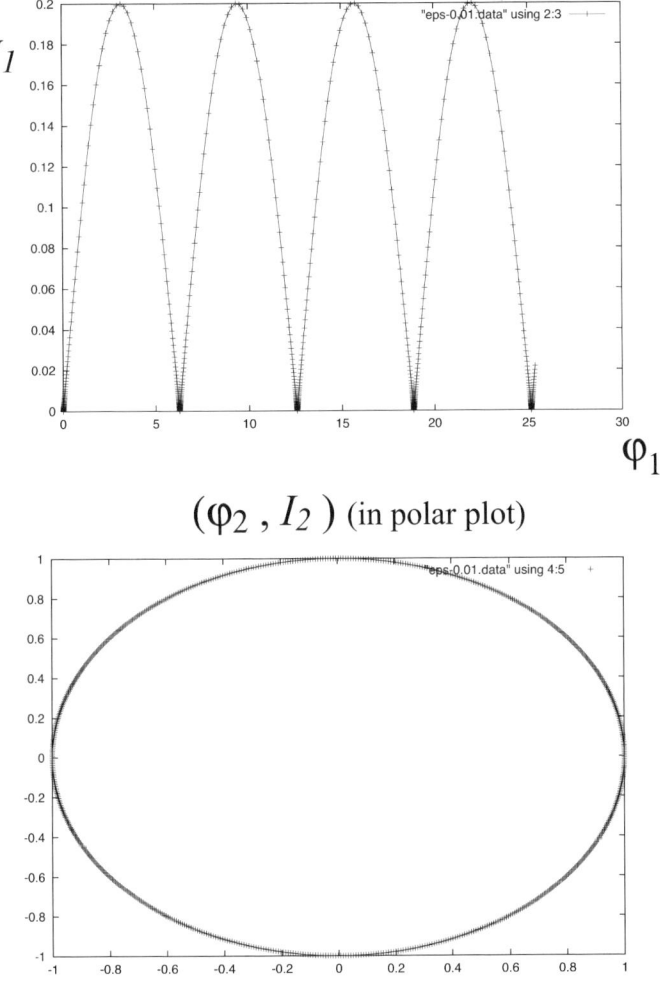

Figure 6. Local phase space: (φ_1, I_1) and (φ_2, I_2) in Eq. (8).

Since μ is quite small, the pendulum (I_1, φ_1) shows approximately periodic motion. I_2 deviates from its initial value when (I_1, φ_1) moves away from the hyperbolic fixed point. When (I_1, φ_1) comes back near the hyperbolic fixed point, I_2 also comes near to the initial value. The difference ΔI_2, [Eq. (12)] is the quantity we are interested in.

Figure 7 shows the dependence of ΔI_2 on ε.

In this figure we can clearly see that the numerical result agrees well with the original estimate (14). That is, "Arnold diffusion" is slow, as expected.

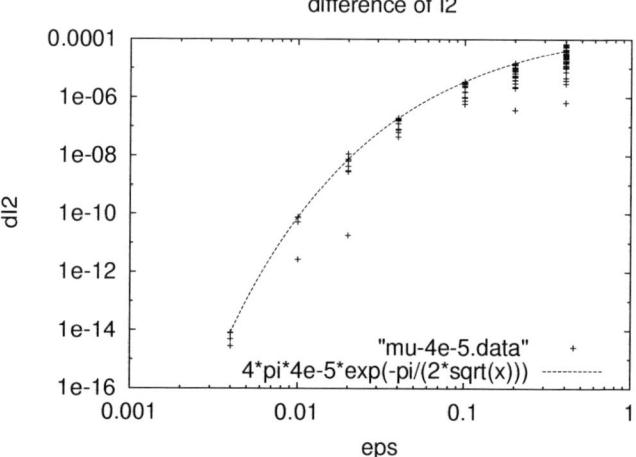

Figure 7. ε dependence of ΔI_2. Dashed line represents the estimation $\Delta I_2 \sim 4\pi\mu\exp(-\pi\omega/2\sqrt{\varepsilon})$ [Eq. (14)].

Now we turn to the next and important question. Although "Arnold diffusion" is called "diffusion," it is not clear whether it is really "diffusive." In other words, whether or not the temporal correlation of ΔI_2 is short and ΔI_2 has Markovian behavior is not confirmed. In Fig. 8 we show the long-time behavior of ΔI_2 for succesive change along one orbit of temporal evolution. Correlation between successive changes $(\Delta I_2, \Delta I_2')$ is shown in Fig. 9.

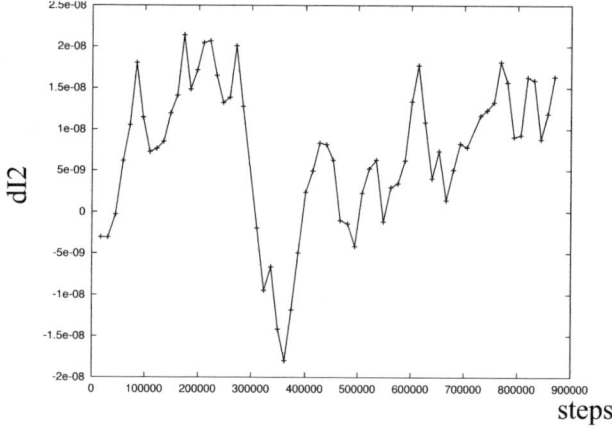

Figure 8. Long-time behavior of ΔI_2. $\varepsilon = 0.02$. This is preliminary data.

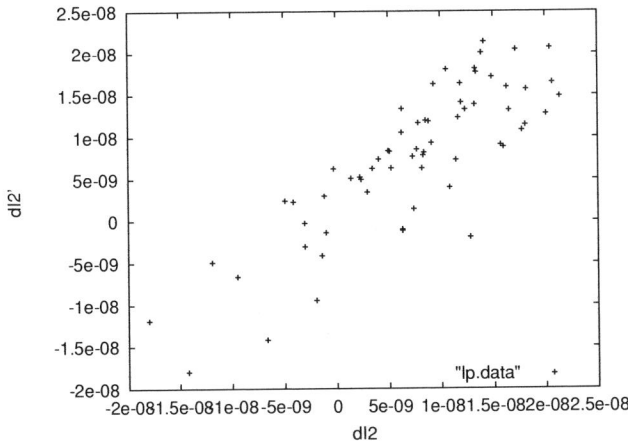

Figure 9. Succesive plot of ΔI_2. Same data as in the previous figure.

The data shown in the above figures, although preliminary and limited in its length or statistics, clearly implies that the action variable I_2 has strong temporal correlation and may not be diffusive.

VI. SUMMARY

In this chapter we numerically examined the Arnold model and found the following. First, by performing extremely high-precision numerical computation using a multi-precision library, "Arnold diffusion" is numerically observed. Next, the difference of the action variable depends singularly on the hyperbolicity ε and is in accordance with the original estimate (14) [1]. The original estimate is given as an inequality and only an upper bound is given, but the the gap between the data and the upper bound appear not so large.

It is confirmed that the singular dependence on ε shown above is actually observed for "not so too small" ε (typically ε ~ 0.1). This is in contrast to some other theorems that are important and rigorous but fail to be applied in actual situations. For example, the KAM theorem is valid only for situations where nonintegrability is extremely small.

The fact that the original estimate (14), which is obtained from first-order perturbation theory with respect to μ, agreed with the numerical results implies that the terms with μ^2 or higher also have singular dependence on ε.

In addition, what is most important is that long-time behavior of ΔI_2 implies that the process has strong temporal correlation and may not be diffusive.

Whether or not "Arnold diffusion" becomes true diffusion after a long time or not is important. Also, if it becomes diffusion, then what is the mechanism of loss of temporal correlation?

We are beginning to understand the real dynamics of global diffusion in the phase space of many-dimensional Hamiltonian systems. From here we are going to travel around the vast world created by the chaotic dynamics of nonlinear systems.

Acknowledgment

I would like to express my sincere thanks to Prof. Stuart Rice for reading the manuscript and making a number of helpful suggestions.

References

1. V. I. Arnold, *Sov. Math. Dokl.* **5**, 581 (1964).
2. B. V. Chirikov, *Phys. Rep.* **52**, 265–379 (1979).
3. P. M. Cincotta, *New Astron. Rev.* **46**, 13–39 (2002).
4. D. M. Leitner and P. G. Wolynes, *Chem. Phys. Lett.* **276**, 289–295 (1997).
5. A. N. Kolmogolov, *Dokl. Akad. Nauk. SSSR* **98**, 527–531 (1954).
6. V. I. Arnold, *Russ. Math. Surv.* **18**, 85 (1963).
7. J. K. Moser, *Nachr. Akad. Wiss. Göttingen. Math. Phys.* **1**, 1 (1962).
8. P. J. Holmes and J. E. Marsden, *J. Math. Phys.* **23**, 669–675 (1982).
9. Z. Xia, *J. Differ. Eq.* **110**, 289–321 (1994).
10. K. Kaneko and R. J. Bagley, *Phys. Lett.* **110A**, 435–440 (1985).
11. C. G. Ragazzo, *Phys. Lett. A* **230**, 183–189 (1997).
12. E. Lega and C. Froeschlé, *Physica D* **95**, 97–106 (1996).
13. R. W. Easton, J. D. Meiss, and G. Roberts, *Physica D* **156**, 201–218 (2001).
14. E. Lega, M. Guzzo, and C. Froeschlé, *Physica D* **182**, 179–187 (2003).
15. A. Cicogna and M. Santroprete, *J. Math. Phys.* **41**, 805–815 (2000).
16. E. Fontich and P. Mart'in, *Nonlinear Anal.* **42**, 1397–1412 (2000).
17. U. Bessi, L. Chierchia, and E. Valdinoci, *J. Math Pure Appl.* **80**, 105–129 (2001).
18. J. Cresson, *J. Diff. Eq.* **187**, 269–292 (2003).
19. S. Honjo, *Adv. Chem. Phys.* **130**, 437–463 (2005).
20. J. Laskar, *Physica D* **67**, 257–281 (1993).
21. G. Haller, *Phys. Lett. A* **200**, 34–42 (1995).
22. G. Haller, *Chaos Near Resonance*, Springer, Berlin, 1999.
23. S. Goto and K. Nozaki, *Prog. Theor. Phys.* **102**, 937–946 (1999).
24. N. N. Nekhoroshev, *Russ. Math. Surv.* **32**, 1–65 (1977).
25. G. Benettin and G. Gallavotti, *J. Stat. Phys.* **44**, 293 (1986).
26. P. Lochak and A. I. Neishtadt, *Chaos* **2**, 495–499 (1992).
27. P. Lochak, *Russ. Math. Surv.* **47**, 57–133 (1992).
28. P. Lochak, *Nonlinearity* **6**, 885–904 (1993).
29. B. Haible, CLN, a Class Library for Numbers. http://clisp.cons.org/~haible/packages-cln.html, (1995,1999).

CHAPTER 22

STRUCTURE OF RESONANCES AND TRANSPORT IN MULTIDIMENSIONAL HAMILTONIAN DYNAMICAL SYSTEMS

SEIICHIRO HONJO and KUNIHIKO KANEKO

Department of Basic Science, Graduate School of Arts and Sciences, University of Tokyo, Komaba, Meguro-ku, Tokyo, 153-8902, Japan

CONTENTS

I. Introduction
II. Model and Basic Quantities
   A. Model
   B. Rotation Number
   C. Diffusion Coefficient
III. Structure of Resonances
   A. Frequency Space and Phase Space
   B. Morphological Change Depending on Parameters
   C. Characterization with Residence Time Distribution
IV. Transport
   A. Transition Diagram
   B. Pathway for Fast Transition
V. Resonance Overlap in Multidimensional Systems
VI. Deflected Diffusion
VII. Summary and Discussion
Acknowledgment
References

I. INTRODUCTION

Understanding in global dynamic behavior in Hamiltonian dynamical systems is a fundamental issue in nonlinear dynamics and statistical physics. Besides the applications in various field such as astronomy, plasma physics, and atomic

---

Geometric Structures of Phase Space in Multidimensional Chaos: A Special Volume of Advances in Chemical Physics, Part B, Volume 130, edited by M. Toda, T Komatsuzaki, T. Konishi, R.S. Berry, and S.A. Rice. Series editor Stuart A. Rice.
ISBN 0-471-71157-8 Copyright © 2005 John Wiley & Sons, Inc.

physics, the study of global stochasticity and diffusion in a multidimensional Hamiltonian dynamical systems is increasingly emphasized in the field of chemical physics, in order to treat chemical reactions as transports in Hamiltonian dynamical systems [1–3].

In nonintegrable Hamiltonian dynamical systems, two mechanisms for instabilities leading to global diffusion in the phase space are well known: Arnold diffusion [4–7] and resonance overlap [5,8]. In a Hamiltonian dynamical system in general, there are resonance conditions $\sum_i m_i \omega_i + M = 0$, where m_i's and M are arbitrary integers and ω_i is the radial frequency of the ith element. The conditions form resonance lines in the phase space, around which the motion is stochastic, giving rise to a layer, while the interwoven resonance layers form the so-called Arnold web. Arnold diffusion is the motion along the resonance layers, and it is a universal behavior in the systems with more than two degrees of freedom. Resonance overlap, on the other hand, derives from destruction of tori that divide each resonance layer, and results in global transport in the phase space, as has been studied in detail in two-dimensional mappings [5,8]. In a system with many degrees of freedom in general, both mechanisms coexist, and it is not so easy to unveil the process of global transport in the phase space. In this chapter, we study this problem by focusing on global structure of resonances.

We introduce a simple model to investigate and calculate a diffusion coefficient as a basic quantity describing transport in Section II, and then we visualize resonances to detect the structure of the Arnold web and overlapped resonances in Section III. With the aid of this representation, to clarify the relevance of Arnold diffusion and diffusion induced by resonance overlap to global transport in the phase space, we compute transition diagrams in the frequency space in Section IV. In Section V, we extend the resonance overlap criterion to multidimensional systems to identify the pathway for fast transport, and in Section VI we revisit the diffusion coefficient to ensure fast transport affecting the global diffusion. A brief summary is given in Section VII.

II. MODEL AND BASIC QUANTITIES

A. Model

As a simple model for Hamiltonian dynamical system with several degrees of freedom, we have chosen Froeschlé map [9–13], given by

$$p_i(n+1) = p_i(n) + K\sin(q_i(n)) + b\sin\left(\sum_{k=1}^{2} q_k(n)\right)$$
$$q_i(n+1) = q_i(n) + p_i(n+1) \qquad (\text{mod } 2\pi) \tag{1}$$

STRUCTURE OF RESONANCES AND TRANSPORT   439

where $i = 1, 2$, $q_i(n)$ is the displacement of ith element, and $p_i(n)$ is its conjugate momentum. Here, K represents nonlinearity of each element and b gives the coupling strength.

The Froeschlé map could be taken as a coupled system consisting of standard maps. The uncoupled standard map with $b = 0$, is studied as a prototype of a Poincaré map for Hamiltonian dynamics with two degrees of freedom. Similarly, the above Froeschlé map could be regarded as a Poincaré map of Hamiltonian dynamical system with three degrees of freedom, and it provides a prototype model for such system.

In Hamiltonian form, the Froeschlé map is expressed by

$$H = \sum_{i=1}^{2} \frac{p_i^2}{2} + \left\{ \sum_{i=1}^{2} K\cos(q_i) + b\cos\left(\sum_{i=1}^{2} q_i\right) \right\} \sum_{n=-\infty}^{\infty} \delta(t-n) \quad (2)$$

which describes a periodically kicked and coupled rotators.

As the Fourier expansion of the periodic delta function, we obtain

$$\sum_{n=-\infty}^{\infty} \delta(t-n) = 1 + 2\sum_{m=1}^{\infty} \cos(2\pi m t) \quad (3)$$

Thus, Hamiltonian (2) means two coupled rotators

$$H = \sum_{i=1}^{2} \frac{p_i^2}{2} + \left\{ \sum_{i=1}^{2} K\cos(q_i) + b\cos\left(\sum_{i=1}^{2} q_i\right) \right\} \quad (4)$$

forced by every color of the frequency with same amplitude.

B. Rotation Number

Here, as the basic frequency to describe resonant state of systems, we define the rotation number

$$\omega_i \equiv \lim_{T\to\infty} \frac{q_i(T) - q_i(0)}{2\pi T} = \lim_{T\to\infty} \sum_{n=1}^{T} \frac{p_i(n)}{2\pi T} \quad (5)$$

to each ith element. Later we are mainly interested in time evolution in the frequency space. Thus, we employ the local rotation numbers computed over finite time length T by

$$\omega_i(jT) \equiv \sum_{n=jT}^{jT+T-1} \frac{p_i(n)}{2\pi T} \quad (6)$$

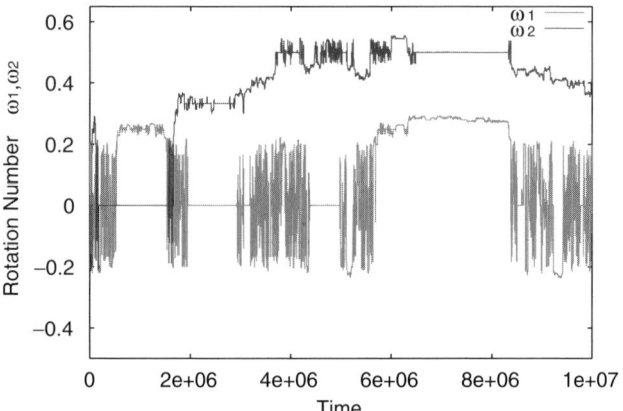

Figure 1. Time series of local rotation numbers. $K = 0.8$, $b = 0.002$. 1×10^7 steps. Residence on rational numbers and intermittent transition among them are observed.

In terms of rotation numbers, the resonance condition of the Froeschlé map is given by $m_1\omega_1 + m_2\omega_2 + M = 0$, where m_i's and M are arbitrary integers, and the order of resonances is defined by $\sum_i |m_i| + |M|$. The value of T must be chosen large enough to enssure the convergence of each rotation number to a certain resonance, but not too large so that transition between the resonance layers is detected. In this case, we choose T typically as 10^3, but change of it within a moderate range does not influence our results to be reported.

Figure 1 shows time series of local rotation numbers. The orbit remains for a while at a certain resonance, and it escapes to another resonance intermittently.

C. Diffusion Coefficient

It is very reasonable to compute the diffusion coefficient in order to study transport in symplectic maps. Diffusion coefficient is defined by

$$D \equiv \lim_{T \to \infty} \left\langle \frac{1}{2} \sum_{i=1}^{2} \frac{(p_i(T) - p_i(0))^2}{T} \right\rangle \quad (7)$$

where $\langle \cdot \rangle$ represents sample average.

At first, we check the convergence of the diffusion coefficient to ensure that there is no anomalous diffusion for the Froeschlé map in long time steps, even with $K < K_c \sim 0.97$, where the last and some other KAM tori of uncoupled standard maps are preserved. In Fig. 2, short-time diffusion coefficients

$$D(T) \equiv \left\langle \frac{1}{2} \sum_{i=1}^{2} \frac{(p_i(T) - p_i(0))^2}{T} \right\rangle \quad (8)$$

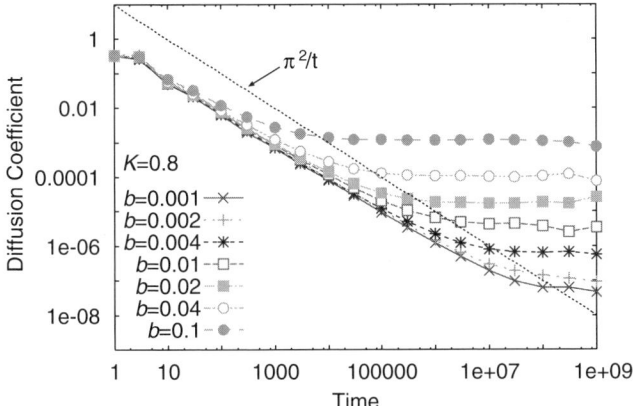

Figure 2. Short time diffusion coefficient. Computed by sequential averages of T steps from single trajectory iterated for 1×10^{10} time steps. $K = 0.8$ and $b = 0.001, 0.002, 0.004, 0.010, 0.020, 0.040, 0.100$. We have no anomalous diffusion in long time steps.

for $K = 0.8$ and various b are plotted. Each diffusion coefficient is computed by sequential averages of T steps from a single trajectory iterated for 1×10^{10} time steps. Thus, fluctuations observed around $T = 1 \times 10^9$ is due to a poor number of samples.

As shown in Fig. 2, we have anomalous diffusion in short time steps, but in long time steps diffusion coefficients converge to certain values to give normal diffusion.

This crossover from anomalous diffusion to normal diffusion corresponds to the 2π periodic property of the Froeschlé map. Diffusion coefficients converge when they satisfy the condition $D = \pi^2/t$, which means $\sqrt{Dt} = \pi$, where the standard deviation of momentum p_i attains π, which is half the length of a side of unit cell of a $2\pi \times 2\pi$ square. Long time correlations are maintained only up to the time when the state jumps out of the unit cell to another cell [14].

Thus, the Froeschlé map is an adequate model to study transport using diffusion coefficients.

Dependence of D on the coupling strength b is shown in Fig. 3, for various values of K.

Random phase approximation $D = (K^2 + b^2)/2$ works well for strong coupling (e.g., $b > 10$) and strong nonlinearity (e.g., $K = 4.0$), that waits for higher-order correction by Fourier path method [11,15,16]. With $K \geq 1.0$, diffusion coefficients approach some constants as $b \to 0$ due to the breakup of the last KAM torus of each standard map, while with $K \leq 0.9$ and smaller b, they are expected to be evaluated by the stochastic pump or three resonance model and their extensions [12,17,18].

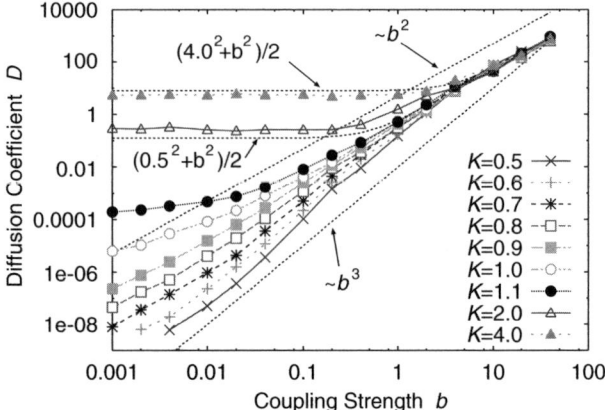

Figure 3. Dependence of diffusion coefficient D on coupling strength b for various K values. D's show power-law dependence ($D \sim b^\alpha$, $2 < \alpha < 3$) for $K \leq 0.9$, while they approach some constants as $b \to 0$ for $K \geq 1.0$. Random phase approximation $D = (K^2 + b^2)/2$ works only for large b or K.

In other parameter regions, with $K \leq 0.9$ and medium b, diffusion coefficients show power-law dependence ($D \sim b^\alpha$, $2 < \alpha < 3$).

To the authors' knowledge, there is no theory and detailed observation about that behavior. Then, in the following sections, we focus on these parameter regions and show how diffusion occurs actually.

III. STRUCTURE OF RESONANCES

A. Frequency Space and Phase Space

To understand dynamics in multidimensional Hamiltonian systems, adequate representation of states is important.

We employ the frequency space to visualize resonances, and besides we use phase space representation to get features of the frequency space.

The idea to investigate structures of resonances in frequency space was, to the authors' knowledge, devised by Martens et al. [19] and then sophisticatedly implemented by Laskar and co-workers [13,20,21]. Instead of using FFT as their methods, we concentrate on a simple toy model and adapt rotation numbers as basic frequencies. By this way, we can easily compute basic frequency so that we can investigate global features of resonances and those dependence on parameters.

To visualize the structures of resonances, we use the density plot in the frequency space to include information on time, as follows: First, we compute the rotation numbers modulo 1 over a finite time interval from trajectories to

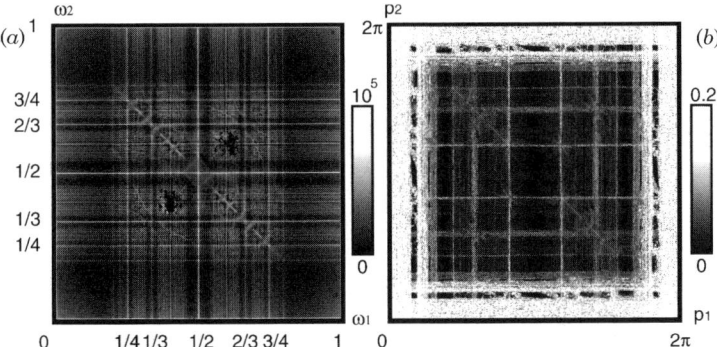

Figure 4. (a) A structure of resonances in the two-dimensional frequency space. The histogram of local rotation numbers, computed by the number of times visited at each local rotation number with 512×512 bins for 10^{10} iterations with $T = 10^3$. The Arnold web is clearly shown. Points are concentrated with higher density on resonances $m_i \omega_i + M = 0$, which construct structures like a web. The coupling resonance $\omega_1 + \omega_2 = 0$ due to the coupling term $b \sin(q_1 + q_2)$ which induces asymmetry. (b) A two-dimensional surface of section of phase space. The section is taken at $q_1 = q_2 = 0$. Maximal Lyapunov exponents averaged over only 256 steps are computed for each initial point in the phase space with increments of $2\pi/512$ (i.e., 512×512 points). Unstable regions correspond to resonance layers. $K = 0.9$ and $b = 0.002$.

describe the structures. Then, the distribution of the rotation number is computed as the histogram of the local rotation numbers. By taking bins of some size over the frequency space [0,1), every visit at each bin is counted, to compute probability density of the local rotation numbers.

The density in the frequency space obtained from a single trajectory described by the gray scale is shown in Fig. 4a with parameters $K = 0.9$ and $b = 0.002$. Higher densities are concentrated on rational numbers, which represent resonance lines.

Vertical and horizontal resonance lines $m_i \omega_i + M = 0$ mean that each element is resonant with external force. Among them, the lowest-order resonances are $(\omega_i + 1 = 0)$'s or $(\omega_i = 1)$'s, and resonance is higher order with larger m_i's. Lines $\sum_i m_i \omega_i + M = 0$ mean that two elements are resonant with each other, thus indicating coupling resonances. In Fig. 4a, the coupling resonance $\omega_1 + \omega_2 = 0$ is clearly visible, while the other lowest-order coupling resonance $\omega_1 - \omega_2 = 0$ is obscure. This is due to the coupling form of Froeschlé map.

A two-dimensional surface of section of phase space with the same parameters as Fig. 4a is shown in Fig. 4b. We measure maximal Lyapunov exponents averaged over a finite time interval and plot each of them to the initial point set on the section of phase space ($q_1 = q_2 = 0$). The brighter region represents the more unstable region, and two vertical lines in the center of Figure 4b

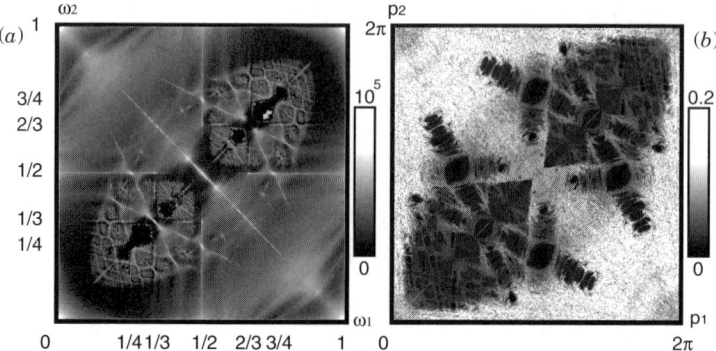

Figure 5. Structure of resonances plotted in the same way as Fig. 4 using parameter values $K = 0.5$ and $b = 0.100$. (a) The frequency space. Coupling resonance $\omega_1 + \omega_2 = 0$ shows the remarkable feature. Regions with higher densities spread wide across this resonance and have a fuzzy structure. (b) A surface of section of phase space. The resonant region around $\omega_1 + \omega_2 = 0$ is overlapped with other resonant region constructing large area. That area corresponds to a region with a fuzzy structure shown in (a).

correspond to $\omega_1 = 1/2$ shown in Fig. 4a. Comparing each part of Fig. 4, we can see that the distinct lines observed in the frequency space represent isolated (non-overlapped) resonances.

The density structure in the frequency space with weaker nonlinearity and stronger coupling is shown in Fig. 5a with parameters $K = 0.5$ and $b = 0.100$. By virtue of strong coupling, coupling resonances of various order are clearly visible. In particular, the lowest-order coupling resonance $\omega_1 + \omega_2 = 0$ shows a remarkable feature. Regions with higher densities spread wide across this resonance and have fuzzy structures. Overlapped resonance layers involving $\omega_1 + \omega_2 = 0$ are observed on the corresponding phase space section shown in Fig. 5b. Thus, fuzzy structures in the frequency space represent overlapped resonances.

As thus far described, in the parameter region we are interested in, coupling resonances have to be seriously taken into account. Then, before we go forward to detailed investigation, we check how global structures are affected by a form of coupling term. To clarify this, we employ a kicked and asymmetric coupled rotators described by

$$H = \sum_{i=1}^{2} \frac{p_i^2}{2} + \left\{ \sum_{i=1}^{2} K \cos(q_i) + b \cos(2q_1 + q_2) \right\} \sum_{n=-\infty}^{\infty} \delta(t - n) \quad (9)$$

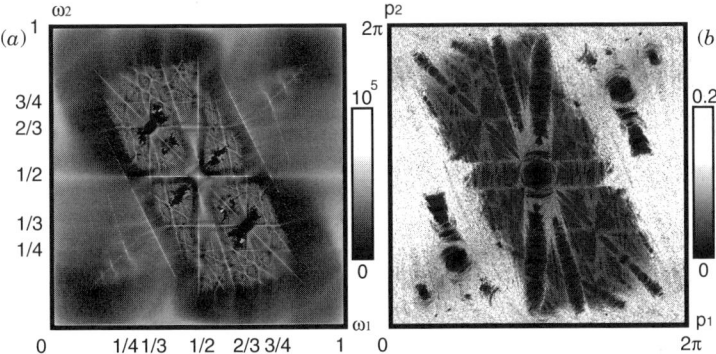

Figure 6. Structures of resonances of the model with coupling term $\sin(2q_1 + q_2)$ in the mapping form. Drawn in the same way as Fig. 4. The coupling resonance $2\omega_1 + \omega_2 = 0$ is enhanced. Overlapped resonant regions are involved as shown in the phase space section (b), and they construct a fuzzy region as shown in the frequency space (a). $K = 0.5$ and $b = 0.40$.

which gives

$$\begin{aligned} p_1(n+1) &= p_1(n) + K\sin(q_1(n)) + 2b\sin(2q_1(n) + q_2(n)) \\ p_2(n+1) &= p_2(n) + K\sin(q_2(n)) + b\sin(2q_1(n) + q_2(n)) \\ q_1(n+1) &= q_1(n) + p_1(n+1) \quad (\text{mod } 2\pi) \\ q_2(n+1) &= q_2(n) + p_2(n+1) \quad (\text{mod } 2\pi) \end{aligned} \quad (10)$$

in a mapping form.

The structures of resonances in the frequency space and on the surface of section of phase space are shown in Fig. 6 with parameters $K = 0.5$ and $b = 0.040$. In this model, the coupling resonance $2\omega_1 + \omega_2 = 0$ directly obtained from the coupling term $\sin(2q_1 + q_2)$, gets involved in resonance overlaps.

B. Morphological Change Depending on Parameters

So far, isolated resonances and overlapped resonances have been shown in the frequency space. Thus, we will investigate how structures of resonance of the Froeschlé map change, depending on parameters in the frequency space.

The orbit does not necessarily diffuse over the whole area of the frequency space within practical computational time, because of slow diffusion under weak nonlinearity and coupling strength. Instead of waiting for a long time steps, we set the initial state in the product of primary stochastic region of each standard maps, and we fix totally iterated time steps.

The structural changes of resonances depending on nonlinearity of standard maps are shown in Fig. 7. By increasing the nonlinear parameter K, we reduce

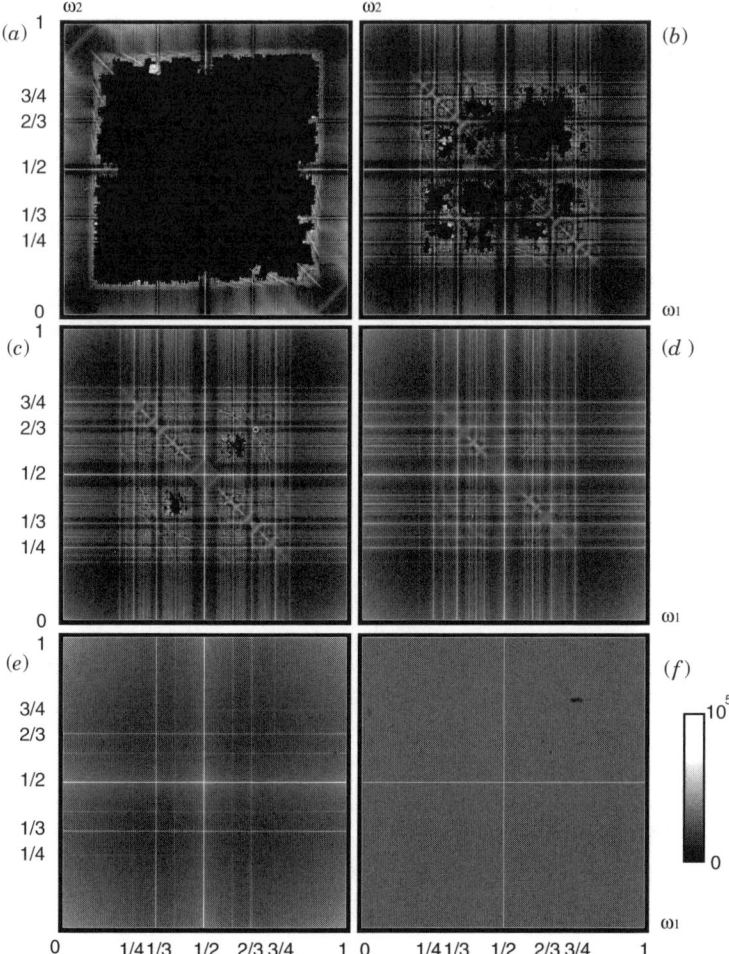

Figure 7. Structural changes of resonances depending on K. $K = 0.5$ (a), 0.8 (b), 0.9 (c), 1.0 (d), 1.2 (e), 2.0 (f). When the parameter K is increased, the size of the unreachable region in finite time steps becomes reduced, while coupling resonance is getting smeared. Higher-order resonances are overlapped more easily. $b = 0.002$ and 1×10^{10} steps.

the size of the region that is not reached from primary resonances. It is observed that higher-order resonances are overlapped more easily.

Next, we investigate the structural changes of resonances depending on coupling strength shown in Fig. 8. Here, when coupling strength is increased, the lowest-order coupling resonance $\omega_1 + \omega_2 = 0$ grows and resonance overlapping involving it is developed, until finally the overlapping region spreads globally.

STRUCTURE OF RESONANCES AND TRANSPORT 447

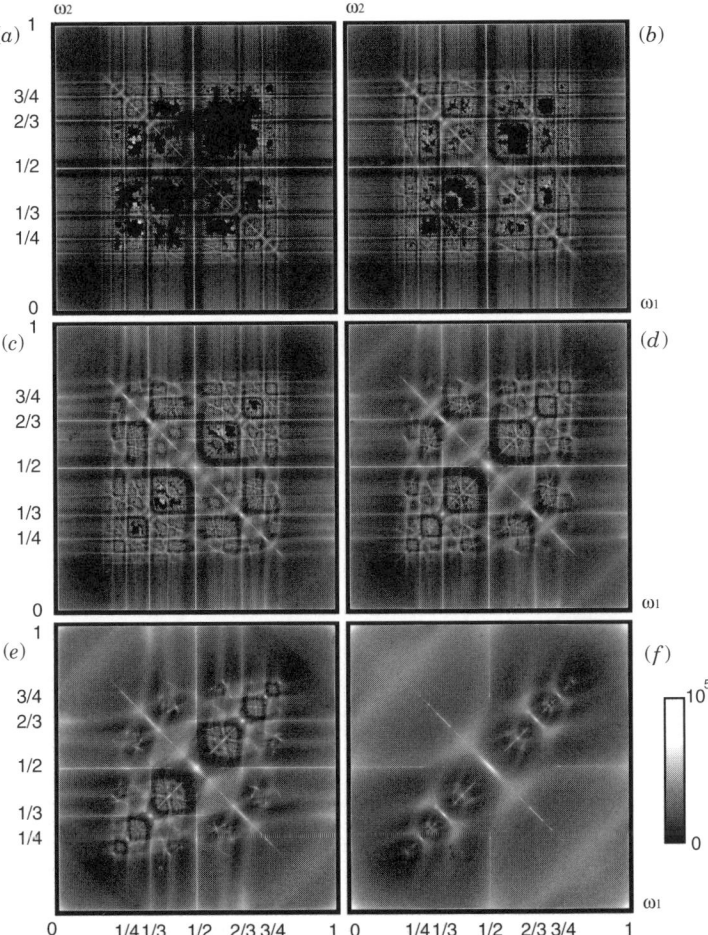

Figure 8. Structural changes of resonances depending on coupling strength b. $b = 0.002$ (a), 0.004 (b), 0.010 (c), 0.020 (d), 0.040 (e), 0.100 (f). When the parameter b is increased, the lowest-order coupling resonance $\omega_1 + \omega_2 = 0$ grows and involves resonance overlapping, up to forming the globally overlapping region. $K = 0.8$ and 1×10^{10} steps.

C. Characterization with Residence Time Distribution

With the representations of these resonance layers in mind, we clarify quantitative differences between isolated resonances and overlapped resonances, by examining residence time distributions $\rho(t)$ at each resonance layer. Since there are fluctuations in local rotation numbers due to the finite time average, we set some threshold W for each resonance condition, so that we compute the

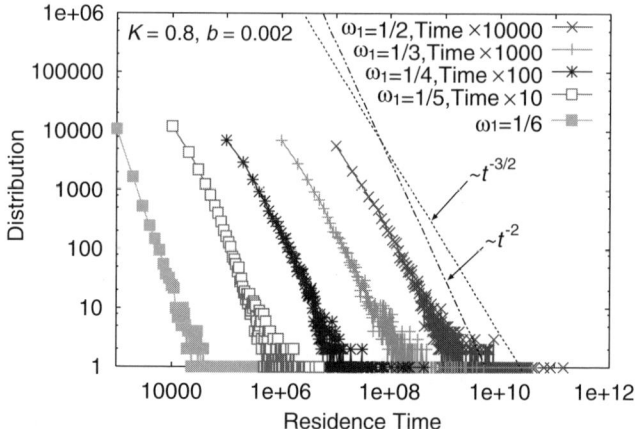

Figure 9. Dependence of residence time distributions on the order of resonances ($\omega_1 = 1/2$, $1/3, 1/4, 1/5, 1/6$). Distributions decay with a power law ($\rho(t) \sim t^{-\alpha}$). The exponent is $\alpha = 3/2$ for lower-order resonances ($\omega_1 = 1/2$), while $\alpha = 2$ for higher-order resonances ($\omega_1 = 1/6$). $K = 0.8$, $b = 0.002$, and $T = 10^{10}$ steps. The data are shifted by some values along the horizontal axis for the purpose of better viewing.

residence time during which $|m_1\omega_1 + m_2\omega_2 + M| < W$ is satisfied. The threshold W is chosen to be around 0.0015, while a moderate change of its value yields almost the same distribution.

First, we investigate the dependence of residence time distributions on the order of resonances, as shown in Fig. 9. For all the resonances with enough residences of orbits to get sufficient statistics, the distribution decays with a power law ($\rho(t) \sim t^{-\alpha}$). The exponent α takes a larger value for higher-order resonances, changing from 3/2 for $\omega_1 = 1/2$ to 2 for $\omega_1 = 1/5$, at least up to our computational time.

Second, we investigate the dependence of residence time distributions on nonlinearity of elements, as shown in Fig. 10. We get power-law decay of the distribution again. The exponent α is $3/2$ for weak nonlinearity, and it approaches 2 for stronger nonlinearity.

Next, we investigate the dependence of residence time distributions on the coupling strength, to get power-law dependence, as shown in Fig. 11. The exponent α is 3/2 for weak coupling, and approaches 2 with increasing coupling strength. Here, we choose $\omega_1 = 1/2$, $K = 0.8$, and $T = 10^{10}$ steps.

These changes of the exponent correspond to the transformation of structures in the frequency space from thin linear layer to fuzzy two-dimensional region. The distribution at the former has the exponent 3/2, while the latter has the exponent 2. Higher-order resonances that are easily overlapped have larger exponents. In addition, the fact that the fraction of resonances with the exponent

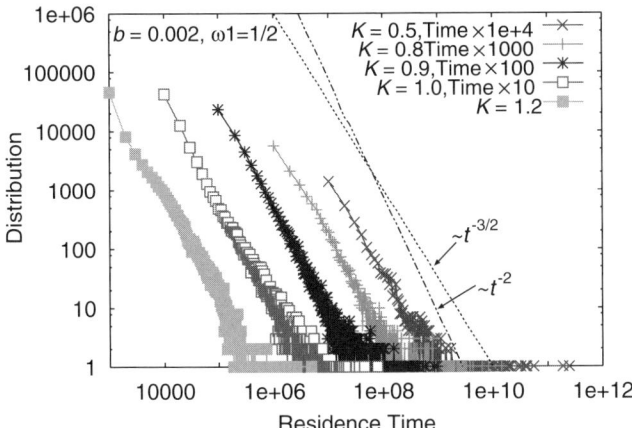

Figure 10. Dependence of residence time distributions on the strength of nonlinearity of each element ($K = 0.5$, 0.8, 0.9, 1.0, 1.2). Distributions decay with a power law ($\rho(t) \sim t^{-\alpha}$). The exponent is $\alpha = 3/2$ with weak nonlinearity, and it approaches $\alpha = 2$ with stronger nonlinearity. $\omega_1 = 1/2$, $b = 0.002$, and $T = 10^{10}$ steps. The data are shifted by some values along the horizontal axis for the purpose of better viewing.

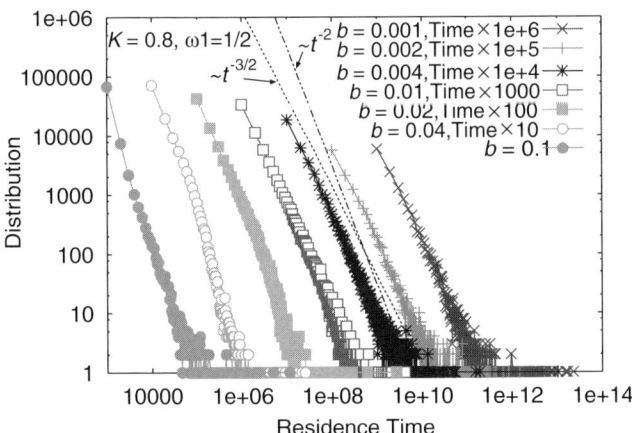

Figure 11. Dependence of residence time distributions on the coupling strength ($b = 0.001$, 0.002, 0.004, 0.010, 0.020, 0.040, 0.100). Distributions decay with a power law ($\rho(t) \sim t^{-\alpha}$). The exponent is $\alpha = 3/2$ with weak coupling, and it approaches $\alpha = 2$ with increasing coupling strength. $\omega_1 = 1/2$, $K = 0.8$, and $T = 10^{10}$ steps. The data are shifted by some values along horizontal axis for the purpose of better viewing.

3/2 increases with the increase of the nonlinearity K or the coupling strength b, and is replaced by those with the exponent 2, is consistent with the observation of overlapping of resonances shown in Figs. 7 and 8.

This power-law distribution is understood by regarding the motion at the resonance layer as Brownian motion. Lifetime of Brownian motion in a finite interval decays with a power law with the exponent $\alpha = 3/2$ in a one-dimensional case and $\alpha = 2$ in a two-dimensional case [22].

The motion along the one-dimensional resonance line called Arnold diffusion is prominent at lower-order resonances when nonlinearity is weak. In fact, the motion with the residence time distribution of the power 3/2 is observed for low-order resonance with weak nonlinearity. On the other hand, overlapped resonances allow the motion across resonances which leads to Brownian motion at a two-dimensional region. Indeed, the distribution with the power 2 is observed at higher-order resonances, and it is more frequently observed with stronger nonlinearity. Hence, one can distinguish clearly the Arnold diffusion from the motion induced by resonance overlaps by the power of the residence time distribution at each resonance condition.

IV. TRANSPORT

Now we discuss the global transport process in the phase space, which consists of the motions across and along the resonance layers, based on the observed geometric structures in the frequency space. For this purpose we compute the transition diagram over resonance lines measured in the frequency space by averaging the momentum over a finite time. Here, we compute the transition matrix among the lower-order resonances $\omega_i = 1/m_i$ and $\omega_i = 1 - (1/m_i)$, where $1 \leq m_i \leq 6$, by coarse-graining the value of ω_i, so that each region contains one junction of the lower-order resonances. The interval of each region and the lower-order resonance in it are summed up in Table I.

TABLE I
Coarse-Graining of the Frequency Space

Interval of Region	Resonance in Region
[0,2/13), [11/13,1)	0 (=1)
[2/13,2/11)	1/6
[2/11,2/9)	1/5
[2/9,2/7)	1/4
[2/7,2/5)	1/3
[2/5,3/5)	1/2
[3/5,5/7)	2/3
[5/7,7/9)	3/4
[7/9,9/11)	4/5
[9/11,11/13)	5/6

As examples, we intensively present the results for the following parameters: nonlinearity $K = 0.5$ and coupling strength $b = 0.040$ and $b = 0.100$, since these two cases, magnitudes of the resonance overlaps involving the lowest-order coupling resonance, are typically different as shown in Figs. 12a,b and

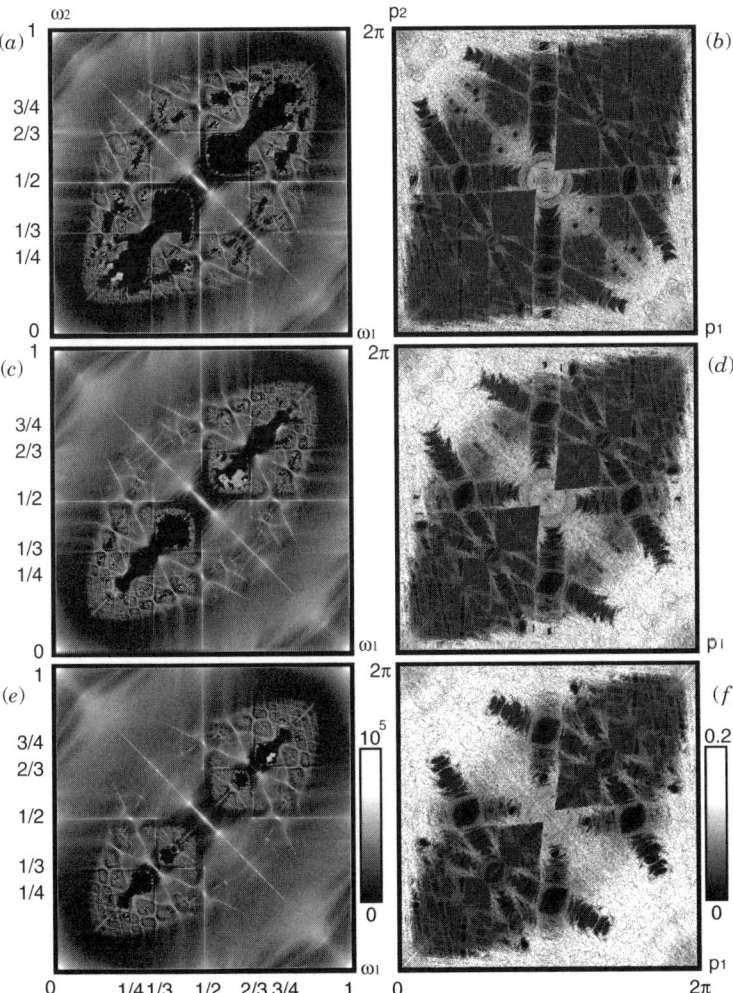

Figure 12. Structures of resonances shown in the frequency space (*left*) and on phase space sections (*right*). Drawn in the same way as Fig. 4. $K = 0.5$, and $b = 0.040$ (a, b), 0.070 (c, d), 0.100 (e, f). With weaker coupling (a, b), overlapped regions around $\omega_i = 0$ and $\omega_i = 1/2$ look isolated, while they are fused with stronger coupling (e, f).

12e,f, although there is no clear transition between these two parameters such as the breakup of the last KAM torus, which occurs in the standard map. Resonance overlaps progress gradually as shown in Fig. 12c,d. Gradual change of the exponent α of the residence time distribution shown in the previous section also indicates this fact.

A. Transition Diagram

At first, we plot transition diagrams for all of the 10^7 transitions that occur during 10^{10} iterations of mapping. They are shown in Figs. 13 and 14 for $b = 0.040$ and $b = 0.100$, respectively.

In the region where resonances are isolated, represented as linear structures in Fig. 12, transition can occur only to the nearest-neighbor regions, while in the overlapped regions shown as fuzzy region in the same figure, long-range transitions occur more freely.

By detailed observation, one can recognize that there are particular directions to transit in the overlapped regions. Transitions across resonances involved in overlaps are enhanced, typically shown as transitions across the lowest-order coupling resonance $\omega_1 + \omega_2 = 0$. When the coupling strength is increased from $b = 0.040$ to $b = 0.100$, the transitions across coupling resonances become increasingly prominent.

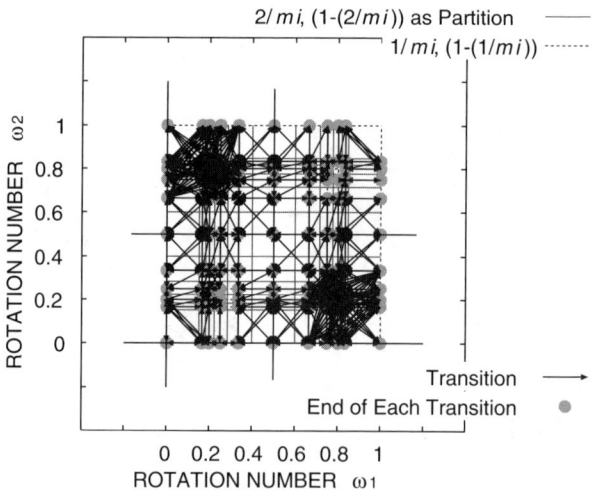

Figure 13. Transition diagram in the frequency space. $K = 0.5$ and $b = 0.040$. 10^7 transitions. In overlapped regions, long-range transitions across resonances are observed. In the region where resonances are isolated, transition can occur only to the nearest-neighbor regions.

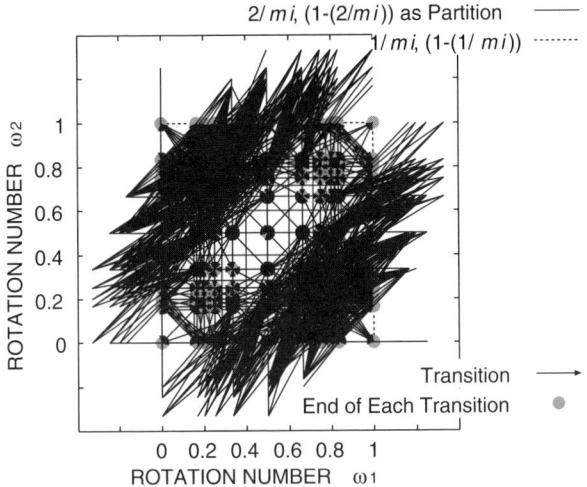

Figure 14. Same as Fig. 13 except parameters $K = 0.5$ and $b = 0.100$. Compared with Fig. 13, longer-range transitions are observed, derived from progress of overlapping.

Next, we plot transition diagrams separately by transition probabilities, for more detailed investigation, as shown in Figs. 15 and 16 for $b = 0.040$ and $b = 0.100$, respectively. We put circles on each end of transitions in addition to directional arrows, so that the region that contains only circles without an arrow means there is no transition from there, even to the nearest-neighbor regions.

In the regions where resonances are isolated, the states remain the same regions with high probabilities; thus, these regions act as the effective barriers for transitions. Note that motions across the overlapped resonances occur with high probabilities. Here, the fact that diffusion across resonances are faster than Arnold diffusion along the resonances, pointed out by Laskar [13], is clearly demonstrated as a global feature of Hamiltonian dynamical systems.

Moreover, regions with resonance overlaps involving coupling resonances inevitably allow for transitions to a variety of directions. In other words, these resonance overlaps play a role as highly connected nodes—that is, "hubs" [23] for the global transport. Thus by visiting the hub parts, the global transport is achieved quickly.

When coupling strength is increased from $b = 0.040$ to $b = 0.100$, the overlapped regions acting as hubs are fused and form larger hubs. Thus, longer-range transitions from $\omega_i = 0$ or 1 to $\omega_i = 1/2$ become able to occur with higher probabilities. While isolated resonances still remain in some regions of phase space, this mechanism of merging hubs, is typical for changing coupling strength b, where the diffusion coefficients increase with b as shown in Fig. 3.

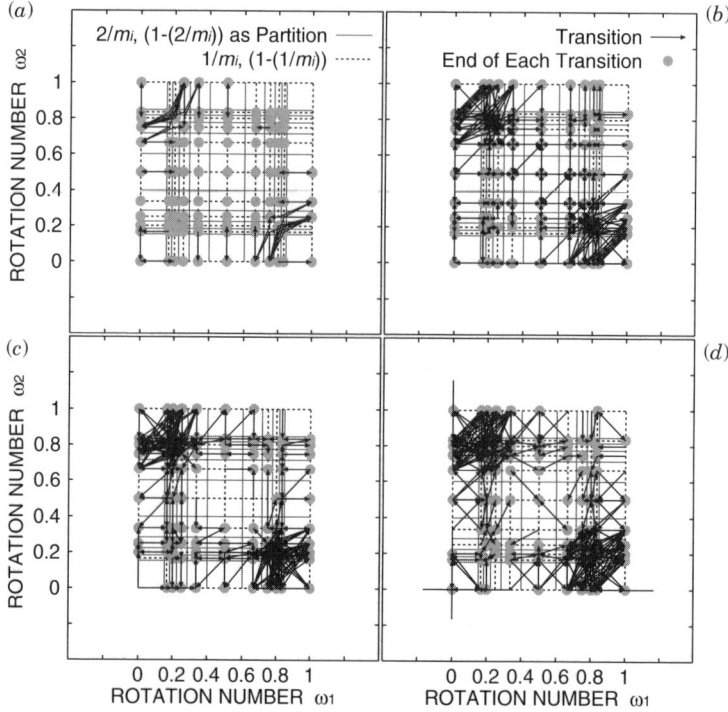

Figure 15. Transition diagrams drawn for each transition probability P. $P \geq 10^{-1}$ (*a*), $10^{-2} \leq P < 10^{-1}$ (*b*), $10^{-3} \leq P < 10^{-2}$ (*c*), and $10^{-4} \leq P < 10^{-3}$ (*d*). Through overlapped regions acting as hubs, long-range transition across resonances are observed with high probabilities. In the region where resonances are isolated, transition does not occur with a high probability. $K = 0.5$, $b = 0.040$.

B. Pathway for Fast Transition

Then, we investigate the existence of a specific pathway for fast transitions through hub regions successively.

As an example, we computed the transition diagram in the course of orbits starting from the region around the golden mean torus $\omega_i \approx 1 - (\sqrt{5} - 1)/2$ and reaching that around the torus with $(\sqrt{5} - 1)/2$. In other words, we study how a transition occurs from a point near one KAM torus to another distant KAM torus, through successive transitions over resonance layers. We have computed the transition diagram from the start state to the goal state, over 64 randomly chosen samples in the the frequency space. The transition diagram depends on each sample, and that with the shortest steps to arrive at the goal state is shown in Figs. 17 and 18 corresponding to Figs. 12a,b and 12e,f,

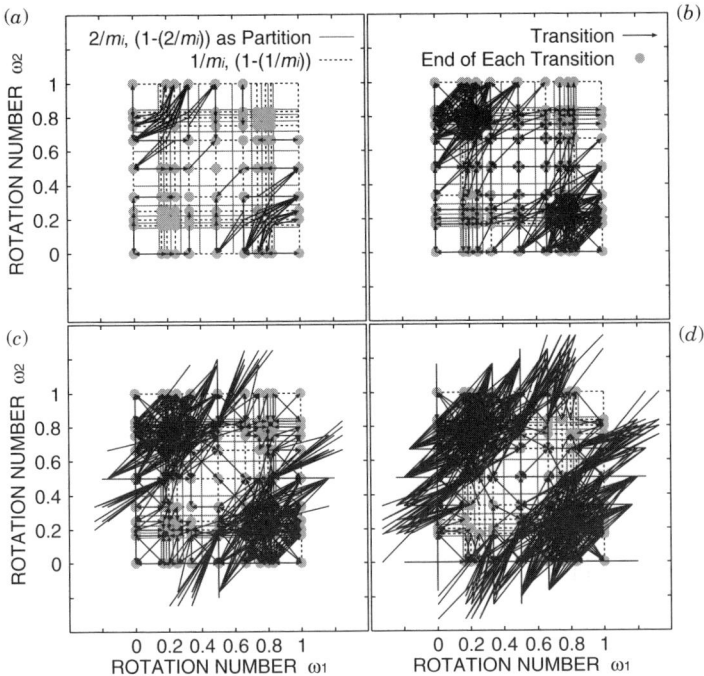

Figure 16. Same as Fig. 15 except parameters $K = 0.5$, and $b = 0.100$. Compared with Fig. 16, overlapped regions are fused and form large hubs, and longer-range transitions occur through overlapped region with higher probabilities.

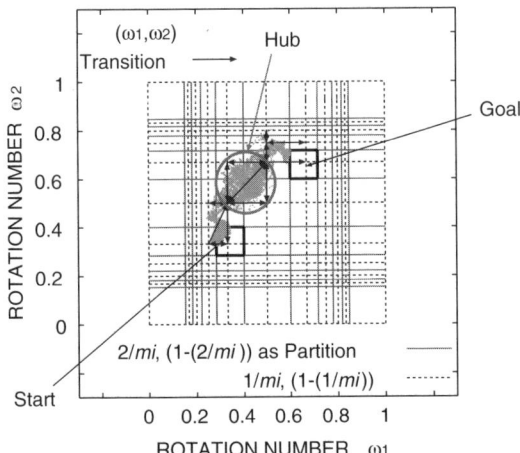

Figure 17. Transition diagram in the frequency space between fixed start and goal. The sample to finish first among randomly chosen 64 samples. It takes $18,702 \times 10^3$ steps to finish. The pathway is through the region that satisfies $1/3 \leq \omega_1 < 1/2$ and $1/2 \leq \omega_2 < 2/3$ shown as the small hub in Fig. 15. $K = 0.5$, $b = 0.040$.

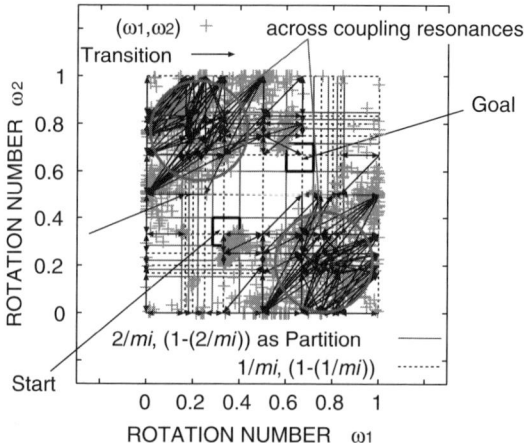

Figure 18. The same plot as in Fig. 17 except for parameters $K = 0.5$ and $b = 0.100$. The sample uses 1416×10^3 steps to reach the goal. The pathway is through the large hub formed by overlapping of the coupling resonance and lower-order resonances.

respectively. Diagrams of other samples with short-time steps for the destination have a feature similar to that of Figs. 17 and 18. It contains transitions through the overlapped resonances across the coupling resonances.

Although resonance overlaps are not so dominant in the phase space, the observed transition diagrams always use these resonance overlaps.

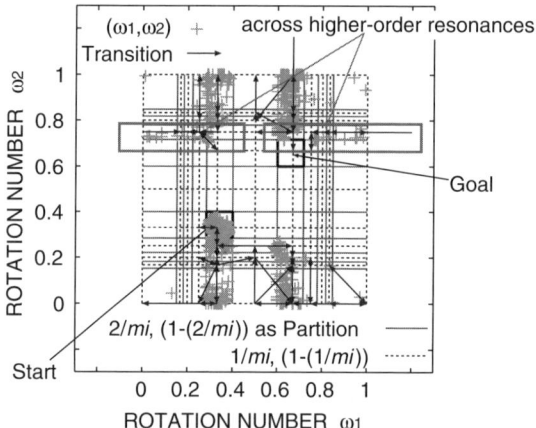

Figure 19. The same plot as in Fig. 17 except for parameters $K = 0.9$ and $b = 0.002$. Samples to finish first use 2187×10^3 steps. The pathway is not on lower-order resonances but through overlapped higher-order resonances (e.g., between $\omega_2 = 5/7$ and $\omega_2 = 3/4$).

As the coupling strength is increased, the number of overlapped resonances that act as such hubs increases, and they become widespread and overlapped each other. Thus, the global transport is faster.

It is, then, natural to ask whether transport actually occurs along resonances with weak coupling in the case that lower-order resonances are not overlapped and no hub region is seen clearly, as shown in Fig. 4. From the diagram shown in Fig. 19, it is not easy to answer the question only by directional arrows of transitions. Focusing on local rotation numbers themselves, however, we conclude that transport occurs between the lower-order resonances, through the overlapped higher-order resonances that exist between the lower-order resonances. Hence, the dominant component of the fast transport is not due to the Arnold diffusion.

V. RESONANCE OVERLAP IN MULTIDIMENSIONAL SYSTEMS

The resonance overlap itself occurs in two-dimensional maps such as the standard map. Then, what is the distinctive characteristic in multidimensional systems? To clarify this point, we extend the resonance overlap criterion [5] to the Froeschlé map we investigate here, and we relate that to the transition diagrams represented in the previous section. In the study of the resonance overlap criterion, it became clear that global diffusion is achieved by overlapping of lower-order resonance layers. Thus, we calculate the width of resonance layers in the increasing order.

To estimate the maximal width of resonance layers of each element, we fix variables in counterpart. Maximal excursion of p_1 is realized at the moment $q_2 = 0$, where the Hamiltonian 2 takes the form

$$H = \frac{p_1^2}{2} + \{(K+b)\cos(q_1)\} \sum_{n=-\infty}^{\infty} \delta(t-n) \quad (11)$$

Then, we can treat this system as the standard map with the nonlinear parameter $K + b$, and it is easy to carry out similar calculations as in Refs. 5 and 8. The maximal half-width of the primary resonances $\omega_i = 0$ or 1, including the separatrix width

$$w = \frac{4(2\pi)^4}{(K+b)^{5/2}} \exp\left(\frac{-\pi^2}{(K+b)^{1/2}}\right) \quad (12)$$

is

$$(\Delta p_i)_1 = \left(1 + \frac{w}{4}\right) 2\sqrt{K+b} \quad (13)$$

Also, the half-width of the resonance layer around $\omega_i = 1/2$ is

$$(\Delta p_i)_2 = \frac{K+b}{2} \tag{14}$$

and that of $\omega_i = 1/3$ is

$$(\Delta p_i)_3 \simeq 9.30 \frac{(K+b)^{3/2}}{(2\pi)^2} \tag{15}$$

To calculate the width of the coupling resonance $\omega_1 + \omega_2 = 0$, we transform the variables by

$$P_1 \equiv \frac{p_1 + p_2}{\sqrt{2}}, \quad P_2 \equiv \frac{-p_1 + p_2}{\sqrt{2}}$$
$$Q_1 \equiv \frac{q_1 + q_2}{\sqrt{2}}, \quad Q_2 \equiv \frac{-q_1 + q_2}{\sqrt{2}} \tag{16}$$

where P_1 is across $\omega_1 + \omega_2 = 0$ and P_2 is along $\omega_1 + \omega_2 = 0$. The Hamiltonian 2 is written as

$$H = \sum_{i=1}^{2} \frac{P_i^2}{2} + \left\{ 2K \cos\left(\frac{Q_1}{\sqrt{2}}\right) \cos\left(\frac{Q_2}{\sqrt{2}}\right) + b \cos(\sqrt{2} Q_1) \right\} \sum_{n=-\infty}^{\infty} \delta(t-n) \tag{17}$$

To employ only the lowest term $b \cos(\sqrt{2} Q_1)$, and with a little algebra, the half-width of the coupling resonance including separatrix width is computed, to get

$$w_c = \frac{4(2\pi)^4}{b^{5/2}} \exp\left(\frac{-\pi^2}{b^{1/2}}\right) \tag{18}$$

and

$$\Delta p_c = \left(1 + \frac{w_c}{4}\right) 2\sqrt{b} \tag{19}$$

Maximal width of these lower-order resonance layers noted here are shown in Figs. 20 and 21, for $b = 0.040$ and $b = 0.100$, respectively. By virtue of the coupling resonances, every resonance layer obviously connects with each other. However, the orbit does not diffuse freely over the whole connected regions, as shown in the previous section. Thus, only overlapping of resonance layers cannot realize a global diffusion in practical time scales, in contrast to the case of two-dimensional mapping.

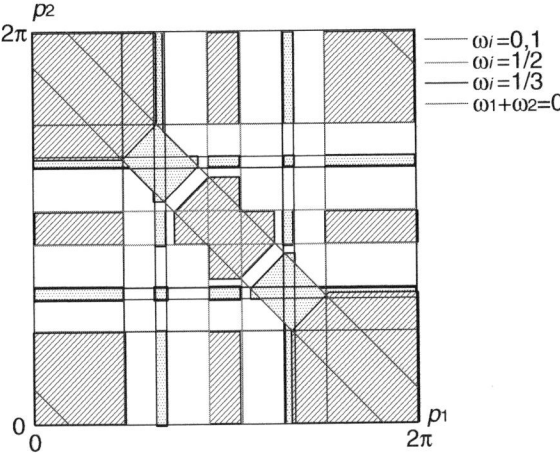

Figure 20. Maximal width of each resonances projected on a two-dimensional momentum space. Hatched regions and dotted regions are accessible only by diffusion across resonances. For hatched regions, resonances $\omega_i = 0, 1, 1/2$ and $\omega_1 + \omega_2 = 0$ are taken into account for these analytic calculations; for dotted regions, resonances $\omega_i = 1/3$ are taken into account. $K = 0.5$ and $b = 0.040$.

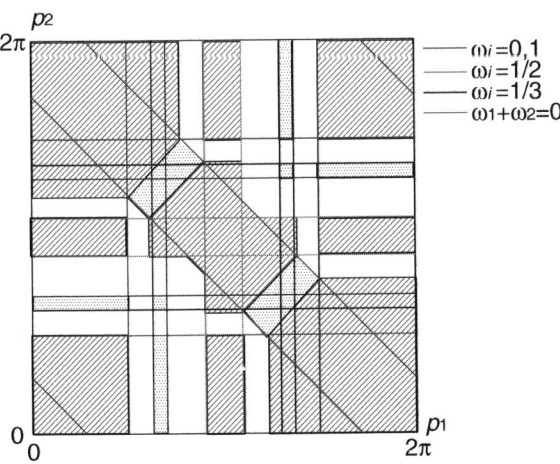

Figure 21. The same drawing as in Fig. 20 except parameters $K = 0.5$ and $b = 0.100$. Transitions from $\omega_i = 0$ or 1 to $\omega_i = 1/2$ are permitted, in contrast to Fig. 20.

By the study of transition diagrams, one can recognize that the motion across resonances is much faster than the motion along resonances. Then, we take account of the difference between diffusion across and along resonances, and we focus on the regions that are accessible by the motion across resonances alone. For the hatched regions shown in Figs. 20 and 21, we employ the coupling resonance $\omega_1 + \omega_2 = 0$ and resonances of each element up to $\omega_i = 1/2$, while for the dotted regions we employ resonances up to $\omega_i = 1/3$. The filled-out regions describe well the transition diagrams shown in the previous section, and it is clear that crossing between a lower-order resonance and a lower-order coupling resonance forms the hub region.

Comparing Fig. 21 with Fig. 20, transitions from $\omega_i = 0$ or 1 to $\omega_i = 1/2$ become possible with increasing coupling strength. If we employ higher-order resonances, it is expected that this transition would be possible artificially for any value of coupling strength, but the study of transition diagrams with different coupling strength values shows that the character of transport is altered, so that the order of resonance must be taken into account.

In multidimensional systems, only overlapping of resonances does not set a good criterion for global transport, where the directions of resonances (which is a missing notion in two-dimensional systems) have a profound effect. Therefore, it is necessary to grasp how lower-order resonances cross each other.

VI. DEFLECTED DIFFUSION

So far we have shown that the transport across the overlapped resonances, is faster than that along the resonances, and existence of hubs enhances transports. However, diffusion is regulated by slow motion, so that we have to check whether the diffusion coefficient is affected by hubs effectively or not.

For this purpose, we compute the diffusion coefficient D_{P_1} and D_{P_2} for P_1 across the lowest-order coupling resonance, and P_2 along it defined by Eq. (16), respectively, and investigate their dependence on the coupling strength as shown in Fig. 22.

As shown, the diffusion coefficient of P_1 is much larger than the other, and the diffusion is anisotropic in this sense.

In the parameter region with large b, D_{P_2} along the coupling resonances is saturated. In terms of P_i's and Q_i's, the Froeschlé map is written as

$$P_1(n+1) = P_1(n) + \sqrt{2}K \sin\left(\frac{Q_1(n)}{\sqrt{2}}\right) \cos\left(\frac{Q_2(n)}{\sqrt{2}}\right) + \sqrt{2}b \sin\left(\sqrt{2}Q_1(n)\right)$$

$$Q_1(n+1) = Q_1(n) + P_1(n+1)$$

$$P_2(n+1) = P_2(n) + \sqrt{2}K \cos\left(\frac{Q_1(n)}{\sqrt{2}}\right) \sin\left(\frac{Q_2(n)}{\sqrt{2}}\right) \qquad (20)$$

$$Q_2(n+1) = Q_2(n) + P_2(n+1)$$

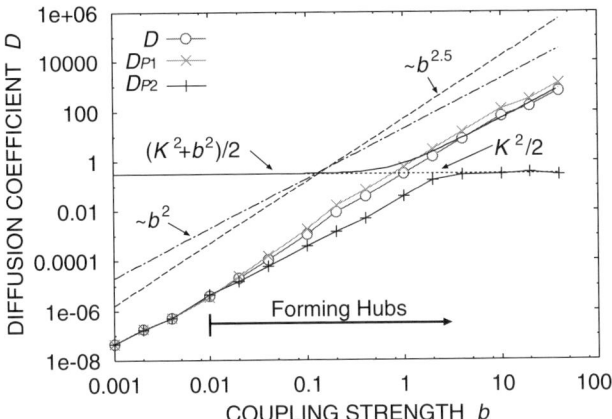

Figure 22. Diffusion coefficients depending on coupling strength b for $K = 0.8$. Diffusion coefficients along and across lowest-order coupling resonance $\omega_1 + \omega_2 = 0$ are also computed. Diffusion across the resonances is faster than diffusion along the resonances, and this difference actually affects the total diffusion coefficient. Random phase approximation gives $D = (K^2 + b^2)/2$ for the total diffusion coefficient and $D = K^2/2$ for that across coupling resonance. Computed from 160 samples with $T = 10^8$.

Thus, only the dynamics of P_1 includes the coupling term by b, while that of P_2 is triggered only by random phase as shown in Eq. (20), without a direct coupling term. Then, random phase approximation, which is appropriate with large b, gives

$$D = \frac{K^2 + b^2}{2}$$
$$D_{P_1} = \frac{K^2}{2} + b^2 \qquad (21)$$
$$D_{P_2} = \frac{K^2}{2}$$

while the data in Fig. 22 show saturation of D_{P_2}.

The difference between D_{P_1} and D_{P_2} is prominent for $b > 0.01$, where the resonance overlaps involving the coupling resonance is remarkable in Fig. 8, showing the dominance of the motion across the hub coupling resonance. This again demonstrates the relevance of the motion across the resonance layer. The usual diffusion coefficient D in Fig. 22 is fitted by the form $D \sim b^\beta$ with $\beta = 2.5$, showing a clear deviation from the $\beta = 2$ for D_2 along the coupling resonance. Indeed, the transport across the coupling resonance at hubs as in Fig. 18 is dominant here, and global diffusion is enhanced by the existence of a hub region, rather than regulated by slow motion.

It is expected that this power law $\beta = 2.5$ could reflect the increase of the fraction of such coupling resonances forming hubs with the coupling b. The exponent β here decreases with K. As K is increased from 0.5 to 0.9, β decreases from 3.5 to 2.2. This dependency suggests that the increase of hub coupling resonances is more relevant because the nonlinearity is weaker.

So far, several estimates of diffusion coefficients have been proposed, as given by the stochastic pump or three-resonance model and their extensions [12,17,18]. These studies, however, assume only the diffusion along the resonances, while our results, in contrast, exhibit the importance of resonance overlap to the global diffusion along resonances.

VII. SUMMARY AND DISCUSSION

In summary, intermingled structure of Arnold web and resonance overlaps are visualized by introducing the representation in the frequency space. It is shown that the motion across the resonances is faster than the motion along the overlapped resonances. By examining the transition over resonance layers, fast transport in the phase space is found to occur mainly through overlapped resonances involving coupling resonances, forming a hub in the transition diagram. There, the simple resonance overlap criterion does not work, because of the difference of the speeds between the motion along and across resonances, and it is necessary to take into account how resonances are crossing. Now it is important to understand heterogeneity induced by hub structures, to study global transport process.

Acknowledgment

This work was partially supported by a Grant-in-Aid for Scientific Research from the Ministry of Education, Science, and Culture of Japan.

References

1. R. E. Gillilan and G. S. Ezra, *J. Chem. Phys.* **94**, 2648–2668 (1991).
2. T. Komatsuzaki and R. S. Berry, *Adv. Chem. Phys.* **123**, 79–152 (2002).
3. M. Toda, *Adv. Chem. Phys.* **123**, 153–198 (2002).
4. V. I. Arnold, *Sov. Math. Dokl.* **5**, 581–585 (1964).
5. B. V. Chirikov, A universal instability of many-dimensional oscillator systems, *Phys. Rep.* **52**, 263–379 (1979).
6. F. Vivaldi, *Rev. Mod. Phys.* **56**, 737–754 (1984).
7. P. M. Cincotta, *New Astron. Rev.* **46**, 13–39 (2002).
8. A. J. Lichtenberg and M. A. Lieberman, *Regular and Chaotic Dynamics*, 2nd ed., Springer-Verlag, Berlin, 1992.

9. C. Froeschlé, *Astrophys. Space Sci.* **14**, 110–117 (1971).
10. K. Kaneko and R. J. Bagley, *Phys. Lett. A* **110**, 435–440 (1985).
11. H. T. Kook and J. D. Meiss, *Phys. Rev. A* **41**, 4143–4150 (1990).
12. B. P. Wood, A. J. Lichtenberg, and M. A. Lieberman, *Phys. Rev. A* **42**, 5885–5893 (1990).
13. J. Laskar, *Physica D* **67**, 257–281 (1993).
14. K. Kaneko and T. Konishi, *Phys. Rev. A* **40**, 6130–6133 (1989).
15. A. B. Rechester and R. B. White, *Phys. Rev. Lett.* **44**, 1586–1589 (1980).
16. A. B. Rechester, M. N. Rosenbluth, and R. B. White, *Phys. Rev. A* **23**, 2664–2672 (1981).
17. B. V. Chirikov and V. V. Vecheslavov, *J. Stat. Phys.* **71**, 243–258 (1993).
18. B. V. Chirikov and V. V. Vecheslavov, *JETP* **85**, 616–624 (1997).
19. C. C. Martens, M. J. Davis, and G. S. Ezra, *Chem. Phys. Lett.* **142**, 519–528 (1987).
20. J. Laskar, *Icarus* **88**, 266–291 (1990).
21. J. Laskar, C. Froeschlé, and A. Celletti, *Physica D* **56**, 253–269 (1992).
22. W. Feller, *An Introduction to Probability Theory and Its Applications*, John Wiley & Sons, New York, 1950.
23. H. Jeong, B. Tombor, R. Albert, Z. N. Oltvai, and A. L. Barabási, *Nature* **407**, 651–654 (2000).

CHAPTER 23

MULTIERGODICITY AND NONSTATIONARITY IN GENERIC HAMILTONIAN DYNAMICS

YOJI AIZAWA

Department of Applied Physics, Faculty of Science and Engineering, Waseda University, Tokyo, 169-8555, Japan

CONTENTS

I. Introduction
II. Multiergodicity and Large Deviation of Stagnant Motions
III. A Universality Conjecture
IV. Survival Time Distribution $P(T)$ in clustering Motions
V. Complexities in Nonstationarity
References

I. INTRODUCTION

Complexity of Hamiltonian dynamics has been studied from various angles, and the knowledge regarding the generic aspect of the complex behaviors has been extended conspicuously: geometrical structures of phase space, kinetic features of long trajectories, ergodic-theoretical characteristics of asymptotic measures, and so forth. However, the understanding of the complexity is still unsatisfactory in the case of generic Hamiltonian systems that are neither completely integrable nor strongly ergodic, because the phase space of generic systems is heterogeneous; hyperbolic and nonhyperbolic regions are intermingled with each other in a very complicated manner [1,2]. We can imagine that the phase space of generic systems consists of two different types of geometry; that is to say, one is for islands of tori and another for chaotic sea [3,4], which are not mutually transitive. In the system with many degrees of freedom, the topological instability leading to the Arnold diffusion makes the problem more difficult.

Geometric Structures of Phase Space in Multidimensional Chaos: A Special Volume of Advances in Chemical Physics, Part B, Volume 130, edited by M. Toda, T Komatsuzaki, T. Konishi, R.S. Berry, and S.A. Rice. Series editor Stuart A. Rice.
ISBN 0-471-71157-8 Copyright © 2005 John Wiley & Sons, Inc.

One of the most striking chaotic phenomena in generic Hamiltonian systems is the onset of nonstationary fluctuations with extremely long-time memories [5–8]. Usually, the nonstationarity is characterized by the infrared catastrophe; that is, the power spectral density function $S(f)$ satisfies $S(f) \sim f^{-\nu}(\nu \geq 1)$ for $f \to 0$.

In the present chapter we discuss the more detailed phenomena induced by the nonstationarity, and we elucidate a universal mechanism hidden behind the statistical laws of nonstationary fluctuations. In Section II some significant ideas, such as "multiergodicity" and "stagnant layer," are explained by using two-dimensional mappings, and the large deviation properties of the power spectral density function are discussed based on numerical simulations. In Section III, a universality conjecture for the stagnant motion is derived from the scaling theoretical approach to the Nekhoroshev theorem [9], and the numerical evidence for the conjecture is explained by using the discrete model of lattice vibrations with many degrees of freedom [6–8]. The process of cluster formation in N-body Hamiltonian systems is discussed in Section IV, where it is shown that the nonstationarity in clustering motions are described in terms of two kinetic laws for the survival time distribution: the Weibull law in the intermediate long-time regime and the log-Weibull law in the intrinsic long-time regime [10,11]. It is emphasized that the log-Weibull law is the same as the universal law discussed in Section III. Observability of nonstationary fluctuations in physical systems are discussed in the final section from the general viewpoint of chaotic dynamics.

II. MULTIERGODICITY AND LARGE DEVIATION OF STAGNANT MOTIONS

Slow motions in phase space are characterized by the local Lyapunov exponent λ_N, which defines the unfolding rate of nearby orbits,

$$\lambda_N = \frac{1}{N} \log \left| \frac{\delta_N}{\delta_0} \right| \qquad (1)$$

where δ_N is the Euclidean distance between two orbits for the time interval $[0, N)$. When the time interval N is finite, λ_N demonstrate the local structure of the restricted region in phase space.

Consider the ensemble of many sample paths with the length $N : [0, N], [N, 2N], \ldots$, and also consider the distribution of their local Lyapunov exponents $P(\lambda)$. Figure 1 shows an example of $P(\lambda)$ and restricted phase portrait of the map (X_n, X_{n+1}) in the case of standard maps, where the distribution $P(\lambda)$ reveals three peaks in general; when the time interval N becomes large enough, the middle peak is gradually abolished, though the first

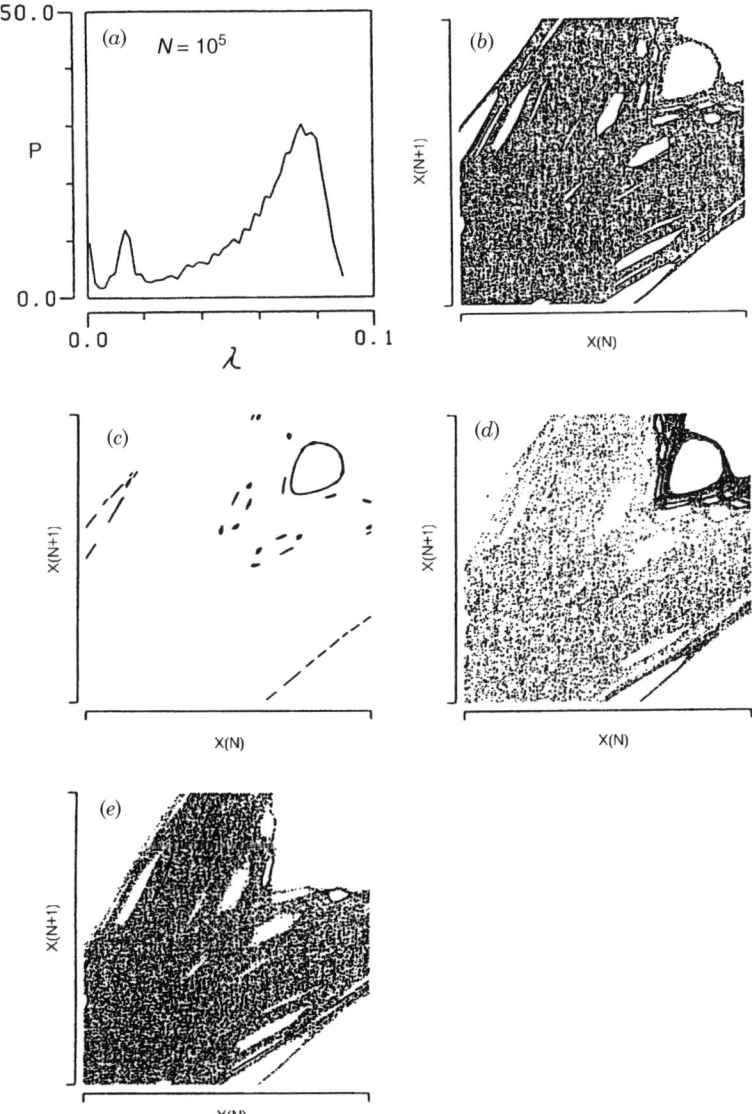

Figure 1. (a) Distribution function $P(\lambda)$. (b) Plots of whole paths. Phase portraits corresponding to three peaks in Fig. 1a are demonstrated in (c), (d), and (e), respectively.

peak around $\lambda \cong 0$ and the third peak (around $\lambda \cong 0.08$) remain stable. Figure 1b is the portrait of whole sample paths, where blank regions represent islands of tori. Figures 1c–1e are restricted phase portraits corresponding to the first, second, and third peaks in Fig. 1a, respectively.

It is especially important that slow motions in generic Hamiltonian dynamics are universally generated from the stagnant motions sticking to the outermost tori, which are demonstrated in Fig. 1c. As uncountable islands of tori are hierarchically distributed in phase space, it is surmised that the number of asymptotic measures corresponding to stagnant motions are also uncountable. Furthermore, the stagnant motion in each sticky region is considered to be uncoupled from other sticky regions in a practical sense, though all the sticky regions are mutually transitive in a topological sense, because it is shown that the mean survival time of the stagnant motion in each sticky region is surmised to be divergent not only from computer simulations but also from the analytical estimation given in the next section. These structures in generic Hamiltonian dynamics are called "multiergodicity" of stagnant motions. The multiergodic aspects can be also reflected in the power spectral density (PSD) function.

Let us define the PSD function $S(\omega : N)$ for the time series x_t with the length N:

$$S(\omega : N) = |\hat{x}(\omega)|^2 \qquad (2)$$

where $\hat{x}$ is the fourier transformation of x_t. The PSD is a random function of frequency $f(= \omega/2\pi)$ depending on the sample path. Figure 2 shows the PSD for

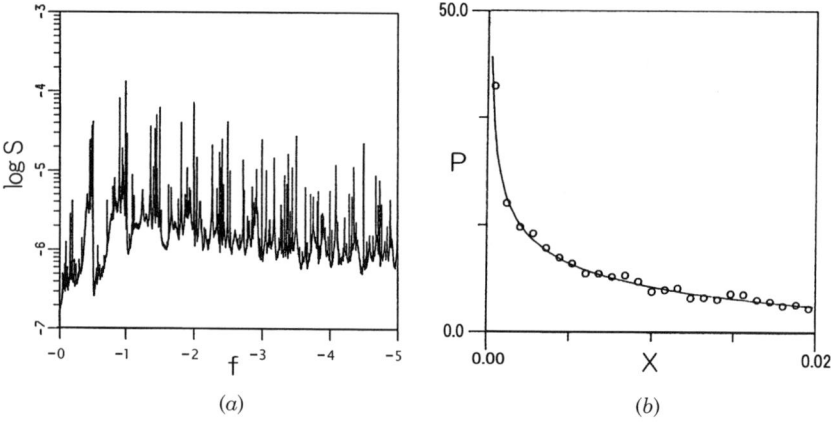

Figure 2. (*a*) Power spectral density. (*b*) Γ-distribution of the PSD function for the standard map at $f = 1/7; \eta = 0.62, \zeta = 0.71$.

the multiergodic case of the standard map, where we observe a lot of singular peaks at resonant frequencies corresponding to stagnant motions. It is known that each singularity reveals algebraic divergence, and the mean PSD is expressed as

$$\langle S(\omega : N) \rangle = \sum A_i |f - f_{Ci}|^{-\nu_i(\pm)} + \langle S(\omega; N) \rangle_C \qquad (3)$$

where f_{Ci}'s stand for resonant frequencies and indices $\nu_i(\pm)$ stand for $f > f_{Ci}$ or $f < f_{Ci}$, respectively, and the nonsingular smooth part $\langle S(\omega : N) \rangle_C$. The multiergodic aspects are characterized by the appearance of an infinite number of resonance peaks. On the other hand, in the simple ergodic case such as the cat map, there appear to be no singularities.

The multiergodic aspects of stagnant motions are also characterized by the large deviation of the PSD function [6]. Some numerical simulations show that the distribution function of PSD functions $P(X)$ ($X = S(\omega : N)$) is generally expressed by Γ-distributions (Fig. 2b),

$$P(X) \propto X^{\eta-1} \exp[-N^\xi X] \qquad (4)$$

In simple ergodic cases, the indices satisfy $\eta \cong \xi \cong 1$ but $\eta < 1$ and $\xi < 1$ in the multiergodic cases, and the index ξ is easily defined from the scaling form of the mean PSD function—that is, $\langle S(\omega : N) \rangle \cong O(N^{-\xi})$. The frequency dependence of the indices seems to characterize the convergence speed of the power spectrum of resonant modes, but the details are still open when we treat the continuous-flow systems with many degrees of freedom.

III. A UNIVERSALITY CONJECTURE

We can expect that the stagnant motions of which statistical properties are discussed in the previous section reflect a universal structure of the interface between torus and chaos. Here we discuss a conjecture concerning the universality in the stagnant layer based on the Nekhoroshev theorem, which proves the onset of a new time scale accompanied by the collapse of tori [9].

Theorem[Nekhoroshev]. Consider nearly integrable systems with n degrees of freedom:

$$H(p, q, \epsilon) = H_0(p) + \epsilon H_1(p, q) \qquad (5)$$

When a torus is destroyed by nonintegrable perturbations $\epsilon H_1(p, q)$, newborn orbits are trapped near the unperturbed torus for an extremely long period T, which characterizes the lifetime of the stagnant motion. The lower bound satisfies

$$T(\epsilon) = \inf(t; |p(t) - p_0(0)| > \epsilon^a)$$
$$\cong \frac{1}{\epsilon}\exp[\epsilon^{-b}] \qquad (6)$$

where $p_0(0)(=p(0))$ is the unperturbed torus. Indices a and b are positive constants determined only by the unperturbed Hamiltonian $H_0(p)$.

We apply the theorem to the very narrow region of the interface between the outermost torus and chaos where stagnant motions are generated. Define the stagnant layer coordinate $\vec{r}$,

$$\vec{r} \cong O(\epsilon^a) \qquad (7)$$

which represents the distance measured from the outermost torus. Lebesgue measure V of the stagnant layer satisfies

$$dV \sim d\vec{r} \cong O(\epsilon^{a'-1}) \qquad (a' = a(n-1)) \qquad (8)$$

Therefore, the distribution of the lifetime of stagnant motions, say $P(T)$, is derived by the following scaling estimation obtained from Eq. (2):

$$\log T \cong \epsilon^{-b}$$
$$d\epsilon/dT \cong \epsilon^{2+b} \exp[-\epsilon^{-b}] \qquad (\epsilon \to 0) \qquad (9)$$

that is,

$$P(T) \equiv \left|\frac{dV}{dT}\right| \cong \frac{1}{T}(\log T)^{-1-a'/b} \qquad (T \to \infty) \qquad (10)$$

The dominant scaling factor $1/T$ comes from the essential singularity in the Nehoroshev bound, and the logarithmic correction comes from the algebraic singularity. If we use a rough approximation, $P(T) \sim 1/T$, the invariant measure $P(\vec{r})$ and the distribution of Lyapunov exponent $P(\lambda)$ satisfy $P(\vec{r}) \cong |\vec{r}|^{-d}$ and $P(\lambda) \cong \lambda^{-\kappa}$ for stagnant motions, where indices d and κ are positive constants determined by a, b, and n. These nonhyperbolic behaviors are similar to the case for infinite measure systems [12]. The most striking point in the above discussion is that the distribution $P(T)$ is equivalent to the log-Weibull distribution, which is discussed in the latter part of this chapter.

Let us see two pieces of numerical evidence to the above-mentioned conjecture; one is the induction phenomena observed in lattice vibration [6–8],

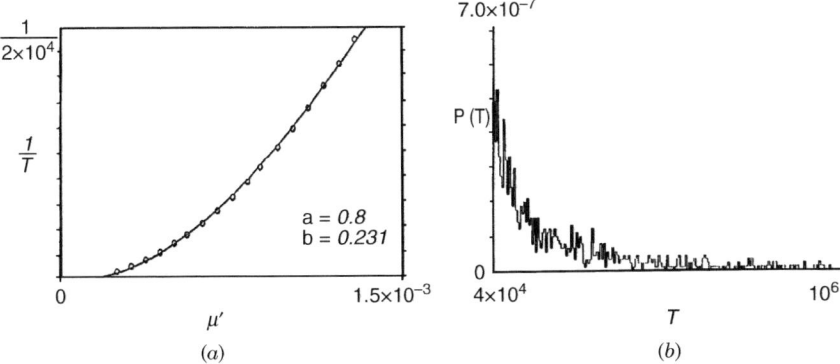

Figure 3. (a) The Nekhoroshev plot by $T \propto \frac{1}{\mu^a}\exp[\mu^{-b}]$, for the induction time of discrete lattice vibrations. (b) The distribution of the induction time T. $P(T)$ is approximately adjusted by $T^{-1.1}$.

which another is the cluster formation of N-body systems, which is discussed in the next section [10,11].

Induction Time Distribution $P(T)$ in Lattice Vibrations—FPU Problem

The mechanism of energy mixing in many-body systems was studied by Fermi, Pasta, and Ulam in 1953, by employing unharmonic lattice vibrations, where they found that the energy of an eigenmode that is excited at an initial time reveals quasi-periodic variation for an extremely long period, and the energy mixing or the energy equipartition does not actually occur in spite of nonlinear couplings among eigenmodes. In 1970, Saito et al. [7] elucidated that a drastic change happens in the mixing process after a long induction time when the nonlinear parameter ϵ exceeds a critical value. The induction time T and its distribution $P(T)$ were studied by using a model with discrete time evolution, and its was numerically proved that the induction time T obeys the Nekhoroshev theorem as shown in Fig. 3, where $\mu = \epsilon - \epsilon_c$ is the effective perturbation to the quasi-periodic motion that appeared during the induction period; that is, in the care of $\epsilon \leq \epsilon_c$ the induction time is divergent [6]. The distribution $P(T)$ is also well-adjusted by Eq. (10).

IV. SURVIVAL TIME DISTRIBUTION $P(T)$ IN CLUSTERING MOTIONS

Cluster formation is one of the universal phenomena in many-body systems, but the dynamical features of clustering particles have not yet been completely understood in terms of Hamiltonian mechanics. Here we study the formation process of liquid droplets in gaseous phase. Let us consider the N-particle system confined in a box, and assume the short-ranged attractive potentials among particles. When the temperature of the total energy is low enough, all the particles collect together to form a liquid cluster (droplet), but when the temperature increases, some particles begin to evaporate into gaseous phase from the cluster, and eventually the liquid–gas phase equilibrium is realized. Even in the equilibrium state, the member particles of the cluster are incessantly changing places with the particles in gaseous phase; and not only the cluster size, but also the shape of the cluster, is always fluctuating. When the cluster size is small, these fluctuations are quite furious; but when the cluster size is large enough, the shape of the cluster is almost globular and the fluctuations become slow and small.

Let us consider the trapping time T of a particle in the inside of the cluster, along with its distribution $P(T)$. Recent numerical simulations for the two-dimensional case show that the distribution $P(T)$ obeys the Weibull distribution $P_W(T)$ in the intermediate long-time regime, but obeys the log-Weibull distribution $P_{LW}(T)$ in the intrinsic long-time regime; their accumulated probabilities $Q_W(T)$ and $Q_{LW}(T)$ corresponding to $P_W(T)$ and $P_{LW}(T)$ are given by

$$Q_W(T) = \int_0^T P_W(X)\,dX = \exp[-AT^{-\alpha}] \tag{11}$$

$$Q_{LW}(T) = \int_0^T P_{LW}(X)\,dX = \exp\left[-B_1\left\{\log\left(\frac{T+B_3}{B_2}\right)\right\}^{-\beta}\right] \tag{12}$$

respectively, with characteristic exponents α and β and adjustable parameters A and B_i's. Figure 4 shows the fitting by $P_W(T)$ (or $Q_W(T)$) in the intermediate long-time regime ($T \leq T_C$), and it shows the log-Weibull fitting by $Q_{LW}(T)$ [10,11]. The log-Weibull regime is systematically prolonged when the total energy E decreases. This is consistent with the phenomenological fact that the cluster is more stabilized as the cluster size increases.

It is a remarkable point that the indices α and β seem to be free from the cluster size. Furthermore, it is worth noting that the intrinsic long-time tail of $P_{LW}(T)$ is the same as the universal scaling form of Eq. (10),

$$P_{LW}(T) \propto \frac{1}{T}(\log T)^{-1-\beta} \tag{13}$$

in the limit of $T \to \infty$. The interrelation between two exponents ($\beta = a'/b$) is still an open question [11].

From the above-mentioned results, it is surmised that the structure of a cluster consists of two geometrical components; one is the central core part with

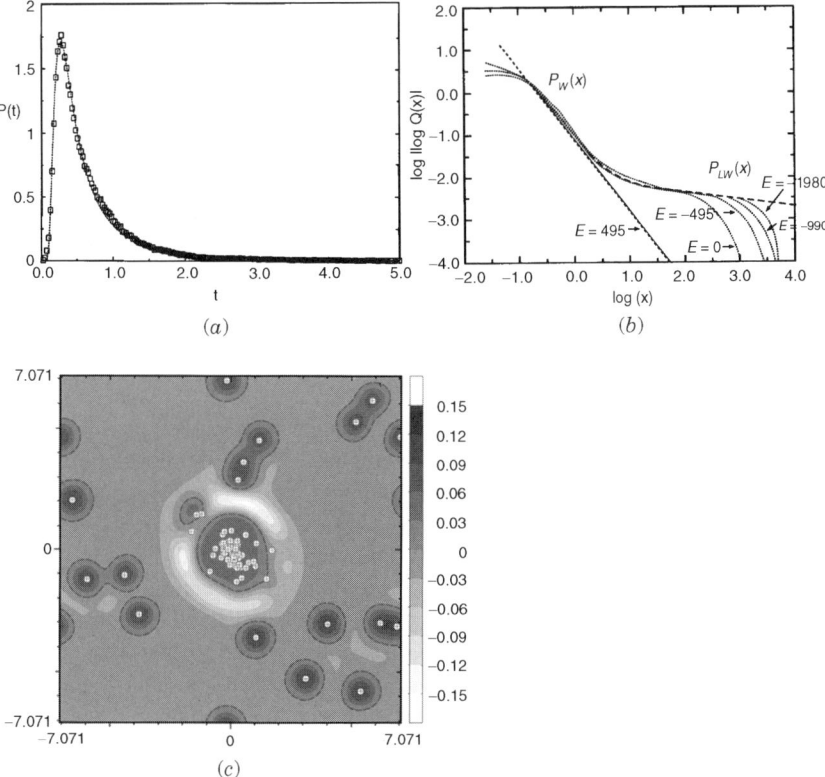

Figure 4. The distribution $P(T)$ of cluster formation. (*a*) The fitting by the Weibull distribution. (*b*) The crossover from the Weibull distribution to the log-Weibull distribution. (*c*) The snapshot of cluster formation in two dimensions. The gray scale demonstrates the value of the Riemannian sectional curvature. (Reprinted figures with permission from Aizawa et al., *Prog. Theor. Phys.* **103**, 519–540 (2000).

relatively high density, and another is the peripheral soft interface that is wrapping the core part. The particles in the core part contribute mainly to the log-Weibull distribution, and the particles in the soft interface contribute mainly to the Weibull distribution. However, it is not easy to define the boundary of a cluster in terms of dynamical system theory, since the soft interface with low density is too ambiguous to distinguish the boundary between the inside and the outside of the cluster.

A clear definition of cluster boundaries, which works successfully at least in numerical simulations, was proposed from the Riemannian geometrization method of Hamiltonian dynamics. The idea was the following: In the inside of the cluster the sectional curvature should be positive because the stability of geodesics must be guaranteed for clustering particles; on the other hand, the sectional curvature should be negative in the outside of the cluster boundary. Figure 4c shows a snapshot of cluster formation, and the value of the sectional curvature is demonstrated by the gray scale. In the core part of the cluster the curvature is very small but positive, and in the soft interface the curvature reveals sharp enhancement. Besides, a drastic transition appears in the sectional curvature near the cluster boundary: from positive curvature to negative curvature. These nonsymmetric structures in the sectional curvature near cluster boundaries seem to contribute to the long-term stabilization of the cluster.

V. COMPLEXITIES IN NONSTATIONARITY

Complexity of generic Hamiltonian dynamics appears in various phases of long-time behaviors. In this chapter we have emphasized that the nonstationarity is a typical phase of the complex behaviors in phase space. The f^{-1} spectrum is the transition point from stationarity to nonstationarity. In the recent studies about the intermittent chaos in a one-dimensional map (modified Bernoulli shift), it was shown that the maximum diversity appears at the transition point [13] and that the dynamical transition process from stationary chaos to nonstationary chaos generally reveals the logarithmic scaling law due to the breakdown of nonadiabatic approximations [14]. Even in generic Hamiltonian systems, we often observe a variety of slow dynamics of which PSD function $S(f) \sim f^{-\nu}$ satisfies $1 < \nu$, depending on the dynamical variables under consideration. So we can surmise that the slow dynamics is quite heterogeneous in generic Hamiltonian systems and that hierarchical structures, called "islands around island," play essential roles in the multiergodic aspects though each stagnant layer satisfies the universal scaling law.

The stability of stagnant motions is a significant problem not only in the theoretical subject but also in the practical measurement. For instance, in the experiments of quartz oscillators the $1/f$ spectral fluctuations are frequently observed, which is considered to be good examples for the nonstationary motions generated in the Hamiltonian dynamics of crystal lattice vibrations [6–8].

The effect of random noise on nonstationary motions has been recently studied to examine the stability of the long-time tails in Hamiltonian dynamics [15], where the survival time distribution for a particle trapped in a potential well is studied under random perturbations. The results were very clear; the Weibull distribution describes the intermediate long-time regime in the same manner shown in the case of cluster formation, but the contribution of the log-Weibull distribution completely disappears in the intrinsic long-time regime and the Gumbel distribution takes the phase of it. The details will be reported in a forthcoming article in relation to the large deviation in nonhyperbolic systems [12], and also to the general aspects in random systems [16]. The Riemannian geometrization is a useful method to describe the complex behaviors in hamiltonian dynamics. In the present chapter we discussed the case for cluster formation, and the cluster boundary can be clearly defined by use of the Riemannian sectional curvature. Complex dynamics, such as in scattering chaos [17,18] and heteroclinic chaos of the mixmaster universe model [19], are successfully analyzed by the Riemannian geometrization method [20]. In the complex processes of chemical reactions, we can also expect that the geometrization method is useful in understanding the clear reaction paths. It seems to be especially important that the positive curvature region generates more complex behaviors than that in the negative curvature region, because the conjugate points of the Jacobi field which induce topological weak instability are densely distributed in the positive curvature region [17]. These will be discussed in another chapter.

References

1. T. Harayama and Y. Aizawa, *Prog. Theor. Phys.* **84**, 23 (1990).
2. S. Kurosaki and Y. Aizawa, *Prog. Theor. Phys.* **98**, 783 (1997).
3. V. K. Mel'nikov, *Trans. Moscow. Soc.* **12**, 1 (1963).
4. V. I. Arnold, *Sov. Math. Dokl.* **5**, 581 (1964).
5. Y. Aizawa, *Prog. Theor. Phys.* **81**, 249 (1989).
6. Y. Aizawa et al., *Prog. Theor. Phys. Suppl.* **98**, 36 (1989).
7. N. Saito et al., *Prog. Theor. Phys. Suppl.* **45**, 209 (1970).
8. Y. Aizawa, *J. Korean Phys. Soc.* **28**, 310 (1995).
9. N. N. Nekhoroshev, *Russ. Math. Surv.* **32**, 1 (1977).
10. Y. Aizawa, K. Sato, and K. Ito, *Prog. Theor. Phys.* **103**, 519 (2000).
11. Y. Aizawa, *Prog. Theor. Phys. Suppl.* **139**, 1 (2000).
12. M. Yuri, *Commun. Math. Phys.* **241**, 453 (2003).
13. Y. Aizawa, *Prog. Theor. Phys. Suppl.* **99**, 149 (1989).
14. T. Akimoto and Y. Aizawa. *Prog. Theor. Phys.* **110**, 849 (2003).
15. Y. Aizawa and R. Imaizumi, *in preparation*, 2004.
16. Y. Kifer, *Stochastics and Dynamics*, 1(1), 1 (2001).
17. Y. Aizawa and T. Miyasaka, *Particles and Field Series*, Vol. 45, Y. H. Ichikawa and T. Tajima, eds., APS, 1990, p. 7.
18. M. Nakato and Y. Aizawa, *Chaos, Solitons & Fractals* **11**, 171 (2000).
19. Y. Aizawa, N. Kogumo and I. Antoniou, *Prog. Theor. Phys.* **98**, 1225 (1997).
20. Y. Aizawa, *J. Phys. Soc. Jpn.* **33**, 1693 (1972).

CHAPTER 24

RELAXATION AND DIFFUSION IN A GLOBALLY COUPLED HAMILTONIAN SYSTEM

YOSHIYUKI Y. YAMAGUCHI

Department of Applied Mathematics and Physics, Kyoto University, 606-8501, Kyoto, Japan

CONTENTS

I. Introduction
II. Model and Initial Condition
III. Relaxation Process
IV. Slow Dynamics
V. Probability Distribution Function of Momenta
VI. Diffusion in Angle Space
    A. Diffusion at Equilibrium
    B. Diffusion in Quasi-Stationary State
    C. Diffusion in Nonstationary State
    D. Dependence on Degrees of Freedom
VII. Summary
References

I. INTRODUCTION

Slow dynamics showing $f^{-\nu}(0 < \nu < 2)$ power spectra is observed in many symplectic systems: a water cluster [1], a ferromagnetic spin system [2,3], and area-preserving mappings [4,5]. To explain the $f^{-\nu}$ spectra, Aizawa and coworkers [6,7] introduced a geometrical model for area preserving mappings. This model assumes an exact self-similar hierarchical structure of phase space consisting of tori and chaotic seas, which is often referred to as "islands around islands" [8]. Motion trapped to KAM tori or Cantori, and hence stagnant motion,

---

Geometric Structures of Phase Space in Multidimensional Chaos: A Special Volume of Advances in Chemical Physics, Part B, Volume 130, edited by M. Toda, T Komatsuzaki, T. Konishi, R.S. Berry, and S.A. Rice. Series editor Stuart A. Rice.
ISBN 0-471-71157-8 Copyright © 2005 John Wiley & Sons, Inc.

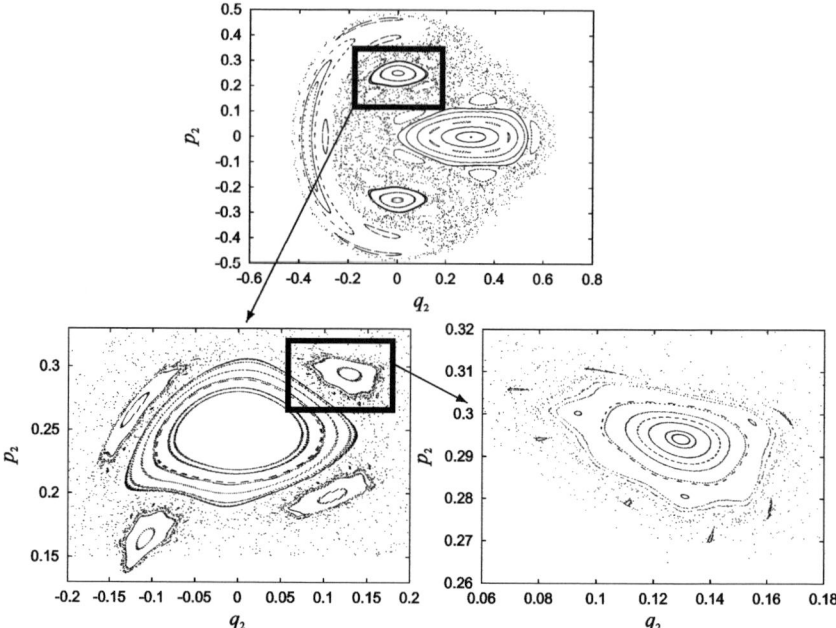

Figure 1. Poincaré sections (q_2, p_2) of the Hénon–Heiles system with $E = 0.125$.

namely slow relaxation, is reproduced. Meiss et al. [9] have successfully proposed a similar model.

The hierarchy of tori is theoretically predicted by the Poincaré–Birkhoff theorem [10] in nearly integrable systems with two degrees of freedom. For instance, the hierarchy in the Hénon–Heiles system represented by the Hamiltonian is shown in Fig. 1.

$$H(q_1, q_2, p_1, p_2) = \tfrac{1}{2}(p_1^2 + p_1^2) + \tfrac{1}{2}(q_1^2 + q_1^2) + q_1^2 q_2 - \tfrac{1}{3} q_2^3$$

Self-similarity is expected to be one of the important concepts to understand statistics and motion in Hamiltonian systems. However, the Poincaré–Birkhoff theorem and the two models introduced by Aizawa et al. and Meiss et al. are based on the two-dimensionality of the phase spaces, and they cannot be directly applied to high-dimensional systems. As far as I know, existence of the self-similarity has not been clearly exhibited, since visualizing the self-similarity is not easy due to the high-dimensionality of Poincaré sections, which has $2N - 2$ dimension for systems with N degrees of freedom.

Then, as the first step of approaching the study of self-similarity, we investigate two features that must appear if self-similarity exists: power-type distribution function of momenta and anomalous diffusion. We are particularly interested in nonequilibrium states since, at thermal equilibrium, a Gaussian distribution and normal diffusion must appear instead of the two anomalies.

In this chapter we consider a system consisting of globally coupled rotators, each of which interacts with the others through a mean field. This model is hence called as Hamiltonian mean field (HMF) model [11,12]. We consider the HMF model for the following two reasons: (i) The mean-filed interaction is a typical long-range interaction [13]. (ii) An effective long-range interaction is easily produced by randomly rewiring the network of interactions [14]. Relaxation to thermal equilibrium has been studied in such long-range Hamiltonian systems [2,11,12,15–17]. Particularly, the HMF model has quasi-equilibrium states that appear before the system goes toward equilibrium [18,19], and the quasi-equilibrium states may show a different nature from that of thermal equilibrium.

In a quasi-equilibrium state, a q-exponential function [20] whose tail is of power type is reported in the HMF model with a certain type of initial condition [15]. The initial condition known as *waterbag* seems special because it lies on a one-dimensional line on two-dimensional μ-space. To check generality, we must compute distribution functions for another type of initial condition that spreads on two-dimensional subspace of the μ-space.

Anomalous diffusion was first investigated in a one-dimensional chaotic map to describe enhanced diffusion in Josephson junctions [21], and it is observed in many systems both numerically [16,18,22–24] and experimentally [25]. Anomalous diffusion is also observed in Hamiltonian dynamical systems. It is explained as due to power-type distribution functions [22,26,27] of trapping and untrapping times of the orbit in the self-similar hierarchy of cylindrical cantori [28].

Latora et al. [18] discussed a relation between the process of relaxation to equilibrium and anomalous diffusion in the HMF model by comparing the time series of the temperature and of the mean-squared displacement of the phases of the rotators. They showed that anomalous diffusion changes to a normal diffusion after a crossover time, and they also showed that the crossover time coincides with the time when the canonical temperature is reached. They also claim that anomalous diffusion occurs in the quasi-stationary states.

The crossover from anomalous to normal diffusion determines the time when the anomalous diffusion finishes. However, it is not clearly pointed out when the anomalous diffusion starts, and hence the study of the relation between the relaxation process and anomalous diffusion is still not complete. Moreover, in Ref. 18, the numerical calculations were performed by using only one type of initial condition—that is, the waterbag initial condition giving a q-exponential distribution [15]—but different types of initial condition may change the conclusion. For instance, in the one-dimensional self-gravitating sheet model, the waterbag initial condition gives a power-type spatial correlation, but a thin width of initial distribution on μ-space breaks the power law [29].

In this chapter, we use a type of initial condition that is different from the waterbag used in Refs. 15 and 18, and we show that (i) probability distribution functions do not have power-law tails in quasi-stationary states and (ii) the diffusion becomes anomalous if and only if the state is neither stationary nor quasi-stationary. In other words, the diffusion is shown to be normal in quasi-stationary states, although a stretched exponential correlation function is present instead of usual exponential correlation. Some scaling laws concerned with degrees of freedom are also exhibited, and the simple scaling laws imply that the results mentioned above holds irrespective of degrees of freedom.

This chapter is organized as follows. The HMF model and initial condition are introduced in Section II. In Section III, we study the relaxation process and divide it into three stages: quasi-stationary, relaxational, and equilibrium stages. Slow dynamics in the HMF model is exhibited in Section IV by using power spectra of the mean field. Single-particle distribution functions of momenta are investigated in Section V to determine time scale as a function of degrees of freedom and to check the existence of a power-law tail. Diffusion process in each stage is investigated in Section VI by using stretched exponential correlation functions of momenta. Dependence on degrees of freedom is also reported in Sections III–VI. The final section, Section VII is devoted to summary.

II. MODEL AND INITIAL CONDITION

In this section we introduce the HMF model and the initial condition that is considered in this chapter. The model consists of N classical and identical rotators confined to move on the unit circle, and the Hamiltonian is composed of a kinetic and a potential part [2,11,12,18,30],

$$H = K + V = \sum_{j=1}^{N} \frac{p_j^2}{2} + \frac{1}{2N} \sum_{i,j=1}^{N} [1 - \cos(\theta_i - \theta_j)] \quad (1)$$

The N particles are globally coupled through the mean field defined as

$$\mathbf{M} = \frac{1}{N} \sum_{j=1}^{N} (\cos\theta_j, \sin\theta_j) = (M\cos\phi, M\sin\phi) \quad (2)$$

where the modulus M ($0 \leq M \leq 1$) represents the magnetization of this system. The magnetization is $M = 1$ in completely ordered phase, and $M \sim O(1/\sqrt{N})$ in completely random phase. We remark that the potential energy V and the kinetic energy K are related to the magnetization M as follows:

$$2V/N = 1 - M^2, \quad 2K/N = 2U - 1 + M^2 \quad (3)$$

where U is the energy per particle; that is, $U = E/N$, and E is the total energy.

The free energy of this system has been obtained in the canonical ensemble [11,12,31], and it has been shown that system (1) has a second-order phase transition at the critical energy $U_c = 0.75$. If the energy U is greater than the critical energy, the largest Lyapunov exponent goes to zero in the thermodynamic limit ($N \to \infty$) [32]. Then, all rotators freely rotate, and diffusion becomes ballistic. On the contrary, if U is small compared to U_c, all rotators are trapped in the potential well and no diffusion occurs. We are therefore interested in a value of the energy which is near, but less than, the critical energy in order to allow some particle diffusion. Hereafter, we set $U = 0.69$ (a value studied also in Refs. 15,18,30, and 31).

The canonical equations of motion for system (1) can be cast in a form that uses the mean field (2) as follows:

$$\frac{d\theta_j}{dt} = p_j, \qquad \frac{dp_j}{dt} = -M(t)\sin(\theta_j - \phi(t)) \qquad (j = 1,\ldots,N) \qquad (4)$$

We numerically integrate Eq. (4) by using fourth-order symplectic integrators [33,34]. The time slice of the integrator is set at $\Delta t = 0.2$ or 0.4, and it suppresses the relative energy error down to $|\Delta E/E| < 5 \times 10^{-7}$.

We have performed the integrations starting from $M(0) \sim O(1/\sqrt{N})$. To prepare these initial conditions numerically, $\theta_j(0)$ is taken from a uniformly random distribution whose support is $[-\pi,\pi]$, and $p_j(0)$ is taken from another uniformly random distribution whose support is $[-\bar{p},\bar{p}]$. The value $\bar{p}$ is chosen to get the energy density U. The total momentum $\sum_{j=1}^{N} p_j$ is an integral of the motion and we initially set it to zero. This initial state corresponds to a local entropy minimum [35], and to a stationary stable solution to the Vlasov–Poisson equation [12], although the system goes toward Gibbs equilibrium due to finite size effects [15]. With respect to the waterbag initial condition giving $M(0) = 1$ chosen in Refs. 15, 18, and 30, the one we choose has the advantage of being a quasi-stationary state from the start.

III. RELAXATION PROCESS

To observe the relaxation process, we use the magnetization $M(t)$. Note that observing $M(t)$ corresponds to observing $2K(t)/N$ by using Eq. (3), and $2K(t)/N$ is the time series of the temperature, since the canonical average of $2K/N$ coincides with the canonical temperature.

Temporal evolutions of $M(t)$ are shown in Fig. 2. In order to suppress fluctuations, we have calculated averages over realizations. Throughout this chapter, unless no comments appear, the numbers of realizations are $n = 1000, 100$, and 8 for $N = 100, 1000$, and $10,000$, respectively. We divide

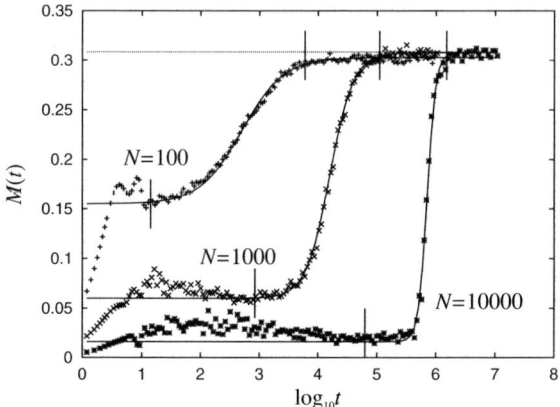

Figure 2. Temporal evolutions of $M(t)$. $U = 0.69$. The horizontal line represents the canonical equilibrium value of M. On each curve, two short vertical lines are marked. The first and the second ones are at the end of Stage I and II, respectively. Solid curves are hyperbolic tangent functions (5). [Reproduced with permission from Y. Y. Yamaguchi, *Phys. Rev. E* **68**, 066210 (2003). Copyright 2004 by the American Physical Society.]

the temporal evolutions into three stages, Stage I, II, and III. In Stage I, the value of magnetization is almost constant but smaller than the canonical value. After Stage I, magnetization rapidly increases toward its equilibrium value M_{eq}, and we call this time interval Stage II. Finally the system reaches equilibrium, during Stage III.

Let us define boundary times between Stages I and II, $t_{I/II}$, and between Stages II and III, $t_{II/III}$, as follows. The magnetization takes the local minimum at t_{min}, and we adopt $t_{I/II} = t_{min}$. We define the other boundary time $t_{II/III}$ as the first passage time that satisfies $M(t) = 0.99 M_{eq}$. Values of the two boundary times are reported in Fig. 3 as functions of degrees of freedom. The local minimum time is proportional to $N^{1.7}$ for $N \geq 100$ with our initial condition $M(0) = 0$, as with another initial condition $M(0) = 1$ [36]. For small N, we cannot neglect the initial time region $t < 6$ in which the level of $M(t)$ goes to $O(1/\sqrt{N})$ coming from the law of large numbers (see Fig. 4d), and hence the power law breaks. The power law recovers by subtracting the initial increasing time 6 from $t_{I/II}$ as shown in Fig. 3; that is, $t_{I/II} - 6 \sim N^{1.7}$ ($N \geq 10$).

A theoretical prediction of $t_{II/III}$, the upper curve in Fig. 3, is obtained by fitting the magnetization $M(t)$ as hyperbolic tangent function,

$$M(t) = [1 + \tanh(a (\log_{10} t - b))] c + d \qquad (5)$$

The parameter d represents the initial level of $M(t)$, and c represent the half-width between initial and equilibrium levels of $M(t)$. The product ac is the slope

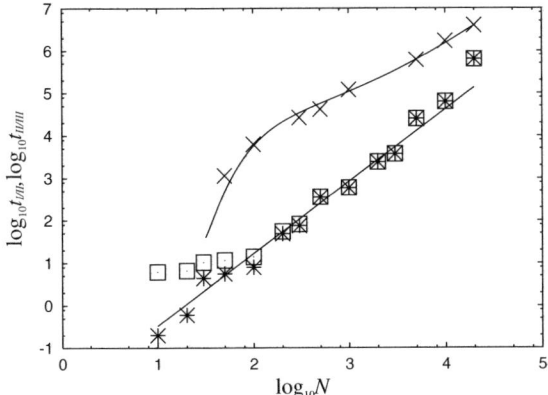

Figure 3. Dependence on degrees of freedom of $t_{I/II}$ (squares) and $t_{II/III}$ (crosses). Asterisks represent $t_{I/II} - 6$. The lower straight line represents the power law $N^{1.7}/150$. The upper curve is a theoretical prediction of the boundary time $t_{II/III}$ using Eqs. (6) and (7) with $M_{th} = 0.99 M_{eq}$. [Reproduced with permission from Y. Y. Yamaguchi, *Phys. Rev. E* **68**, 066210 (2003). Copyright 2004 by the American Physical Society.]

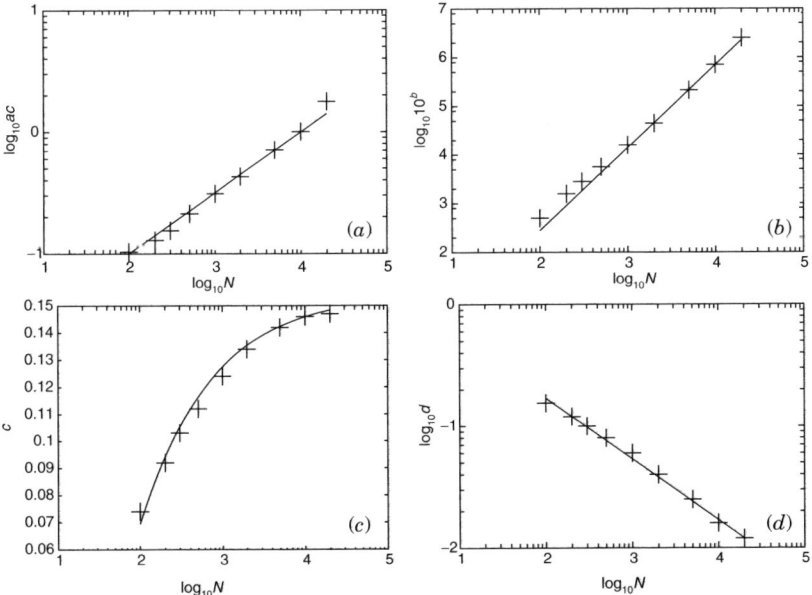

Figure 4. Four parameters a, b, c, and d are reported as functions of degrees of freedom. (a) Log–log plot of ac. (b) Log–log plot of 10^b. (c) Linear-log plot of c. (d) Log–log plot of d. Solid curves are scaling functions described in Eq. (6). [Reproduced with permission from Y. Y. Yamaguchi, *Phys. Rev. E* **68**, 066210 (2003). Copyright 2004 by the American Physical Society.]

at $\log_{10} t = b$; that is, $ac = \mathrm{d}M/\mathrm{d}(\log_{10} t)|_{\log_{10} t = b}$, and 10^b is the time scale. As shown in Fig. 4, these four parameters are fitted as

$$a(N) = \frac{\sqrt{N}}{100\, c(N)}, \qquad 10^{b(N)} = \tfrac{1}{9} N^{1.7}$$
$$c(N) = \frac{(M_{\mathrm{eq}} - d(N))}{2}, \qquad d(N) = \frac{1.7}{\sqrt{N}} \tag{6}$$

By using the scaling law (6), we can predict when $M(t)$ reaches a given threshold level, M_{th}, as a function of N. Let t_{th} be the threshold time, which satisfies $M(t_{\mathrm{th}}) = M_{\mathrm{th}}$; then t_{th} is expressed as

$$t_{\mathrm{th}} = 10^b \left(\frac{M_{\mathrm{th}} - d}{M_{\mathrm{eq}} - M_{\mathrm{th}}} \right)^{\frac{\ln 10}{2a}} \tag{7}$$

In Fig. 3, t_{th} is reported for $M_{\mathrm{th}} = 0.99 M_{\mathrm{eq}}$, and the prediction is in good agreement with numerical results. We remark that, roughly speaking, $t_{\mathrm{II/III}}$ is asymptotically proportional to $N^{1.7}$.

The system seems quasi-stationary in Stage I. The existence of quasi-stationary states for a sufficiently long time has been questioned in Ref. 36. We will positively answer to the question by observing dependence on τ of the correlation function $C_p(t; \tau)$ in Section VI.B.

IV. SLOW DYNAMICS

In Stages I, II, and III, we exhibit power spectra of the magnetization $M(t)$, and we show that the HMF model has slow dynamics described by $f^{-\nu}$ spectra. The power spectra are shown in Fig. 5 for $N = 100, 1000$, and $10,000$, and they are of power types in long-time regions. The periods in which the power spectra are computed are arranged in Table I. Values of the exponent ν are estimated as $\nu = 0.6, 1.4$, and 1.4 for $N = 10,000$ in Stages I, II, and III, respectively.

A noticeable dependence on degrees of freedom is a shift of the (local) peak frequency in Stage II. The shift is guessed as a reflection of time scale that grows as N increases, as shown in Section III.

V. PROBABILITY DISTRIBUTION FUNCTION OF MOMENTA

In Section III we have found that time scale is proportional to $N^{1.7}$ through temporal evolution of magnetization $M(t)$—in other words, in angle space $(\theta_1, \ldots, \theta_N)$. In this section we check validity of the time scale by observing

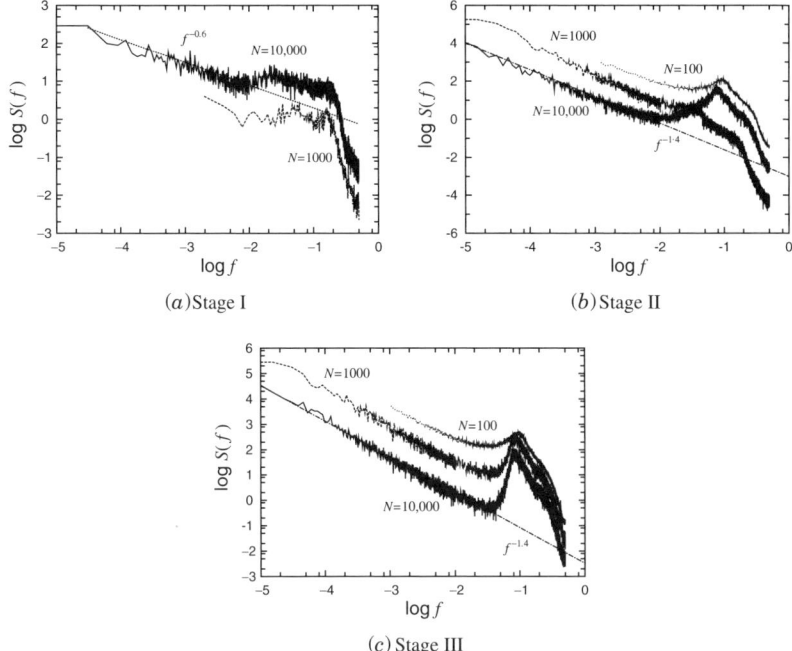

Figure 5. Power spectra of the magnetization $M(t)$ for Stages I, II, and III.

temporal evolution of single-particle probability distributions of momenta $f(p,t)$, namely in momentum space $(p_1, \ldots, p_N)$. The distribution functions with scaled time, $f(p, t/N^{1.7})$, are exhibited in Fig. 6 for $N = 1000, 2000, 3000$ and $10,000$, and they are in good agreement with each other irrespective of degrees of freedom. Moreover, trivial scaling t/N [37] does not work as shown in Fig. 7. We can therefore conclude that the time scale of relaxation process is proportional to $N^{1.7}$ both in angle and in momentum spaces.

Now let us check whether the distribution functions have power-law tails or not in quasi-stationary states. The time shown in Fig. 6d is $t/N^{1.7} = 0.0048$,

TABLE I
Periods in Which Power Spectra, Shown in Fig. 5, Are Computed

N	Stage I	Stage II	Stage III
100	—	1000–5096	10,000–26,384
1000	0–512	10,000–75,536	120,000–185,536
10,000	0–16,384	500,000–631,072	1,500,000–1,631,072

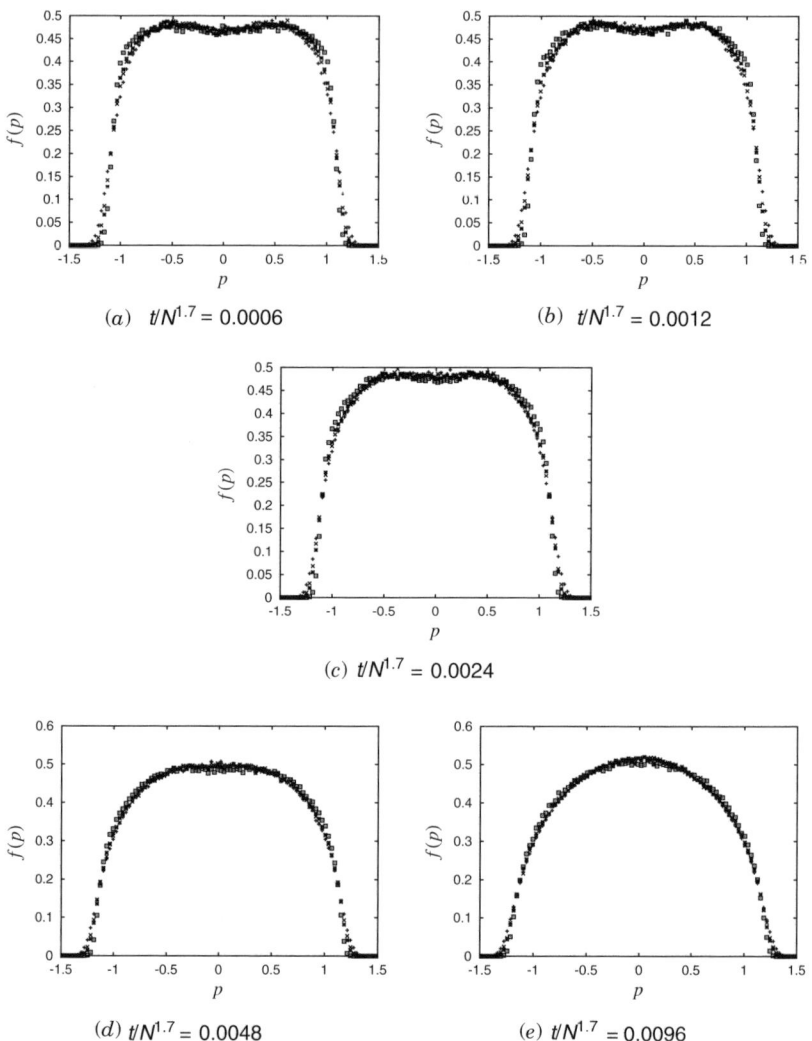

Figure 6. Probability distribution functions of momenta for scaled time.

which reads $t = 600$ for $N = 1000$, and hence the system is in quasi-stationary Stage I (see Fig. 2). A log–log plot of Fig. 6d is shown in Fig. 8, and the distribution sharply decreases. Hence we can numerically exclude the presence of power-law tails even though the system is in a quasi-stationary state.

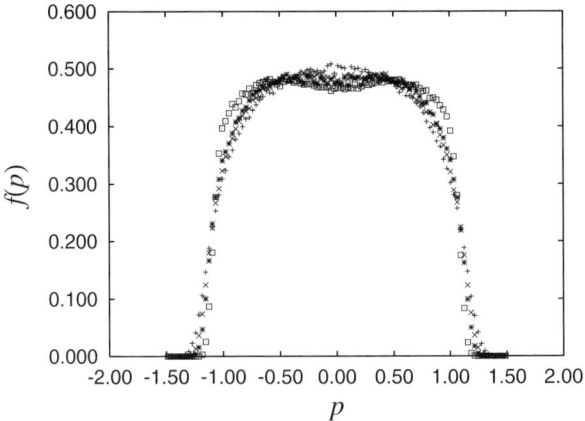

Figure 7. Same with Fig. 6, but time is scaled as $t/N = 0.6$. $N = 1000\,(+)$, $2000\,(\times)$, $3000\,(*)$, and $10{,}000\,(\square)$.

The authors of Ref. 15 observe distribution functions at the same time (i.e., $t =$ constant) for several degrees of freedom, and they conclude that distribution function of momenta goes to a q-exponential function in the limit $N \to \infty$. We remark that the distribution functions should be taken at the same scaled time (i.e., $t/N^{1.7} =$ constant) if one considers the thermodynamic limit.

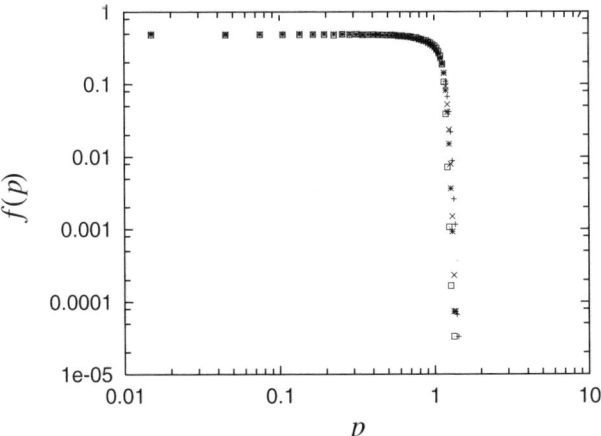

Figure 8. Log–log plot of Fig. 6d.

VI. DIFFUSION IN ANGLE SPACE

To observe the diffusion process, we introduce the mean-square displacement of phases $\sigma_\theta^2(t)$ defined as

$$\sigma_\theta^2(t) = \frac{1}{N}\sum_{j=1}^{N}[\theta_j(t) - \theta_j(0)]^2 = \left\langle [\theta_j(t) - \theta_j(0)]^2 \right\rangle_N \qquad (8)$$

The symbol $\langle \cdot \rangle_N$ represents the average over all the N rotators. The quantity $\sigma_\theta^2(t)$ typically scales as $\sigma_\theta^2(t) \sim t^\alpha$, and the diffusion is anomalous when $\alpha \neq 1, 2$, while it is normal when $\alpha = 1$ and ballistic for $\alpha = 2$. The quantity $\sigma_\theta^2(t)$ can be rewritten by using the correlation function of momenta $C_p(t;\tau)$ as

$$\begin{aligned}\sigma_\theta^2(t) &= \int_0^t dt_1 \int_0^t dt_2 \, \langle p_j(t_1) p_j(t_2) \rangle_N \\ &= 2\int_0^t ds \int_0^{t-s} d\tau \, C_p(s;\tau)\end{aligned} \qquad (9)$$

where $C_p(t;\tau)$ is defined as

$$C_p(t;\tau) = \langle p_j(t+\tau) p_j(\tau) \rangle_N \qquad (10)$$

Moreover, if the system is stationary and $C_p(t;\tau)$ does not depend on τ accordingly,

$$C_p(t;\tau) = C_p(t;0) \qquad (\forall \tau > 0) \qquad (11)$$

then Eq. (9) is simplified as

$$\sigma_\theta^2(t) = 2\int_0^t (t-s) \, C_p(s;0) \, ds \qquad (12)$$

As described in Eq. (9), the mean-square displacement $\sigma_\theta^2(t)$ is obtained from the correlation function of momenta $C_p(t;\tau)$, and hence we study diffusion process by observing the correlation function. We start from the simplest stage, Stage III, because we may use the simple expression (12). Next, we progress to Stage I, where we expect that the system is quasi-stationary and that we may use Eq. (12) again. In nonstationary stage, Stage II, we check whether diffusion is of a power type. Finally we investigate dependence on degrees of freedom for some important parameters.

A. Diffusion at Equilibrium

Assuming that the system has reached equilibrium at $t = t_{\text{eq}}$, we observe $C_p(t; t_{\text{eq}})$ and $\sigma_\theta^2(t; t_{\text{eq}})$, where

$$\sigma_\theta^2(t; t_{\text{eq}}) = \left\langle [\theta_j(t + t_{\text{eq}}) - \theta_j(t_{\text{eq}})]^2 \right\rangle_N \tag{13}$$

At equilibrium we may assume that the system is stationary,

$$C_p(t; t_{\text{eq}} + \tau) = C_p(t; t_{\text{eq}}) \qquad (\forall \tau > 0) \tag{14}$$

and hence

$$\sigma_\theta^2(t; t_{\text{eq}}) = 2 \int_0^t (t - s) C_p(s; t_{\text{eq}}) \, ds \tag{15}$$

Now let us consider the correlation function for $N = 1000$. We adopt $t_{\text{eq}} = 2^{20} \simeq 10^6$, which is long enough to reach equilibrium imaging from Fig. 2. The correlation function $C_p(t; t_{\text{eq}})$ is reported in Fig. 9a, and it is well approximated by the stretched exponential function [38]

$$C_p(t; t_{\text{eq}}) = 0.47 \exp[-(t/410)^{0.32}] \tag{16}$$

rather than by a pure exponential [see the inset of Fig. 9a, which is a log-linear plot of $C_p(t; t_{\text{eq}})$].

We remark that a stretched exponential function $\exp[-x^\beta]$ with a small exponent $|\beta| \ll 1$ is indistinguishable from a power-type function in the region $|\beta \ln x| \ll 1$:

$$\exp[-x^\beta] = \exp[-\exp(\beta \ln x)] \sim \exp[-1 - \beta \ln x] = x^{-\beta}/e$$

However, the fitting function (16) well agree with numerical result even around $|0.32 \ln(t/410)| = 1$, whose two solutions are $t \simeq 18, 9330$. We therefore adopt a stretched exponential function as an approximation of $C_p(t; t_{\text{eq}})$.

By using the fitting function (16) and Eq. (15), we numerically reproduce $\sigma_\theta^2(t; t_{\text{eq}})$, and the reproduced curve well approximates the numerical result as shown in Fig. 9b. Note that $\sigma_\theta^2(t; t_{\text{eq}})$ is proportional to t^2 in the limit of $t \to 0$, since $C_p(s; t_{\text{eq}})$ in Eq. (15) goes to the constant $C_p(0; t_{\text{eq}})$. On the other hand, in the limit of $t \to \infty$, $\sigma_\theta^2(t; t_{\text{eq}})$ is proportional to t, because both $C_p(s; t_{\text{eq}})$ and $sC_p(s; t_{\text{eq}})$ are almost zeros in the long-time region, and hence their integrals become constants. The crossover from t^2 to t is also observed if we assume an exponential correlation function, and hence we conclude that diffusion at equilibrium is normal as expected although a stretched exponential is present.

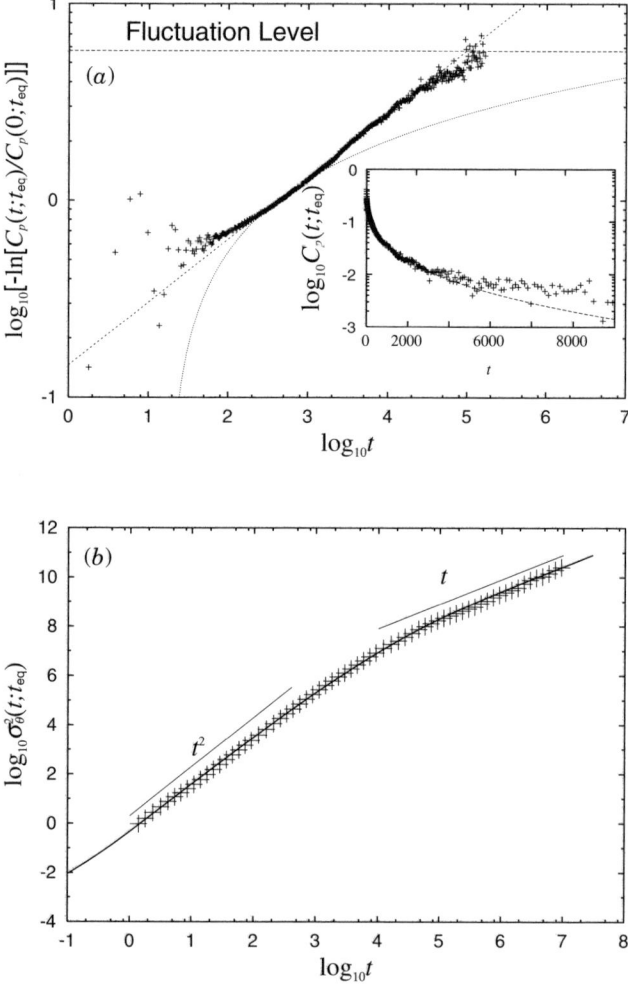

Figure 9. (a) Double log–log plot of normalized correlation function $C_p(t;t_{eq})/C_p(0;t_{eq})$ at equilibrium with $t_{eq} = 2^{20}$. $N = 1000$. We take an average over $n = 100$ realizations. The straight line and the curve represent the stretched exponential function (16) and the power-type function $(t/410)^{-0.32}/e$, respectively. The upper horizontal line is fluctuation level $O(1/\sqrt{Nn})$. The inset shows a log-linear plot with the stretched exponential function. (b) Log–log plot of $\sigma_\theta^2(t;t_{eq})$ with the approximate function produced by Eqs. (15) and (16). [Reproduced with permission from Y. Y. Yamaguchi, *Phys. Rev. E* **68**, 066210 (2003). Copyright 2004 by the American Physical Society.]

B. Diffusion in Quasi-Stationary State

Except for Stage III, we cannot expect stationarity to hold Eq. (11) any more. However, from the temporal evolutions of $M(t)$ (Fig. 2), we may expect quasi-stationarity in Stage I,

$$C_p(t;\tau) = C_p(t;0) + \epsilon(t;\tau) \qquad (17)$$

where τ belongs to Stage I and $\epsilon(t)$ is suitably small.

The correlation function $C_p(t;\tau)$ for various values of τ is reported in Figs. 10a and 10b for $N = 1000$, and the relative error of correlation function, defined as

$$R(\tau) = \max_t \frac{|\epsilon(t;\tau)|}{C_p(0;0)} \qquad (18)$$

is also reported in Fig. 10c as a function of τ. The error $R(\tau)$ is suppressed up to the end of Stage I, and hence we conclude that the system is quasi-stationary in Stage I. We believe that the quasi-stationary states correspond to stationary stable states of the Vlasov equation [39]. We remark that $R(\tau)$ is constant in Stage III again due to stationarity at equilibrium.

It seems natural that we regard $C_p(t;\tau)$ as a series of stretched exponential functions of t rather than power-type functions, since this function fits $C_p(t;\tau)$ in more than two decades of time (power-law fits of the correlation functions hold in one decade). Moreover, at equilibrium, $C_p(t;t_{\rm eq})$ is also a stretched exponential rather than a pure exponential, as shown in Fig. 9.

In the quasi-stationary region, Stage I, the mean-square displacement $\sigma_\theta^2(t)$ can be derived by the correlation function $C_p(t;0)$, which is reported in Fig. 11 for $N = 100, 1000$ and $10,000$. We approximate $C_p(t;0)$ by a stretched exponential function as

$$\begin{aligned} N = 100: &\quad C_p(t;0) = 0.38\,\exp[-(t/20)^{0.68}] \\ N = 1000: &\quad C_p(t;0) = 0.38\,\exp[-(t/180)^{0.91}] \\ N = 10,000: &\quad C_p(t;0) = 0.38\,\exp[-(t/2200)^{0.90}] \end{aligned} \qquad (19)$$

The prefactor 0.38 comes from $C_p(0;0) = 2K(0)/N$.

Using the approximate functions (19) and Eq. (12), we are able to reproduce $\sigma_\theta^2(t)$, as shown in Fig. 12. The approximation is good in Stage I—that is, in the quasi-stationary time region—irrespective of the value of N. Consequently, there is no anomaly in diffusion in Stage I, since the diffusion is explained by stretched exponential correlation function.

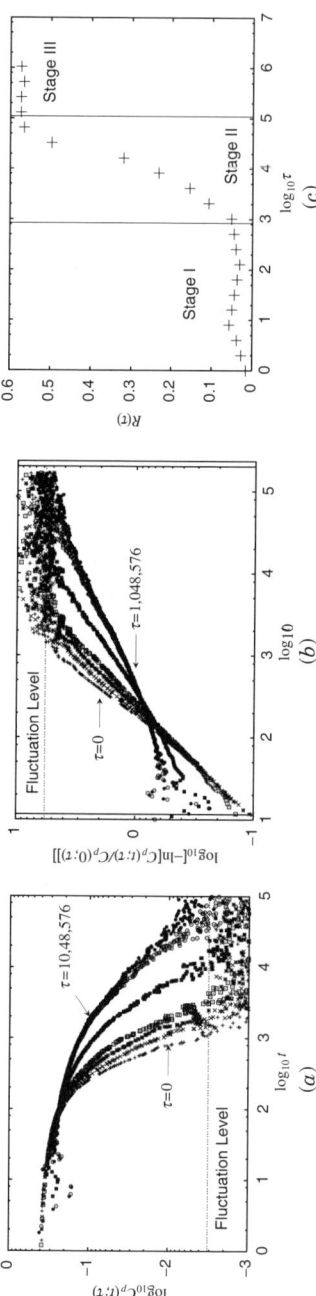

Figure 10. Correlation function of momenta $C_p(t;\tau)$ for various values of $\tau = 0, 1024, 2048, 4096, 16,384, 65,536$ and $1,048,576$ from left to right. (a) Log–log plot. (b) Double log–log plot. (c) Relative error, Eq. (18), of $C_p(t;\tau)$ from $C_p(t;0)$. [Reproduced with permission from Y. Y. Yamaguchi, *Phys. Rev. E* **68**, 066210 (2003). Copyright 2004 by the American Physical Society.]

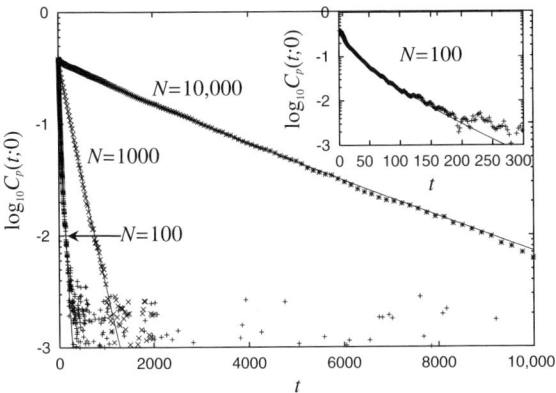

Figure 11. Correlation function of momenta at $\tau = 0$, that is, $C_p(t; 0)$. The inset is magnification of the horizontal axis around $t = 0$ for $N = 100$. These numerical results are approximated by solid curves that are stretched exponential functions in Eq. (19). [Reproduced with permission from Y. Y. Yamaguchi, *Phys. Rev. E* **68**, 066210 (2003). Copyright 2004 by the American Physical Society.]

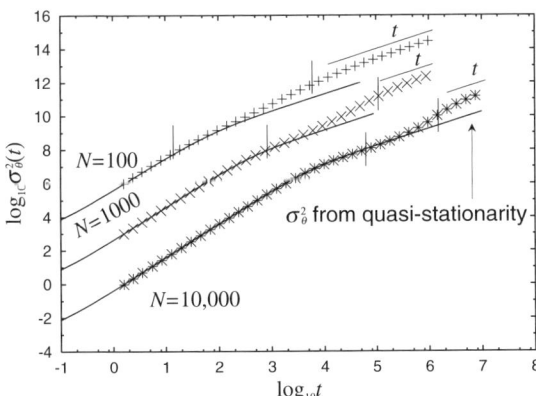

Figure 12. Time series of the mean-square displacement of the phases $\sigma_\theta^2(t)$. $N = 100, 1000$, and $10,000$ from top to bottom. The vertical axis is the original scale only for $N = 10,000$, and is multiplied by 10^3 and 10^6 for $N = 1000$ and 100, respectively, just for a graphical reason. In Stage I where the system is quasi-stationary, the numerical results are approximated by solid curves that are obtained from Eq. (12) using functions in Eq. (19). After the system reaches equilibrium, diffusion becomes normal. Anomaly in diffusion is observed only in Stage II. The two short vertical lines on each curve show the end of Stages I and II, which correspond to the ones found in Fig. 2. [Reproduced with permission from Y. Y. Yamaguchi, *Phys. Rev. E* **68**, 066210 (2003). Copyright 2004 by the American Physical Society.]

C. Diffusion in Nonstationary State

After the quasi-stationary region, diffusion becomes anomalous, which is faster than normal diffusion, in Stage II. If we fit $\sigma_\theta^2(t)$ by a power-type function t^α in Stage II, the exponent α is estimated as 1.54, 1.59, and 1.74 for $N = 100$, 1000 and 10,000, respectively. The values of exponent tend to increase as N increases as reported for the system having the so-called two-dimensional egg-crate potential [16]. On the other hand, the duration in which diffusion is anomalous becomes shorter and shorter in logarithmic time scale as N increases, in accordance with the sharper change of $M(t)$. Moreover, σ_θ^2/t^α is not constant, but has a wave in Stage II (see Fig. 13). Hence we guess that the anomaly in diffusion is not anomalous diffusion taking a power-type function, but a transient anomaly due to nonstationarity of Stage II.

Let us proceed to investigate the origin of anomaly in diffusion. We focus on the behavior for $N = 1000$. The mean-square displacement $\sigma_\theta^2(t)$ is perfectly determined by the correlation function $C_p(t;\tau)$ using Eq. (9), once we assume that $C_p(t;\tau)$ is a series of stretched exponential functions. We introduce three parameters, $C_p(0;\tau)$, $t_{\text{corr}}(\tau)$, and $\beta(\tau)$, to describe the stretched exponential function as

$$C_p(t;\tau) = C_p(0;\tau) \exp[-\{t/t_{\text{corr}}(\tau)\}^{\beta(\tau)}] \tag{20}$$

We investigate which of the three parameters is the most important to yield anomaly in diffusion.

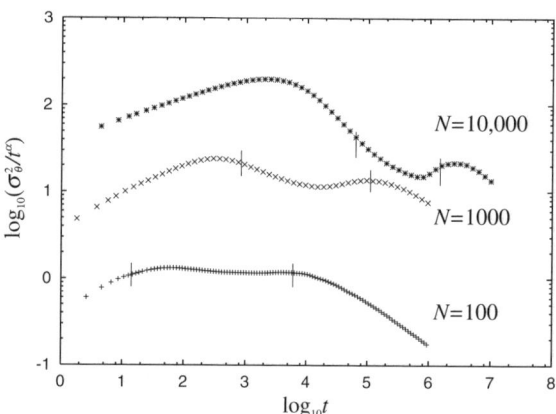

Figure 13. Log–log plot of $\sigma_\theta^2(t)/t^\alpha$. The exponent α is estimated as 1.54, 1.59, and 1.74 for $N = 100, 1000$ and $10,000$, respectively. The two short vertical lines on each curve show the end of Stages I and II. In Stage II, σ_θ^2/t^α is not constant. The vertical axis is the original scale only for $N = 100$, and it is multiplied by 10 and 100 for $N = 1000$ and $N = 10,000$, respectively, for a graphical reason. [Reproduced with permission from Y. Y. Yamaguchi, *Phys. Rev. E* **68**, 066210 (2003). Copyright 2004 by the American Physical Society.]

The strategy is as follows. We reproduce $d\sigma_\theta^2(t)/dt$ by using the three parameters and the formula

$$\frac{d\sigma_\theta^2}{dt}(t) = 2\int_0^t d\tau\, C_p(0;\tau) \exp[-\{(t-\tau)/t_{\text{corr}}(\tau)\}^{\beta(\tau)}] \tag{21}$$

We consider the first derivative of σ_θ^2 instead of σ_θ^2 itself, because the former requires only single integration while the latter requires double integrations [Eq. (9)]. We first omit the dependence on τ of the parameter $C_p(0;\tau)$ and fix it to a constant value to observe how it affects the anomaly in diffusion. We then fix the other two parameters $t_{\text{corr}}(\tau)$ and $\beta(\tau)$ to determine their effect on the mean-square displacement.

From the numerical results of $C_p(t;\tau)$, Fig. 10b, we determine the values of three parameters $C_p(0;\tau)$, $t_{\text{corr}}(\tau)$, and $\beta(\tau)$ at some value of τ by using the least-squares method. The discrete values of the parameters are not enough to reproduce $d\sigma_\theta^2(t)/dt$ accurately, and then we approximate the parameters by hyperbolic tangent functions as follows:

$$C_p(0;\tau) = 0.046\,[1 + \tanh(2.5(\log_{10}\tau - 4.35))] + 0.385$$

$$t_{\text{corr}}(\tau) = 80\,[1 + \tanh(1.5(\log_{10}\tau - 3.4))] + 170 \tag{22}$$

$$\beta(\tau) = 0.31\,[1 + \tanh(1.5(\log_{10}\tau - 3.8))] + 0.29$$

The hyperbolic tangent functions are in good agreement with numerical results, as shown in Fig. 14. To confirm the validity of the approximation, we reproduced

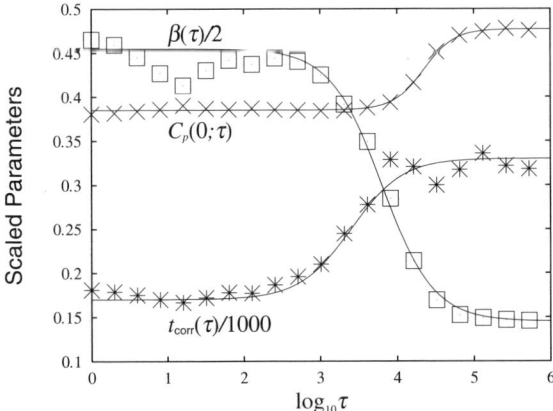

Figure 14. The three parameters $C_p(0;\tau)$, $t_{\text{corr}}(\tau)$ and $\beta(\tau)$ as functions of τ. The latter two parameters $t_{\text{corr}}(\tau)$ and $\beta(\tau)$ are multiplied by $1/1000$ and $1/2$, respectively for a graphical reason. Solid curves are hyperbolic tangent functions described in Eq. (22). [Reproduced with permission from Y. Y. Yamaguchi, *Phys. Rev. E* **68**, 066210 (2003). Copyright 2004 by the American Physical Society.]

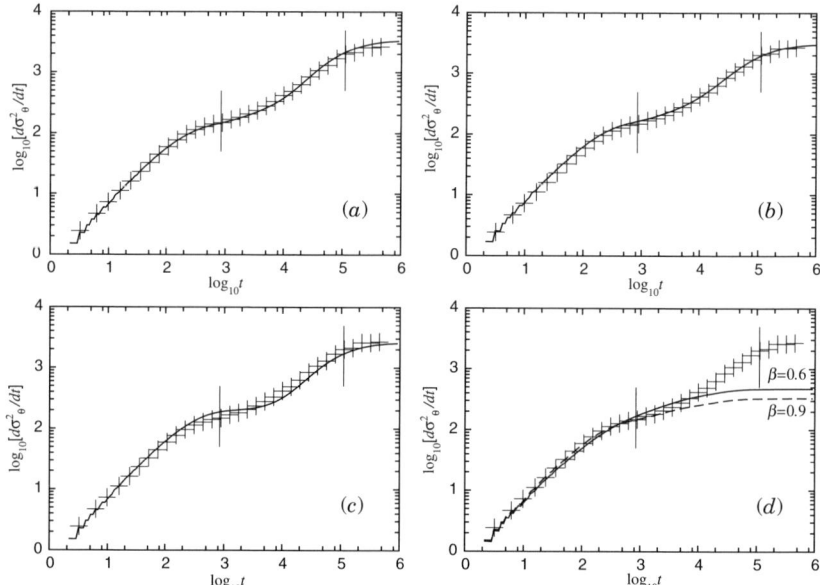

Figure 15. Time derivative of the mean-square displacement, $d\sigma_\theta^2(t)/dt$. (a) Numerical results (crosses) and reproduced one (solid curve) using Eq. (21) and the approximate functions of the three parameters in Eqs. (22). (b, c, d), $C_p(0;\tau)$, $t_{\text{corr}}(\tau)$, and $\beta(\tau)$ are kept constant, respectively. In part d, two constants for $\beta(\tau)$ have been tested. Solid and dashed curves represent $\beta = 0.6$ and 0.9, respectively. The short vertical lines mark the end of Stages I and II. [Reproduced with permission from Y. Y. Yamaguchi, *Phys. Rev. E* **68**, 066210 (2003). Copyright 2004 by the American Physical Society.]

$d\sigma_\theta^2/dt$ using Eqs. (21) and (22), and the reproduced one is in good agreement with numerical results, as shown in Fig. 15a.

If we fix $C_p(0;\tau)$ at its middle value, 0.431, we find that the dependence on τ of $C_p(0;\tau)$ does not affect significantly $d\sigma_\theta^2/dt$, as shown in Fig. 15b. By fixing $t_{\text{corr}}(\tau)$ at its middle value, 250, we obtain the same conclusion for $t_{\text{corr}}(\tau)$ as for $C_p(0;\tau)$, particularly in Stage II (see Fig. 15c). On the contrary, if we fix $\beta(\tau)$ at 0.6 or 0.9, we observe no anomaly in diffusion as shown in Fig. 15d, because $d\sigma_\theta^2(t)/dt$ is proportional to t and is constant in short- and long-time regions respectively, and the same behavior is obtained at equilibrium (see Fig. 9b). Consequently, among the three parameters, $\beta(\tau)$ plays a crucial role to produce anomaly in diffusion.

D. Dependence on Degrees of Freedom

In Stage I (and III), we fit $C_p(t;0)$ (resp. $C_p(t;t_{\text{eq}})$) by a stretched exponential function, which has three parameters: $C_p(0;0)$, $t_{\text{corr}}(0)$ and $\beta(0)$ (resp. $C_p(0;t_{\text{eq}})$,

RELAXATION AND DIFFUSION IN A GLOBALLY COUPLED 497

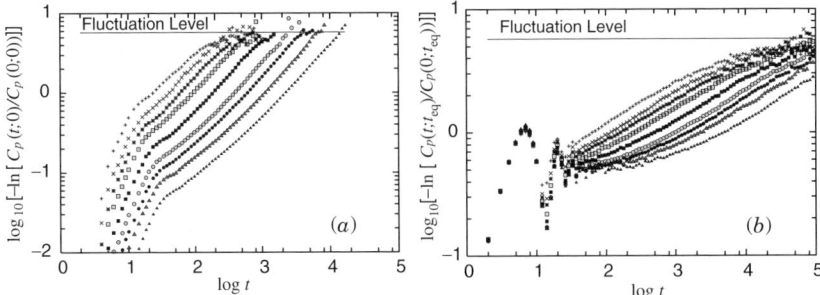

Figure 16. Double log–log plots of correlation functions for various values of degrees of freedom N. (a) $C_p(t;0)$ (Stage I). (b) $C_p(t;t_{eq})$ (Stage III). Both in (a) and in (b), $N = 100(1000)$, 200(500), 300(300), 500(200), 1000(100), 2000(100), 3000(50), 5000(10) and 10000(10) from top to bottom, where the inside of parentheses represent numbers of realizations for $C_p(t;t_{eq})$. For $C_p(t;0)$, the number is 1000 for $N = 1000$ and is 100 for the others. [Reproduced with permission from Y. Y. Yamaguchi, *Phys. Rev. E* **68**, 066210 (2003). Copyright 2004 by the American Physical Society.]

$t_{corr}(t_{eq})$ and $\beta(t_{eq})$). In order to obtain scaling laws for the parameters, we show them as functions of degrees of freedom N. The parameters $C_p(0;0)$ and $C_p(0;t_{eq})$ represent temperature at $t = 0$ and at equilibrium, respectively, and hence they do not depend on N. We therefore focus on the other four parameters, $t_{corr}(0)$, $\beta(0)$, $t_{corr}(t_{eq})$ and $\beta(t_{eq})$. The correlation functions, $C_p(t;0)$ and $C_p(t;t_{eq})$, are shown in Fig. 16, and values of the four parameters are reported as functions of N in Fig. 17.

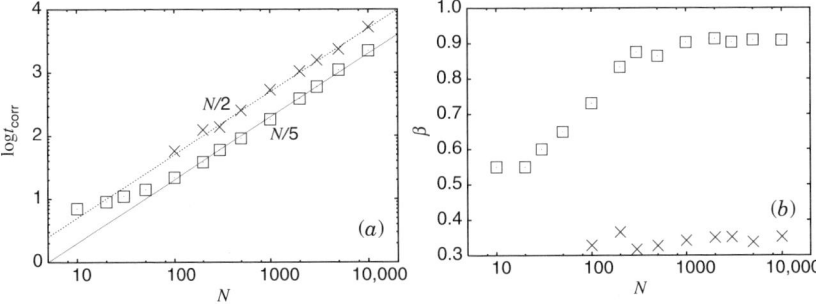

Figure 17. Two parameters of correlation function as functions of degrees of freedom, $t_{corr}(N)$ (a) and $\beta(N)$ (b). In both parts a and b, squares represent values of the parameters for $C_p(t;0)$ (Stage I), and crosses represent those for $C_p(t;t_{eq})$ (Stage III). [Reproduced with permission from Y. Y. Yamaguchi, *Phys. Rev. E* **68**, 066210 (2003). Copyright 2004 by the American Physical Society.]

For large N, $N \geq 200$, the correlation times are proportional to N, that is $t_{\mathrm{corr}}(0) = N/5$ and $t_{\mathrm{corr}}(t_{\mathrm{eq}}) = N/2$, and the stretching exponents $\beta(0)$ and $\beta(t_{\mathrm{eq}})$ are almost constants. We expect that these scaling lows for the four quantities are kept even in the thermodynamic limit, although they break for small N where $t_{\mathrm{corr}}(0)$ is larger than $N/5$ and $\beta(0)$ is smaller than the constant. The duration of Stage I, $t_{\mathrm{I/II}}$, is around 23 for $N = 100$, and hence $t_{\mathrm{corr}}(0)$ and $\beta(0)$ are estimated mainly not in Stage I, but in Stage II from Fig. 16a. In Stage II, $t_{\mathrm{corr}}(\tau)$ and $\beta(\tau)$ are increasing and decreasing functions of τ, respectively (see Fig. 14), and hence $t_{\mathrm{corr}}(0)$ and $\beta(0)$ are larger and smaller than expected values, respectively.

VII. SUMMARY

In a Hamiltonian system having mean field interaction, referred to as the HMF model, we have investigated two features that must reflect self-similar hierarchy of phase space: power-type distribution and anomalous diffusion. They have been reported in the same model for one type of initial condition, and we used a different type of initial condition to check generality.

We have studied the relaxation process through magnetization, and relaxation process is divided into three stages: quasi-stationary, relaxational, and equilibrium. In each stage, the power spectrum of the magnetization takes a power-type tail in a long-time region, and the power-type tail implies that the system has slow dynamics.

The length of the first quasi-stationary stage and relaxation time reaching the third equilibrium stage increases as a nontrivial power-type function of degrees of freedom N, namely $\tau \sim N^{1.7}$. This time scale is also confirmed by observing temporal evolution of single-particle distribution of momenta. The distributions taken at the same scaled time (i.e., $t/N^{1.7}$ = constant) are well-superposed irrespective of values of N, while a trivial time scale $\tau \sim N$ is numerically excluded.

At the quasi-stationary stage, we have checked whether the distribution has a power-law tail or not. The tail of distributions sharply decreases in a log–log plot, and hence we numerically exclude the presence of power-law tails even when the system is in a quasi-stationary state.

We showed that diffusion becomes anomalous only in the relaxational stage, where magnetization is increasing and goes toward the canonical value. This result does not depend on the number of degrees of freedom, at least from $N = 100$ to $10,000$. The interval where the anomaly in diffusion appears becomes shorter and shorter in logarithmic time scale as N increases corresponding to a sharper change of magnetization. Moreover, a detailed investigation exhibits the absence of power-type diffusion even in the relaxational stage. Consequently, we guess that anomaly in diffusion is not anomalous power-type diffusion but instead a transient anomaly due to nonstationarity.

Diffusion is obtained by integrating the correlation function of momenta $C_p(t;\tau)$, and the correlation function is approximated by a series of stretched exponential functions $C_p(t;\tau) = C_p(0;\tau)\exp[-(t/t_{\text{corr}}(\tau))^{\beta(\tau)}]$. Among the three parameters $C_p(0;\tau)$, $t_{\text{corr}}(\tau)$, and $\beta(\tau)$, the stretching exponent $\beta(\tau)$ plays a crucial role to yield anomaly in diffusion. If we assume that $\beta(\tau)$ is a constant, we never observe anomaly in diffusion. This result is consistent with the fact that anomaly in diffusion does not appear in a (quasi-)stationary state, because the correlation functions $C_p(t;\tau)$ and, accordingly, $\beta(\tau)$ are almost invariant with respect to τ.

Both in quasi-stationary and in equilibrium stages, t_{corr} is proportional to N, and β is almost constant. These simple scaling laws imply that fitting by stretched exponential functions is valid irrespective of degrees of freedom.

We have not understood the theoretical reason for the appearance of a stretched exponential function, and the origin of slow dynamics is described by power-law tails of power spectra. Diversity of time scale might be a useful concept to solve the problems. Each rotator of the system is described as a pendulum with moving separatrix, and it may have various time scales determined by distance from the separatrix. If we assume that several time scales with exponential correlation function, $\exp(-t/t_{\text{corr}})$, are present, and we assume the probability distribution function of t_{corr}, $P(t_{\text{corr}})$, then we obtain a stretched exponential function $\int P(t_{\text{corr}})\exp(-t/t_{\text{corr}})dt_{\text{corr}}$ by choosing suitable forms for $P(t_{\text{corr}})$ [40,41]. Moreover, according to the Boltzmann–Jeans conjecture [42–46], the rate of energy transfer between two subsystems becomes exponentially small as a function of λ, which describes the difference between time scales of the two subsystems. We expect that the small transfer rate implies the slow dynamics showing the $f^{-\nu}$ spectrum.

References

1. A. Baba, Y. Hirata, S. Saito, and I. Ohmine, *J. Chem. Phys.* **106**, 3329 (1997) and references therein.
2. Y. Y. Yamaguchi, *Prog. Theor. Phys.* **95**, 717 (1996).
3. Y. Y. Yamaguchi, *Int. J. Bifucation and Chaos* **7**, 839 (1997).
4. C. F. F. Karney, *Physica D* **8**, 360 (1983).
5. B. V. Chirikov and D. L. Shepelyansky, *Physica D* **13**, 395 (1984).
6. Y. Aizawa, *Prog. Theor. Phys.* **71**, 1419 (1984).
7. Y. Aizawa, Y. Kikuchi. T. Haruyama, K. Yamamoto, M. Ota, and K. Tanaka, *Prog. Theor. Phys. Suppl.* **98**, 36 (1989).
8. J. D. Meiss, *Phys. Rev. A* **3**, 2375 (1986).
9. J. D. Meiss and E. Ott, *Physica D* **20**, 387 (1986).
10. A. J. Lichtenberg and M. A. Lieberman, *Regular and Chaotic Dynamics*, 2nd ed., Springer-Verlag, Berlin, 1992, pp. 183–185.

11. S. Ruffo, in *Transport and Plasma Physics*, edited by S. Benkadda et al., eds. World Scientific, Singapore, 1994, p. 114.
12. M. Antoni and S. Ruffo, *Phys. Rev. E* **52**, 2361 (1995).
13. T. Dauxois, S. Ruffo, E. Arimondo, and M. Wilkens, eds., *Dynamics and Thermodynamics in Systems with Long-Range Interactions*, Lecture Notes in Physics 602, Springer, New York, 2002.
14. B. J. Kim, H. Hong, P. Holme, G. S. Jeon, P.Minnhagen, and M. Y. Choi, *Phys. Rev. E* **64**, 056135 (2001).
15. V. Latora, A. Rapisarda, and C. Tsallis, *Physica A* **305**, 129 (2002).
16. A. Torcini and M. Antoni, *Phys. Rev. E* **59**, 2746 (1999).
17. T. Tsuchiya, N. Gouda, and T. Konishi, *Phys. Rev. E* **53**, 2210 (1996).
18. V. Latora, A. Rapisarda, and S. Ruffo, *Phys. Rev. Lett.* **83**, 2104 (1999).
19. Y. Y. Yamaguchi, *Phys. Rev. E* **68**, 066210 (2003).
20. V. Latora, A. Rapisarda, and C. Tsallis, *Phys. Rev. E* **64**, 056134 (2001).
21. T. Geisel, J.Nierwetberg, and A. Zacherl, *Phys. Rev. Lett.* **54**, 616 (1985).
22. J. Klafter and G. Zumofen, *Phys. Rev. E* **49**, 4873 (1994).
23. V. Latora, A. Rapisarda, and S. Ruffo, *Phys. Rev. Lett.* **80**, 692 (1998).
24. K. Kaneko and T. Konishi, *Phys. Rev. A* **40**, 6130 (1989).
25. T. H. Solomon, E. R. Weeks, and H. L. Swinney, *Phys. Rev. Lett.* **71**, 3975 (1993).
26. J. D. Meiss and E. Ott, *Phys. Rev. Lett.* **55**, 2741 (1985); *Physica* **13D**, 395 (1984).
27. G. M. Zaslavsky, D. Stevens, and H. Weitzner, *Phys. Rev. E* **48**, 1683 (1993).
28. T. Geisel, in *Lévy Flights and Related Topics in Physics*, M. F. Schlesinger et al., eds., Springer-Verlag, New York, 1995, p. 153 (see also references therein).
29. H. Koyama and T. Konishi, *Phys. Lett. A* **279**, 226 (2001).
30. M. A. Montemurro, F. Tamarit, and C. Anteneodo, *Phys. Rev. E* **67**, 031106 (2003).
31. V. Latora, A. Rapisarda, and S. Ruffo, *Physica D* **131**, 38 (1999).
32. M.-C. Firpo, *Phys. Rev. E* **57**, 6599 (1998).
33. H. Yoshida, *Phys. Lett. A* **150**, 262 (1990); *Celestial Mech. Dynam. Astron.* **56**, 27 (1993).
34. R. I. McLachlan and P. Atela, *Nonlinearity* **5**, 541 (1992).
35. M. Antoni, H. Hinrichsen, and S. Ruffo, *Chaos, Solitons and Fractals* **13**, 393 (2002).
36. D. H. Zanette, and M. A. Montemurro, *Phys. Rev. E* **67**, 031105 (2003).
37. T. Dauxois, V. Latora, A. Rapisarda, S. Ruffo, and A. Torcini, The Hamiltonian Mean Field Model: From Dynamics to Statistical Mechanics and Back, in Ref. [13]; also cond-mat/0208456.
38. J. C. Phillips, *Rep. Prog. Phys.* **59**, 1133 (1996).
39. Y. Y. Yamaguchi, J. Barré, F. Bouchet, T. Dauxiois, and S. Ruffo, *Physica A*, **00**, 000 (2005).
40. R. G. Palmer, D. L. Stein, E. Abrahams, and P. W. Anderson, *Phys. Rev. Lett.* **53**, 958 (1984).
41. R. A. Pelcovits and D. Mukamel, *Phys. Rev. B* **28**, 5374 (1983).
42. L. Boltzmann, *Nature* **51**, 413 (1895).
43. J. H. Jeans, *Philos. Mag.* **6**, 279 (1903); **10**, 91 (1905).
44. L. Landau and E. Teller, *Physik. Z. Sowjetunion* **11**, 18 (1936).
45. G. Benettin, L Galgani and A. Giorgilli, *Commun. Math. Phys.* **121**, 557 (1989).
46. O. Baldin and G. Benettin, *J. Stat. Phys.* **62**, 201 (1991).

CHAPTER 25

FINITE-TIME LYAPUNOV EXPONENTS IN MANY-DIMENSIONAL DYNAMICAL SYSTEMS

TERUAKI OKUSHIMA

Department of Physics, Tokyo Metropolitan University, Minami-Ohsawa, Hachioji, Tokyo, 192-0397, Japan

CONTENTS

I. Introduction
II. Finite-Time Lyapunov Exponents and Vectors
III. QR Method and Its Correcting Procedure
   A. The Standard QR Method
   B. Finite-Time Error in QR Method
   C. Correcting Procedure for the QR Results
   D. Summary
IV. Lyapunov Instability in Many-Dimensional Hamiltonian Systems
   A. Order of Motion and Lyapunov Instability
   B. Coexistence of Qualitatively Different Local Instabilities
   C. Correction in Many-Dimensional Systems
   D. Summary
V. Conclusion
Acknowledgments
References

I. INTRODUCTION

The aim of this chapter is twofold. One is to give a new method for computing *finite-time* Lyapunov exponents and vectors in many-dimensional dynamical systems, and the other is to discuss the Lyapunov instability of a ϕ^4 model with this method.

Geometric Structures of Phase Space in Multidimensional Chaos: A Special Volume of Advances in Chemical Physics, Part B, Volume 130, edited by M. Toda, T Komatsuzaki, T. Konishi, R.S. Berry, and S.A. Rice. Series editor Stuart A. Rice.
ISBN 0-471-71157-8 Copyright © 2005 John Wiley & Sons, Inc.

The spectrum of Lyapunov exponents provides fundamental and quantitative characterization of a dynamical system. Lyapunov exponents of a reference trajectory measure the exponential rates of principal divergences of the initially neighboring trajectories [1]. Motion with at least one positive Lyapunov exponent has strong sensitivity to small perturbations of the initial conditions, and is said to be chaotic. In contrast, the principal divergences in regular motion, such as quasi-periodic motion, are at most linear in time, and then all the Lyapunov exponents are vanishing. The Lyapunov exponents have been studied both theoretically and experimentally in a wide range of systems [2–5], to elucidate the connections to the physical phenomena of importance, such as transports in phase spaces and nonequilibrium relaxation [6,7].

The existence of Lyapunov exponents is proved, under a general condition, by the multiplicative ergodic theorem of Oseledec [8]. However, the convergence of the exponents is found to be quite slow (algebraically) in time for a generic dynamical system [9], due to its nonhyperbolicity.

In a nonhyperbolic system, chaotic and regular motions coexist, which induces large variations in the magnitudes of instability along a reference trajectory [10]. These variations are related to the alternations, *along a trajectory*, of chaotic and quasi-regular (laminar) motions in two-dimensional discrete-time dynamical systems [11], and further, that of random and cluster motions in high-dimensional Hamiltonian systems [12]. The variations of the instability can be quantified by *finite-time* Lyapunov exponents, which measure the exponential rates of principal divergences during *finite-time* intervals. Statistical scaling analysis of the distribution of finite-time exponents have revealed the presence of invariant structure in phase space, such as homoclinic tangencies [13]. Moreover, recent understandings of shadowability (i.e., computability of chaotic systems) [14,15], mixing process in two-dimensional incompressible flow, entropy production in advection-diffusion equation, and dynamo phenomena [16] have been widely developed with the essential use of finite-time Lyapunov exponents and vectors.

When a dynamical system is nonhyperbolic, there exist time intervals where part of the finite-time Lyapunov exponents accumulate around zero. Hence the spectra of the exponents are (quasi-)degenerate. These degenerate spectra impede our ability to obtain accurate numerical values of finite-time Lyapunov exponents using the existing numerical methods, namely, the QR method and the SVD method [9,17]:

1. The QR methods, based on the matrix factorization of QR decomposition [18], are effective and thus are widely used algorithms for computing the Lyapunov exponents [1,19–22]. However, for the *finite-time* Lyapunov exponents, these methods introduce errors that decrease only algebraically

in time [9]. Goldhirsch, Sulem, and Orszag have derived a correction for the standard QR method [9,23]. This correction is rather effective when the spectrum is nondegenerate, but insufficient to compute the accurate values of *finite-time* exponents when the spectrum is (quasi-) degenerate.

2. On the other hand, the SVD methods, based on the matrix factorization of singular value decomposition (SVD) [18], are algorithms, capable of computing accurate values of finite-time exponents [9,17]. However, the SVD methods not only have a practical disadvantage of requiring quite large computational costs, relative to QR method, but also have a severe limitation of being applicable only to continuous-time dynamical systems with nondegenerate Lyapunov spectra [17].

This chapter is organized as follows: Section II provides the basic definitions of Lyapunov quantities. In Section III, after summarizing the difficulty of the QR method, we construct a new method for computing accurate values of finite-time Lyapunov exponents, by generalizing the correction given by Goldhirsch et al. to higher-order corrections. We present the detail of our method, as well as discuss its efficiency with numerical confirmation. In Section IV, we apply this method to a many-dimensional ϕ^4 model and clearly show that qualitatively different local instabilities coexist along a trajectory and that the observation enables us to accurately determine lifetimes of ordered motions. The usefulness of our method in many-dimensional systems is confirmed there. Section V summarizes this chapter.

II. FINITE-TIME LYAPUNOV EXPONENTS AND VECTORS

Let us consider continuous- or discrete-time dynamical systems in n-dimensional phase space $x = (x^1, x^2, \ldots, x^n)$, whose equations of motion are, respectively,

$$\frac{dx^j(t)}{dt} = f^j(x(t)) \quad \text{or} \quad x^j(t) = F^j(x(t-1)) \quad (1)$$

for $j = 1, 2, \ldots, n$. We write the solution of Eq. (1) starting from x_0 at $t = 0$, as $x(t, x_0)$. Then, the stability matrix from a time t_i to t_f along a reference trajectory $x(t, x_0)$ is given by the $n \times n$ Jacobian matrix $M(t_f, t_i; x_0)$ whose j–k element is defined as

$$M(t_f, t_i; x_0)_{j,k} = \frac{\partial x^j(t_f, x_0)}{\partial x^k(t_i, x_0)} \quad (2)$$

Here, an infinitesimal perturbation $v = (v^1, v^2, \ldots, v^n)$ at $t = t_i$ is transformed to $M(t_f, t_i; x_0) v$ at $t = t_f$. It is noteworthy that $M(t_f, t_i; x_0)$ satisfies the relation

$$M(t_f, t_i; x_0) = M(t_f, 0; x_0) M(t_i, 0; x_0)^{-1} \qquad (3)$$

The time-evolution equations of the stability matrix $M(t, 0)$ are the variational equation of Eq. (1), given by

$$\frac{dM(t, 0)}{dt} = Df \cdot M(t, 0) \quad \text{or} \quad M(t, 0) = DF \cdot M(t - 1, 0) \qquad (4)$$

where Df and DF denote the $n \times n$ Jacobian matrices of f and F evaluated at $x = x(t, x_0)$, respectively. The initial condition for Eq. (4) is $M(t = 0, 0) = I$, where I is the $n \times n$ identity matrix.

Now we introduce the finite-time Lyapunov exponents and corresponding vectors, utilizing the SVD of the stability matrix. The SVD of any $n \times n$ matrix, say A, is a matrix factorization [18]

$$A = U D V^T \qquad (5)$$

where D is a diagonal matrix, and U and V are orthogonal matrices. These SVD matrices are constructed as follows: The symmetric matrix $A^T A$ has nonnegative eigenvalues $\{\sigma_j\}$ and the orthonormalized eigenvectors $\{v_j\}$. Defining u_j as

$$u_j = \frac{A v_j}{\mu_j} \qquad (6)$$

with $\mu_j = \sqrt{\sigma_j}$ for $\sigma_j \neq 0$ and setting the other vectors u_j so that $\{u_j\}$ forms a orthonormalized bases set, we have $A = U D V^T$ with

$$D = \text{diag}(\mu_1, \mu_2, \ldots, \mu_n) \qquad (7)$$
$$U = [u_1, u_2, \ldots, u_n] \qquad (8)$$
$$V = [v_1, v_2, \ldots, v_n] \qquad (9)$$

where $[u_1, u_2, \ldots, u_n]$ denotes the matrix whose ith column vector is u_i.

Any matrix, and thus the stability matrix $M(t_f, t_i)$, is also factorized in the following SVD form:

$$\begin{aligned} M(t_f, t_i) &= U(t_f, t_i) D(t_f, t_i) V(t_f, t_i)^T \\ &= \sum_{j=1}^{n} u_j(t_f, t_i) \mu_j(t_f, t_i) v_j(t_f, t_i)^T \end{aligned} \qquad (10)$$

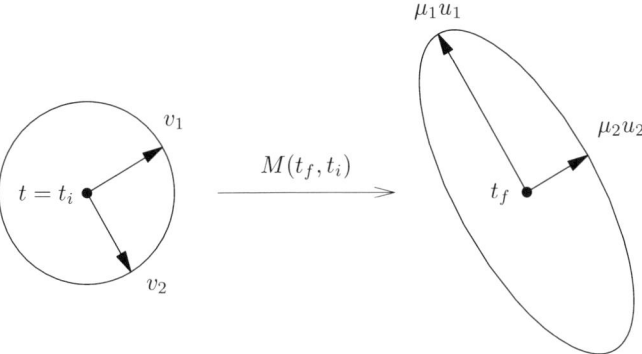

Figure 1. Graphic illustration of the singular value decomposition, Eq. (10), in a two-dimensional system.

where μ_j are, without loss of generality, assumed to be ordered as

$$\mu_1 \geq \mu_2 \geq \cdots \geq \mu_n \tag{11}$$

The meaning of the singular value decomposition, Eq. (10), is depicted in Fig. 1, which shows that the tangent vector v_j at $t = t_i$ is transformed to the tangent vector u_j multiplied by a scalar μ_j at $t = t_f$.

From this decomposition, the finite-time Lyapunov exponents in the time interval from t_i to t_f are given by

$$\lambda_j(t_f, t_i) = \frac{\log \mu_j(t_f, t_i)}{(t_f - t_i)} \quad (j = 1, 2, \ldots, n) \tag{12}$$

The corresponding finite-time and left finite-time Lyapunov vectors are given by $v_j(t_f, t_i)$ and $u_j(t_f, t_i)$ in Eq. (10), respectively.

The ordinary (i.e., infinite-time) Lyapunov exponents and vectors are given by the $t_f \to \infty$ limits of the associating finite-time Lyapunov quantities:

$$\lambda_j = \lim_{t_f \to \infty} \lambda_j(t_f, t_i) \tag{13}$$

$$v_j(t_i) = \lim_{t_f \to \infty} v_j(t_f, t_i) \tag{14}$$

Here, the t_i-independency of the exponents is shown by the multiplicative ergodic theory of Oseledec [8].

III. QR METHOD AND ITS CORRECTING PROCEDURE

A. The Standard QR Method

The QR methods [1,19] are based on the relation

$$\lambda_j = \lim_{t \to \infty} \frac{\log R(t,0)_{j,j}}{t} \tag{15}$$

where $R(t,0)$ is the following matrix given by the QR factorization of the stability matrix:

$$M(t,0) = Q(t,0) R(t,0) \tag{16}$$

where the upper-triangular matrix $R(t,0)$ with nonnegative diagonal elements, as well as the orthogonal matrix $Q(t,0)$, is uniquely determined.

For a multidimensional chaotic system, the condition number of $M(t,0)$, which is defined by $\|M\| \times \|M^{-1}\|$ with $\|M\| := \max\{|Mx|/|x| : x \neq 0\}$ [18], becomes exponentially large for large t, which introduces a large amount of errors into the direct evaluation of Eq. (16) (orthonormalizing the column vectors in M).

To evade this numerical difficulty, the standard QR method evaluates Eq. (16) as follows [1,19]:

1. By dividing time into intervals τ ($t_k = k\tau$ for $k = 1, 2, \ldots$), $M(t,0)$ is represented as

$$M(t,0) = T_n T_{n-1} \ldots T_1 \quad \text{for } t = n\tau \tag{17}$$

with

$$T_k = M(t_k, t_{k-1}) \tag{18}$$

2. Then, with utilizing QR decomposition repeatedly, Q_k and R_k ($k = 1, 2, \ldots$) are introduced as follows:

$$\begin{aligned} T_1 &= Q_1 R_1 \\ T_k Q_{k-1} &= Q_k R_k \quad (k \geq 2) \end{aligned} \tag{19}$$

These matrices satisfy

$$M(t,0) = Q_n R_n R_{n-1} \ldots R_1 \tag{20}$$

and thus

$$R(t,0) = R_n R_{n-1} \ldots R_1 \tag{21}$$

3. The standard QR method evaluates the Lyapunov exponents as

$$\lambda_j = \lim_{t \to \infty} \frac{\sum_{k=1}^{n} \log(R_k)_{j,j}}{t} \qquad (22)$$

since $\log(R(t,0))_{j,j} = \sum_{k=1}^{n} \log(R_k)_{j,j}$ for upper-tridiagonal R_k.

B. Finite-Time Error in QR Method

We now introduce a normalized form of the stability matrix, which gives approximate values of finite-time Lyapunov exponents and vectors:

$$M(t_f, t_i) = U e^d r V^T \qquad (23)$$

where the dependencies, on t_i and t_f, of the matrices in the right-hand side are omitted for notational simplicity. Here, U and V are orthogonal, d is diagonal, and r is an upper-triangular matrix whose diagonal elements are normalized to unity. These matrices are determined as follows:

1. Since $M(t_f, t_i) = M(t_f, 0) M(t_i, 0)^{-1}$ and $M(t_n, 0) = Q_n R_n R_{n-1} \ldots R_1$, the stability matrix $M(t_f, t_i)$ is represented by

$$\begin{aligned} M(t_f, t_i) &= Q_f R_f R_{f-1} \ldots R_2 R_1 \cdot R_1^{-1} R_2^{-1} \ldots R_i^{-1} Q_i^T \\ &= Q_f R_f R_{f-1} \ldots R_{i+1} Q_i^T \end{aligned} \qquad (24)$$

where $t_i = i\tau$, $t_f = f\tau$. Therefore, U and V are chosen as

$$U = Q_f, \qquad V = Q_i \qquad (25)$$

2. The remaining matrices d and r are, respectively, given by

$$d_{j,j} = \log(R_{i+1})_{j,j} + \log(R_{i+2})_{j,j} + \cdots + \log(R_f)_{j,j} \qquad (26)$$

$$r = e^{-d} R_f R_{f-1} \ldots R_{i+2} R_{i+1} \qquad (27)$$

Practically, in order to obviate the numeric overflow or underflow, r is computed as

$$r = r_{f-i} \ldots r_2 r_1 \qquad (28)$$

where

$$r_1 = e^{-d_1} R_{i+1} \qquad (29)$$

$$r_k = e^{-d_{k+1}} R_{i+k} e^{d_k} \qquad (k \geq 2) \qquad (30)$$

and $(d_k)_{j,j} = \log(R_{i+1})_{j,j} + \log(R_{i+2})_{j,j} + \cdots + \log(R_{i+k})_{j,j}$. The estimations of U, d, and V are straightforward.

Note that if all off-diagonal elements of r are negligibly small compared to the diagonal elements, that is,

$$|r_{i,j}| \ll 1 \quad \text{for} \quad 1 \leq i < j \leq n \tag{31}$$

then the jth finite-time Lyapunov exponent and (left) vector are given by $d_{j,j}/(t_f - t_i)$ and the jth column vectors of $(V)U$, respectively. Therefore, we define the QR-approximate exponents $\lambda_j^{QR}(t_f, t_i)$ as

$$\lambda_j^{QR}(t_f, t_i) = \frac{d_{j,j}}{(t_f - t_i)} \tag{32}$$

In general, however, r is far from diagonal, and thus λ_j^{QR}, U, and V are not accurate approximations of corresponding Lyapunov quantities.

To see the discrepancy between the exact and the approximate exponents, we have computed these values using the standard map (also referred to as the Chirikov–Taylor map):

$$\begin{aligned} y(t) &= y(t-1) - K\sin(x(t-1)) \\ x(t) &= x(t-1) + y(t) \end{aligned} \tag{33}$$

The exact exponents λ_j are directly computed by diagonalizing the symmetric matrix $M^T M$ with high-precision computation to evade its roundoff error. Figure 2 plots the error in the smallest exponents, $|\lambda_2^{QR}(t,0) - \lambda_2(t,0)|$, against t. We can see that until $t \simeq 30$ the error rapidly decreases. After the initial dropping stage, it decreases slowly as $\sim 1/t$ (inset). This quite slow convergence shows that λ_j^{QR} is not a sufficiently accurate approximation of the finite-time Lyapunov exponent. Note here that behavior of λ_1 is similar to that of λ_2.

C. Correcting Procedure for the QR Results

Now we present our novel method for computing accurate values of finite-time Lyapunov exponents and vectors, by correcting the finite-time error in λ^{QR} [24].

To this end, we construct a sequence of refinements $U_{(k)}$, $d_{(k)}$, $r_{(k)}$, and $V_{(k)}$ ($k = 0, 1, 2, \ldots$) satisfying

$$r_{(k)} \to \text{identity matrix} \quad \text{as} \quad k \to \infty \tag{34}$$

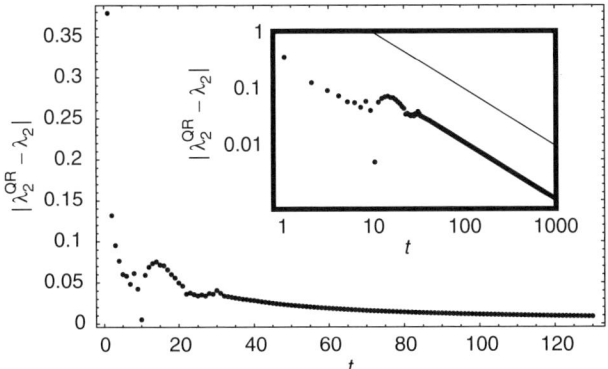

Figure 2. The finite-time errors in QR method, $|\lambda_2^{QR}(t,0) - \lambda_2(t,0)|$, are plotted against t for $K = 1.5$, $(x(0), y(0)) = (1.1\pi, 0)$. Semilog plot for $t < 130$ and log–log plot for $t < 1000$ (**inset**). In the inset, the solid line $10/t$ is shown for eye guidance.

with the normalization condition

$$M = \begin{cases} U_{(k)} e^{d_{(k)}} r_{(k)} V_{(k)}^T & \text{for even } k \\ U_{(k)} r_{(k)}^T e^{d_{(k)}} V_{(k)}^T & \text{for odd } k \end{cases} \quad (35)$$

Here $U_{(k)}$ and $V_{(k)}$ are orthogonal, $d_{(k)}$ is diagonal, and $r_{(k)}$ is an upper-triangular matrix whose diagonal elements are normalized to unity.

The construction of these matrices is as follows. Starting from $U_{(0)} = U$, $d_{(0)} = d$, $r_{(0)} = r$, and $V_{(0)} = V$, we generate the successors $d_{(k)}, r_{(k)}$ ($k \geq 1$) by

$$\begin{aligned} r_{(k-1)}^T &= \mathcal{Q}_{(k)} \mathcal{R}_{(k)} \quad \text{(QRD)} \\ r_{(k)} &= e^{-d_{(k-1)}} \mathcal{D}_{(k)}^{-1} \mathcal{R}_{(k)} e^{d_{(k-1)}} \\ d_{(k)} &= \log(\mathcal{D}_{(k)}) + d_{(k-1)} \end{aligned} \quad (36)$$

where $\mathcal{D}_{(k)}$ is the diagonal matrix equal to the diagonal part of $\mathcal{R}_{(k)}$. The matrices $U_{(k)}, V_{(k)}$ are given by

$$U_{(k)} = U_{(0)} \mathcal{Q}_{(2)} \mathcal{Q}_{(4)} \cdots \mathcal{Q}_{(2\lfloor k/2 \rfloor)} \quad (37)$$

$$V_{(k)} = V_{(0)} \mathcal{Q}_{(1)} \mathcal{Q}_{(3)} \cdots \mathcal{Q}_{(2\lfloor (k-1)/2 \rfloor + 1)} \quad (38)$$

where $\lfloor x \rfloor$ denotes the largest integer not greater than x. As shown in Fig. 3, this procedure is intrinsically regarded as the diagonalization of the symmetric

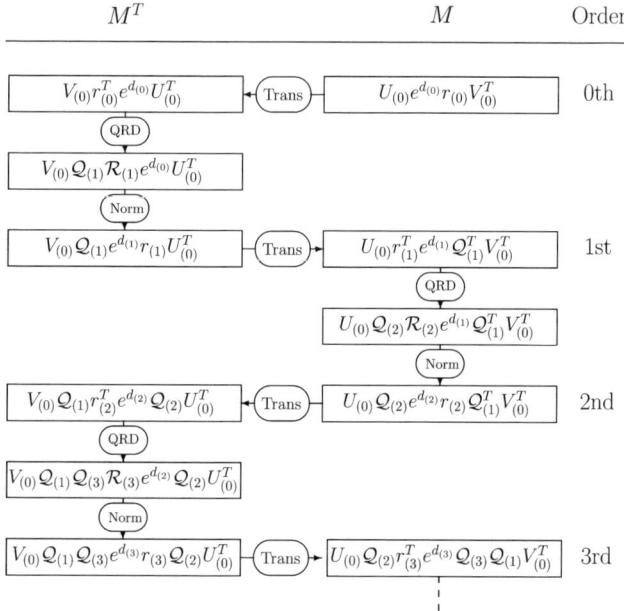

Figure 3. Diagrammatic representation of the correcting procedure Eq. (36), where "QRD" denotes QR decomposition of the matrix $r_{(k)}$ and "Norm" denotes the normalization procedure: $\mathcal{R}_k e^{d_{(k-1)}} \to e^{d_{(k)}} r_{(k)}$ (see text).

matrix $M^T M$ via the LR method [18], except for including the normalization for evading numerical disasters, arising from the large condition number of the stability matrix. From the general property of the LR method, we can see that this iterative procedure converges to the exact SVD exponentially as $k \to \infty$ [18]. Using $U_{(k)}$, $V_{(k)}$, and $d_{(k)}$, we define the kth corrected Lyapunov exponent as

$$\lambda_j^{(k)}(t_f, t_i) = \frac{d_{(k)j,j}}{(t_f - t_i)} \tag{39}$$

and the kth corrected (left) Lyapunov vectors as the jth column vector of $(U_{(k)})V_{(k)}$. Note that the first corrected exponent $\lambda_j^{(1)}(t, 0) = d_{j,j}/t + \log(\mathcal{D}_{(1)j,j})$ is equal to the correction derived by Goldhirsch et al. for nondegenerate spectra systems in Refs. 9 and 23. Namely, our correcting procedure is a generalization of the correction term proposed by them.

Here we numerically test our method using the standard map with the same parameters as in Fig. 2. The kth corrected errors in the smallest exponents, $|\lambda_2^{(k)}(t,0) - \lambda_2(t,0)|$, are plotted in Fig. 4a as a function of t for $k = 0, 1, 2,$

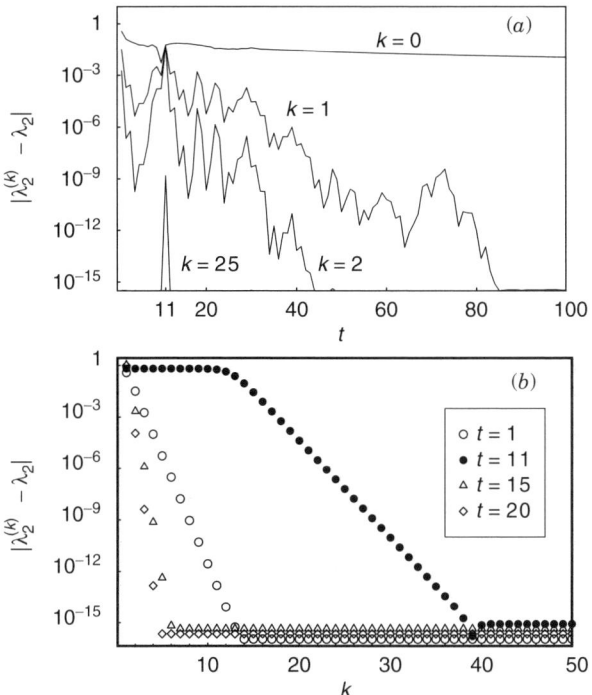

Figure 4. The kth corrected errors in the smallest exponents are plotted: (a) $|\lambda_2^{(k)}(t,0) - \lambda_2(t,0)|$ versus t for $k = 0, 1, 2$, and 25. (b) $|\lambda_2^{(k)}(t,0) - \lambda_2(t,0)|$ versus k for $t = 1, 11$, and 20.

and 25.[1] We can see that, for all t computed, the errors rapidly decrease as k increases, and we can also see that the convergencies have different rates dependent on t. For example, the slowest convergence is observed at $t = 11$. To see the detail of the convergencies, we plot the errors as a function of k in Fig. 4b. There are intervals of k in which the errors decrease exponentially, up to precisions close to the floating number precision (16 digits). At $t = 11$, the initial step k_0 at which the interval starts is larger ($k_0 \sim 15$) than that of $t \neq 11$ ($k_0 \sim 1$), because the standard QR method fails to describe the sudden changes in directions of Lyapunov vectors. This is the major reason for the slowest convergence observed at $t = 11$ in Fig. 4b. These observations show that the higher corrections ($k \geq 2$) are generally important for our high-accuracy computation of finite-time Lyapunov exponents.

---

[1] The behavior of λ_1 here is also similar to that of λ_2.

D. Summary

In this section, we have developed a numerical algorithm for computing accurate values of finite-time Lyapunov exponents and vectors, based on the standard QR method, by constructing a new correcting procedure. This correcting procedure is regarded as a generalized LR method. As a result, the correcting process exponentially converges to the exact Lyapunov quantities, and is, in contrast to the existing method, applicable for general dynamical systems, even when their Lyapunov spectra are (quasi-)degenerate or they are nonhyperbolic systems. This method is, as a whole, very efficient, because of the rapid convergence, and because the correcting procedure is only called when the exact quantities are necessary. Another virtue of this method is a practical advantage of being easy to implement [see Eqs. (19) and (36)].

IV. LYAPUNOV INSTABILITY IN MANY-DIMENSIONAL HAMILTONIAN SYSTEMS

In this section, we study the singular values of an many-dimensional oscillator system, as the first step toward the elucidation of the dynamical origin of ordered motions, and show a clear evidence of coexistence of qualitatively different local instabilities along a trajectory, which enables us to accurately determine lifetimes of ordered motions.

A. Order of Motion and Lyapunov Instability

Let us consider a $(2\Lambda + 1)$-degrees-of-freedom oscillator system (Λ is a nonnegative integer), whose Hamiltonian is given by a ϕ^4-interaction model truncated in reciprocal space (ϕ^4 MTRS):

$$H = \sum_{j=-\Lambda}^{\Lambda} \left(\tfrac{1}{2} p_j p_{-j} + \frac{\omega_j}{2} q_j q_{-j} \right) + \frac{\lambda}{4} \sum{}' q_{j_1} q_{j_2} q_{j_3} q_{j_4} \tag{40}$$

where $\omega_j = \sqrt{1 + j^2}$ and λ is a nonlinearity parameter. Here, all modes $j = -\Lambda, -\Lambda + 1, \ldots, \Lambda$ satisfy the reality conditions $q_j = q^*_{-j}$; $p_j = p^*_{-j}$ and

$$\sum{}' = \sum_{j_1, j_2, j_3, j_4 = -\Lambda}^{\Lambda} \delta_{j_1 + j_2 + j_3 + j_4, 0} \tag{41}$$

For the ϕ^4 MTRS with $\lambda = 1.0$ and $N(= 2\Lambda + 1) = 5$, we compute the finite-time Lyapunov exponents $\lambda_j(t, t - \Delta T)$ with $\Delta T = 1000$, via the corrected QR method developed in the previous section. The initial condition

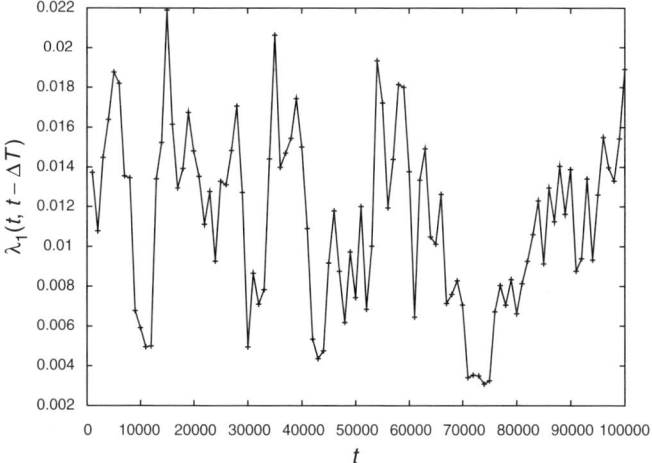

Figure 5. Plot of $\lambda_1(t, t - \Delta T)$ against t with $\Delta T = 1000$ for ϕ^4 MTRS.

has total energy 200 where the action variables are set to have ratio

$$I_{-2} : I_{-1} : I_0 : I_1 : I_2 = 2 : 0 : 1 : 3 : 1 \tag{42}$$

and the angular variables are given by random numbers form uniform distribution in $[0, 2\pi)$. (The following arguments are independent of the random number realizations.)

Figure 5 shows that the leading exponents are positive and thus the ϕ^4 MTRS is chaotic. In addition, they varies depending on the time intervals: We see relatively small instability regions, for example, around the intervals 9000–12,000 and 70,000–75,000. The variations of finite-time Lyapunov exponents have been related to the alternations between qualitatively different motions, such as (a) chaotic and quasi-regular, laminar motions in two-dimensional systems [11] and (b) random and cluster motions in high-dimensional systems [12], and they have been utilized for detecting these ordered motions.

Let us confirm the correspondence between the relative magnitude of instability and the orders of motions with this ϕ^4 MTRS. Figure 6 shows stroboscopic data of $j = 0$ mode variables (x_0, p_0) with sampling period $1/\omega_0$. The left panel presents the data in the time interval $70,000 \leq t \leq 75,500$, where the leading finite-time Lyapunov exponent is relatively small. A coherent, quasi-periodic motion is clearly seen in these variables. Since almost all energy concentrates in the zeroth mode, the other mode variables $j \neq 0$ are hardly excited in this coherent motion. In contrast, the right panel in Fig. 6 presents the data in the time interval $20,000 \leq t \leq 25,500$, where the leading exponent is

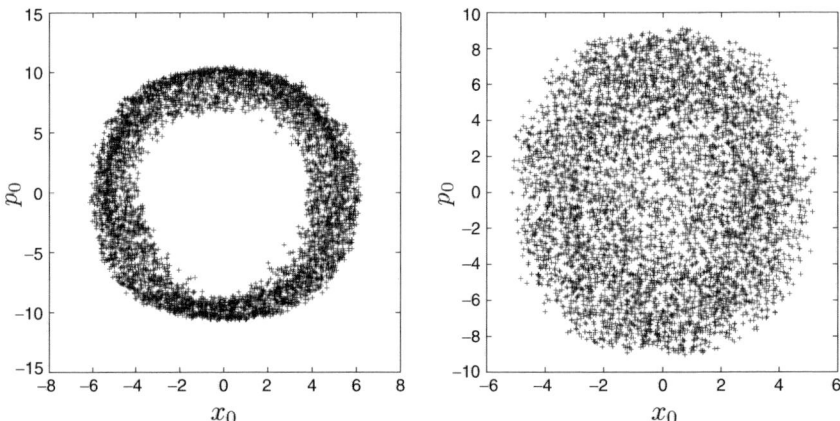

Figure 6. Plot of (x_0, p_0) with sampling period $1/\omega_0$ for the time interval $70,000 \leq t \leq 75,500$ (**left panel**) and for $20,000 \leq t \leq 25,500$ (**right panel**).

relatively large. In this case, the other modes, as well as the zeroth mode, show irregular, chaotic motions in energetically allowed regions.

As in the pioneering works, such as Ref. 12, we have confirmed that this many-dimensional model is a nonhyperbolic dynamical system, which has chaotic trajectories, along which motions change intermittently form irregular to ordered, and vice versa. However, it is not fully understood why we can find ordered motions just by monitoring relatively small local Lyapunov instability regions, although their magnitude itself has no relation to the orders of motions in principle.

B. Coexistence of Qualitatively Different Local Instabilities

When a continuous-time Hamiltonian system has an extra, independent conserved quantity, there exist a singular value increasing linearly in time and another singular value decreasing inversely proportional to time. Thus, when a system has an extra, independent quantity that is quasi-conserved during an finite time interval, such as ordered motion in Fig. 6 (left), there must exist a singular value increasing linearly in time and another singular value decreasing inversely proportional to time in the time interval.

Examples of typical plots of singular values $\mu_1(t, t_i)(= e^{\lambda_j(t, t_i)(t-t_i)})$ for ordered and irregular motions ($\Lambda = 2$, $\lambda = (32\pi)^{-1}$) are, respectively, the thick lines in Figs. 7a and 7b. These figures show that the local instability has a qualitative difference that corresponds to the orders of motions: μ_1 increases linearly in time, $t - t_i$, for (a) quasi-periodic ordered motions and exponentially for (b) irregular motions.

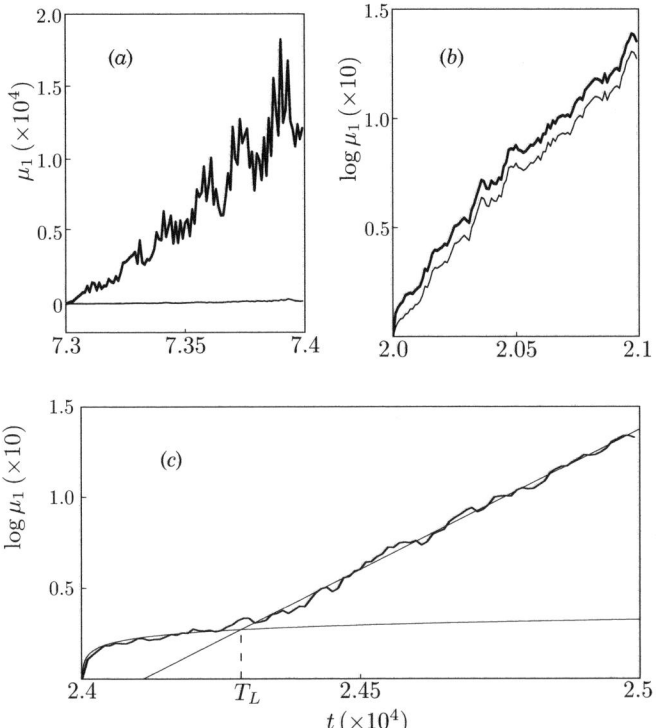

Figure 7. $\mu_1(t, t_i)$ against t, $t_i < t < t_i + 1000$, for an initial condition $(H(p(0), q(0)) = 300)$: (a) $t_i = 73,000$ and (b) $t_i = 20,000$. The thick and the thin lines are our corrected results and the approximate results of the standard QR method, respectively. The thick line in (c) is μ_1 for $t_i = 24,000$. The thin lines are the fitted lines in the early linear and in the latter exponential stage, respectively. $T_L \sim 24,300$ is the time at the intersection of these fitted lines.

Moreover, thanks to the great accuracy of our method, lifetimes of ordered motions are preciously determined as follows. Figure 7c shows a typical plot of μ_1 that changes from linear to exponential increase. The crossover time T_L in Fig. 7c corresponds to the change from an ordered to an irregular motion. Thus the lifetime of the ordered motion is accurately given by T_L. Note that this method for computation of lifetimes has advantages over other more naive methods, such as direct presentation of trajectories with suitable variables, because it is applicable even when the values of T_L is too small to recognize their ordered motions or the choice of projected variables to visualize them is nontrivial.

Here we discuss the availability of conventional method for finding ordered motions. Suppose that a Hamilton system with N degrees of freedom has a number M ($M < N$) of quasi-conserved quantities I_j ($j = 1, 2, \ldots, M$). Then, each singular values that correspond to the quasi-conserved quantities I_j vary linearly and inverse linearly in time, and thus both of the associated finite-time Lyapunov exponents approach zero. The remaining singular values change exponentially and then the associated finite-time Lyapunov exponents approach nonzero finite values. Hence, for quasi-conserved quantities with sufficiently long lifetimes, their existence are shown by the finite-time Lyapunov exponents around zero. This is the case in which the conventional method, via relatively small instabilities in time series or the spectra, for detecting ordered motions are justified. In contrast, the singular value method presented here can find ordered motions, even when quasi-conserved quantities do not have long lifetimes enough to distinguish themselvs from nonzero value.

C. Correction in Many-Dimensional Systems

So far, the singular values have been computed by using our method with the relation $\mu_j = \exp(\lambda_j(t - t_i))$, where the accuracy is confirmed by comparing them to the result of the high-precision diagonalization of $M^T M$, as in Section III.C. Here, we show the indispensability of our novel corrections for acquiring the quantitative and qualitative properties.

The approximate μ_1 computed via the standard QR method are plotted for comparison, with the thin lines in Figs. 7a and 7b. These figures show that the standard QR method dose not have enough accuracy in both cases: (a) For ordered, quasi-periodic motions, the QR-approximate μ_1 gives much smaller result than the accurate result and does note give the original, linear increase. Hence, due to the absence of the exponential instability, the standard QR method fails to reproduce the *qualitative* change of μ_1. (b) For irregular chaotic motions, the approximate $\log \mu_1$ grows linearly but is smaller by an approximately constant gap compared to the accurate result. Namely, the standard QR method fails to describe the quantitatively accurate change of $\log \mu_1$ in this case.

These also show that our corrections are necessary to obtain lifetimes of ordered motions, because, without these corrections, the qualitative changes, and thus the clear crossover times, generally disappear.

Thus, we have again confirmed that our corrections are necessary for accurate computation of local Lyapunov instability in this multidimensional dynamical system.

D. Summary

In this section, we have applied our novel method to the ϕ^4 model and clarified the following: (1) We give a new scheme for detecting ordered motions with the linear increase of singular values of stability matrixes. From the crossover times of local instabilities that change from linear to exponential increases, lifetimes of the associated ordered motions are determined accurately. (2) We then elucidated the condition that the conventional method for finding ordered motions are available. (3) The corrections are shown to be indispensable also for many-dimensional dynamical systems, in which magnitudes and directions of stable, unstable, and marginal Lyapunov instabilities may fluctuate.

V. CONCLUSION

In this chapter, we have developed a numerical algorithm for computing accurate values of finite-time Lyapunov exponents and vectors, by constructing a correcting procedure to the standard QR method. This procedure is a generalized LR method. As a result, the corrected results exponentially converge to the exact Lyapunov quantities for generic multidimensional dynamical systems including nonhyperbolic systems with (quasi-)degenerate Lyapunov spectra. This method is easy to implement [see Eqs. (19) and (36)] and very efficient, because of the exponential convergence and because the correcting procedure is called only when the exact quantities are necessary. We have demonstrated the efficiency of our method by applying it both to the standard map and to a multidimensional system consisting of oscillators. In the application to the latter system, alternations in qualitatively different local instabilities have been found along a trajectory. From crossover times of local instabilities that change from linear to exponential increases, lifetimes of the associated ordered motions are determined accurately.

We expect that this correcting procedure can be applicable for other numerical methods, such as the symplectic method [21,22], and that faster convergence may be accomplished by introducing shifts [18] into the correcting process. These variants would be useful for exploring instability, especially, in high-dimensional dynamical systems. We hope that these methods will help us to develop understandings of generic multidimensional nonhyperbolic chaotic systems.

Acknowledgments

I express gratitude to A. Shudo and A. Tanaka for continuous encouragement and enlightening discussions. The support given by the following sponsors is deeply appreciated: the Japan–U.S. Cooperative Science Program of Japan Society of the Promotion of Science, Inoue Foundation for Science, and Yukawa Institute for Theoretical Physics at Kyoto University.

References

1. J.-P. Eckmann and D. Ruelle, *Rev. Mod. Phys.* **57**, 617 (1985).
2. M. Sano and Y. Sawada, *Phys. Rev. Lett.* **55**, 1082 (1985).
3. T. D. Sauer, J. A. Tempkin, and J. A. Yorke, *Phys. Rev. Lett.* **81**, 4341 (1998).
4. T. D. Sauer and J. A. Yorke, *Phys. Rev. Lett.* **83**, 1331 (1999).

5. H. D. I. Abarbenel, *Analysis of Observed Chaotic Data*, Springer-Verlag, New York, 1995.
6. A. J. Lichtenberg and M. A. Lieberman, *Regular and Stochastic Motion*, Springer-Verlag, New York, 1983.
7. E. Ott, *Chaos in Dynamical Systems*, Cambridge University Press, Cambridge, 1993.
8. V. I. Oseledec, *Moscow Math. Soc.* **19**, 197 (1968).
9. I. Goldhirsch, P.-L. Sulem, and S. A. Orszag, *Physica D* **27**, 311 (1987).
10. Y. Aizawa, Y. Kikuchi, T. Harayama, K. Yamamoto, M. Ota, and K. Tanaka, *Prog. Theor. Phys. Suppl.* **98**, 36 (1989).
11. M. A. Sepúlveda, R. Badii, and E. Pollak, *Phys. Rev. Lett.* **63**, 1226 (1989).
12. T. Konishi and K. Kaneko, *J. Phys. A* **25**, 6283 (1991).
13. T. Morita, H. Hata, H. Mori, T. Horita, and K. Tomita, *Prog. Theor. Phys.* **79**, 296 (1988).
14. S. Dawson, C. Grebogi, T. Sauer, and J. A. Yorke, *Phys. Rev. Lett.* **73**, 1927 (1994); T. Sauer, C. Grebogi, and J. A. Yorke, *Phys. Rev. Lett.* **79**, 59 (1997).
15. W. B. Hayes, *Phys. Rev. Lett.* **90**, 54104 (2003).
16. X. Z. Tang and A. H. Boozer, *Physica D* **95**, 283 (1996); M. M. Alvarez, F. J. Muzzio, S. Cerbelli, A. Adrover, and M. Giona, *Phys. Rev. Lett.* **81**, 3395 (1998); M. Giona and A. Adrover, *Phys. Rev. Lett.* **81**, 3864 (1998); A. Adrover and M. Giona, *Phys. Rev. E* **60**, 347 (1999); J.-L. Thiffeault and A. H. Boozer, *Chaos* **11**, 16 (2001); M. Giona, S. Cerbelli, and A. Adrover, *Phys. Rev. Lett.* **88**, 024501 (2002).
17. K. Geist, U. Parlitz, and W. Lauterborn, *Prog. Theor. Phys.* **83**, 875 (1990).
18. G. H. Golub and C. F. Van Loan, *Matrix Computations*, 3rd ed., Johns Hopkins University Press, Baltimore, 1996.
19. G. Benettin, L. Galgani, and J. M. Strelcyn, *Phys. Rev. A* **14**, 2338 (1976); I. Shimada and T. Nagashima, *Prog. Theor. Phys.* **61**, 1605 (1979).
20. F. Christiansen and H. H. Ruth, *Nonlinearity* **10**, 1063 (1997).
21. S. Habib and R. D. Ryne, *Phys. Rev. Lett.* **74**, 70 (1995); G. Rangarajan, S. Habib, and R. D. Ryne, *Phys. Rev. Lett.* **80**, 3747 (1998).
22. M. H. Partovi, *Phys. Rev. Lett.* **82**, 3424 (1999).
23. J.-L. Thiffeault, *Physica (Amsterdam)* **172D**, 139 (2002).
24. T. Okushima, *Phys. Rev. Lett.* **91**, 254101 (2003).

CHAPTER 26

THE ROLE OF CHAOS FOR INERT AND REACTING TRANSPORT

MASSIMO CENCINI and ANGELO VULPIANI

Dipartimento di Fisica, Università di Roma "la Sapienza" and Center for Statistical Mechanics and Complexity INFM UdR Roma 1, Piazzale Aldo Moro 5, I-00185 Roma, Italy

DAVIDE VERGNI

Istituto Applicazioni del Calcolo, CNR Viale del Policlinico 137, I-00161 Roma, Italy

To know that you know when you do know
and to know that you do not know when you do not know:
that is knowledge.

—Confucius [Analects, 2:17]

CONTENTS

I. Introduction
II. Transport of Inert Substances
   A. Standard and Anomalous Diffusion
   B. About the Meaning of Anomalous
   C. Strong Anomalous Diffusion in Chaotic Flows
III. Transport of Reacting Substances
   A. Fronts in Cellular Flows
   B. Slow and Fast Reaction Regimes
   C. Geometrical Optics Limit
   D. On the Relevance of Chaos in Front Propagation
IV. Summary
Acknowledgments
References

Geometric Structures of Phase Space in Multidimensional Chaos: A Special Volume of Advances in Chemical Physics, Part B, Volume 130, edited by M. Toda, T Komatsuzaki, T. Konishi, R.S. Berry, and S.A. Rice. Series editor Stuart A. Rice.
ISBN 0-471-71157-8 Copyright © 2005 John Wiley & Sons, Inc.

I. INTRODUCTION

The dynamics of fields advected by a velocity field is a problem of considerable practical and theoretical interest in many disciplines as astrophysics, geophysics, chemical engineering, and disordered media [1, 2]. Though typically the advected field backreacts on the flow modifying the velocity field, often the advected field can be considered as passive—that is, transported without dynamical effects on the velocity field. This is the case we shall consider in this contribution. Under this simplifying assumption, the most general equation describing the evolution of the concentrations of N species, $\theta_i(x,t)$, is

$$\partial_t \theta_i + \boldsymbol{u} \cdot \boldsymbol{\nabla} \theta_i = D_i \Delta \theta_i + \frac{1}{\tau_i} f_i(\theta_1, \ldots, \theta_N) \tag{1}$$

where on the left-hand side the second term accounts for the transport by an incompressible velocity field; on the right-hand side the first term represents molecular diffusion (D_i is the diffusion constant for the ith species), and the second one models possible chemical or biological processes (of characteristic time scale τ_i) taking place among the different substances.

When $f_i(\theta_1, \ldots, \theta_N) = 0$ we speak of inert transport while if $f_i(\theta_1, \ldots, \theta_N) \neq 0$ of reacting transport.

In typical transport problems, one considers a scalar field θ, representing, for example, the concentration of some substance that evolves according to the advection–diffusion equation

$$\partial_t \theta + \boldsymbol{u} \cdot \boldsymbol{\nabla} \theta = D \Delta \theta \tag{2}$$

being D the molecular diffusivity; often a source term is added in the right-hand side. Given the field $\boldsymbol{u}$, the main goal is to understand the dynamical and statistical properties of the field θ. Remarkably, in the last few years, much progress has been achieved in this direction, so that now we have a satisfactory understanding of the statistics of passive fields. The interested reader may consult the review [3] where an exhaustive discussion on passive fields in turbulent flows can be found.

Transport problems of fields can be investigated in terms of particles motion, which is an interesting aspect per se. This is clear if we note that Eq. (2) is nothing but the Fokker–Planck equation of the stochastic process describing the motion of test particles:

$$\dot{\boldsymbol{x}}(t) = \boldsymbol{u}(\boldsymbol{x}(t), t) + \sqrt{2D}\, \boldsymbol{\eta}(t) \tag{3}$$

where $\boldsymbol{u}(\boldsymbol{x}(t),t)$ is the Eulerian velocity field at the particle position $\boldsymbol{x}(t)$, and $\boldsymbol{\eta}$ is a Gaussian white noise with zero mean and $\langle \eta_i(t)\eta_j(t')\rangle = \delta_{ij}\delta(t-t')$.

After the seminal works of Arnold and Aref, it is now well recognized that particle motion can be highly nontrivial even in simple laminar velocity fields due to the so-called Lagrangian chaos [4]. For instance, the dispersion properties are greatly enhanced by the combined effects of molecular diffusion and advection [1, 4]. At large times and scales (with respect to the typical time and length scales of u), the test particle motion can be typically described as a Brownian process with an enhanced diffusion coefficient [5]; that is, $\langle (x_i(t) - x_i(0))^2 \rangle \simeq 2 D_{ii}^E t$ where the eddy diffusion coefficient, $D_{ii}^E > D$, accounts for the effects of u. For a large class of velocity fields, there are now well-established techniques for computing D^E (see, e.g., Refs. 5 and 6). An asymptotic diffusive behavior of particles implies that the coarse-grained concentration of the scalar field obeys the Fick equation:

$$\partial_t \langle \theta \rangle = D_{ij}^E \partial^2_{x_i x_j} \langle \theta \rangle, \quad i,j = 1, \ldots, d \qquad (4)$$

where d is the space dimension, and the average $\langle \theta \rangle$ is over a volume of linear dimension larger than the typical velocity length scale.

Under certain conditions, anomalous diffusion may take place—that is, $\langle (x(t) - x(0))^2 \rangle \sim t^{2\nu}$ with $\nu \neq 1/2$ [2]—and the macroscopic description, Eq. (4), does not hold anymore. In the following we discuss in details the necessary conditions to observe anomalous diffusion. In particular, we consider the case of incompressible velocity fields where either standard diffusion ($\nu = 1/2$) or superdiffusion ($\nu > 1/2$) may appear [5].

The problem of reacting transport is different due to the presence of the production term $f_i(\theta_1, \ldots, \theta_N)$ [see Eq. (1)] that makes it to be nonlinear. Here we consider the simplest nontrivial case of Eq. (1): a unique scalar field $\theta(x,t)$ evolving according to the advection–reaction–diffusion equation

$$\partial_t \theta + u \cdot \nabla \theta = D \Delta \theta + \frac{1}{\tau} f(\theta) \qquad (5)$$

where θ represents the fractional concentration of a reacting substance. The glossary is as follows: $\theta = 0$ indicates fresh material that still has nto reacted; $0 < \theta < 1$ means coexistence of fresh material and products; $\theta = 1$ means that the reaction is over [7]. In biological applications, θ is the concentration of a population of organisms that are transported by the flow and grow/die according to the dynamics $f(\theta)$ [8].

In this chapter, we shall restrict our to the study of a specific class of production terms, $f(\theta)$, known as Fisher–Kolmogorov–Petrovsky–Piskunov (FKPP) type [9, 10]. Specifically we consider $f(\theta)$'s that are convex functions ($f''(\theta) < 0$) with $f(0) = f(1) = 0$ and $f'(0) = 1$. A typical example is $f(\theta) = \theta(1 - \theta)$. For more general $f(\theta)$ see Refs. 7 and 8. With this choice,

$\theta = 0$ and 1 are the unstable and stable steady states of the dynamics, respectively. At a phenomenological level, provided that at the initial time a small portion of the system is with $\theta \neq 0$, one observes a front connecting the unstable and stable states propagating through the space.

The problem of front propagation has been extensively studied in many different fields [11,12] such as chemical reaction fronts [7], flames propagation in gases [8], and population dynamics of biological communities [11,12]. In many of these systems the reaction takes place in moving media (i.e., fluids), so that it is important to understand how the flow affects the front propagation.

In the absence of stirring, the front propagation speed reaches an asymptotic value, $v_0 = 2\sqrt{D/\tau}$, and the thickness of the reaction zone is $\xi = 8\sqrt{D\tau}$ [9,10]. While in the presence of a velocity field the front propagates usually with an average speed v_f greater than v_0 [13–15]. Moreover, if $f(\theta)$ is not convex, under special conditions, the flow may stop ("quench") the reaction [16].

The front velocity v_f is the result of the interplay among the flow characteristics (i.e., intensity U and length scale L), the diffusivity D, and the production time scale τ. In this chapter we shall study the problem of front propagation in the case of cellular flows. In particular, introducing the Damköhler number $Da = L/(U\tau)$ (the ratio of advective to reactive time scales) and the Péclet number $Pe = UL/D$ (the ratio of diffusive to advective time scales), we shall discuss how the front speed can be expressed as a nondimensional function such as $v_f/v_0 = \phi(Da, Pe)$. A crucial role in determining $\phi(Da, Pe)$ is played by the renormalization of the diffusion coefficient and chemical time scale [13] induced by the advection.

We also consider an important limit case—that is, the so-called geometrical optics limit—which is realized for $(D, \tau) \to 0$ maintaining D/τ constant [17]. In this limit, one has a nonzero bare front speed, v_0, while the front thickness ξ goes to zero; that is, the front is sharp. Physically speaking, this limit corresponds to situations in which ξ is very small compared with the other length scales of the problem. In this case we provide a simple prediction for the front speed as a nondimensional function $v_f/v_0 = \psi(U/v_0)$.

Other interesting questions concern the modification of the front geometry as a consequence of advection. In particular, one may ask if the presence of Lagrangian chaos has a role in the front dynamics. We shall briefly discuss this problem in the framework of the geometrical optics limit.

The material is organized as follows. In Section II we discuss particles motion in (both regular and chaotic) laminar flows. In Section III after a brief discussion on some general results that do not depend on the specific properties of the velocity field, we shall analyze front propagation in cellular flows.

II. TRANSPORT OF INERT SUBSTANCES

The dynamical and statistical properties of advected passive fields are tightly related to those of test particles. A straightforward way to understand this point is to reconsider the transport equation [Eq. (2)] and solve it in terms of particle trajectories (i.e., the so-called characteristics) that obey the following dynamics:

$$\dot{x}(s;t) = u(x(s;t),s) + \sqrt{2D}\,\eta(s), \qquad x(t;t) = x \tag{6}$$

that is, Eq. (3) where we explicitly fixed the final position to be x. Along the path described by Eq. (6), the transport equation [Eq. (2)] becomes simply $\dot{\theta} = 0$; that is, the concentration field is conserved along the particles trajectories. This means that to compute the field θ at point x and time t given the initial condition, $\theta(x,0)$ one has to consider all paths $x(s;t)$ with final condition x, that is,

$$\theta(x,t) = \langle \theta(x(0;t),0) \rangle \tag{7}$$

where the average is performed over all the paths (i.e., over the realization of noise η). This approach has a straightforward generalization when a source term is added to the right-hand side of Eq. (2), and it is at the basis of our understanding of scalar fields [3, 18]. For this reason, in the next subsections we shall focus mainly on the statistical and dynamical properties of particle trajectories in a variety of flow conditions. In particular, we shall discuss in detail the conditions responsible for the long-time large-scale properties of the particles motion.

A. Standard and Anomalous Diffusion

Investigating the diffusive properties of single-particle motion allows us to predict the characteristics of the macroscopic motion of concentration fields [cf. Eq. (4)]. In this framework it is important to identify the conditions that may lead to anomalous diffusion that implies as a consequence the failure of the Fickian description of transport; that is, Eq. (4) does not hold anymore. From Eq. (3) it is easy to obtain the following relation [19]:

$$\langle (x_i(t) - x_i(0))^2 \rangle = \int_0^t dt_1 \int_0^t dt_2 \langle v_i(x(t_1))\, v_i(x(t_2)) \rangle \simeq 2t \int_0^t d\tau\, C_{ii}(\tau) \tag{8}$$

where

$$C_{ij}(\tau) = \langle v_i(x(\tau))\, v_j(x(0)) \rangle \tag{9}$$

is the correlation function of the Lagrangian velocity, $v = \dot{x}$.

From Eq. (8) it is not difficult to understand that anomalous diffusion can occur only when one or both of the following conditions are violated:

(i) Finite variance of the velocity: $\langle v^2 \rangle < \infty$.
(ii) Fast decay of the Lagrangian velocities correlation function: $\int_0^t d\tau \, C_{ii}(\tau) < \infty$.

If both $\langle v^2 \rangle < \infty$ and $\int_0^t d\tau \, C_{ii}(\tau) < \infty$, then one has standard diffusion and the effective diffusion coefficients are given by

$$D_{ii}^E = \lim_{t \to \infty} \frac{1}{2t} \langle (x_i(t) - x_i(0))^2 \rangle = \int_0^\infty d\tau \, C_{ii}(\tau) \qquad (10)$$

Let us now mention two examples in which the above conditions are violated and anomalous diffusion takes place. It is worth remarking that here the term anomalous diffusion refers to asymptotic nonstandard diffusion. Sometimes in the literature the term is used also for long (but nonasymptotic) transient behaviors.

Violation of (i) can be obtained in the so-called Lévy flight model [20]. The simplest instance is the discrete (in time) one-dimensional case, where the particle position $x(t+1)$ at the time $t+1$ is obtained from $x(t)$ as follows:

$$x(t+1) = x(t) + U(t) \qquad (11)$$

and $U(t)$'s are independent variables identically distributed according to a α-Lévy-stable distribution, $P_\alpha(U)$, that is,

$$\int dU e^{ikU} P_\alpha(U) \propto e^{-c|k|^\alpha} \quad \text{and} \quad P_\alpha(U) \sim U^{-(1+\alpha)} \quad \text{for } |U| \gg 1 \qquad (12)$$

with $0 < \alpha \leq 2$. An easy computation gives $\langle x(t)^q \rangle = C_q \, t^{q/\alpha}$ if $q < \alpha$, and $\langle x(t)^q \rangle = \infty$ if $q \geq \alpha$. Though $\langle x^2 \rangle = \infty$ for any $\alpha < 2$, one can consider the Lévy flight as a sort of anomalous diffusion in the sense that $x_{\text{typical}} \sim t^{1/\alpha} \gg t^{1/2}$. However, in our opinion, the Lévy flight model has a doubtful importance for physical systems due to the very unrealistic property of infinite variance.

Physically more interesting is the Lévy walk model [21], which is still described by Eq. (11) but now $U(t)$ is a random variable with finite variance but nontrivial time correlations—for example, such that $C_{ii}(\tau) \sim \tau^{-\beta}$ with $\beta < 1$. In other words, condition (ii) is violated. This can be obtained as follows. Let us assume that $U(t)$ takes the values $\pm u_0$, and maintains its value for a duration T that is random and with probability density $\psi(T) \sim T^{-(\alpha+1)}$. Then standard diffusion is realized for $\alpha > 2$, while anomalous (super)diffusion takes place for $\alpha < 2$:

$$\langle x(t)^2 \rangle \sim t^{2\nu}, \quad \nu = \begin{cases} 1/2, & \alpha > 2 \\ (3-\alpha)/2, & 1 < \alpha < 2 \\ 1, & \alpha < 1 \end{cases} \quad (13)$$

Besides the above simplified models, more interesting is the understanding of the anomalous diffusion in incompressible velocity fields or deterministic maps. In this direction, Avellaneda, Majda, and Vergassola [22, 23] obtained a very important and general result about the character of the asymptotic diffusion in an incompressible velocity field $\boldsymbol{u}(\boldsymbol{x})$. If the molecular diffusivity D is nonzero and the infrared contribution to the velocity field are weak enough, namely,

$$\int d\mathbf{k} \, \frac{\langle |\hat{\boldsymbol{u}}(\mathbf{k})|^2 \rangle}{k^2} < \infty \quad (14)$$

then one has standard diffusion; that is, the effective diffusion coefficients D_{ii}^E's in Eq. (10) are finite. The average $\langle \cdot \rangle$ indicates the time average, and $\hat{\boldsymbol{u}}$ is the Fourier transform of the velocity. Then there are two possible causes for the superdiffusion:

(a) $D > 0$ and, in order to violate Eq. (14), $\boldsymbol{u}$ with strong spatial correlation;
(b) $D = 0$ and strong correlation between $\mathbf{v}(\boldsymbol{x}(t))$ and $\mathbf{v}(\boldsymbol{x}(t+\tau))$ at large τ.

One of the few nontrivial systems for which the presence of anomalous diffusion can be proved rigorously is the $2d$ random shear flow $\boldsymbol{u} = (u(y), 0)$, where $u(y)$ is a random function [6] such that

$$u(y) = \int_{-\infty}^{\infty} dk \, e^{iky} \, \hat{u}(k), \quad \langle \hat{u}(k) \, \hat{u}(k') \rangle = S(k) \, \delta(k - k') \quad (15)$$

$S(k)$ is the power spectrum and the average $\langle \cdot \rangle$ is taken over the field realizations. Matheron and De Marsily [24] showed that superdiffusion in the x-direction occurs if $\int dk \, S(k) k^{-2} = \infty$. On the contrary, if this integral is finite one has standard diffusion with an effective diffusivity $D_{11}^E \gg D$. Consider now $S(k) \sim k^\zeta$ for $k \mapsto 0$, it is easy to realize that if $\zeta > 1$ standard diffusion takes place, while, if $-1 \leq \zeta \leq 1$ [5]

$$\langle |x(t) - x(0)|^2 \rangle \sim t^{2\nu} \quad \nu = \frac{3-\zeta}{4} \geq \tfrac{1}{2} \quad (16)$$

The physical interpretation of the condition $\int dk \, S(k) \, k^{-2} = \infty$ is as follows. Dimensionally $\int dk \, S(k) \, k^{-2} \sim \langle u^2 \rangle L^2$ where L is the typical length of the function $u(y)$, i.e. the typical distance between two sequent zeros of $u(y)$. If $\langle u^2 \rangle < \infty$ and $\int dk\, S(k)\, k^{-2} < \infty$ the diffusion process is basically similar to that one characterized by a velocity field given by a sequence of strips of size L and velocity $\pm \sqrt{\langle u^2 \rangle}$, i.e. the transversal Taylor diffusion in channels [25]. The origin of the anomalous diffusion is then due to the fact that a test particle travels in a given direction for a very long time before changing direction and so on.

B. About the Meaning of Anomalous

Let us now discuss in more general terms the anomalous diffusion problem considering moments of arbitrary order of the particle's displacement. Two cases are possible [26]: (a) *weak* anomalous diffusion when a unique exponent is involved,

$$\langle |\, x(t) - x(0)\, |^q \rangle \sim t^{q\nu}, \qquad \forall q > 0 \text{ and } \nu > \tfrac{1}{2} \qquad (17)$$

and (b) *strong* anomalous diffusion when

$$\langle |\, x(t) - x(0)\, |^q \rangle \sim t^{q\,\nu(q)}, \qquad \nu(q) \neq \text{const}, \nu(2) > \tfrac{1}{2} \qquad (18)$$

and $\nu(q)$ is a nondecreasing function of q.

In terms of the probability $P(\Delta x, t)$ of observing a displacement $\Delta x = x(t) - x(0)$ at time t, weak anomalous diffusion amounts to the scaling property:

$$P(\Delta x, t) = t^{-\nu} F(\Delta x\, t^{-\nu}) \qquad (19)$$

where the function F is not necessarily the Gaussian one. On the contrary, strong anomalous diffusion is not compatible with the scaling, Eq. (19).

In the case of weak anomalous diffusion, it is natural to conjecture

$$F(z) \propto e^{-c|z|^\alpha} \qquad (20)$$

where in general α is not determined by ν. However, an argument á la Flory due to Fisher [27] suggests that

$$P(\Delta x, t) \sim t^{-\nu} \exp\left[-c \left(\frac{|\Delta x|}{t^\nu} \right)^{\frac{1}{1-\nu}} \right] \qquad (21)$$

which means $\alpha = \frac{1}{1-\nu}$. Remarkably the random shear flow is in agreement with Fisher's prediction; indeed for $\zeta = 0$ (i.e., $\nu = 3/4$), one has $F(a) \sim e^{-c|a|^4}$ for $|a| \gg 1$ [28]. While for the properties of dispersion the detailed functional dependence of $P(\Delta x, t)$ is not particularly important, it has a nontrivial role in determining the propagation properties in reactive systems [29].

C. Strong Anomalous Diffusion in Chaotic Flows

If Eq. (14) holds then anomalous diffusion may appear only for $D = 0$ and very strong Lagrangian velocity correlations. The latter condition can be realized—for example, in time periodic velocity fields where the Lagrangian phase space has a complicated self-similar structure of islands and cantori [30]. Here superdiffusion is due to the almost trapping, for arbitrarily long time, of the ballistic trajectories close to the cantori, which are organized in complicated self-similar structures.

In this framework an interesting example is the Lagrangian motion in velocity field given by a simple model for Rayleigh–Bénard convection [31], which is given by the stream function:

$$\psi(x, y, t) = \psi_0 \sin\left[\frac{2\pi}{L}(x + B \sin \omega t)\right] \sin\left[\frac{2\pi}{L} y\right] \qquad (22)$$

where the velocity is given by $u = (\partial_y \psi, -\partial_x \psi)$, $\psi_0 = UL/2\pi$ (L is the periodicity of the cell; here we use $L = 2\pi$), and U is the velocity intensity. The even oscillatory instability is accounted for by the term $B \sin \omega t$, representing the lateral oscillation of the rolls [31]. At fixed B, the control parameter for particle diffusion is $\epsilon \equiv \omega L^2 / \psi_0$—that is, the ratio between the lateral roll oscillation frequency (ω) and the characteristic circulation frequency (ψ_0/L^2) inside the cell. Different regimes take place depending on ϵ. For instance, at $\epsilon \sim 1$ the synchronization between the circulation in the cells and their global oscillation causes a very efficient way of jumping from cell to cell. This mechanism, similar to stochastic resonance, makes the effective diffusivity as a function of the frequency ω very structured [26] (see Fig. 1). Moreover, for vanishing molecular diffusivity, anomalous superdiffusion takes place in a narrow window of ω values around the peaks, that is,

$$\langle (x(t) - x(0))^2 \rangle \propto t^{2\nu(2)} \quad \text{with } \nu(2) > 1/2 \qquad (23)$$

[see Fig. 2 (left)]. The presence of genuine anomalous diffusion is confirmed by the fact that effective diffusivity diverges as $D_{11}^E \sim D^{-\beta}$ with $\beta > 0$ [see Fig. 2 (right)], as suggested in Ref. 5.

The remarkable property of the flow [Eq. (22)] is that moments of the particle displacement display a strong anomalous behavior [Eq. (18)], indeed Fig. 3 (left) shows that $q\nu(q)$'s is a nontrivial function of q. In particular, the curve $q\nu(q)$ versus q displays a nonlinear behavior. A closer inspection shows that two linear regions are present: the first one up to $q \sim 2$, the second elsewhere.

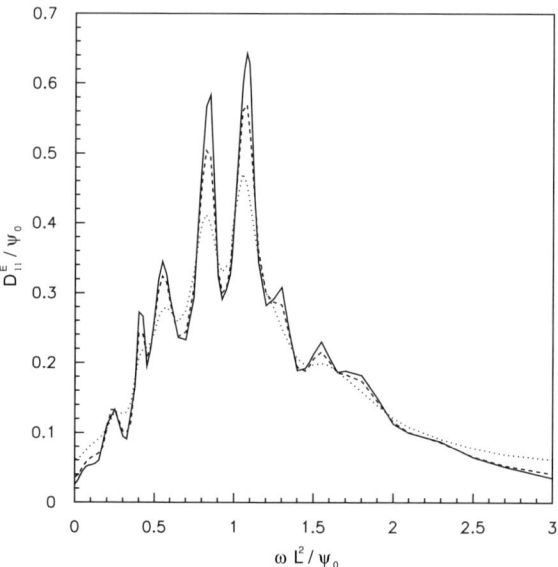

Figure 1. The renormalized diffusivity D_{11}^E/ψ_0 versus the frequency $\omega L^2/\psi_0$ for different values of the molecular diffusivity D/ψ_0. $D/\psi_0 = 3 \times 10^{-3}$ (dotted curve); $D/\psi_0 = 1 \times 10^{-3}$ (broken curve); $D/\psi_0 = 5 \times 10^{-4}$ (full curve).

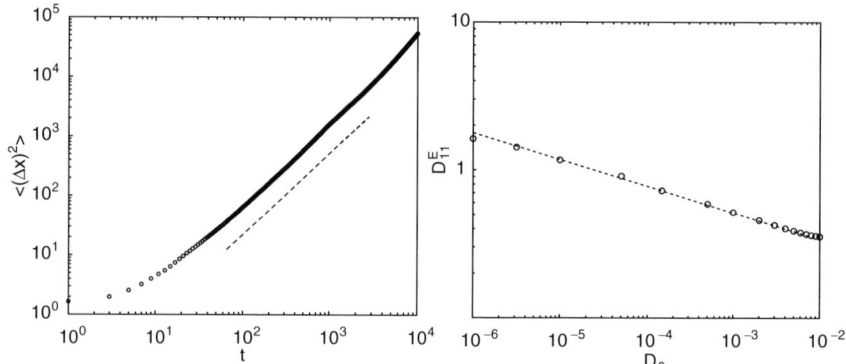

Figure 2. (**Left**) Mean-squared displacement versus the time for the flow [Eq. (22)] with $D = 0$, and $\omega = 1.1$. Lengths and times are shown in units of L and L^2/ψ_0, respectively. The best-fit (dashed) line corresponds to $2\nu(2) = 1.3$. (**Right**) The diffusion coefficient D_{11}^E as a function of D for the frequency of the roll oscillation $\omega = 1.1$. The diffusivities are reported in units of ψ_0. The best-fit (dashed) line has the slope $-\beta = -0.18$.

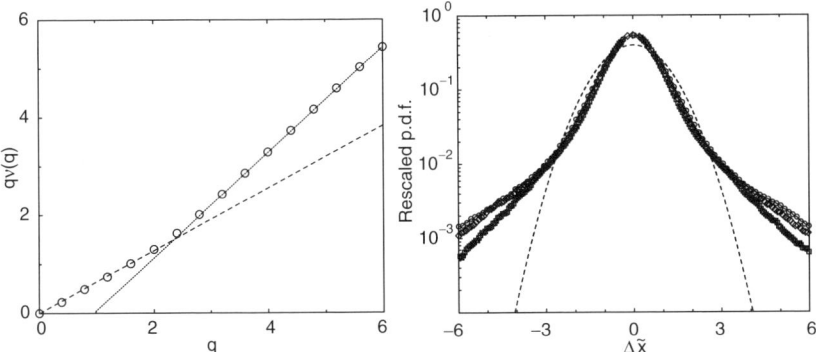

Figure 3. (**Left**) The measured scaling exponents $q\,v(q)$'s (joined by dot-dashed straight lines) of the moments of the displacement Δx, as a function of the order q. The dashed line corresponds to $0.65\,q$ while the dotted line corresponds to $q - 1.04$. (**Right**) The normalized probability distribution function $P(\Delta x(t)/\tilde{\sigma})$ versus $\Delta \tilde{x} \equiv \Delta x/\tilde{\sigma}$ ($\tilde{\sigma} = \exp\langle \ln|\Delta x(t)|\rangle$) for the three times $t_1 = 500$ (circles), $t_2 = 2t_1$ (diamonds) and $t_3 = 2t_2$ (squares). The dashed line represents the Gaussian function.

The two linear regions are associated to two different mechanisms in the diffusion process. For small q's—that is, for the core of the probability distribution function $P(\Delta x, t)$—only one exponent ($v_1 \equiv v(q) \simeq 0.65$ for $q < 2$) fully characterizes the diffusion process. This means that the typical (i.e., nonrare) events obey a (weak) anomalous diffusion process. Roughly speaking, one can say that at scale l the characteristic time $\tau(l)$ behaves as $\tau(\ell) \sim l^{1/v_1}$. On the other hand, for $q > 2$ the behavior $q\,v(q) \sim q - \text{const}$ suggests that the large deviations are essentially associated with ballistic transport, $\tau(\ell) \sim l$, due to the synchronization between the circulation in the cells and their global oscillation.

Strong anomalous diffusion is also highlighted by the normalized probability densities $P(\Delta x, t)$ at different times that do not collapse onto a unique curve [see Fig. 3 (right)], suggesting that the scaling property, Eq. (19), does not hold.

Strong anomalous diffusion appears also in other systems, as, for instance, in the standard map [32]:

$$J_{t+1} = J_t + K \sin(\theta_t)$$
$$\theta_{t+1} = \theta_t + J_{t+1} \qquad \text{mod } 2\pi \qquad (24)$$

For specific values of K, one has the coexistence of many accelerator modes (i.e., ballistic trajectories), and one observes anomalous diffusion in the action variable:

$$\langle [J_t - J_0]^2 \rangle \sim t^{2\nu(2)}, \quad t \gg 1 \quad \text{with} \quad \nu(2) \neq 1/2 \qquad (25)$$

The sticking of the chaotic orbits to stable islands leads to the appearance of blocks of long-range correlation in the sequences of the J_t variable that are responsible for its anomalous behavior [33].

These results along with those in Ref. 34 suggest that the phenomenon of anomalous diffusion in chaotic systems is possible but rare, taking place only for specific values of the parameters. Small changes typically restore standard diffusion $\nu(2) = 1/2$.

III. TRANSPORT OF REACTING SUBSTANCES

We consider the case of a unique scalar reactive field $\theta(x, t)$. This model is appropriate in aqueous autocatalytic premixed reactions, as well as in gaseous combustion with a large flow intensity but low value of gas expansion across the flame [7]. The field θ evolves according to the advection–reaction–diffusion (ARD) equation:

$$\partial_t \theta + \boldsymbol{u} \cdot \boldsymbol{\nabla} \theta = D\Delta\theta + \frac{1}{\tau} f(\theta) \qquad (26)$$

where $f(\theta)$ accounts for the reaction and τ is the reaction characteristic time [7, 13–15]. As mentioned in the introduction, we take $f(\theta) = \theta(1 - \theta)$. The results we are going to describe do not depend on the specific form of $f(\theta)$, provided that $f(\theta)$ is convex ($f''(\theta) < 0$), positive in the interval $[0, 1]$, and vanishing at its extremes, and $f'(0) = 1$. This corresponds to the FKPP type of reaction [9, 10].

In Section II we saw the link between the solution of Eq. (2) and Lagrangian trajectories. There exist a similar relation also for Eq. (26) [35]:

$$\theta(x, t) = \left\langle \theta(x(0), 0) \exp\left(\frac{1}{\tau} \int_0^t \frac{f(\theta(x(s;t), s))}{\theta(x(s;t), s)} ds\right) \right\rangle \qquad (27)$$

where the average is performed over all the trajectories $x(s; t)$ that started in $x(0)$ and ended in $x(t; t) = x$ [as in Eq. (6)].

Using the maximum principle [35] and noting that $f(\theta)/\theta \leq f'(0)$, because of the convexity of $f(\theta)$, one can write an upper bound for θ in terms of the solution, θ_L, of the linearized ARD:

$$\partial_t \theta_L + \boldsymbol{u} \cdot \boldsymbol{\nabla} \theta_L = D\Delta\theta_L + \frac{f'(0)}{\tau} \theta_L \qquad (28)$$

In fact, if $\theta(x,0) \leq \theta_L(x,0)$, one has [35]

$$\theta(x,t) \leq \theta_L(x,t) \tag{29}$$

at all times. From Eqs. (27–29), one obtains

$$\theta(x,t) \leq \theta_L(x,t) = \langle \theta(x(0;t),0) \rangle \exp\left(\frac{f'(0)}{\tau} t\right) \tag{30}$$

where $\langle \theta(x(0;t),0) \rangle$ is as in Eq. (7) with initial condition $\theta(x,0)$ (that we assume localized around $x = 0$).

In Section II.A we saw that under general conditions (i.e., spatial and temporal short-range correlations), Eq. (2) has the same asymptotic behavior of a Fick equation. As a consequence, we have

$$\langle \theta(x(0;t),0) \rangle \sim \frac{1}{\sqrt{4\pi D_{11}^E t}} \exp\left(-\frac{x^2}{4D_{11}^E t}\right) \tag{31}$$

Eqs. (30) and (31) imply that, along the x-direction, the field θ is exponentially small until a time t of the order of $x/\sqrt{4D_{11}^E f'(0)/\tau}$. Therefore we have an upper bound for v_f:

$$v_f \leq 2\sqrt{D_{11}^E f'(0)/\tau} \tag{32}$$

The preceding discussion shows that, if standard diffusion holds, then there is a front propagating with a constant speed; that is, the solvable case $u = 0$ is recovered with a renormalized diffusion constant. Nevertheless, the analytical determination of v_f, for a given velocity field, u, is a rather difficult problem even for simple laminar fields [13–15,17].

A. Fronts in Cellular Flows

The bound in Eq. (32) is very general and holds for generic incompressible flows and production terms. Here we numerically investigate the properties of front propagation in the particular case of the cellular flow, Eq. (22). This flow is interesting because, at variance with shear flows, all the streamlines are closed and, therefore, the front propagation is determined by the mechanisms of contamination of one cell to the other [13,15]. As first we consider the time-independent case; that is, $B = 0$ in Eq. (22). Since we are interested in the propagation in the x-direction, we take periodic boundary conditions in y and an infinite extent along the x-axis with boundary conditions $\theta(-\infty, y; t) = 1$ and $\theta(+\infty, y; t) = 0$.

We define the velocity as [14]

$$v_f(t) = \frac{1}{L}\int_0^L dy \int_{-\infty}^{\infty} dx\, \partial_t \theta(x,y;t) \qquad (33)$$

that is, the so-called bulk burning rate which coincides with the front speed when the latter exists, but it is well-defined even when the front itself is not well-defined. The asymptotic (average) front speed is determined by

$$v_f = \lim_{T\to\infty} \frac{1}{T}\int_0^T dt\, v_f(t)$$

In our discussion, we always suppose that the diffusion time scale is the slowest one and thus $Pe \gg 1$ and $Da \cdot Pe \gg 1$.

At large scales and long times the effects of the velocity field can be modeled in terms of a reaction–diffusion process with renormalized coefficients [13],

$$\partial_t \theta = D^E \Delta \theta + \frac{1}{\tau_{\text{eff}}} F(\theta) \qquad (34)$$

The renormalized diffusivity D^E accounts for the process of diffusion from cell to cell as a result of the nontrivial interaction of advection and molecular diffusion [6]. The renormalized reaction time τ_{eff} is the time it takes for a single cell to be filled by inert material, and it depends on the interaction of advection and production. F indicates the functional form of the renormalized chemistry. Therefore, the limiting speed of the front in the moving medium is given by $v_{\text{eff}} \sim \sqrt{D^E/\tau_{\text{eff}}}$ [13,14]. The problem is now reduced to derive the expressions for the renormalized parameters by means of physical considerations.

Now using as interpretative framework the above-described macroscopic model, we present the results of numerical simulations for slow ($Da \ll 1$) and fast ($Da \gg 1$) reaction.

B. Slow and Fast Reaction Regimes

First we need the renormalized diffusion coefficient and reaction time. The eddy diffusivity depends on Pe and D as [6]

$$\frac{D^E}{D} \sim Pe^{1/2}, \qquad Pe \gg 1 \qquad (35)$$

The renormalized characteristic time can be estimated as follows. At small Da, the reaction is significantly slower than the advection, and consequently the region where the reaction takes place extends over several cells (distributed

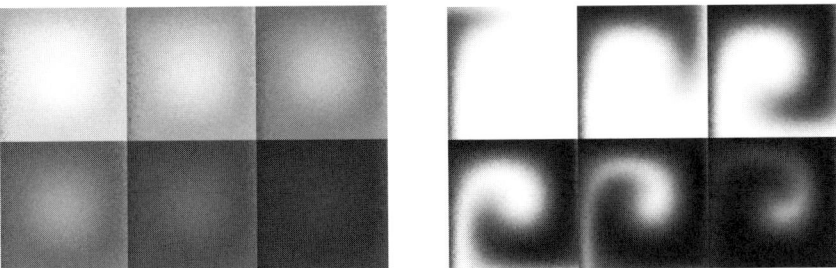

Figure 4. (**Left**) Six snapshots of the field θ within the same cell, at six successive times with a delay τ/6 (from left to right, top to bottom), as a result of the numerical integration of Eq. (26). Here $Da \simeq 0.4$ and $Pe \simeq 315$. Black stands for $\theta = 1$, white for $\theta = 0$. (**Right**) The same but for $Da = 4$ and $Pe = 315$, τ is now replaced by $\tau_{\text{eff}} \sim L/U$. Note that a spiral wave invades the interior of the cell, with a speed comparable to U.

front). Therefore, in the slow reaction regime, $Da \ll 1$, a single cell is first invaded by a mixture of reactants and products (on the fast advective time scale), and subsequently a complete reaction is achieved on the slower time scale $\tau_{\text{eff}} \simeq \tau$ (see Fig. 4). In the case of fast reaction, $Da \gg 1$, two sharply separated phases emerge inside the cell and the filling process is characterized by an inward spiral motion of the outer, stable phase (see Fig. 4) at a speed proportional to U. Therefore we have

$$\frac{\tau_{\text{eff}}}{\tau} \sim \begin{cases} 1, & Da \ll 1 \\ Da, & Da \gg 1 \end{cases} \qquad (36)$$

From Eqs. (35)–(36) and recalling that $v_f \sim \sqrt{D^E/\tau_{\text{eff}}}$, one obtains [13,14]

$$\frac{v_f}{v_0} \sim \begin{cases} Pe^{1/4}, & Da \ll 1, Pe \gg 1 \\ Pe^{1/4} Da^{-1/2}, & Da \gg 1, Pe \gg 1 \end{cases} \qquad (37)$$

The case of $Pe \ll 1$ is less interesting because the dynamics is dominated by diffusion.

In terms of the typical velocity of the cellular flow, we have $v_f \propto U^{1/4}$ for slow reaction ($U \gg L/\tau$, or equivalently $Da \ll 1$) whereas $v_f \propto U^{3/4}$ for fast reaction ($U \ll L/\tau$, or $Da \gg 1$). The scaling $v_f \propto U^{1/4}$ for slow reaction (i.e., fast advection) is a consequence of $D^E \propto D Pe^{1/2}$ [6] in the homogenization limit [13,14], and it has been obtained in Refs. 13 and 15. Numerical simulations of Eq. (26) confirm these predictions (see Fig. 5); that is, one has the two scaling laws $v_f \propto U^{1/4}$ and $v_f \propto U^{3/4}$ at fast and slow advection, respectively [13].

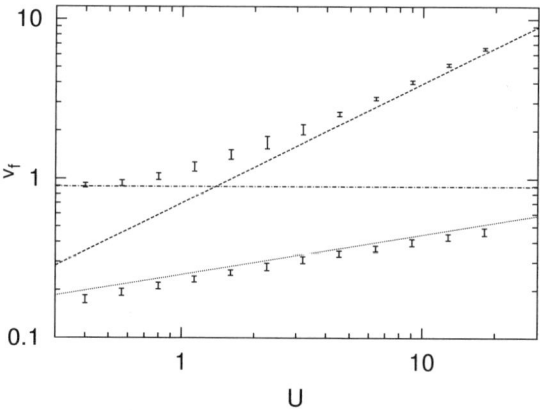

Figure 5. The front speed v_f as a function of U, the typical flow velocity. The lower curve shows data at $\tau = 20.0$ (fast advection). The upper curve shows data at $\tau = 0.2$ (slow advection). For comparison, the scalings $U^{1/4}$ and $U^{3/4}$ are shown as dotted and dashed lines, respectively. The horizontal line indicates v_0, the front velocity without advection, for $\tau = 0.2$.

C. Geometrical Optics Limit

Physically speaking, the geometrical optics regime corresponds to the limit when the front thickness and the reaction time are much smaller than the length and time scales of the velocity field. Mathematically speaking, this means the limit $(D, \tau) \to 0$ maintaining D/τ constant [17,36], so that the bare front speed, v_0, is finite while the front thickness ξ goes to zero.

The sharp interface separating the reactants from the products is modeled by the G-equation, Eq. (38) [8,36]:

$$\frac{\partial G}{\partial t} + \boldsymbol{u} \cdot \boldsymbol{\nabla} G = v_0 |\boldsymbol{\nabla} G| \qquad (38)$$

The front is defined by a constant level surface of the scalar function $G(r, t)$.

In cellular flows, the front border is wrinkled by the velocity field during propagation and its length increases until pockets of fresh material develop [37,38] (see Fig. 6). After this, the front propagates periodically in space and time with an enhanced speed $v_f > v_0$.

Let us now consider the flow [Eq. (22)] with $B = 0$—that is, the stationary case. The problem addressed here is the dependence of the effective speed v_f on the flow intensity, U, and the bare velocity, v_0, that is expected of the form [39]

$$\frac{v_f}{v_0} = \psi\left(\frac{U}{v_0}\right) \qquad (39)$$

where $\psi(\mathscr{U})$ is a function that depends on the flow details.

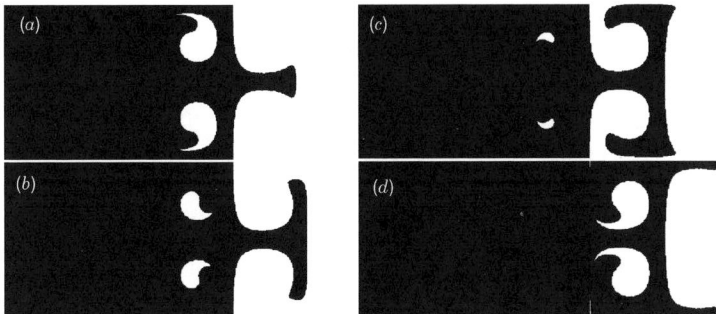

Figure 6. Snapshot of the front shape with time step $T/8$ (from (a) to (d)), where T is the period of the front dynamics, for $v_0 = 0.5$, $U = 4.0$ and $L = 2\pi$. Unburnt (burnt) material is indicated in white (black).

As far as we know, apart from very simple shear flows (for which $\psi(\mathcal{U}) = 1 + \mathcal{U}$ [40]), there are no methods to compute $\psi(\mathcal{U})$ from first principles. Mainly one has to resort to numerical simulations and phenomenological arguments.

For turbulent flows, by means of dynamical renormalization group techniques, Yakhot [40] proposed

$$\frac{v_\mathrm{f}}{v_0} = e^{(U/v_\mathrm{f})^\alpha} \qquad (40)$$

with $\alpha = 2$; now U indicates the root mean squared average velocity (see also Refs. 42 and 43. Therefore, from Eq. (40) one has that $v_\mathrm{f} \to U/\sqrt{\ln(U)}$ for $U \to \infty$.

For the cellular flow under investigation, albeit the exact form of the function $\psi(\mathcal{U})$ is not known, a simple argument can be given for an upper and a lower bound by mapping the front dynamics onto a one-dimensional problem. The starting point is the following observation. In the optical regime, since $\theta(x, y)$ is a two-valued function ($\theta = 1$ and $\theta = 0$), we can track the farther edge of the interface between product and material $(x_M(t), y_M(t))$, which is defined as the rightmost point (in the x-direction) for which $\theta(x_M, y_M; t) = 1$. Then we can define a velocity

$$\tilde{v}_f = \lim_{t \to \infty} \frac{x_M(t)}{t} \qquad (41)$$

which is equivalent to the definition in Eq. (33). After a transient, in the unit cell $[0, 2\pi]$ (we describe the case $L = 2\pi$) the point $(x_M(t), y_M(t))$ moves to the right along the separatrices of the streamfunction [Eq. (22)], so that $y_M(t)$ is essentially close to the values 0 or π. Along this path, one can reduce the edge dynamics to the $1d$-problem

$$\frac{dx_M}{dt} = v_0 + U\beta |\sin(x_M(t))| \qquad (42)$$

where the second term of the right-hand side is the horizontal component of the velocity field. We have neglected the y-dependence, replacing it with a constant β that takes into account the average effect of the vertical component of the velocity field along the path followed by (x_M, y_M). By solving Eq. (42) in the interval $x_M \in (0, 2\pi)$, one obtains the time, T, needed for x_M to reach the end of the cell. The front speed, as the speed of the edge particle, is then given by $v_f = 2\pi/T$. The final result is

$$\frac{v_f}{v_0} = \psi_\beta(\mathscr{U}) = \frac{\pi\sqrt{(\mathscr{U}\beta)^2 - 1}}{2\ln\left(\mathscr{U}\beta + \sqrt{(\mathscr{U}\beta)^2 - 1}\right)} \qquad (43)$$

Note that Eq. (43) is valid only for $\mathscr{U}\beta \geq 1$.

We have taken $\beta = 1$ for the upper bound and $\beta = 1/2$ (which is the average of $|\cos(y)|$ between 0 and π) for the lower bound. We have also computed the average of $|\cos(y_M(t))|$ in a period of its evolution obtaining $\beta \approx 0.89$, which gives indeed a very good approximation of the measured curve (see Fig. 7).

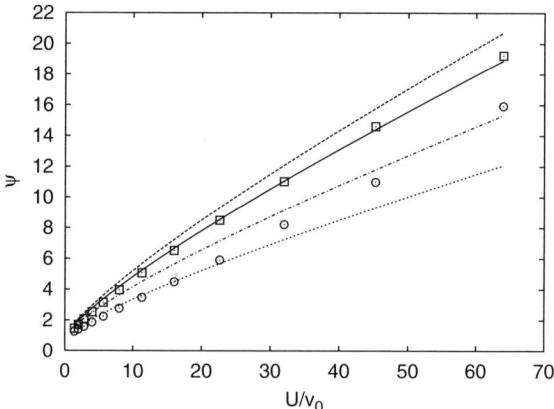

Figure 7. The measured $\psi(U/v_0)$ as a function of U/v_0 (squares), the Yakhot formula [Eq. (40)] with $\alpha = 2$ (circles), and the function ψ_β for $\beta = 1, 1/2$ (dashed and dotted lines) and for $\beta = 0.89$ (solid line). The dashed–dotted line is the bound given in Eq. (44).

This agreement is an indication that the average of $|\cos(y_M(t))|$ depends on U and v_0 very weakly (as we checked numerically). Previous studies [42,43] reported an essentially linear dependence of the front speed on the flow intensity—that is, $v_f \propto U$ for large U, which is not too far but different from our result. A rigorous lower bound has been obtained in Ref. 44 by using the G-equation:

$$v_f \geq U/(\ln(1 + U/v_0)) \tag{44}$$

which is also shown in Fig. 7.

From Eq. (43), asymptotically (i.e., for $U \gg v_0$) one has $v_f \sim U/\ln(U)$, which corresponds to (40) for $\alpha = 1$. Expressions such as Eq. (40) have been proposed for flows with many scales (e.g., turbulent flows), and different values of α have been reported [41–44]. The fact that also the simple one-scale vortical flow investigated here displays such a behavior may be incidental. However, we believe that it can be due to physical reasons. Indeed, the large-scale features of the flow—for example, the absence of open channels (like for the shear flow)—can be more important than the detailed multiscale properties of the flow [37].

D. On the Relevance of Chaos in Front Propagation

An interesting problem is the thin front dynamics in the presence of Lagrangian chaos generated by the time periodic streamfunction [Eq. (22)]. We are mainly interested in addressing the two following issues. First, since trajectories starting near the roll separatrices typically have a positive Lyapunov exponent, it is natural to wonder about the role of Lagrangian chaos on front propagation. Second, as we have shown in Section II.C, we know that for the time-dependent streamfunction [Eq. (22)] the transport properties are strongly enhanced; therefore it is worth investigating if similar effects are reflected also in the front speed.

In order to define the instantaneous front length, $\mathscr{L}(t)$, we introduce the variable $\sigma_\epsilon(x, y; t)$ which assumes the value 0 if θ is constant inside a circle of radius ϵ centered in (x, y) at time t, otherwise $\sigma_\epsilon(x, y; t) = 1$ (i.e., $\sigma_\epsilon(x, y; t) = 1$ only if the ϵ-ball centered in (x, y) contains a portion of the front). The front length is then defined by

$$\mathscr{L}(t) = \lim_{\epsilon \to 0} \frac{1}{\epsilon} \int_{-\infty}^{\infty} dx \int_0^L dy\, \sigma_\epsilon(x, y; t) \tag{45}$$

A direct consequence of Lagrangian chaos is the exponential growth of passive scalar gradients and material lines [1, 4]: A (passive) material line of initial length ℓ_0 for large times grows as

$$\ell(t) \sim \ell_0 e^{\Lambda(1)t} \tag{46}$$

where $\Lambda(1)$ is the generalized Lyapunov exponent, $\Lambda(1) \geq \lambda$ [45].

Figure 8. Snapshots at two successive times, $t = 3.6$ and 7.5, of the evolution of passive (**top**) and reactive line of material for two values of v_0 (**middle** $v_0 = 0.7$ and **bottom** $v_0 = 2.1$) for $U = 1.9$, $B = 1.1$ and $\omega = 1.1U$. The initial condition is a straight vertical line.

In the presence of molecular diffusivity, the exponential growth of $\ell(t)$ stops due to diffusion, and chaos has just a transient effect [46]. For reacting scalars, something very similar happens. Let us compare the evolution of a material line in the passive and reactive cases (see Fig. 8). While in the passive case (without molecular diffusivity) structures on smaller and smaller scales develop due to stretching and folding, in the reactive systems after a number of folding events structures on smaller scales are inhibited as a consequence of the Huygens dynamics: The interface between the two phases merges. This phenomenon is responsible for the formation of *pockets*. Of course, "merging" is more and more efficient as v_0 increases (compare the middle and the bottom pictures of Fig. 8).

In Fig. 9 we show the time evolution of the line length, $\mathscr{L}(t)$, as a function of t for the passive and reactive material at different values of v_0. While at small times both the passive and reactive scalar lines grow exponentially with a rate close to $\Lambda(1)$, at large time $t > t^*$ (where t^* is a transient time depending on v_0) the reacting ones stop due to merging. Asymptotically, the front length varies periodically with an average value depending on v_0. The time t^* can be estimated as follows: Two initially separated parts of the line (e.g., originally at distance ℓ_0) become closer and closer, roughly as $\sim \ell_0 \exp(-\Lambda(1)t)$. When such a distance becomes of the order of $v_0 t$, merging takes place and, hence, to leading order

$$t^* \propto \frac{1}{\Lambda(1)} \ln\left(\frac{\Lambda(1)\ell_o}{v_0}\right) \qquad (47)$$

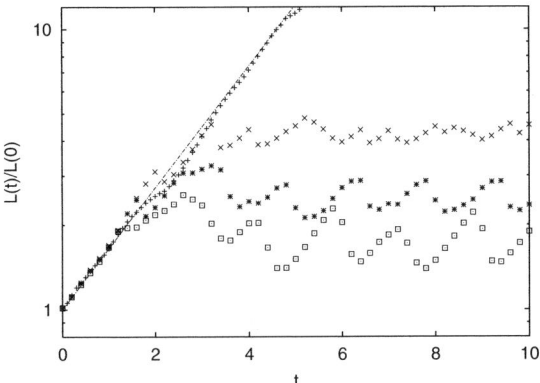

Figure 9. $\mathscr{L}(t)/\mathscr{L}(0)$ as a function of time for $U = 1.9$, $B = 1.1$, and $\omega = 1.1U$ for the passive (+) and reactive case: from top $v_0 = 0.3\,(\times)$, $0.5\,(*)$, $0.7\,(\text{square})$. The straight line indicates the curve $\exp(\Lambda(1)t)$ with $\Lambda(1) \approx 0.5$, which has been directly measured.

In the asymptotic state ($t > t^*$) both the spatial and temporal structures of the flow become periodic.

Let us now switch to the effects of Lagrangian chaos on the asymptotic dynamics of front propagation. An immediate consequence of Eq. (47) is that the asymptotic front length (45) behaves as $L_f \sim v_0^{-1}$ for values of v_0 small enough. Indeed,

$$L_f \sim L e^{\Lambda(1)t^*} \sim \frac{L^2 \Lambda(1)}{v_0} \qquad (48)$$

which is in fairly good agreement with the simulations [37].

It is worth remarking that even if the scaling [Eq. (48)] holds when chaos is present, in general it is not peculiar of chaotic flows. For instance, for the shear flow ($u_x = U\sin(y)$, $u_y = 0$) one has $v_f = U + v_0$. On the other hand, since $v_f \sim L_f v_0$ [8], even if the shear flow is not chaotic, $L_f \sim 1/v_0$ for $U/v_0 \gg 1$. From the previous discussion, it seems that the front length dependence on v_0 is not an unambiguous effect of chaos on the asymptotic dynamics. But, if we look at Fig. 8, the spatial "complexity" of the front in the presence of Lagrangian chaos is evident.

Let us now introduce an indicator to quantitatively evaluate this qualitative observation. Let us use W_f to represent the size of the region in which burnt and unburnt material coexist. Introducing a measure, $\mu(x)$, in that region we can defined W_f as the standard deviation of $\mu(x)$ [37]:

$$\mu(x) = \frac{|\partial_x \tilde{\theta}(x)|}{\int dx |\partial_x \tilde{\theta}(x)|} \qquad (49)$$

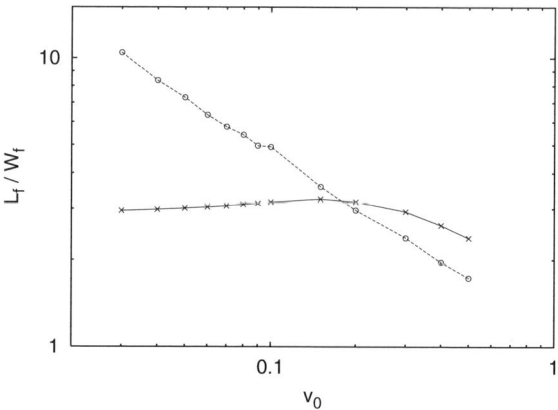

Figure 10. L_f/W_f as a function of v_0 for the time dependent ($\circ$) cases with parameters $U = 1.9$, $B = 1.1$, $\omega = 1.1U$ and the time-independent cases ($\times$) with $U = 1.9$.

where $\tilde{\theta}(x) = 1/L \int_0^L \theta(x,y)\, dy$, that is,

$$W_f = \left[\int x^2 \mu(x)\, dx - \left(\int x \mu(x) dx \right)^2 \right]^{1/2} \tag{50}$$

For a simple shear flow, W_f and L_f display the same kind of dependence on v_0 (actually they are proportional). In generic chaotic flows there is an increasing of the front length, while chaotic mixing induces a decrease of W_f. This is indeed what one observes in Fig. 10, where we show the ratio L_f/W_f both for the nonchaotic and the chaotic flow. For the latter this ratio diverges for very small v_0 values as a signature of chaos. From a physical point of view the ratio L_f/W_f is an indicator of the spatial complexity of the front. Indeed it indicates the degree of wrinkling of the front with respect to the size of the region in which the front is present. Loosely speaking, we can say that the *temporal* complexity of Lagrangian trajectories converts in the *spatial* complexity of the front.

IV. SUMMARY

We reviewed some aspect of passive transport in fluid flows. As far as inert substances are concerned, we described the problem of transport from a Lagrangian point of view focusing on single-particle properties. In particular, the conditions for having asymptotic standard or anomalous diffusion have been discussed in details. As for the problem of reacting substances, we study the

THE ROLE OF CHAOS FOR INERT AND REACTING TRANSPORT   541

dependence of the front speed on the flow characteristics, considering the case of reaction that are slow or fast with respect to the typical time scales of the advection. Moreover, we have shown that Lagrangian chaos has a minor role on the front speed and mostly acts as a source of complexity for the front structures.

Acknowledgments

We thank M. Abel, G. Boffetta, P. Castiglione, A. Celani, G. Lacorata, C. Lopez, R. Mancinelli, P. Muratore-Ginanneschi, A. Mazzino, R.A. Pasmanter, A. Torcini, M. Vergassola, and E. Zambianchi for fruitful collaborations and interesting discussions during the last few years.

References

1. H. K. Moffatt, *Rep. Prog. Phys.* **46**, 621 (1983).
2. J. P. Bouchaud and A. Georges, *Phys. Rep.* **195**, 127 (1990).
3. G. Falkovich, K. Gawędzki, and M. Vergassola, *Rev. Mod. Phys.* **73**, 913 (2001).
4. A. Crisanti, M. Falcioni, G. Paladin, and A. Vulpiani, *Riv. Nuovo Cim.* **14**, 1 (1991).
5. L. Biferale, A. Crisanti, M. Vergassola, and A. Vulpiani, *Phys. Fluids* **7**, 2725 (1995).
6. A. J. Majda and P. R. Kramer, *Phys. Rep.* **314**, 237 (1999).
7. J. Xin, *SIAM Rev.* **42**, 161 (2000).
8. N. Peters, *Turbulent Combustion*, Cambridge University Press, Cambridge, 2000.
9. A. N. Kolmogorov, I. G. Petrovskii, and N. S. Piskunov, *Moscow Univ. Bull. Math.* **1**, 1 (1937).
10. R. A. Fisher, *Ann. Eugenics* **7**, 355 (1937).
11. E. R. Abraham, *Nature* **391**, 577 (1998).
12. J. Ross, S. C. Müller, and C. Vidal, *Science* **240**, 460 (1988).
13. M. Abel, A. Celani, D. Vergni, and A. Vulpiani, *Phys. Rev. E* **64**, 046307 (2001).
14. P. Constantin, A. Kiselev, A. Oberman, and L. Ryzhik, *Arch. Rational Mech.* **154**, 53 (2000).
15. B. Audoly, H. Beresytcki, and Y. Pomeau, *C. R. Acad. Sci.* **328**, Série II b, 255 (2000).
16. N. Vladimirova, P. Constantin, A. Kiselev, O. Ruchayskiy, and L. Ryzhik, *Comb. Theory Model* **7**, 487 (2003).
17. A. R. Kerstein, W. T. Ashurst, and F. A. Williams, *Phys. Rev. A* **37**, 2728 (1988).
18. U. Frisch, A. Mazzino, and M. Vergassola, *Phys. Rev. Lett.* **80**, 5532 (1998).
19. G. I. Taylor, *Proc. Lond. Math. Soc. Ser. 2* **20**, 196 (1921).
20. E. Montroll and M. Schlesinger, in *Studies in Statistical Mechanics*, Vol. 11, E. W. Montroll and J. L. Lebowitz, eds., North-Holland, Amsterdam, 1984, p.1.
21. M. F. Schlesinger, B. West, and J. Klafter, *Phys. Rev. Lett.* **58**, 1100 (1987).
22. M. Avellaneda and A. Majda, *Commun. Math. Phys.* **138**, 339 (1991).
23. M. Avellaneda and M. Vergassola, *Phys. Rev. E* **52**, 3249 (1995).
24. G. Matheron and G. De Marsily, *Water Resouces Res.* **16**, 901 (1980).
25. G. I. Taylor, *Proc. R. Soc. A* **219**, 186 (1953); *Proc. R. Soc. A* **225**, 473 (1954).
26. P. Castiglione, A. Mazzino, P. Muratore-Ginanneschi, and A. Vulpiani, *Physica D* **134**, 75 (1999).

27. M. E. Fisher, *J. Chem. Phys.* **44**, 616 (1966).
28. J.-P. Bouchaud, A. Georges, J. Koplik, A. Provat, and S. Redner, *Phys. Rev. Lett.* **64**, 2503 (1990).
29. R. Mancinelli, D. Vergni, and A. Vulpiani, *Eur. Phys. Lett.* **60**, 532 (2002); *Physica D* **185**, 175 (2003).
30. G. M. Zaslavsky, D. Stevens, and H. Weitzener, *Phys. Rev. E* **48**, 1683 (1993).
31. T. H. Solomon and J. P. Gollub, *Phys. Rev. A* **38**, 6280 (1988).
32. A. J. Lichtenberg and M. A. Lieberman, *Physica D* **33**, 211 (1988).
33. Y. H. Ichikawa, T. Kamimura, and T. Hatori, *Physica D* **29**, 247 (1987).
34. P. Leboeuf, *Physica D* **116**, 8 (1998).
35. M. Freidlin, *Functional Integration and Partial Differential Equations*, Princeton University Press, Princeton, NJ, 1985.
36. P. F. Embid, A. J. Majda, and P. E. Souganidis, *Phys. Fluids* **7** (8), 2052 (1995).
37. M. Cencini, A. Torcini, D. Vergni, and A. Vulpiani, *Phys. Fluids* **15**, 679 (2003).
38. W. T. Ashurst and G. I. Shivanshinsky, *Comb. Sci. Tech.* **80**, 159 (1991).
39. P. D. Ronney, in *Modeling in Combustion Science*, J. Buckmaster and T. Takeno, eds., Springer-Verlag Lecture Notes in Physics, Springer-Verlag, Berlin, 1994, pp. 3–22.
40. B. Khouider, A. Bourlioux, and A. J. Majda, *Comb. Th. Model.* **5** 295 (2001).
41. V. Yakhot, *Comb. Sci. Tech.* **60**, 191 (1988).
42. R. C. Aldredge, in *Modeling in Combustion Science*, J. Buckmaster and T. Takeno, eds., Springer-Verlag Lecture Notes in Physics, Springer-Verlag, Berlin, 1994, pp. 23–35.
43. R. C. Aldredge, *Comb. and Flame* **106**, 29 (1996).
44. A. Oberman, Ph.D. thesis, University of Chicago, 2001.
45. G. Boffetta, M. Cencini, M. Falcioni, and A. Vulpiani, *Phys. Rep.* **356**, 367 (2002).
46. A. K. Pattanayak, *Physica D* **148**, 1 (2001).

CHAPTER 27

ON RECURSIVE PRODUCTION AND EVOLVABILITY OF CELLS: CATALYTIC REACTION NETWORK APPROACH

KUNIHIKO KANEKO

Department of Basic Science, College of Arts and Sciences, University of Tokyo, Komaba, Meguro-ku, Tokyo, 153-8902, Japan

CONTENTS

I. Basic Question for Recursive Production of a Cell as Reaction Dynamics of Catalytic Network
   A. Q1: Origin of Heredity
   B. Q2: Recursiveness and Evolvability with Diverse Chemicals
II. Brief Historical Survey
   A. Eigen's Hypercycle
   B. Dyson's Loose Reproduction System
III. Constructive Biology
   A. Standpoint of Constructive Biology
   B. Modeling Strategy for the Chemical Reaction Networks
IV. Minority Control Hypothesis for the Origin of Genetic Information
   A. Model
   B. Results
   C. Minority-Controlled State
      1. Preservation of Minority Molecule
      2. Control of the Growth Speed
      3. Control of Chemical Composition by the Minority Molecule
      4. Evolvability
   D. Experiment
   E. Discussion
      1. Heredity from a Kinetic Viewpoint
      2. Some Remarks
V. Recursive Production in an Autocatalytic Network
   A. Model
   B. Results

Geometric Structures of Phase Space in Multidimensional Chaos: A Special Volume of Advances in Chemical Physics, Part B, Volume 130, edited by M. Toda, T Komatsuzaki, T. Konishi, R.S. Berry, and S.A. Rice. Series editor Stuart A. Rice.
ISBN 0-471-71157-8 Copyright © 2005 John Wiley & Sons, Inc.

1. Phases
2. Dependence of Phases on the Basic Parameters
3. Maintenance of Recursive Production
4. Switching
C. Evolution
D. Statistical Law
VI. Summary
Acknowledgments
References

I. BASIC QUESTION FOR RECURSIVE PRODUCTION OF A CELL AS REACTION DYNAMICS OF CATALYTIC NETWORK

Question: A cell consists of several replicating molecules that mutually help the synthesis and keep some synchronization for replication. At least a membrane that partly separates a cell from the outside has to be synthesized, keeping some degree of synchronization with the replication of other internal chemicals. How is such recursive production maintained, while keeping diversity of chemicals? Furthermore, this recursive production is not complete, and there appears a slow "mutational" change over generations, which leads to evolution. How is evolvability compatible with recursive production [1]?

A. Q1: Origin of Heredity

In a cell, among many chemicals, only some chemicals (e.g., DNA) are regarded to carry genetic information. Why do only some specific molecules play the role to carry the genetic information? How has such separation of roles in molecules between genetic information and metabolism progressed? Is it a necessary course of a system with internal degrees and reproduction?

In a cell, however, a variety of chemicals form a complex reaction network to synthesize themselves. Then how can such a cell with a huge number of components and complex reaction network sustain reproduction, keeping similar chemical compositions?

To consider this problem, we start from a simple prototype cell that consists of mutually catalyzing molecule species whose growth in number leads to division of the protocell [2]. In this protocell, the molecules that carry the genetic information are not initially specified. The first question we discuss here is how heredity to maintain production of the protocell emerges. Related with the question, we ask if there appears some specific molecules to carry information for heredity, to realize continual reproduction of such protocell. We note that in the present cells, it is generally believed that information is encoded in DNA, which controls the behavior of a cell.

Here, we do not necessarily take a "genocentric" standpoint, in the sense that a gene determines the course of a cell. In fact, even in these cells, proteins and DNA both influence their replication process. Still, it cannot be denied that there exists a difference between DNA and protein molecules with regard to the role as information carrier. In spite of this mutual dependence, why is a DNA molecule usually regarded as the carrier of heredity? Is there any general rule that some specific molecules play the role of carrier of genetic information so that the recursive production of cells continues?

Now, the origin of genetic information in a replicating system is an important theoretical topic that should be studied, not necessarily as a property of certain molecules, but as a general property of replicating systems. To investigate this problem, we need to clarify what "information" really means. In considering information, one often tends to be interested in how several messages are encoded on a molecule. In fact, a heteropolymer such as DNA would be suited to encode many bits of information. One might point out that with this reason DNA molecules would be selected as an information carrier. Although this "combinatorial" capacity of an information carrier is important, what we are interested here is a basic property that has to be satisfied prior to that—that is, origin of just "1-bit" information.

As Shannon beautifully demonstrated, information means selection of one branch from several possibilities [3,4]. Assume that there are two possibilities in an event, each of which can occur with the probability $1/2$. In this case, when one of these possibilities turns out to be true, then this choice of a branch is regarded to have 1-bit information. In this sense, if a specific cell state is selected from several possible states, this selection process has information, and a molecule to control such process carries information.

Now, a molecule that carries the information is postulated to play the role to control for the choice of cellular state. Furthermore, to play the role to carry the information for heredity, the molecules must be transmitted to next generations relatively faithfully. These two features—that is, control and preservation—are nothing but the problem of heredity.

Let us reconsider what "heredity" really means. The heredity causes a high correlation in phenotype between ancestor and offspring. Then, for a molecule to carry heredity, we identify the following two features as necessary.

1. If this molecule is removed or replaced by a mutant, there is a strong influence on the behavior of the cell. We refer to this as the "control property."
2. Such molecules are preserved well over generations. The number of such molecules exhibits smaller fluctuations than that of other molecules, and their chemical structure (such as polymer sequence) is preserved over a long time span, even under potential changes by fluctuations through the

synthesis of these molecules. We refer to this as the "preservation property."

These two conditions are regarded as a fundamental condition for a molecule to establish the heredity. Now, the problem of "information" at a minimal level (i.e., 1-bit information) is nothing but the problem of the origin of heredity. As the origin of heredity, we study how a molecule starts to have the above two properties in a protocell. In other words, we study how 1-bit information starts to be encoded on a single molecule in a replicating cell system. After we answer this basic question, we will then discuss how a protocell with the heredity in the above sense attains incentive to evolve genetic information in today's sense.

To sum up, the first question we address here is restated as follows. Consider a protocell with mutually catalyzing molecules. Then, under what conditions does recursive production continue to maintain catalytic activities? How are recursiveness and diversity in chemicals compatible? How is evolvability of such protocells possible? To answer these questions, are molecules carrying heredity necessary? Under what conditions does one molecule species begin to satisfy the conditions 1 and 2 so that the molecule carries heredity? We show, under rather general conditions in our model of mutually catalyzing system, that a symmetry breaking between the two kinds of molecules takes place, and through replication and selection, one kind of molecule comes to satisfy the conditions 1 and 2.

B. Q2: Recursiveness and Evolvability with Diverse Chemicals

In a cell, the total number of molecules is limited. If there are a huge number of chemical species that catalyze each other, the number of some molecules species may go to zero. Then molecules that are catalyzed by them no longer are synthesized. Then, other molecules that are catalyzed by them cannot be synthesized, either. In this manner, the chemical compositions may vary drastically, and the cell may lose reproduction activity.

Of course, a cell state is not constant, and a cell may not keep on dividing forever. Still, a cell state is sustained to some degree to keep producing similar offspring cells. We refer to such a condition for reproduction of cells as "recursive production" or "recursiveness." The question we address here is whether there are some conditions on distribution of chemicals or structure of reaction network for recursive production.

There are two directions of study. One is with regard to the static aspect of reaction network structure (e.g., topology). The other is the number distribution of chemical species and their dynamics. Of course, one needs to combine the two aspects to fully understand the condition for recursive production of a cell.

Currently there is much interest in the reaction network structure. For example, Jeong et al. [5] studied the metabolic reaction network, without going into details of the topology. Write down all (known) metabolic reaction equations. Here, the rate of reactions is disregarded, and we are only concerned with whether or not such a reaction equation exists in a cell. Then compute how many times a specific molecule species appears in such reaction equations. If this number is large, the molecule species is related with many biochemical reactions. For example, H_2O has a large number of connections, since in many reactions it appears either in the left-hand or right-hand side of the equation. ATP has a relatively high number of connections, too. From these data the histogram $P(n)$ is obtained, as the number of molecules species that appears n times in the equations. From the data, it is shown that $P(n)$ decays with some power of n as $n^{-\alpha}$ [5].

So far, the discussion is limited only to topological structure of the network. In the reaction network dynamics, the numbers of molecules are distributed. On each "node" of the network, the abundance of the corresponding molecule species is assigned. Accordingly, some path is "thick" where such reactions occur frequently. Such abundance, as well as their fluctuations and dynamics, has to be investigated.

In a cell, the number of each molecule changes in time through reaction; the number, on average, is increased for the cell replication. For this growth to progress effectively, some positive feedback process underlying the replication process should exist, which, then, may lead to amplification of the number fluctuations in molecules. With such large fluctuations and complexity in the reaction network, how is recursive production of cells sustained? Is there any universal statistics in the number distribution of molecules?

II. BRIEF HISTORICAL SURVEY

A. Eigen's Hypercycle

Of course, the problem raised in the previous section has been addressed in the study on the origin of life, or origin of replicating system. Here we are not necessarily interested in "what happened in the past," but instead, we intend to unveil the universal logic of cell. Still, it is relevant to review the earlier studies.

To consider the origin of replication system, one needs to discuss how genetic information is faithfully transferred to the next generation. Mills et al. [6] set up an experiment of RNA replication, by using a solution of RNA and enzyme. In this experiment, some enzymes are supplied from outside, and in this sense it is not an autonomous replication system. Still, his group found that RNA molecules with proper sequences are reproduced under some error.

Following this experimental study of Spiegelman on replication of RNA, Eigen's group started theoretical study on the replication of molecules [7]. The replication process of polymer in biochemical reaction is generally carried out with the aid of enzymes. The enzyme is given by a polymer, while its catalytic activity strongly depends on its sequence. For most sequences of the polymers, the catalytic activity is very small, but few of them may have high catalytic activity. Depending on the sequence, some polymer has a much higher catalytic activity, and the replication rate of polymers depends on the sequence. As a theoretical argument, consider replication of polymers whose replication rate depends on its sequence. Now, assume that a "good" sequence has replication rate α times larger than its mutant with a substitution of a monomer from the original sequence. Here, the replication progresses under some error. Without fine machinery for error correction, this error is not negligible. Assume that in each replication process, a monomer is substituted by another monomer with the rate μ. Then the probability that a polymer consisting of N monomers can produce itself is given by $(1 - \mu)^N \approx exp(-N\mu)$, assuming that μ is small.

Now, let us examine if the good polymer can continue replication, maintaining its sequence, so that the information of this sequence is transferred. The condition that the good sequence dominates in populations in the ensemble of polymers is given by

$$N < \ln(\alpha)/\mu \qquad (1)$$

Here, $\ln(\alpha)$ is typically $O(1)$, while the error rate in the replication of monomer is estimated to be around 0.01–0.1, in the usual polymer replication process. Then the above condition gives $N < 100$ or so. In other words, information using a polymer with a sequence longer than this threshold N is hardly sustained. This problem was first posed by Eigen and is called "error catastrophe" [7]. On the other hand, information for the replication for a minimal life system must require much more information. Of course, the error rate could be reduced once some machinery for faithful replication as in the present life emerges. However, such machinery requires much more information to be transmitted by the polymer.

Summing up: For replication to progress, catalysts are necessary, and information on a polymer to replicate itself must be preserved. However, error rate in replication must have been high at a primitive stage of life; accordingly, it is recognized that the information to carry catalytic activity will be lost within few generations. In other words, faithful replication system requires larger information, while a larger information requires faithful replication system. Thus there appears catch-22-type paradox.

To resolve this problem of inevitable loss of catalytic activities through replication errors, Eigen and Schuster proposed the hypercycle [7], where replicating chemicals catalyze each other and form a cycle: "A catalyzes the synthesis of B, B catalyzes the synthesis of C, C catalyzes the synthesis of A." In this case, each chemical mutually amplifies the synthesis of the corresponding chemical species in this cycle. There occur a variety of mutations to each species, but this mutant is not generally catalyzed in some other species in the cycle. Then, such a mutant is not catalyzed by C. This is also

CATALYTIC REACTION NETWORK APPROACH   549

understood by writing out the rate equation for the increase of the population. In this hypercycle the population increase is given by the product of the populations of molecules such as $N_A \times N_B$, $N_B \times N_C$, and $N_C \times N_A$, while the growth of the population of the mutants is linear to each population N_A, N_B, and N_C. In the previous estimate for error catastrophe, both the good and mutant sequences increase linearly to the number. Then the number of variety of mutants dominates. In the present case, once the population of the good sequence in the hypercycle is dominant, it is maintained against possible emergence of mutants. With this hypercycle, the original problem of error accumulation is avoided.

Since the proposal of the hypercycle, population dynamics of molecules for such catalytic networks have been developed. However, the hypercycle itself turned out to be weak against parasitic molecules—that is, those which replicate, catalyzed by a molecule in the cycle, but do not catalyze those in the cycle. In contrast to the previous mutant, the growth rate of the population of these molecules is again the product of the populations of two species, and such parasitic molecules can invade.

Although the hypercycle itself may be weak against parasitic molecules (i.e., those which are catalyzed but do not catalyze others), it is then discussed that compartmentalization by a cell structure may suppress the invasion of parasitic molecules [7] or that the reaction–diffusion system at spatially extended system resolves this parasite problem [8]. As chemistry of lipid, it is not so surprising that a compartment structure is formed. Still, as the origin of life, this means that more complexity and diversity in chemicals are required other than a set of information-carrying molecules (e.g., RNA).

B. Dyson's Loose Reproduction System

If initially there are a variety of chemicals that form a complex network of mutual catalyzation, this system may be robust against the invasion of parasitic molecules. Such an idea resembles the stability of an ecosystem, where a complex network of several species may resist to invasion of external species. Hence we need to study if replication of complex reaction network can be sustained. In this case, from the beginning, there are many molecular species that mutually catalyze, allowing for the existence of many parasitic molecules. Here, complete replication of the system is probably difficult. Then the question we have to address is whether such a complex network can maintain molecules that catalyze the synthesis of the network species. This question was addressed by Dyson [9], as a possibility of a loose reproduction system.

Dyson, noting the experiment of Oparin on the formation of a cell-like structure, considered a collection of molecules with proteins and others. These molecules cannot replicate themselves like DNA or RNA. They, on the other hand, can have enzyme activities and can catalyze the synthesis of other molecules, albeit not faithful reproduction. Still, they may keep similar compositions. Although accurate replication of such a variety of chemicals is not possible, chemicals, as a set, may continue reproducing themselves loosely while keeping catalytic activity. Indeed, the accurate replication must be difficult at the early stage of life, but loose reproduction could be easier. However, whether this collection of molecules can keep catalytic activity through reproduction is not evident.

Dyson obtained a condition for the sustainment of catalytic activities in these collections of molecules, by taking an abstract model. For simplicity he classified molecules into two states depending on whether they have catalytic activity or not. Furthermore, he assumed that the ratio of the synthesis of catalytic molecules is amplified as the fraction of catalytic molecules is larger; that is, a positive feedback process is assumed. This model is mapped to a kind of Ising model. With the aid of mean-field analysis in statistical physics, he showed that the catalytic activities can be sustained depending on the number of molecules and their species. Although his model is abstract, the result he obtained probably can be applied to any system with a set of catalytic molecules, be it protein, lipids, or other polymers.

It is important to study whether such loose reproduction as a set is possible in a mutually catalytic reaction network (also see, e.g., Refs. 10 and 11). If this is possible and if these chemicals also include molecules forming a membrane for compartmentalization, then reproduction of a primitive cell will become possible. In fact, from the chemical nature of lipid molecules, it is not so surprising that a compartment structure is formed.

Still, in this reproduction system, any particular molecules carrying information for reproduction do not exist, in contrast to the present cell that has specific molecules (DNA) for it. As for a transition from early loose reproduction to later accurate replication with genetic information, Dyson did not give an explicit answer. He only referred to "genetic takeover" that was originally proposed by Cairns-Smith [12], who discussed that a precise replication system by nucleic acids took over the original loose reproduction system by clay. Indeed, Dyson wrote that his idea is based on "Cairns-Smith theory minus clay." However, the logic for this "takeover" is not unveiled.

Considering these theoretical studies so far, it is important to study how recursive production of a cell is possible, with the appearance of some molecules to play a specific role for heredity.

III. CONSTRUCTIVE BIOLOGY

A. Standpoint of Constructive Biology

Before describing our theoretical model and explaining the numerical results, it is relevant to briefly summarize our basic standpoint in the study of biology, referred to as "constructive biology" [1,13].[1] Here we are interested not in details of specific biological function but in universal features of a biological system. Accordingly, we need to study some features that are not influenced by the details of complicated biological processes. The present organisms, however, include detailed elaborated processes that are captured through the history of evolution. Then, for our purpose, it is desirable to set up a minimal biological system, to understand universal logic that organisms necessarily should obey. Hence, the approach that should be taken will be "constructive" in nature. This constructive approach is carried out both experimentally and theoretically.

Our "constructive biology" consists of the following steps of studies: (i) Construct a model system by combining procedures; (ii) clarify a universal class of phenomena through the constructed model(s); (iii) reveal the universal logic underlying the class of phenomena and extract logic that the life process should obey; (iv) provide a new look at data on the present organisms from our discovered logic.

There are three levels to perform these steps: (1) gedanken experiment (logic) (2) computer model, and (3) real experiment. The first one is theoretical study, revealing a logic that underlies universal features in life processes, essential in understanding the logic of "what is life." Still, the life system has a complex relationship among many parts, which constitute the characteristic feature as a whole, which then influences the process of each part. We have not gained sufficient theoretical intuition with regard to such a complex system. Then it is also relevant to make computer experiments and heuristically find some logic that cannot be easily reached by logical reasoning only. This is the second approach mentioned above—that is, construction of an artificial world in a computer. Here we combine well-defined simple procedures, to extract a general logic therein [2,14–17].

Still, in a system with potentially huge degrees of freedom such as life, the construction in a computer may miss some essential factors. Hence, we need the third experimental approach—that is, construction in a laboratory. In this case again, one constructs a possible biology world in a laboratory, by combining several procedures. For example, this experimental constructive biology has been pursued by Yomo and co-workers (see, e.g., Refs. 18–21 at the levels of biochemical reaction, cell, and ensembles of cells).

Taking this standpoint of constructive biology, we have been working problems listed in Table I both theoretically and experimentally. The first two items in the table are related with the construction of a replicating system with compartment, raised in the questions in Section I. Of course, this problem is essential to consider the origin of a cellular life. However, we do not intend to

---

[1]One can skip this subsection, if one is not much interested in the general standpoint in the study of biology.

TABLE I
Examples of Constructive Biology Under Current Investigation

Construction of	Experiment	Theory	Question to be Addressed
Replicating system	In vitro replicating system with several enzymes	Minority control	Origin of information
Cell system	Replicating liposome with internal reaction network	Dynamic bottleneck in autocatalytic reaction system	Evolvability and recursiveness for growth
Multicellular system	Interaction-induced differentiation of an ensemble of cells	Isologous diversification in inter- and intradynamics	Robustness in development
Developmental process (I)	Controlled differentiation from undifferntiated cells	Emergence of differentiation rule	Irreversibility in development
Developmental process (II)	Activin-controlled construction of tissues formation	Self-consistency between pattern and dynamics	Origin of positional information
Generation	Germ-line segregation from ensemble of cells	Higher-level recursiveness	Origin of recursive individuality
Evolution	Interaction-dependent evolution of E Coli	Symbiotic sympatric speciation	Genetic fixation of phenotypic differentiation

reproduce what has occurred in the earth. We do not try to guess the environmental condition of the past earth. Rather we try to construct such a replication system from a complex reaction network under a condition previously set up by us. For example, by constructing a protocell, in the present chapter, we ask the condition for the heredity, or universal features of the reaction dynamics, to support the recursive production of cells.

The third to sixth items are related to the construction of multicellular organisms with developmental process. When cells are aggregated, they start to form differentiation of roles; and then from a single cell, a robust developmental process to form an organized structure of differentiated cells is generated. This developmental process to form a cell aggregate is transferred to the next generation. An experimental construction of multicellular organisms (with cell differentiation) from bacteria is one target. Here again, we do not try to imitate the process of the present multicellular organisms. For example, by putting bacteria cells into some artificial condition, we study whether the cells can differentiate into distinct types or form some robust distribution of cells. Also, in vitro construction of morphogenesis from undifferentiated cells has been possible by putting cells into some given conditions [22]. With these studies, we can establish a viewpoint of universal dynamics underlying development rather than the conventional picture as a finely tuned-up process for it [15,16].

The seventh item is construction of evolution—in particular, the speciation process—which is how a species splits into two distinct groups different both in phenotype and genotype [17].

To carry out this plan experimentally, we need a system to design a life system controlled as we like. Such controlled experiments are now possible by recent advances in technology, such as flow cytometry, imaging techniques, and microarray to measure gene expressions, while advances in nanotechnology provide a powerful tool in constructing a system to regulate and observe behaviors of a single cell or multiple cells, in a well-controlled situation.

Here this construction is interesting by itself, but our goal is not the construction itself. Rather, we try to extract general features that a life system should satisfy, and we set up general questions. For example, as posed in Section I, we set up a question as to whether there are some "information molecules" that control the replication system. Then we answer the question by setting up a theory. For each item, we set up general questions, make model simulations, and set up a general theory to answer the question. This theoretical part is carried out in tight collaboration with the experiment.

To close this subsection, we give a brief remark on the study of the so-called artificial life (AL). Indeed, our approach may have something in common with AL [23]. In the AL study, people intended to construct life-as-it-could-be, not restricted to the present organisms. Originally, in the study of AL, they have been interested in the logic of life that all possible biological systems should obey, be it on this earth or in other conditions in the universe.

Indeed, there are some important studies on the origin of replicating structure from the side of computation (e.g., Ref. 24). However, the conventional AL study often tended to imitate life, and could not propose basic concepts to understand "what is life." Also, the conventional AL study was often biased into the study in a computer. It often assumes a combination of logical processes with manipulation of symbols like the study of artificial intelligence.

Our approach is distinct from the conventional artificial life study in the two points. First, we do not take such symbol-based approach, but rather we use dynamical systems approach. Second, tight collaboration between experiment and theory is essential. Note, however, that this collaboration is not of the type to "fit the data" by some theoretical expression, but rather at a conceptual level. We will see an example of such collaboration in Section IV.

B. Modeling Strategy for the Chemical Reaction Networks

Now, we discuss a standpoint in a modeling cell, based on the standpoint of the last section. Then, what type of a model is best suited for a cell to answer the question in Section I? With all the current biochemical knowledge, we can say that one could write down several types of intended models. Due to the complexity of a cell, there is a tendency of building a complicated model in trying to capture the essence of a cell. However, doing so only makes one difficult to extract new concepts, although simulation of the model may produce similar phenomena as those in living cells. Therefore, to avoid such failures, it may be more appropriate to start with a simple model that encompasses only the essential factors of living cells. Simple models may not produce all the observed natural phenomena, but are comprehensive enough to bring us new thoughts on the course of events occurring in nature.

In setting up a theoretical model here, we do not impose many conditions to imitate the life process. Rather we impose the postulates as minimally as possible, and we study universal properties in such system. For example, as a minimal condition for a cell, we consider a system consisting of chemicals separated by a membrane. The chemicals are synthesized through catalytic reactions, and accordingly the amount of chemicals increases, including the membrane component. Because the volume of this system is larger, the surface tension for the membrane can no longer sustain the system, and it will divide. After the division, these protocell systems should interact with each other, since they share resource chemicals. Under such minimum setup as will be discussed later, we study the condition for the recursive growth of a cell, as well as differentiation of the cell.

Let us start from simple argument for a biochemical process that a cell that grows must at least satisfy. In a cell, there are a huge number of chemicals that catalyze each other and form a complex network. These molecules are spatially arranged in a cell, and in some problems such spatial arrangement is very important, while for some others the discussion on just the composition of chemicals in a cell is sufficient to determine the state of a cell. Hence, for the starting point we disregard the spatial structure within a cell, and we consider just the composition of chemicals in a cell. Hence, if there are k chemical species in a cell, the cell state is characterized by the number of molecules of each species as $N_1, N_2, \ldots, N_k$. These molecules change their number through reaction among these molecules. Since most reactions are catalyzed by some other molecules, the reaction dynamics consist of a catalytic reaction network.

Through membrane, some chemicals may flow in, which are successively transformed to other chemicals through this catalytic reaction network. For a cell to grow recursively, a set of chemicals has to be synthesized for the next generation. When the number of molecules is large enough, the membrane is no longer sustained, as already mentioned. Then, when the number of molecules is larger than some value, it is expected to divide. Hence, the basic picture for a simple toy cell is shown in Fig. 1.

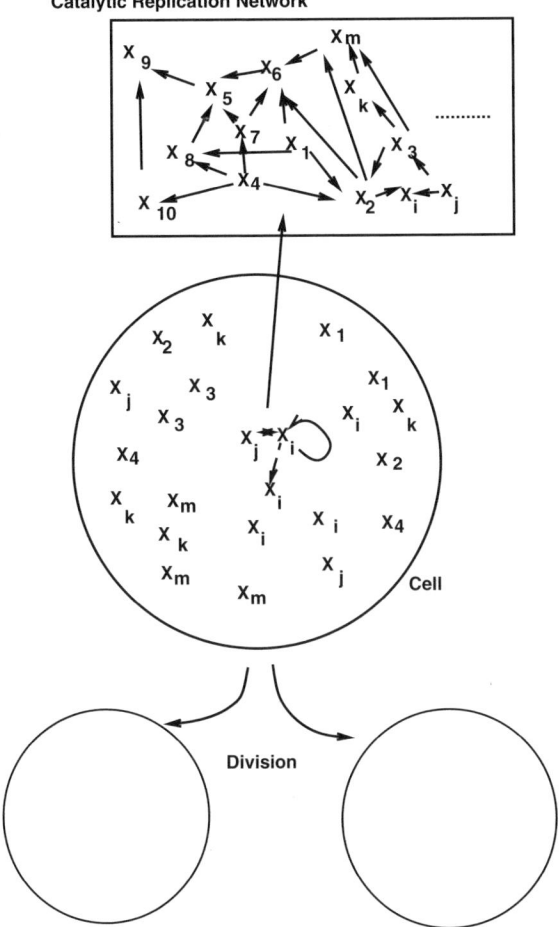

Figure 1. Schematic representation of our modeling strategy of a cell.

Of course, it is impossible to include all possible chemicals in a model. Because our constructive biology is aimed at neither making a complicated realistic model for a cell nor imitating a specific cellular function, we set up a minimal model with reaction network, to answer the questions raised in Section I. Now, there are several levels of modeling, depending on what question we are trying to answer.

0. By taking reversible two-body reactions, including all levels of reactions, ranging from metabolites, proteins, nucleic acids, and so forth. For example, to

answer the general question, "How is a nonequilibrium condition sustained in a cell?", such a level of model is desirable [25].

1. Assuming that some reaction processes are fast, they can be adiabatically eliminated. Also, most fast reversible reactions can be eliminated by assuming that they are already balanced. Then we need to discuss only the concentration (number) of molecular species, which changes relatively slowly. For example, by assuming that an enzyme is synthesized and decomposed fast, the concentrations can be eliminated to give catalytic reaction network dynamics consisting of the reactions with

$$X_i + X_j \to X_\ell + X_j \tag{2}$$

where X_j catalyzes the reaction [15,26]. If the catalysis progresses through several steps, this process is replaced by

$$X_i + mX_j \to X_\ell + mX_j \tag{3}$$

leading to higher-order catalysis [16].

For a cell to grow, some resource chemicals must be supplied through a membrane. Through the above catalytic reaction network, the resource chemicals are transformed to others; as a result, the cell grows. Indeed, this class of model is adopted to study the condition for cell growth, to unveil universal statistics for such cells, and also as a model for cell differentiation.

2. Model focusing on the dynamics of replicating units (e.g., hypercycle). For a cell to grow effectively, there should be some positive feedback process to amplify the number of each molecular species. Such a positive feedback process leads to an autocatalytic process to synthesize each molecular species. For reproduction of a cell, (almost) all molecule species are somehow synthesized. Then, it would be possible to take a replication reaction from the beginning as a model. For example, consider a reaction

$$S + X + Y \to X' + Y : S' + X' \to 2X$$

Then as a total, the reaction is represented as

$$S + S' + X + Y \to 2X + Y$$

Assuming the resources S and S' are constantly supplied, we can consider the replication reaction

$$X + Y \to 2X + Y \tag{4}$$

catalyzed by Y. At this level, we can take a unit of replicator and consider a replication reaction network. This model was first discussed in the hypercycle by Eigen and Schuster (see Section II.A).

3. Coarse-grained (phenomenological) level. Some other reduced model is adopted for the study of gene expression or signal transduction network. The modeling at this level is relevant when attempting to understand specific function of a cell.

In the present chapter we mainly use the modeling of level 2. This class of model can be obtained by reducing from the level 1 model, by restricting our interest only to take into account of replicating units. In this sense, the model is a bit simpler than the level 1 model. On the other hand, it may not be suitable to discuss the condition for cell growth, since at the level 2 model, the supply of resource chemicals is automatically assumed, and one cannot discuss how transported chemicals are transformed into others. In the present chapter, we briefly refer to the level 1 model only at the end of Section V.D, to demonstrate the universality of our result, but for details see the original articles [15,26] on level 1 modeling.

To sum up, we envision a (proto)cell containing molecules. With a supply of chemicals available to the cell, these molecules replicate through catalytic reactions, so that their numbers within a cell increase. When the total number of molecules exceeds a given threshold, the cell divides into two, with each daughter cell inheriting half of the molecules of the mother, chosen randomly. Regarding the choice of chemical species and the reaction, we have a discussion later for specific models (see Fig. 1 for a schematic representation).

IV. MINORITY CONTROL HYPOTHESIS FOR THE ORIGIN OF GENETIC INFORMATION

In the present section we propose an answer to the question raised in Section I.A, by taking a simple model of a cell with replicating molecules, proposing a novel concept on minority control, and providing corresponding experimental results.

A. Model

As discussed in Section III.B, we start from consideration of a prototype of cell, consisting of molecules that catalyze each other. As the reaction progresses, the number of molecules in this protocell will increase. Then, this cell will be divided when its volume (the total number of molecules) is beyond some threshold. Then the molecules split into two "daughter cells." Then our question in Section I is restated as follows: How are the chemical compositions transferred to the offspring cells? Do some specific molecules start to carry heredity in the sense of control and preservation, so that the reproduction continues?

Before considering the specific model, it may be relevant to recall the difference of roles between DNA (or RNA) and protein. According to the present understanding of molecular biology [27], changes undergone by DNA molecules are believed to exercise stronger influences on the behavior of cells than other chemicals. Also, a DNA molecule is transferred to offspring cells relatively accurately, compared with other constitutes of the cell. Hence a DNA molecule satisfies (at least) the "preservation" and "control" properties 1 and 2 in Section I.A.

In addition, a DNA molecule is stable, and the time scale for the change of DNA (e.g., its replication process as well as its decomposition process) is much slower. Because of this relatively slow replication, the number of DNA molecules is smaller than the number of protein molecules. At each generation of cells, single replication of each DNA molecule typically occurs, while other molecules undergo more replications (and decompositions).

With these natures of DNA in mind, while without assuming the detailed biochemical properties of DNA, we seek a general condition for the differentiation of the roles of molecules in a cell and study the origin of the control and preservation of some specific molecules.

Now, we consider a very simple protocell system [2], consisting of two species of replicating molecules that catalyze each other (see Fig. 2), assuming that only two kinds of molecules X and Y exist in this protocell, and they catalyze each other for the synthesis of the molecules.

$$X + Y \to 2X + Y; Y + X \to 2Y + X \qquad (5)$$

Here, this "catalytic reaction" is not necessarily a single reaction. In general, there can be several intermediate processes for each "reaction." The model simply states that there are two molecules that help the synthesis of the other, directly or indirectly. In general, the catalytic activities as well as the synthesis speeds differ by types of molecules. Without losing generality, one can assume that X is synthesized faster than Y.

With this synthesis of molecules, the total number of molecules in the protocell will increase, until it divides into two. As long as the molecules catalyze each other, this synthesis continues, as well as the division (reproduction) of protocell. However, some structural changes in molecules can occur through replication ("replication error"). These structural changes in each kind of molecule may result in the loss of catalytic activity. Indeed, the molecules with catalytic activity are not so common. On the other hand, molecules without catalytic activity can grow their number, if they are catalyzed by other catalytic molecules. Then, as discussed in Section II.A, the maintenance of reproduction is not so easy.

Summarizing the above discussion, we consider the following model, as a first step in answering the question posed in Section I.A [2].

CATALYTIC REACTION NETWORK APPROACH 559

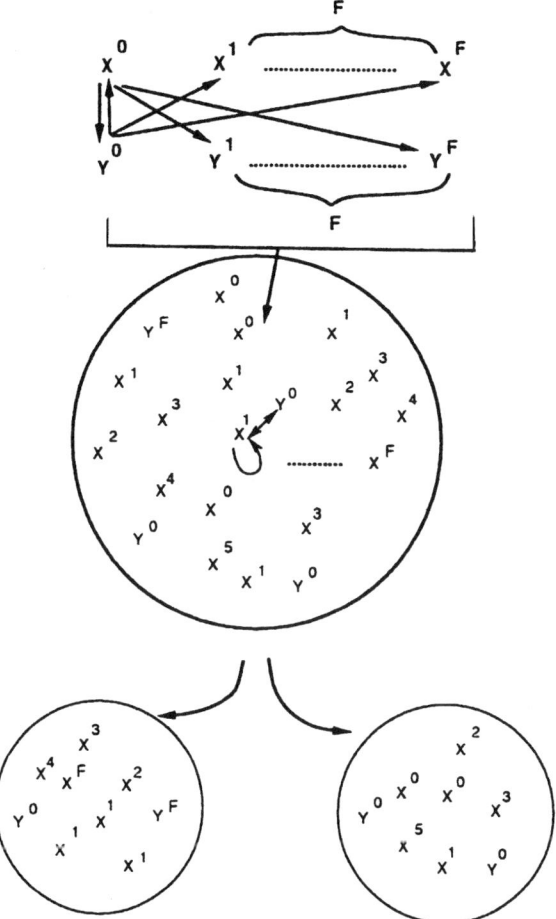

Figure 2. Schematic representation of our model.

(i) There are two species of molecules, X and Y, which are mutually catalyzing.

(ii) For each species, there are active and inactive ("I") types. Consider that the active molecule type is rather rare. There are F types of inactive molecules per active type. For most simulations, we consider the case in which there is only one type of active molecule for each species.

Active types are denoted as X^0 and Y^0, while there are inactive types X^I and Y^I with $I = 1, 2, \ldots, F$. The active type has the ability to catalyze the replication of both types of the other species of molecules. The catalytic reactions for replication are assumed to take the form

$$X^J + Y^0 \to 2X^J + Y^0 \quad (\text{for } J = 0, 1, \ldots, F)$$

and

$$Y^J + X^0 \to 2Y^J + X^0 \quad (\text{for } J = 0, 1, \ldots, F)$$

(iii) The rates of synthesis (or catalytic activity) of the molecules X and Y differ. We stipulate that the rate of the above replication process for Y, γ_y, is much smaller than that for X, γ_x. This difference in the rates may also be caused by a difference in catalytic activities between the two molecular species.

(iv) In the replication process, there may occur structural changes that alter the activity of molecules. Therefore the type (active or inactive) of a daughter molecule can differ from that of the mother. The rate of such structural change is given by μ, which is not necessarily small, due to thermodynamic fluctuations. This change can consist of the alternation of a sequence in a polymer or other conformational change, and they may be regarded as replication "error." Note that the probability for the loss of activity is F times greater than for its gain, since there are F times more types of inactive molecules than active molecules. Hence, there are processes described by

$$X^I \to X^0 \quad \text{and} \quad Y^I \to Y^0 \quad (\text{with rate } \mu)$$
$$X^0 \to X^I \quad \text{and} \quad Y^0 \to Y^I \quad (\text{with rate } \mu \text{ for each})$$

resulting from structural change.

(v) When the total number of molecules in a protocell exceeds a given value $2N$, it divides into two, and the chemicals therein are distributed into the two daughter cells randomly, with N molecules going to each. Subsequently, the total number of molecules in each daughter cell increases from N to $2N$, at which point these divide.

(vii) To include competition, we assume that there is a constant total number M_{tot} of protocells, so that one protocell, randomly chosen, is removed whenever a (different) protocell divides into two.

With the above-described process, we have basically four sets of parameters: the ratio of synthesis rates γ_y/γ_x, the error rate μ, the fraction of active molecules $1/F$, and the number of molecules N. (The number M_{tot} is not important, as long as it is not too small.)

We carried out simulation of this model, according to the following procedure. First, a pair of molecules is chosen randomly. If these molecules are of different species, then if the X molecule is active, a new Y molecule is produced with the probability γ_y, and if the Y molecule is active, a new X molecule is produced with the probability γ_x. Such replications occur with the error rates given above. All the simulations were thus carried out stochastically, in this manner.

We consider a stochastic model rather than the corresponding rate equation, which is valid for large N, since we are interested in the case with relatively small N. This follows from the fact that in a cell, often the number of molecules of a given species is not large, and thus the continuum limit implied in the rate equation approach is not necessarily justified [28].

Furthermore, it has recently been found that the discrete nature of a molecule population leads to qualitatively different behavior than in the continuum case in a simple autocatalytic reaction network [29]. In a simple autocatalytic reaction system with a small number of molecules, a novel steady state is found when the number of molecules is small, which is not described by a continuum rate equation of chemical concentrations. This novel state is first found by stochastic particle simulations. The mechanism is now understood in terms of fluctuation and discreteness in molecular numbers. Indeed, some state with extinction of specific molecule species shows a qualitatively different behavior from that with very low concentration of the molecule. This difference leads to a transition to a novel state, referred to as discreteness-induced transition. This phase transition appears by decreasing the system size or flow to the system, and it is analyzed from the stochastic process, where a single-molecule switch changes the distributions of molecules drastically.

Reference 29 gives examples in which a discreteness in molecule number leads to a novel phase that is not observed from a continuous rate equation of chemical reaction. In a cell, since the number of some molecular species is very small, we need to seriously consider the possibility that the discreteness in molecule numbers may lead to a novel behavior distinct from the continuum description.

B. Results

If N is very large, the above-described stochastic model can be replaced by a continuous model given by the rate equation. Let us represent the total number of inactive molecules for each of X and Y as

$$N_x^I = \sum_{j=1}^{F} N_x^j, \quad N_y^I = \sum_{j=1}^{F} N_y^j$$

Then the growth dynamics of the number of molecules N_x^J and N_y^J is described by the rate equations, using the total number of molecules N^t:

$$dN_x^j/dt = \gamma_x N_x^j N_y^0/N^t; dN_y^j/dt = \gamma_y N_x^0 N_y^j/N^t \qquad (6)$$

From these equations, under repeated divisions, it is expected that the relations $\frac{N_x^0}{N_y^0} = \frac{\gamma_x}{\gamma_y}$, $\frac{N_x^0}{N_x^I} = \frac{1}{F}$, and $\frac{N_y^0}{N_y^I} = \frac{1}{F}$ are eventually satisfied. Indeed, even with our stochastic simulation, this number distribution is approached as N is increased.

However, when N is small, there appears a significant deviation from the above distribution [2], under the presence of the selection process. In Fig. 3, we have plotted the average numbers $\langle N_x^0 \rangle$, $\langle N_x^I \rangle$, $\langle N_y^0 \rangle$, and $\langle N_y^I \rangle$. Here, each molecule number is computed for a cell just prior to the division, when the total number of molecules is $2N$, while the average $\langle \ldots \rangle$ is taken over all cells that divided throughout the simulation. (Accordingly, a cell removed without division does not contribute to the average.) As shown in the figure, there appears a state satisfying $\langle N_y^0 \rangle \approx 2 - 10$, $\langle N_y^I \rangle \approx 0$. Since $F \gg 1$, such a state with $\frac{\langle N_y^0 \rangle}{\langle N_y^I \rangle} > 1$ is not expected from the rate equation, Eq. (6). Indeed, for the X species, the number of inactive molecules is much larger than the number of active ones. Hence, we have found a novel state that can be realized due to the smallness of the number of molecules and the selection process.

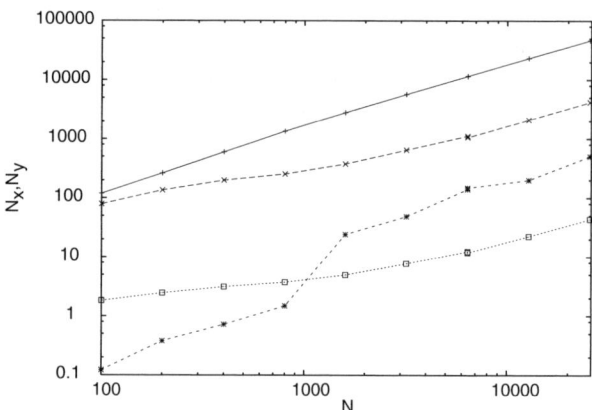

Figure 3. Dependence of $\langle N_x^0 \rangle (\times)$, $\langle N_x^I \rangle (+)$, $\langle N_y^0 \rangle$ ($\square$), and $\langle N_y^I \rangle (*)$ on N. The parameters were fixed as $\gamma_x = 1$, $\gamma_y = 0.01$, and $\mu = 0.05$. Plotted are the averages of N_x^0, N_x^I, N_y^0, and N_y^I at the division event, and thus their sum is $2N$. We use $M_{tot} = 100$, and the sampling for the averages were taken over 10^5 to 3×10^5 steps, where the number of divisions ranges from 10^4 to 10^5, depending on the parameters. Reproduced with permission from Ref. 2.

CATALYTIC REACTION NETWORK APPROACH   563

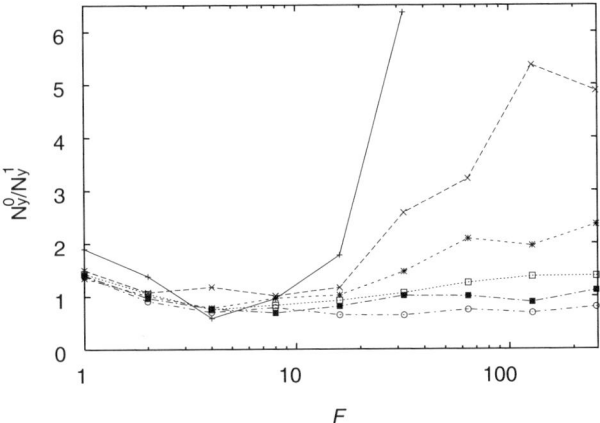

Figure 4. Dependence of the active-to-inactive ratio, $\langle N_y^0 \rangle / \langle N_y^I \rangle$, on F. The parameters were fixed as $\gamma_x = 1$, $\gamma_y = 0.01$, $\mu = 0.05$, and $F = 128$. Plots for $\gamma_y = 0.005$ ($\diamond$), 0.01 (+), 0.015 ($\square$), 0.02 (×), 0.025 ($\triangle$), and 0.03 (∗) are overlaid. Plotted are the averages of N_x^0, N_x^I, N_y^0, and N_y^I at the division event. Reproduced with permission from Ref. 2.

For the dependence of $\{\langle N_x^0 \rangle, \langle N_x^I \rangle, \langle N_y^0 \rangle, \langle N_y^I \rangle\}$ on these parameters, see also figures in Ref. 2. From these numerical results, it is shown that the above-mentioned state with $\langle N_y^0 \rangle \approx 2\text{--}10$, $\langle N_y^I \rangle < 1$ is reached and sustained when γ_y/γ_x is small and F is sufficiently large. In fact, for most dividing cells, N_y^I is exactly 0, while there appear a few cells with $N_y^I > 1$ from time to time. It should be noted that the state with almost no inactive Y molecules appears in the case of larger F—that is, in the case of a larger possible variety of inactive molecules. This suppression of Y^I for large F contrasts with the behavior found in the continuum limit (the rate equation). In Fig. 4, we have plotted $\langle N_y^0 \rangle / \langle N_y^I \rangle$ as a function of F. Up to some value of F, the proportion of active Y molecules decreases, in agreement with the naive expectation provided by Eq. (6), but this proportion increases with further increase of F, in the case that γ_y/γ_x is small ($\lesssim 0.02$) and N is small.

This behavior of the molecular populations can be understood from the viewpoint of selection: In a system with mutual catalysis, both X^0 and Y^0 are necessary for the replication of protocells to continue. The number of Y molecules is rather small, since their synthesis speed is much slower than that of X molecules. Indeed, the fixed point distribution given by the continuum limit equations possesses a rather small N_y^0. However, in a system with mutual catalysis, both X^0 and Y^0 must be present for replication of protocells to continue. Note that for the replication of X molecules to continue, at least a single active Y molecule is necessary. Hence, if N_y^0 vanishes, only the replication of inactive Y molecules occurs, and divisions from this cell cannot proceed indefinitely, because the number of X^0 molecules is cut in half at each division. Furthermore, for a cell with $N_y^0 = 1$, only one of its daughter cells can have an active Y molecule. Summing up, under the presence of selection, protocells with $N_y^0 > 1$ are selected.

On the other hand, the total number of Y molecules is limited to small values, due to their slow synthesis speed. This implies that a cell that suppresses the number of Y^I molecules to be as small as possible is preferable under selection, so that there is a room for Y^0 molecules. Hence, a state with almost no Y^I molecules and a few Y^0 molecules, once realized through fluctuations, is expected to be selected through competition for survival (see Fig. 5 for schematic representation).

Of course, the probability for such rare fluctuations decrease quite rapidly as the total molecule number increases, and for sufficiently large numbers, the continuum description of the rate equation is valid. Clearly then, a state of the type described above is selected only when the total number of molecules within a protocell is not too large. In fact, a state with very small N_Y^I appears only if the total number N is smaller than some threshold value depending on F and γ_y. In other words, too large a cell is not favorable, because the fluctuation is too small to produce such a rare state.

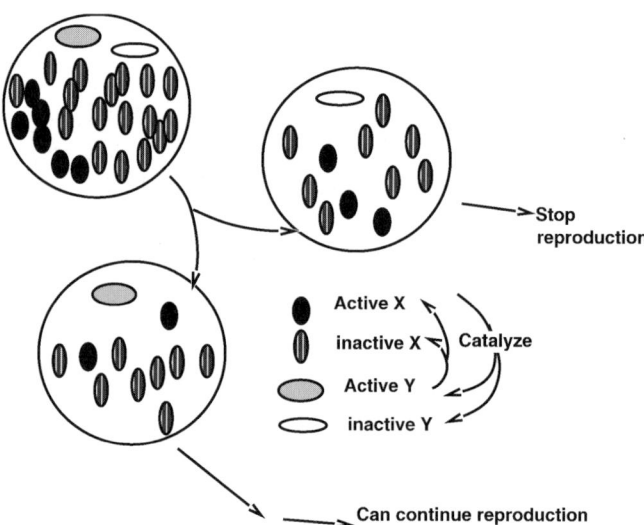

Figure 5. Schematic representation of our logic. Once an active molecule of each molecule species is lost, the reproduction does not continue.

C. Minority-Controlled State

We showed that in a mutually catalyzing replication system, the selected state is one in which the number of inactive molecules of the slower replicating species, Y, is drastically suppressed. In this section, we first show that the fluctuations of the number of active Y molecules is smaller than those of active X molecules in this state. Next, we show that the molecular species Y (the minority species) becomes dominant in determining the growth speed of the protocell system. Then, considering a model with several active molecule types, the control of chemical composition through specificity symmetry breaking is discussed.

1. Preservation of Minority Molecule

First, we computed the time evolution of the number of active X and Y molecules, to see if the selection process acts more strongly to control the number of one or the other. We computed N_x^0 and N_y^0 at every division to obtain the histograms of cells with given numbers of active molecules.

The fluctuations in the value of N_y^0 are found to be much smaller than those of N_x^0. The selection process discriminates more strongly between different concentrations of active Y molecules than between those of active X molecules. Hence the active Y molecules are well-preserved with relatively smaller fluctuations in the number.

2. Control of the Growth Speed

Now, it is expected that the growth speed of our protocell has a stronger dependence on the number of active Y molecules than the number of active X molecules. We have found that the division time is a much more rapidly decreasing function of N_y^0 than of N_x^0. Even a slight change in the number of active Y molecules has a strong influence on the division time of the cell. Of course, the growth rate also depends on N_x^0, but this dependence is much weaker. Hence, the growth speed is controlled mainly by the number of active Y molecules.

3. Control of Chemical Composition by the Minority Molecule

As another demonstration of control, we study a model in which there is more specific catalysis of molecule synthesis. Here, instead of single active molecule types for X and Y, we consider a system with k types of active X and Y molecules, X^{0i} and Y^{0i} ($i = 1, 2, \ldots, k$). In this model, each active molecule type catalyzes the synthesis of only a few types ($m < k$) of the other species of molecules. Here we assume that both X and Y molecules have the same "specificity" (i.e., the same value of m) and study how this symmetry is broken.

As already shown, when N, γ_y, and F satisfy the conditions necessary for realization of a state in which N_y^I is sufficiently small, the surviving cell type contains only a few active Y molecules, while the number of inactive ones vanishes or is very small. Our simulations show that in the present model with several active molecule types, only a single type of active Y molecule remains after a sufficiently long time. We call this "surviving type," i_r ($1 \leq i_r \leq k$). Contrastingly, at least m types of X^0 species that can be catalyzed by the remaining Y^{0i_r} molecular species remain. Accordingly, for a cell that survived after a sufficiently long time, a single type of Y^{0i_r} molecule catalyzes the synthesis of (at least) m kinds of X molecular species, while the multiple types of X molecules catalyze this single type of Y^{0i} molecules. Thus, the original symmetry regarding the catalytic specificity is broken as a result of the difference between the synthesis speeds.

Due to autocatalytic reactions, there is a tendency for further increase of the molecules that are in the majority. This leads to competition for replication between molecular types of the same species. Since the total number of Y molecules is small, this competition leads to all-or-none behavior for the survival of molecules. As a result, only a single type of species Y remains, while for species X the numbers of molecules of different types are statistically distributed as guaranteed by the uniform replication error rate.

Although X and Y molecules catalyze each other, a change in the type of the remaining active Y molecule has a much stronger influence on X than a change in the types of the active X molecules on Y, since the number of Y molecules is much smaller.

With the results so far, we can conclude that the Y molecules—that is, the minority species—control the behavior of the system, and are preserved well over many generations. We therefore call this state the minority-controlled (MC) state.

4. Evolvability

An important characteristic of the MC state is evolvability. Consider a variety of active molecules $0i$, with different catalytic activities. Then the synthesis rates γ_x and γ_y depend on the activities of the catalyzing molecules. Thus, γ_x can be written in terms of the molecule's inherent growth rate, g_x, and the activity, $e_y(i)$, of the corresponding catalyzing molecule Y^{0i}:

$$\gamma_x = g_x \times e_y(i), \qquad \gamma_y = g_y \times e_x(i)$$

Since such a biochemical reaction is entirely facilitated by catalytic activity, a change of e_y or e_x—for example, by the structural change of polymers—is more important. Given the occurrence of such a change to molecules, those with greater catalytic activities will be selected through evolution, leading to the selection of larger e_y and e_x. As an example to demonstrate this point, we have extended the model to include k kinds of active molecules with different catalytic activities. Then, molecules with greater catalytic activities are selected through competition.

Since only a few molecules of the Y species exist in the MC state, a structural change to them strongly influences the catalytic activity of the protocell. On the other hand, a change to X molecules has a weaker influence, on average, since the deviation of the *average* catalytic activity caused by such a change is smaller, as can be deduced from the law of large numbers. Hence the MC state is important for a protocell to realize evolvability.

D. Experiment

Recently, there have been some experiments to construct minimal replicating systems *in vitro*. As an experiment corresponding to this problem, we describe an *in vitro* replication system, constructed by Yomo's group [18].

In general, proteins are synthesized from the information on DNA through RNA, while DNA are synthesized through the action of proteins. As a set of chemicals, they autonomously replicate themselves. Now simplifying this replication process, Matsuura et al. [18] constructed a replication system consisting of DNA and DNA polymerase (i.e., an enzyme for the synthesis of DNA), and so forth. This DNA polymerase is synthesized by the corresponding gene in the DNA, while it works as the catalyst for the corresponding DNA. Through this mutual catalytic process the chemicals replicate themselves.

As for the amplification of DNA, the polymerase chain reaction (PCR) is widely used and is a standard tool for molecular biology. In this case, however, enzymes that are necessary for the replication of DNA must be supplied externally. In this sense, it is not a self-contained autonomous replication system. In the experiment by Yomo's group, while they use PCR as one step of experimental procedures, the enzyme (DNA polymerase) for DNA synthesis is also replicated *in vitro* within the system. Of course, some (raw) material, such as amino acid or ATP, has to be supplied, but otherwise the chemicals are replicated by themselves (see Fig. 6 for the experimental procedure).

In this experiment, there is mutual synthetic process between gene and enzymes. Roughly speaking, the polymerase in the experiment corresponds to X in our model, while the polymerase gene corresponds to Y.

Now, at each step of replication, about 2^{30}–2^{40} DNA molecules are replicated. Here, of course, there are some errors. These errors can occur in the synthesis of enzyme and also in the synthesis of DNA. With these errors, there appear DNA molecules with different sequences. Now a pool of DNA molecules with a variety of sequences is obtained as a first generation.

From this pool, the DNA and enzymes are split into several tubes. Then, materials with ATP and amino acids are supplied, and the replication process is repeated (see Fig. 6). In other words, the "test tube" here plays the role of "cell

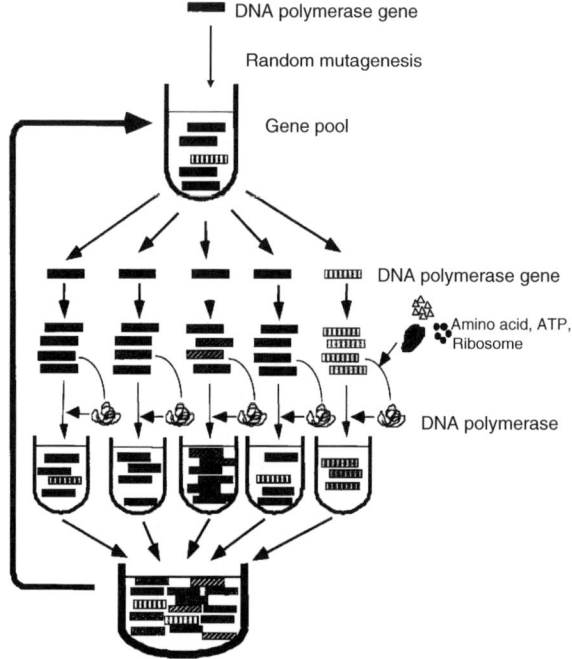

Figure 6. Illustration of *in vitro* autonomous replication system consisting of DNA and DNA polymerase. See text and Ref. 18 for details. Provided with the courtesy of Yomo et al.

compartmentalization." Instead of autonomous cell division, the split into several tubes is operated externally.

In this experiment, instead of changing the synthesis speed γ_y or N in the model, one can control the number of genes, by changing the condition how the pool is split into several test tubes.

Indeed, they studied the two distinct cases: (a) split to tubes containing a single DNA in each and (b) split to tubes containing 100 DNA molecules. Recall that in the theory, the evolvability by minority control is predicted. Hence, the behavior between the two cases may be drastically different.

First, we describe the case with a single DNA in each tube. Here, the pool of chemicals is split into 10 tubes, each of which has a single DNA molecule, and the replication process already described progresses in each tube. Here, the sequence of DNA molecules could be different by tube, since there is replication error. Then the activity of DNA polymerase by each tube is also different, and the number of DNA molecules synthesized in each tube is different. In other words, some DNA molecules can produce more offspring, but others cannot.

active to inactive molecules in the model), only when the population of DNA polymerase genes is small and competition of replicating systems is applied. When the number of genes (corresponding to Y) is small, the information containing in the DNA polymerase genes is preserved. This is made possible by the maintenance of rare fluctuations, as found in our theory. The system has evolvability only if the amount of DNA in the system is small. Otherwise, the system gradually loses its activity to replicate itself. These experimental results are consistent with the minority control theory described.

E. Discussion

1. Heredity from a Kinetic Viewpoint

In this section we have shown that in a mutually catalyzing system, molecules Y with the slower synthesis speed and minority in number tend to act as the carrier of heredity. Through the selection under reproduction, a state in which there are a few active Y and almost zero inactive Y molecules is selected. This state is termed the "minority-controlled state." Between the two molecular species, there appears separation of roles, between that with a larger number and that with a greater catalytic activity. The former has a variety of chemicals and reaction paths, while the latter works as a basis for the heredity, in the sense of the two properties mentioned in Sections I.A and IV.C, "preservation" and "control." We now discuss these properties in more detail.

Preservation Property. A state that can be reached only through very rare fluctuations is selected, and it is preserved over many generations, even though the realization of such a state is very rare when we consider the fluctuations around the rate equation obtained in the continuum limit.

Control Property. A change in the number of Y molecules has a stronger influence on the growth rate of a cell than a change in the number of X molecules. Also, a change in the catalytic activity of the Y molecules has a strong influence on the growth of the cell. The catalytic activity of the Y molecules acts as a control parameter of the system.

Once this minority controlled state is established, the following scenario for the evolution of genetic information is expected. First, a new selection pressure is now possible to emerge, to evolve a machinery to ensure that the minority molecule makes it into the offspring cells, since otherwise the reproduction of the cell is highly damaged. Hence a machinery to guarantee the faithful transmission of the minority molecule should evolve. Now, the origin of heredity is established. Here, for this heredity, any specific metabolic or genetic contents transmitted faithfully is not necessary. It can appear from the loose reproduction system that Dyson considered (as in Section II.B). This heredity evolves just as a result of kinetic phenomenon and is a rather general phenomenon in a reproducing protocell consisting of mutually catalytic molecules.

This faithful transmission of minority molecule provides a basis for critical information for reproduction of the protocells. Since this minority molecule is protected to be transmitted, other chemicals that are synthesized in connection with it are probable to be transmitted, albeit not always faithfully. Hence there appears a further evolutionary incentive to package life-critical information into the minority molecule. Now more information ("many bits" of information) are encoded on the minority molecule. Then, the molecules work as a carrier of genetic information in the today's sense. With this evolution having more molecules catalyzed by the minority molecule, it is then easier to further develop the machinery to better take care of minority molecules, since this minority molecule is essential to many reactions for the synthesis of many other molecules.

Hence the evolution of faithful transmission of minority molecules and of coding of more information reinforce each other. At this point one can expect a separation of metabolism and genetic information.

To sum up, how a single molecule starts to reign the heredity is understood from a kinetic viewpoint. We first show the minority-controlled state as a rather general consequence of kinetic process of mutually catalytic molecules. This provides a basis for heredity. Taking advantage of the evolvability of minority-controlled state, then, preservation mechanism of the minority molecule evolves, which allows for more information encoded on it, leading to separation of genetic information and metabolism. In this sense, the minority molecular species with slower synthesis speed, leading to the preservation of rare states and control of the behavior of the system, acts as an information carrier. The important point of our theory is that heredity arises prior to any metabolic information that needs to be inherited.

2. Some Remarks

In Section II, we described two standpoints on the origin of life—that is, genetic information first or complex metabolism first. We pointed out some difficulty at each standpoint. In the former picture, there was a problem on the stability against parasites, while the latter cannot solve how genetic information took over the original loose reproduction system. The minority control gives a new look to these problems.

The first problem in Section II.A was the appearance of parasitic molecules to destroy the hypercycle—that is, the mutually catalytic reaction cycle. If only the replication process of molecules is concerned, it is not so easy to resolve the problem. Here we consider the dual level of replication—that is, molecular and cellular replication.

In the present theory for the origin of information, existence of a cell unit that reproduces itself is required. Two levels of reproduction, both molecules and cells, are assumed here. Hence a cell with parasitic molecules cannot grow, and it is selected out. Relevance of this type of two-level reproduction to avoid molecular parasites has been discussed [8,30,31]. Here, relevance of cellular compartment to the *origin of genetic information* is more important.

This two-level selection works effectively, with the aid of minority control of specific molecules for a cell. Indeed, surviving cells satisfy the minority control. With the selection pressure for reproduction of cells, there appears a state that is not expected by the rate equation for reaction of molecules, where the number of inactive Y molecules that are parasitic to the catalytic reaction is suppressed. Furthermore, resistance against parasitic (inactive) Y molecules is established by this minority-controlled state.

This minority control also resolves the question on the genetic takeover, the problem in the "metabolism first" standpoint (in Section II.B). Among several molecules, specific molecule species that are minority in population controls the behavior of a cell and is well preserved. The possible scenario mentioned in the beginning of this section gives one plausible answer regarding how genetic takeover progresses.

The differentiation of role between the molecules looks like "symmetry breaking." When initially two states are equally possible, and later only one of them is selected, it is said that the symmetry is broken. In the differentiation of roles of molecules studied here, however, the molecules have different characters as to the replication speed from the beginning. Here a difference in one character (i.e., the replication speed) is "transformed" into the difference in the control behavior and in the role as a carrier of heredity. In other words, a characteristic with already broken symmetry is transformed into a different type of symmetry breaking. This kind of transformation of one character's difference to another is often seen in biology, as we have already discussed in the study of morphogenesis and sympatric speciation [16,17].

V. RECURSIVE PRODUCTION IN AN AUTOCATALYTIC NETWORK

Now we come to the second question raised in Section I. In the model of the previous section, we considered a system consisting of two kinds of molecules. In a cell, however, a variety of chemicals form a complex reaction network to synthesize themselves. Here we study a model with a large number of chemical species to discuss how a cell with such large number of components and complex reaction network can sustain reproduction, keeping similar chemical compositions [32–34].

A. Model

To unveil general features of a system with mutually catalyzing molecules, we study a system with a variety of chemicals (k molecular species), forming a mutually catalyzing network. The molecules replicate through catalytic reactions, so that their numbers within a cell increase (see Fig. 1 again for schematic representation of the model).

We envision a (proto)cell containing k molecular species with some of the species possibly having a zero population. A chemical species can catalyze the synthesis of some other chemical species as

$$[i] + [j] \rightarrow [i] + 2[j] \qquad (7)$$

with $i, j = 1, \ldots, k$ according to a randomly chosen reaction network, where the reaction is set at far-from-equilibrium, In Eq. (7) the molecule i works as a catalyst for the synthesis of the molecule j, while the reverse reaction is neglected, as discussed in the hypercycle model. For each chemical the rate for the path of catalytic reaction in Eq. (7) is given by ρ; that is, each species has about $k\rho$ possible reactions. The rate is kept fixed throughout each simulation. Considering catalytic reaction dynamics, the reverse reaction process is neglected, and reactions $i \leftrightarrow j$ are not included. (Here we investigate the case without direct mutual connections; the path $i \rightarrow j$ is excluded when there was a path $j \rightarrow i$, although this condition is not essential for the results to be discussed.) Furthermore, each molecular species i has a randomly chosen catalytic ability $c_i \in [0, 1]$ (i.e., the above reaction occurs with the rate c_i). Assuming an environment with an ample supply of chemicals available to the cell, the molecules then replicate, leading to an increase in their numbers within a cell.

Again, when the total number of molecules exceeds a given threshold (here we used $2N$), the cell is assumed to divide into two, with each daughter cell inheriting half of the molecules of the mother cell, chosen randomly.

During the replication process, structural changes (e.g., the alternation of a sequence in a polymer) may occur that alter the catalytic activities of the molecules. Therefore, the activities of the replicated molecule species can differ from those of the mother species. The rate of such structural changes is given by the replication "error rate" μ. As a simplest case, we assume that this "error" leads to all other molecular species with equal probability (i.e., with the rate $\mu/(k-1)$) and could thus regard it as a background fluctuation. In reality, of course, even after a structural change, the replicated molecule will keep some similarity with the original molecule, and a replicated species with the "error" would be within a limited class of molecule species. Hence, this equal rate of transition to other molecular species is a drastic simplification. Some simulations where the errors in replication only lead to a limited range of molecule species, however, show that the simplification does not affect the basic conclusions presented here. Hence we use the simplest case for most simulations.

In statistical physics, people study mostly the case where the total number of molecules N is very large, at least much much larger than a number of molecule species k. In this case, the continuum description is relevant. When N/k is rather small, some molecular species can often fluctuate around 0, where the discreteness 0, 1, 2, ... will be important, as already discussed. In order to take the importance of the discreteness in the molecule numbers into account, we adopted a stochastic rather than the usual differential equations approach, by taking a variety of possible chemicals where N and k are of a comparable order.

The model is simulated as follows: At each step, a pair of molecules, say, i and j, is chosen randomly. If there is a reaction path between species i and j, and i (j) catalyzes j (i), one molecule of the species j (i) is added with probability c_i (c_j), respectively. The molecule is then changed to another randomly chosen species with the probability of the replication error rate μ. When the total number of molecules exceeds a given threshold (denoted as N), the cell divides into two such that each daughter cell inherits half ($N/2$) of the molecules of the mother cell, chosen randomly [2].

Again, to include competition, we assume that there is a constant total number M_{tot} of protocells, so that one protocell, randomly chosen, is removed whenever a (different) protocell divides into two. However, the result here does not depend on M_{tot} so much. We choose mostly $M_{tot} = 1$ in the results below, but the simulation with $M_{tot} = 100$ gives essentially the same behavior.

B. Results

1. Phases

Our main concern here is the dynamics of these molecule numbers N_i of the species i in relationship with the condition of the recursive growth of the (proto)cell. In our model there are four basic parameters: the total number of molecules N, the total number of molecular species k, the mutation rate μ, and the reaction path rate ρ. By carrying out simulations of this model (choosing a variety of parameter values N, k, μ, and ρ) and also by taking various random networks, we have found that the behaviors are classified into the following three phases [32,33]:

1. Fast switching states without recursiveness.
2. Achievement of recursive production with similar chemical compositions.
3. Switch over several quasi-recursive states.

In phase 1, there is no clear recursive production and the dominant molecular species changes by generation frequently. Even though each generation has some

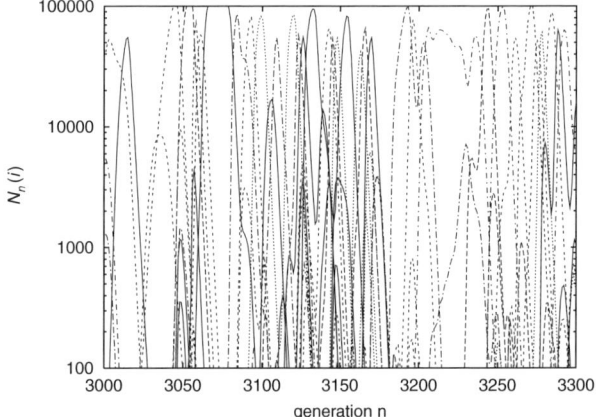

Figure 8. The number of molecules $N_n(i)$ for the species i plotted as a function of generation n of cells—that is, at each successive division event n. Only the population of species which exceed 100 during this time span are plotted. A random network with $k = 500$ and $\rho = 0.2$. Dominant species change successively in generation.

dominating species with regard to the molecule numbers, the dominating species change every few generations. At one generation, some chemical species are dominant, but only a few generations later. Information regarding the previously dominating species is totally lost often to the point that its population drops to zero (see Fig. 8). Here no stable mutual catalytic relationships are formed among molecules. Hence, the time required for reproduction of a cell is quite large, and it is much larger than in case 2.

In phase 2, a recursive state is established, and the chemical composition is stabilized such that it is not altered much by the division process (see Fig. 9). Generally, all the observed recursive states consist of 5–12 species, except for those species with one or two molecule numbers, which exist only as a result of replication errors. These 5–12 chemicals mutually catalyze, by forming a catalytic network as in Fig. 10, which will be discussed later. The member of these 5–12 species do not change by generations, and the chemical compositions are transferred to the offspring cells. Once reached, this state is preserved throughout whole simulations, lasting over more than 10,000 generations.

The recursive state observed here is not necessarily a fixed point with regard to the population dynamics of the chemical concentrations. In some case, the chemical concentrations oscillate in time, but the nature of the oscillation is not altered by the process of cell division.

For example, in the recursive state depicted in Fig. 9a, 11 species remain in existence throughout the simulation. As shown, three species have much higher

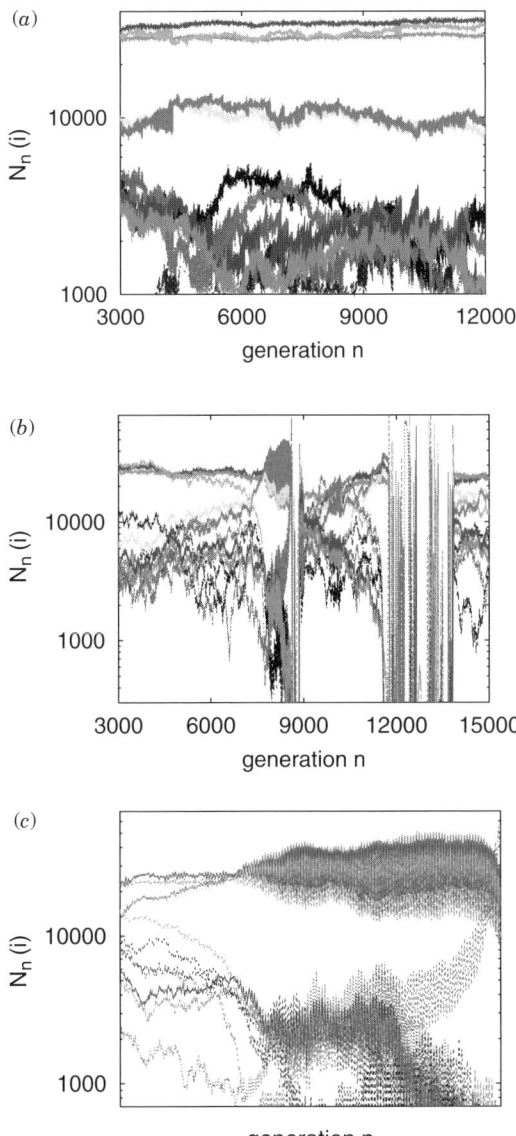

Figure 9. The number of molecules $N_n(i)$ for the species i plotted as a function of generation n of cells—that is, at each successive division event n. Results from a random network with $k = 200$ and $\rho = 0.1$ were adopted, with $N = 64{,}000$ and $\mu = 0.01$ (*a*) and $\mu = 0.1$ (*b*). Only some species (whose population get large at some generation) are plotted. In (a), a recursive production state is established, while in (b), a few quasi-recursive states are visited successively. (*c*) Expansion of (b) around the time step 100,000.

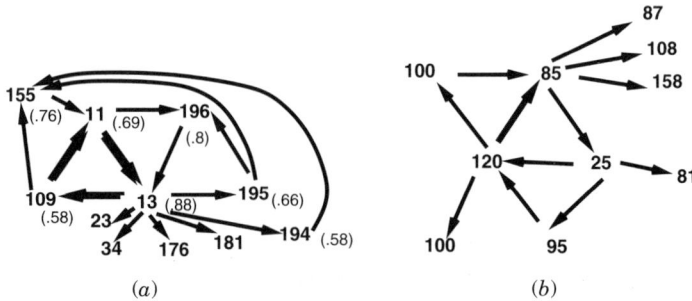

Figure 10. The catalytic network of the dominant species that constitute the recursive state. The catalytic reaction is plotted by an arrow $i \to j$, as the replication of the species j with the catalytic species i. The numbers in parentheses denote c_i of the species. Only the species that continue to exist with the population larger than 10 is plotted. (Note that many other species can exist at each generation, through the replication error.) (a) Corresponding to the recursive state of Fig. 9a, where the three species connected by thick arrows are the top three species in Fig. 9a. (b) Another example of a network observed in a different set of simulations with $k = 200$ and $\rho = 0.1$, but with a different reaction network from Fig. 9.

populations than others, which form a hypercycle as $109 \to 11 \to 13 \to 109$. (The numbers 11, 13, ... are indices of chemical species, initially assigned arbitrarily.) The hypercycle sustains the replication of the molecules and is called "core hypercycle." The catalytic activities of the species satisfy $c_{13} > c_{109} > c_{11}$, and accordingly the respective populations satisfy $N_{11} > N_{109} > N_{13}$.

In phase 3, after one recursive state lasts over many generations (typically a thousand generations), a fast switching state appears until a new (quasi-)-recursive state appears. As shown in Fig. 9b, for example, each (quasi-)recursive state is similar to that in phase 2, but in this case its lifetime is finite, and it is replaced by the fast switching state as in phase 1. Then the same or different (quasi-)recursive state is reached again, which lasts until the next switching occurs. In the example of Fig. 9b (see also Fig. 9c for its expansion), around the 12,000th generation, the core network is taken over by parasites to enter fast switching state which in turn gives way for a new quasi-recursive state around the 14,000th generation.

In the example of Fig. 9b, there is another type of switching, as shown around the 85,000th generation, as shown in Fig. 9c with magnification. Here, the quasi-recursive state is still stable, but the core hypercycle consisting of dominant species changes. As in Fig. 9c, a switch occurs from an initial core hypercycle (109,11,13) to the next core hypercycle (11,13,195,155) around the 8500th generation.

This latter switching is the competition among core networks, while the former drastic switch is due to the invasion of parasitic molecules, which is most commonly observed. The mechanism of this switching is discussed again in Section V.B.4.

2. Dependence of Phases on the Basic Parameters

Although the behavior of the system depends on the choice of the network, there is a general trend with regard to the phase change, from phase 1 to phase 3 and then to phase 2 with the increase of N, or with the decrease of k, as schematically shown in Fig. 11. By choosing a variety of networks, however, we find a clear dependence of the fraction of the networks on the parameters, leading to a rough sketch of the phase diagram. Generally, the fraction of phase 2 increases and the fraction of phase 1 decreases also with the decrease of ρ or μ. For example, the fraction of phase 1 (or (3)) gets larger as k is decreased from $k \lesssim 300$ for $N = 50,000$ (with $\rho = 0.1$ and $\mu = 0.01$), while dependence on ρ will be discussed below.

For a quantitative investigation, it is useful to classify the phases by the similarity of the chemical compositions between two cell division events [34]. To check the similarity, we first define a k-dimensional vector $\vec{V}_n = (p_n(1), \ldots, p_n(k))$ with $p_n(i) = N_n(i)/N$. Then, we measure the similarity between ℓ successive generations with the help of the inner product as

$$H_\ell = \vec{V}_n \cdot \vec{V}_{n+\ell} / (|V_n||V_{n+\ell}|) \tag{8}$$

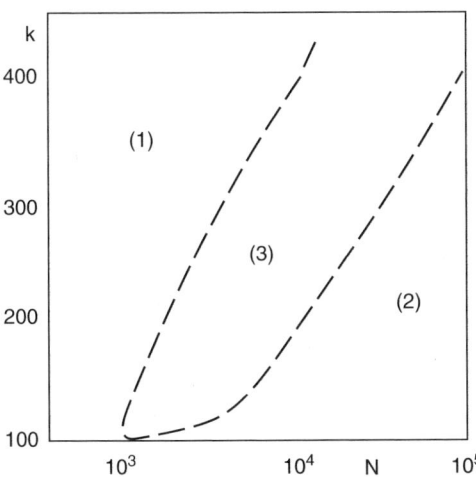

Figure 11. Schematic representation of the phase diagram of the three phases, plotted as a function of the total number of molecules N and the total possible number of molecular species k.

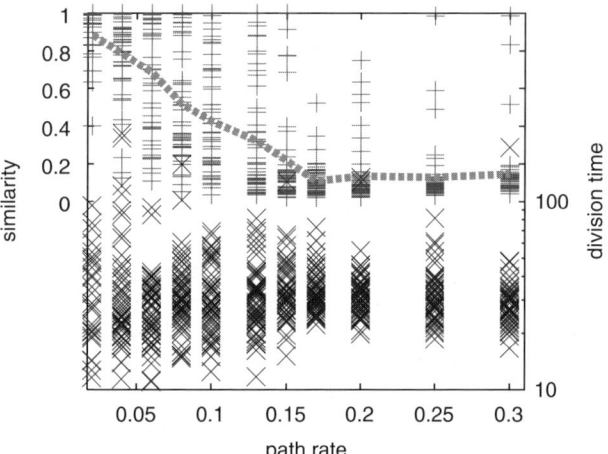

Figure 12. The average similarity $\overline{H_{20}}$ (+) and the average division time (×) plotted as a function of the path rate ρ. For each ρ, data from 50 randomly chosen networks are plotted. The average is taken over 600 division events. The dotted line indicates the average of $\overline{H_{20}}$ over the 50 networks for each ρ. For ρ > 0.2, networks over 98% have $H < 0.4$, and they show fast switching, while for ρ = 0.08, about 95% belong to phase 2 or 3. At ρ = 0.02, 25 out of 50 networks cannot support cell growth, and four cannot at ρ = 0.04. (Reprinted Figure 3 with permission from K. Kaneko, *Phys. Rev. E* **68**, 031909 (2003). Copyright (2003) by the American Physical Society.)

In Fig. 12, the average similarity $\overline{H_{20}}$ and the average division time are plotted for 50 randomly chosen reaction networks as a function of the path probability ρ. Roughly speaking, the networks with $\overline{H_{20}} > 0.9$ belong to phase 2, and those with $\overline{H_{20}} < 0.4$ belong to (1), empirically. Hence, for ρ > 0.2, phase 1 is observed for nearly all the networks (e.g., 48/50), while for lower path rates, the fraction of phase 2 or phase 3 increases. The value ρ ~ 0.2 gives the phase boundary in this case.

Generally speaking, a positive correlation between the growth speed of a cell and the similarity H exists. In Fig. 12, the division time is also plotted, where each point with a lower division time corresponds to that with a high similarity H. The network with higher similarity (i.e., in phase 2) gives a higher growth speed. Indeed, the recursive states maintain higher growth speeds since they effectively suppress parasitic molecules. In Fig. 12, by decreasing path rates, the variations in the division speeds of the networks become larger, and some networks that reach recursive states have higher division speeds than do networks with larger ρ. On the other hand, when the path rate is too low, the protocells generally cannot grow since the probability to have mutually catalytic connections in the network is nearly zero. Indeed, there exists an optimal path rate seems (e.g., around 0.05 for $k = 200$, $N = 12{,}800$ as in Fig. 12) for having a network with high growth speeds. Consequently, under competition for growth, protocells having such optimal networks will be evolved as will be discussed in Section V.C.

Besides the correlation between the growth speed and similarity, the correlation with the diversity of the molecules also exists. Protocells with higher growth speed and similarity in phase 2 have higher chemical diversity also. In phase 1, one (or a very few) molecular species is dominant in the population, while about 10 species have higher population in phase 2 with higher growth speed.

3. Maintenance of Recursive Production

How is the recursive production sustained in phase 2? We have already discussed the danger of parasitic molecules that have lower catalytic activities and are catalyzed by molecules with higher catalytic activities. As discussed in Section II.A, such parasitic molecules can invade the hypercycle. Indeed, under the structural changes and fluctuations, the recursive production state could be destabilized. To answer the question on the stability of recursive states, we have examined several reaction networks. The unveiled logic for the maintenance of recursive state is summarized as follows.

a. Stabilization by Intermingled Hypercycle Network. The 5–12 species in the recursive state form a mutually catalytic network—for example, as in Fig. 10. This network has a *core hypercycle network*, as shown by thick arrows in Fig. 10a. As shown by Fig. 13, such a core hypercycle has a mutually catalytic relationship: "A catalyzes B, B catalyzes C, and C catalyzes A." However, they are also connected with other hypercycle networks such as $G \to D \to B \to G$, $D \to C \to E \to D$, and so forth. The hypercylces are intermingled to form a network. Coexistence of a core hypercycle and other attached hypercycles are common to the recursive states we have found in our model.

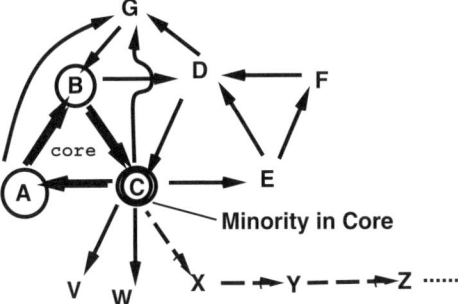

Figure 13. An example of mutually catalytic network in our model. The core network for the recursive state is shown by circles, while parasitic molecules (X, Y, ...) connected by broken arrows are suppressed at a (quasi-)recursive state.

This intermingled hypercycle network (IHN) leads to stability against parasites and fluctuations. Assume that there appears a parasitic molecule to one species in the member of IHN (say X as a parasite to C in Fig. 13). Species X may decrease the number of species C. If there were only a single hypercycle $A \to B \to C \to A$, the population of all the members A, B, C would be easily decreased by this invasion of parasitic molecules, resulting in the collapse of the hypercycle. In the present case, however, other parts of the network (say, that consisting of A, B, G, D in Fig. 13) compensate the decrease of the population of C by the parasite, so that the population of A and B are not so much decreased. Then, through the catalysis of species B, the replication of molecule C progresses, so that the population of C is recovered. Hence the complexity in the hypercycle network leads to stability against the attack of parasite molecules.

Next, IHN is also relevant to the stability against fluctuations. It is known that the population dynamics of a simple hypercycle often leads to a heteroclinic cycle [35], where the population of one (or a few) member approaches 0 and then is recovered. For a continuum model, such a heteroclinic cycle can continue forever, but in a stochastic model, due to fluctuations, the number of the corresponding molecular species is totally extinct sometimes. Once this molecular species goes extinct completely, its recovery by replication error would require a very long time. Hence, to achieve stability against fluctuations, a state with the heteroclinic cycle dynamics or any oscillation in which some of the population goes very low should be avoided. Indeed, by forming IHN, such oscillatory instability is often avoided or reduced. Due to coexistence of several hypercycle processes, instability in each hypercycle cancels out, leading to fixed-point dynamics or oscillation with a smaller amplitude. Thus the danger that the population of some molecules in the hypercycle goes to zero by fluctuations is reduced.

Stability of coexistence of many species is discussed as "homeochaos" [36], while stable reproduction in reaction network is also seen in Ref. 37.

b. Minority in the Core Hypercycle. Now we study more closely the population dynamics in a core hypercycle. Here, the number of molecules N_j of molecular species j is in the inverse order of their catalytic activity $c(j)$; that is, $N_A > N_B > N_C$ for $c_A < c_B < c_C$. Because a molecule with higher catalytic activity helps the synthesis of others to a greater extent, this inverse relationship is expected. Indeed, the population sizes of just three species A, B, C, with the catalytic relationship $A \to B \to C \to A$ are estimated by taking the continuum limit $N \to \infty$ and obtaining a fixed-point solution of the rate equation for the concentrations of the chemicals as discussed in Ref. 7. From a straightforward calculation we have $N_A : N_B : N_C = c_A^{-1} : c_B^{-1} : c_C^{-1}$.

Here, the C molecule is catalyzed by a molecular species with higher activities but larger populations (A). Hence, the parasitic molecule species cannot easily invade to disrupt this mutually catalytic network. Since the minority molecule (C) is catalyzed by the majority molecule (A) [with the aid of another molecule (B)], a large fluctuation in molecule numbers is required to destroy this network.

The stability in the minority molecule is also accelerated by the complexity in IHN. If the catalytic activity of C is highest, the recursive state here is mainly achieved by catalysis of the molecule C. On the other hand, this also implies that C is the minority in the core network. (The population of the molecule C is usually larger than D, E, etc., in Fig. 13, however.) Hence the attack to C molecule is most relevant to destroy this recursive state. In the IHN, this minority molecular species is involved in several hypercycles as in C in Fig. 13. On the one hand, this demonstrates the prediction in Section IV.E that more species are catalyzed by the minority molecules, while on the other hand it leads to the suppression of the fluctuation in the number of minority molecules, as will be discussed in Section V.D. With the decrease of the fluctuation, the probability that the minority molecules is extinct is reduced, so that the recursive state is hardly destroyed.

c. Localization in a Random Network. The present system belongs to a class of system with reaction and diffusion, since the structural change by replication error leads to the diffusion within the network space. With random connection in the catalytic network, the present system is nothing but a reaction–diffusion in a random network. Generally, such a problem is related with the Anderson localization, where concentrations are localized within some part of the network, depending on the degree of the connectivity in the network and the strength of the diffusion coupling. From this viewpoint, the formation of IHN, localized only within a limited species in the global network, may be understood as an example of such localization. It will be interesting to study the stability of the recursive production in terms of the localization transition in the reaction network [38].

4. Switching

Next, we discuss the mechanism of switching. In phase 3, the recursive production state is destabilized, when the population of parasitic molecules increase. For example, the number of the molecule C may be decreased due to fluctuations, while the number of some parasitic molecules (X) that are not originally in the catalytic network but are catalyzed by C may increase. Frequency of such fluctuation increases as the total population of molecules in a cell becomes smaller. If such fluctuation appears, the other molecular species in the original network loses the main source of molecules that catalyze their synthesis, successively. Then the new parasitic molecule X occupies a large portion of populations. However, the molecule's main catalyst (C) soon

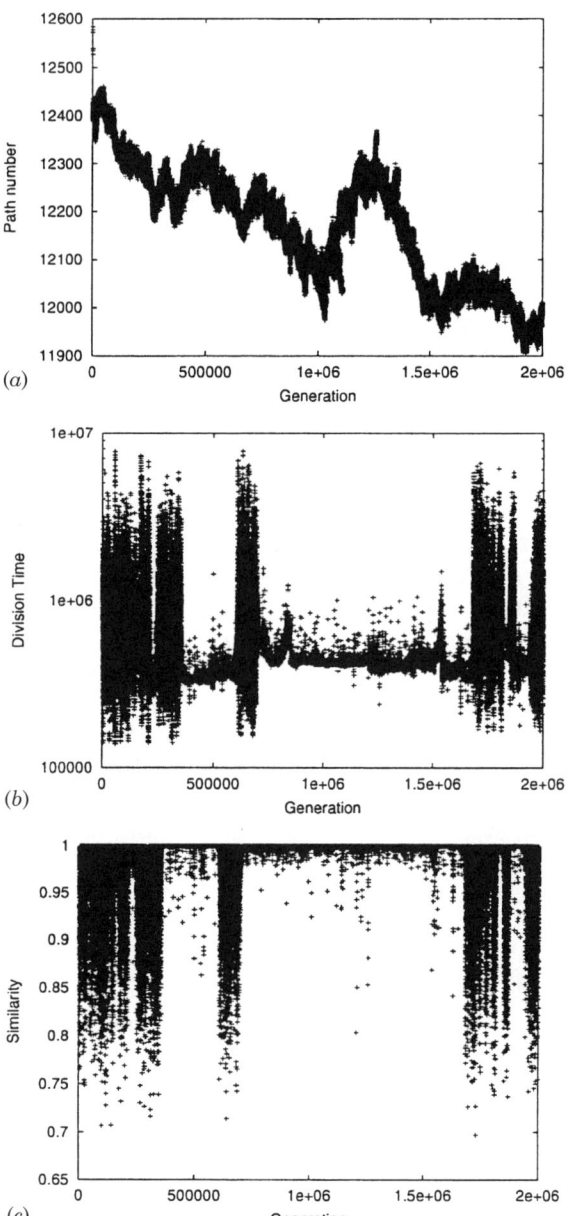

disappears, the synthesis of X is stopped, and this species X is taken over by some molecules Y that are catalyzed by X (see the broken arrows in Fig. 13). Then, within a few generations, dominant species changes and recursive production does not continue. Indeed, this is what occurred in phase 1. Then the parasitic molecule X is taken over by some other Y. This take over by parasites continues successively, until a new (or same) recursive state with hypercycle network is formed. Hence the fluctuation in the minority molecule in the core network is relevant to the switching process.

C. Evolution

Model A

The next question we have to address is whether the recursive production state is achieved through evolution. To check this problem, we have extended our model to further include a "mutational" change of network at each division event (model A). To be specific, at each division event we add or delete randomly (with equal probability) a few reaction paths, whose connection $i \rightarrow j$ is again chosen randomly. Here to see the evolution of catalytic activity, the index of the species is ordered with the value of catalytic activity; that is, the index j is ordered so that c_j monotonically increases with j. Since the mutational change is assumed to be random, a new path is added or deleted independent of the catalytic activity. In the simulation displayed here, there are five mutations of the network path at every generation. We have carried out numerical experiments of this model, to see if the path rate of the network stays around the state supporting the recursive production.

An example of the time series of path rates at each generation is shown in Fig. 14, along with the time series of the division time as well as chemical diversity. Corresponding to this time series, the change of dominant species is plotted over generations in Fig. 15.

As shown, the recursive state is achieved, and it is maintained over many generations until it switches to other states. At each reproduction, there are changes in the reaction paths here. In spite of such mutations, the recursive production state is sustained over many generations. In each recursive

◄─────────

Figure 14. Evolution of path rates, recursiveness, and division time, plotted versus generation. The total number of species k is 500, where c_i is chosen as $100^{-(k-i)/k}$, so that it ranges from 0.01 to 1.0 equally in logarithmic scale. The number of molecules N in a cell is set to 50,000, so that the cell divided when the total molecule number is 100,000. The initial path rate is set at $\rho = 0.1$—that is, 125,000 paths totally. At every division, five paths are "mutated"; that is, with equal probability, five paths are added or eliminated randomly. Totally there are $M_{tot} = 100$, so that one of 100 cells are eliminated when one cell is divided into two. (*a*) The total path number. The path rate is obtained by dividing the number by k^2. (*b*) The division time—that is, the required steps for a cell divide. (*c*) The similarity $H^1(i)$, defined in Section V.B.

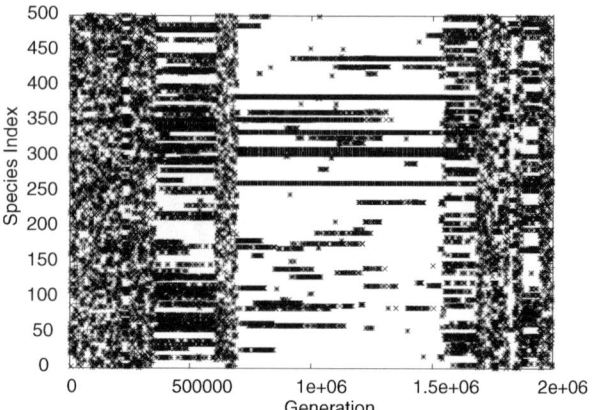

Figure 15. Evolution of cell. Those species i with $N(i) > 100$ are plotted with the vertical axis as the species index i and with the longitudinal axis as the generation. The data are from the results of the simulation for Fig. 14.

production state, the path rate remains rather low. Here, such a network that supports the recursive production is selected and is maintained. Note that many molecules are catalyzed by the minority species in the core hypercycle network. In this sense, a prototype of the evolution to package the information into the minority molecule that is suggested in Section IV.E is observed here.

An example of the network of dominant species is given in Fig. 16. Here intermingled hypercycle networks (IHN) are formed so that recursive production is formed. Again, there is a core hypercycle, and other hypercycles are connected with it. The surviving molecule species have a large connectivity

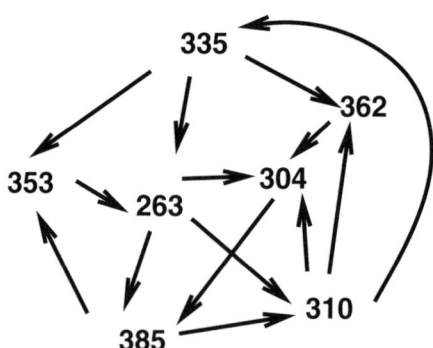

Figure 16. The catalytic network of the species that constitute the recursive state around 10^6th generation of Fig. 14 or 15.

in reaction paths, much larger than expected from a random network of the reaction path rate here. As in Fig 16. the IHN here forms a highly connected network, even though the average path rate remains small (as shown in Fig. 14, the path per species is about 0.1 or lower). The paths forming the IHN are preserved over long generations, while a few paths are sometimes eliminated. Here, coexistence of several parallel paths among species is important to give the robustness of the recursive state against mutation that may delete one of the paths. As in the dynamics of phase 3, the recursive production state is destabilized finally with the mutation of reaction paths, while after some generations, other recursive networks are formed through the mutation of the network.

To sum up, phase 3 gives a basis for evolvability, since a novel, (quasi-)-recursive state with different chemical compositions is visited successively.

Model B

So far, we have assumed that the structural change in the replication can occur equally to any other molecular species. Of course, this is a simplification, and the replication error occurs only to limited types of molecular species that have similarity to the original. To see this point, we have studied another model (model B) with some modifications from the original model of Section V.A.

Here, the catalytic activity is set as $c_i = i/k$; that is, the activity is monotonically increasing with the species index. Then, instead of global change to any molecular species by replication error, we modify the rule so that the change occurs only within a given range $i_0(\ll k)$; that is, when the molecule species j is synthesized, with the error rate μ, the molecule $j + j'$ with j' a random number over $[-i_0, i_0]$ is synthesized.

In model B, we have not included any change of the network. The network is fixed in the beginning, and it is not changed through the simulation. Instead, by local change of structural error, the range of existing species evolves by generations. Here we take species only with $i < i_{\text{ini}}$ in the initial condition, and examine if the evolution to a network with higher catalytic activities (i.e., with much larger i) progresses or not. In other words, we examine if the indices i in the network increase successively or not. An example is shown in Fig. 17, where the catalytic activity increases through successively switching to one (quasi-)-recursive state (consisting of species within the width of the order $2i_0$), to another.

Here the switching occurs as in phase 3. With the pressure for selection of the protocells, cells with a new (quasi-)recursive state are selected that consist of molecules with higher catalytic activities (i.e., with larger indices of species). Again each recursive state consists of IHN, and the species with the highest catalytic activity in the core hypercycle is minority in population. Once the population of such species is decreased by fluctuations, there occurs a switch to

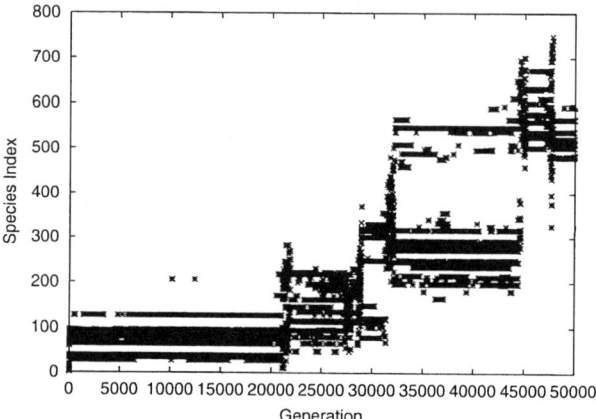

Figure 17. Evolution of species in a cell. Those species i with $N(i) > 100$ are plotted with the vertical axis as the species index i and with the longitudinal axis as the generation. The total number of species k is 5000, where c_i is chosen as $c_i = i/k$, so that it ranges from 0.0002 to 1.0 equally distributed. The number of molecules in a cell is set at 8000, so that the cell divides when the total molecule number is 16,000. The path rate is set at $\rho = 0.1$. The replication error for the species occurs within the range of species $[i - 100, i + 100]$, instead of global selection from all species. Totally there are $M_{tot} = 10$ cells, so that one of 10 cells is eliminated when a cell is divided into two.

a new state that has higher catalytic activities, and the species indices successively increase. Hence, evolution from a rather primitive cell consisting of low catalytic activities to that with higher activities is possible, by taking advantage of minority molecules.

Note that this switching cannot occur if the total number of molecules N is small. When the number is too small, the mutation of paths to destroy the recursive state hardly occurs. On the other hand, if the total number of molecules is too large, it is harder to establish a recursive state, due to a larger possibility to change the network. Hence, there is optimal value of the number of molecules in a protocell to realize the recursive production as well as the evolution.

D. Statistical Law

To close the present section, we investigate the fluctuations of the molecule numbers of each of the species, by coming back to the original model studied in Section V.B, without evolution of reaction paths. The characteristics of the fluctuations of the number of each molecular species over the generations can have a significant impact on the recursive production of a cell, since the number of each molecular species is not very large. In order to quantitatively characterize the sizes of these fluctuations, we have measured the distribution $P(N_i)$ for each molecular species i, by sampling over division events.

Our numerical results are summarized as follows:

I. For the fast switching states, the distribution $P(N_i)$ satisfies the power law

$$P(N_i) \approx N_i^{-\alpha} \tag{9}$$

with $1 < \alpha < 2$, as shown in Fig. 18a. The exponent α depends on the parameters, and it approaches 2 as alternation of dominant species is more

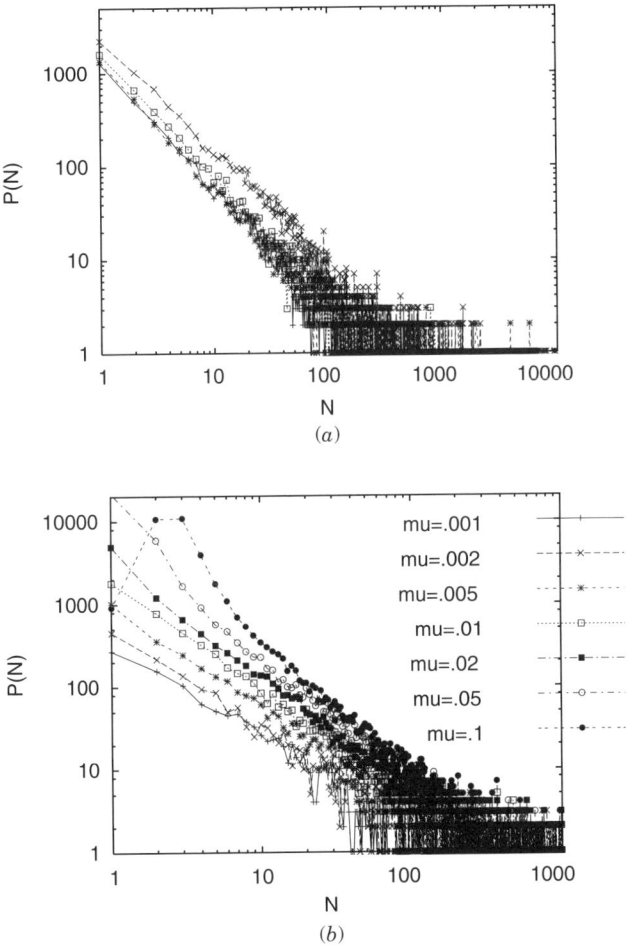

Figure 18. The number distribution of the molecules corresponding to the network in Fig. 7 (fast switching states). (a) The distribution is sampled from 100,000 division events. Plotted for four molecule species among 500. Log–log plot. (b) Change of the distribution with the change of the error rate μ, for a specific molecular species.

frequent. For example, as shown in Fig. 18b, the exponent α increases from 1 to 2, with the increase of the error rate μ.

II. For recursive states, the fluctuations in the core network (i.e., 13, 11, 109 in Fig. 9a or 10a) are typically small (and are roughly fit by Gaussian distribution). On the other hand, for species that are peripheral to but catalyzed by the core hypercycle, the number distribution is closer to log-normal distributions

$$P(N_i) \approx \exp\left[-\frac{(\log N_i - \overline{\log N_i})^2}{2\sigma}\right] \quad (10)$$

as shown in Fig. 19.

Even though the distribution does not agree well with the log-normal distribution, at least the distribution is roughly symmetric after taking the logarithm (i.e., because the 0th approximation the distribution is not normal but log-normal). The origin of the log-normal distributions here can be understood by the following rough argument: For a replicating system, the growth of the molecule number N_m of the species m is given by

$$dN_m/dt = AN_m \quad (11)$$

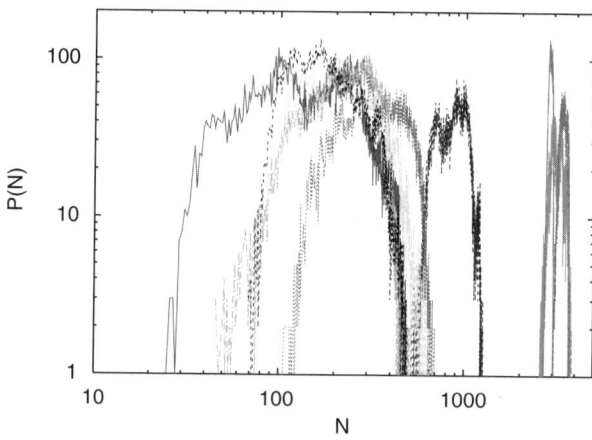

Figure 19. The number distribution of the molecules corresponding to the network in Fig. 9a or 10a. The distribution is sampled from 1000 division events. From right to left, the plotted species are 11, 109, 13, 155, 176, 181, 195, 196, and 23. Log-log plot. (Reprinted with permission from K. Kaneko, *Phys. Rev. E* **68**, 031909 (2003). Copyright (2003) by the American Physical Society.)

where A is the average effect of all the molecules that catalyze m. We can then obtain the estimate

$$d \log N_m / dt = \bar{a} + \eta(t) \qquad (12)$$

by replacing A with its temporal average $\bar{a}$ plus fluctuations $\eta(t)$ around it. If $\eta(t)$ is approximated by a Gaussian noise, the log-normal distribution for $P(N_m)$ is suggested. This argument is valid if $\bar{a} > 0$. As such, this equation diverges with time, but here the cell divides into two before the divergence becomes significant. Although the asymptotic distribution as $N \to \infty$ is not available then, the argument on the distribution form is valid as long as N is sufficiently large.

For the fast switching state, the growth of each molecular species is close to zero on the average. In this case N_m in the Langevin equation [Eq. (12)] can approach 0, and we need to consider the equation by seriously taking into account the absorbing boundary condition at $N_m = 0$. By taking into account the normalization of the probability, the stationary solution for the Fokker–Planck equation corresponding to Eq. (12) for $\bar{a} \le 0$ is given by

$$P(N) \propto N^{-(1+v)} \qquad (13)$$

with

$$v = |\bar{a}|/(\overline{a^2} - \bar{a}^2) \qquad (14)$$

(see, e.g., Refs. 39 and 40). Change of the exponent α against the error rate in Fig. 18b will be understood as the change of the ratio of variance to the mean of a.

If several molecules mutually catalyze each other, however, one would expect that the fluctuations will not increase as in the Brownian motion as in Eq. (12). For example, consider that the number of one species in the core cycle increases due to the fluctuation. Then it relatively decreases the number of molecules of the other species in the core network, resulting in the suppression of the catalytic reaction to replicate the increased species. Then the catalytic molecule of the original molecule species decreases. Hence the fluctuations in the core hypercycle is reduced.

Another reason for the reduction of fluctuation of the species in the core cycle is high connectivity in the IHN. The chemicals of the core part have catalytic paths with a large number of molecular species. Hence many processes work in parallel to the synthesis of the core species. Then, fluctuations due to other chemical concentrations are added in parallel. Thus, the fluctuations can come close to Gaussian distribution (recall the central limit theorem).

Note also that for some networks, the distributions of the molecule numbers in the recursive sates may sometimes be intermediate between log-normal and Gaussian and may occasionally even have double peaks.

By studying a variety of networks, the observed distributions of the molecule numbers can be summarized as follows:

- (1) Distribution close to Gaussian form, with relatively small variances in the core (hypercycle) of the network.
- (2) Distribution close to log-normal, with larger fluctuations for a peripheral part of the network.
- (3) Power-law distributions for parasitic molecules that appear intermittently.

To quantitatively study the magnitude of variance in the IHN for the recursive production, we have also plotted the variance $\overline{(N_i - \overline{N_i})^2}$ (the overbar denotes the average of the distribution $P(N_i)$). As can be seen in Fig. 20, the variance in the core network is small, especially for the minority species (i.e., 13). For molecule species that do not belong to the core hypercycle, the variance scaled by the average increases as the average decreases. Suppression of the relative fluctuation in the core hypercycle comes from the direct feedback of the population change of the molecular species in the core, as well as multiple parallel reaction paths, as already mentioned.

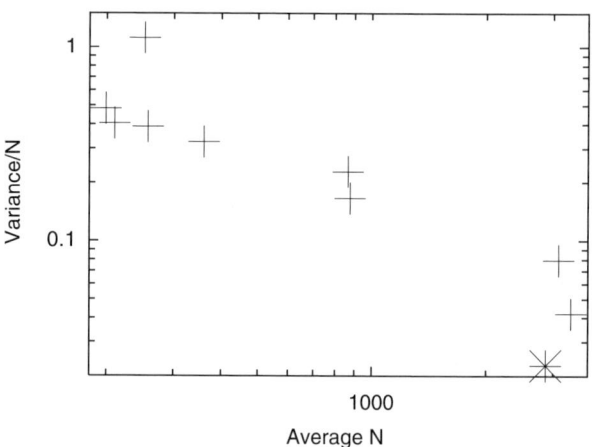

Figure 20. Scaled variance—that is, the variance of the molecule number divided by its average—plotted against the average. From the largest to the smallest, the species 11 (the largest $\overline{N_i}$), 109 (the second largest), 13, 155, 194, 176, 195, 181, 196, 23, and 34 (smallest $\overline{N_i}$) are plotted. Computed from the data in Fig. 19. The asterisk denotes the species 13, which has largest catalytic activity here and the minority in the hypercycle core. Reprinted Figure 5 with permission from K. Kaneko, *Phys. Rev.* E **68**, 031909 (2003). Copyright (2003) by the American Physical Society.)

CATALYTIC REACTION NETWORK APPROACH

Remark: Universal Statistics

Quite recently, Furusawa, Kaneko, and co-workers [26,41] have studied several models of minimal cell consisting of catalytic reaction networks, without assuming the replication process itself. In other words, the molecules are successively synthesized from nutrition chemicals transported from the membrane, where the level 1 model of Section III.B is adopted. They have found a universal statistical law of chemicals for a cell that grows recursively.

(i) The number of molecules of each chemical species over all cells generally obeys the log-normal distribution. This distribution is universally observed for a state with recursive production. Existence of such log-normal distributions is also experimentally verified [26]. Ubiquity of log-normal distribution in the level 2 model described in this section is thus supported in the level 1 model.

(ii) A power law in the average abundances of chemicals exists. This is statistics against a huge number of molecular species. When the abundances of all chemical species are ordered according to the magnitude, the abundances of chemicals are inversely proportional to the rank of the magnitude. Such a law was originally found in the linguistics by Zipf [42]. This Zipf's law on chemical abundances [26] is found to be universal when a cell optimizes the efficiency and faithfulness of self-reproduction. It is a universal statistics when the cell shows a recursive growth under fluctuations in the molecule numbers. Furthermore, using data from gene expression databases on various organisms and tissues, the abundances of expressed genes are found to exhibit this law. Thus, the universal statistics are also supported experimentally. It is shown that this power law of gene expression is maintained by a hierarchical organization of catalytic reactions. Major chemical species are synthesized, catalyzed by chemicals with a little less abundant chemicals. The latter chemicals are synthesized by chemicals with much less abundance, and this hierarchy of catalytic reactions continues until it reaches the minor chemical species.

Remark: Search for the Deviation from Universal Statistics

So far we have observed ubiquity of log-normal distribution, in several models. The fluctuations in such distribution are generally very large. This is in contrast to our naive impression that a process in a cell system must be well-controlled.

Then, is there some relevance of such large fluctuations to biology? Quite recently, we have extended the idea of the fluctuation–dissipation theorem in statistical physics to evolution and have proposed a linear relationship (or high correlation) between (genetic) evolution speed and (phenotypic) fluctuations. This proposition turns out to be supported by experimental data on the evolution of *E. coli* to enhance the fluorescence in its proteins [43]. Hence the fluctuations are quite important biologically.

The log-normal distribution is also rather universal in the present cell, as demonstrated in the distribution of some proteins, measured by the degree of fluorescence [41]. Now, is this universality the final statement for "cell statistical mechanics"? We have to be cautious here, since too universal laws may not be so relevant to biological function. In fact, chemicals that obey the log-normal distribution may have too large fluctuations to control some function. Some other mechanism to suppress the fluctuation may work in a cell.

Indeed, the minority control suggests the possibility of such control to suppress the fluctuation, as discussed in Section IV.E. For a recursive production system, some mechanism to decrease the fluctuation in minority molecule may be evolved.

At least there can be two possibilities to decrease the fluctuation leading to deviation from log-normal distribution.

The first one is some negative feedback process. In general, the negative feedback can suppress the response as well as the fluctuation. Still, it is not a trivial question how chemical reaction can give rise to suppression of fluctuation, since to realize the negative feedback in chemical reaction, production of some molecules is necessary, which may further add fluctuations.

The second possible mechanism is the use of multiple parallel reaction paths. If several processes work sequentially, the fluctuations would generally be increased. When reaction processes work in parallel for some species, the population change of such a molecule is influenced by several fluctuation terms added in parallel. If a synthesis (or decomposition) of some chemical species is a result of the average of these processes working in parallel, the fluctuation around this average can be decreased by the law of large numbers. Indeed, the minority in the core network that has higher reaction paths has relatively lower fluctuation as in Fig. 20. Suppression of fluctuation by multiple parallel paths may be a strategy adopted in a cell. Note that this is also consistent with the scenario that more and more molecules are related with the minority species as discussed in Section IV.E. With the increase of the paths connected with the minority molecules, the fluctuation of minority molecules is reduced, which further reinforces the minority control mechanism. Hence the increase of the reaction paths connected with the minority molecule species through evolution, the decrease of the fluctuation in the population of minority molecules, and the enhancement of minority control reinforce each other. In this regard, the search for molecules that deviate from log-normal distribution should be important in the future.

In physics, we are often interested in some quantities that deviate from Gaussian (normal) distribution, since the deviation is exceptional. Indeed, in physics, search for power-law distributions or log-normal distributions has been popular over a few decades. On the other hand, a biological unit can grow and reproduce, to increase the number. For such a system, the components within have to be synthesized, so that the amplification process is common. Then, the fluctuation is also amplified. In such a system, the power-law or log-normal distributions are quite common, as already discussed here and as is also shown in several models and experiments [26,41]. In this case, the Gaussian (normal) distribution is not so common (normal). Then exceptional molecules that obey the normal distribution with regard to their concentration may be more important.

Also, the ubiquity of log-normal distribution we found is true for a state with recursive production. If a cell is not in a stationary growth state but in a transient process switching from one steady state to another, the universal statistics can be violated. Search for such a violation will be important both experimentally and theoretically.

VI. SUMMARY

We have studied a problem of recursive production and evolution of a cell, by adopting a simple protocell system. This protocell consists of catalytic reaction network with replicating molecules. The basic concepts we have proposed through several simulations are as follows:

1. *Minority Control.* In a cell system with mutually catalytic molecules, replicating molecules with a smaller size in population are shown to control the behavior of the total cell system. This minority-controlled state is achieved by preserving rare fluctuations with regard to the molecule number. The molecule species, minority in its number, works as a carrier of heredity, in the sense that it is preserved well with suppressed number fluctuations and that it controls the behavior of a cell relatively strongly. Since molecules that are replicated by this minority species are also preserved, more molecules will be synthesized with the help of this species. In addition, reaction paths to stabilize the replication of this minority molecules is expected to evolve. Hence, the replication of more and more molecule species is packaged into the synthesis of this minority molecule, which also ensures the transmission of the minority molecule. Thus, the minority molecular species gives a basis for "genetic information." Hence evolution from a loose reproduction system to a faithful replication system with genes is understood from a kinetic viewpoint of chemical reaction.

2. *Recursiveness of Production in an Intermingled Hypercycle Network.* Next, a protocell model consisting of a variety of mutually catalyzing molecular species is investigated. When the numbers of molecules in a cell is not too small and the number of possible species is not too large in a cell, recursive production of a cell is achieved. This recursive production state consists of 5–12

dominant molecule species, which form an intermingled hypercycle network (IHN). Within this IHN, there is a core hypercycle, while parallel multiple reaction paths in the IHN are important to ensure the stability of the state against invasion of parasitic molecules and against fluctuations in the molecule number.

3. *Itinerant Dynamics Over Recursive Production States.* When the fluctuation in molecule number is not small enough, there appears switches over (quasi-)recursive production states. A given quasi-recursive state is destabilized by being taken over by some parasitic molecules. Then, the dominant molecule species change frequently by generations, where the growth speed of a cell is suppressed. After this transient, the fast change of chemical compositions is reduced so that a quasi-recursive production of a cell is sustained again. Each switching occurs with the loss of chemical diversity. Note that in high-dimensional dynamical systems, such switching over quasi-stable states through unstable transient dynamics is studied as chaotic itinerancy [44–46], where the loss of degrees of freedom is also observed in the process of switching.

Destabilization of a recursive state in the present model occurs through the decrease of the population of the minority molecules in the core hypercycle. As this molecular species is taken over by parasitic molecules, the switching starts to occur. In this sense, the process in the switching is not random, but is restricted to specific routes within the phase space of chemical composition, as in the chaotic itinerancy. It is interesting to study the present switching over the recursive state by generalizing the concept of chaotic itinerancy [46], to include stochastic process.

4. *Evolution Through Itinerant Dynamics.* By considering change in the available reaction paths to the model, this hypercycle network evolves to recursive production states. Following the itinerant dynamics above, each recursive state is destabilized, but later another recursive state is evolved. Through these successive visits of recursive states, a cell can evolve to have a chemical network supporting a higher growth speed. Since the minority species in the hypercycle network is relevant to this switch, minority molecules are shown to be important to evolution.

5. *Universal Statistics and Control of Fluctuations.* Statistics of the number fluctuations of each molecular species is studied. We have found (i) power-law distribution of fast switching molecules, (ii) suppression of fluctuation in the core hypercycle species, and (iii) ubiquity of log-normal distribution for most other molecular species. The origin of log-normal distribution is generally due to multiplicative stochastic process in the catalytic reaction dynamics, as is confirmed in several other reaction network models. On the other hand, suppression of the number fluctuations of the core hypercycle is due to high connections in reaction paths with other molecules. In particular, reduced is the number fluctuations of the minority molecular species that has high catalytic connections with others. This suppression of fluctuation further reinforces the

minority control for the reproduction of a cell. The deviation from ubiquitous log-normal distribution thus appears, which may be important in control of cell function.

In the present chapter, we have not discussed cell–cell interaction and have restricted our study only to a production process of a single cell. Of course, cells start to interact with each other, as the cell density is increased through the cell division. Indeed, including the cell–cell interaction to the present cell model with reaction network, cell differentiation and morphogenesis of a cell aggregate are studied [15,16]. Through instability of intracellular dynamics with cell–cell interaction, cell differentiation, irreversible loss of plasticity in cells, and robust pattern formation process appear as a general course of development with the increase of the cell number. Relevance of minority control and deviation from universal statistics to such multicellular developmental processes will be an important issue to be studied in the future.

Acknowledgments

The author is grateful to T. Yomo, C. Furusawa, W. Fontana, Y. Togashi, A. Awazu, and K. Fujimoto for discussions. The work is partially supported by Grant-in-Aids for Scientific Research from the Ministry of Education, Science, and Culture of Japan (11CE2006).

References

1. K. Kaneko *What is Life?: A Complex Systems Approach*, University of Tokyo Press, Tokyo, 2003 (in Japanese).
2. K. Kaneko and T. Yomo, *J. Theor. Biol.* **214**, 563–576 (2002).
3. C. Shannon and W. Weaver, *The Mathematical Theory of Communication*, University of Ilinois Press, Champaign, IL, 1949.
4. L. Brillouin, *Science and Information Theory*, Academic Press, New York, 1969.
5. H. Jeong et al., *Nature* **407**, 651 (2000); H. Jeong, S. P. Mason, and A.-L. Barabási, *Nature* **411**, 41 (2001).
6. D. R. Mills, R. L. Peterson, and S. Spiegelman, *Proc. Natl. Acad. Sci. USA* **58**, 217; (1967) D. R. Mills, F. R. Kramer, and S. Spiegelman, *Science* **180**, 916 (1973).
7. M. Eigen and P. Schuster, *The Hypercycle*, Springer, Berlin, 1979.
8. M. Boerlijst and P. Hogeweg, *Physica* **48D**, 17 (1991); P. Hogeweg, *Physica* **75D**, 275–291 (1994).
9. F. Dyson, *Origins of Life*, Cambridge University Press, Cambridge, 1985.
10. S. A. Kauffman, *The Origin of Order*, Oxford University Press, New York, 1993.
11. R. Bagley, J. D. Farmer, and S. Kauffmans, in *Artificial Life*, C. Langton, ed., 1989, pp. 93–140.
12. A. G. Cairns-Smith, *Clay Minerals and the Origin of Life*, Cambridge University Press, New York, 1982.
13. K. Kaneko, in *Function and Regulation of Cellular Systems*, A. Deutsch et al., eds., Birkhauser, Boston, 2003.
14. K. Kaneko, *Complexity* **3**, 53–60 (1998).
15. K. Kaneko and T. Yomo, *Physica* **75D**, 89–102 (1994); *Bull. Math. Biol.* **59**, 139 (1997); *J. Theor. Biol.* **199**, 243–256 (1999).

16. C. Furusawa and K. Kaneko, *Bull. Math. Biol.* **60**, 659–687 (1998); *Phys. Rev. Lett.* **84**, 6130–6133 (2000); *J. Theor. Biol.* **209**, 395–416 (2001); *Anat. Rec.* **268**, 327–342 (2002); *J. Theor. Biol.* **224**, 413–435 (2003).
17. K. Kaneko and T. Yomo, *Proc. R. Soc. B* **267**, 2367–2373 (2000); K. Kaneko, *Popul. Ecol.* **44**, 71–85 (2002).
18. T. Matsuura, T. Yomo, M. Yamaguchi, N. Shibuya., E. P. Ko-Mitamura, Y. Shima, and I. Urabe, *Proc. Natl. Acad. Sci. USA* **99**, 7514–7517 (2002).
19. E. Ko, T. Yomo, and I. Urabe, *Physica* **75D**, 81–88 (1993).
20. A. Kashiwagi, W. Noumachi, M. Katsuno, M. T. Alam, I. Urabe, and T. J. Yomo, *Mol. Evol.* **52**, 502–509 (2001).
21. A. Kashiwagi, I. Urabe, K. Kaneko, and T. Yomo, submitted 2003.
22. T. Ariizumi and M. Asashima, *Int. J. Dev. Biol.* **45**, 273–279 (2001).
23. C. Langton, eds., *Artificial Life* Addison-Wesley, Reading, MA, 1989.
24. W. Fontana and L. W. Buss, *Bull. Math. Biol.* **56**, 1–64 (1994).
25. A. Awazu and K. Kaneko, preprint 2003.
26. C. Furusawa and K. Kaneko, *Phys. Rev. Lett.* **90**, 088102 (2003).
27. B. Alberts, D. Bray, J. Lewis, M. Raff, K. Roberts, and J. D. Watson, *The Molecular Biology of the Cell*, 1983, 1989, 1994, 2002.
28. B. Hess and A. Mikhailov, *Science* **264**, 223 (1994); A. Mikhailov and B. Hess, *J. Theor. Biol.* **176**, 185–192 (1995).
29. Y. Togashi and K. Kaneko, *Phys. Rev. Lett.* **86**, 2459 (2001); *J. Phys. Soc. Japan* **72**, 62–68 (2003); preprint 2003.
30. E. Szathmary and J. Maynard Smith, *J. Theor. Biol.* **187**, 555–571 (1997).
31. M. Eigen, *Steps Towards Life*, Oxford University Press, New York, 1992.
32. K. Kaneko, *J. Biol. Phys.* **28**, 781 (2002); *Adv. Complex Syst.* **6**, 79–92 (2003).
33. K. Kaneko, *Phys. Rev. E* **68**, 031909 (2003).
34. D. Segré, D. Ben-Eli, and D. Lancet, *Proc. Natl. Acad. Sci. USA* **97**, 4112 (2000); D. Segré et al., *J. Theor. Biol.* **213**, 481 (2001); D. Segré and D. Lancet, *EMBO Rep.* **1**, 217 (2000).
35. J. Hofbauer and K. Sigmund, *Evolutionary Games and Population Dynamics*, Cambridge University Press, Cambridge, 1998.
36. K. Kaneko and T. Ikegami, *Physica D* **56**, 406–429 (1992).
37. T. Ikegami and T. Hashimoto, *Artifi. Life* **2**, 305–318 (1996).
38. H. Takagi and K. Kaneko, preprint 2003.
39. A. S. Mikhailov and V. Calenbuhr, *From Cells to Societies*, Springer, Berlin, 2002.
40. D. Sornette, *Critical Phenomena in Natural Science*, Springer, Berlin, 2002.
41. C. Furusawa, T. Suzuki, A. Kashiwagi, T. Yomo, and K. Kaneko, Ubiquity of Log-Normal Distribution in Gene Expression, preprint.
42. G. K. Zipf, *Human Behavior and the Principle of Least Effort*, Addison-Wesley, Cambridge, 1949.
43. K. Sato, Y. Ito, T. Yomo, and K. Kaneko, *Proc. Natl. Acad. Sci. USA* **100**, 14086–14090 (2003).
44. K. Kaneko, *Physica D* **41**, 137–172 (1990).
45. I. Tsuda, *Neural Networks* **5**, 313 (1992).
46. K. Kaneko and I. Tsuda, ed., Focus issue on "Chaotic Itinerancy," *Chaos* **13**, 926 (2003).

AUTHOR INDEX

Numbers in parentheses are reference numbers and indicate that the author's work is referred to although his name is not mentioned in the text. Numbers in *italic* show the pages on which the complete references are listed. Letter in **boldface** indicates the volume.

Abarbanel, H. D.-I.: **B:**285(77), **B:**288(77), **B:**289(80), **B:**294(77), **B:***312*; **B:**502(5), **B:**506(5), **B:***518*
Abe, H., **A:**330(37), **A:***35*
Abel, M., **B:**522(13), **B:**531-533(13), **B:***541*
Abou-Chacra, R., **B:**212(80), **B:***253*
Abraham, E. R., **B:**522(11), **B:***541*
Abrahams, E., **B:**499(40), **B:***500*
Adachi, S., **A:**410(34), **A:***434*
Adams, J. E.: **B:**74-75(47), **B:***85*; **B:**89(10), **B:**108(20), **B:***127*
Adrover, A., **B:**502(16), **B:***518*
Agekian, T. A., **A:**330(36), **A:***335*
Aizawa, Y.: **B:**270(63), **B:**273(63,67-68,71), **B:**274(71), **B:***312*; **B:**359(20), **B:***371*; **B:**381(25-26), **B:**385(26,38), **B:***419*; **B:**465(1-2), **B:**466(5-6,8,10-11), **B:**469(6), **B:**470(6,8), **B:**471-472(10-11), **B:**474(6,8,13-15), **B:**475(17-20), **B:***475*; **B:**477(6-7), **B:***499*; **B:**502(10), **B:***518*
Ajayan, P. M., **B:**156(2-3), **B:***176*
Akimoto, T., **B:**474(14), **B:***475*
Akiyama, R., **B:**200(43), **B:***203*
Albert, R., **B:**453(23), **B:***463*
Albert, S., **A:**289(27), **A:***302*
Alberts, B.: **B:**180(12), **B:***201*; **B:**558(27), **B:***598*
Aldredge, R. C., **B:**537(42-43), **B:***542*
Alexander, S., **B:**208(35-36), **B:**230(35), **B:**232-233(35), **B:**241(35-36), **B:**246(36), **B:**250(35-36), **B:***252*

Alhassid, Y., **B:**7(5), **B:***23*
Allan, D. W., **B:**270(61), **B:***312*
Allen, J. P., **B:**228(133), **B:***255*
Allen, L., **A:**436(5), **A:**443(5), **A:***457*
Allen, P. B.: **B:**190(32), **B:**195(32), **B:**198(32), **B:***202*; **B:**208(40), **B:**223(109), **B:**238(146), **B:**240(146), **B:**243(146), **B:**250(40), **B:***252*, **B:***254-255*
Alligood, K. T., **B:**307(99), **B:***313*
Almeida, M. A., **A:**307(19), **A:***334*
Almlof, J., **A:**402(6), **A:***433*
Almöf, J. E., **B:**131-132(34), **B:***153*
Alvarez, M. M., **B:**502(16), **B:***518*
Alvarez-Ramírez. M., **A:**323(30), **A:***335*
Amar, F. G.: **B:**35(26), **B:**46(26), **B:**68(26), **B:***84*; **B:**130(6), **B:**139(6), **B:***152*; **B:**156(5), **B:***176*
Amitrano, C.: **A:**146(26-27), **A:***169*; **A:**178(58,61), **A:***214*; **B:**13(11), **B:**17(11,13), **B:***23*; **B:**90(25), **B:**106(25), **B:***127*; **B:**130(13), **B:**139(13), **B:**143(13), **B:***153*
Anderson, J., **B:**406(77), **B:***421*
Anderson, P. W.: **B:**212(80), **B:***253*; **B:**499(40), **B:***500*
Andreoni, W., **B:**180(18), **B:**201(18), **B:***202*
Anfinrud, P. A., **B:**200(36), **B:***202*
Angelescu, D. E., **B:**221(102), **B:***254*
Anosova, J. P., **A:**330(36), **A:***335*
Anteneodo, C., **B:**480-481(30), **B:***500*

Geometric Structures of Phase Space in Multidimensional Chaos: A Special Volume of Advances in Chemical Physics, Part B, Volume 130, edited by M. Toda, T Komatsuzaki, T. Konishi, R.S. Berry, and S.A. Rice. Series editor Stuart A. Rice.
ISBN 0-471-71157-8   Copyright © 2005 John Wiley & Sons, Inc.

Antoni, M., **B:**479(12,16), **B:**480(12), **B:**481(12,35), **B:**494(16), **B:***500*
Antoniou, D., **B:**206(14), **B:***251*
Antoniou, I., **B:**475(19), **B:***475*
Apkarian, V. A., **B:**186(28), **B:***202*
Aquilanti, V., **B:**89(13-15), **B:**96-97(13), **B:**111(13), **B:**114-115(14), **B:***127*
Ariizumi, T., **B:**553(22), **B:***598*
Arimondo, E., **B:**479(13), **B:***500*
Arnold, V. I.: **A:**173-174(13), **A:**184-186(13), **A:***213*; **A:**221(20-21), **A:***263*; **A:**340-341(14), **A:**358(14), **A:**372(14), **A:***398*; **B:**424(1), **B:**425(6), **B:**427(1), **B:**429(1), **B:**435(1), **B:***436*; **B:**438(4), **B:***462*; **B:**465(4), **B:***475*
Asashima, M., **B:**553(22), **B:***598*
Ashurst, W. T., **B:**534(38), **B:***542*
Atela, P., **B:**481(34), **B:***500*
Aubry, S., **B:**379(16), **B:**383(31-32), **B:***419*
Audoly, B., **B:**522(15), **B:**531(15), **B:***541*
Aurell, E., **B:**301(92-93), **B:***313*
Austin, R. H., **B:**207(26-27), **B:**209(26-27), **B:***252*
Avellaneda, M., **B:**525(22-23), **B:***541*
Awazu, A., **B:**556(25), **B:***598*
Azzam, T., **A:**287-288(41), **A:**291(41), **A:**293-294(41), **A:**296-299(41), **A:**310(41), **A:***303*

Baba, A., **B:**477(1), **B:***499*
Babikov, D., **A:**257(85), **A:***265*
Bachuber, K., **B:**406(79), **B:***421*
Bacic, Z.: **A:**198(106), **A:***215*; **A:**280(30), **A:***303*
Bader, J. S., **B:**180(4), **B:**183-185(4), **B:***201*
Badii, R., **B:**502(11), **B:**513(11), **B:***518*
Baer, T., **A:**5(6), **A:***140*, **B:**215(85), **B:**220(91), **B:***253-254*
Bagley, R. J.: **B:**427(10), **B:***436*; **B:**438(10), **B:***463*; **B:**550(11), **B:***597*
Bai, Y., **B:**180(14), **B:***201*
Bai, Z.-I., **A:**308(23), **A:**318(23), **A:***334*
Baldan, O., **B:**378(12), **B:**398(12), **B:**401(12), **B:***418*
Baldin, O., **B:**499(46), **B:***500*
Balk, M. W., **B:**220(97), **B:***254*
Ball, K. D.: **A:**178(66), **A:***214*, **B:**26(1), **B:***83*
Bandrauk, A. D., **A:**198(136), **A:***216*
Barbáasi, A. L., **B:**453(23), **B:***463*

Barbara, P. F., **A:**402(6), **A:***433*, **B:**131-132(34), **B:***153*
Barbieri, M., **B:**329(12), **B:***351*
Barkai, E., **B:**228(123), **B:***254*
Barré, J., **B:**489(39), **B:***500*
Barrick, D., **B:**200(35), **B:***202*
Barthel, J., **B:**406(79), **B:***421*
Batista, V. S., **A:**456(30), **A:***459*
Beck, C., **A:**287(33-34), **A:**284(34), **A:**287(35), **A:**293(34-35), **A:**298-299(35), **A:**301(35), **A:***303*, **B:**364(13), **B:***371*
Beck, T. L.: **A:**178(53), **A:***214*, **B:**5(4), **B:***23*; **B:**45(31), **B:**53(31,34), **B:**54(34), **B:**56(31), **B:***84*; **B:**90(22), **B:***127*; **B:**130(5,8-10,21), **B:**135(5), **B:**139(5,8-10,21), **B:**143(10), **B:***152-153*; **B:**156(5), **B:***176*; **B:**209(54), **B:***253*
Becker, O. M., **B:**261(22), **B:**262(28), **B:**267(28), **B:***311*
Begleiter, H., **B:**338(19), **B:**341(19), **B:***351*
Bembenek, S. D., **B:**391(52), **B:***420*
Ben-Avraham, D., **B:**208(33), **B:**230(33), **B:***252*
Benderskii, V. A., **A:**402(6), **A:***433*
Ben-Eli, D., **B:**573(34), **B:**579(34), **B:***598*
Benettin, G.: **B:**31(22), **B:**46(22), **B:***84*; **B:**378(12-13), **B:**381(24), **B:**396(66), **B:**397(70), **B:**398(12-13,73), **B:**399(13), **B:**401(12-13,73), **B:***418-420*; **B:**429(25), **B:***436*; **B:**499(45-46), **B:***500*; **B:**502(19), **B:**506(19), **B:***518*
Benito, R. M., **A:**167(72), **A:***170*
Benjamin, I., **B:**7(5), **B:***23*
Ben-Shaul, A., **B:**72(41), **B:***84*
Bensimon, D.: **B:**5(10), **A:**18(10), **A:**30(10), **A:***140*; **A:**177(41), **A:***214*
Bentley, J. A.: **A:**198(114,120), **A:***216*; **A:**280(29), **A:***303*
Beratan, D. N., **B:**201(45), **B:***203*
Berblinger, M., **B:**43(28), **B:***84*
Berendsen, H. J. C., **B:**267(60), **B:**273(70), **B:**309(60), **B:***312*
Beresytcki, H., **B:**522(15), **B:**531(15), **B:***541*
Berge, P., **A:**293(49), **A:***303*
Bergmann, K., **A:**436(8), **A:***457*
Berne, B. J.: **A:**144(12), **A:***168*; **A:**228(40), **A:***264*; **B:**180(1,4), **B:**183-185(4), **B:***201*; **B:**205(2), **B:***251*
Bernstein, R. B., **B:**27-29(2), **B:**71-73(2), **B:**75(2), **B:**79-80(2), **B:***83*

Berry, M. V.: **A**:106(71), **A**:107(73), **A**:*142*; **A**:410(32), **A**:*434*; **B**:4(1), **B**:*23*; **B**:56(36), **B**:*84*
Berry, R. S.: **A**:7(26), **A**:137(26), **A**:*140*; **A**:146(25-31), **A**:*169*; **A**:173(5), **A**:178(5,53-54,57-58,61-63,65-66,69-74), **A**:179(69-73), **A**:186(63,69), **A**:*213–215*; **A**:218-219(4), **A**:229(4), **A**:234(4,50), **A**:245(50), **A**:*263–264*; **A**:338-339(2), **A**:341(2), **A**:352(2), **A**:392(42), **A**:*398–399*; **B**:4(3), **B**:5(4), **B**:10(8-9), **B**:13(9,11), **B**:16(9), **B**:17(11-12), **B**:21(16-18), **B**:22(20-21), **B**:*23–24*; **B**:26(1), **B**:35(26), **B**:45(31), **B**:46(26), **B**:52(31), **B**:53-54(34), **B**:56(31), **B**:66(39-40), **B**:68(26), **B**:*83–84*; **B**:90(22,25-27), **B**:106(25), **B**:*127*; **B**:130(5-16,26), **B**:135(5), **B**:139(5-16,26), **B**:143(10,12-13), **B**:*152–153*; **B**:156(5-6), **B**:170(6), **B**:*176*; **B**:209(53-55), **B**:*253*; **B**:260(13-20), **B**:261(21), **B**:262(29), **B**:263(13-20), **B**:265(13-20), **B**:266(29,56), **B**:284(56), **B**:300(13-20), **B**:*310–311*; **B**:388(46), **B**:*420*; **B**:438(2), **B**:*462*
Bessi, U., **B**:427(17), **B**:*436*
Bestiale, S., **B**:130(28), **B**:139(28), **B**:*153*
Beswic, J. A., **A**:63(43), **A**:*141*
Bethardy, G. A., **A**:278(28), **A**:*302*, **B**:210(77), **B**:*253*
Bhalla, K., **B**:180(13), **B**:*201*
Bhatia, P., **A**:228(42), **A**:*264*
Bies, W. E., **A**:403(11), **A**:*433*
Biferale, L., **B**:521(5), **B**:525(5), **B**:*541*
Bigwood, R., **A**:137(99), **A**:*141*, **B**:206-207(20), **B**:209(20,50), **B**:212(20), **B**:214(20), **B**:217(20), **B**:*252*
Bihary, Z., **B**:186(28), **B**:*202*
Birkhoff, G. D., **A**:278(19), **A**:*302*
Bizzarri, A. R.: **B**:228(119), **B**:*254*; **B**:265(52), **B**:*311*
Blencowe, M. P., **B**:221(103), **B**:*254*
Blumel, R., **A**:339(13), **A**:358(13), **A**:*398*
Blumen, A., **B**:264(51), **B**:*311*
Blumenfeld, R., **B**:228(116), **B**:*254*
Boczko, E. M., **B**:266-257(59), **B**:*311*
Boerlijst, M., **B**:549(8), **B**:573(8), **B**:*597*
Boffetta, G.: **B**:301(92-94), **B**:*313*; **B**:537(45), **B**:*542*
Bogomolny, E. B., **A**:107(74), **A**:*142*

Boguski, M. S., **B**:348(21), **B**:*351*
Bohigas, O., **A**:403(11), **A**:*433*, **B**:210(72), **B**:*253*
Bohr, N., **A**:306(4), **A**:*334*
Bolhuis, P. G., **A**:232(43), **A**:*264*
Boltzmann, H., **B**:378(10), **B**:401(10), **B**:*418*
Boltzmann, L., **B**:499(42), **B**:*500*
Bonnet, L., **A**:249(75), **A**:*265*
Boozer, A. H., **B**:502(16), **B**:*518*
Borg, I., **B**:318-319(5), **B**:*351*
Borkovec, M.: **A**:144(12), **A**:*168*; **A**:228(37), **A**:*264*; **B**:180(1), **B**:*201*; **B**:205(2), **B**:*251*
Bossy-Wetzel, E., **B**:180(13), **B**:*201*
Botina, J., **A**:436(13), **A**:437-439(13), **A**:456(29), **A**:*457–458*
Botstein, D., **B**:348(21), **B**:*351*
Bouchaud, J.-P.: **B**:228(120), **B**:*254*; **B**:520-521(2), **B**:526(28), **B**:*541–542*
Bouchet, F., **B**:489(39), **B**:*500*
Boudon, V., **A**:248(66), **A**:*265*
Bountis, T., **A**:403(19), **A**:410(19), **A**:*434*
Bourlioux, A., **B**:535(40), **B**:*542*
Bovin, J. O., **B**:156(2), **B**:*176*
Bowman, J. M.: **A**:198(113-119), **A**:*216*; **A**:280(29), **A**:287(37,39), **A**:288(39), **A**:293(39), **A**:*303*
Boxer, S. G., **B**:200(37), **B**:*203*
Braier, P. A., **B**:90(27), **B**:*127*
Braunstein, D., **B**:228(115), **B**:*254*
Bray, D.: **B**:180(12), **B**:*201*; **B**:558(27), **B**:*598*
Brayer, G. D., **B**:180(17), **B**:*202*
Breen, J. J., **A**:61(41-42), **A**:*141*
Briant, C. L., **B**:130(1), **B**:139(1), **B**:*152*
Brillouin, L., **A**:288(43), **A**:*303*, **B**:545(4), **B**:*597*
Brodier, O., **A**:403(11), **A**:*433*
Brody, T. A., **B**:210(63), **B**:214(63), **B**:*253*
Brooks, B. R.: **B**:190(33), **B**:*202*; **B**:228-229(126), **B**:*255*
Brooks, C. L. III: **B**:190(33), **B**:*202*; **B**:228(128), **B**:*255*; **B**:266-257(59), **B**:*311*
Broucke, R. A., **A**:330(38), **A**:*335*
Browaeys, A., **A**:403(13), **A**:*433*
Brown, D., **A**:280(30), **A**:*303*
Brown, P. O., **B**:348(21), **B**:*351*
Brown, R.: **B**:289(80), **B**:*312*; **B**:364(16), **B**:*371*
Brown, R. C., **B**:128(91), **A**:*142*
Bruccoleri, R. E., **B**:190(33), **B**:*202*

Brumer, P.: **A**:8(31), **A**:20(35), **A**:27(35), **A**:114(83), **A**:128(97), **A**:134(97), **A**:140(103), **A**:*141–142*; **A**:436(10), **A**:456(27,30), **A**:*457–458*; **B**:43(27), **B**:*84*
Brunet, J. P., **B**:89(10), **B**:*127*
Bryngelson, J. D., **B**:263-264(38), **B**:*311*
Bu, L.: **B**:180(9,16), **B**:197(9), **B**:*201–202*; **B**:228(137), **B**:*255*
Buch, V., **B**:186(28), **B**:*202*
Buchheim, M. A., **B**:332(15), **B**:*351*
Buldum, A., **B**:221(107), **B**:*254*
Bunimovich, L., **B**:387(41-42), **B**:*419*
Bunker, P. R., **A**:198(86), **A**:*215*
Burghardt, I., **A**:218(3), **A**:228(3), **A**:232(45), **A**:237(55), **A**:244(55), **A**:*263–265*
Burleigh, D., **B**:210(78), **B**:*253*
Burton, J. J., **B**:130(1), **B**:139(1), **B**:*152*
Bushnell, G. W., **B**:180(17), **B**:*202*
Buss, L. W., **B**:553(24), **B**:*598*

Cahill, D. G., **B**:221(99), **B**:*254*
Cai, J., **B**:180(13), **B**:*201*
Cairns-Smith, A. G., **B**:550(12), **B**:*597*
Calenbuhr, V., **B**:591(39), **B**:*598*
Campbell, D. M., **B**:207(31), **B**:219(31), **B**:249(31), **B**:*252*
Campolieti, G., **A**:114(83), **A**:*142*
Cannistraro, S.: **B**:228(119,134), **B**:*254–255*; **B**:265(52), **B**:*311*
Cao, L., **B**:290(81-82), **B**:291(81-82), **B**:300(82), **B**:*312*
Cao, W., **B**:180(14), **B**:*201*
Carcía, A. E., **B**:180(15), **B**:*202*
Cardenas, A. E., **B**:180(15), **B**:*202*
Careri, G., **B**:221(155), **B**:248(155), **B**:*255*
Carlini, P.: **B**:228(119), **B**:*254*; **B**:265(52), **B**:*311*
Carrington, T., **A**:124(89), **A**:*142*
Carry, J. R., **A**:146(46), **A**:*169*
Carter, D., **A**:20(34), **A**:*141*
Carter, S.: **A**:96(65), **A**:*142*; **A**:198(110), **A**:199-200(142), **A**:202(142), **A**:*215–216*
Casati, G., **A**:7(28), **A**:128(28), **A**:*141*
Casdagli, M., **B**:285(75), **B**:286(79), **B**:300(75), **B**:307-309(75), **B**:*312*
Castiglione, P., **B**:526-527(26), **B**:*541*
Cavalli, S., **B**:89(13-15), **B**:96-97(13), **B**:111(13), **B**:114-115(14), **B**:*127*
Cederbaum, L. S., **B**:210(64-65,71), **B**:*253*

Celani, A., **B**:522(13), **B**:531-533(13), **B**:*541*
Celleti, A., **B**:442(21), **B**:*463*
Cencini, M., **B**:534(37), **B**:537(37,45), **B**:539(37), **B**:*542*
Cerbelli, S., **B**:502(16), **B**:*518*
Cerjan, C. J.: **B**:132(35), **B**:*153*; **B**:171(14), **B**:*177*
Chakravarty, C.: **B**:31(20), **B**:39(20), **B**:45(20), **B**:*84*; **B**:209(57), **B**:*253*
Champion, P. M., **B**:180(14), **B**:200(35), **B**:*201–202*
Chan, H. S., **B**:90(32), **B**:*128*
Chan, W.-T., **A**:198(129), **A**:*216*
Chandler, D., **B**:205(7), **B**:207(7), **B**:209(7), **B**:217(7,17), **B**:218(7), **B**:*251*, **B**:*253*
Chandra, A. K., **A**:198(131-132), **A**:*216*
Chang, S.-J., **A**:442(23), **A**:*457*
Chang, Y.-T., **A**:402(6), **A**:*433*
Chapman, R. L., **B**:332(15), **B**:*351*
Chapuisat, X., **A**:198(133-134), **A**:*216*, **B**:89(8-10), **B**:*127*
Chatfield, D. C., **A**:174(23), **A**:178(23), **A**:*213*
Chawanya, T., **B**:301(90), **B**:*313*
Chelkowski, S., **A**:198(136), **A**:*216*
Chenciner, A., **A**:219(15), **A**:*263*
Chernov, N. I., **B**:387(43), **B**:*419*
Chesnavich, W. J., **A**:249(73), **A**:*265*
Chiang, J., **A**:96(100), **A**:*142*
Chierchia, L.: **B**:396(66), **B**:*420*; **B**:427(17), **B**:*436*
Child, M. S.: **A**:176(37), **A**:*213*; **A**:223(35), **A**:228(39), **A**:230(39), **A**:232(39), **A**:*264*
Chin, J. K., **B**:180(10), **B**:199(10), **B**:*201*
Chirikov, B. V.: **A**:7(28), **A**:128(28), **A**:*141*; **A**:340(16), **A**:372(16), **A**:*398*; **B**:383(27-28), **B**:397(69), **B**:*419–420*; **B**:424(2), **B**:427(2), **B**:430(2), **B**:*436*; **B**:438(5), **B**:441(17-18), **B**:457(5), **B**:*462–463*; **B**:477(5), **B**:*499*
Chiti, F., **B**:264(46), **B**:*311*
Cho, M.: **B**:229(139), **B**:*255*; **B**:391(54), **B**:*420*
Choi, D., **B**:396(64), **B**:*420*
Choi, M. Y., **B**:479(14), **B**:*500*
Choi, N. N., **A**:308(24), **A**:*335*
Chow, T. S., **B**:208(34), **B**:233(142), **B**:*252*, **B**:*255*
Christiansen, F., **B**:502(20), **B**:*518*
Christoffel, K. M., **A**:198(119), **A**:*216*
Chuang, I. L., **A**:436(1), **A**:*457*
Cicogna, A., **B**:427(15), **B**:*436*

Cincotta, P. M.: **B**:424(3), **B**:427(3), **B**:*436*;
B:438(7), **B**:*462*
Ciraci, S., **B**:221(107), **B**:*254*
Clark, J. W., **A**:436(11), **A**:*457*
Cline, J. I., **A**:63(44-46), **A**:65(48), **A**:*141*
Cohen, E., **B**:354(1-2), **B**:*370*
Cohen, E. D. D., **B**:364(13), **B**:*371*
Coltrin, M. E., **A**:402(6), **A**:*433*
Connor, J. N. L., **A**:249(74), **A**:*265*
Conroy, J. A., **B**:324(9), **B**:*351*
Constantin, P., **B**:522(14,16), **B**:531-533(14),
B:*541*
Costley, J., **B**:106(34), **B**:*128*
Coy, S. L., **A**:278(22), **A**:*302*, **B**:210(66), **B**:*253*
Creagh, S. C.: **A**:218(7), **A**:*263*;
A:403(11-12,14), **A**:404(14), **A**:*433*;
B:414(83), **B**:*421*
Cresson, J., **B**:427(18), **B**:*436*
Crisanti, A.: **B**:301(92-94), **B**:*313*; **B**:521(4-5),
B:525(5), **B**:537(4), **B**:*541*
Crooks, G. E., **B**:354(4), **B**:*370*
Cross, M. C., **B**:221(102), **B**:*254*
Cross, P. C.: **A**:271(9), **A**:*302*, **B**:89(16),
B:93(16), **B**:95(16), **B**:106(16), **B**:*127*;
B:187(30), **B**:*202*
Csaszar, A. G., **A**:269(7), **A**:274-275(7),
A:277(7), **A**:*302*
Curioni, A., **B**:180(18), **B**:201(18), **B**:*202*
Cushman, R., **B**:248(63), **A**:*265*
Cvitanovic, P., **B**:380(18), **B**:*419*

Dahleh, M. A., **A**:436(12), **A**:*457*
D'Alessandro, M., **B**:395(61), **B**:*420*
Dana, I., **A**:177(41), **A**:*214*
D'Aquino, A., **B**:395(61), **B**:*420*
Dateo, C. E.: **A**:198(114,116), **A**:*216*;
A:280(29), **A**:*303*
Daubechies, I., **B**:316(2), **B**:*351*
Dauxois, T., **B**:479(13), **B**:485(37), **B**:489(39),
B:*500*
Davidson, G. S., **B**:348(20), **B**:*351*
Davis, D. J., **A**:31(36), **A**:*141*
Davis, H. L.: **B**:90(22), **B**:*127*; **B**:130(7),
B:139(7), **B**:*152*
Davis, M. J.: **A**:5(11), **A**:6(12), **A**:20(33),
A:31(36), **A**:35(11), **A**:39(12),
A:59-60(11-12), **A**:65(11), **A**:*140–141*;
A:145(17), **A**:166(17), **A**:167(71),
A:*169–170*; **A**:177(42-44),
A:198(91,94-95), **A**:*214–215*; **A**:228(29),

A:*264*; **A**:339(9), **A**:*398*; **A**:404(21),
A:428(37), **A**:*434*; **B**:205(9),
B:209(9,58,62), **B**:210(77), **B**:*251*, **B**:*253*;
B:260(10), **B**:*310*; **B**:442(19), **B**:*463*
Davis, P., **A**:388(38), **A**:*399*
Dawson, S., **B**:502(14), **B**:*518*
De Almeida, A. M., **A**:7(25), **A**:75(25), **A**:*140*
De Carvalho, A., **B**:380(20), **B**:*419*
Decatur, S. M., **B**:200(37), **B**:*203*
Decius, J. C.: **A**:271(9), **A**:*302*, **B**:89(16),
B:93(16), **B**:95(16), **B**:106(16), **B**:*127*;
B:187(30), **B**:*202*
Delagado, J., **A**:323(30), **A**:*335*
DeLeon, N.: **A**:7(23-25), **A**:75(23-25),
A:83-84(56), **A**:87(23,57), **A**:88(59),
A:90(59), **A**:94(23,63), **A**:*140–141*;
A:146(35-37), **A**:153(35-37), **A**:159(36),
A:*169*; **A**:178(56), **A**:198(127),
A:*214–216*; **A**:228(40), **A**:232(48),
A:250(48), **A**:*264*; **B**:48(33), **B**:82(33),
B:*84*
DeLorenzi, G., **A**:173(9), **A**:*213*
Delos, J. B., **A**:278(21), **A**:*302*
DeLuca, J., **B**:395(62), **B**:*420*
Delwiche, C. F., **B**:332(15), **B**:*351*
De Marsily, G., **B**:525(24), **B**:*541*
De Micheli, E., **A**:249(77), **A**:*265*
Demidov, A. A., **B**:180(14), **B**:200(35),
B:*201–202*
Demikhovskii, V. Ya., **A**:131(95-96), **A**:*142*,
B:209(59), **B**:*253*
Demtröder, W., **B**:210(64), **B**:*253*
Deprit, A., **A**:178-179(75), **A**:194(75), **A**:*215*
De Sousa Dias, M. E. R., **A**:248(71), **A**:*265*
Devaney, D. L., **A**:311(27), **A**:317(27), **A**:*335*
De Vivie-Riedl, R., **A**:436(3), **A**:*457*
DeVore, R. A., **B**:316(2), **B**:*351*
Dewey, T. G., **B**:228(130), **B**:237(130),
B:*255*
Diacu, F., **A**:307(18), **A**:309(25), **A**:330(25),
A:*334–335*
Dian, B. C., **B**:249(157), **B**:*255*
Dickerson, R. E., **B**:180(11), **B**:*201*
Diener, M., **A**:389(41), **A**:*399*
Dill, K. A., **B**:90(32), **B**:*128*
DiNola, A., **B**:267(60), **B**:309(60), **B**:*312*
Dion, C. M., **A**:198(136), **A**:*216*
Dlott, D. D., **B**:200(37), **B**:*203*
Dobbyn, A. J., **A**:280(31), **A**:*303*
Dobson, C. M., **B**:264(46), **B**:*311*

Doll, J. D.: **A**:406(28), **A**:410(28), **A**:*434*, **B**:130(21-22), **B**:139(21-22), **B**:*153*; **B**:209(56), **B**:*253*
Donoho, D. L., **B**:316(2), **B**:*351*
Dorsey, N. E., **B**:225-227(113), **B**:247(113), **B**:*254*
Doster, W., **B**:228(115), **B**:*254*
Doye, J. P. K.: **B**:90(27), **B**:*127*; **B**:130(25), **B**:139(25), **B**:*153*; **B**:266-267(57), **B**:284(57), **B**:*311*
Dragt, A. J.: **A**:167(74-75), **A**:*170*; **A**:277(16-18), **A**:*302*
Drews, A. R., **B**:228(134), **B**:*255*
Drobits, J. C., **A**:66(52), **A**:*141*
Duke, K., **B**:348(20), **B**:*351*
Dumont, R. S., **B**:43(27), **B**:*84*
Dunning, T. H., **A**:402(6), **A**:*433*
Duppen, K., **B**:200(42), **B**:*203*
Düren, R., **A**:287(38), **A**:298-299(38), **A**:301(38), **A**:*303*
Dyre, J. C., **B**:392(56), **B**:*420*
Dyson, F., **B**:549(9), **B**:*597*

Easton, R. W., **B**:427(13), **B**:*436*
Eberly, J. H., **A**:436(5), **A**:443(5), **A**:*457*
Eckart, C., **B**:89(19), **B**:93(19), **B**:106(19), **B**:*127*
Eckhardt, B., **A**:173(8), **A**:*213*
Eckmann, J.-P., **B**:502(1), **B**:*517*
Edler, J., **B**:200(44), **B**:*203*
Efstathiou, K., **A**:269(4), **A**:*302*
Egorov, S. A., **B**:180(6), **B**:183-185(6), **B**:*201*
Eguchi, J., **B**:207(30), **B**:219(30), **B**:249(30), **B**:*252*
Egydo de Carvalho, R., **A**:403(11), **A**:*433*
Eigen, M., **B**:547-548(7), **B**:573(31), **B**:*597-598*
Einstein, Albert: **A**:288(42), **A**:*303*; **A**:306(5), **A**:*334*
Eisen, M. B., **B**:348(21), **B**:*351*
Eisenberg, D., **B**:378(6), **B**:405(6), **B**:*418*
Eizinger, A., **B**:348(20), **B**:*351*
Elber, R.: **B**:180(15), **B**:186(28), **B**:*202*; **B**:223(110), **B**:228(110,117,131-132), **B**:235(131-132), **B**:238(144), **B**:257(131), **B**:*254-255*
Ellegaard, C., **B**:210(75), **B**:*253*
Ellinger, Y., **A**:198(130), **A**:*216*
Elliott, S. R., **B**:208(42), **B**:250(42), **B**:*252*

Elmaci, N., **B**:262(29), **B**:266(29,56), **B**:284(56), **B**:*311*
Elran, Y., **A**:262(86), **A**:*265*
Embid, P. F., **B**:534(36), **B**:*542*
Engholm, J. R., **B**:207(25), **B**:209(25), **B**:239(25), **B**:*252*
Englander, S. W., **B**:180(14), **B**:*201*
Entin-Wohlman, O., **B**:208(36-37), **B**:241(36-37), **B**:246(36-37), **B**:250(36-37), **B**:*252*
Era, M., **B**:180(19), **B**:*202*
Erskin, J., **B**:354(12), **B**:*371*
Esaki, S., **B**:265(53), **B**:299(53), **B**:*311*
Estebaranz, J. M., **A**:167(72), **A**:*170*
Etters, R. D., **B**:130(2-4), **B**:139(2-4), **B**:*152*
Eubank, S., **B**:286(79), **B**:*312*
Evans, D., **B**:354(1), **B**:*370*
Evans, M. G.: **A**:144(4), **A**:*168*; **A**:176(28), **A**:*213*; **B**:258(3), **B**:*310*
Evard, D. D., **A**:63(44-46), **A**:65(48), **A**:*141*
Eyring, H.: **A**:144(3), **A**:*168*; **A**:172(3), **A**:176(3,27), **A**:*213*; **B**:258(2), **B**:*310*
Ezra, G. S., **A**:*264*; **A**:6(18), **A**:20(33), **A**:48(18), **A**:*140-141*; **A**:145(20), **A**:146(48), **A**:166(20), **A**:167(71), **A**:*169-170*; **A**:178(55), **A**:*214*; **A**:307(12), **A**:*334*; **A**:339(11), **A**:358(11), **A**:*398*; **B**:205(10), **B**:209(10,58), **B**:*251*, **B**:*258*; **B**:438(1), **B**:442(19), **B**:*462-463*

Fabian, J.: **B**:190(32), **B**:195(32), **B**:198(32), **B**:*202*; **B**:208(40), **B**:238(145-146), **B**:240(145-146), **B**:243(145-146), **B**:250(40), **B**:*252, 255*
Fair, J. R., **A**:178(64), **A**:*214*; **B**:48(33), **B**:82(33), **B**:*84*
Falcioni, M., **B**:521(4), **B**:537(4,45), **B**:*541-542*
Falkovich, G., **B**:520(3), **B**:523(3), **B**:*541*
Fano, U., **A**:307(13), **A**:*334*
Farantos, S.: **A**:198(99-104), **A**:*215*; **A**:228(42), **A**:*264*; **A**:278(25), **A**:287(35-36,38), **A**:292(45), **A**:293(35-36,38), **A**:298(35-36,38), **A**:299(35,38), **A**:301(35,38), **A**:*302-303*
Farmer, J. D.: **B**:286(79), **B**:*312*; **B**:550(11), **B**:*597*

Farrelly, D.: **A**:173(6,10,18), **A**:177(6,18), **A**:180-181(6,18), **A**:198(85), **A**:*213,* **A**:*215*; **A**:219(13), **A**:234(13,51), **A**:248(65), **A**:*263-265*
Faure, A., **A**:245(58-59), **A**:248(58), **A**:256(58,84), **A**:257-258(59), **A**:*265*
Fayer, M. D.: **B**:180(5), **B**:183-184(5), **B**:189-190(5), **B**:200(35,37,43), **B**:*201-203*; **B**:207(28), **B**:209(28), **B**:239(28), **B**:*252*
Federov, D. G., **B**:201(47), **B**:*203*
Fein, A. E.: **B**:190(31), **B**:*202*; **B**:238(143), **B**:*255*
Feldman, J. L., **B**:208(40), **B**:250(40), **B**:*252*
Felker, P. M., **B**:209(43), **B**:220(43,94-95), **B**:*252,* **B**:*254*
Feller, W., **B**:450(22), **B**:*463*
Fenichel, N., **A**:338(4-7), **A**:340(5), **A**:341(7), **A**:349(7), **A**:*398*
Fermi, E., **B**:376(1), **B**:393(1), **B**:*418*
Ferrer, S., **A**:238(57), **A**:*265*
Fersht, A., **B**:263(35), **B**:*311*
Field, R. W.: **A**:282(32), **A**:287(34-35), **A**:288(34), **A**:293(34-35), **A**:298-299(35), **A**:301(35), **A**:*303,* **B**:210(66), **B**:*253*
Finn, J. M.: **A**:167(74-75), **A**:*170*; **A**:277(16-17), **A**:*302*
Firpo, M.-C., **B**:481(32), **B**:*500*
Fisher, M. E., **B**:526(27), **B**:*542*
Fisher, R. A., **B**:521-522(10), **B**:*541*
Flach, S., **B**:395(60), **B**:*420*
Fleming, G. R.: **B**:220(97), **B**:229(139), **B**:*254-255*; **B**:391(54), **B**:*420*
Fleming, P. R., **A**:198(105), **A**:*215*
Fleurat-Lesard, P., **A**:257(85), **A**:*265*
Floquet, G., **A**:407(29), **A**:430(29), **A**:*434*
Flores, J., **B**:210(63), **B**:214(63), **B**:*253*
Floudas, C. A., **A**:198(138), **A**:*216*
Flynn, C. P., **A**:173(9), **A**:*213,* **B**:228(133), **B**:*255*
Fontana, W., **B**:553(24), **B**:*598*
Fontich, E., **B**:427(16), **B**:*436*
Ford, J., **B**:376(2), **B**:393(2), **B**:*418*
Forest, E., **A**:277(18), **A**:*302*
Forst, W., **A**:176(30), **A**:*213*
Founargiotakis, M, **A**:198(99,103), **A**:*215*
Francisco, J. E.: **B**:27-28(7), **B**:71-72(7), **B**:75(7), **B**:79-80(7), **B**:*84*; **B**:90(29), **B**:*127*; **B**:180(1), **B**:*201*

Franks, F., **B**:405(7), **B**:*418*
Frantz, D. D., **B**:130(22), **B**:139(22), **B**:*153*
Fraser, A. M., **B**:292-293(83), **B**:*312*
Frauenfelder, H., **B**:228(114-115), **B**:*254*
Fredj, E., **B**:186(28), **B**:*202*
Freed, K. F., **B**:210(69), **B**:*253*
Freeman, D. L., **B**:130(21-22), **B**:139(21-22), **B**:*153*
Freidlin, M., **B**:530-531(35), **B**:*542*
French, J. B., **B**:210(63), **B**:214(63), **B**:*253*
Fried, L. E., **A**:146(48), **A**:*169*
Friedman, R. S., **A**:174(23), **A**:178(23), **A**:*213*
Frisch, U., **B**:523(18), **B**:*541*
Froeschlé, C.: **B**:427(12,14), **B**:*436*; **B**:438(9), **B**:442(21), **B**:*463*
Frost, W., **A**:5(5), **A**:*140*
Froyland, G., **B**:290-291(82), **B**:300(82), **B**:*312*
Fuchigami, S., **B**:200(40), **B**:*203*
Fuchikami, N., **B**:396(64), **B**:*420*
Fujimoto, H., **B**:74(46), **B**:*85*
Fujimura, Y., **A**:198(136-137), **A**:*216*
Fujisaki, H., **B**:436(7), **A**:443(25), **A**:456(31-32), **A**:*457-458*
Fukui, K.: **B**:74(46), **B**:*85*; **B**:130(30), **B**:*153*
Fukuzawa, K., **B**:201(47), **B**:*203*
Fulton, N. G., **A**:220(19), **A**:*263*
Funatsu, T., **B**:265(55), **B**:299(55), **B**:*311*
Furusawa, C., **B**:551(16), **B**:553(16), **B**:556(16,26), **B**:557(26), **B**:573(16), **B**:593(26,41), **B**:594(41), **B**:595(26,41), **B**:597(16), **B**:*598*

Gadre, S. R., **A**:198(121), **A**:*216*
Galbraith, H. W., **B**:89(17), **B**:106(17), **B**:*127*
Galgani, L.: **B**:31(22), **B**:46(22), **B**:*84*; **B**:378(12-13), **B**:396(66), **B**:397(70,72), **B**:398(12-13), **B**:399(13), **B**:401(12-13), **B**:402(72), **B**:*418-420*; **B**:499(45), **B**:*500*; **B**:502(19), **B**:506(19), **B**:*518*
Gallavotti, G.: **B**:354(2), **B**:*370*; **B**:396(66), **B**:397(70), **B**:*420*; **B**:429(25), **B**:*436*
Gandhi, S. R., **A**:436(6), **A**:*457*
Garbary, D. J., **B**:332(15), **B**:*351*
García, A. E.: **B**:228(116), **B**:*254*; **B**:264(40), **B**:267-268(42), **B**:300(40), **B**:*311*
Garrett, B. C.: **A**:144(10,14), **A**:163(10), **A**:*168-169*; **A**:174(23), **A**:178(23), **A**:*213*; **A**:218(2), **A**:*263*; **A**:402(6), **A**:*433*; **B**:258(6), **B**:300(6), **B**:*310*

Gaspard, P.: **A**:22(35), **A**:27(35), **A**:*141*;
  A:218(3), **A**:228(3), **A**:232(45-46),
  A:237(55), **A**:244(55), **A**:*263–265*;
  A:307(11), **A**:330(11), **A**:*334*;
  A:436-437(18), **A**:*457*; **B**:83(51), **B**:*85*;
  B:386-387(40), **B**:*419*
Gatti, F., **A**:198(133-134), **A**:*216*
Gawedzki, K., **B**:520(3), **B**:523(3), **B**:*541*
Gazdy, B.: **A**:198(113-114,116,118), **A**:*216*;
  A:280(29), **A**:*303*
Geisel, T.: **A**:128(92), **A**:*142*, **B**:383-384(37),
  B:*419*; **B**:479(21,28), **B**:*500*
Geist, K., **B**:502-503(17), **B**:*518*
Gelbart, W. M., **B**:210(69), **B**:*253*
George, T. G., **A**:406(28), **A**:410(28), **A**:*434*
Georges, A.: **B**:228(120), **B**:*254*; **B**:520-521(2),
  B:526(28), **B**:*541–542*
Gerber, R. B.: **B**:186(28), **B**:*202*; **B**:228(117),
  B:238(144), **B**:*254–255*
Geva, E., **B**:186(26), **B**:*202*
Gibson, J., **B**:286(79), **B**:*312*
Gillilan, R. E., **A**:*264*; **A**:6(18), **A**:48(18),
  A:*140*; **A**:145(20), **A**:166(20), **A**:*169*;
  A:178(55), **A**:*214*; **A**:339(11), **A**:358(11),
  A:*398*; **B**:205(10), **B**:209(10), **B**:*251*;
  B:438(1), **B**:*462*
Giona, M., **B**:502(16), **B**:*518*
Giorgilli, A.: **B**:378(12-13), **B**:396(66),
  B:397(70-72), **B**:398(12-13), **B**:399(13),
  B:401(12-13,71), **B**:402(72), **B**:*418–420*;
  B:499(45), **B**:*500*
Girifalco, L. A., **B**:159(12), **B**:160(13),
  B:167(13), **B**:*176–177*
Giugliarelli, G., **B**:228(134), **B**:*255*
Glansdorff, P., **B**:354(9), **B**:*370*
Glotzer, S. C., **B**:392(56), **B**:*420*
Go, J., **B**:210(77), **B**:*253*
Go, N.: **B**:223-224(112), **B**:228-229(112),
  B:*254*; **B**:261(24-25), **B**:264(37),
  B:301(24-25), **B**:*311*
Goda, N., **A**:340(21), **A**:388(21),
  A:*398*
Gogonea, V., **B**:201(46), **B**:*203*
Gojobori, T., **B**:328(11), **B**:*351*
Goldhirsch, I., **B**:502-503(9), **B**:*518*
Goldstein, H., **A**:219(16), **A**:*263*,
  B:115(36), **B**:*128*
Gollub, J. P., **B**:527(31), **B**:*542*
Golub, G. H., **B**:502-504(18), **B**:506(18),
  B:510(18), **B**:517(18), **B**:*518*

Gomez Llorente, J. M.: **A**:198(100-102,104),
  A:*215*; **A**:278(25), **A**:*302*
Gong, J. B.: **A**:128(97), **A**:134(97), **A**:140(103),
  A:*142*; **A**:147(55), **A**:153(55), **A**:*170*;
  A:218(1), **A**:228(1), **A**:*263*; **A**:456(27),
  A:*457*
Gonzalez, L., **A**:198(137), **A**:*216*
Goriely, A., **A**:403(20), **A**:*434*
Goswami, D., **A**:436(6), **A**:*457*
Goto, S., **B**:427(23), **B**:*436*
Gotze, W., **B**:391(51), **B**:*420*
Gouda, N., **B**:479(17), **B**:*500*
Gower, J. C., **B**:262(26-27), **B**:*311*
Graff, S. M., **A**:357(31), **A**:362(31), **A**:*399*
Grammaticos, B., **A**:403(19), **A**:410(19), **A**:*434*
Grandy, W. T., **B**:38(14), **B**:72(14), **B**:*84*
Grassberger, P., **B**:7(6-7), **B**:*23*
Gray, H. B., **B**:201(45), **B**:*203*
Gray, S. K.: **A**:5(11), **A**:6(12), **A**:35(11),
  A:39(12), **A**:59-60(11-12),
  A:65(11,47,49,51,55), **A**:66(49),
  A:124(90), **A**:*140–142*; **A**:145(17-18),
  A:166(17), **A**:*169*; **A**:177(43), **A**:178(52),
  A:198(90-93,95), **A**:*214–215*; **A**:228(29),
  A:*264*; **A**:339(9), **A**:*398*; **A**:428(37),
  A:*434*; **B**:205(9), **B**:209(9), **B**:*251*;
  B:260(10-11), **B**:*310*
Grebenshchikov, S. Yu., **A**:287(38), **A**:293(52),
  A:298-299(38), **A**:301(38), **A**:*303*
Grebogi, C.: **A**:339(13), **A**:358(13), **A**:*398*;
  A:456(28), **A**:*457*; **B**:364(16), **B**:*371*;
  B:502(14), **B**:*518*
Green, D. R., **B**:180(13), **B**:*201*
Gregurick, S. K., **B**:186(28), **B**:*202*
Grice, M. E., **A**:249(73), **A**:*265*
Groenen, P., **B**:318-319(5), **B**:*351*
Gruebele, M., **A**:137(99-100), **A**:*142*,
  B:206(20), **B**:207(20,22),
  B:209(20,22,50-51), **B**:210(22),
  B:212(20), **B**:214(20), **B**:216(22),
  B:217(20,89-90), **B**:*252*, **B**:*254*
Guckenheimer, I., **A**:293(51), **A**:*303*
Guhr, T., **B**:210(73-74), **B**:*253*
Guichardet, A., **B**:88(4), **B**:*127*
Guijarro, J. I., **B**:264(46), **B**:*311*
Guilini, D., **A**:140(101), **A**:*142*
Guiraldenq, P., **B**:160(13), **B**:167(13), **B**:*177*
Guo, Z., **B**:266-257(59), **B**:*311*
Gustavson, F. G., **A**:278(20), **A**:*302*
Gutmann, M., **A**:61(41-42), **A**:*141*

Gutzwiller, M. C.: **A**:7(27), **A**:*141*; **A**:176(34), **A**:*213*; **A**:293(50), **A**:*303*; **A**:306(1,9-10), **A**:320(9), **A**:*334*; **A**:402(9-10), **A**:*433*; **A**:436(16), **A**:442(16), **A**:*457*; **B**:83(52), **B**:*85*
Guzzo, M.: **B**:381(24), **B**:*419*; **B**:427(14), **B**:*436*

Haak, J. R., **B**:267(60), **B**:309(60), **B**:*312*
Haake, F.: **A**:7(29), **A**:*141*; **A**:436(17), **A**:438-439(17), **A**:*457*
Habib, S., **A**:140(102), **A**:*142*, **B**:502(21), **B**:517(21), **B**:*518*
Haffer, H., **A**:403(13), **A**:*433*
Hahn, O.: **A**:198(100-102, **A**:104), **A**:*215*; **A**:278(25), **A**:*302*
Haible, B., **B**:431(29), **B**:*436*
Halberstadt, N., **A**:65(48), **A**:*141*
Halicioglu, T., **B**:55(35), **B**:*84*
Hall, T., **B**:380(20), **B**:*419*
Haller, G., **A**:347(28), **A**:351(28), **A**:372(28), **A**:*399*, **B**:427(21-22), **B**:*436*
Halonen, L. O.: **A**:96(65), **A**:*142*; **A**:199-200(142), **A**:202(142), **A**:*216*
Hamada, D., **B**:264(46), **B**:*311*
Hamm, P.: **B**:200(44), **B**:*203*; **B**:207(24), **B**:209(24), **B**:*252*
Han, S., **B**:180(14), **B**:*201*
Handy, N. C.: **A**:198(110), **A**:*215*, **B**:74-75(47), **B**:*85*; **B**:89(10), **B**:108(20), **B**:*127*
Hänggi, P., **A**:228(37), **A**:*264*, **B**:221(106), **B**:*254*
Harayama, T.: **B**:273(67), **B**:*312*; **B**:465(1), **B**:*475*; **B**:502(10), **B**:*518*
Harding, L., **A**:198(135), **A**:*216*
Hariharan, A., **A**:436(6), **A**:*457*
Harris, G. J., **A**:269(7-8), **A**:274(7-8), **A**:275(7), **A**:277(7-8), **A**:*302*
Harthcock, M. A., **A**:91(61-62), **A**:*141*
Haruyama, T., **B**:477(7), **B**:*499*
Harvey, S. C., **B**:228(129), **B**:*255*
Hase, W. L.: **A**:5(6), **A**:*140*; **A**:168(78), **A**:*170*; **A**:198(82), **A**:*215*; **A**:293(52), **A**:*303*; **B**:27-28(7), **B**:71-72(7), **B**:75(7); **B**:79-80(7), **B**:*84*; **B**:90(29), **B**:*127*; **B**:180(1), **B**:*201*; **B**:215(85), **B**:*253*; **B**:259(7), **B**:*310*
Hasegawa, H.: **B**:200(40), **B**:*203*; **B**:326-327(10), **B**:*351*; **B**:354(10-11),

B:356(10), **B**:358(11), **B**:359(10,19), **B**:368(21), **B**:*370–371*
Hasha, D. L., **B**:207(30), **B**:219(30), **B**:249(30), **B**:*252*
Hashimoto, N., **B**:34(25), **B**:46(32), **B**:56(32), **B**:58(32), **B**:*84*
Hashimoto, T., **B**:582(37), **B**:*598*
Hasselblatt, B., **B**:378(15), **B**:*418*
Hata, H., **B**:502(13), **B**:*518*
Hatano, T., **B**:354(8), **B**:*370*
Hatori, T., **B**:530(33), **B**:*542*
Hauschildt, J., **A**:287(38), **A**:298-299(38), **A**:301(38), **A**:*303*
Havlin, S., **B**:208(33), **B**:230(33), **B**:*252*
Hayasaka, K., **B**:328(11), **B**:*351*
Hayes, W. B., **B**:502(15), **B**:*518*
Head-Gordon, T., **B**:411(82), **B**:*421*
Heckenberg, N. R., **A**:403(13), **A**:*433*
Heidrich, D., **A**:198(109), **A**:*215*
Helgaker, T., **B**:131-132(33), **B**:*153*
Heller, E. J.: **A**:107(72), **A**:129(93), **A**:*142*; **A**:403(11), **A**:404(21), **A**:*433–434*; **B**:209(62), **B**:210(68), **B**:*253*
Helmerson, K., **A**:403(13), **A**:*433*
Henderson, J. R., **A**:220(19), **A**:*263*
Henrard, J., **A**:248(67), **A**:*265*
Henrichsen, H., **B**:481(35), **B**:*500*
Henriksen, E. A., **B**:221(101), **B**:*254*
Hensinger, W. K., **A**:403(13), **A**:*433*
Herbst, E., **A**:256(82-83), **A**:*265*
Hernandez, R.: **A**:146(51-52), **A**:148(60), **A**:*169–170*; **A**:178(49,51), **A**:*214*; **B**:21(19), **B**:*23*; **B**:210(67), **B**:*253*
Hershkovitz, E., **A**:228(38), **A**:*264*
Hess, B., **B**:561(28), **B**:*598*
Hetzenauer, H., **B**:406(79), **B**:*421*
Higo, J., **A**:165(66-67), **A**:*170*, **B**:302(96), **B**:*313*
Hilger, Adam, **B**:376(3), **B**:382-383(3), **B**:*418*
Hill, J. R.: **B**:200(37), **B**:*203*; **B**:207(28), **B**:209(28), **B**:239(28), **B**:*252*
Hinde, R. J.: **A**:146(28), **A**:*169*; **A**:178(57,63), **A**:186(63), **A**:*214*; **A**:234(50), **A**:245(50), **A**:*264*; **B**:10(8-9), **B**:13(9), **B**:16(9), **B**:*23*; **B**:66(39-40), **B**:*84*; **B**:89(21), **B**:90(25), **B**:106(25), **B**:108(21), **B**:*127*; **B**:130(14-15), **B**:139(14-15), **B**:*153*; **B**:209(57), **B**:*253*

Hirano, T., **A:**198(140), **A:***216*
Hirata, F., **B:**261(24-25), **B:**301(24-25), **B:***311*
Hirata, Y., **B:**477(1), **B:***499*
Hirooka, H., **B:**381(25), **B:***419*
Hirsch, M., **A:**198(109), **A:***215*
Hirsch, M. W., **A:**338(8), **A:**347(8), **A:***398*, **A:**338(39), **A:***399*
Hirschfelder, J. O., **A:**176(29), **A:***213*
Hoare, M. R.: **B:**29(16), **B:**31(16), **B:***84*; **B:**136(39), **B:***153*
Hochstrasser, R. M., **B:**207(24), **B:**209(24), **B:**220(94), **B:**248(152), **B:***252*, **B:***254-255*
Hofbauer, J., **B:**582(35), **B:***598*
Hoff, W., **B:**207(26), **B:**209(26), **B:***252*
Hoffman, D. K., **B:**131(32), **B:**132(32), **B:***153*
Hogeweg, P., **B:**549(8), **B:**573(8), **B:***597*
Hoki, K., **A:**198(137), **A:***216*
Holbrook, K. A., **A:**176(31), **A:***213*
Holme, P., **B:**479(14), **B:***500*
Holme, T. A., **A:**198(96), **A:***215*
Holmes, P.: **A:**293(50-51), **A:***303*; **A:**309(25), **A:**330(25), **A:***335*
Holmes, P. J., **B:**427(8), **B:***436*
Honeycutt, J. D., **B:**266-267(58), **B:***311*
Hong, H., **B:**479(14), **B:***500*
Hong, M. K., **B:**228(115), **B:***254*
Honjo, S., **A:**372(33), **A:**378(33), **A:***399*, **B:**427(19), **B:***436*
Hoover, W. G., **B:**130(28), **B:**139(28), **B:***153*
Horai, S., **B:**328(11), **B:***351*
Horiguchi, H., **B:**72-73(42), **B:***84*
Horita, T., **B:**502(13), **B:***518*
Horiuti, J., **A:**176(38), **A:***213*
Hoshino, K.: **A:**147(59), **A:***170*; **A:**340(23), **A:***399*; **B:**285(74), **B:**301(95), **B:***312-313*
Hotta, K., **B:**83(56), **B:***85*
Huang, C. M., **A:**198(120), **A:***216*
Huang, G. M., **A:**436(11), **A:***457*
Hudson, J. S. Jr., **B:**348(21), **B:***351*
Hudspeth, E., **B:**220(92), **B:***254*
Hufnagel, L., **A:**136(98), **A:***142*
Hummer, G.: **B:**180(15), **B:***202*; **B:**228(116), **B:***254*; **B:**264(40), **B:**300(40), **B:***311*
Hutchinson, J. S., **A:**178(64), **A:**198(96,105), **A:***214-215*, **B:**48(33), **B:**82(33), **B:***84*

Hynes, J. T.: **A:**144(11), **A:**163(11), **A:**164(63), **A:***168*, **A:***170*, **B:**180(7), **B:**187(29), **B:**200(29), **B:***201-202*; **B:**216(86), **B:***254*

Iben, I. E. T., **B:**228(115), **B:***254*
Ibrado, A. M., **B:**180(13), **B:***201*
Ichihashi, T., **B:**156(2), **B:***176*
Ichikawa, Y. H., **B:**530(33), **B:***542*
Ichiki, K., **B:**378(14), **B:**409(14), **B:**411-412(14), **B:***418*
Iijima, S., **B:**156(2), **B:***176*
Ikeda, K. S.: **A:**340(18), **A:**388(18), **A:***398*; **A:**403(15-18), **A:**404(22-25), **A:**405(23-25,27), **A:**406(17,25), **A:**407(22,24,27), **A:**408(22), **A:**409(15,22), **A:**410(22,25,35), **A:**412(22), **A:**413(24-25), **A:**417-418(25), **A:**423(25), **A:**425(22-23), **A:**428(16-18, **A:**24), **A:**429(24), **A:**431(25), **A:***433-434*; **B:**156(8-9), **B:**159-160(11), **B:**167(8,11), **B:**168(11), **B:***176*
Ikegami, T., **B:**582(36-37), **B:***598*
Imaizumi, R., **B:**474(15), **B:***475*
Imig, O., **A:**198(109), **A:***215*
Inadomi, Y., **B:**201(47), **B:***203*
Inoue, A., **B:**83(53-54), **B:***85*
Ionascu, D., **B:**200(35), **B:***202*
Ionov, S., **B:**180(1), **B:***201*
Isaacson, A. D., **A:**402(6), **A:***433*
Ishii, K., **B:**229(138), **B:**242(138), **B:***255*
Ishii, Y.: **A:**403(18), **A:**428(18), **A:***434*; **B:**265(55), **B:**299(55), **B:***311*; **B:**380(19), **B:***419*
Ishikawa, H., **A:**287(34-35), **A:**288(34), **A:**293(34-35), **A:**298-299(35), **A:**301(35), **A:***303*
Ishioka, S., **B:**396(64), **B:***420*
Ito, K., **B:**466(10), **B:**471-472(10), **B:***475*
Ito, Y., **B:**594(43), **B:***598*
Iung, C., **B:**210(79), **B:***253*
Iwai, T., **B:**88(5-6), **B:***127*
Iwane, A. H., **B:**265(53), **B:**299(53), **B:***311*
Iyer, V. R., **B:**348(21), **B:***351*
Izrailev, F. M.: **A:**131(95-96), **A:***142*; **A:**442(22), **A:***457*; **B:**209(59), **B:***253*; **B:**397(69), **B:***420*

Jackson, T. A., **B:**200(36), **B:***202*
Jacucci, G., **A:**173(9), **A:***213*

Jaffé, C.: **A:**7(22), **A:**21(22), **A:***140*; **A:**144(15), **A:**147(15,54,58), **A:**148(15,54), **A:**163(15), **A:***169–170*; **A:**173(6-7,10,18-19,21), **A:**177(6,18-19,45-47), **A:**180-181(6,18), **A:**182(78), **A:**212(19,21), **A:***213–215*; **A:**219(8-9,13), **A:**221(8), **A:**233(9), **A:**234(8,13,51), **A:**235(9), **A:**237(9), **A:**250(8), **A:***263–264*; **A:**338-339(3), **A:**341(3), **A:**352(3), **A:**398; **A:**428(38), **A:***434*
Jaffe, R. J., **B:**72-73(43), **B:***85*
Jagannathan, A., **B:**208(37), **B:**241(37), **B:**246(37), **B:**250(37), **B:**252
Jalnapurkar, S. M., **A:**248(72), **A:***265*
Janda, K. C., **A:**63(44-46), **A:**65(48), **A:***141*
Jang, S.: **A:**6(19), **A:**87(58), **A:**88(60), **A:**97-98(19), **A:**104(67), **A:**108(75), **A:**124(88), **A:***140–142*; **A:**168(76), **A:***170*; **A:**198(122), **A:***216*
Jansen, T. I. C., **B:**200(42), **B:***203*
Jarzynski, C., **B:**354(3), **B:**355(17), **B:**356(3), **B:***370–371*
Jaynes, E. T., **B:**28(12), **B:**72(12), **B:***84*
Jean, J. M., **B:**209(44), **B:**252
Jeans, J. H.: **B:**378(11), **B:**401(11), **B:***418*; **B:**499(43), **B:***500*
Jellinek, J.: **B:**45(31), **B:**53(31,34), **B:**54(34), **B:**56(31), **B:***84*; **B:**90(22), **B:***127*; **B:**130(5,7-8), **B:**135(5), **B:**139(5,7-8), **B:***152*; **B:**156(5), **B:***176*
Jena, P., **A:**178(66), **A:***214*
Jensen, H. J. A., **B:**131-132(33), **B:***153*
Jeon, G. S., **B:**479(14), **B:***500*
Jeong, H.: **B:**453(23), **B:***463*; **B:**546-547(5), **B:***597*
Jiang, M., **B:**348(20), **B:***351*
Jimenez, R., **B:**180(10), **B:**199(10), **B:***201*
Johan, T., **A:**245(59), **A:**257-258(59), **A:***265*
Johnson, B. R., **A:**245(61), **A:***265*
Johnson, J. B., **A:**228(115), **B:***254*
Johnson, K. E., **A:**61-62(39-40), **A:***141*
Johnson, R. W., **B:**73(44), **B:***85*
Jonas, D. M., **A:**278(24), **A:***302*
Jonas, J., **B:**207(30), **B:**219(30), **B:**249(30), **B:***252*
Jones, C. K. R. T., **A:**347(26), **A:**349(26), **A:***399*
Jones, D. P., **B:**180(13), **B:***201*
Joos, E., **A:**140(101), **A:***142*
Jordahl, O. M., **A:**272(11), **A:**277(11), **A:***302*

Jordan, K. D., **B:**171(14), **B:***177*
Jørgensen, P., **B:**131-132(33), **B:***153*
Jortner, J., **A:**63(43), **A:***141*
Jost, R., **A:**287-288(39), **A:**293(39), **A:***303*
Joyeux, M.: **A:**198(141), **A:***216*; **A:**219(10), **A:**223(34), **A:***263–264*; **A:**269(1-6), **A:**274(1), **A:**278(3), **A:**287(34-36,39), **A:**288(34,39), **A:**293(34-36,39), **A:**298(35-36), **A:**299(35), **A:**301(35), **A:***302–303*
Judd, K., **B:**290-291(82), **B:**300(82), **B:***312*
Judson, R. S., **A:**436(14), **A:***457*
Jun, B., **B:**264(45), **B:***311*
Jung, C., **A:**249(76,78), **A:***265*
Jungwirth, P., **B:**186(28), **B:***202*
Justum, Y., **A:**198(133-134), **A:***216*

Kadanoff, L. P.: **A:**5(10), **A:**18(10), **A:**30(10), **A:***140*; **A:**177(41), **A:***214*
Kaelberer, J., **B:**130(2-4), **B:**139(2-4), **B:***152*
Kamimura, T., **B:**530(33), **B:***542*
Kamino, T., **B:**156-157(7), **B:***176*
Kan, I., **A:**339(13), **A:**358(13), **A:***398*
Kaneko, K.: **A:**308(21), **A:***334*; **A:**340(19), **A:**372(33), **A:**388(19), **A:**378(33), **A:***398–399*; **B:**301(89-90), **B:***312–313*; **B:**427(10), **B:***436*; **B:**438(10), **B:**441(14), **B:***463*; **B:**479(24), **B:***500*; **B:**502(12), **B:**513-514(12), **B:***518*; **B:**544(1-2), **B:**551(1-2,13-17,21), **B:**553(15-17), **B:**556(15-16,25-26), **B:**557(15,26), **B:**558(2), **B:**561(29), **B:**562-563(2), **B:**573(16-17,32-33), **B:**575(2,32), **B:**582(36), **B:**583(38), **B:**592(33), **B:**593(26,41), **B:**594(41,43), **B:**595(26,41), **B:**596(44,46), **B:**597(15-16), **B:***597–598*
Kantz, H.: **B:**285(78), **B:**294(78), **B:***312*; **B:**395(63), **B:***420*
Kantz, T. S., **B:**332(15), **B:***351*
Kaper, T. J., **B:**347(27), **B:**349(27), **A:***399*
Kaplan, I., **A:**403(11), **A:***433*
Karney, C. F. F.: **B:**383(29), **B:***419*; **B:**477(4), **B:***499*
Karp, G., **B:**180(12), **B:***201*
Karplus, M.: **B:**180(15), **B:**190(33), **B:***202*; **B:**228(126,128,132), **B:**229(126), **B:**235(132), **B:***255*; **B:**262(28), **B:**263-264(36), **B:**267(28), **B:***311*
Kashiwagi, A., **B:**551(21), **B:**593-595(41), **B:***598*

Kashiwagi, H., **B**:180(19), **B**:*202*
Kassel, L. S.: **A**:5(2), **A**:*140*; **A**:144(2), **A**:*168*
Kataoka, M., **B**:264(46), **B**:*311*
Kato, H., **B**:74(46), **B**:*85*
Kato, S., **B**:72-73(43), **B**:74(46), **B**:*85*
Katok, A., **B**:378(15), **B**:379(34), **B**:383(34), **B**:*418–419*
Kauffman, S. A., **B**:550(10), **B**:*597*
Kauffmans, S., **B**:550(11), **B**:*597*
Kawai, S.: **A**:147(58), **A**:*170*; **A**:338-339(3), **A**:341(3), **A**:352(3), **A**:*398*
Kawamura, H., **B**:396(64), **B**:*420*
Kay, K. G.: **A**:111(77), **A**:114(82), **A**:*142*; **A**:262(86), **A**:*265*; **B**:209(47), **B**:*252*
Keck, J. C.: **A**:144(8), **A**:163(8), **A**:*168*; **A**:173(15), **A**:176(15), **A**:*213*; **B**:258(5), **B**:300(5), **B**:*310*
Keilin, D., **B**:180(11), **B**:*201*
Keller, H. M., **A**:280(31), **A**:*303*
Keller, J., **A**:288(44), **A**:*303*
Keller, J. B., **A**:402(4), **A**:*433*
Kellman, M. E., **A**:287-288(34), **A**:293(34), **A**:*303*
Kelner, J., **B**:223(109), **B**:*254*
Kemble, E. C., **A**:272(12), **A**:277(12), **A**:*302*
Kendrick, B. K., **A**:257(85), **A**:*265*
Kenkre, V. M., **B**:180(5), **B**:183-184(5), **B**:189-190(5), **B**:*201*
Kennel, M. B., **B**:289(80), **B**:*312*
Kenny, J. E., **A**:61-62(40), **A**:*141*
Keplinski, P., **B**:221(108), **B**:*254*
Kerstein, A. R., **B**:522(17), **B**:531(17), **B**:534(17), **B**:*541*
Keshavamurthy, S., **A**:178(50), **A**:*214*
Keske, J. C., **B**:205(8), **B**:209(8), **B**:*251*
Ketzmerick, R., **A**:136(98), **A**:*142*
Khalatnikov, I. M., **A**:402(4), **A**:*433*
Kholodenko, Y., **B**:248(152), **B**:*255*
Khouider, B., **B**:535(40), **B**:*542*
Kidera, A.: **B**:181(21), **B**:195(21), **B**:*202*; **B**:238(150-151), **B**:*255*
Kiefer, C., **A**:140(101), **A**:*142*
Kiefer, J. H., **A**:198(135), **A**:*216*
Kiefhaber, T., **B**:254(43), **B**:*311*
Kifer, Y., **B**:475(16), **B**:*475*
Kikuchi, Y.: **B**:273(67), **B**:*312*; **B**:477(7), **B**:*499*; **B**:502(10), **B**:*518*
Kim, B. J., **B**:479(14), **B**:*500*
Kim, C. N., **B**:180(13), **B**:*201*
Kim, H., **B**:186(27), **B**:*202*
Kim, S. B.: **A**:146(32), **A**:*169*; **A**:174(24), **A**:178(24), **A**:*213*
Kim, S. K., **B**:348(20), **B**:*351*
Kimball, G. E., **A**:172(3), **A**:176(3), **A**:*213*
Kindt, J. T., **B**:406(80), **B**:*421*
Kinsey, J. L., **B**:27-28(4), **B**:71-73(4), **B**:75(4), **B**:79-80(4), **B**:*83*
Kiraly, M., **B**:348(20), **B**:*351*
Kirczenow, G., **B**:221(104), **B**:*254*
Kiselev, A., **B**:522(14,16), **B**:531-533(14), **B**:*541*
Kishino, H., **B**:326-327(10), **B**:*351*
Kitagawa, T., **B**:200(35), **B**:*202*
Kitamura, K., **B**:265(53), **B**:299(53), **B**:*311*
Kitao, A., **B**:261(24-25), **B**:301(24-25), **B**:*311*
Kitao, O., **B**:390(48), **B**:*420*
Kitaura, K., **B**:201(47), **B**:*203*
Klafter, J.: **B**:228(122-123), **B**:*254*; **B**:264(48,50-51), **B**:*311*; **B**:479(22), **B**:*500*; **B**:524(21), **B**:*541*
Klee, S., **A**:278(27), **A**:*302*
Klemperer, W.: **A**:198(87-90), **A**:*215*; **A**:278(22), **A**:*302*
Klippenstein, S. J.: **A**:144(14), **A**:*169*; **A**:218(2), **A**:*263*
Kluck, R. M., **B**:180(13), **B**:*201*
Knoll, J., **A**:402(7), **A**:*433*
Knowles, P. J., **A**:269(7), **A**:274-275(7), **A**:277(7), **A**:*302*
Ko, E., **B**:551(19), **B**:*598*
Kobayashi, T., **B**:156(9), **B**:159-160(11), **B**:167-168(11), **B**:*176*
Kocher, T. D., **B**:324(9), **B**:*351*
Koeppl, G. W., **A**:176(39), **A**:*214*
Koga, N., **B**:285(74), **B**:*312*
Kogumo, N., **B**:475(19), **B**:*475*
Koizumi, H., **B**:156(1), **B**:*176*
Kolmogorov, A. N.: **B**:425(5), **B**:*436*; **B**:521-522(9), **B**:*541*
Komatsu, M., **B**:156-157(7), **B**:*176*
Komatsuzaki, T.: **A**:7(26), **A**:137(26), **A**:*140*; **A**:146(38-45,49-50), **A**:147(41,45,59), **A**:148(38-39,41-45), **A**:151(44-45), **A**:152-153(45), **A**:161(44), **A**:162(38-45,49), **A**:166(42-43), **A**:167(42-44), **A**:*169–170*; **A**:173(5), **A**:178(5,67-74), **A**:179(69-73), **A**:186(69), **A**:*213–215*; **A**:218-219(4), **A**:229(4), **A**:234(4,50), **A**:235(52), **A**:245(50,62), **A**:247(62), **A**:*263–265*; **A**:338-339(2),

AUTHOR INDEX

A:340(23), A:341(2), A:352(2),
A:*398-399*; B:21(14-18), B:22(20-21),
B:*23-24*; B:90(26), B:*127*; B:209(53),
B:*253*; B:260(13-20), B:261(21),
B:263(13-20), B:264(41), B:265(13-20),
B:266-267(41), B:274(73), B:285(74),
B:294(41), B:300(13-20), B:301(73,95),
B:*310-313*; B:438(2), B:*462*
Komeji, Y., B:201(47), B:*203*
Ko-Mitamura, E. P., B:551(18), B:567(18), B:570(18), B:*598*
Komornicki, A., B:72-73(43), B:*85*
Komuro, M., A:340(20), A:388(20), A:*398*
Kondepudi, D., B:354(9), B:*370*
Kondorskiy, A., A:456(31), A:*459*
Konishi, T.: A:340(21), A:342(25), A:377(25), A:388(21), A:*398-399*, B:385(39), B:*419*; B:441(14), B:*463*; B:479(17,24,29), B:*500*; B:502(12), B:513-514(12), B:*518*
Kook, H. T., B:438(11), B:441(11), B:*463*
Koon, W. S.: A:168(77), A:*170*; A:248(68), A:*265*; A:340(22), A:*399*
Koplik, J., B:526(28), B:*542*
Köppel, H., B:210(64-65,71), B:*253*
Koput, J., A:287(33,35,38), A:293(35,38), A:298-299(35,38), A:301(35,38), A:*303*
Kosloff, R., A:436(3), A:*457*
Kostov, K. S., B:264(41), B:266-267(41), B:294(41), B:*311*
Kovács, Z., A:224(26), A:228(41), A:249(41), A:250(26), A:*264*
Koyama, H., B:479(29), B:*500*
Kozin, I. N., A:248(64,69-70), A:*265*
Kozlov, V. V., A:173-174(13), A:184-186(13), A:*213*
Kramer, F. R., B:547(6), B:*597*
Kramer, P. R., B:521(6), B:*541*
Kramers, H. A., A:144(6), A:163(6), A:*168*
Krumhansl, J. A., B:228(116), B:*254*
Kruskal, J. B., B:317(3), B:319(3), B:*351*
Kubachewski, O., B:157(10), B:*176*
Kubo, R., B:46(32), B:56(32), B:58(32), B:*84*
Kugimiya, T., A:165(66), A:*170*
Kuharski, R. A., B:205(7), B:207(7), B:209(7), B:217(7,17), B:218(7), B:*251*, B:*254*
Kulkami, S. A., A:198(121), A:*216*
Kumar, A. T. N., B:200(35), B:*202*
Kumeda, Y., A:198(140), A:*216*
Kundkar, L. R., B:220(96), B:*254*
Kunz, R. E., B:26(1), B:*83*

Kupperman, A., A:245(60), A:*265*, B:89(11), B:115(11), B:*127*
Kupsch, J., A:140(101), A:*142*
Kurkal, V., A:198(126), A:*216*
Kurosaki, S., B:465(2), B:*475*
Kurtz, S. R., B:228(133), B:*255*
Kuzmin, M., B:209(48), B:*252*
Kwok, A., B:207(28), B:209(28), B:239(28), B:*252*

Labastie, P., B:90(23), B:*127*
Lai, Y. C., A:339(13), A:358(13), A:*398*
Laird, B. B., B:391(52), B:*420*
Lambert, W. R., B:209(43), B:220(43,94), B:*252*, B:*254*
Lan, B. L., A:198(115,117), A:*216*
La Nave, E., B:391(55), B:*420*
Lancet, D., B:573(34), B:579(34), B:*598*
Landau, L., B:499(44), B:*500*
Landolfi, M., B:394(57), B:*420*
Langmuir, I., A:306(6), A:*334*
Langton, C., B:553(23), B:*598*
Lanne, J., A:91(61-62), A:100(66), A:*141-142*
Lara, M., A:238(57), A:*265*
Larregaray, P., A:249(75), A:*265*
Lashkari, D., B:348(21), B:*351*
Laskar, J.: A:427(20), B:*436*; B:438(13), B:442(13,20-21), B:453(13), B:*463*
Lasker, L., A:372(34), A:378(34), A:*399*
Latora, V., B:479(15,18,20,23), B:480(15,18), B:481(15,18,31), B:485(37), B:487(15), B:*500*
Lauterborn, W., B:502-503(17), B:*518*
Lauvergnat, D., A:198(134), A:*216*
LaViolette, R. A., B:250(159), B:405(75), B:*255*, B:*420*
Leboeuf, P., A:219(12), A:*263*, B:530(34), B:*542*
Lebowitz, J. L., B:354(5), B:*370*
Le Daeron, P. Y., B:383(32), B:*419*
Lee, F., A:248(65), A:*265*
Lee, H. W., A:105(68), A:*142*
Lee, J. C. F., B:348(21), B:*351*
Lee, M.-H., A:308(24), A:*335*
Lee, S. Y., A:403(11), A:*433*
Lee, T. J., A:198(114,116), A:*216*
Lee, T. K., A:280(29), A:*303*
Lega, E., B:427(12,14), B:*436*
Lehman, E. L., B:322(8), B:*351*

Lehmann, K. K.: **A:**198(87-90), **A:***215*;
    A:278(22,26), **A:***302*; **B:**209(45), **B:***252*
Leitner, D. M.: **A:**131(94), **A:***142*; **A:**178(53),
    A:*214*; **B:**4(3), **B:**5(4), **B:***23*;
    B:130(10,22,26), **B:**139(10,22,26),
    B:143(10), **B:***152–153*; **B:**180(2,15),
    B:181(22), **B:**190(22), **B:**193(22),
    B:195(21-22), **B:**197(22), **B:***201–202*;
    B:205(3-4,6), **B:**206(4,16-19,21),
    B:207(3-4,6,19,21),
    B:209(16-21,54-57,60-61),
    B:210(16,21,65,70-71,76), **B:**212(20),
    B:213(16,18-19), **B:**214(16-17,20),
    B:217(20), **B:**220(3-4,6), **B:**221(105,107),
    B:223(111), **B:**229(111), **B:**233(141),
    B:238(147-149), **B:**240(111,147-149),
    B:241(111), **B:**242(147),
    B:243(111,147-149), **B:**246(147-148),
    B:247(111), **B:**248(156), **B:**249(3-4,6),
    B:250(111), **B:***251–255*; **B:**424(4),
    B:427(4), **B:***436*
Le Quéré, F., **A:**124(90), **A:***142*
Lessen, D., **A:**276(14), **A:**278(14), **A:***302*
Lester, M. I., **A:**66(52-54), **A:***141*
Letokhov, V., **B:**180(1), **B:***201*
Levin, S. A., **B:**538(47), **B:***542*
Levine, B., **B:**205(6), **B:**207(6), **B:**220(6),
    B:249(6), **B:***251*
Levine, D. J., **B:**214(82), **B:***253*
Levine, R. D.: **A:**116(87), **A:***142*, **B:**7(5), **B:***23*;
    B:27(2-5), **B:**28(2-5,13), **B:**29(2),
    B:71(2-5), **B:**72(2-5,13), **B:**73(2-5),
    B:75(2-5), **B:**79(2-5), **B:**80(2-4),
    B:*83–84*
Levitt, M., **B:**228-229(127), **B:***255*
Levy, D. H., **A:**61-62(39-40), **A:***141*
Levy, R. M., **B:**261(23), **B:***311*
Lewis, J.: **B:**180(12), **B:***201*; **B:**558(27), **B:***598*
Lewis, L. A., **B:**332(15), **B:***351*
Leyvraz, F., **A:**403(11), **A:***433*
Li, C.-B., **B:**354(10), **B:**356(10), **B:**359(10),
    B:*370*
Li, F.-Y., **B:**26(1), **B:***83*
Lian, T. Q., **B:**248(152), **B:***255*
Liao, Jie-Lou, **A:**218(5), **A:***263*
Lichtenberg, A. J.: **A:**5(7), **A:**8(7), **A:***140*;
    A:146(47), **A:***169*; **A:**176(33), **A:***213*;
    A:340(17), **A:**372(17), **A:***398*; **B:**31(21),
    B:45(21), **B:**68(21), **B:***84*; **B:**376(4),
    B:392(4), **B:**395(62), **B:***418*, **B:***420*;
    B:438(8,12), **B:**441(12), **B:**457(8),
    B:463(12), **B:***462–463*; **B:**478(10), **B:***499*;
    B:502(5), **B:***518*; **B:**529(32), **B:***542*
Lidar, D. A., **B:**228(117), **B:***254*
Lieberman, M. A.: **A:**5(7), **A:**8(7), **A:***140*;
    A:146(47), **A:***169*; **A:**176(33), **A:***213*;
    A:340(17), **A:**372(17), **A:***398*; **B:**31(21),
    B:45(21), **B:**68(21), **B:***84*; **B:**376(4),
    B:392(4), **B:***418*; **B:**438(12), **B:**441(12),
    B:463(12), **B:***463*; **B:**478(10), **B:***499*;
    B:502(5), **B:***518*; **B:**529(32), **B:***542*
Light, J. C.: **A:**124(89), **A:***142*; **A:**280(30),
    A:*303*; **B:**4(2), **B:***23*
Lim, M.: **B:**200(36), **B:***202*; **B:**207(24),
    B:209(24), **B:***252*
Ling, S.: **A:**83-84(56), **A:***141*; **A:**146(37),
    A:153(37), **A:***169*; **A:**198(127-128), **A:***216*
Lipp, C., **A:**250(78), **A:***265*
Litke, A., **B:**338(19), **B:**341(19), **B:***351*
Littlejohn, R. G., **A:**220(17), **A:**248(17),
    A:250(17), **A:***263*, **B:**88(7), **B:**89(7,12-14),
    B:91(7), **B:**93-95(7), **B:**96(12-13),
    B:97(13), **B:**108(7), **B:**111(13), **B:***127*
Litvak-Hinenzon, A., **A:**180(77), **A:**182(77),
    A:*215*
Liu, X., **B:**180(13), **B:***201*
Livi, R., **B:**395(58,63), **B:***420*
Lo, M. W.: **A:**168(77), **A:***170*; **A:**173(10),
    A:*213*; **A:**219(13), **A:**234(13), **A:**248(68),
    A:*263*, **A:***265*; **A:**340(22), **A:***399*
Lochak, P.: **B:**397(70), **B:***420*; **B:**429(26-28),
    B:*436*
Locke, B., **B:**248(152), **B:***255*
Lockwood, S. F., **B:**324(9), **B:***351*
Logan, D. E.: **B:**180(2), **B:***201*; **B:**206(15),
    B:210(15), **B:**212-213(15), **B:***251*
Longarte, A., **B:**249(157), **B:***255*
López-Castillo, A., **A:**307(19), **A:***334*
Loring, R. F., **B:**200(43), **B:***203*
Losada, J. C., **A:**167(72), **A:***170*
Louck, J. D., **B:**89(17), **B:**106(17), **B:***127*
Loudon, R., **A:**429(39), **A:***434*
Louie, G. V., **B:**180(17), **B:***202*
Lovejoy, E. R.: **A:**146(32-33), **A:***169*;
    A:174(24-25), **A:**178(24-25), **A:***213*
Luck, S., **B:**228(115), **B:***254*
Lund, J., **B:**348(20), **B:***351*
Luthey-Schulten, Z. A., **B:**90(31), **B:***128*

Macdonald, R. G., **A:**278(28), **A:***302*
MacElroy, R. D.: **B:**250(159), **B:***255*;
    B:405(75), **B:***420*

MacKay, R. S.: **A**:5(8-9), **A**:30(8-9), **A**:*140*;
  A:177(41), **A**:193(81), **A**:*214–215*,
  B:376(3), **B**:382(3), **B**:383(3,35),
  B:*418–419*
MacKerrel, A. D.: **B**:190(33), **B**:*202*; **B**:261(22),
  B:*311*
Madsen, D., **B**:217(89), **B**:*254*
Maiti, B., **A**:228(42), **A**:*264*
Maitra, N. T., **A**:129(93), **A**:*142*
Majda, A. J., **B**:521(6), **B**:525(22), **B**:534(36),
  B:535(40), **B**:*541–542*
Makarov, D. E., **A**:402(6), **A**:*433*
Maki, A., **A**:278(27), **A**:*302*
Malyshev, A. I., **A**:131(95-96), **A**:*142*,
  B:209(59), **B**:*253*
Mancinelli, R., **B**:527(29), **B**:*542*
Manneville, P., **B**:359(18), **B**:*371*
Manz, J., **A**:198(137), **A**:*216*, **B**:106(35), **B**:*128*
Maradudin, A. A.: **B**:190(31), **B**:*202*;
  B:238(143), **B**:*255*
Marcelin, A., **A**:176(26), **A**:*213*
Marcus, R. A.: **A**:5(3-4), **A**:*140*; **A**:144(7,16),
  A:146(16), **A**:*168–169*; **A**:178(60), **A**:*214*;
  A:402(6), **A**:*433*; **B**:209(49), **B**:220(96),
  B:*252*, **B**:*254*
Marinari, E. G., **B**:273(69), **B**:*312*
Marks, L. D., **B**:156(2-3), **B**:*176*
Marsden, J. E.: **A**:168(77), **A**:*170*; **A**:173(10),
  A:*213*; **A**:219(13), **A**:221(22), **A**:234(13),
  A:248(22,68,72), **A**:*263–265*; **A**:340(22),
  A:*399*; **B**:427(8), **B**:*436*
Marston, C. C., **A**:*215*; **A**:7(23,25),
  A:75(23,25), **A**:87(23), **A**:94(23,63),
  A:*140–141*; **A**:232(48), **A**:250(48), **A**:*264*
Martens, C. C.: **A**:20(33), **A**:*141*; **A**:167(71),
  A:*170*; **B**:209(58), **B**:*253*; **B**:442(19),
  B:*463*
Martin, C. H., **B**:206(12), **B**:*251*
Mart'in, P., **B**:427(16), **B**:*436*
Martínez, T. J., **B**:205(6), **B**:207(6), **B**:220(6),
  B:249(6), **B**:*251*
Martinoli, A., **B**:397(72), **B**:402(97), **B**:*420*
Marvulle, V., **A**:403(11), **A**:*433*
Maslov, V. P., **A**:402(4), **A**:*433*
Mather, J. N., **B**:383(33), **B**:*419*
Matheron, G., **B**:525(24), **B**:*541*
Matsuda, H., **B**:229(138), **B**:242(138), **B**:*255*
Matsumoto, K., **A**:340(18), **A**:388(18), **A**:*398*
Matsumoto, M., **B**:390(49-50), **B**:*420*
Matsunaga, Y.: **A**:147(59), **A**:*170*; **A**:340(23),
  A:*399*; **B**:264(41), **B**:266-267(41),
  B:274(73), **B**:285(74), **B**:294(41),
  B:301(73,95), **B**:*311–313*
Matsuura, T., **B**:551(18), **B**:567(18),
  B:570(18), **B**:*598*
Mayne, L. C., **B**:180(14), **B**:*201*
Mayoral, E., **B**:364(15), **B**:*371*
Mazzino, A., **B**:523(18), **B**:526-527(26), **B**:*541*
McCammon, J. A.: **B**:180(15), **B**:*202*;
  B:228(118,129), **B**:*254–255*; **B**:261(23),
  B:*311*
McCoy, A. B., **A**:198(111-112), **A**:*215–216*
McDonald, J. D., **B**:214(81), **B**:*253*
McGehee, R., **A**:308(22), **A**:310(22),
  A:315(22), **A**:320(22), **A**:322(22), **A**:*334*
McKaye, K. R., **B**:324(9), **B**:*351*
McKenzie, C., **A**:403(13), **A**:*433*
McLachlan, R. I., **B**:481(34), **B**:*500*
McLafferty, F. J., **A**:173(16), **A**:176(16), **A**:*213*
McWhorter, D. A., **B**:220(92-93), **B**:*254*
Mees, A., **B**:290-291(82), **B**:300(82), **B**:*312*
Mehta, M. A., **A**:*215*; **A**:7(25), **A**:75(24-25),
  A:88(59), **A**:90(59), **A**:*140–141*;
  A:146(35), **A**:153(35), **A**:*169*; **A**:232(48),
  A:250(48), **A**:*264*; **B**:48(33), **B**:82(33),
  B:*84*; **B**:210(63), **B**:214(63), **B**:*253*
Meiss, J. D.: **A**:5(8-9), **A**:30(8-9), **A**:*140*;
  A:177(41), **A**:*214*; **B**:376(3), **B**:382(3),
  B:383(3,35-36), **B**:*418–419*; **B**:427(13),
  B:*436*; **B**:438(11), **B**:441(11), **B**:*463*;
  B:477(8), **B**:478(9), **B**:479(26),
  B:*499–500*
Melinger, J. S., **A**:436(6), **A**:*457*
Mellau, G. C., **A**:289(27), **A**:*302*
Mello, P. A., **B**:210(63), **B**:214(63), **B**:*253*
Mel'nikov, V. K., **B**:465(3), **B**:*475*
Merchant, K. A., **B**:200(43), **B**:*203*
Merz, K. M., **B**:201(46), **B**:*203*
Metzler, R., **B**:228(122-123), **B**:*254*
Meyer, K. R., **A**:221(24), **A**:234(24), **A**:237(24),
  A:241(24), **A**:250(24), **A**:*264*
Meyer, N., **A**:250(78), **A**:*265*
Mezey, P. G., **B**:133(36-37), **B**:*153*
Mikami, T., **B**:186-187(24), **B**:189(24),
  B:193(24), **B**:*202*
Mikhailov, A., **B**:561(28), **B**:591(39), **B**:*598*
Mikkola, S., **A**:323(31-32), **A**:*335*
Milburn, G. J., **A**:403(13), **A**:*433*
Miller, M. A.: **B**:56(37), **B**:*84*; **B**:90(27), **B**:*127*;
  B:130(19), **B**:139(19), **B**:*153*;
  B:266-267(57), **B**:284(57), **B**:301(95),
  B:*311*, **B**:*313*

Miller, R. J. D., **B**:206(13), **B**:209(13), **B**:*251*
Miller, W. H.: **A**:54(37), **A**:57(38), **A**:111(78), **A**:112(80), **A**:114(81), **A**:115(80,84-86), **A**:*141–142*; **A**:144(9,13), **A**:146(51), **A**:148(9,60), **A**:*168–170*; **A**:172-173(2), **A**:178(48-50), **A**:193(80), **A**:198(83,90), **A**:*213–215*; **A**:218(6), **A**:232(6), **A**:*263–264*; **A**:402(1-2,6,8), **A**:404(1-2), **A**:406(28), **A**:409(2), **A**:410(28), **A**:*433–434*; **B**:74-75(47), **B**:*85*; **B**:89(10), **B**:90(30), **B**:108(20), **B**:*127*; **B**:132(35), **B**:*153*; **B**:171(14), **B**:*177*; **B**:186(25), **B**:*202*; **B**:210(67), **B**:*253*
Mills, D. R., **B**:547(6), **B**:*597*
Mills, I. M., **A**:198(110), **A**:*215*
Mil'Nikov, G. V., **A**:227(36), **A**:*264*
Milnor, J., **B**:380(17), **B**:*419*
Minami, Y., **A**:198(140), **A**:*216*
Minnhagen, P., **B**:479(14), **B**:*500*
Mishler, B. D., **B**:332(15), **B**:*351*
Mitchell, K. A., **B**:89(13-14), **B**:96-97(13), **B**:111(13), **B**:114-115(14), **B**:*127*
Mitome, M., **B**:156(2), **B**:*176*
Miyadera, T., **A**:443(25), **A**:456(32), **A**:*457–458*
Miyasaka, T., **B**:475(17), **B**:*475*
Miyashita, O.: **B**:181(21), **B**:195(21), **B**:*202*; **B**:238(150-151), **B**:*255*
Mizutani, Y., **B**:200(35), **B**:*202*
Mladenovic, M., **A**:198(106), **A**:*215*
Moffatt, H. K., **B**:520-521(1), **B**:537(1), **B**:*541*
Montemurro, M. A., **B**:480-481(30), **B**:482(36), **B**:*500*
Montgomery, J. A., **B**:205(7), **B**:207(7), **B**:209(7), **B**:217-218(7), **B**:*251*
Montgomery, R., **A**:219(15), **A**:*263*, **B**:88(1-2), **B**:*126*
Montroll, E. W.: **B**:228(121), **B**:*254*; **B**:524(20), **B**:*541*
Moore, C. B.: **A**:146(32-33), **A**:*169*; **A**:174(24-25), **A**:178(24-25), **A**:*213*; **B**:210(67), **B**:*253*
Moore, G. R., **B**:180(11), **B**:*201*
Moore, T., **B**:348(21), **B**:*351*
Mordasini, T., **B**:180(18), **B**:201(18), **B**:*202*
Mori, H.: **B**:156-157(7), **B**:*176*; **B**:502(13), **B**:*518*
Morita, T., **B**:502(13), **B**:*518*
Moritsugu, K.: **B**:181(21), **B**:195(21), **B**:*202*; **B**:238(150-151), **B**:*255*

Morokuma, K., **B**:72-73(43), **B**:*85*
Morriss, G., **B**:354(1), **B**:*370*
Mortenson, P. N., **B**:266-267(57), **B**:284(57), **B**:*311*
Moser, J. M., **B**:425(7), **B**:*436*
Moser, J. K.: **A**:306(3), **A**:*334*; **A**:358(32), **A**:*399*
Mouchet, A., **A**:219(12), **A**:*263*
Mount, K. E., **A**:410(32), **A**:*434*, **B**:56(36), **B**:*84*
Moyal, J. E., **A**:106(70), **A**:111(70), **A**:*142*
Mudipalli, P. S., **A**:198(135), **A**:*216*
Mukamel, D., **B**:499(41), **B**:*500*
Müller, S. C., **B**:522(12), **B**:*541*
Muller-Groeling, A., **B**:210(74), **B**:*253*
Muratore-Ginanneschi, P., **B**:526-527(26), **B**:*541*
Murdock, J., **A**:237(54), **A**:*264*
Murray, N. V., **A**:177(41), **A**:*214*
Murrell, J. N.: **A**:96(65), **A**:*142*; **A**:199-200(142), **A**:202(142), **A**:*216*
Mutschke, G., **B**:395(60), **B**:*420*
Muzzio, F. J., **B**:502(16), **B**:*518*

Nagaoka, M.: **A**:146(49-50), **A**:162(49), **A**:*169*; **A**:178(67-68), **A**:*214*; **A**:235(52), **A**:*264*; **B**:21(14-15), **B**:*23*; **B**:90(26), **B**:*127*
Nagaya, K., **A**:436(7), **A**:*457*
Nakamura, H.: **A**:165(66), **A**:*170*; **A**:227(36), **A**:*264*; **A**:436(7), **A**:456(31), **A**:*457–458*
Nakamura, K., **B**:83(51), **B**:*85*
Nakanishi, K., **B**:390(48), **B**:*420*
Nakano, T., **B**:201(47), **B**:*203*
Nakato, M., **B**:475(18), **B**:*475*
Nakayama, T., **B**:208(32), **B**:*252*
Nara, S., **A**:388(38), **A**:*399*
Nauts, A., **A**:198(134), **A**:*216*, **B**:89(8,10), **B**:*127*
Navarro, J. F., **A**:248(67), **A**:*265*
Nayak, S. K.: **A**:178(66), **A**:*214*, **B**:31(20), **B**:39(20), **B**:45(20), **B**:*84*; **B**:130(27), **B**:139(27), **B**:*153*
Neishtadt, A. I.: **A**:173-174(13), **A**:184-186(13), **A**:*213*, **B**:429(26), **B**:*436*
Nekhoroshev, N. N.: **B**:381(23), **B**:*419*; **B**:429(24), **B**:*436*; **B**:466(9), **B**:469(9), **B**:*475*
Nemoto, T., **B**:201(47), **B**:*203*
Nesbet, R. K., **B**:27-28(6), **B**:71-72(8), **B**:75(8), **B**:79(8), **B**:*84*

Nesbitt, D. J., **A**:282(32), **A**:*303*
Neumann, M., **B**:406(76), **B**:*421*
Newmeyer, D. D., **B**:180(13), **B**:*201*
Newton, R. G., **A**:407(30), **A**:*434*
Nguyen, P. H., **B**:200(39), **B**:*203*
Nielse, J. K., **B**:391(53), **B**:*420*
Nielsen, M. A., **A**:436(1), **A**:*457*
Nierwetberg, J., **B**:479(21), **B**:*500*
Nilsson, L., **B**:190(33), **B**:*202*
Nishikawa, T., **B**:223-224(112), **B**:228-229(112), **B**:*254*
Nitzan, A., **B**:221(106), **B**:*254*
Noid, D. W., **A**:65(51), **A**:*141*
Noid, W. G., **B**:200(43), **B**:*203*
Noli, C., **A**:249(74), **A**:*265*
Nord, R. S., **B**:131(32), **B**:132(32), **B**:*153*
Nordholm, K. S. J., **B**:209(46), **B**:*252*
Nordholm, S., **B**:205(5), **B**:207(5), **B**:215(5), **B**:*251*
Northrup, F. J., **A**:278(28), **A**:*302*
Northrup, S. H.: **B**:180(15), **B**:*202*; **B**:216(86), **B**:*254*
Nozaki, K., **B**:427(23), **B**:*436*
Nyman, G., **B**:186(27), **B**:*202*
Nymeyer, H., **B**:267-268(42), **B**:*311*

Oberman, A., **B**:522(14), **B**:531-533(14), **B**:537(44), **B**:*541*–*542*
Ohmine, I.: **B**:229(139), **B**:250(158), **B**:*255*; **B**:262(30-31), **B**:265(30-31), **B**:273(30), **B**:300(30-31), **B**:*311*; **B**:378(8-9), **B**:388(8-9), **B**:389(47), **B**:390(8-9,47-50), **B**:391(54), **B**:397(47), **B**:401(47), **B**:406(78,81), **B**:408(8-9), **B**:409(9), **B**:*418*, **B**:*420*–*421*; **B**:477(1), **B**:*499*
Ohtaki, Y., **B**:354(10), **B**:356(10), **B**:359(10,19), **B**:368(21), **B**:*370*–*371*
Ohtsuki, Y., **A**:198(137), **A**:*216*
Okabe, T., **B**:274(72), **B**:*312*
Okazaki, S., **B**:186-187(24), **B**:189(24), **B**:193(24), **B**:*202*
Okubo, A., **B**:538(47), **B**:*542*
Okumura, K., **B**:200(41), **B**:*203*
Okushima, T., **A**:393(44), **B**:508(24), **A**:399, **B**:*518*
Olafson, B. D., **B**:190(33), **B**:*202*
Olson, W. K., **B**:261(23), **B**:*311*
Oltvai, Z. N., **B**:453(23), **B**:*463*
Onishi, T., **A**:403(16-17), **A**:406(17), **A**:428(16-17), **A**:*433*

Onuchic, J. N.: **B**:90(31), **B**:*128*; **B**:201(45), **B**:*203*; **B**:205(1), **B**:206(1), **B**:*251*; **B**:267-268(42), **B**:*311*
Oono, Y.: **B**:317(4), **B**:*351*; **B**:354(6), **B**:*370*
Ooyama, N., **B**:381(25), **B**:*419*
Orbach, R. L., **B**:208(32,35-39), **B**:230(35), **B**:232-233(35), **B**:241(35-39), **B**:246(36-37), **B**:250(35-37), **B**:*252*
Ormos, P., **B**:228(115), **B**:*254*
Orszag, S. A., **B**:502-503(9), **B**:*518*
Oseledec, V. I., **B**:502(8), **B**:505(8), **B**:*518*
Oskay, W. H., **A**:403(13), **A**:*433*
Ota, M.: **B**:273(67), **B**:*312*; **B**:477(7), **B**:*499*; **B**:502(10), **B**:*518*
Otsuka, K., **A**:340(18), **A**:388(18), **A**:*398*
Ott, E.: **A**:393(45), **A**:*399*; **A**:456(28), **A**:*457*; **B**:364(16), **B**:*371*; **B**:383(36), **B**:*419*; **B**:478(9), **B**:479(26), **B**:*499*–*500*; **B**:502(6), **B**:*518*
Otto, M. F., **A**:136(98), **A**:*142*
Oxtoby, D. W., **B**:180(3), **B**:183-184(3), **B**:*201*
Ozorio De Almeida, A. M., **A**:*215*; **A**:146(34), **A**:153(34), **A**:159(34), **A**:161(34), **A**:*169*; **A**:224(25), **A**:232(48), **A**:250(48), **A**:*264*

Pack, R. T., **A**:257(85), **A**:*265*
Pal, P.: **B**:29(16), **B**:31(16), **B**:*84*; **B**:136(39), **B**:*153*
Palacián, J.: **B**:147(54,58), **A**:148(54), **A**:*170*; **A**:173(21), **A**:178-179(76), **A**:194(76), **A**:197(76), **A**:212(21), **A**:*213*, **A**:*215*; **A**:219(9,11), **A**:233(9), **A**:235(9), **A**:237(9), **A**:238(57), **A**:*263*, **A**:*265*; **A**:338-339(3), **A**:341(3), **A**:352(3), **A**:*398*
Paladin, G.: **B**:310(91-93), **B**:*313*; **B**:521(4), **B**:537(4), **B**:*541*
Palao, J. P., **A**:436(3), **A**:*457*
Palmer, R. G., **B**:499(40), **B**:*500*
Pandey, A., **B**:210(63), **B**:214(63), **B**:*253*
Paniconi, M., **B**:354(6), **B**:*370*
Paparella, F., **B**:300(88), **B**:301(94), **B**:*312*–*313*
Parisi, G., **B**:273(69), **B**:*312*
Park, J., **B**:248(156), **B**:*255*
Park, K., **B**:181(20), **B**:185-186(20), **B**:*202*
Parlitz, U., **B**:502-503(17), **B**:*518*
Partovi, M. H., **B**:502(22), **B**:517(22), **B**:*518*
Pasta, J., **B**:376(1), **B**:393(1), **B**:*418*
Patashinskii, A. Z., **A**:402(4), **A**:*433*
Pate, B. H., **B**:205(8), **B**:209(8,45), **B**:220(92-93), **B**:*251*–*252*, **B**:*254*

Pattanayak, A. K., **B**:538(46), **B**:*542*
Patterson, C. W., **A**:198(97), **A**:*215*
Pavlichenkov, I. M., **A**:248(69), **A**:*265*
Pear, M. P., **B**:180(15), **B**:*202*
Pearman, R., **B**:217(89-90), **B**:*254*
Pearson, E. M., **B**:55(35), **B**:*84*
Pechukas, P.: **A**:173(16-17), **A**:176(16,35-36), **A**:177(35-36), **A**:186(17), **A**:193-194(35), **A**:*213–214*; **A**:228(39), **A**:230(39), **A**:232(39), **A**:*264*; **B**:106(34), **B**:*128*
Peirce, A. P., **A**:436(12), **A**:*457*
Pelcovits, R. A., **B**:499(41), **B**:*500*
Peng, T.-I., **B**:180(13), **B**:*201*
Percival, I. C.: **A**:5(8-9), **A**:30(8-9), **A**:*140*; **A**:177(41), **A**:*214*; **B**:383(30,35), **B**:*419*
Pérez-Chavela, E., **A**:307(18), **A**:*334*
Perry, D. S., **B**:210(77), **B**:*253*
Persch, G., **B**:210(64), **B**:*253*
Peters, N., **B**:521-522(8), **B**:534(8), **B**:539(8), **B**:*541*
Peterson, K. A.: **A**:287(37,40), **A**:299(40), **A**:*303*, **B**:200(37), **B**:*203*; **B**:207(25), **B**:209(25), **B**:239(25), **B**:*252*
Peterson, R. L., **B**:547(6), **B**:*597*
Petrovskii, I. G., **B**:521-522(9), **B**:*541*
Pettigrew, G. W., **B**:180(11), **B**:*201*
Pettini, M., **B**:394(57), **B**:395(58-59), **B**:*420*
Pettit, B. M., **B**:228(128), **B**:*255*
Phillips, J. C., **B**:489(38), **B**:*500*
Phillips, W. D., **A**:403(13), **A**:*433*
Phillpot, S. R., **B**:221(108), **B**:*254*
Piskunov, N. S., **B**:521-522(9), **B**:*541*
Platt, N., **B**:300(87-88), **B**:*312*
Plotkin, S. S., **A**:164(65), **A**:*170*, **B**:264(44), **B**:302(97), **B**:*311*, **B**:*313*
Pohorill, A.: **B**:250(159), **B**:*255*; **B**:405(75), **B**:*420*
Poincaré, H.: **A**:176(32), **A**:*213*; **A**:306(2), **A**:310(2), **A**:*334*
Pokrovskii, V. L., **A**:402(4), **A**:*433*
Polanyi, M.: **A**:144(4), **A**:*168*; **A**:176(27-28), **A**:*213*; **B**:258(3), **B**:*310*
Polik, W. F., **B**:210(67), **B**:*253*
Pollak, E., **A**:164(64), **A**:*170* **A**:173(17), **A**:176(37), **A**:177(40), **A**:186(17), **A**:*213–214*; **A**:218(5), **A**:228(39), **A**:230(39), **A**:232(39), **A**:*263–264*; **B**:73(45), **B**:*85*; **B**:106(35), **B**:*128*; **B**:502(11), **B**:513(11), **B**:*518*

Polyanski, O. L.: **A**:220(19), **A**:*263*; **A**:269(7-8), **A**:274(7-8), **A**:275(7), **A**:277(7), **A**:*302*
Pomeau, Y.: **A**:293(49), **A**:*303*, **B**:359(18), **B**:*371*; **B**:522(15), **B**:531(15), **B**:*541*
Porjesz, B., **B**:338(19), **B**:341(19), **B**:*351*
Porter, C. E., **B**:210(63), **B**:214(63), **B**:*253*
Posch, H. A., **B**:130(28), **B**:139(28), **B**:*153*
Pöschel, J., **B**:397(70), **B**:*420*
Postma, J. P. M., **B**:267(60), **B**:309(60), **B**:*312*
Potts, A. R., **B**:220(91), **B**:*254*
Poulsen, J. A., **B**:186(27), **B**:*202*
Pratt, L. R.: **B**:250(159), **B**:*255*; **B**:405(75), **B**:*420*
Prigogine, I., **B**:354(9), **B**:*370*
Pritchard, H. O., **A**:198(129), **A**:*216*
Procaccia, I., **B**:7(6-7), **B**:*23*
Provat, A., **B**:526(28), **B**:*542*
Provenzale, A., **B**:300(88), **B**:301(94), **B**:*312–313*
Proykova, A., **B**:26(1), **B**:*83*
Pryer, K. M., **B**:329(14), **B**:*351*
Pugh, C. C., **A**:338(8), **A**:347(8), **A**:*398*

Qi, P. X., **B**:180(14), **B**:*201*
Qian, J., **B**:209(44), **B**:*252*
Quapp, W.: **A**:198(108-109), **A**:*215*; **A**:278(27), **A**:*302*
Quenneville, J., **B**:205(6), **B**:207(6), **B**:220(6), **B**:249(6), **B**:*251*

Rabii, F., **B**:205(7), **B**:207(7), **B**:209(7), **B**:217-218(7), **B**:*251*
Rabitz, H., **A**:436(2,12-15), **A**:437-439(13), **A**:454(15), **A**:456(29), **A**:*457–458*
Radons, G., **A**:128(92), **A**:*142*, **B**:383-384(37), **B**:*419*
Raff, M.: **B**:180(12), **B**:*201*; **B**:558(27), **B**:*598*
Ragazzo, C. G., **B**:427(11), **B**:*436*
Rahman, A.: **B**:135(38), **B**:*153*; **B**:388(44), **B**:*419*
Rahman, N., **A**:456(29), **A**:*458*
Raizen, M. G., **A**:403(13), **A**:*433*
Rajaram, S., **B**:322(7), **B**:*351*
Ramakrishna, V., **A**:436(2), **A**:*457*
Ramani, A., **A**:403(19), **A**:410(19), **A**:*434*
Ramaswany, R.: **B**:31(20), **B**:39(20), **B**:45(20), **B**:*84*; **B**:130(27), **B**:139(27), **B**:*153*
Ramaswarmy, R., **B**:389-390(47), **B**:397(47), **B**:401(47), **B**:*420*

Rammal, R., **B:**230-231(140), **B:**233(140), **B:***255*
Ramsperger, H. C.: **A:**5(1), **A:***140*; **A:**144(1), **A:***168*
Rao, V. S., **A:**198(131), **A:***216*
Rapisarda, A., **B:**479(15,18,20,23), **B:**480(15,18), **B:**481(15,18,31), **B:**485(37), **B:**487(15), **B:***500*
Ratiu, T., **A:**221(22), **A:**248(22), **A:***264*
Ratner, M. A.: **B:**182(23), **B:**186(28), **B:***202*; **B:**238(144), **B:***255*
Rayez, J. C., **A:**249(75), **A:***265*
Rechester, A. B., **B:**441(15-16), **B:***463*
Rector, K. D., **B:**207(28), **B:**209(28), **B:**239(28), **B:***252*
Redmon, L. T., **A:**277(15), **A:***302*
Redner, S., **B:**526(28), **B:***542*
Regan, J. J., **B:**201(45), **B:***203*
Rego, L. G. C., **B:**221(104), **B:***254*
Reichl, L. E., **B:**83(51), **B:***85*
Reid, B. P., **A:**65(48), **A:***141*
Reinhardt, P., **B:**140(40), **B:***153*
Reinhardt, W. P.: **A:**166(68-69), **A:***170*; **A:**198(85), **A:***215*; **B:**187(29), **B:**200(29), **B:***202*
Reinsch, M., **A:**220(17), **A:**248(17), **A:**250(17), **A:***263*, **B:**88(7), **B:**89(7,12,14), **B:**91(7), **B:**93-95(7), **B:**108(7), **B:**114(14), **B:**115(12,14), **B:***127*
Rella, C. W.: **B:**200(37), **B:***203*; **B:**207(25,28), **B:**209(25,28), **B:**239(25,28), **B:***252*
Remacle, F., **A:**116(87), **A:***142*
Rencher, A. C., **B:**316(1), **B:**319(1), **B:**330(1), **B:***351*
Renzaglia, K. S., **B:**332(15), **B:***351*
Rey, R., **B:**180(7), **B:***201*
Rice, O. K.: **A:**5(1,3), **A:***140*; **A:**144(1), **A:***168*; **B:**221(98), **B:***254*
Rice, S. A.: **A:**6(12-16,19), **A:**8(30), **A:**20(32), **A:**22(35), **A:**27(35), **A:**39(12), **A:**41(13-14), **A:**59-60(12), **A:**65(51), **A:**66(16,55), **A:**70(15), **A:**85(15), **A:**87(58), **A:**95(64), **A:**99(60), **A:**97-98(19), **A:**104(67), **A:**108(75), **A:**124(88), **A:***140–142*; **A:**145(18-19), **A:**147(55), **A:**153(55), **A:**167(76), **A:***169–170*; **A:**178(52,59), **A:**198(91-93,122-126), **A:***214–216*; **A:**218(1), **A:**222(28), **A:**228(1,28), **A:**232(46), **A:***263–264*; **A:**307(11),
A:330(11), **A:***334*; **A:**436(4,9), **A:**437-438(4), **A:***457*; **B:**205(9), **B:**209(9,46), **B:**210(68-69), **B:***251–253*; **B:**260(11-12), **B:***310*
Richter, K.: **A:**219(14), **A:**244(14), **A:***263*; **A:**306(8), **A:**307(12,17), **A:***334*; **A:**308(20), **A:***334*
Rick, S. W., **B:**130(22), **B:**139(22), **B:***153*
Risser, S. M., **B:**201(45), **B:***203*
Rist, C., **A:**256(84), **A:***265*
Roberts, G., **B:**427(13), **B:***436*
Roberts, K.: **B:**180(12), **B:***201*; **B:**558(27), **B:***598*
Roberts, R. M., **A:**248(64,70-71), **A:***265*
Robinson, P. J., **A:**176(31), **A:***213*, **B:**215(85), **B:***253*
Robledo, A., **B:**364(15), **B:***371*
Rogaski, C. A., **A:**278(23), **A:***302*
Roitberg, A.: **B:**186(28), **B:***202*; **B:**238(144), **B:***255*
Rolston, S. L., **A:**403(13), **A:***433*
Romanini, D., **A:**278(26), **A:***302*
Römelt, J., **B:**106(35), **B:***128*
Romesberg, F., **B:**180(10), **B:**199(10), **B:***201*
Rom-Kedar, V., **A:**180(77), **A:**182(77), **A:***215*
Ronney, P. D., **B:**534(39), **B:***542*
Rosca, F., **B:**180(14), **B:**200(35), **B:***201–202*
Rose, J. P., **B:**266(56), **B:**284(56), **B:***311*
Rosenbluth, M. N., **B:**441(16), **B:***463*
Ross, B. D., **B:**207(29), **B:**219(29), **B:**249(29), **B:***252*
Ross, D. T., **B:**348(21), **B:***351*
Ross, J., **B:**522(12), **B:***541*
Ross, S. D.: **A:**168(77), **A:***170*; **A:**173(10), **A:**198(86), **A:***213*, **A:***215*; **A:**219(13), **A:**234(13), **A:**248(68), **A:***263*, **A:***265*; **A:**340(22), **A:***399*
Rossky, P. J., **B:**186(27), **B:***202*
Rost, J.-M., **A:**306(8), **A:**307(15), **A:**330(34), **A:***334–335*
Rost, S., **A:**219(14), **A:**244(14), **A:***263*
Roukes, M. L., **B:**221(100-102), **B:***254*
Rousseau, D. L., **B:**180(14), **B:***201*
Roux, B.: **A:**190(33), **A:***202*; **B:**261(22), **B:***311*
Rubinsztein-Dunlop, H., **A:**403(13), **A:***433*
Rubner, J., **A:**128(92), **A:***142*
Ruchayskiy, O., **B:**522(16), **B:***541*
Ruedenberg, K., **B:**131(32), **B:**132(32), **B:***153*
Ruelle, D.: **B:**273(69), **B:***312*; **B:**502(1), **B:***517*
Ruf, B. A., **A:**402(6), **A:***433*

Ruffo, S.: **B**:395(58,62-63), **B**:*420*;
  B:479(11-13,18,23), **B**:480(11-12,18),
  B:481(11-12,18,31,35), **B**:485(37),
  B:489(39), **B**:*500*
Rupley, J. A., **B**:221(155), **B**:248(154-155),
  B:*255*
Rüssmann, H., **B**:396(65), **B**:*420*
Ruth, H. H., **B**:502(20), **B**:*518*
Ryne, R. D., **B**:502(21), **B**:517(21), **B**:*518*
Ryzhik, L., **B**:522(14,16), **B**:531-533(14),
  B:*541*

Sadeghi, R., **A**:232(44), **A**:*264*
Sadovskii, D. A.: **A**:248(66), **A**:*265*;
  A:269(4-5), **A**:*302*
Sagnella, D. E.: **B**:180(8), **B**:200(36),
  B:*201–202*; **B**:228(135-136),
  B:248(135-136), **B**:*255*
Saito, N.: **B**:381(25), **B**:*419*; **B**:466(7),
  B:470-471(7), **B**:474(7), **B**:*475*
Saito, S.: **B**:229(139), **B**:*255*; **B**:262(31,34),
  B:265(31), **B**:300(31), **B**:*311*; **B**:378(14),
  B:390(49), **B**:391(54), **B**:406(78,81),
  B:409(14), **B**:411-412(14), **B**:*418,*
  B:*420–421*; **B**:477(1), **B**:*499*
Sakurai, J. J., **B**:182(23), **B**:*202*
Sander, C., **B**:228-229(127), **B**:*255*
San Juan, J. F., **A**:238(57), **A**:*265*
Sannami, A., **B**:380(22), **B**:*419*
Sano, M., **B**:502(2), **B**:*517*
Sano, M. M., **A**:322-325(29), **A**:328(29),
  A:*335*
Santoprete, M., **B**:427(15), **B**:*436*
Sasa, S., **B**:354(8), **B**:*370*
Sasai, M., **A**:165(66), **A**:*170*, **B**:389-390(47),
  B:397(47), **B**:401(47), **B**:*420*
Sastry, S., **B**:392(56), **B**:*420*
Sathyamurthy, N., **A**:228(42), **A**:*264*
Sato, F., **B**:180(19), **B**:*202*
Sato, K.: **B**:466(10), **B**:471-472(10), **B**:*475*;
  B:594(43), **B**:*598*
Sauer, T. D.: **B**:285(75), **B**:300(75),
  B:307(75,99), **B**:308-309(75), **B**:*312–313*;
  B:502(3-4,14), **B**:*517–518*
Sawada, S., **B**:156(4,8-9), **B**:159-160(11),
  B:165(4), **B**:167(8,11), **B**:*176*
Sawada, Y., **B**:502(2), **B**:*517*
Scala, A., **B**:391(55), **B**:*420*
Schaefer, H. F. III, **A**:198(90), **A**:*215*

Schatz, C., **B**:182(23), **B**:*202*
Schelling, P. K., **B**:221(108), **B**:*254*
Scher, H., **B**:228(121), **B**:*254*
Scherer, G. J., **A**:198(87-90), **A**:*215*
Schinke, R.: **A**:257(85), **A**:*265*; **A**:280(31),
  A:287(33-36,38,41), **A**:288(34,41),
  A:291(41), **A**:293(34-36,38,41,52),
  A:294(41), **A**:296-297(41),
  A:298(35-36,38,41), **A**:299(35,38,41),
  A:301(35,38,41), **A**:*303*
Schlagheck, P., **A**:403(11), **A**:*433*
Schlesinger, M. F., **B**:524(20-21), **B**:*541*
Schlier, C., **B**:43(28), **B**:*84*
Schmelcher, P., **B**:209(61), **B**:*253*
Schmidt, P. P., **A**:198(139), **A**:*216*
Schofield, S. A.: **B**:180(2), **B**:*201*; **B**:207(23),
  B:209-210(23), **B**:214(23,83-84),
  B:216(84), **B**:*252–253*
Scholz, H. J., **A**:249(76), **A**:*265*
Schranz, H. W., **B**:200(38), **B**:*203*
Schreiber, T., **B**:285(78), **B**:294(78),
  B:*312*
Schroder, T. B., **B**:392(56), **B**:*420*
Schubart, J., **A**:324(33), **A**:*335*
Schuler, G., **B**:348(21), **B**:*351*
Schulman, L. S., **A**:402(5), **A**:*433*
Schulte, A., **B**:228(115), **B**:*254*
Schulten, K., **B**:206(12), **B**:248(153), **B**:*251,*
  B:*255*
Schultz, S. L., **B**:209(44), **B**:*252*
Schuster, P., **B**:547-548(7), **B**:*597*
Schuttenmaer, C. A., **B**:406(80), **B**:*421*
Schwab, K., **B**:221(101), **B**:*254*
Schwartz, S. D., **A**:111(78), **A**:*142*, **B**:206(14),
  B:*251*
Schweiters, C. D., **A**:456(29), **A**:*458*
Schwenke, D. W., **A**:174(23), **A**:178(23),
  A:*213*
Schwettman, H. A., **B**:207(25,28),
  B:209(25,28), **B**:239(25,28), **B**:*252*
Sciortino, F., **B**:391(55), **B**:*420*
Scoles, G., **B**:209(45), **B**:*252*
Scott, S. K., **A**:389(40), **A**:*399*
Seckler, B., **A**:402(4), **A**:*433*
Segal, D., **B**:221(106), **B**:*254*
Segré, D., **B**:573(34), **B**:579(34), **B**:*598*
Seideman, T., **A**:144(12), **A**:*168*
Sekimoto, K., **B**:354-355(7), **B**:*370*
Sekine, S., **B**:402(74), **B**:*420*

Seko, C.: **B**:27(8,10), **B**:28(11), **B**:30(19), **B**:33(19), **B**:36(19), **B**:39(19), **B**:43(29-30), **B**:45(19), **B**:47(29), **B**:49(10), **B**:51(10), **B**:52(30), **B**:56(19), **B**:58(30), **B**:60(19,30), **B**:61-63(11), **B**:66(19), **B**:68(19), **B**:67(11), **B**:*84*; **B**:90(24), **B**:*127*; **B**:130(17-18,20,29), **B**:139(17-18,20,29), **B**:140(17), **B**:143(17), **B**:*153*; **B**:270-271(62), **B**:*312*
Seligman, T. H., **A**:250(78), **A**:*265*
Sepúlveda, M. A., **B**:502(11), **B**:513(11), **B**:*518*
Serva, M., **B**:310(91), **B**:*313*
Shafin, W., **A**:61-62(40), **A**:*141*
Shah, S. P., **A**:198(124-125), **A**:*216*
Shalon, D., **B**:348(21), **B**:*351*
Shannon, C., **B**:545(3), **B**:*597*
Shanz, H., **A**:136(98), **A**:*142*
Shapere, A., **B**:88(3), **B**:*127*
Shapiro, M.: **A**:8(31), **A**:*141*; **A**:436(10), **A**:*457*
Shatalov, V. E., **A**:410(31), **A**:*434*
Shavitt, I., **A**:277(15), **A**:*302*
Sheeran, M., **B**:180(14), **B**:*201*
Shen, D., **A**:198(129), **A**:*216*
Shen, J., **B**:180(15), **B**:*202*
Shen, T. Y., **B**:228(118), **B**:*254*
Sheng, P., **B**:208(41), **B**:250(41), **B**:*252*
Shepard, R. N., **B**:317(3), **B**:319(3), **B**:*351*
Shepelyanski, D. L.: **B**:383(27), **B**:*419*; **B**:477(5), **B**:*499*
Shewmon, P. G., **B**:160(13), **B**:167(13), **B**:*177*
Shi, K.-J., **A**:442(23), **A**:*457*
Shi, Q., **B**:186(26), **B**:*202*
Shibata, T., **B**:301(89-90), **B**:*312-313*
Shibuya, N., **B**:551(18), **B**:567(18), **B**:570(18), **B**:*598*
Shida, N.: **B**:402(6), **A**:*433*, **B**:30(17), **B**:*84*; **B**:131-132(34), **B**:*153*; **B**:259(8), **B**:*310*
Shiga, M., **B**:186-187(24), **B**:189(24), **B**:193(24), **B**:*202*
Shima, Y., **B**:551(18), **B**:567(18), **B**:570(18), **B**:*598*
Shimizu, Y., **B**:156(8-9), **B**:159-160(11), **B**:167(8,11), **B**:168(11), **B**:*176*
Shimono, M., **B**:*311*
Shinbrot, T., **A**:456(28), **A**:*457*
Shirai, H., **A**:165(66), **A**:*170*
Shirts, R. B.: **A**:166(68-69), **A**:*170*; **A**:198(97-98), **A**:*215*
Shivanshinsky, G. I., **B**:534(28), **B**:*542*

Shizume, K., **A**:140(102), **A**:*142*
Shleesinger, M. F., **B**:264(48-49), **B**:*311*
Shore, B. W., **A**:436(8), **A**:*457*
Shore, J. E., **B**:73(44), **B**:*85*
Shrake, A., **B**:217(88), **B**:*254*
Shtilerman, M., **B**:180(14), **B**:*201*
Shub, M., **A**:338(8), **A**:347(8), **A**:*398*
Shudo, A.: **A**:403(15-18), **A**:406(17), **A**:409(15), **A**:410(35), **A**:428(16-18), **A**:*433-434*, **B**:262(34), **B**:*311*; **B**:378(14), **B**:409(14), **B**:411-412(14), **B**:*418*
Sibert, E. L. III: **A**:198(111-112,141), **A**:*215-216*; **A**:269(2), **A**:274(13), **A**:*302*, **B**:187(29), **B**:200(29), **B**:*202*; **B**:210(78), **B**:*253*
Siegel, S. L, **A**:306(3), **A**:*334*
Sigmund, K., **B**:582(35), **B**:*598*
Simó, C., **B**:376(5), **B**:*418*
Simó, Carles, **A**:173(20), **A**:175(20), **A**:*213*
Simonovic, N., **A**:330(34), **A**:*335*
Sinai, Ya. G., **B**:387(41), **B**:*419*
Singer, S. J.: **B**:205(7), **B**:207(7), **B**:209(7), **B**:217-218(7), **B**:*251*; **B**:217(87), **B**:*254*
Sivakumer, N., **A**:63(46), **A**:65(48), **A**:*141*
Sjodin, T., **B**:180(14), **B**:200(35), **B**:*201-202*
Skene, J. M., **A**:66(52-54), **A**:*141*
Skinner, J. L., **B**:180(6), **B**:181(20), **B**:183-184(6), **B**:185(6,20), **B**:186(20), **B**:*201-202*
Skodje, R. T.: **A**:232(44), **A**:*264*; **A**:402(6), **A**:*433*
Skog, J. E., **B**:329(14), **B**:*351*
Skokov, S., **A**:287(37,39), **A**:288(39), **A**:293(39), **A**:*303*
Sligar, S. G.: **B**:200(35), **B**:*202*; **B**:228(114), **B**:*254*
Smale, S., **A**:388(39), **A**:*399*
Smith, A. M., **A**:278(22), **A**:*302*
Smith, A. R., **A**:329(14), **B**:*351*
Smith, D. J., **B**:156(2), **B**:*176*
Smith, J. Maynard, **B**:573(30), **B**:*598*
Smith, K. S., **A**:198(97-98), **A**:*215*
Snijder, J. G., **B**:200(42), **B**:*203*
Sokolov, I. M., **B**:264(51), **B**:*311*
Solina, S. A. B., **B**:210(66), **B**:*253*
Solomon, T. H.: **B**:479(25), **B**:*500*; **B**:527(31), **B**:*542*
Someda, K., **B**:200(40), **B**:*203*

Song, K.: **A**:168(78), **A**:*170*; **A**:249(73), **A**:*265*; **B**:259(7), **B**:*310*
Sornette, D., **B**:591(40), **B**:*598*
Sosnick, T. R., **B**:180(14), **B**:*201*
Souganidis, P. E., **B**:534(36), **B**:*542*
Sparpaglione, M., **B**:395(58), **B**:*420*
Spiegel, E. A., **B**:300(87-88), **B**:*312*
Spiegelman, S., **B**:547(6), **B**:*597*
Spohn, H., **B**:354(5), **B**:*370*
Srinivasan, A. R., **B**:261(23), **B**:*311*
Stamatescu, I. O., **A**:140(101), **A**:*142*
Stamatiadis, S.: **A**:228(42), **A**:*264*; **A**:287(38), **A**:293(38), **A**:298-299(38), **A**:301(38), **A**:*303*
Stanley, H. E., **B**:391(53,55), **B**:*420*
Stapleton, H. J., **B**:228(133-134), **B**:*255*
Starr, F. W., **B**:391(53,55), **B**:*420*
States, D. J., **B**:190(33), **B**:*202*
Staudt, L. M., **B**:348(21), **B**:*351*
Stauffer, J. R., **B**:324(9), **B**:*351*
Steck, D. A., **A**:403(13), **A**:*433*
Stein, D. L., **B**:499(40), **B**:*500*
Steinbach, P. J., **B**:228(115), **B**:*254*
Steinfeld, J. I.: **B**:27-28(7), **B**:71-72(7), **B**:75(7), **B**:79-80(7), **B**:*84*; **B**:90(29), **B**:*127*; **B**:180(1), **B**:*201*
Stern, P. S., **B**:228-229(127), **B**:*255*
Sternin, B. Y., **A**:410(31), **A**:*434*
Stevens, D.: **B**:479(27), **B**:*500*; **B**:527(30), **B**:*542*
Stewart, G. M., **B**:214(81), **B**:*253*
Stillinger, D., **B**:130(23), **B**:139(23), **B**:*153*
Stillinger, F. H.: **B**:30(18), **B**:35(18), **B**:*84*; **B**:90(28), **B**:*127*; **B**:130(23), **B**:131(31), **B**:139(23), **B**:*153*; **B**:262(32-33), **B**:*311*; **B**:388(44-45), **B**:411(82), **B**:*419*, **B**:*421*
Stinson, D. G., **B**:228(133), **B**:*255*
Stock, G., **B**:200(39), **B**:*203*
Stokes, G. G., **A**:410(31), **A**:*434*
Strang, G., **B**:93(33), **B**:*128*
Stratt, R. M.: **B**:229(139), **B**:*255*; **B**:391(54), **B**:*420*
Straub, E., **B**:180(1), **B**:200(36), **B**:*201-202*
Straub, J. E.: **A**:144(12), **A**:*168*, **B**:180(8-9,16), **B**:197(9), **B**:*201-202*; **B**:205(2), **B**:228(135-137), **B**:248(135-136), **B**:*251*, **B**:*255*
Strelcyn, J. M.: **B**:31(22), **B**:46(22), **B**:*84*; **B**:396(66), **B**:*420*; **B**:502(19), **B**:506(19), **B**:*518*

Stuart, J. M., **B**:348(20), **B**:*351*
Stuchebruckhov, A. A., **B**:209(48-49), **B**:*252*
Stuchebrukhov, A., **B**:180(1), **B**:*201*
Stuchi, T. J., **A**:307(19), **A**:*334*
Stumpf, M., **A**:280(31), **A**:*303*
Suárez, D., **B**:201(46), **B**:*203*
Suen, J., **B**:354(12), **B**:*371*
Sugano, S., **B**:156(1,4), **B**:*176*
Sugny, D.: **A**:198(141), **A**:*216*; **A**:219(10), **A**:*263*; **A**:269(1-3), **A**:274(1), **A**:278(3), **A**:287-288(34), **A**:293(34), **A**:*302–303*
Sulem, P.-L., **B**:502-503(9), **B**:*518*
Sun, L., **A**:168(78), **A**:*170*, **B**:259(7), **B**:*310*
Sun, X., **A**:115(84-86), **A**:*142*
Sutcliffe, B. T., **B**:89(18), **B**:106(18), **B**:*127*
Suzuki, T., **B**:593-595(41), **B**:*598*
Swaminathan, S., **B**:190(33), **B**:*202*
Swimm, R. T., **A**:278(21), **A**:*302*
Swinney, H. L.: **B**:292-293(83), **B**:*312*; **B**:479(25), **B**:*500*
Syage, J. A., **B**:220(94), **B**:*254*
Szalay, V., **A**:198(107), **A**:*215*
Szathmary, E., **B**:573(30), **B**:*598*

Tabor, M.: **A**:288(45), **A**:293(45), **A**:*303*; **A**:403(20), **A**:*434*; **B**:83(51), **B**:*85*
Tachibana, A., **B**:88(5), **B**:*127*
Taddei, N., **B**:264(46), **B**:*311*
Taguchi, Y.-H., **B**:317(4), **B**:*351*
Tai, K., **B**:228(118), **B**:*254*
Takada, S., **B**:264(39), **B**:285(74), **B**:299(39), **B**:*311–312*
Takagi, H., **B**:583(38), **B**:*598*
Takahashi, K.: **A**:403(16-17), **A**:404(22-25), **A**:405(23-25,27), **A**:406(17,25), **A**:407(22, **A**:24,27), **A**:408-409(22), **A**:410(22,25), **A**:412(22), **A**:413(24-25), **A**:417-418(25), **A**:423(25), **A**:425(22-23), **A**:428(16-17,24), **A**:429(24), **A**:431(25), **A**:*433–434*; **A**:443(24), **A**:*457*
Takahashi, S., **B**:83(56), **B**:*85*
Takami, T., **A**:437(21), **A**:443(25-26), **A**:449(26), **A**:*457*
Takano, K., **A**:198(140), **A**:*216*
Takano, M., **B**:272(64-65), **B**:274(65), **B**:*312*
Takatsuka, K.: **A**:220(18), **A**:244-245(18), **A**:*263*; **A**:392(43), **A**:*399*; **B**:27(8,10), **B**:28(11,15), **B**:30(19), **B**:32(23-24), **B**:33(19), **B**:34(25), **B**:36(19), **B**:39(19), **B**:43(29), **B**:44(30), **B**:45(19), **B**:47(29),

AUTHOR INDEX

B:49(10), B:51(10), B:52(30), B:56(19),
B:58(30), B:60(19,30), B:61-63(11),
B:64(38), B:66(19), B:68(19), B:67(11),
B:83(48-50,53-56), B:*84–85*; B:90(24),
B:*127*; B:130(17-18,20,29),
B:139(17-18,20,29), B:140(17),
B:143(17), B:*153*; B:259(9), B:*310*;
B:270-271(62), B:*312*
Takayanagi, K., B:156(2), B:*176*
Takeda, K., B:156-157(7), B:*176*
Takens, F., B:285(76), B:302(76), B:*312*
Taketsugu, T., A:198(140), A:*216*
Talbi, D., A:198(130), A:*216*
Talkner, P., A:228(37), A:*264*
Tamarit, F., B:480-481(30), B:*500*
Tanaka, A., A:456(32), A:*459*
Tanaka, H.: B:265(53), B:299(53), B:*311*;
  B:378(8-9), B:388(8-9), B:390(8-9),
  B:408(8-9), B:409(9), B:*418*; B:502(10),
  B:*518*
Tanaka, K.: B:270(63), B:273(63,67), B:*312*;
  B:477(7), B:*499*
Tang, H.: A:6(19), A:97-98(19), A:104(67),
  A:*140*, A:*142*; A:198(122), A:*216*
Tang, K. T., A:252(82), A:*265*
Tang, X. Z., B:502(16), B:*518*
Tanikawa, K., A:323(31-32), A:330(37), A:*335*
Tanimura, Y., B:200(41), B:*203*
Tanishiro, Y., B:156(2), B:*176*
Tanner, G.: A:219(14), A:244(14), A:*263*;
  A:306(8), A:307(12,14,17), A:308(24),
  A:330(14), A:333(24), A:*334–335*
Tannor, D. J., A:436(9), A:*457*
Taraskin, S. N., B:208(42), B:250(42), B:*252*
Tarn, T. J., A:436(11), A:*457*
Taylor, G. I., B:523(19) 526(25), B:*541*
Taylor, H. S.: A:198(100-102,104), A:*215*;
  A:278(25), A:*302*
Teller, E., B:499(44), B:*500*
Tempkin, J. A., B:502(3), B:*517*
Tenenbaum, A., B:395(61), B:*420*
Tennyson, J.: A:198(99), A:*215*; A:220(19),
  A:248(64,70), A:*263*, A:*265*;
  A:269(5,7-8), A:274(7-8), A:275(7),
  A:277(7), A:*302*
Teramoto, H., B:83(50), B:*85*
Teranishi, Y., A:436(7), A:456(31), A:*457–458*
Tersigni, S. H., A:20(32), A:*141*
Tesch, C. M., A:436(3), A:*457*
Tesch, M., B:248(153), B:*255*

Thayer, B. D., B:228(134), B:*255*
Theuer, H., A:436(8), A:*457*
Thiffeault, J.-L., B:502(16), B:503(23), B:*518*
Thirumalai, D.: B:228(117,135), B:248(135),
  B:*254–255*; B:266-267(58), B:*311*
Thommen, F., A:63(44-45), A:*141*
Thompson, D. E., B:200(43), B:*203*
Thouless, D. J., B:212(80), B:*253*
Thurston, W., B:380(17), B:*419*
Tiller, W. A., B:55(35), B:*84*
Tinghe, T. S., B:221(100), B:*254*
Tiyapan, A., A:177(46-47), A:*214*
Toda, M.: A:6(17), A:21(17), A:138(17), A:*140*;
  A:145(22-24), A:147(56), A:*169–170*;
  A:234(49), A:245(62), A:247(62),
  A:*264–265*; A:338(1), A:339(12),
  A:340(1), A:358(12), A:372(1),
  A:378(37), A:381(12), A:*398–399*;
  B:46(32), B:56(32), B:58(32), B:*84*;
  B:205(11), B:209(11), B:*251*; B:274(73),
  B:301(73), B:*312*; B:438(3), B:*462*
Toennies, J. P., A:252(82), A:*265*
Togashi, Y., B:561(29), B:*598*
Tokmakoff, A., B:180(5), B:183-184(5),
  B:189-190(5), B:*201*
Toller, M., A:173(9), A:*213*
Tombor, B., B:453(23), B:*463*
Tomita, K., B:502(13), B:*518*
Tomsovic, S., A:403(11), A:*433*,
  B:210(72), B:*253*
Topper, R. Q.: A:75(24), A:88(59), A:90(59),
  A:*140–141*; A:146(35), A:153(35), A:*169*;
  B:48(33), B:82(33), B:*84*
Torcini, A.: B:479(16), B:485(37), B:494(16),
  B:*500*; B:534(37), B:537(37), B:539(37),
  B:*542*
Toulouse, G., B:230-231(140), B:233(140),
  B:*255*
Trautmann, D., A:250(78), A:*265*
Trent, J. M., B:348(21), B:*351*
Tresser, C., B:300(87-88), B:*312*
Tribus, M., B:28(13), B:72(13), B:*84*
Tromp, J. W., A:111(78), A:*142*
Trosset, M. W., B:*351*
True, N. S., B:207(29), B:219(29), B:249(29),
  B:*252*
Truhlar, D. G.: A:144(10,14), A:163(10,61),
  A:*168–170*; A:174(23), A:178(23), A:*213*;
  A:218(2), A:*263*; A:402(6), A:*433*;
  B:258(6), B:300(6,86), B:*310*, B:*312*

Tsai, C. J., **B:**171(14), **B:***177*
Tsallis, C.: **B:**299-300(85), **B:***312*; **B:**364(14), **B:***371*; **B:**479(15,20), **B:**480-481(15), **B:**487(15), **B:***500*
Tsuchiya, T., **A:**340(21), **A:**388(21), **A:***398*, **B:**479(17), **B:***500*
Tsuda, I., **B:**596(45-46), **B:***598*
Tufillaro, N. B., **B:**299-300(84), **B:***312*
Tyng, V., **A:**287-288(34), **A:**293(34), **A:***303*

Uchimaru, T., **A:**198(132), **A:***216*
Ueno, Y., **B:**201(47), **B:***203*
Ulam, S., **B:**376(1), **B:**393(1), **B:***418*
Ullmo, D., **A:**403(11), **A:***433*, **B:**210(72), **B:***253*
Ullo, J. J., **B:**406(77), **B:***421*
Umeda, H., **A:**198(136-137), **A:***216*
Umehara, N., **A:**330(37), **A:***335*
Upcroft, B., **A:**403(13), **A:***433*
Urabe, I., **B:**551(18-21), **B:**567(18), **B:**570(18), **B:***598*
Uzer, T.: **A:**7(22), **A:**21(22), **A:***140*; **A:**144(15), **A:**147(15,54,58), **A:**148(15,54), **A:**163(15), **A:***169–170*; **A:**173(6-7,10,18-19,21), **A:**177(6,18), **A:**180-181(6,18), **A:**182(78), **A:**212(19,21), **A:***213,* **A:***215*; **A:**219(8-9,13), **A:**221(8), **A:**233(9), **A:**234(8,13,51), **A:**235(9), **A:**237(9), **A:**248(64), **A:**250(8), **A:***263–265*; **A:**338-339(3), **A:**341(3), **A:**352(3), **A:***398*; **A:**428(38), **A:***434*; **B:**180(1), **B:***201*; **B:**209(52), **B:***252*

Valdinoci, E., **B:**427(17), **B:***436*
Valiron, P.: **A:**245(58), **A:**248(58), **A:**256(58), **A:***265*; **A:**256(84), **A:***265*
Van der Meer, A. G. F., **B:**207(27), **B:**209(27), **B:***252*
Van der Meer, L., **B:**207(26), **B:**209(26), **B:***252*
Van der Vaart, A., **B:**201(46), **B:***203*
van der Zwan, G., **A:**164(63), **A:***170*
Van Dishoek, E. F., **A:**256(81), **A:***265*
Van erp, T. S., **A:**232(43), **A:***264*
Van Gunsteren, W. F., **B:**267(60), **B:**273(70), **B:**309(60), **B:***312*
van Hecke, Ch., **A:**248(66), **A:***265*
Van Loan, C. F., **B:**502-504(18), **B:**506(18), **B:**510(18), **B:**517(18), **B:***518*
Van Mourik, T., **A:**269(7), **A:**274-275(7), **A:**277(7), **A:***302*

Van Vleck, J. H.: **A:**272(10), **A:**277(10), **A:***302*; **A:**306(7), **A:***334*
Vanzini, S., **B:**397(72), **B:**402(97), **B:***420*
Vecheslavov, V. V., **B:**441(17-18), **B:***463*
Vekhter, B., **B:**266(56), **B:**284(56), **B:***311*
Vela-Arevalo, L. V., **A:**167(73), **A:***170*
Vergassola, M., **B:**520(3), **B:**521(5), **B:**522(18), **B:**523(3), **B:**525(5,23), **B:***541*
Vergni, D., **B:**522(13), **B:**527(29), **B:**531-533(13), **B:**534(37), **B:**537(37), **B:**539(37), **B:***541–542*
Vetterli, M., **B:**316(2), **B:***351*
Viano, G. A., **A:**249(77), **A:***265*
Viartola, A., **A:**238(57), **A:***265*
Vidal, C., **A:**293(49), **A:***303*, **B:**522(12), **B:***541*
Vijay, A., **A:**198(131), **A:***216*
Villa, J., **A:**163(61), **A:***170*, **B:**300(86), **B:***312*
Vittot, M., **B:**396(67), **B:***420*
Vivaldi, F., **B:**438(6), **B:***462*
Vladimirova, N., **B:**522(16), **B:***541*
Von Hardenberg, J. G., **B:**300(88), **B:***312*
Von Zeipel, H., **A:**310(26), **A:***335*
Voros, A., **A:**410(33), **A:***434*
Vulpiani, A.: **B:**310(91-94), **B:***313*; **B:**395(58), **B:***420*; **B:**521(4-5), **B:**522(13), **B:**525(5), **B:**526(26), **B:**527(26,29), **B:**531-533(13), **B:**534(37), **B:**537(4,37), **B:**539(37,45), **B:***541–542*

Wadi, H., **A:**222(33), **A:**232(47), **A:***264*
Wagner, A. F.: **A:**198(135), **A:***216*; **A:**402(6), **A:***433*
Wagner, C., **B:**254(43), **B:***311*
Wagner, G. C., **B:**228(134), **B:***255*
Waite, B. A., **A:**198(83-84), **A:***215*
Wales, D. J.: **A:**146(25), **A:***169*; **A:**178(54,57), **A:***214*; **A:**234(50), **A:**245(50), **A:***264*; **B:**10(8), **B:**12(10), **B:***23*; **B:**26(1), **B:**56(37), **B:**66(39), **B:***83–84*; **B:**90(25,27), **B:**106(25), **B:***127*; **B:**130(11-12,14,19,24-25), **B:**139(11-12,14,19,24-25), **B:**143(12), **B:***152–153*; **B:**209(57), **B:**250(158), **B:***253,* **B:***255*; **B:**258(1), **B:**266(57), **B:**267(1,57), **B:**284(57), **B:**301(95), **B:***310–311,* **B:**313
Walker, D. M., **B:**299-300(84), **B:***312*
Walker, R. B., **A:**257(85), **A:***265*
Wallenberg, R., **B:**156(2), **B:***176*

Walsh, T. R., **B:**266-267(57), **B:**284(57), **B:***311*
Walter, J., **A:**172(3), **A:**176(3), **A:***213*
Wang, H., **A:**115(85-86), **A:***142*
Wang, Q., **B:**380(21), **B:**394(21), **B:***419*
Wang, W.: **B:**180(14), **B:***201*; **B:**338(19), **B:**341(19), **B:***351*
Wang, X., **B:**180(13), **B:***201*
Wannier, G. H., **A:**307(16), **A:***334*
Warren, W. S., **A:**436(6), **A:***457*
Watanabe, M.: **B:**140(40), **B:***153*; **B:**261(22), **B:***311*
Waterland, R. L., **A:**66(53), **A:***141*
Watson, J. D., **B:**558(27), **B:***598*
Wayne, E., **B:**397(68), **B:***420*
Wazawa, T., **B:**265(55), **B:**299(55), **B:***311*
Weaver, D. L., **B:**264(45), **B:***311*
Weaver, W., **B:**545(3), **B:***597*
Weber, T. A.: **B:**30(18), **B:**35(18), **B:***84*; **B:**90(28), **B:***127*; **B:**131(31), **B:***153*; **B:**262(32 33), **B:***311*; **B:**388(45), **B:***419*
Weeks, E. R., **B:**479(25), **B:***500*
Weiss, J., **A:**287(38), **A:**298-299(38), **A:**301(38), **A:***303*
Weiss, S., **B:**265(54), **B:**299(54), **B:***311*
Weissman, M. C., **B:**273(66), **B:***312*
Weitzner, H.: **B:**479(27), **B:***500*; **B:**527(30), **B:***542*
Weizer, V. G., **B:**159(12), **B:***176*
West, B., **B:**524(21), **B:***541*
Westerberg, K. M., **A:**198(138), **A:***216*
Weyl, H., **A:**105(69), **A:***142*
Wharton, D., **B:**200(35), **B:***202*
Wharton, L., **A:**61-62(39), **A:***141*
Whelan, N. D., **A:**403(11,14), **A:**404(14), **A:***433*
Whetten, R. L., **B:**90(23), **B:***127*
White, R. B., **B:**441(15-16), **B:***463*
Whiteley, T. W. J., **A:**249(74), **A:***265*
Whitnell, R. M., **A:**280(30), **A:***303*, **B:**209(55-56), **B:***253*
Whitney, H., **B:**302(98), **B:**304-305(98), **B:***313*
Wiberg, K. B., **B:**217(88), **B:***254*
Wiedenmüller, H. A., **B:**210(73-74), **B:***253*
Wiesenfeld, L.: **A:**7(22), **A:**21(22), **A:***140*; **A:**144(15), **A:**147(15,57), **A:**148(15), **A:**163(15), **A:**167(70), **A:***169–170*; **A:**173(19), **A:**212(19), **A:***213*; **A:**219(8), **A:**221(8,26), **A:**222(33), **A:**228(38,41), **A:**232(41,47), **A:**234(8), **A:**236(53), **A:**237(53), **A:**238(56), **A:**245(58-59,62), **A:**247(62), **A:**248(56,58), **A:**249(41),

A:250(8,26), **A:**256(58), **A:**257-258(59), **A:***263–265*; **A:**428(38), **A:***434*
Wiggins, S.: **A:**7(20-22), **A:**20(20-21), **A:**22(21), **A:***140*; **A:**144(15), **A:**145(21), **A:**147(15,53-54), **A:**148(15,54), **A:**163(15,21,53), **A:**167(73), **A:***169–170*; **A:**173(14,19,21), **A:**174(22), **A:**179(14,22), **A:**183(22), **A:**187(22), **A:**203(22), **A:**212(19,21), **A:***213*; **A:**219(8-9), **A:**221(8,23,27), **A:**233(9), **A:**234(8), **A:**235(9), **A:**237(9,23), **A:**238(56), **A:**248(23,56), **A:**250(8), **A:***263–265*; **A:**292(46-47), **A:***303*; **A:**333(39), **A:***335*; **A:**339(10,14), **A:**341(24), **A:**243(29), **A:**351(30), **A:**358(14,24), **A:**362(24), **A:**364(24), **A:***398–399*; **A:**428(38), **A:**432(40), **A:***434*
Wight, C. A., **A:**402(6), **A:***433*
Wigner, E., **B:**258(4), **B:***310*
Wigner, E. P.: **A:**144(5), **A:***168*; **A:**172(4), **A:**173(4,11-12), **A:**176(4,11-12,29), **A:***213*
Wilczek, F., **B:**88(3), **B:***127*
Wilkens, M., **B:**479(13), **B:***500*
Wilkinson, M., **A:**402(3), **A:**404(3), **A:***433*
Willberg, D. M., **A:**61(41-42), **A:***141*
Wilson, E. B.: **A:**271(9), **A:***302*, **B:**89(16), **B:**93(16), **B:**95(16), **B:**106(16), **B:***127*; **B:**187(30), **B:***202*
Wilson, M. A.: **B:**250(159), **B:***255*; **B:**405(75), **B:***420*
Windey, P., **B:**273(69), **B:***312*
Winkler, J. E., **B:**201(45), **B:***203*
Winnewisser, G., **A:**256(83), **A:***265*
Wintgen, D., **A:**307(12,14,17), **A:**308(20), **A:**330(14), **A:***334*
Withnell, R. M., **B:**4(2-3), **B:***23*
Wodtke, A. M., **A:**278(23-24), **A:***302*
Wolf, R. J., **A:**198(82), **A:***215*
Wolynes, P. G.: **A:**131(94), **A:***142*; **A:**164(65), **A:***170*; **B:**90(31), **B:***128*; **B:**180(2,15), **B:***201–202*; **B:**205(1,3,6), **B:**206(1,15,20-21), **B:**207(3,6,19-20,23), **B:**209(16-21,23,60), **B:**210(16,21,23), **B:**212(15,20), **B:**213(15-19), **B:**214(16-17,20,23,83-84), **B:**216(84), **B:**217(20), **B:**220(3,6), **B:**221(105), **B:**228(114), **B:**249(3,6), **B:***251–254*; **B:**263-264(38), **B:**302(97), **B:***311*, **B:***313*; **B:**378(8), **B:**388(8), **B:**390(8), **B:**408(8), **B:***418*; **B:**424(4), **B:**427(4), **B:***436*

Won, Y., **B:**190(33), **B:***202*
Wong, C. F., **B:**180(15), **B:***202*
Wong, S. M., **B:**210(63), **B:**214(63), **B:***253*
Wood, B. P., **B:**438(12), **B:**441(12), **B:**463(12), **B:***463*
Wooten, F., **B:**208(40), **B:**250(40), **B:***252*
Worlock, J. M., **B:**221(100-101), **B:***254*
Wörner, H. J., **A:**128(97), **A:**134(97), **A:***142*
Wozny, C. E., **A:**65(47,49), **A:**66(49), **A:***141*
Wright, K. R., **A:**178(64), **A:***214*, **B:**48(33), **B:**82(33), **B:***84*
Wyatt, R. E.: **A:**128(91), **A:***142*; **A:**198(120), **A:***216*; **B:**180(2), **B:***201*; **B:**210(79), **B:**214(84), **B:**216(84), **B:***253*
Wylie, B. N., **B:**348(20), **B:***351*

Xia, Z.: **A:**311(28), **A:***335*; **A:**373(35-36), **A:**377-378(35), **A:***399*; **B:**427(9), **B:***436*
Xie, A., **B:**207(26-27), **B:**209(26-27), **B:**228(115), **B:***252*, **B:***254*
Xie, X. S., **B:**228(125), **B:***255*
Xin, J., **B:**521-522(7), **B:**530(7), **B:***541*
Xu, D., **B:**206(12), **B:***251*

Yagihara, S., **A:**165(67), **A:***170*, **B:**302(96), **B:***313*
Yakhot, V., **B:**537(41), **B:***542*
Yakubo, K., **B:**208(32), **B:***252*
Yamada, H., **B:**274(72), **B:***312*
Yamaguchi, M., **B:**551(18), **B:**567(18), **B:**570(18), **B:***598*
Yamaguchi, Y., **A:**198(90), **A:***215*
Yamaguchi, Y. Y.: **B:**385(39), **B:***419*; **B:**477(2-3), **B:**479(2,19), **B:**480(2), **B:**485(39), **B:***499-500*
Yamamoto, K.: **B:**273(67), **B:***312*; **B:**477(7), **B:***499*; **B:**502(10), **B:***518*
Yamamoto, T.: **A:**111(79), **A:***142*; **A:**308(21), **A:***334*
Yanagida, T., **B:**265(53,55), **B:**299(53,55), **B:***311*
Yanao, T.: **A:**220(18), **A:**244-245(18), **A:***263*, **B:**28(11), **B:**61-63(11), **B:**64(38), **B:**67(11), **B:**83(48-49), **B:***84-85*; **B:**90(24), **B:***127*
Yang, H., **B:**228(124-125), **B:***255*
Yang, J., **B:**180(13), **B:***201*
Yang, P.-H., **B:**248(154), **B:***255*
Yang, X., **A:**278(23-24), **A:***302*

Yanguas, P.: **A:**147(54,58), **A:**148(54), **A:***170*; **A:**173(21), **A:**178-179(76), **A:**194(76), **A:**197(76), **A:**212(21), **A:***213*, **A:***215*; **A:**219(9,11), **A:**233(9), **A:**235(9), **A:**237(9), **A:**238(57), **A:***263-264*; **A:**338-339(3), **A:**341(3), **A:**352(3), **A:***398*
Yasuda, H., **B:**156-157(7), **B:***176*
Ye, X., **B:**180(14), **B:**200(35), **B:***201-202*
Yeh, S.-R., **B:**180(14), **B:***201*
Yip, S., **B:**406(77), **B:***421*
Yokomizo, T., **A:**165(67), **A:***170*, **B:**302(96), **B:***313*
Yokoyama, K., **B:**317(4), **B:**334(18), **B:***351*
Yomo, T., **B:**544(2), **B:**551(2,13-15,17-21), **B:**553(15,17), **B:**556-557(15), **B:**558(2), **B:**562-563(2), **B:**567(18), **B:**570(18), **B:**573(17), **B:**575(2), **B:**593(41), **B:**594(41,43), **B:**595(41), **B:**597(15), **B:***597-598*
Yonetani, T., **B:**200(37), **B:***202*
Yorke, J. A.: **A:**456(28), **A:***457*, **B:**285(75), **B:**300(75), **B:**307(75,99), **B:**308-309(75), **B:***312-313*; **B:**502(3-4,14), **B:***517-518*
Yoshida, H., **B:**481(33), **B:***500*
Yoshida, T., **B:**265(55), **B:**299(55), **B:***311*
Yoshihiro, T., **B:**180(19), **B:***202*
Yoshimoto, A., **A:**404(24), **A:**405(24,26), **A:**407(24,26), **A:**413(24), **A:**428-429(24,26), **A:***434*
Young, L.-S., **B:**380(21), **B:**394(21), **B:***419*
Young, R. D., **B:**228(115), **B:***254*
Yu, X.: **B:**180(15), **B:**181(22), **B:**190(22), **B:**193(22), **B:**195(21-22), **B:**197(22), **B:***202*; **B:**223(111), **B:**229(111), **B:**233(141), **B:**240-241(111), **B:**243(111), **B:**247(111), **B:**248(156), **B:**250(111), **B:***254-255*
Yuri, M., **B:**470(12), **B:**475(12), **B:***475*

Zacharl, A., **B:**383-384(37), **B:***419*
Zacherl, A., **B:**479(21), **B:***500*
Zanette, D. H., **B:**482(36), **B:***500*
Zaslavsky, G. M.: **B:**264(48), **B:***311*; **B:**479(27), **B:***500*; **B:**527(30), **B:***542*
Zechman, F. W., **B:**332(15), **B:***351*
Zeh, H. D., **A:**140(101), **A:***142*
Zewail, A. H., **A:**61(41-42), **A:***141*, **B:**209(43), **B:**220(43,94-96), **B:***252*, **B:***254*

Zhang, D. H., **A**:65(50), **A**:94(64), **A**:96(100), **A**:*141–142*
Zhang, J. Z. H., **A**:65(50), **A**:*141*
Zhang, X. L., **B**:338(19), **B**:341(19), **B**:*351*
Zhang, Z. Q., **B**:208(41), **B**:250(41), **B**:*252*
Zhao, M., **B**:260(12), **B**:*310*
Zhao, M. S.: **A**:6(13-16,19),
 A:8(30), **A**:41(13-14), **A**:66(16),
 A:70(15), **A**:85(15), **A**:88(60),
 A:97-98(19), **A**:104(67), **A**:108(75),
 A:124(88), **A**:*140–142*; **A**:145(19),
 A:147(55), **A**:153(55), **A**:168(76),
 A:*169–170*; **A**:178(59), **A**:198(122-123),
 A:*214,* **A**:*216*; **A**:218(1), **A**:222(28),
 A:228(1,28), **A**:*263–264*; **A**:436-438(4),
 A:*457*
Zheng, C., **B**:180(15), **B**:*202*
Zhilinskii, B. I., **A**:248(66), **A**:*265*
Zhou, M., **B**:208(41), **B**:250(41), **B**:*252*
Zhu, W., **A**:436(13,15), **A**:437-439(13),
 A:454(15), **A**:*457*
Zimmerman, T., **B**:210(64), **B**:*253*
Zipf, G. K., **B**:593(42), **B**:*598*
Zumhofen, G., **B**:479(22), **B**:*500*
Zumofen, G., **B**:264(50), **B**:*311*
Zurek, W. H., **A**:140(102), **A**:*142*
Zwanzig, R., **A**:164(62), **A**:*170*
Zwier, T. S., **B**:249(157), **B**:*255*

SUBJECT INDEX

Letter in **boldface** indicates the volume.

Ab initio calculations:
  intramolecular dynamics, floppy molecules, canonical perturbation theory, **A**:272–278
  resonantly coupled isomerizing/dissociating systems, high energy bifurcations, **A**:299–301
Activation mechanism, rapid alloying, microcluster dynamics:
  nano-sized clusters, **B**:159
  radial and surface diffusion, **B**:168–170
  saddle point energy distribution, **B**:174–175
Adams-Moulton predictor-corrector method, onset dynamics, argon clusters, **B**:135–136
Adiabatic delocalization, isomerizing systems, intramolecular dynamics, *vs.* nonadiabatic, **A**:278–286
Adiabatic invariants:
  fluctuation-dissipation theorem, excess heat production:
    anomalous variance, **B**:361–368
    Hamiltonian chaotic systems, **B**:363–367
  slow relaxation, internal degrees of freedom, Hamiltonian systems, **B**:401–403
Adiabatic solution, multidimensional barrier tunneling:
  low-frequency approximation, **A**:418–422
  Melnikov method, **A**:417–418

Advection-reaction-diffusion equation (ARD), chaotic transitions, inert and reactive substances, **B**:521–522
FKPP reaction, **B**:530–531
Allan variance, multibasin protein landscapes, chaotic transition, regularity:
  nonstationarity in energy fluctuations, **B**:270–285
  temperature dependence in dimensionality of folding dynamics, **B**:293–299
Alternative Rice, Ramsperger, Kassel and Marcus (ARRKM) rate theory:
unimolecular reaction:
  Gray-Rice-Davis predissociation theory, **A**:39–41
  helium-iodine predissociation, **A**:60–61
  quantization, **A**:108–111
  rigorous quantum rate *vs.*, **A**:111–114
  unimolecular reaction rate, isomerization theory, **A**:70–75
Amplitude measurements:
  fringed tunneling models, multiple trajectories, **A**:424–425
  phase-space transition states, Melnikov integral, **A**:367–371
Angle-action variables, unimolecular reaction rate theory:
  KAM theorem, **A**:13–14
  phase-space structure, **A**:11–12
  quantum transport, cantori systems, **A**:129–131

Geometric Structures of Phase Space in Multidimensional Chaos: A Special Volume of Advances in Chemical Physics, Part B, Volume 130, edited by M. Toda, T Komatsuzaki, T. Konishi, R.S. Berry, and S.A. Rice. Series editor Stuart A. Rice.
ISBN 0-471-71157-8   Copyright © 2005 John Wiley & Sons, Inc.

SUBJECT INDEX

Angle space diffusion, globally-coupled Hamiltonian systems, relaxation and diffusion, Hamilton mean field model, **B:**488–498
  equilibrium diffusion, **B:**489–490
  nonstationary state, **B:**494–496
  quasi-stationary state, **B:**491–493
Angular momentum, phase-space transition state geometry, **A:**247–261
  astrophysics applications, **A:**256–261
  inelastic scattering, **A:**257–261
  rotating frame dynamics, **A:**248–256
  relative equilibrium, **A:**249–251
  van der Waals complex, **A:**251–256
  three degrees of freedom, Hénon-Heiles potential, **A:**238–244
  triatomic dynamics, zero angular momentum, **A:**244–247
Anharmonicity:
  heat transfer, quantum energy flow, protein vibrational states, **B:**238–241
  resonantly coupled isomerizing/dissociating systems, **A:**286–287
Anomalous diffusion:
  chaotic transitions, inert substances, **B:**523–527
  strong anomalous diffusion, **B:**527–530
  weak vs. strong diffusion, **B:**526–527
  globally-coupled Hamiltonian systems, relaxation and diffusion, Hamilton mean field (HMF) model, **B:**479–480
  heat transfer, quantum energy flow, **B:**227–238
  multidimensional Hamiltonian systems, resonance and transport, **B:**440–442
  slow relaxation, internal degrees of freedom, Hamiltonian systems, mixed-phase space, **B:**379–387
  unimolecular reaction rate, faster-than-classical dynamics, **A:**134–137
Anomalous time series, multichannel chemical isomerization, multi-basin potential, structural transitions, **B:**31–34
Anomalous variance, nonergodic adiabatic invariant, fluctuation-dissipation theorem, excess heat production, **B:**361–368
  Hamiltonian chaotic systems, **B:**363–367
  microcanonical distribution, **B:**361–363

Anti-harmonic oscillator equation, phase-space transition state geometry, one degree of freedom model, linearized Hamiltonian, **A:**225–227
Antiproton-proton-antiproton system, collinear eZe configuration, **A:**330
A priori distribution:
  multichannel chemical isomerization: linear surprisal theory, **B:**80
  microcanonical temperature, **B:**58–62
  multichannel isomerization, **B:**28
"Apt" coordinates, Wigner's transition state dynamics, rank-one saddle phase-space structure, **A:**184–186
Argon clusters:
  multibasin landscapes, chaotic transition, regularity, **B:**265–266
  onset dynamics, phase transition:
    analytic techniques, **B:**131–136
    configuration entropy, **B:**140
    configuration space, **B:**140–142
    gradient extremal, **B:**131–133
    Lindemann's criterion, **B:**138–139
    additional potentials, **B:**148–151
    Lyapunov and KS entropy, **B:**142–143
    partitioned cell dynamics, **B:**145–146
    potential energy surface, cell petition, **B:**133–134
    potential function, MD simulation, and temperature, **B:**135–136
    reaction paths, **B:**136–138
    stationary points, **B:**143–145
    structural characteristics, **B:**136
    watershed, **B:**146–148
  power spectra and phase-space dimensions, **B:**5–11
  slow relaxation, internal degrees of freedom, Hamiltonian systems, **B:**389–392
Argon-iodine molecules, unimolecular reaction rate theory, predissociation, **A:**61–63
Arnold diffusion
  atomic clusters, **B:**23
  multidimensional Hamiltonian Systems, resonance and transport, **B:**438, **B:**462
  multidimensional phase space:
    slow dynamics, **B:**427–430
    multi-precision numerical method, **B:**430–435

SUBJECT INDEX

unimolecular reaction rate theory:
  phase-space structure, bottlenecks, many-dimensional systems, **A:**20
  quantum transport, classically chaotic systems, **A:**131–134
  web model, phase-space transition states:
    Melnikov integral derivation, **A:**341–342, **A:**371–377
    normally hyperbolic invariant manifold connections, **A:**340
    tangency and, **A:**378–385
Arnold model: *see* Arnold diffusion
Arrhenius relation:
  isomerization:
    atomic clusters, **B:**27–28
    multichannel dynamics, **B:**27–28
  multichannel isomerization:
    isomer lifetimes, **B:**53–70
      basic principles, **B:**60–62
      canonical temperature, **B:**70
      definition, **B:**57–60
      density of states evaluation, **B:**55–56
      exponential relation, average lifetimes, **B:**62–70
        liquid-like ergodicity and nonergodicity, **B:**67–69
        M_7 single exponential form case study, **B:**65–67
        multiexponential form, **B:**62–63
        single exponential form, **B:**64–65
      lifetime averaging law, **B:**53–55
      local temperatures, **B:**60
      microcanonical temperature, numerical calculation, **B:**60–62
      rapid alloying, microcluster dynamics, radial and surface diffusion, **B:**167–170
Artificial life (AL) research, recursive cell production and evolution, catalytic reaction network, **B:**553–557
Astrophysics applications, phase-space transition state geometry, angular momentum, **A:**256–261
Asymptotic conditions:
  intramolecular dynamics, floppy molecules, canonical perturbation theory, **A:**275–278
  phase-space transition states:
    angular momentum, astrophysics applications, **A:**261
    tangency, **A:**378–385

Atomic clusters:
  basic properties, **B:**3–4
  Kolmogorov entropy, **B:**5–11
  level spacing distributions, **B:**4–5
  Lyapunov exponents, **B:**5–11
  molecular internal space, isomerization dynamics, **B:**90–91
  nonlinear canonical transformation, **B:**21–22
  phase-space dimensions, **B:**5–11
  power spectra, **B:**5–11
  regularity, chaos and ergodicity characteristics, **B:**11–20
Autocatalytic network, recursive cell production and evolution, **B:**573–595
  core hypercycle minority, **B:**582–583
  evolution models, **B:**585–588
  intermingled hypercycle network stabilization, **B:**581–582
  molecular models, **B:**574–575
  phase states, **B:**575–581
  random network localization, **B:**583
  statistical law, **B:**588–595
    deviation from universal statistics, **B:**593–595
    universal statistics, **B:**593
  switching mechanism, **B:**584–585
Autocorrelation, fluctuation-dissipation theorem, excess heat production, **B:**357–359
Averaged kinetic temperature controlling method (AKTCM), rapid alloying, microcluster dynamics:
  nano-sized clusters, **B:**159
  size effect, **B:**161–164
Average mutual information, multibasin landscapes, chaotic transition, regularity, state-space structures, **B:**292–294

Backward reactive trajectories, Wigner's transition state dynamics, rank-one saddle phase-space structure:
  normally hyperbolic invariant manifolds, **A:**188–190
  transition state principles, **A:**190–191
Ballistic pathway, chaotic transition, regularity, two-basin landscapes, Kramers-Grote-Hynes theory, **A:**164–165

Basin transition:
  molecular internal space, democratic centrifugal force, **B:**106
  multichannel chemical isomerization, microcanonical temperature, **B:**57–62
Bath interaction:
  chaotic transition, regularity, two-basin landscapes, Kramers-Grote-Hynes theory, **A:**164–165
  vibrational energy relaxation, force-force-correlation function approximations, **B:**187–190
  Wigner's transition state dynamics, theoretical background, **A:**174–175
Belousov-Zhabotinsky (BZ) reaction, phase-space transition states, hyperbolicity breakdown, **A:**389–392
Bending energy fluctuation, multibasin landscapes, chaotic transition, regularity, nonstationarity in, **B:**278–282
Berendsen algorithm, multibasin landscapes, chaotic transition, regularity, constant-temperature molecular dynamics, **B:**309–310
Bernoulli system, multibasin landscapes, chaotic transition, regularity, nonstationarity in energy fluctuations, **B:**274–280
Bifurcation mechanisms:
  multichannel isomerization, inter-basin mixing, reaction tubes, **B:**47–50
  phase-space transition states:
    hyperbolicity breakdown, **A:**391–392
    multidimensional chaos crisis, **A:**392–395
    resonantly coupled isomerizing/dissociating systems:
      high energy bifurcations, **A:**296–301
      saddle-node bifurcations, **A:**287–296
Binary clusters, rapid alloying, microcluster dynamics:
  floppy surface atoms and PES reaction paths, **B:**170–171
  heat of solution, **B:**164–165
  molecular dynamics simulation, **B:**160–161
  nano-sized clusters, **B:**157–160
  procedural characteristics, **B:**157–158
  simulation model, **B:**158–160
  radial and surface diffusion, **B:**167–170
  reaction path numerical simulation, **B:**171–173
  saddle point energy distribution, **B:**173–175
  size effect, **B:**161–164
  solid phase, **B:**165–166
Binary collisions, Coulomb three-body problem, zero angular momentum, **A:**315–319
Birkhoff-Gustavson procedure, intramolecular dynamics, floppy molecules, canonical perturbation theory, **A:**278
Birkhoff normal form, phase-space transition states, normally hyperbolic invariant manifolds, **A:**339–340
BLN protein models, multibasin landscapes, chaotic transition, regularity:
  basic properties, **B:**267–270
  nonstationarity in energy fluctuations, **B:**270–285
  temperature dependence in dimensionality of folding dynamics, **B:**296–299
Body-fixed frame, molecular internal space, kinematics, **B:**88–90
Bohr-Sommerfeld (BS) quantization condition, three-body problem, **A:**306–309
Boltzmann constant, multichannel isomerization:
  canonical temperature, **B:**70
  density of state evaluation, microcanonical temperature, **B:**55–56
Boltzmann equilibrium distribution:
  fluctuation-dissipation theorem, excess heat production:
    second law from, **B:**369–370
    superstatistical equilibrium distributions, **B:**360–361
    thermodynamics, **B:**355–356
  multichannel isomerization, linear surprisal theory, **B:**71–74
Boltzmann-Jeans conjecture, slow relaxation, internal degrees of freedom, Hamiltonian systems, **B:**401–403
  hypothesis validity, **B:**405–412
Bond energy fluctuation, multibasin landscapes, chaotic transition, regularity, nonstationarity in, **B:**277–278
Born-Oppenheimer approximation, phase-space transition states, singular perturbation theory *vs.*, **A:**342–345
Bottlenecks:
  chaotic transition, regularity, two-basin landscapes, normally hyperbolic invariant manifolds (NHIM), **A:**166–168

SUBJECT INDEX

multibasin landscapes, chaotic transition, regularity, **B:**259–260
protein structures, **B:**265–266
slow relaxation, internal degrees of freedom, Hamiltonian systems, **B:**407–412
unimolecular reaction rate theory:
Davis-Gray predissociation analysis, **A:**30–39
dissociation dynamics, Hamiltonian equations, **A:**123
helium-iodine predissociation, **A:**60–61
phase-space structure:
few-dimensional systems, **A:**18–19
many-dimensional systems, **A:**19–20
quantum scars, **A:**108
quantum transport, cantori systems, **A:**129–131
Zhao-Rice approximation, **A:**41–54
intramolecular dividing surface, **A:**46–48
isomerization theory, **A:**71–75
separatrix crossing rate, **A:**50–54
zeroth-order rate constant, **A:**48–54
Bound-bound coupling potentials, dissociation dynamics, unimolecular reactions, Hamiltonian equations, **A:**116–123
Bound-continuum coupling potentials, dissociation dynamics, unimolecular reactions, Hamiltonian equations, **A:**116–123
Box-counting dimensions, multibasin landscapes, chaotic transition, regularity, embedding theorems, **B:**307–309
Brain wave analysis, nonmetric multidimensional scaling algorithm, **B:**338–341
Branching structure of paths:
fringed tunneling models, **A:**415–417
multiple trajectories, **A:**423–425
multidimensional barrier tunneling:
Melnikov method, **A:**417–418
M-set critical point, **A:**414–415
phase-space transition states:
chaotic itinerancy, **A:**385–389
tangency, **A:**377–385
Brownian motion:
chaotic transitions, inert and reactive substances, **B:**522
multidimensional Hamiltonian systems, resonance and transport structures, **B:**450

Canonical correlation analysis (CCA), nonmetric multidimensional scaling algorithm and:
fern genotype/phenotype, **B:**330–332
green autotrophs, **B:**332–333
Canonical ensemble, globally-coupled Hamiltonian systems, relaxation and diffusion, Hamilton mean field (HMF) model, **B:**480–481
Canonical perturbation theory (CPT), floppy molecules, intramolecular dynamics, **A:**269–278
Canonical temperature, multichannel isomerization, **B:**70
Canonical transformation:
atomic clusters, nonlinear transformation, **B:**21–22
chaotic transition, regularity, two-basin landscapes, saddle regions, dynamical regularity, **A:**148–151
unimolecular reaction rate theory:
phase-space structure, molecular dynamics, **A:**9–10
reaction path analysis, **A:**57–59
Cantori systems:
slow relaxation, internal degrees of freedom, mixed-phase space, **B:**383–387
unimolecular reaction rate, quantum transport, classically chaotic systems, **A:**128–129
Whisker mapping, **A:**129–131
Cao's algorithm, multibasin landscapes, chaotic transition, regularity:
state-space structures, **B:**291–292
temperature dependence in dimensionality of folding dynamics, **B:**296–299
Capped octahedron (COCT) structures, multichannel isomerization:
inter-basin mixing, **B:**46–57
time scales, **B:**50–51
liquid-like phase, **B:**42–44
microcanonical temperature, single exponential form, **B:**66–67
multi-basin potential, **B:**28–30
anomalous time series, **B:**32–34
nonstatistical behavior, low-energy phase, **B:**44–45
Carbon monoxide myoglobin (MbCO):
heat transfer, quantum energy flow, protein vibrational states, anharmonic decay, **B:**238–241

Carbon monoxide myoglobin
  (MbCO) (*Continued*)
  heat transport, quantum energy flow,
    B:250–251
  vibrational energy relaxation,
    cytochrome *c*, CD stretching mode,
    B:200
Carbon-oxygen "wagging," unimolecular
  reaction rate, isomerization,
  cyclobutanone, **A:**100–104
Cartesian coordinates:
  phase-space transition state geometry, three
    degrees of freedom, Hénon-Heiles
    potential, **A:**238–244
  vibrational energy relaxation, cytochrome *c*,
    CD stretching mode, coupling constants,
    B:193–195
  Wigner's transition state dynamics,
    hydrogen cyanide isomerization,
    A:207–210
Cascading saddlebacks, resonantly coupled
  isomerizing/dissociating systems, high
  energy bifurcations, **A:**298–301
Catalytic reaction network, recursive cell
  production and evolution:
  autocatalytic network, **B:**573–595
  core hypercycle minority, **B:**582–583
  evolution models, **B:**585–588
  intermingled hypercycle network
    stabilization, **B:**581–582
  molecular models, **B:**574–575
  phase states, **B:**575–581
  random network localization, **B:**583
  statistical law, **B:**588–595
    deviation from universal statistics,
      B:593–595
    universal statistics, **B:**593
  switching mechanism, **B:**584–585
  constructive biology, **B:**550–557
  chemical reaction networks modeling,
    B:553–557
  diverse chemicals, **B:**546–547
  Dyson's loose reproduction system,
    B:549–550
  Eigen's hypercycle, **B:**547–549
  heredity origins, **B:**544–546
  minority control hypothesis, **B:**557–573
    evolvability, **B:**566–567
    experimental protocol, **B:**567–571
    growth speed, **B:**565
  intermingled hypercycle network
    production, **B:**595–596
  itinerant dynamics, **B:**596
  kinetic theory, heredity and, **B:**571–572
  model parameters, **B:**557–561
  molecule chemical composition,
    B:565–566
  molecule preservation, **B:**565
  stochastic results, **B:**561–564
  universal statistics and fluctuation control,
    B:596–597
Cayley tree topology, unimolecular reaction
  kinetics, quantum energy flow, local
  random matrix theory, **B:**212–214
CD stretching mode, vibrational energy
  relaxation:
  basic principles, **B:**180–181
  cytochrome *c*, **B:**190–200
  carbon monoxide myoglobin (MbCO),
    B:200
  classical calculation, **B:**197
  coupling constants calculation, **B:**192–195
  full width at half maximum spectra,
    B:199–200
  lifetime parameter assignment, **B:**195–196
  quantum calculation, **B:**197–199
  system and bath characteristics,
    B:190–192
  force-force-correlation function
    approximations, **B:**187–190
Celestial mechanics, *n*-body problem in,
  A:309–312
Cell-cell interaction, recursive cell production
  and evolution, catalytic reaction
  networks, **B:**597
Cell partition, onset dynamics, argon clusters,
  potential energy surfaces, **B:**133–134
Cellular flow fronts, chaotic transitions, inert and
  reactive substances, **B:**531–532
Center manifold:
  phase-space transition state geometry, *n*
    degrees of freedom structures, **A:**237
  Wigner's transition state dynamics, stationary
    points, **A:**179
Centrifugal forces, phase-space transition state
  geometry, angular momentum,
  astrophysics applications, **A:**256–261
Chain rule, Wigner's transition state dynamics,
  Lie transformation, normal-form
  coordinates, **A:**198

SUBJECT INDEX

Chaotic itinerancy, phase-space transition states,
  A:385–389
  Melnikov integral, **A:**363–371
  normally hyperbolic invariant manifold
    connections, **A:**340
Chaotic transitions. *See also* Quantum chaos
  atomic clusters, local characteristics,
    B:11–20
  inert substances:
    standard and anomalous diffusion,
      B:523–527
    strong anomalous diffusion, **B:**527–530
  multibasin landscapes, regularity in:
    Berendsen algorithm, constant-temperature
      molecular dynamics, **B:**309–310
    embedded techniques:
      basic principles, **B:**302–309
      phase-space reconstruction,
        B:285–288
    energy nonstationarity, protein landscapes,
      B:270–285
      bending energy fluctuation, **B:**278–282
      bond energy flucation, **B:**277–278
      torsional angle energy fluctuation,
        B:282–285
    folding dynamic dimensionality,
      temperature dependency, **B:**294–299
    global/local collective coordinates,
      B:261–262
    liquid water, **B:**262–263
    minimalistic 46-bead protein models,
      B:266–270
    phase space transport geometry,
      B:260–261
    proteins, **B:**263–266
    state-space structure, **B:**285–299
      average mutual information,
        B:292–294
      false nearest neighbors, **B:**288–292
      phase-space reconstruction, embedding
        of, **B:**285–288
  multidimensional barrier tunneling:
    global dynamics, **A:**402–406
    quantum mapping, **A:**428
  multidimensional phase space slow dynamics,
    global motion, **B:**425–427
  phase-space transition states:
    angular momentum, astrophysics
      applications, **A:**261
    Arnold model, **A:**371–377

Melnikov integral, **A:**368–371
multidimensional chaos crisis, **A:**392–395
normally hyperbolic invariant manifold
  connections, **A:**339–340
reacting substances:
  front propagation, **B:**537–540
  fronts in cellular flows, **B:**531–532
  geometric optics limit, **B:**534–537
  slow and fast reaction regimes, **B:**532–534
regularity, in two-basin landscapes:
  Kramers-Grote-Hynes theory, **A:**163–165
  phase space geometrics, **A:**151–163
  reactive island theory, **A:**153–163
  saddle crossing stochasticity, **A:**165–166
  saddle regions, dynamical regularity,
    A:147–151
three-body problem, **A:**306–309
unimolecular reaction rate, **A:**128–137
  Arnold diffusion suppression, **A:**131–134
  Cantori model, **A:**129–131
  faster-than-classical anomalous diffusion,
    A:134–137
CHARMM potential, vibrational energy
  relaxation:
  cytochrome *c*, CD stretching mode,
    B:190–192
  coupling constants, **B:**193–195
Chemical molecules, catalytic reaction network,
  recursive cell production and evolution,
    B:546–547
  autocatalytic phases, **B:**575–581
  modeling strategy, **B:**553–557
Chirikov-Taylor map, finite-time Lyapunov
  exponents, multidimensional
  Hamiltonian dynamical systems, **B:**508
Christoffel symbols, molecular internal space,
  Eckart subspace dynamics, **B:**109
Clade techniques, nonmetric multidimensional
  scaling algorithm and, molecular
  taxonomy, **B:**326–329
Classical autocorrelation function, vibrational
  energy relaxation:
  cytochrome *c*, CD stretching mode, **B:**197
  quantum correction factor and, **B:**185–186
Clusters
  heat transfer, quantum energy flow,
    B:221–248
  unimolecular reaction kinetics:
    energy diffusion, **B:**222–223
    proteins, **B:**241–248

Clusters (*Continued*)
protein vibrational energy:
anharmonic decay, **B**:237–241
anomalous subdiffusion,
B:227–237
water clusters, **B**:223–227
nonmetric multidimensional scaling
algorithm, **B**:318–320
protein family, **B**:342–343
survival time distribution, Hamiltonian
system multiergodicity, **B**:471–474
Coarse-grained representation:
optimal control theory, Zhu-Botina-Rabitz
formula, **A**:450–453
quantum chaos systems:
Rabi state and frequency, **A**:446–448
random vector transition, **A**:449
rotating-wave approximation, **A**:440–449
transition element, **A**:450–453
Coexisting phase, rapid alloying, binary clusters,
B:156–157
Coherent control, quantum chaos, **A**:436
Cohomology equation, intramolecular
dynamics, floppy molecules, canonical
perturbation theory, **A**:272–278
Collective coordinates:
molecular internal space, four-body systems
isomerization, **B**:118–121
multibasin landscapes, chaotic transition,
regularity, local-global postulation,
B:260–266
Collinear electron-electron-nucleus (eeZ)
configuration, Coulomb three-body
problem, zero angular momentum,
A:312–319
Collinear electron-nucleus-electron (eZe)
configuration:
Coulomb three-body problem:
mass ratio effect, **A**:319–330
antiproton-proton-antiproton system,
A:330
triple collision manifold, **A**:320–323
triple collision orbits, **A**:323–329
zero angular momentum, **A**:312–319
Collinear transition state, molecular internal
space, democratic centrifugal force,
B:104–106
Compact clusters, rapid alloying, microcluster
dynamics, reaction path enumeration,
B:172–173

Configuration entropy, onset dynamics, argon
clusters, **B**:140
Conjugate action-angle-like coordinates,
resonantly coupled isomerizing/
dissociating systems, polyad folding and
saddle-node bifurcation, **A**:290–296
Constrained dynamics, molecular internal space,
gauge field reaction rates, **B**:109–110
Constructive biology, recursive cell production
and evolution, catalytic reaction
network, **B**:550–557
chemical reaction networks modeling,
B:553–557
Controlled random matrix, optimal control
theory, quantum chaos systems,
A:438–439
Control property, recursive cell production and
evolution, catalytic reaction network,
heredity kinetics, **B**:571–573
Control schemes, quantum chaos, **A**:436–437
Core hypercycle network, recursive cell
production and evolution:
catalytic reaction networks, **B**:581–583
statistical laws, **B**:591–595
evolution models, **B**:586–588
minority molecules, **B**:582–583
Coriolis coupling, unimolecular reaction rate
theory, reaction path analysis, **A**:57–59
Correction factor, atomic clusters, nonlinear
canonical transformation, **B**:21–22
Correlation dimension, atomic clusters, power
spectra and phase-space, **B**:7–11
Correlation function:
globally coupled Hamiltonian systems,
relaxation and diffusion:
equilibrium diffusion, **B**:489–490
quasi-stationary state, **B**:491–493
multichannel isomerization, inter-basin
mixing, **B**:52–53
Coulomb three-body problem:
Arnold web model, **A**:378
celestial mechanics, **A**:309–312
collinear eZe case, mass ratio effect, **A**:319–330
antiproton-proton-antiproton system,
A:330
triple collision manifold, **A**:319–323
triple collision orbits, **A**:323–329
free-fall case, **A**:330–332
two-dimensional case, zero angular
momentum, **A**:312–319

SUBJECT INDEX

Coupling coefficients, vibrational energy relaxation, cytochrome *c*, CD stretching mode, **B**:192–195
Coupling states:
 intramolecular dynamics, adiabatic *vs.* nonadiabatic delocalization, **A**:284–286
 resonantly coupled isomerizing/dissociating systems, polyad folding and saddle-node bifurcation, **A**:287–296
C++ programming, multidimensional phase space, slow dynamics, multi-precision numerical method, **B**:431–435
Critical point:
 fringed tunneling models:
  global structure of branches, **A**:415–417
  multiple trajectories, **A**:423–425
  perturbation strength, **A**:425–426
 multidimensional barrier tunneling:
  Melnikov method, **A**:417–418
  M-set local structure, **A**:414–415
  periodic perturbation effects, **A**:413–414
Curvilinear reaction coordinates, multichannel isomerization, linear surprisal theory, **B**:74–76
Cyclobutanone, unimolecular reaction, isomerization, **A**:100–104
Cyclohexane ring inversion, unimolecular reaction kinetics, **B**:216–221
Cylindrical manifolds:
 chaotic transition, regularity, two-basin landscapes, reactive island theory, **A**:154–163
 multibasin landscapes, chaotic transition, regularity, phase-space transition states, **B**:259–260
 phase-space transition state geometry, two degrees of freedom, **A**:232–234
 unimolecular reaction rate, reactive island theory (RIT), **A**:76–80
 Wigner's transition state dynamics, rank-one saddle phase-space structure, normally hyperbolic invariant manifolds, **A**:188–190
Cytochrome *c*, CD stretching mode, vibrational energy relaxation, **B**:190–200
 carbon monoxide myoglobin (MbCO), **B**:200
 classical calculation, **B**:197
 coupling constants calculation, **B**:192–195

full width at half maximum spectra, **B**:199–200
 lifetime parameter assignment, **B**:195–196
 quantum calculation, **B**:197–199
 system and bath characteristics, **B**:190–192

Damköhler number, chaotic transitions, inert and reactive substances, **B**:522
Data mining, nonmetric multidimensional scaling:
 algorithm characteristics, **B**:320–321
 basic principles, **B**:317–320
 brain wave analysis, **B**:338–341
 embedded point estimation, **B**:322–323
 gene expression, temporal patterns, **B**:348–349
 microarray data, **B**:343–349
 molecular/morphology comparisons, **B**:329–333
 ferns, **B**:329–332
 green autotrophs, **B**:332–333
 molecular taxonomy, **B**:324–329
 pointwise criterion, **B**:321–322
 protein family, **B**:342–343
 soil bacteria biodiversity, **B**:334–338
Davis-Gray rate theory:
 chaotic transition, regularity, two-basin landscapes, **A**:145–147
 phase-space transition states, Melnikov integral, **A**:358
 unimolecular reaction:
  Gray-Rice isomerization theory, **A**:69–70
  predissociation theory, **A**:30–39
Deactivation rate, unimolecular reaction kinetics, LRMT dynamical corrections to RRKM theory, **B**:215–216
Debye frequency:
 rapid alloying, microcluster dynamics, size effect, **B**:162–164
 slow relaxation, internal degrees of freedom, Hamiltonian systems, **B**:408–412
Deflected diffusion, multidimensional Hamiltonian dynamical systems, resonance and transport structures, **B**:460–462
Delocalization, isomerizing systems, intramolecular dynamics, adiabatic *vs.* nonadiabatic, **A**:278–286

Delta function, vibrational energy relaxation, cytochrome c, CD stretching mode, **B**:195–196
Democratic centrifugal force (DCF), molecular internal space:
 atomic cluster isomerization dynamics, **B**:90–91
 classical equations and metric force, **B**:98–99
 four-body systems, PAHC equations of motion, **B**:117–118
 gauge field suppression, **B**:111–113
 kinematics, **B**:89–90
 mass-balance asymmetry and trapping trajectories, **B**:103–106
 motion trapping, transition state, **B**:121–123
Density functional theory (DFT), vibrational energy relaxation, basic principles, **B**:180–181
Density of states (DOS):
 dissociation dynamics, unimolecular reactions, Hamiltonian equations, **A**:123
 heat transfer, quantum energy flow, clusters and macromolecules, **B**:221–248
 multibasin landscapes, chaotic transition, regularity, minimalistic 46-bead protein models, **B**:268–270
 multichannel isomerization, microcanonical temperature, **B**:55–56
 multiexponential lifetime averaging, **B**:63
 phase-space transition state geometry, angular momentum, astrophysics applications, **A**:256–261
 vibrational energy relaxation, cytochrome c, CD stretching mode, **B**:190–192
Density probability, resonantly coupled isomerizing/dissociating systems, polyad folding and saddle-node bifurcation, **A**:293–296
Deterministic diffusion, optimal control theory, controlled quantum kicked rotor, **A**:439–443
Deterministic diffusion, slow relaxation, internal degrees of freedom, mixed-phase space, **B**:387
Devil's staircase trajectory, atomic clusters, regularity, chaos, and ergodicity, **B**:16–20

Diffusion coefficient:
 globally-coupled Hamiltonian systems, relaxation and diffusion, angle space diffusion:
  equilibrium diffusion, **B**:489–490
  Hamilton mean field model, **B**:488–498
  nonstationary state, **B**:494–496
  quasi-stationary state, **B**:491–493
 heat transfer, quantum energy flow, proteins, **B**:242–248
 multidimensional Hamiltonian dynamical systems, resonance and transport structures, **B**:440–442
 deflected diffusion, **B**:460–462
 rapid alloying, microcluster dynamics:
  nano-sized clusters, **B**:157–158
  radial and surface atoms, **B**:167–170
Diffusive stochastic processes, chaotic transition, regularity, two-basin landscapes, saddle regions, dynamical regularity, **A**:149–151
Dimensionality:
 atomic clusters, phase-space transition states, **B**:7–11
 multibasin landscapes, chaotic transition, regularity, temperature dependency in folding dynamics, **B**:293–299
 phase-space transition states:
  Hamiltonian dynamics, **A**:221–223
  tangency and, **A**:381–385
 slow relaxation, internal degrees of freedom, mixed-phase space, **B**:386–387
Diophantine condition, slow relaxation, mixed-phase space, **B**:380–387
Dirac coordinate eigenstate, unimolecular reaction rate, semiclassical approximation, rigorous quantum rate, **A**:115–116
Discrete variable representation (DVR), unimolecular reaction rate, wave packet dynamics, **A**:124–128
Dispersion relation, heat transfer, quantum energy flow, anomalous subdiffusion, **B**:231–238
Dissociation dynamics, unimolecular reactions, Hamiltonian equations, **A**:116–123
DNA molecules, catalytic reaction network, recursive cell production and evolution, minority control hypothesis, **B**:557–573

SUBJECT INDEX 637

Double-well systems:
  chaotic transition, regularity, two-basin landscapes, reactive island theory, **A:**153–163
  multidimensional barrier tunneling, **A:**404–406
  slow relaxation, internal degrees of freedom, Hamiltonian systems, **B:**414–418
  unimolecular reaction rate:
    Gray-Rice isomerization theory *vs.* reactive island theory, **A:**82–84
    isomerization in, **A:**84–88
Dunham expansion model, resonantly coupled isomerizing/dissociating systems:
  anharmonicity, **A:**286–287
  polyad folding and saddle-node bifurcation, **A:**287–296
Dynamical heterogeneity, slow relaxation, internal degrees of freedom, Hamiltonian systems, molecular systems, **B:**392
Dynamical localization, unimolecular reaction rate, quantum transport, classically chaotic systems, **A:**128–129
Dynamical propensity rule, chaotic transition, regularity, two-basin landscapes, saddle regions, dynamical regularity, **A:**151
Dyson's loose reproduction system, recursive cell production and evolution, catalytic reaction networks, **B:**549–550
  kinetic heredity, **B:**571–573

Eckart frame, molecular internal space:
  gauge-dependent expression, rotation-vibration energy, **B:**93
  gauge field isomerization suppression, **B:**106–113
    applications, **B:**110–113
    Eckart subspace dynamics, **B:**107–109
    four-body systems, **B:**123–125
    quantitive role, **B:**106–107
    reaction rate effects, **B:**109–110
    gauge field suppression, subspace parameterization, **B:**110–113
  kinematics, **B:**89–90
Eckart type potential:
  fringed tunneling, semiclassical method, **A:**406–407
  multidimensional barrier tunneling, periodic perturbation effects, **A:**412–414
Edge-bridging saddle, atomic clusters, regularity, chaos, and ergodicity, **B:**16–20
Edge running (ER), rapid alloying, microcluster dynamics:
  saddle point energy distribution, **B:**175
Ehrenfest adiabatic state, isomerizing systems, intramolecular dynamics, **A:**278–286
Eigen's hypercycle, recursive cell production and evolution, catalytic reaction networks, **B:**547–549
Eigenstates, intramolecular dynamics, adiabatic *vs.* nonadiabatic delocalization, **A:**278–286
Eigenvalues:
  atomic clusters, basic principles, **B:**4
  molecular internal space, three-atom clusters, isomerization dynamics, **B:**96–97
  phase-space transition state geometry:
    one degree of freedom model, linearized Hamiltonian, **A:**224–227
    three degrees of freedom, Hénon-Heiles potential, **A:**240–244
  Wigner's transition state dynamics:
    hydrogen cyanide stationary point geometry, **A:**203–205
    saddle region energy landscapes, stability analysis, **A:**181–182
Eigenvector components, unimolecular reaction kinetics, quantum energy flow, **B:**213–214
Eigenvector-following algorithm, onset dynamics, argon clusters, gradient external path, **B:**132–133
Einstein-Brillouin-Keller (EBK) quantization rule:
  resonantly coupled isomerizing/dissociating systems, polyad folding and saddle-node bifurcation, **A:**288–296
  three-body problem, **A:**306–309
    collinear eZe configuration, **A:**330
Einstein-Shannon entropy, fluctuation-dissipation theorem, excess heat production, Boltzmann equilibrium distribution, **B:**355–356
Elastic bounce phenomenon, *n*-body problem in celestial mechanics, **A:**310–312

Embedding theorems:
multibasin landscapes, chaotic transition, regularity, **B:**305–309
false nearest-neighbor, **B:**289–292
future research issues, **B:**300–302
state-space structures, **B:**285–288
temperature dependence in dimensionality of folding dynamics, **B:**296–299
theoretical principles, **B:**302–309
nonmetric multidimensional scaling algorithm, **B:**320–321
brain wave analysis, **B:**338–341
fern genotype/phenotype, **B:**330–332
illustrative case, **B:**323–324
microarray data, **B:**345–349
molecular taxonomy case, **B:**324–329
point estimation, **B:**322–323
protein structures, **B:**342–344
soil bacteria biodiversity, **B:**334–338
Energy barrier tunneling, global dynamics, **A:**404–406
Energy diffusion, heat transfer, quantum energy flow:
anomalous subdiffusion, **B:**232–238
clusters and macromolecules, **B:**222–223
Energy gain equation, multidimensional barrier tunneling, low-frequency approximation, **A:**420–422
Energy gap measurement, phase-space transition states, Melnikov integral, **A:**364–371
Energy level spacings, atomic clusters, distribution in, **B:**4–5
Energy nonstationarity, multibasin landscapes, chaotic transition, regularity, protein structures, **B:**270–285
bending energy fluctuation, **B:**278–282
bond energy flucation, **B:**277–278
torsional angle energy fluctuation, **B:**282–285
Energy surfaces, Wigner's transition state dynamics:
Lie transformation, normal-form coordinates, **A:**197–198
saddle regions, phase-space analysis, **A:**182–183
Entropy deficiency:
fluctuation-dissipation theorem, excess heat production, **B:**354–355
multichannel isomerization, linear surprisal theory, **B:**73–74

Equi-energy surface gaps, phase-space transition states:
Arnold model, **A:**371–377
Melnikov integral, **A:**364–371
multidimensional chaos crisis, **A:**393–395
Equilibrium diffusion, globally-coupled Hamiltonian systems, relaxation and diffusion, **B:**489–490
Ergodicity:
atomic clusters:
local characteristics, **B:**11–20
power spectra and phase-space dimensions, **B:**7–11
Hamiltonian system multiergodicity and nonstationarity:
complex behaviors, **B:**474–475
stagnant motion deviation, **B:**466–469
survival time distribution, clustering motions, **B:**471–474
universality conjecture, **B:**469–471
multichannel isomerization:
global mixing, **B:**47
lifetime averaging uniformity, **B:**35–36
liquid-like dynamics, **B:**67–70
slow relaxation, internal degrees of freedom, Hamiltonian systems, **B:**375–378
nearly integrable systems, **B:**393–398
Error rate, recursive cell production and evolution, autocatalytic network, **B:**574–595
Euclidean distance:
Hamiltonian system multiergodicity, stagnant motion deviation, **B:**466–469
heat transfer, quantum energy flow, anomalous subdiffusion, **B:**230–238
molecular internal space, Eckart subspace dynamics, **B:**107–109
multibasin landscapes, chaotic transition, regularity, state-space structural embedding, **B:**289–292
Euler angles, molecular internal space, gauge-dependent expression, rotation-vibration energy, **B:**91–93
Euler-Lagrangian equations, molecular internal space, gauge-invariant energy, **B:**94
Evolution models, recursive cell production and evolution:
autocatalytic reaction networks, **B:**585–588
catalytic reaction network, itinerant dynamics, **B:**596

Evolvability, recursive cell production and evolution, catalytic reaction networks, **B:**566–567

Excess heat production, fluctuation-dissipation theorem:
Boltzmann equilibrium distribution, **B:**355–359
hysteresis loop area, **B:**358–359
thermodynamics, **B:**355–356
Jarzynski's nonequilibrium work relation, **B:**368–369
nonergodic adiabatic invariant, anomalous variance, **B:**361–368
Hamiltonian chaotic systems, **B:**363–367
microcanonical distribution, **B:**361–363
second law, **B:**369–370
superstatistical equilibrium distributions, **B:**359–361

Exponential decay, multichannel chemical isomerization, lifetime averaging uniformity, **B:**39–42

Exponential relation, microcanonical temperature and average lifetimes:
linear surprisal theory, **B:**74–80
multichannel isomerization, **B:**62–70
liquid-like ergodicity and nonergodicity, **B:**67–69
M_7 single exponential form case study, **B:**65–67
multiexponential form, **B:**62–63
single exponential form, **B:**64–65

"Falling cat" phenomenon, molecular internal space:
Eckart subspace dynamics, **B:**108–109
kinematics, **B:**88–90

False nearest-neighbor (FNN), multibasin landscapes, chaotic transition, regularity, state-space structural reconstruction, **B:**288–292

Faster-than-classical quantum anomalous diffusion, unimolecular reaction rate, **A:**134–137

Fast reaction regimes, chaotic transitions, inert and reactive substances, **B:**532–534

Fast switching states, recursive cell production and evolution, catalytic reaction networks, autocatalytic phases, **B:**578–581

Fast transition pathways, multidimensional Hamiltonian, resonance and transport structures, **B:**454–457

Feenberg renormalized perturbation, unimolecular reaction kinetics, quantum energy flow, **B:**212–214

Fenichel normal form, phase-space transition states:
chaotic itinerancy, **A:**388–389
Lie perturbation theory, **A:**354–358
Melnikov integral, **A:**361–371
normally hyperbolic invariant manifolds, **A:**338–340, **A:**345–352

Fermi-Pasta-Ulam (FPU) computer experiments:
Hamiltonian system, multiergodicity, **B:**471
slow relaxation, internal degrees of freedom, Hamiltonian systems:
hypothesis validity, **B:**406–412
mixed-phase space, **B:**381–387
molecular systems, **B:**387–392
nearly integrable systems, **B:**392–398

Fermi resonance, resonantly coupled isomerizing/dissociating systems:
anharmonicity, **A:**286–287
high energy bifurcations, **A:**297–301
polyad folding and saddle-node bifurcation, **A:**287–296

Fermi's golden rule:
heat transfer, quantum energy flow, protein vibrational states, anharmonic decay, **B:**238–241
unimolecular reaction kinetics, quantum energy flow, **B:**214
vibrational energy relaxation, **B:**180
general formula, **B:**183–184
perturbation expansion, **B:**182

Fern genotype/phenotype, nonmetric multidimensional scaling algorithm and, **B:**329–332

Few-dimensional systems, unimolecular reaction rate theory, phase-space structure, bottlenecks, **A:**18–19

Fick equation, chaotic transitions, inert and reactive substances, **B:**522

FKPP reaction, **B:**531

Final-state interaction, multichannel isomerization, linear surprisal theory, **B:**78

Finger formation pattern, unimolecular reaction rate, wave packet dynamics, **A:**126–128
Finite-horizon configuration, slow relaxation, internal degrees of freedom, mixed-phase space, **B:**387
Finite-time Lyapunov exponents: multidimensional Hamiltonian dynamical systems:
  instability properties, **B:**512–517
  correction, **B:**516
  order of motion, **B:**512–514
  qualitative different instabilities, **B:**514–516
  QR method corrections, **B:**506–512
  correction procedure, **B:**508–511
  finite-time error, **B:**507–508
  standard method, **B:**506–507
  vectors, **B:**503–505
phase-space transition states, multidimensional chaos crisis, **A:**393–395
First-order kinetics, chaotic transition, regularity, two basin landscapes, reactive island theory, **A:**158–163
Fisher-Kolmogorov-Petrovsky-Piskunov (FKPP) reaction, chaotic transitions, inert and reactive substances, **B:**521–522
  ARD equation, **B:**530–531
Five-body systems, celestial mechanics, **A:**311–312
Floater hopping (FH), rapid alloys:
  binary clusters, **B:**174–175
  microcluster dynamics, **B:**176
Floaters, rapid alloying, binary clusters:
  floppy surface atom reaction paths, **B:**170–171
  reaction path enumeration, **B:**172–173
  saddle point energy distribution, **B:**173–174
Floppy molecules, canonical perturbation theory, **A:**269–278
Floppy surface atoms, rapid alloying, microcluster dynamics, **B:**170–171
Floquet solution, fringed tunneling, semiclassical method, **A:**407
Fluctuation-dissipation theorem:
  excess heat production:
    Boltzmann equilibrium distribution, **B:**355–359

hysteresis loop area, **B:**358–359
thermodynamics, **B:**355–356
Jarzynski's nonequilibrium work relation, **B:**368–369
nonergodic adiabatic invariant, anomalous variance, **B:**361–368
  Hamiltonian chaotic systems, **B:**363–367
  microcanonical distribution, **B:**361–363
  second law, **B:**369–370
  superstatistical equilibrium distributions, **B:**359–361
recursive cell production and evolution, catalytic reaction networks, **B:**593–595
Flux dynamics, Wigner's transition state dynamics, rank-one saddle phase-space structure, transition state pathway, **A:**193–194
Fokker-Planck equation:
  chaotic transitions, inert and reactive substances, **B:**520–522
  recursive cell production and evolution, catalytic reaction network, **B:**591–595
Folding dynamics, multibasin landscapes, chaotic transition, regularity, temperature dependence in dimensionality of, **B:**293–299
Foliation, phase-space transition states, normally hyperbolic invariant manifolds, **A:**349–352
Force-force-correlation function approximations, vibrational energy relaxation, **B:**186–190
  first term contribution, **B:**187–190
  Taylor expansion, **B:**186–187
Forward reactive trajectories, Wigner's transition state dynamics, rank-one saddle phase-space structure:
  normally hyperbolic invariant manifolds, **A:**188–190
  transition state principles, **A:**191
Four-body systems, molecular internal space, **B:**113–125
  collective coordinates, isomerization mechanism, **B:**118–121
  gauge field isomerization suppression, Eckart frame, **B:**123–125
  principal-axis hyperspherical coordinates, equations of motion, **B:**114–118

transition state trapped motion, DCF effects,
  B:121–123
Four-dimensional degrees of freedom:
  phase-space transition state geometry,
    A:243–244
  phase-space transition states:
    Arnold web model, **A:**375–377
    tangency and, **A:**381–385
Four-dimensional free rotor, unimolecular
  fragmentation mapping, Morse-like
  kicking field, **A:**27–30
Fourier transform:
  chaotic transitions, inert and reactive
    substances, standard and anomalous
    diffusion, **B:**525–527
  intramolecular dynamics, floppy molecules,
    canonical perturbation theory,
    A:271–278
  multidimensional Hamiltonian systems,
    resonance and transport, **B:**439
  vibrational energy relaxation:
    basic principles, **B:**180
    cytochrome c, CD stretching mode, **B:**197
    symmetrized autocorrelation function,
      B:184–185
Fractal distribution:
  heat transfer, quantum energy flow,
    anomalous subdiffusion, **B:**230–238
  multichannel isomerization, inter-basin
    mixing, turning points, **B:**51–53
"Fraction" modes, heat transfer, quantum
  energy flow, proteins, **B:**241–248
Free-fall problem, Coulomb three-body
  structures, **A:**330–332
"Freezing" energy, multichannel chemical
  isomerization, multi-basin potential,
  B:30–31
Frequency-dependent energy diffusion, heat
  transfer, quantum energy flow, proteins,
  B:243–248
Frequency filter, heat transfer, quantum energy
  flow, **B:**225–227
Frequency space, multidimensional
  Hamiltonian, resonance and structure,
  B:442–445
Fringed tunneling models:
  global branch structures, **A:**415–417
  multiple characteristic trajectories,
    A:422–425
  perturbation strength, **A:**425–426

semiclassical technique, **A:**406–407
branch global structures, **A:**415–417
complex-domain method, **A:**407–410
two-dimensional barriers, **A:**428–431
  example, **A:**428–431
Froeschlé mapping, multidimensional
  Hamiltonian, resonance and transport,
  B:438–439
diffusion coefficient, **B:**440–442
frequency and phase space, **B:**443–445
resonance overlap, **B:**457–460
Front propagation, chaotic transitions, inert and
  reactive substances:
  thin front dynamics, **B:**537–540
Fukui's criterion, onset dynamics, argon
  clusters, **B:**130–131
Full width at half maximum (FWHM),
  vibrational energy relaxation,
  cytochrome c, CD stretching mode,
  B:199–200
Fully developed chaotic regimes, chaotic
  transition, regularity, two-basin
  landscapes, **A:**146–147
  saddle crossing stochasticity, **A:**166
"Funnel-like" phenomenon, multibasin
  landscapes, chaotic transition, regularity,
  minimalistic 46-bead protein models,
  B:267–270

Γ periodic orbit, phase-space transition state
  geometry, two degrees of freedom,
  A:230–234
Gap problem, phase-space transition states:
  Arnold web model, **A:**377
  Melnikov integral, **A:**397–398
Gaspard-Rice mapping:
  quantum suppression of Arnold diffusion,
    A:134
  unimolecular fragmentation, Morse-like
    kicking field:
    four-dimensional rotor, **A:**27–30
    two-dimensional free particles, **A:**24–27
Gauge field isomerization suppression,
  molecular internal space, **B:**106–113
  applications, **B:**110–113
  Eckart subspace dynamics, **B:**107–109
  four-body systems, Eckart frame,
    B:123–125
  quantitive role, **B:**106–107
  reaction rate effects, **B:**109–110

Gauge-invariant expression, molecular internal space:
  classical equations and metric force, **B**:97–99
  metrics and kinetic energy, **B**:93–94
Gauge-theoretical formalism, molecular internal space:
  kinematics, **B**:88–90
  n-body systems:
    gauge-invariant kinetic energy expression, **B**:93–94
    rotation-vibration kinetic energy, **B**:91–93
Gaussian distribution, recursive cell production and evolution, catalytic reaction networks, **B**:591–595
Gaussian Orthogonal Ensemble (GOE):
  atomic clusters, energy level distribution, **B**:4–5
  optimal control theory, perfect control solution, **A**:454–456
  quantum chaos systems, random matrix system, **A**:438–439
Gaussian random vectors, quantum chaos systems, optimal control theory, **A**:439–441
Gaussian stochastic force, multibasin landscapes, chaotic transition, regularity, Berendsen algorithm, constant-temperature MD, **B**:309–310
Gaussian wavepacket, quantum chaos systems, **A**:437
Gene expression:
  catalytic reaction network, recursive cell production and evolution, **B**:545–546
  intermingled hypercycle network production, **B**:595–596
  itinerant dynamics, **B**:596
  minority control hypothesis, **B**:557–573
    evolvability, **B**:566–567
    experimental protocol, **B**:567–571
    growth speed, **B**:565
    kinetic theory, heredity and, **B**:571–573
    model parameters, **B**:557–561
    molecule chemical composition, **B**:565–566
    molecule preservation, **B**:565
    stochastic results, **B**:561–564
  nonmetric multidimensional scaling algorithm and:
    microarray data, **B**:345–348
    temporal patterns, **B**:348–349

Genotype/phenotype analysis, nonmetric multidimensional scaling algorithm and, **B**:329–333
  ferns, **B**:329–332
  green autotrophs, **B**:332–333
Geometrical optics limit, chaotic transitions, inert and reactive substances:
  G-equation, **B**:534–537
Geometrics of phase-space:
  angular momentum, **A**:247–261
    astrophysics applications, **A**:256–261
    inelastic scattering, **A**:257–261
    rotating frame dynamics, **A**:248–256
    relative equilibrium, **A**:249–251
    van der Waals complex, **A**:251–256
  atomic clusters:
    regularity, chaos, and ergodicity, **B**:14–20
  chaotic transition, regularity, two-basin landscapes, **A**:151–163
  fringed tunneling, multiple trajectories, **A**:422–425
  Hamiltonian dynamics, **A**:219–223
    dimensions, **A**:221–223
    general equations, **A**:219–221
  multichannel isomerization, **B**:27–28
  inter-basin mixing, **B**:45–53
    Markov-type isomers, **B**:45–47
    reaction tube bifurcation, **B**:47–50
    time scale, **B**:50–51
    turning point fractal dimension, **B**:51–53
  n degrees of freedom, **A**:234–247
    Hénon-Heiles potential, **A**:237–244
    normally hyperbolic invariant manifolds, dimensions and, **A**:234–237
    geometric characteristics, **A**:236–237
    linear regime, **A**:235–236
    tri-body dynamics, zero angular momentum, **A**:244–247
  one degree of freedom, **A**:223–228
    linear case, linearization, **A**:223–227
    nonlinearities, **A**:227
  two degrees of freedom, **A**:228–234
    linear theory, **A**:229–230
    periodic orbit dividing serfaces, **A**:230–234
  Wigner's transition state dynamics, **A**:174–175
  stationary points, **A**:179

Geometric structure:
G-equation, chaotic transitions, inert and reactive substance, geometrical optics limit, **B:**534–537
Gibbs free energy:
binary clusters, microcluster dynamics, rapid alloying, **B:**156–157
globally-coupled Hamiltonian systems, relaxation and diffusion, Hamilton mean field (HMF) model, **B:**481
Global dynamics:
fringed tunneling branches, **A:**415–417
Hamiltonian systems relaxation and diffusion:
angle space diffusion, **B:**488–498
equilibrium diffusion, **B:**489–490
nonstationary state, **B:**494–496
quasi-stationary state, **B:**491–493
model and initial condition, **B:**480–481
probability distribution function of momenta, **B:**484–487
relaxation process, **B:**481–484
slow dynamics, **B:**484
multidimensional barrier tunneling, **A:**402–406
phase-space transition states:
Arnold diffusion model, **A:**371–377
chaotic itinerancy, **A:**385–389
Lie perturbation, **A:**352–358
many degrees of freedom, **B:**425–427
Melnikov integral, **A:**358–371
examples, **A:**395–398
multidimensional chaos, **A:**392–395
normally hyperbolic invariant manifolds, **A:**338–340, **A:**345–352
breakdown, **A:**389–392
singular perturbation theory, **A:**342–345
skeleton bifurcation, **A:**340–341
tangency principles, **A:**341–342, **A:**377–385
Global mixing, multichannel isomerization, **B:**46–47
Gō-like BLN protein models, multibasin landscapes, chaotic transition, regularity:
basic properties, **B:**267–270
nonstationarity in energy fluctuations, **B:**270–285
temperature dependence in dimensionality of folding dynamics, **B:**295–299

G-protein genes, nonmetric multidimensional scaling algorithm and, microarray data, **B:**345–348
Gradient external path (GEP), onset dynamics, argon clusters, **B:**131–133
Gram-Schmidt orthogonalization, unimolecular reaction rate theory, reaction path analysis, **A:**56–59
Gray-Rice-Davis rate theory, unimolecular reaction:
predissociation, ARRKM theory, **A:**39–41
Gray-Rice isomerization theory, unimolecular reaction rate, **A:**66–70
cyclobutanone, **A:**102–104
reactive island theory (RIT) vs., **A:**80–84
Zhao-Rice approximation, **A:**73–75
Green autotrophs, nonmetric multidimensional scaling algorithm and, **B:**332–333
Green fluorescent protein (GFP), heat transfer, quantum energy flow:
anomalous subdiffusion, **B:**228–238
proteins, **B:**242–248
Green function, fringed tunneling, semiclassical method, **A:**407
Growth speed control, recursive cell production and evolution, catalytic reaction networks, **B:**565
Gumbel distribution, Hamiltonian system, multiergodicity, **B:**474–475
Gutzwiller trace formula, three-body problem, **A:**306–309
Gyration radii, molecular internal space:
democratic centrifugal force, **B:**104–106
four-body systems, PAHC equations of motion, **B:**115–118
three-atom clusters, isomerization dynamics, **B:**96–97
topographical mapping, **B:**99–103
Gyration space, molecular internal space:
classical equations and metric force, **B:**97–99
topographical mapping, **B:**101–103

Hamiltonian systems:
atomic clusters, basic principles, **B:**4
chaotic systems, fluctuation-dissipation theorem, excess heat production, adiabatic invariant, **B:**363–367
Coulomb three-body problem, zero angular momentum, **A:**312–319

Hamiltonian systems (*Continued*)
  fringed tunneling, semiclassical method,
    A:406–407
  globally coupled Hamiltonian systems,
    relaxation and diffusion:
    angle space diffusion, **B:**488–498
    equilibrium diffusion, **B:**489–490
    nonstationary state, **B:**494–496
    quasi-stationary state, **B:**491–493
    model and initial condition, **B:**480–481
    probability distribution function of
      momenta, **B:**484–487
    relaxation process, **B:**481–484
    slow dynamics, **B:**484
  intramolecular dynamics:
    floppy molecules, canonical perturbation
      theory, **A:**269–278
    resonantly coupled isomerizing/
      dissociating systems:
      high energy bifurcations, **A:**297–301
      polyad folding and saddle-node
        bifurcation, **A:**288–296
  multichannel chemical isomerization,
    multi-basin potential, **B:**28–30
  multidimensional dynamical transport and
    resonance:
    basic principles, **B:**450–452
    deflected diffusion, **B:**460–462
    diffusion coefficient, **B:**440–442
    fast transition pathway, **B:**454–457
    frequency and phase space, **B:**442–445
    model components, **B:**438–439
    morphological change, **B:**445–447
    overlap, **B:**457–460
    residence time distribution, **B:**447–450
    rotation number, **B:**439–440
    transition diagram, **B:**452–454
  multiergodicity and nonstationarity:
    complex behaviors, **B:**474–475
    stagnant motion deviation, **B:**466–469
    survival time distribution, clustering
      motions, **B:**471–474
    universality conjecture, **B:**469–471
  phase-space transition states, **A:**219–223
    angular momentum:
    astrophysics inelastic scattering,
      A:257–261
    rotating frame dynamics, **A:**249–256
    Arnold web model, **A:**372–377
    dimensions, **A:**221–223

  general equations, **A:**219–221
  global reaction dynamics, **A:**341–342
  Lie perturbation theory, **A:**353–358
  Melnikov integral, **A:**358–371
  tangency, **A:**378–385
  three degrees of freedom, Hénon-Heiles
    potential, **A:**237–244
  triatomic dynamics, zero angular
    momentum, **A:**244–247
  two degrees of freedom, periodic orbit
    dividing surfaces, **A:**230–234
  quantum chaos systems:
    kicked rotor system, **A:**442–443
    random matrix system, **A:**438–439
  slow relaxation, internal degrees of freedom:
    FPU models, **B:**398–403
    mixed-phase space systems, anomalous
      transport, **B:**379–387
    molecular systems, **B:**387–392
    nearly integrable picture, applicability,
      B:392–398
    validity of hypothesis, **B:**403–412
  unimolecular reaction rate:
    dissociation techniques, **A:**116–123
    isomerization:
      cyclobutanone molecule, **A:**100–104
      HCN to CHN, **A:**96–100
    normally hyperbolic invariant manifolds
      (NHIM), **A:**7–8
    phase-space structure:
      canonical transformation, **A:**9–10
      normally hyperbolic invariant manifolds
        (NHIM), **A:**20–22
    quantum energy flow, local random matrix
      theory, **B:**212–214
    quantum suppression of Arnold diffusion,
      A:131–134
    Zhao-Rice approximation, **A:**41–54
  Wigner's transition state dynamics:
    Lie transformation normalization,
      A:194–198
    rank-one saddle phase-space structure,
      A:183–184
Hamilton-Jacobi equation:
  unimolecular reaction rate theory, phase-
    space structure, **A:**9–10
  action/angle variables, **A:**11–12
  Wigner's transition state dynamics, hydrogen
    cyanide isomerization model,
    A:200–202

SUBJECT INDEX

Hamilton mean field (HMF) model, globally-coupled Hamiltonian systems, relaxation and diffusion:
 initial condition, **B:**480–481
 probability distribution function of momenta, **B:**484–487
 relaxation process, **B:**481–484
Harmonic approximation, unimolecular reaction rate, isomerization theory, cyclobutanone, **A:**103–104
Harmonic oscillator:
 heat transfer, quantum energy flow, anomalous subdiffusion, **B:**229–238
 intramolecular dynamics, floppy molecules, canonical perturbation theory, **A:**273–278
 onset dynamics, argon clusters, configuration entropy, **B:**140
 phase-space transition states:
  Lie perturbation theory, **A:**353–358
  one degree of freedom model, **A:**225–227
 slow relaxation, internal degrees of freedom, Hamiltonian systems, **B:**399–403
 vibrational energy relaxation:
  force-force-correlation function approximations, **B:**187–190
  quantum correction factor and, **B:**186
Harthcock-Lane potential energy surface, unimolecular reaction rate, 3-phospholene isomerization, **A:**91–96
Hausdorff dimension, atomic clusters, power spectra and phase-space dimensions, **B:**5–11
Heat of solution, rapid alloying, microcluster dynamics, numerical results, **B:**164–165
Heat reservoir, rapid alloying, microcluster dynamics, nano-sized clusters, **B:**158–160
Heat transfer:
 clusters and macromolecules, quantum energy flow, unimolecular reaction kinetics, **B:**221–248
 energy diffusion, **B:**222–223
 proteins, **B:**241–248
 protein vibrational energy:
  anharmonic decay, **B:**237–241
  anomalous subdiffusion, **B:**227–237
 water clusters, **B:**223–227
 quantum energy flow, **B:**206

Heaviside step function, multichannel isomerization, inter-basin mixing, **B:**52–53
Heisenberg representation, unimolecular reaction rate, quantized ARRKM theory vs. rigorous quantum rate, **A:**112–114
Helium-chloride (HeCl$_2$), unimolecular reaction rate, predissociation theory, **A:**63–66
Helium-iodine-chloride, unimolecular reaction rate, predissociation theory, **A:**66
Helium-iodine (HeI$_2$), unimolecular reaction:
 Gray-Rice-Davis ARRKM theory, **A:**39–41
 predissociation:
  Davis-Gray analysis, **A:**35–39
  intramolecular bottleneck, **A:**60–61
  Morse potentials, **A:**59–60
  rate constants, **A:**60
Hénon-Heiles potential:
 globally coupled Hamiltonian systems, **B:**478–480
 multichannel isomerization, **B:**83
 phase-space transition state geometry:
  angular momentum, **A:**248
  three degrees of freedom, **A:**237–244
Heredity, recursive cell production and evolution, catalytic reaction network, **B:**544–546
 kinetic theory, **B:**571–573
Hessian matrices:
 multichannel isomerization, microcanonical temperature, single exponential form, **B:**66–67
 onset dynamics, argon clusters:
  cell partition, potential energy surfaces, **B:**133–134
  gradient external path, **B:**132–133
 phase-space transition state geometry, one degree of freedom model, linearized Hamiltonian, **A:**224–227
 phase-space transition states, Lie perturbation theory, **A:**353–358
 vibrational energy relaxation, cytochrome c, CD stretching mode, **B:**190–192
 coupling coefficients, **B:**192–195
Heteroclinicity, multidimensional barrier tunneling:
 global dynamics, **A:**405–406

Heterogeneous scenario, slow relaxation,
  internal degrees of freedom, Hamiltonian
  systems:
    FPU models, **B:**398–403
    molecular systems, **B:**392
Hilbert space, unimolecular reaction rate theory:
    faster-than-classical anomalous diffusion,
      A:134–137
    quantum transport, cantori systems,
      A:129–131
Homeochaos, catalytic reaction networks, **B:**582
Homoclinic tangency:
    chaotic transition, regularity, two-basin
      landscapes:
        reactive island theory, **A:**155–163
        transition state theory, **A:**145–147
    phase-space transition states, Melnikov
      integral, **A:**361–371
    unimolecular fragmentation mapping, Morse-
      like kicking field, two-dimensional free
      particle, **A:**24–27
    unimolecular reaction rate:
      normally hyperbolic invariant manifolds
        (NHIM), **A:**21–22
      reactive island theory (RIT), **A:**78–80
Homogeneous scenario, slow relaxation, internal
  degrees of freedom, Hamiltonian
  systems, molecular systems, **B:**392
Homology equation, Wigner's transition state
  dynamics, Lie transformation
  normalization, **A:**195–197
Hopping mechanism, heat transport, quantum
  energy flow, **B:**250–251
Horseshoe dynamics. *See* Chaos theory
Husimi representation, optimal control theory,
  kicked rotor system, **A:**443
Huygens dynamics, chaotic transitions, inert and
  reactive substances, front propagation,
  B:538–540
Hydrodynamics, singular perturbation theory,
  A:345
Hydrogen-bond network, slow relaxation,
  internal degrees of freedom, Hamiltonian
  systems, molecular systems, **B:**388–392
Hydrogen cyanide (HCN):
  intramolecular dynamics:
    adiabatic *vs.* nonadiabatic delocalization,
      A:278–286
    floppy molecules, canonical perturbation
      theory, **A:**269–278

  canonical transformations, **A:**301–302
  isomerization to CHN, unimolecular reaction
    rate theory:
      potential energy surfaces, **A:**96–100
      Wigner's transition state dynamics,
        isomerization, **A:**198–212
      Hamiltonian equation, **A:**200–202
      model system, **A:**199–200
      nonreactive degrees of freedom
        quantization, **A:**211–212
      normal form transformation, **A:**205–207
      stationary flow points, **A:**202–205
      visualization techniques, **A:**207–210
Hydrogen isocyanide (HNC):
  hydrogen cyanide isomerization to,
    A:198–212
  intramolecular dynamics:
    adiabatic *vs.* nonadiabatic delocalization,
      A:278–286
    floppy molecules, canonical perturbation
      theory, **A:**269–278
Hyperbolicity:
  Coulomb three-body problem, collinear eZe
    configuration, triple collision orbits,
    A:325–329
  phase-space transition states:
    breakdown of, **A:**341, **A:**389–392
    chaotic itinerancy, **A:**388–389
    Melnikov integral, **A:**366–371
    n degrees of freedom structures, **A:**237
    normally hyperbolic invariant manifolds,
      A:345–352
  slow relaxation, internal degrees of freedom,
    Hamiltonian systems, anomalous
    transport, mixed-phase space,
    B:379–387
Hypercycle structures, recursive cell production
  and evolution, catalytic reaction
  networks, **B:**548–549
  autocatalytic phases, **B:**578–581
  evolution models, **B:**585–588
  intermingled hypercycle network,
    B:581–583
  kinetics of heredity, **B:**571–573
Hypothesis validity, slow relaxation, internal
  degrees of freedom, Hamiltonian
  systems, **B:**403–412
Hysteresis loop, fluctuation-dissipation
  theorem, excess heat production,
  B:358–359

1/f-noise spectral density, multibasin landscapes, chaotic transition, regularity, nonstationarity in energy fluctuations, **B:**272–280
Incident waves, fringed tunneling models, two-dimensional barrier systems, **A:**429–431
Induction time distribution, Hamiltonian system, multiergodicity, **B:**471
Inelastic scattering, phase-space transition state geometry, angular momentum, **A:**257–261
Inequality conditions, phase-space transition states, Arnold web model, **A:**375–377
Inert substances, chaotic transitions:
standard and anomalous diffusion, **B:**523–527
strong anomalous diffusion, **B:**527–530
Infinite-horizon configuration, slow relaxation, internal degrees of freedom, mixed-phase space, **B:**387
Inherent structure (IS) surfaces, slow relaxation, internal degrees of freedom, Hamiltonian systems, molecular systems, **B:**388–392
Inhomogeneity, slow relaxation, internal degrees of freedom, Hamiltonian systems, molecular systems, **B:**390–392
Initial value representation (IVR), unimolecular reaction rate, semiclassical approximation, rigorous quantum rate, **A:**114–116
Input-boundary condition, multidimensional barrier tunneling, low-frequency approximation, **A:**422
Instability properties, multidimensional Hamiltonian dynamical systems, finite-time Lyapunov exponents, **B:**512–517
correction, **B:**516
order of motion, **B:**512–514
qualitative different instabilities, **B:**514–516
Instantaneous normal mode (INM) analysis, slow relaxation, internal degrees of freedom, Hamiltonian systems:
hypothesis validity, **B:**404–412
molecular systems, **B:**391–392
Instanton trajectory, multidimensional tunneling, **A:**402–406

Integrable systems, global motion, many degrees of freedom, **B:**425–427
Integration paths, classical trajectories, multidimensional barrier tunneling:
periodic perturbation effects, **A:**412–414
static barriers, **A:**410–412
Interatomic distances, Wigner's transition state dynamics, hydrogen cyanide isomerization model, **A:**199–200
Hamiltonian equations, **A:**200–202
Inter-basin mixing, multichannel isomerization, **B:**45–53
reaction tube bifurcation, **B:**47–50
time scale, **B:**50–51
turning point fractal dimension, **B:**51–53
Interlayer mixing (IM), rapid alloying, microcluster dynamics:
future research issues, **B:**176
saddle point energy distribution, **B:**175
Intermediate, semi-chaotic regime, chaotic transition, regularity, two-basin landscapes, saddle regions, dynamical regularity, **A:**149–151
Intermingled hypercycle network (IHN), recursive cell production and evolution, catalytic reaction networks, **B:**581–583
evolution models, **B:**585–588
Intramolecular energy transfer:
chaotic transition, regularity, two-basin landscapes, dynamical bottlenecks, **A:**166–168
dissociation dynamics:
resonantly coupled systems, **A:**286–301
high-energy bifurcations, **A:**296–301
polyad folding and saddle-node bifurcations, **A:**287–296
unimolecular reactions, Hamiltonian equations, **A:**117–123
isomerization pathways:
nearly separable systems, **A:**269–286
adiabatic *vs.* nonadiabatic delocalization, **A:**278–286
canonical perturbation theory, floppy molecules, **A:**269–278
resonantly coupled systems, **A:**286–301
high-energy bifurcations, **A:**296–301
polyad folding and saddle-node bifurcations, **A:**287–296

Intramolecular energy transfer (*Continued*)
   unimolecular reaction rate theory:
      Davis-Gray predissociation analysis,
         A:34–39
      phase-space quantum scars, **A:**108
      3-phospholene isomerization, **A:**94–96
      quantum transport, cantori systems,
         A:129–131
      Zhao-Rice approximation:
         bottleneck dividing surface, **A:**46–48
         isomerization theory, **A:**72–75
         zeroth-order rate constant, bottleneck
            crossing, **A:**48–50
Intramolecular vibrational energy relaxation
   (IVR):
   heat transport, **B:**249–251
   phase-space transition states:
      Arnold web model, **A:**372–377
      normally hyperbolic invariant manifold
         connections, **A:**340
   unimolecular reaction kinetics:
      cyclohexane ring inversion, **B:**217–221
      LRMT dynamical corrections to RRKM
         theory, **B:**216
      quantum energy flow, **B:**208–221
      vibrational energy relaxation, **B:**200–201
Invariant measure, unimolecular reaction rate
   theory, phase-space structure, **A:**10–11
Invariant structures:
   chaotic transition, regularity, two-basin
      landscapes, saddle crossing stochasticity,
      A:165–166
   multidimensional barrier tunneling, global
      dynamics, **A:**404–406
   phase-space transition states,
      multidimensional chaos crisis,
      A:393–395
Inverse harmonic potential, phase-space
   transition states, Lie perturbation theory,
   A:353–358
In vitro experiments, recursive cell production
   and evolution, catalytic reaction
   networks, **B:**567–571
Isomerization:
   atomic clusters:
      molecular internal space, **B:**90–91
      passage-times, **B:**27
      chaotic transition, regularity, two-basin
         landscapes, phase space geometrics,
         A:152–163

intramolecular dynamics:
   nearly separable systems, **A:**269–286
   adiabatic *vs.* nonadiabatic delocalization,
      A:278–286
   canonical perturbation theory, floppy
      molecules, **A:**269–278
   resonantly coupled systems, **A:**286–301
   high-energy bifurcations, **A:**296–301
   polyad folding and saddle-node
      bifurcations, **A:**287–296
multichannel chemical clusters:
   inter-basin mixing, **B:**45–53
   reaction tube bifurcation, **B:**47–50
   time scale, **B:**50–51
   turning point fractal dimension,
      B:51–53
linear surprisal theory, **B:**70–81
   chemical reaction dynamics temperature,
      B:70–71
   maximum entropy principle, **B:**72–74
   variational structure, **B:**74–80
   nonequilibrium stationary flow,
      B:74–76
   population ratio, **B:**76–78
   prior distribution, **B:**78–80
liquid-like state behavior, **B:**34–45
   lifetime averaging uniformity, basin
      transition, **B:**35–45
   accumulated residence time/
      ergodicity, **B:**35–36
   exponential decay expansion
      uniformity, **B:**39–42
   non-RRKM behaviors, **B:**42–45
   nonstatistical low-energy behavior,
      B:44–45
   passage time and uniformity,
      B:36–39
   short-time behavior, **B:**42–44
   unimolecular dissociation via transition
      state, **B:**34–35
Markov-type appearance, **B:**45–47
memory-losing dynamic geometry,
   B:45–53
microcanonical temperature, **B:**53–70
   Arrhenius-like relation, **B:**60–62
   canonical temperature, **B:**70
   definition, **B:**57–60
   density of states evaluation, **B:**55–56
   exponential relation, average lifetimes,
      B:62–70

liquid-like ergodicity and nonergodicity, **B:**67–69
  M_7 single exponential form case study, **B:**65–67
  multiexponential form, **B:**62–63
  single exponential form, **B:**64–65
  lifetime averaging law, **B:**53–55
  local temperatures, **B:**60
multi-basin potential, **B:**28–34
  anomalous time series, **B:**31–34
  M_7-like system, **B:**28–30
  solid-liquid transition, **B:**30–31
phase-space transition state geometry, nonzero angular momentum, **A:**247–248
unimolecular reaction kinetics, **A:**66–104
  cyclobutanone, **A:**100–104
  double-well system, **A:**84–88
  Gray-Rice theory, **A:**66–70
  reactive island theory *vs.*, **A:**80–84
  HCN → CNH, **A:**96–100
  3-phospholene, **A:**91–96
  reactive island theory, **A:**75–80
  Gray-Rice theory *vs.*, **A:**80–84
  triple-well system, **A:**88–91
  Zhao-Rice approximation, **A:**70–75
Wigner's transition state dynamics, hydrogen cyanide, **A:**198–212
  Hamiltonian equation, **A:**200–202
  model system, **A:**199–200
  nonreactive degrees of freedom quantization, **A:**211–212
  normal form transformation, **A:**205–207
  stationary flow points, **A:**202–205
  visualization techniques, **A:**207–210
Itinerant dynamics, recursive cell production and evolution, catalytic reaction networks, **B:**596

Jacobi vectors:
  finite-time Lyapunov exponents, multidimensional Hamiltonian dynamical systems, **B:**503–505
  intramolecular dynamics, floppy molecules, canonical perturbation theory, **A:**269–278
  molecular internal space:
    Eckart subspace dynamics, **B:**107–109

  four-body systems, PAHC equations of motion, **B:**114–118
  gauge-dependent expression, rotation-vibration energy, **B:**92–93
  three-atom clusters, isomerization dynamics, **B:**96–97
  Wigner's transition state dynamics, hydrogen cyanide isomerization model, **A:**199–200
Jarzynski's nonequilibrium work relations, fluctuation-dissipation theorem, excess heat production, **B:**368–369

K entropy, multichannel isomerization, microcanonical temperature, single exponential form, **B:**66–67
Keplerian forces, phase-space transition state geometry, triatomic dynamics, zero angular momentum, **A:**244–247
Kicked rotor systems, optimal control theory, quantum chaos systems, **A:**439–446
Kicking potential:
  unimolecular fragmentation mapping, Morse-like kicking field:
    four-dimensional free rotor, **A:**27–30
    two-dimensional free particle, **A:**22–27
  unimolecular reaction rate, faster-than-classical anomalous diffusion, **A:**134–137
Kinematics, molecular internal space, **B:**88–90
Kinetic energy:
  heat transfer, quantum energy flow:
    anomalous subdiffusion, **B:**232–238
    clusters and macromolecules, **B:**222–223
  molecular internal space, gauge-invariant expression, **B:**93–94
  recursive cell production and evolution, catalytic reaction network, **B:**571–573
Kolmogorov-Arnold-Moser (KAM) theorem:
  multidimensional Hamiltonian systems, resonance and transport:
    diffusion coefficient, **B:**440–442
    fast transition pathways, **B:**454–457
  multidimensional phase space slow dynamics:
    Arnold model, **B:**429–430
    global motion, **B:**425–427
  phase-space transition states:
    Lie perturbation theory, **A:**357–358
    Melnikov integral, **A:**362–371

Kolmogorov-Arnold-Moser (KAM)
theorem (*Continued*)
slow relaxation, internal degrees of freedom,
Hamiltonian systems, **B:**376–378
mixed-phase space, **B:**380–381
nearly integrable systems, **B:**394–398
unimolecular reaction rate theory:
isomerization, cyclobutanone,
A:102–104
phase-space structure, **A:**12–14
bottlenecks:
few-dimensional systems, **A:**18–19
many-dimensional systems, **A:**19–20
Poincaré surface of section, **A:**14–16
stability analysis, **A:**17–18
Kolmogorov entropy:
atomic clusters:
basic principles, **B:**3–4
power spectra and phase-space dimensions,
B:5–11
regularity, chaos, and ergodicity, **B:**11–20
chaotic transition, regularity, two-basin
landscapes, **A:**146–147
Wigner's transition state dynamics,
A:178–179
Kolmogorov-Sinai (KS) entropy:
fluctuation-dissipation theorem, excess heat
production:
long-period limit, **B:**358–359
multibasin landscapes, chaotic transition,
regularity, nonstationarity in energy
fluctuations, **B:**274–280
onset dynamics, argon clusters:
results from, **B:**142–143
slow relaxation, internal degrees of freedom
Komatsuzaki-Berry technique, unimolecular
reaction, **A:**138–140
Kramers-Grote-Hynes theory, chaotic transition,
regularity, two-basin landscapes,
A:163–165
Kronecker delta, molecular internal space,
gauge-dependent expression, rotation-
vibration energy, **B:**92–93

Ladder operators, intramolecular dynamics,
floppy molecules, canonical perturbation
theory, **A:**271–278
Lagrangian equations:
chaotic transitions, inert and reactive
substances, **B:**521–522
front propagation, **B:**537–540
standard and anomalous diffusion,
B:523–527
strong anomalous diffusion, **B:**527–530
molecular internal space:
classical equations and metric force,
B:98–99
democratic centrifugal force, **B:**104–106
four-body systems, PAHC equations of
motion, **B:**115–118
gauge field suppression, **B:**111–113
gauge-invariant energy, **B:**94
onset dynamics, argon clusters, gradient
external path, **B:**132–133
Lagrangian singularity, phase-space transition
states, tangency and, **A:**381–385
Lagrangian time integral, unimolecular reaction
rate, semiclassical approximation,
rigorous quantum rate, **A:**115–116
Landau-Teller-Zwanzig (LTZ) formula,
vibrational energy relaxation, quantum
correction factor and, **B:**186
Langevin capture theory, phase-space transition
state geometry, angular momentum,
astrophysics applications, **A:**256–261
Langevin dynamics simulations, multibasin
landscapes, chaotic transition, regularity,
nonstationarity in energy fluctuations,
B:273–280
Langevin equation:
chaotic transition, regularity, two-basin
landscapes, Kramers-Grote-Hynes
theory, **A:**163–165
multibasin landscapes, chaotic transition,
regularity, Berendsen algorithm,
constant-temperature MD, **B:**309–310
recursive cell production and evolution,
catalytic reaction network, statistical law,
B:591–595
Langmuir orbit:
Coulomb three-body problem, free-fall
problem, **A:**332
three-body problem, **A:**307–309
Laplace transform, multichannel isomerization,
B:34–45
Lattice vibrations, Hamiltonian system
multiergodicity, **B:**741
LBL bead sequence, multibasin landscapes,
chaotic transition, regularity,
nonstationarity in, **B:**280–282

SUBJECT INDEX 651

Lebesgue measure, unimolecular fragmentation mapping, Morse-like kicking field, two-dimensional free particle, **A:**23–27
Lennard-Jones potential:
atomic reactions:
power spectra and phase-space, **B:**10–11
regularity, chaos, and ergodicity, **B:**13–20
transformation mechanisms, **B:**21–22
cluster dynamics, chaotic transition, regularity, two-basin landscapes, **A:**146–147
molecular internal space, atomic cluster isomerization dynamics, **B:**91
multibasin landscapes, chaotic transition, regularity, nonstationarity in energy fluctuations, **B:**274–280
multichannel chemical isomerization:
liquid-like state, **B:**35
microcanonical temperature, **B:**58–62
multi-basin potential, **B:**28–30
solid-liquid tansition, **B:**30–31
onset dynamics, argon clusters, **B:**135–136
phase-space transition state geometry, angular momentum, astrophysics applications, **A:**256–261
rapid alloying, binary clusters, **B:**156–157
slow relaxation, internal degrees of freedom, Hamiltonian systems, molecular systems, **B:**392
Level shift mechanism, dissociation dynamics, unimolecular reactions, Hamiltonian equations, **A:**117–123
Lévy flight model:
atomic clusters, **B:**23
chaotic transitions, inert and reactive substances, **B:**524–527
Lévy walk model, chaotic transitions, inert and reactive substances, **B:**524–527
Liapunov exponents. *See* Lyapunov exponents
Lie canonical perturbation theory, atomic clusters, nonlinear transformation, **B:**21–22
Lie-Deprit transforms, Wigner's transition state dynamics, theoretical background, **A:**179
Lie perturbation theory, phase-space transition states, **A:**352–358
normally hyperbolic invariant manifold connections, **A:**339–340
Lie transformation algorithm:
canonical perturbation theory:

chaotic transition, regularity, two-basin landscapes, **A:**146–147
unimolecular reaction rate theory, phase-space structure, **A:**10
chaotic transition, regularity, two-basin landscapes:
normally hyperbolic invariant manifolds, **A:**167–168
reactive island theory, **A:**162–163
intramolecular dynamics, floppy molecules, canonical perturbation theory, **A:**277–278
Wigner's transition state dynamics, **A:**178–179
basic principles, **A:**194–197
hydrogen cyanide isomerization, **A:**205–207
visualization techniques, **A:**207–210
normal-form coordinates, dynamics, **A:**197–198
normalization, **A:**194–198
Lie triangle, Wigner's transition state dynamics, Lie transformation normalization, **A:**195–197
Lifetime averaging uniformity, multichannel isomerization:
liquid-like states, **B:**34–45
accumulated residence time and ergodicity, **B:**35–36
exponential decay expression, **B:**39–42
non-RRKM behaviors, **B:**42–45
nonstatistical low energy behavior, **B:**44–45
short-time behavior, **B:**42–44
passage time and uniformity, **B:**36–39
microcanonical temperature/Arrhenius relation, **B:**53–55
exponential relation, **B:**62–70
Lifetime parameter, vibrational energy relaxation, cytochrome *c*, CD stretching mode, **B:**195–196
Limit cycles, phase-space transition states, hyperbolicity breakdown, **A:**390–392
Lindemann's criterion (index):
multichannel isomerization:
lifetime averaging, liquid-like dynamics, **B:**69–70
multi-basin potential, solid-liquid tansition, **B:**30–31
onset dynamics, argon clusters:

Lindemann's criterion (index) (*Continued*)
argon$_5$ clusters, **B:**148–151
configuration entropy, **B:**140–142
phase transition results, **B:**138–139
rapid alloying, microcluster dynamics, solid phase transition, **B:**165–166
unimolecular reaction kinetics, LRMT dynamical corrections to RRKM theory, **B:**215–216
Linear harmonic energy, resonantly coupled isomerizing/dissociating systems, polyad folding and saddle-node bifurcation, **A:**288–296
Linearization, phase-space transition state geometry:
  n degrees of freedom structures, normally hyperbolic invariant manifolds, **A:**235–236
  one degree of freedom model, **A:**223–227
  three degrees of freedom, Hénon-Heiles potential, **A:**240–244
Linear surprisal theory, multichannel isomerization, **B:**70–81
  chemical reaction dynamics, temperature, **B:**70–71
  maximum entropy principle, **B:**72–74
  recent developments in, **B:**80
  variational structure, **B:**74–80
    nonequilibrium stationary flow, **B:**74–76
    population ratio, **B:**76–78
    prior distribution, **B:**78–80
Liouville density function, unimolecular reaction rate theory, Wigner function and, **A:**106
Liouville equation, fluctuation-dissipation theorem, excess heat production:
  Boltzmann equilibrium distribution, **B:**355–356
  microcanonical distribution, **B:**362–363
Lippman-Schwinger equation, multichannel isomerization, linear surprisal theory, **B:**74–76
Liquid-phase dynamics:
  multichannel chemical isomerization, **B:**34–45
  lifetime averaging uniformity, basin transition, **B:**35–45

accumulated residence time/ergodicity, **B:**35–36
exponential decay expansion uniformity, **B:**39–42
non-RRKM behaviors, **B:**42–45
nonstatistical low-energy behavior, **B:**44–45
passage time and uniformity, **B:**36–39
short-time behavior, **B:**42–44
onset dynamics, argon clusters, **B:**139
Liquid water models:
  multibasin landscapes, chaotic transition, regularity, **B:**262–263
  slow relaxation, internal degrees of freedom, Hamiltonian systems:
    molecular systems, **B:**388–392
    nearly integrable systems, **B:**393–398
Lissajous figure, chaotic transition, regularity, two-basin landscapes, reactive island theory, **A:**162–163
Local equilibrium assumption, chaotic transition, regularity, two-basin landscapes, **A:**144–147
Local Random Matrix Theory (LRMT):
  unimolecular reaction kinetics:
    cyclohexane ring inversion, **B:**217–221
    dynamical corrections to RRKM theory, **B:**215–216
    quantum energy flow, **B:**209–214
Locking phenomenon, molecular internal space:
  four-body systems, collective coordinates, **B:**118–121
  topographical mapping, **B:**100–103
Log-normal distribution, recursive cell production and evolution, catalytic reaction networks, **B:**591–595
Long-period limit, fluctuation-dissipation theorem, excess heat production, 358–359
Long-time correlation decay, slow relaxation, internal degrees of freedom, Hamiltonian systems, molecular systems, **B:**388–392
Lorentz gas model, slow relaxation, internal degrees of freedom:
  mixed-phase space, **B:**386–387
Lorentzian function:
  slow relaxation, internal degrees of freedom, Hamiltonian systems, **B:**402–403
  hypothesis validity, **B:**406–412

SUBJECT INDEX

vibrational energy relaxation, cytochrome c,
 CD stretching mode, **B:**197
Lorenz equation, multibasin landscapes, chaotic
 transition, regularity:
 average mutual information, **B:**293–294
 state-space structural embedding,
  B:286–288
Low-energy states, multichannel isomerization,
 nonstatistical behavior, liquid-like phase,
 B:44–45
Low-frequency approximation,
 multidimensional barrier tunneling:
 adiabatic solution, **A:**418–422
 Melnikov method, **A:**431–432
 theoretical analysis, **A:**417–418
LR method, finite-time Lyapunov exponents,
 multidimensional Hamiltonian
 dynamical systems, **B:**510–512
L-set, fringed tunneling models, global structure
 of branches, **A:**415–417
Lyapunov exponents. *See also* Maximum
 Lyapunov exponent (MLE)
 atomic clusters:
  basic principles, **B:**3–4
  power spectra and phase-space dimensions,
   B:5–11
  regularity, chaos, and ergodicity, **B:**11–20
 chaotic transitions:
  inert and reactive substances, front
   propagation, **B:**537–540
  two-basin landscapes, **A:**146–147
 finite-time exponents, multidimensional
  Hamiltonian dynamical systems:
  instability properties, **B:**512–517
  correction, **B:**516
  order of motion, **B:**512–514
  qualitative different instabilities,
   B:514–516
  QR method corrections, **B:**506–512
   correction procedure, **B:**508–511
   finite-time error, **B:**507–508
   standard method, **B:**506–507
   vectors, **B:**503–505
 Hamiltonian systems, multiergodicity,
  B:466–469
 universality conjecture, **B:**470–471
 multibasin landscapes, chaotic transition,
  regularity:
  nonstationarity in energy fluctuations,
   B:274–280

multichannel isomerization:
 global mixing, **B:**46–47
 inter-basin mixing, reaction tube
  bifurcation, **B:**49–50
multidimensional Hamiltonian, resonance and
 transport, frequency and phase space,
 B:443–445
onset dynamics, argon clusters:
 results from, **B:**142–143
 phase-space transition states:
  hyperbolicity breakdown, **A:**391–392
  multidimensional chaos crisis,
   A:393–395
  normally hyperbolic invariant manifolds,
   A:338–340, **A:**347–352
 slow relaxation, internal degrees of freedom,
  Hamiltonian systems:
  anomalous transport, mixed-phase space,
   B:379–387
  nearly integrable systems, **B:**394–398
 unimolecular reaction rate theory:
  phase-space quantum scars, **A:**107–108
  phase-space structure, stability analysis,
   A:18
 Wigner's transition state dynamics,
  A:178–179

Macromolecules, heat transfer, quantum energy
 flow, **B:**221–248
 unimolecular reaction kinetics:
  energy diffusion, **B:**222–223
  proteins, **B:**241–248
  protein vibrational energy:
   anharmonic decay, **B:**237–241
   anomalous subdiffusion, **B:**227–237
   water clusters, **B:**223–227
Magnetization, globally-coupled Hamiltonian
 systems, relaxation and diffusion:
 Hamilton mean field model, **B:**481–484
 probability distribution function,
  B:484–487
Many-dimensional systems, unimolecular
 reaction rate theory:
 phase-space structure, bottlenecks,
  A:19–20
Mapping models, unimolecular fragmentation,
 A:22–30
Morse-like kicking field:
 four-dimensional free rotor, **A:**27–30
 two-dimensional free particles, **A:**22–27

Mapping time, chaotic transition, regularity, two-basin landscapes, reactive island theory, **A:**159–163
Maradudin-Fein formula, CD stretching mode, **B:**190
Markov-type stochastic process:
 fluctuation-dissipation theorem, excess heat production, **B:**354–355
 Boltzmann equilibrium distribution, **B:**355–356
 multichannel isomerization:
  inter-basin mixing, **B:**45–47
  lifetime averaging unformity, **B:**40–42
 multidimensional phase space, slow dynamics, Arnold diffusion, **B:**434–435
 slow relaxation, internal degrees of freedom, mixed-phase space, **B:**383–387
Mass ratio effect, Coulomb three-body problem, collinear electron-nucleus-electron (eZe) configuration, **A:**319–330
 antiproton-proton-antiproton system, **A:**330
 triple collision manifold, **A:**320–323
 triple collision orbits, **A:**323–329
Mathematical reduction, phase-space transition state geometry, angular momentum, rotating frame dynamics, **A:**248–256
Maximum entropy principle (MEP), multichannel isomerization, linear surprisal theory, **B:**72–74
Maximum Lyapunov exponent, multichannel chemical isomerization, multi-basin potential, **B:**30–31
McGehee's blow-up technique:
 collinear eZe configuration, mass ratio effect, triple collision manifold, **A:**320–323
 n-body problem in celestial mechanics, **A:**309–312
 three-body problem, theoretical background, **A:**308–309
Mean square displacement (MSD), globally-coupled Hamiltonian systems, relaxation and diffusion:
 angle space diffusion, **B:**488–498
 multibasin landscapes, chaotic transition, regularity, protein structures, **B:**264–266
 quasi-stationary state, **B:**491–493
Melnikov-Arnold integral, multidimensional phase space slow dynamics:
 global motion, **B:**427
 slow relaxation, **B:**429–430
Melnikov function:
 fringed tunneling models, perturbation strength, **A:**426
 multidimensional barrier tunneling:
  low-frequency approximation, **A:**420–422
  theoretical analysis, **A:**417–418
  unstable periodic orbits, **A:**431–432
 phase-space transition states:
  Arnold web model, **A:**371–377
  examples using, **A:**395–398
  global reaction dynamics, **A:**341–342
  normally hyperbolic invariant manifold connections, **A:**340, **A:**358–371
  tangency and, **A:**378–385
"Melting" energy, multichannel chemical isomerization, multi-basin potential, **B:**30–31
Memory-losing dynamics:
 multibasin landscapes, chaotic transition, regularity, **B:**259–260
 temperature dependence in dimensionality of folding dynamics, **B:**295–299
 multichannel isomerization:
  inter-basin mixing, **B:**45–53
  reaction tube bifurcation, **B:**47–50
  time scale, **B:**50–51
  turning point fractal dimension, **B:**51–53
Meso time scale, rapid alloying:
 binary clusters, **B:**156–157
 microcluster dynamics, **B:**176
Metric force, molecular internal space:
 internal motion equations, **B:**97–99
 three-atom cluster isomerization, collective coordinates, **B:**94–106
 democratic centrifugal force, mass-balance asymmetry and trapping trajectories, **B:**103–106
 internal motion equations, **B:**97–99
 principal-axis hyperspherical coordinates, **B:**94–97
 topographical mapping, PAHC, **B:**99–103
Metric tensor, molecular internal space, gauge-invariant energy, **B:**93–94
Microarray data, nonmetric multidimensional scaling algorithm and, **B:**343–349
 gene expression patterns, **B:**348–349
 information content, **B:**345–348

Microcanonical distribution, fluctuation-
  dissipation theorem, anomalous
  variance, nonergodic adiabatic invariant,
  B:361–363
Microcanonical ensembles, multichannel
  isomerization:
  inter-basin mixing, **B:**45–53
  reaction tube bifurcation, **B:**47–50
  time scale, **B:**50–51
  turning point fractal dimension, **B:**51–53
  linear surprisal theory, **B:**70–81
  chemical reaction dynamics temperature,
    B:70–71
  maximum entropy principle, **B:**72–74
  recent developments in, **B:**80
  variational structure, **B:**74–80
  nonequilibrium stationary flow, **B:**74–76
  population ratio, **B:**76–78
  prior distribution, **B:**78–80
  liquid-like state behavior, **B:**34–45
  lifetime averaging uniformity, basin
    transition, **B:**35–45
  accumulated residence time/ergodicity,
    B:35–36
  exponential decay expansion uniformity,
    B:39–42
  non-RRKM behaviors, **B:**42–45
  nonstatistical low-energy behavior,
    B:44–45
  passage time and uniformity, **B:**36–39
  short-time behavior, **B:**42–44
  unimolecular dissociation via transition
    state, **B:**34–35
  Markov-type appearance, **B:**45–47
  memory-losing dynamic geometry,
    B:45–53
  microcanonical temperature, **B:**53–70
  Arrhenius-like relation, **B:**60–62
  canonical temperature, **B:**70
  definition, **B:**57–60
  density of states evaluation, **B:**55–56
  exponential relation, average lifetimes,
    B:62–70
  liquid-like ergodicity and nonergodicity,
    B:67–69
  M_7 single exponential form case study,
    B:65–67
  multiexponential form, **B:**62–63
  single exponential form, **B:**64–65
  lifetime averaging law, **B:**53–55

local temperatures, **B:**60
multi-basin potential, **B:**28–34
  anomalous time series, **B:**31–34
  M_7-like system, **B:**28–30
  solid-liquid transition, **B:**30–31
Microcanonical rate constant, unimolecular
  reaction, quantized ARRKM theory vs.
  rigorous quantum rate, **A:**112–114
Microcanonical temperature:
  multichannel chemical isomerization,
    B:53–70
  Arrhenius-like relation, **B:**60–62
  canonical temperature, **B:**70
  definition, **B:**57–60
  density of states evaluation, **B:**55–56
  exponential relation, average lifetimes,
    B:62–70
  liquid-like ergodicity and nonergodicity,
    B:67–69
  M_7 single exponential form case study,
    B:65–67
  multiexponential form, **B:**62–63
  single exponential form, **B:**64–65
  lifetime averaging law, **B:**53–55
  local temperatures, **B:**60
  multichannel isomerization, future research
    issues, **B:**82–83
Microscopic mechanisms, rapid alloying,
  microcluster dynamics, **B:**170–175
  floppy surface atoms and PES reaction paths,
    B:170–171
  reaction path enumeration, **B:**171–173
  saddle point energy distribution,
    B:173–175
Miller-Handy-Adams reaction, unimolecular
  reaction rate theory, reaction path
  analysis, **A:**54–59
Minimalistic 46-bead protein models, multibasin
  landscapes, chaotic transition, regularity,
  B:266–270
Minimum energy path (MEP), intramolecular
  dynamics, floppy molecules, canonical
  perturbation theory, **A:**269–278
Minority control hypothesis, recursive cell
  production and evolution, catalytic
  reaction network, **B:**557–573
  core hypercycles, **B:**582–583
  evolvability, **B:**566–567
  experimental protocol, **B:**567–571
  growth speed, **B:**565

Minority control hypothesis, recursive cell production and evolution, catalytic reaction network (*Continued*)
intermingled hypercycle network production, B:595–596
itinerant dynamics, B:596
kinetic theory, heredity and, B:571–572
model parameters, B:557–561
molecule chemical composition, B:565–566
molecule preservation, B:565
stochastic results, B:561–564
Mixed phase space, Hamiltonian systems, anomalous transport, B:379–387
Model truncated in reciprocal space (MTRS), multidimensional Hamiltonian dynamical systems, finite-time Lyapunov exponent instability, B:512–517
MOIL program, heat transfer, quantum energy flow, anomalous subdiffusion, B:228–238
Molecular diffusivity, chaotic transitions, inert and reactive substances, standard and anomalous diffusion, B:524–527
Molecular dynamics (MD):
multibasin landscapes, chaotic transition, regularity:
Berendsen algorithm, B:309–310
liquid water, B:262–263
local/global collective coordinates, B:260–266
minimalistic 46-bead protein models, B:267–270
nonstationarity in energy fluctuations, B:273–280
onset dynamics, argon clusters, B:135–136
configuration entropy, B:140
phase-space structure, unimolecular reaction rate theory, A:9–22
action/angle variables, A:11–12
canonical transformation, A:9–10
few-dimensional system bottlenecks, A:18–19
invariant measure, A:10–11
KAM theorem, A:12–14
many-dimensional system bottlenecks, A:19–20
normally hyperbolic invariant manifold, A:20–22
Poincaré surface of section, A:14–16

stability analysis, A:17–18
rapid alloying, binary clusters:
numerical results, B:160–161
research background, B:156–157
rapid alloying, microcluster dynamics:
radial and surface diffusion, B:169–170
saddle point energy distribution, B:175
slow relaxation, internal degrees of freedom, Hamiltonian systems, B:377–378
anomalous transport, mixed-phase space, B:379–387
molecular systems, B:387–392
nearly integrable system, B:392–398
Molecular internal space:
atomic cluster isomerization, B:90–91
four-body systems, B:113–125
collective coordinates, isomerization mechanism, B:118–121
gauge field isomerization suppression, Eckart frame, B:123–125
principal-axis hyperspherical coordinates, equations of motion, B:114–118
transition state trapped motion, DCF effects, B:121–123
future research issues, B:125–126
gauge field isomerization suppression, B:106–113
applications, B:110–113
Eckart subspace dynamics, B:107–109
quantitive role, B:106–107
reaction rate effects, B:109–110
gauge-theoretical formalism, *n*-body systems:
gauge-invariant kinetic energy expression, B:93–94
rotation-vibration kinetic energy, B:91–93
kinematics, B:88–90
metric force collective coordinates, three-atom cluster isomerization, B:94–106
democratic centrifugal force, mass-balance asymmetry and trapping trajectories, B:103–106
internal motion equations, B:97–99
principal-axis hyperspherical coordinates, B:94–97
topographical mapping, PAHC, B:99–103
Molecular systems:
recursive cell production and evolution, catalytic reaction networks:
chemical composition control, B:565–566

growth speed control, **B:**565
minority molecule preservation, **B:**565
statistical laws, **B:**593–595
slow relaxation, internal degrees of freedom, **B:**387–392
Molecular taxonomy, nonmetric multidimensional scaling algorithm and, **B:**324–329
Molecule optimal dynamic coordinates (MODC), multibasin landscapes, chaotic transition, regularity, protein structures, **B:**264–266
Monkey saddle points, onset dynamics, argon clusters, **B:**152
Monodromy matrix, unimolecular reaction rate theory, phase-space structure, stability analysis, **A:**17–18
Morphological changes, multidimensional Hamiltonian resonance and transport structures, **B:**445–447
Morse-like kicking field, unimolecular fragmentation mapping:
  four-dimensional free rotor, **A:**27–30
  two-dimensional free particle, **A:**22–27
Morse potential:
  molecular internal space:
    atomic cluster isomerization dynamics, **B:**90–91
    four-body systems, **B:**113–125
  multichannel chemical isomerization, microcanonical temperature, **B:**58–62
  multichannel isomerization, inter-basin mixing, **B:**45–43, **B:**52–53
  phase-space transition states:
    angular momentum, astrophysics applications, **A:**256–261
    Melnikov integral, **A:**371
    triatomic dynamics, zero angular momentum, **A:**245–247
  rapid alloying, microcluster dynamics: **B:**175–176
  nano-sized clusters, **B:**159–160
  unimolecular reaction rate theory, Zhao-Rice approximation, **A:**42–54
Motion trapping, molecular internal space, democratic centrifugal force, **B:**121–123
MRRKM. *See* Zhao-Rice approximation
M-set:
  fringed tunneling models:
    global structure of branches, **A:**415–417

multiple trajectories, **A:**423–425
perturbation strength, **A:**425–426
semiclassical techniques, **A:**409–410
multidimensional barrier tunneling:
  critical point local structure, **A:**414–415
  Melnikov method, theoretical background, **A:**417–418
Multibasin potential:
chaotic transition, regularity on:
  Berendsen algorithm, constant-temperature molecular dynamics, **B:**309–310
  embedded techniques:
    basic principles, **B:**302–309
    phase-space reconstruction, **B:**285–288
  energy nonstationarity, protein landscapes, **B:**270–285
    bending energy fluctuation, **B:**278–282
    bond energy flucation, **B:**277–278
    torsional angle energy fluctuation, **B:**282–285
  folding dynamic dimensionality, temperature dependency, **B:**294–299
  global/local collective coordinates, **B:**261–262
  liquid water, **B:**262–263
  minimalistic 46-bead protein models, **B:**266–270
  phase space transport geometry, **B:**260–261
  proteins, **B:**263–266
  state-space structure, **B:**285–299
    average mutual information, **B:**292–294
    false nearest neighbors, **B:**288–292
    phase-space reconstruction, embedding of, **B:**285–288
  multichannel chemical isomerization, **B:**28–34
    anomalous time series, **B:**31–34
    lifetime averaging uniformity, **B:**35–45
    M_7-like system, **B:**28–30
    solid-liquid transition, **B:**30–31
  slow relaxation, internal degrees of freedom, Hamiltonian systems, **B:**413–418
Multicellular organisms, recursive cell production and evolution, catalytic reaction network, **B:**552–557
Multichannel chemical isomerization:
  inter-basin mixing, **B:**45–53
  reaction tube bifurcation, **B:**47–50
  time scale, **B:**50–51

Multichannel chemical
  isomerization (Continued)
  turning point fractal dimension, B:51–53
  linear surprisal theory, B:70–81
  chemical reaction dynamics temperature,
    B:70–71
  maximum entropy principle, B:72–74
  recent developments in, B:80
  variational structure, B:74–80
  nonequilibrium stationary flow, B:74–76
  population ratio, B:76–78
  prior distribution, B:78–80
  liquid-like state behavior, B:34–45
  lifetime averaging uniformity, basin
    transition, B:35–45
  accumulated residence time/ergodicity,
    B:35–36
  exponential decay expansion uniformity,
    B:39–42
  non-RRKM behaviors, B:42–45
  nonstatistical low-energy behavior,
    B:44–45
  passage time and uniformity, B:36–39
  short-time behavior, B:42–44
  unimolecular dissociation via transition
    state, B:34–35
  Markov-type appearance, B:45–47
  memory-losing dynamic geometry, B:45–53
  microcanonical temperature, B:53–70
    Arrhenius-like relation, B:60–62
    canonical temperature, B:70
    definition, B:57–60
    density of states evaluation, B:55–56
    exponential relation, average lifetimes,
      B:62–70
    liquid-like ergodicity and nonergodicity,
      B:67–69
    M_7 single exponential form case study,
      B:65–67
    multiexponential form, B:62–63
    single exponential form, B:64–65
  lifetime averaging law, B:53–55
  local temperatures, B:60
  multi-basin potential, B:28–34
    anomalous time series, B:31–34
    M_7-like system, B:28–30
    solid-liquid transition, B:30–31
Multidimensional systems:
  barrier tunneling:
    fringed tunneling models, A:406–407

  global branch structures, A:415–417
  multiple characteristic trajectories,
    A:422–425
  two dimensional barriers, A:428–431
  global dynamics, A:402–406
  Melnikov integral, A:431–432
  perturbation strength, A:425–426
  semiclassical method, A:407–410
  M-set structure at critical point,
    A:414–415
  periodic perturbation effects, A:412–414
  static barrier, A:410–412
    classical solution, A:410
    trajectory singularities and integration
      paths, A:410–412
  theoretical analyses:
    low-frequency approximation,
      A:418–422
    overview, A:417–418
Hamiltonian dynamical systems:
  finite-time Lyapunov exponents:
    instability properties, B:512–517
      correction, B:516
      order of motion, B:512–514
      qualitatively different instabilities,
        B:514–516
    QR method corrections, B:506–512
      correction procedure, B:508–511
      finite-time error, B:507–508
      standard method, B:506–507
      vectors, B:503–505
  resonance structure:
    deflected diffusion, B:460–462
    diffusion coefficient, B:440–442
    frequency and phase space, B:442–445
    model components, B:438–439
    morphological change, B:445–447
    overlap, B:457–460
    residence time distribution, B:447–450
    rotation number, B:439–440
  transport structure:
    basic principles, B:450–452
    deflected diffusion, B:460–462
    diffusion coefficient, B:440–442
    fast transition pathway, B:454–457
    model components, B:438–439
    rotation number, B:439–440
    transition diagram, B:452–454
  phase space systems (See also Nonmetric
    multidimensional scaling (nMDS))

SUBJECT INDEX

chaos crisis in, **A:**392–395
multibasin landscapes, chaotic transition, regularity:
  nonstationarity in energy fluctuations, **B:**285
  state-space structural embedding, **B:**285–288
normally hyperbolic invariant manifold connections, **A:**339–340
onset dynamics, argon clusters, cell partition, potential energy surfaces, **B:**134
slow relaxation dynamics:
  Arnold model, **B:**427–430
  future research issues, **B:**435–436
  numerical method and results, **B:**430–435
  global motion, many degrees of freedom, **B:**425–427
  research background, **B:**423–424
Wigner's transition state dynamics, **A:**173–175
Multiergodicity. *See* Ergodicity
Multiexponential lifetime averaging, multichannel isomerization, microcanonical temperature and, **B:**62–63
Multiple parallel reaction paths, recursive cell production and evolution, catalytic reaction networks, **B:**594–595
Multi-precision method, multidimensional phase space slow dynamics, Arnold model, **B:**430–435
Multivariate analysis (MVA), nonmetric multidimensional scaling:
  brain wave analysis, **B:**341
  data mining applications, **B:**316–320
  microarray data, **B:**347–349
  soil bacteria biodiversity, **B:**336–338
Murrell, Carter, and Halonen (MCH) potential, Wigner's transition state dynamics, hydrogen cyanide isomerization model, **A:**199–211

Nano-sized clusters, rapid alloying, microcluster dynamics, **B:**157–160
  procedural characteristics, **B:**157–158
  simulation model, **B:**158–160
Navier-Stokes equation, singular perturbation theory, **A:**345

N-body problem in celestial mechanics, basic principles, **A:**309–312
Nearest-neighbor level spacings:
  atomic clusters, energy level distribution, **B:**4–5
  rapid alloying, microcluster dynamics, numerical results, **B:**164–165
Nearly integrable system:
  global motion, many degrees of freedom, **B:**425–427
  slow relaxation, internal degrees of freedom, Hamiltonian systems, **B:**392–398
  slow relaxation, mixed-phase space, **B:**380–387
Nearly separable isomerizing systems, intramolecular dynamics, **A:**269–286
  adiabatic *vs.* nonadiabatic delocalization, **A:**278–286
  canonical perturbation theory, floppy molecules, **A:**269–278
Negative feedback process, recursive cell production and evolution, catalytic reaction networks, **B:**594–595
Negative heat of solution, rapid alloying, microcluster dynamics:
  nano-sized clusters, **B:**158–160
  numerical results, **B:**164–165
"Negative" temperature, multichannel isomerization, linear surprisal theory, 79–80
Nekhoroshev theorem:
  Hamiltonian system multiergodicity and nonstationarity, **B:**466
  universality conjecture, **B:**469–471
  slow relaxation, internal degrees of freedom, Hamiltonian systems, **B:**376–378, **B:**400–403
  mixed-phase space, **B:**381–387
  nearly integrable systems, **B:**394–398
Neon-chloride (NeCl$_2$), unimolecular reaction rate, predissociation theory, **A:**63–66
Neon-iodine-chloride, unimolecular reaction rate, predissociation theory, **A:**66
Neon-iodine molecules, unimolecular reaction rate theory, predissociation, **A:**61–63
Newton-Raphson algorithm, onset dynamics, argon clusters, gradient external path, **B:**132–133

SUBJECT INDEX

Nonadiabatic delocalization, isomerizing systems, intramolecular dynamics, *vs.* adiabatic, **A:**278–286

Noncompact clusters, rapid alloying, microcluster dynamics, reaction path enumeration, **B:**173

Nonequilibrium stationary flow, multichannel isomerization, linear surprisal theory, **B:**74–76

Nonergodicity:
adiabatic invariant, anomalous variance, fluctuation-dissipation theorem, excess heat production, **B:**361–368
Hamiltonian chaotic systems, **B:**363–367
microcanonical distribution, **B:**361–363
multichannel isomerization, lifetime averaging, liquid-like dynamics, **B:**67–70

Nonhyperbolic systems, slow relaxation, internal degrees of freedom, Hamiltonian systems, **B:**379–387

Nonlinear dynamics:
atomic clusters, **B:**21–22
phase-space transition states:
Arnold model, **A:**371–377
one degree of freedom model, **A:**227
three degrees of freedom, Hénon-Heiles potential, **A:**241–244
Wigner's transition state dynamics, rank-one saddle phase-space structure, normally hyperbolic invariant manifolds, **A:**187–190

Nonmetric multidimensional scaling (nMDS), data mining using:
algorithm characteristics, **B:**320–321
basic principles, **B:**317–320
brain wave analysis, **B:**338–341
embedded point estimation, **B:**322–323
gene expression, temporal patterns, **B:**348–349
microarray data, **B:**343–349
molecular/morphology comparisons, **B:**329–333
ferns, **B:**329–332
green autotrophs, **B:**332–333
molecular taxonomy, **B:**324–329
pointwise criterion, **B:**321–322
protein family, **B:**342–343
soil bacteria biodiversity, **B:**334–338
theoretical background, **B:**316–317

Nonreactive trajectories:
phase-space transition state geometry, one degree of freedom model, **A:**225–227
Wigner's transition state dynamics:
hydrogen cyanide isomerization model, quantization, **A:**211–212
rank-one saddle phase-space structure, normally hyperbolic invariant manifolds, **A:**188–190

Nonstationarity:
globally-coupled Hamiltonian systems, relaxation and diffusion, angle diffusion, **B:**494–496
Hamiltonian systems:
complex behaviors, **B:**474–475
research background, **B:**465–466
stagnant motion deviation, **B:**466–469
survival time distribution, clustering motions, **B:**471–474
universality conjecture, **B:**469–471
multibasin landscapes, chaotic transition, regularity, protein structures, **B:**270–285
bending energy fluctuation, **B:**278–282
bond energy fluctuation, **B:**277–278
torsional angle energy fluctuation, **B:**282–285
slow relaxation, internal degrees of freedom, mixed-phase space, **B:**385–387

Nonstatistical low-energy behavior, multichannel isomerization, liquid-like phase, **B:**44–45

Nonzero angular momentum, phase-space transition state geometry, basic principles, **A:**247–261

No-recrossing rule, transition state theory, Wigner's dynamical perspective, **A:**173–175

No-return assumption, chaotic transition, regularity, two-basin landscapes, **A:**144–147

Normal-Form theory, Wigner's transition state dynamics:
hydrogen cyanide isomerization, Lie transformation, **A:**205–207
Lie transformation normalization, **A:**197–198
rank-one saddle phase-space structure:
"apt" coordinates, **A:**184–186
Hamiltonian equations, **A:**184
normally hyperbolic invariant stable/unstable manifolds, **A:**186–190

transition state structure location,
  A:191–193
Normally hyperbolic invariant manifolds
  (NHIM):
  chaotic transition, regularity, two-basin
    landscapes:
    reactive island theory, **A**:163
    skeleton structure, **A**:166–168
    transition state theory, **A**:147
  Coulomb three-body problem:
    collinear eZe configuration, triple collision
      orbits, **A**:325–329
    future research issues, **A**:333–334
  multibasin landscapes, chaotic transition,
    regularity, **B**:259–260
  phase-space transition states:
    angular momentum, rotating van der Waals
      complex, **A**:252–256
    Arnold web model, **A**:373–377
    breakdown of hyperbolicity, **A**:341
    chaotic itinerancy, **A**:385–389
    global chemical reactions, **A**:338–340,
      A:345–352
    breakdown, **A**:389–392
    Hamiltonian dynamics, **A**:222–223
    hyperbolicity breakdown, **A**:389–392
    Lie perturbation theory, **A**:352–358
    Melnikov integral, **A**:340, **A**:358–371
    multidimensional chaos crisis, **A**:392–395
    n degrees of freedom structures,
      A:234–237
    three degrees of freedom, Hénon-Heiles
      potential, **A**:241–244
    triatomic dynamics, zero angular
      momentum, **A**:244–247
    tangency, **A**:377–385
  unimolecular reaction rate theory:
    phase-space structure, **A**:20–22
    reactive island theory (RIT), **A**:76–80
  Wigner's transition state dynamics:
    hydrogen cyanide isomerization model,
      A:200
    Lie transformation, **A**:205–207
    visualization techniques, **A**:207–210
    Lie transformation, stable/unstable
      manifolds, **A**:197–198
    rank-one saddle phase-space structure,
      A:183–194
    "apt" coordinates, Normal Form theory,
      A:184–186

n-degree-of-freedom Hamiltonian,
  A:183–184
  stable/unstable manifolds, **A**:186–190
  transition state defined, **A**:190–191
  transition state flux, **A**:193–194
  transition state search technique,
    A:191–193
  theoretical background, **A**:174–175,
    A:179
Nucleotide sequencing, nonmetric
  multidimensional scaling algorithm and,
  molecular taxonomy, **B**:324–329
Number distribution of chemicals, recursive cell
  production and evolution:
  autocatalytic network, **B**:575–581
  catalytic reaction network, **B**:547

Octahedral structure, atomic clusters, regularity,
  chaos, and ergodicity, **B**:14–20
1.5-dimensional scattering barrier:
  fringed tunneling models, **A**:430–431
  multidimensional barrier tunneling, global
    dynamics, **A**:404–406
One degree of freedom model, phase-space
  transition states, **A**:223–228
  linear case/linearization, **A**:223–227
  Melnikov integral, **A**:358–371
  nonlinearities, **A**:227
One-dimensional attractor, multibasin
  landscapes, chaotic transition,
  regularity, embedding theorems,
  B:302–309
Onset dynamics, argon cluster phase transition:
  analytic techniques, **B**:131–136
  configuration entropy, **B**:140
  configuration space, **B**:140–142
  gradient extremal, **B**:131–133
  Lindemann's criterion, **B**:138–139
  additional potentials, **B**:148–151
  Lyapunov and KS entropy, **B**:142–143
  partitioned cell dynamics, **B**:145–146
  potential energy surface, cell partition,
    B:133–134
  potential function, MD simulation, and
    temperature, **B**:135–136
  reaction paths, **B**:136–138
  research background, **B**:129–131
  stationary points, **B**:143–145
  structural characteristics, **B**:136
  watershed, **B**:146–148

Optimal control theory (OCT), quantum chaos systems, **A**:437–443
 analytic expression, **A**:449–456
 controlled kicked rotor, **A**:439–440
 controlled random matrix, **A**:438–439
 perfect control solution, **A**:453–456
Order of motion, multidimensional Hamiltonian dynamical systems, finite-time Lyapunov exponent instability, **B**:512–517
Ordinary differential equation (ODE):
 phase-space transition state geometry: Hamiltonian dynamics, **A**:220–221
 one degree of fredom model, nonlinear regime, **A**:227
 phase-space transition states, singular perturbation theory, **A**:343–345
Oscillatory shifts, phase-space transition states: Melnikov integral, **A**:366–371
 tangency and, **A**:383–385
Overlap mechanisms, multidimensional Hamiltonian resonance, **B**:457–460

Painlevé analysis:
 multidimensional barrier tunneling, global dynamics, **A**:403–406
 n-body problem in celestial mechanics, **A**:310–312
Pairwise resonances, unimolecular reaction rate theory, Zhao-Rice approximation, bottleneck dividing surface, **A**:48
Parallel sliding, multibasin landscapes, chaotic transition, regularity, nonstationarity in energy fluctuations, **B**:284–285
Parasitic molecules, recursive cell production and evolution, catalytic reaction network, **B**:583–585
Partitioned cell dynamics, onset dynamics, argon clusters, **B**:145–146
Passage time, multichannel chemical isomerization, lifetime averaging uniformity, **B**:36–39
Path integral techniques:
 multidimensional barrier tunneling, M-set critical point, **A**:414–415
 three-body problem, **A**:306–309
Péclet number, chaotic transitions, inert and reactive substances, research background, **B**:522

Pentagonal bipyrimidal (PBP) structures, multichannel isomerization:
 inter-basin mixing, **B**:46–57
 microcanonical temperature, single exponential form, **B**:65–67
 multi-basin potential, **B**:28–30
 anomalous time series, **B**:31–34
 nonstatistical behavior, low energy phase, **B**:45
Periodic orbit dividing surfaces (PODS):
 phase-space transition state geometry, two degrees of freedom, **A**:230–234
 resonantly coupled isomerizing/dissociating systems:
  high energy bifurcations, **A**:297–301
  polyad folding and saddle-node bifurcation, **A**:292–296
 Wigner's transition state dynamics:
  rank-one saddle phase-space structure, normally hyperbolic invariant manifolds, **A**:186–190
  theoretical background, **A**:176–179
Periodic orbit instability, multidimensional barrier tunneling, Melnikov method, **A**:431–432
Periodic perturbation, multidimensional barrier tunneling, semiclassical results, **A**:412–414
Perpendicular coordinates:
 intramolecular dynamics, **A**:268–269
 resonantly coupled isomerizing/dissociating systems, polyad folding and saddle-node bifurcation, **A**:287–296
Perturbation expansion, vibrational energy relaxation, **B**:181–182
Perturbation theory. *See also* Singular perturbation theory
 dissociation dynamics, unimolecular reactions, Hamiltonian equations, **A**:118–123
 fringed tunneling:
  semiclassical method, **A**:406–407
  strength characteristics, **A**:425–426
 phase-space transition states:
  geometry research, **A**:266
  Melnikov integral, **A**:362–371
  normally hyperbolic invariant manifolds, **A**:348–352
  theoretical background, **A**:218–219

SUBJECT INDEX 663

time scales, **A:**341–342
slow relaxation, internal degrees of freedom, Hamiltonian systems, **B:**400–403
future research issues, **B:**413–418
mixed-phase space, anomalous transport, **B:**380–387
Wigner's transition state dynamics, rank-one saddle phase-space structure, normally hyperbolic invariant manifolds, **A:**187–190
Phase categories, recursive cell production and evolution, catalytic reaction networks, **B:**575–581
Phase-space transition states. *See also* Multidimensional phase space systems
atomic clusters, **B:**5–11
nonlinear canonical transformation, **B:**21–22
regularity, chaos, and ergodicity, **B:**11–20
geometry of:
angular momentum, **A:**247–261
astrophysics applications, **A:**256–261
inelastic scattering, **A:**257–261
rotating frame dynamics, **A:**248–256
relative equilibrium, **A:**249–251
van der Waals complex, **A:**251–256
Hamiltonian dynamics, **A:**219–223
dimensions, **A:**221–223
general equations, **A:**219–221
n degrees of freedom, **A:**234–247
Hénon-Heiles potential, **A:**237–244
normally hyperbolic invariant manifolds, dimensions and, **A:**234–237
geometric characteristics, **A:**236–237
linear regime, **A:**235–236
tri-body dynamics, zero angular momentum, **A:**244–247
one degree of freedom, **A:**223–228
linear case, linearization, **A:**223–227
nonlinearities, **A:**227
theoretical background, **A:**218–219
two degrees of freedom, **A:**228–234
linear theory, **A:**229–230
periodic orbit dividing surfaces, **A:**230–234
global chemical reactions:
Arnold model, **A:**371–377
chaotic itinerancy, **A:**385–389
Lie perturbation, **A:**352–358

Melnikov integral, **A:**358–371
examples, **A:**395–398
multidimensional chaos, **A:**392–395
normally hyperbolic invariant manifolds, **A:**338–340, **A:**345–352
breakdown, **A:**389–392
singular perturbation theory, **A:**342–345
skeleton bifurcation, **A:**340–341
tangency principles, **A:**341–342, **A:**377–385
molecular dynamics:
chaotic transition, regularity, two-basin landscapes:
geometrics, **A:**151–163
reactive island theory, **A:**162–163
saddle regions, dynamical regularity, **A:**148–151
unimolecular reaction rate theory, **A:**9–22
action/angle variables, **A:**11–12
ARRKM quantization, **A:**108–111
canonical transformation, **A:**9–10
faster-than-classical quantum anomalous diffusion, **A:**134–137
few-dimensional system bottlenecks, **A:**18–19
invariant measure, **A:**10–11
KAM theorem, **A:**12–14
many-dimensional system bottlenecks, **A:**19–20
normally hyperbolic invariant manifold, **A:**20–22
Poincaré surface of section, **A:**14–16
quantized ARRKM theory *vs.* rigorous quantum rate, **A:**111–114
quantum scars, **A:**106–108
stability analysis, **A:**17–18
wave packet dynamics, **A:**124–128
Wigner function and Weyl's rule, **A:**104–106
multibasin landscapes, chaotic transition, regularity:
geometrical structure, **B:**259–260
state-space structure, **B:**285–299
average mutual information, **B:**292–294
false nearest neighbors, **B:**288–292
phase-space reconstruction, embedding of, **B:**285–288
multichannel chemical isomerization, multi-basin potential, **B:**32–34

Phase-space transition states. (*Continued*)
  multidimensional Hamiltonian resonance and
    structure, **B:**442–445
  onset dynamics, argon clusters:
    analytic techniques, **B:**131–136
    configuration entropy, **B:**140
    configuration space, **B:**140–142
    gradient extremal, **B:**131–133
    Lindemann's criterion, **B:**138–139
    additional potentials, **B:**148–151
    Lyapunov and KS entropy,
      B:142–143
    partitioned cell dynamics, **B:**145–146
    potential energy surface, cell partition,
      B:133–134
    potential function, MD simulation, and
      temperature, **B:**135–136
    reaction paths, **B:**136–138
    stationary points, **B:**143–145
    structural characteristics, **B:**136
    watershed, **B:**146–148
  slow relaxation, internal degrees of freedom,
    Hamiltonian systems, mixed-phase
    space, **B:**382–387
  Wigner's transition state dynamics,
    A:173–175
  hydrogen cyanide isomerization model,
    A:199–200
  rank-one saddle structure,
    A:183–194
  "apt" coordinates, Normal Form theory,
    A:184–186
  n-degree-of-freedom Hamiltonian,
    A:183–184
  normally hyperbolic invariant manifolds
    (NHIMs), stable/unstable manifolds,
    A:186–190
  transition state defined, **A:**190–191
  transition state flux, **A:**193–194
  transition state search technique,
    A:191–193
  saddle region energy landscapes,
    coordinate space *vs.*, **A:**180–181
Phase-space volume, multichannel chemical
  isomerization, microcanonical
  temperature, **B:**57–62
3-Phospholene, unimolecular reaction rate,
  isomerization, **A:**91–96
Photofragment excitation, Wigner's transition
  state dynamics, **A:**178

Planar dynamics, phase-space transition
  state geometry, angular momentum,
  rotating van der Waals complex,
  A:251 256
Planck's constant, unimolecular reaction rate
  theory:
  faster-than-classical anomalous diffusion,
    A:134–137
  quantum suppression of Arnold diffusion,
    A:132–134
Pocket formation, chaotic transitions, inert and
  reactive substances, front propagation,
  B:538–540
Poincaré-Birkhoff theorem:
  globally coupled Hamiltonian systems,
    B:478–480
  slow relaxation, internal degrees of freedom,
    Hamiltonian systems, mixed-phase
    space, **B:**382–387
Poincaré mapping, multidimensional
  Hamiltonian, resonance and transport,
  B:439
Poincaré surface of section (PSS):
  chaotic transition, regularity, two-basin
    landscapes:
    reactive island theory, **A:**154–163
    transition state theory, **A:**145–147
  n-body problem in celestial mechanics,
    A:309–312
  phase-space transition state geometry,
    Hamiltonian dynamics, **A:**222–223
  slow relaxation, internal degrees of freedom,
    Hamiltonian systems, mixed-phase
    space, **B:**381–387
three-body problem:
  collinear eZe configuration, triple collision
    orbits, **A:**323–329
  research background, **A:**305–309
  unimolecular reaction rate theory:
    Davis-Gray predissociation analysis,
      A:31–39
    double-well isomerization, **A:**85–88
    faster-than-classical anomalous diffusion,
      A:1353–137
    Gray-Rice isomerization theory, **A:**69–70
    isomerization:
      cyclobutanone, **A:**102–104
      HCN to CHN, **A:**97–100
      3-phospholene, **A:**94–96
      phase-space structure, **A:**14–16

SUBJECT INDEX

Zhao-Rice approximation, separatrix crossing rate, **A**:53–54
Pointwise criterion, nonmetric multidimensional scaling algorithm, **B**:321–322
Poisson brackets:
 intramolecular dynamics, floppy molecules, canonical perturbation theory, **A**:277–278
 phase-space transition states:
  Lie perturbation theory, **A**:354–358
  Melnikov integral, **A**:365–371
  Wigner's transition state dynamics, Lie transformation, normal-form coordinates, **A**:198
Poisson distribution, atomic clusters, energy level distribution, **B**:4–5
Polyad folding, resonantly coupled isomerizing/dissociating systems, **A**:287–296
Polymerase chain reaction (PCR), recursive cell production and evolution, catalytic reaction networks, **B**:567–571
Population ratio, multichannel isomerization, linear surprisal theory, **B**:76–78
Porter-Thomas distribution, unimolecular reaction kinetics, quantum energy flow, **B**:214
Positronium negative ions, collinear eZe configuration, **A**:330
Potential energy surfaces (PES):
 Coulomb three-body problem, zero angular momentum, **A**:313–319
 intramolecular dynamics, **A**:268–269
  adiabatic *vs.* nonadiabatic delocalization, **A**:280–286
  nearly separable isomerizing systems, floppy molecules, canonical perturbation theory, **A**:269–278
  resonantly coupled isomerizing/dissociating systems, high energy bifurcations, **A**:298–301
 molecular internal space:
  atomic cluster isomerization dynamics, **B**:90–91
  democratic centrifugal force, **B**:104–106
  topographical mapping, **B**:101–103
 multibasin landscapes, chaotic transition, regularity:
  basic principles, **B**:259–260
  liquid water, **B**:262–263

 local/global collective coordinates, **B**:260–266
 minimalistic 46-bead protein models, **B**:267–270
 nonstationarity in energy fluctuations, **B**:272–280
 multichannel chemical isomerization, microcanonical temperature, **B**:57–62
 onset dynamics, argon clusters:
  argon$_5$ clusters, **B**:148–151
  cell partition, **B**:133–134
  configuration entropy, **B**:140–142
  stationary points, **B**:143–145
  structural analysis, **B**:136
 phase-space transition state geometry, three degrees of freedom, Hénon-Heiles potential, **A**:241–244
 rapid alloying, microcluster dynamics, floppy surface atom reaction paths, **B**:170–171
 slow relaxation, internal degrees of freedom, **B**:416–418
 unimolecular reaction rate, isomerization:
  cyclobutanone, **A**:101–104
  3-phospholene, **A**:91–96
 Wigner's transition state dynamics:
  hydrogen cyanide isomerization, stationary point geometry, **A**:202–205
  saddle region energy landscapes, **A**:180–183
  theoretical background, **A**:178–179
Potential function, onset dynamics, argon clusters, **B**:135–136
Power-law distribution, recursive cell production and evolution, catalytic reaction networks, **B**:592–595
Power spectra:
 atomic clusters, **B**:5–11
 slow relaxation, internal degrees of freedom:
  hypothesis validity, **B**:408–412
  mixed-phase space, **B**:383–387
Power spectral density function (PSD), Hamiltonian system multigodicity, stagnant motion deviation, **B**:468–469
 complex behaviors, **B**:474–475
Predissociation theory, unimolecular reaction rate, **A**:30–66
 Davis-Gray analysis, **A**:30–39
 Gray-Rice-Davis ARRKM theory, **A**:39–41
 HeCl$_2$ and NeCl$_2$, **A**:63–66

Predissociation theory, unimolecular reaction
  rate (*Continued*)
  HeICl and NeICl, **A:**66
  reaction path analysis, **A:**54–59
  separatrix rate crossing constant, **A:**50–54
  van der Waals molecules, **A:**59–63
    HeI$_2$, **A:**59–61
    NeI$_2$ and ArI$_2$, **A:**61–63
  wave packet dynamics, **A:**126–128
  Zhao-Rice approximation (MRRKM),
    A:41–54
    intramolecular bottleneck:
      approximate dividing surface, **A:**46–48
      zeroth-order rate constant calculation for
        crossing, **A:**48–50
      separatrix approximate dividing surface,
        A:45–46
Preservation property, recursive cell production
  and evolution, catalytic reaction
  network, heredity kinetics, **B:**571–573
Primitive melting, atomic clusters, power spectra
  and phase-space dimensions, **B:**7–11
Principal-axis hyperspherical coordinates
  (PAHC), molecular internal space:
  classical equations and metric force, **B:**97–99
  four-body systems, **B:**113–125
    collective coordinates, **B:**118–121
    equations of motion, **B:**114–118
    gauge field suppression, **B:**110–113
    isomerization topography mapping, **B:**99–103
    kinematics, **B:**89–90
  three-atom clusters, isomerization dynamics,
    B:94–97
Principal component analysis (PCA):
  data mining and, **B:**316–317
  multibasin landscapes, chaotic transition,
    regularity:
    future research issues, **B:**300–302
    local/global collective coordinates,
      B:260–266
    temperature dependence in dimensionality
      of folding dynamics, **B:**293–299
    nonmetric multidimensional scaling
      algorithm and:
      brain wave analysis, **B:**338–341
      fern genotype/phenotype, **B:**330–332
      green autotrophs, **B:**332–333
      microarray data, **B:**345–349
      protein family, **B:**342–343
      soil bacteria biodiversity, **B:**334–338

Prior distributions, multichannel isomerization,
  linear surprisal theory, **B:**78–80
Probability distribution function of momenta,
  globally-coupled Hamiltonian systems,
  relaxation and diffusion, Hamilton mean
  field model, **B:**484–487
Prompt states, survival probability, **A:**119–120
Protein folding, chaotic transition, regularity,
  two-basin landscapes, Kramers-
  Grote-Hynes theory, **A:**164–165
Protein structures:
  heat transfer, quantum energy flow,
    B:241–248
  anharmonic decay, vibrational states,
    B:238–241
  anomalous subdiffusion of vibrational
    energy, **B:**227–238
  multibasin landscapes, chaotic transition,
    regularity, **B:**263–266
  energy nonstationarity, **B:**270–285
    bending energy fluctuation,
      B:278–282
    bond energy fluctuation, **B:**277–278
    torsional angle energy fluctuation,
      B:282–285
  minimalistic 46-bead protein models,
    B:266–270
  nonmetric multidimensional scaling,
    B:342–343
Pruning front, slow relaxation, internal degrees
  of freedom, Hamiltonian systems,
  B:380–387
Pseudo-Arnold diffusion, phase-space transition
  states, Arnold web model, **A:**377
Pseudometrics, molecular internal space:
  Eckart subspace dynamics, **B:**108–109
  four-body systems, PAHC equations of
    motion, **B:**116–118
  gauge field effects, **B:**106–107
Pseudo-potential energy curves, intramolecular
  dynamics:
  adiabatic *vs.* nonadiabatic delocalization,
    A:279–286
  floppy molecules, canonical perturbation
    theory, **A:**276–278
Pulse-timing control, quantum chaos, **A:**436

QR methods, finite-time Lyapunov exponents,
  multidimensional Hamiltonian
  dynamical systems:

SUBJECT INDEX 667

corrections, **B**:506–512
  correction procedure, **B**:508–511
  finite-time error, **B**:507–508
  standard method, **B**:506–507
Quadratic corrections, resonantly coupled isomerizing/dissociating systems, polyad folding and saddle-node bifurcation, **A**:288–296
Qualitative analysis, multidimensional Hamiltonian dynamical systems, finite-time Lyapunov exponent instability, **B**:514–516
Quantitative analysis, molecular internal space, gauge field effects, **B**:106–107
Quantum calculation, vibrational energy relaxation, cytochrome c, CD stretching mode, **B**:197–198
Quantum chaos systems:
  coarse-grained representation:
    procedures, **A**:448–449
    Rabi state and frequency, **A**:446–448
    random vector transition, **A**:449
    rotating-wave approximation, **A**:440–449
    transition element, **A**:450–453
  optimal control theory, **A**:437–443
    analytic expression, **A**:449–456
    controlled kicked rotor, **A**:439–440
    controlled random matrix, **A**:438–439
    perfect control solution, **A**:453–456
Quantum correction factor (QCF), vibrational energy relaxation:
  basic principles, **B**:180–181
  comparison with other methods, **B**:185–186
  cytochrome c, CD stretching mode, **B**:199–200
Quantum energy flow:
  theoretical background, **B**:206–208
  unimolecular reaction kinetics:
    cyclohexane ring inversion, **B**:216–221
    heat transfer in clusters and macromolecules, **B**:221–248
    energy diffusion, **B**:222–223
    proteins, **B**:241–248
    protein vibrational energy:
      anharmonic decay, **B**:237–241
      anomalous subdiffusion, **B**:227–237
    water clusters, **B**:223–227
  localization and rate influence, **B**:208–221

dynamical corrections to RRKM from LRMT, **B**:215–216
local random matrix theory, **B**:209–214
Rice-Ramsperger-Kassel-Marcus theory, **B**:214–215
theoretical background, **B**:206
Quantum flux-flux autocorrelation function, unimolecular reaction, quantized ARRKM theory $vs.$ rigorous quantum rate, **A**:111–114
Quantum mapping, multidimensional barrier tunneling, **A**:428
Quantum mechanics:
  intramolecular dynamics, adiabatic $vs.$ nonadiabatic delocalization, **A**:279–286
  phase-space transition states, theoretical background, **A**:218–219
  resonantly coupled isomerizing/dissociating systems, polyad folding and saddle-node bifurcation, **A**:288–296
  three-body problem, **A**:305–309
  unimolecular reaction rate, **A**:104–128
    ARRKM theory, **A**:108–111
    Hamiltonian approach, **A**:116–123
    phase space quantum scars, **A**:106–108
    predissociation theory, helium chloride/neon-chloride molecules, **A**:65–66
    quantum transport, classically chaotic systems, **A**:128–137
      Arnold diffusion suppression, **A**:131–134
      Cantori model, **A**:129–131
      faster-than-classical anomalous diffusion, **A**:134–137
    rigorous quantum rate:
      $vs.$ quantized ARRKM, **A**:111–114
      semiclassical approximation, **A**:114–116
    wave packet dynamics, **A**:123–128
    Wigner function and Weyl's rule, **A**:104–106
"Quantum scars," unimolecular reaction rate, phase-space structures, **A**:106–108
Quasi-classical rate theory, unimolecular reaction rate, semiclassical approximation, rigorous quantum rate, **A**:115–116
Quasi-elasticity, phase-space transition state geometry, angular momentum, astrophysics applications, **A**:260–261

Quasi-equilibrium, globally-coupled
  Hamiltonian systems, relaxation and
  diffusion, Hamilton mean field (HMF)
  model, **B:**479–480
Quasi-periodic modes, fluctuation-dissipation
  theorem, excess heat production,
  adiabatic invariant, **B:**366–367
Quasi-regular regimes, chaotic transition,
  regularity, two-basin landscapes,
  A:146–147
  normally hyperbolic invariant manifolds,
    A:167–168
  saddle crossing stochasticity, **A:**165–166
  saddle regions, dynamical regularity,
    A:149–151
Quasi-stationarity:
  fluctuation-dissipation theorem, excess heat
    production, **B:**356
  microcanonical distribution, **B:**362–363
  superstatistical equilibrium distributions,
    B:360–361
  globally coupled Hamiltonian systems,
    relaxation and diffusion, Hamilton mean
    field model, **B:**484
  angle diffusion, **B:**491–493
Quenching technique, multichannel chemical
  isomerization:
  multi-basin potential, **B:**30
  passage time and uniformity, **B:**36–39

Rabi state and frequency:
  coarse-grained representation:
    optimal control theory, **A:**450–453
    perfect control solution, **A:**453–456
    quantum chaos systems, **A:**446–448
    random vector transition, **A:**449
  optimal control theory, future research, **A:**456
Radial diffusion, rapid alloying, microcluster
  dynamics, **B:**167–170
Raman scattering, slow relaxation, internal
  degrees of freedom, Hamiltonian
  systems, hypothesis validity, **B:**407–412
Random matrix system, optimal control theory,
  quantum chaos systems, **A:**438–439
Random network localization, recursive cell
  production and evolution, catalytic
  reaction network, **B:**583
Random noise, Hamiltonian system
  multiergodicity, stagnant motion
  deviation, **B:**474–475

Random phase approximation, multidimensional
  Hamiltonian resonance and structure,
  B:441–442
Random sampling techniques, multichannel
  isomerization, lifetime averaging, liquid-
  like dynamics, **B:**68–70
Random vectors, optimal control theory,
  quantum chaos systems, smooth
  transition, **A:**449
Rank-one saddle phase-space structure,
  Wigner's transition state dynamics,
  A:183–194
  "apt" coordinates, Normal Form theory,
    A:184–186
  n-degree-of-freedom Hamiltonian,
    A:183–184
  normally hyperbolic invariant manifolds
    (NHIMs), stable/unstable manifolds,
    A:186–190
  transition states:
    defined, **A:**190–191
    flux, **A:**193–194
    search technique, **A:**191–193
Rapid alloying (RA), binary clusters,
  microcluster dynamics:
  floppy surface atoms and PES reaction paths,
    B:170–171
  future research issues, **B:**175–176
  heat of solution, **B:**164–165
  molecular dynamics simulation, **B:**160–161
  nano-sized clusters, **B:**157–160
  procedural characteristics, **B:**157–158
  simulation model, **B:**158–160
  radial and surface diffusion, **B:**167–170
  reaction path numerical simulation, **B:**171–
    173
  saddle point energy distribution, **B:**173–175
  size effect, **B:**161–164
  solid phase, **B:**165–166
Rate equations, recursive cell production and
  evolution, catalytic reaction networks,
  minority control hypothesis, **B:**562–573
Rayleigh-Bénard convection, chaotic transitions,
  inert and reactive substances, strong
  anomalous diffusion, **B:**527–530
Reacting substances, chaotic transitions:
  front propagation, **B:**537–540
  fronts in cellular flows, **B:**531–532
  geometric optics limit, **B:**534–537
  slow and fast reaction regimes, **B:**532–534

SUBJECT INDEX

Reaction coordinates:
atomic clusters:
regularity, chaos, and ergodicity, **B:**20
separability and regularity of, **B:**22
fringed tunneling models, two-dimensional barrier systems, **A:**429–431
intramolecular dynamics, **A:**268–269
multibasin landscapes, chaotic transition, regularity, **B:**260
multichannel isomerization, linear surprisal theory, nonequilibrium stationary flow, **B:**74–76
onset dynamics, argon clusters, **B:**136–138
configuration entropy, **B:**140–142
phase-space transition states:
Melnikov integral, **A:**358–371
multidimensional chaos crisis, **A:**392–395
normally hyperbolic invariant manifold connections, **A:**339–340
one degree of freedom model, **A:**225–227
rapid alloying, microcluster dynamics:
floppy surface atoms, **B:**170–171
numerical enumeration of, **B:**171–173
resonantly coupled isomerizing/dissociating systems, polyad folding and saddle-node bifurcation, **A:**287–296
unimolecular reaction rate theory, predissociation, **A:**54–59
Reaction network structure, recursive cell production and evolution, diverse chemical molecules, **B:**546–547
"Reaction path Hamiltonian," molecular internal space, kinematics, **B:**89–90
Reaction tubes, multichannel isomerization, inter-basin mixing and bifurcation of, **B:**47–50
Reactive island theory (RIT):
chaotic transition, regularity, two-basin landscapes, phase space geometrics, **A:**153–163
unimolecular reaction rate:
Gray-Rice isomerization theory vs., **A:**80–84
isomerization, **A:**75–80
double-well systems, **A:**85–88
3-phospholene, **A:**95–96
triple-well systems, **A:**90–91
research background, **A:**7–8
Reactive trajectories, Wigner's transition state dynamics, rank-one saddle phase-space structure, normally hyperbolic invariant manifolds, **A:**188–190
Recrossing trajectories:
chaotic transition, regularity, two-basin landscapes:
Kramers-Grote-Hynes theory, **A:**163–165
phase space geometrics, **A:**151–163
reactive island theory, **A:**158–163
saddle regions, dynamical regularity, **A:**149–151
phase-space transition states, theoretical background, **A:**218–219
Recurrence phenomenon, slow relaxation, internal degrees of freedom, Hamiltonian systems, nearly integrable systems, **B:**395–398
Recursive cell production and evolution, catalytic reaction network:
autocatalytic network, **B:**573–595
core hypercycle minority, **B:**582–583
evolution models, **B:**585–588
intermingled hypercycle network stabilization, **B:**581–582
molecular models, **B:**574–575
phase states, **B:**575–581
random network localization, **B:**583
statistical law, **B:**588–595
deviation from universal statistics, **B:**593–595
universal statistics, **B:**593
switching mechanism, **B:**584–585
constructive biology, **B:**550–557
chemical reaction networks modeling, **B:**553–557
diverse chemicals, **B:**546–547
Dyson's loose reproduction system, **B:**549–550
Eigen's hypercycle, **B:**547–549
heredity origins, **B:**544–546
minority control hypothesis, **B:**557–573
evolvability, **B:**566–567
experimental protocol, **B:**567–571
growth speed, **B:**565
intermingled hypercycle network production, **B:**595–596
itinerant dynamics, **B:**596
kinetic theory, heredity and, **B:**571–572
model parameters, **B:**557–561

Recursive cell production and evolution,
catalytic reaction network (*Continued*)
molecule chemical composition,
B:565–566
molecule preservation, **B:**565
stochastic results, **B:**561–564
universal statistics and fluctuation control,
B:596–597
Regularity characteristics, atomic clusters,
B:11–20
Relative equilibria (RE), phase-space transition
state geometry, angular momentum:
astrophysics inelastic scattering, **A:**257–261
rotating frame dynamics, **A:**249–251
rotating van der Waals complex, **A:**251–256
Relative stability, phase-space transition state
geometry, angular momentum, rotating
frame dynamics, **A:**248–256
Renormalized diffusion, chaotic transitions,
inert and reactive substances, **B:**532
Replication mechanisms, recursive cell
production and evolution:
autocatalytic network, **B:**574–595
catalytic reaction networks, **B:**547–549
minority control hypothesis, **B:**560–573
in vitro experiments, **B:**569–571
Repulsion of energy levels:
atomic clusters, spacing distribution,
B:4–5
heat transfer, quantum energy flow, protein
vibrational states, anharmonic decay,
B:240–241
Residence time distribution:
multichannel chemical isomerization, lifetime
averaging uniformity, **B:**35–36
multidimensional Hamiltonian resonance and
transport structures, **B:**447–450
Resonance condition:
multidimensional Hamiltonian dynamical
systems:
deflected diffusion, **B:**460–462
diffusion coefficient, **B:**440–442
frequency and phase space, **B:**442–445
model components, **B:**438–439
morphological change, **B:**445–447
overlap, **B:**457–460
residence time distribution, **B:**447–450
rotation number, **B:**439–440
phase-space transition states, tangency and,
A:378–385

resonantly coupled isomerizing/dissociating
systems, polyad folding and saddle-node
bifurcation, **A:**295–296
Resonantly coupled isomerizing/dissociating
systems, intramolecular dynamics,
A:286–301
high-energy bifurcations, **A:**296–301
polyad folding and saddle-node bifurcations,
A:287–296
Rice-Ramsperger-Kassel-Marcus (RRKM) rate
theory. *See also* Alternative Rice,
Ramsperger, Kassel and Marcus rate
theory; Zhao-Rice approximation
(MRRKM)
atomic clusters, research background,
B:27–28
multichannel isomerization:
liquid-like state, **B:**34–45
unimolecular reaction:
ARRKM quantization, **A:**108–111
isomerization:
cyclobutanone, **A:**103–104
double-well systems, **A:**87–88
HCN to CHN, **A:**98–100
3-phospholene, **A:**95–96
predissociation, Davis-Gray analysis,
A:35–39
unimolecular reaction kinetics:
cyclohexane ring inversion, **B:**218–221
LRMT dynamical corrections,
B:215–216
quantum energy flow, **B:**214–215
theoretical background, **B:**206
unimolecular reaction rate, predissociation
theory, helium-iodine, **A:**60–61
Rice-Ramsperger-Kassel (RRK) rate theory,
unimolecular reaction, research
background, **A:**5–8
Riemann geometrization, Hamiltonian system
multiergodicity:
complex behaviors, **B:**475
survival time distribution, **B:**473–474
Riemann sheets:
fringed tunneling models, perturbation
strength, **A:**425–426
multidimensional barrier tunneling, static
barriers, **A:**412
Rigorous quantum rate, unimolecular reaction:
quantized ARRKM theory and, **A:**111–114
semiclassical approximation, **A:**114–116

SUBJECT INDEX

Ring inversion rates, unimolecular reaction kinetics, **B**:218–221
Ring puckering coordinate, unimolecular reaction rate, isomerization:
  cyclobutanone, **A**:100–104
  3-phospholene, **A**:91–96
Robustness parameters, chaotic transition, regularity, two-basin landscapes, normally hyperbolic invariant manifolds (NHIM), **A**:166–168
Rotating frame dynamics, phase-space transition state geometry:
  angular momentum, **A**:248–256
  relative equilibrium, **A**:249–251
  van der Waals complex, **A**:251–256
Rotating-wave approximation (RWA), coarse-grained representation, quantum chaos systems, **A**:443, **A**:446
Rotation numbers, multidimensional Hamiltonian, resonance and transport, **B**:439–440
  frequency and phase space, **B**:442–445
Rotation-vibration energy, molecular internal space, gauge-dependent expression, **B**:91–93
Rotor systems:
  faster-than-classical anomalous diffusion, **A**:134–137
  unimolecular fragmentation mapping, Morse-like kicking field, **A**:27–30
Rugged multibasin dynamics, chaotic transition, regularity, two-basin landscapes, **A**:167–168
Rydberg atoms, Wigner's transition state dynamics, saddle region energy landscapes, **A**:180–183

Saddle crossings, chaotic transition, regularity, two-basin landscapes, stochasticity parameters, **A**:165–166
Saddle indexes, phase-space transition states:
  Lie perturbation theory, **A**:352–358
  normally hyperbolic invariant manifolds, **A**:338–340
Saddle-node bifurcations:
  intramolecular dynamics, future research, **A**:301–302
  resonantly coupled isomerizing/dissociating systems, **A**:287–296
  high energy bifurcations, **A**:296–301

Saddle point approximation, fringed tunneling models, semiclassical techniques, **A**:408–410
Saddle regions:
  atomic clusters:
    power spectra and phase-space, **B**:8–11
    regularity, chaos, and ergodicity, **B**:15–20
  chaotic transition, dynamical regularity, two-basin landscapes, **A**:147–151
  molecular internal space, four-body systems, **B**:113–125
  multibasin landscapes, chaotic transition, regularity, **B**:258–260
  phase-space transition states, **B**:259–260
  protein structures, **B**:264–266
  multichannel chemical isomerization, multi-basin potential, **B**:28–30
  onset dynamics, argon clusters, **B**:136
  phase-space transition states, chaotic itinerancy, **A**:386–389
  rapid alloying, microcluster dynamics, energy distribution, **B**:173–174
  Wigner's transition state dynamics:
    energy landscapes, **A**:180–183
    phase space:
      $vs.$ coordinate space, **A**:180–181
      energy landscapes, **A**:182–183
    rank-one saddle phase-space structure, **A**:183–194
    "apt" coordinates, Normal Form theory, **A**:184–186
    n-degree-of-freedom Hamiltonian, **A**:183–184
    normally hyperbolic invariant manifolds (NHIMs), stable/unstable manifolds, **A**:186–190
    transition states:
      defined, **A**:190–191
      flux, **A**:193–194
      search technique, **A**:191–193
      stability, **A**:181–182
    theoretical background, **A**:178–179
Scaling laws:
  globally-coupled Hamiltonian systems, relaxation and diffusion:
    future research issues, **B**:499
    Hamilton mean field model, **B**:484
    Hamiltonian system multiergodicity, universality conjecture, **B**:470–471

Scaling laws (*Continued*)
multibasin landscapes, chaotic transition, regularity, embedding theorems, **B**:307–309
recursive cell production and evolution, catalytic reaction networks, **B**:592–595
Scaling transformation, Coulomb three-body problem:
free-fall problem, **A**:330–332
zero angular momentum, **A**:314–319
Schrödinger equation:
intramolecular dynamics, adiabatic *vs.* nonadiabatic delocalization, **A**:282–286
multidimensional barrier tunneling, global dynamics, **A**:403–406
optimal control theory, quantum chaos systems, **A**:438
Rabi state and frequency, **A**:447–448
unimolecular reaction rate, wave packet dynamics, **A**:123–128
Schubart orbits, Coulomb three-body problem, collinear eZe configuration, **A**:324–329
"Scissors" barriers, Wigner's transition state dynamics:
rank-one saddle phase-space structure, normally hyperbolic invariant manifolds, **A**:188–190
Self-similar island chain:
globally coupled Hamiltonian systems, **B**:478–480
slow relaxation, internal degrees of freedom:
future research issues, **B**:413–418
mixed-phase space, **B**:382–387
Semi-chaotic regime, chaotic transition, regularity, two-basin landscapes, **A**:146–147
saddle crossing stochasticity, **A**:165–166
"Semiclassical eigenfunction hypothesis," unimolecular reaction rate theory, Wigner function, **A**:106–108
Semiclassical techniques:
multidimensional barrier tunneling, **A**:407–410
fringed tunneling models, **A**:406–407
complex-domain method, **A**:407–410
global branch structures, **A**:415–417
multiple characteristic trajectories, **A**:422–425
two-dimensional barriers, **A**:428–431

global dynamics, **A**:402–406
Melnikov integral, **A**:431–432
M-set structure at critical point, **A**:414–415
periodic perturbation effects, **A**:412–414
perturbation strength, **A**:425–426
static barrier, **A**:410–412
classical solution, **A**:410
trajectory singularities and integration paths, **A**:410–412
theoretical analyses:
low-frequency approximation, **A**:418–422
overview, **A**:417–418
three-body problem, **A**:306–309
unimolecular reaction rate, **A**:104–128
ARRKM theory, **A**:108–111
Hamiltonian approach, **A**:116–123
phase space quantum scars, **A**:106–108
predissociation theory, helium chloride/neon-chloride molecules, **A**:65–66
rigorous quantum rate, **A**:114–116
semiclassical approximation, **A**:114–116
vs. quantized ARRKM, **A**:111–114
wave packet dynamics, **A**:123–128
Wigner function and Weyl's rule, **A**:104–106
Semiclassical techniques, multichannel isomerization, future research, **B**:83
Semi-global dynamics, chaotic transition, regularity, two-basin landscapes, reactive island theory, **A**:156–163
Separatrix construction:
multidimensional Hamiltonian resonance and transport, resonance overlap, **B**:458–460
phase-space transition states:
Hamiltonian dynamics, **A**:222–223
Melnikov integral, **A**:359–371
examples using, **A**:395–398
one degree of freedom model, **A**:226–227
tangency and, **A**:384–385
unimolecular reaction rate theory:
Gray-Rice-Davis ARRKM theory, **A**:39–41
Gray-Rice isomerization theory, **A**:67–70
isomerization, cyclobutanone, **A**:102–104
normally hyperbolic invariant manifolds (NHIM), **A**:7–8

quantum suppression of Arnold diffusion,
    A:132–134
Zhao-Rice approximation:
    crossing rate constant, **A**:50–54
    dividing surface approximation,
        A:45–46
    isomerization theory, **A**:70–75
Shannon entropy, multichannel isomerization:
    lifetime averaging, liquid-like dynamics,
        B:68–70
    linear surprisal theory, **B**:72–74
Shepard plots, nonmetric multidimensional
    scaling algorithm and, molecular
    taxonomy, **B**:326–329
Short-period limit, fluctuation-dissipation
    theorem, excess heat production, **B**:359
Short-time behavior, multichannel
    isomerization:
    liquid like phase, **B**:42–44
    microcanonical temperature/Arrhenius
        relation, **B**:53–55
Single exponential lifetime averaging,
    multichannel isomerization,
    microcanonical temperature, **B**:64–65
Singularities, multidimensional barrier
    tunneling:
    M-set critical point, **A**:414–415
    static barriers, **A**:410–412
Singular perturbation theory, phase-space
    transition states, **A**:342–345
Singular value decomposition (SVD) theorem:
    finite-time Lyapunov exponents,
        multidimensional Hamiltonian
        dynamical systems:
        vectors, **B**:504–505
    molecular internal space:
        four-body systems, PAHC equations of
            motion, **B**:114–118
        three-atom clusters, isomerization
            dynamics, **B**:95–97
Size effect:
    globally-coupled Hamiltonian systems,
        relaxation and diffusion, Hamilton mean
        field (HMF) model, **B**:481
    rapid alloying, microcluster dynamics:
        nano-sized clusters, **B**:158
        numerical results, **B**:161–164
Skeleton reaction profile:
    chaotic transition, regularity, two-basin
        landscapes, **A**:166–168

multibasin landscapes, chaotic transition,
    regularity, phase-space transition states,
    B:260
phase-space transition states:
    bifurcation, **A**:340–341
    chaotic itinerancy, **A**:385–389
    normally hyperbolic invariant manifold
        connections, **A**:341
    tangency of intersections,
        A:377–385
SKEW structures, multichannel isomerization:
    inter-basin mixing, **B**:46––47
    time scales, **B**:50–51
    liquid-like phase, **B**:43–44
    microcanonical temperature, single
        exponential form, **B**:66–67
    nonstatistical behavior, low energy phase,
        B:44–45
Slow relaxation dynamics:
    chaotic transitions, inert and reactive
        substances, **B**:532–534
    Hamiltonian systems:
        internal degrees of freedom:
            FPU models, **B**:398–403
            future research issues, **B**:412–418
            mixed-phase space systems, anomalous
                transport, **B**:379–387
            molecular systems, **B**:387–392
            nearly integrable picture, applicability,
                B:392–398
            theoretical background, **B**:375–378
            validity of hypothesis, **B**:403–412
        multiergodicity and nonstationarity,
            B:466–469
    multidimensional phase space systems:
        Arnold model, **B**:427–430
        numerical method and results,
            B:430–435
        global motion, many degrees of freedom,
            B:425–427
Smale horseshoe, unimolecular fragmentation
    mapping, Morse-like kicking field,
    two-dimensional free particle,
    A:24–27
Smallness parameter, multidimensional barrier
    tunneling, low-frequency approximation,
    A:419–422
Smoothing conditions, phase-space transition
    states, normally hyperbolic invariant
    manifolds, **A**:348–352

Soil bacteria biodiversity, nonmetric multidimensional scaling algorithm and, **B:**334–338

Solid-liquid transition, multichannel chemical isomerization, multi-basin potential, **B:**30–31

Solid-phase transition, rapid alloying, microcluster dynamics:
nano-sized clusters, **B:**158
numerical results, **B:**165–166

Space-fixed frame, molecular internal space, gauge-dependent expression, rotation-vibration energy, **B:**92–93

SPC potential, slow relaxation, internal degrees of freedom, Hamiltonian systems, hypothesis validity, **B:**403–412

Stability analysis:
Coulomb three-body problem, zero angular momentum, **A:**317–319
phase-space transition state geometry, triatomic dynamics, zero angular momentum, **A:**245–247
unimolecular reaction rate theory, phase-space structure, **A:**17–18
Wigner's transition state dynamics, saddle region energy landscapes, **A:**181–182

Stable/unstable manifolds. *See also* Normally hyperbolic invariant manifolds (NHIM)
Coulomb three-body problem, zero angular momentum, **A:**317–319
multidimensional phase space, slow dynamics, Arnold model, **B:**427–430
phase-space transition states:
angular momentum, rotating van der Waals complex, **A:**254–256
Arnold web model, **A:**373–377
chaotic itinerancy, **A:**386–389
Melnikov integral, **A:**361–372
multidimensional chaos crisis, **A:**392–395
normally hyperbolic invariant manifolds, **A:**347–352
connections, **A:**339–340
one degree of freedom model, **A:**227
tangency and branching, **A:**377–385
rank-one saddle phase-space structure, **A:**186–190
Wigner's transition state dynamics, Lie transformation, **A:**197–198

Stagnant motion deviation. *See also* Nonstationarity

Hamiltonian system multiergodicity and nonstationarity, **B:**466–469
complexities, **B:**474–475
universality conjecture, **B:**469–471

Standard diffusion, chaotic transitions, inert substances, **B:**523–527

State-space structure, multibasin landscapes, chaotic transition, regularity, **B:**285–299
average mutual information, **B:**292–294
false nearest neighbors, **B:**288–292
phase-space reconstruction, embedding of, **B:**285–288

Static barrier, multidimensional barrier tunneling, **A:**410–412
classical solution, **A:**410
low-frequency approximation, **A:**419–422
trajectory singularities and integration paths, **A:**410–412

Stationary points:
onset dynamics, argon clusters, **B:**143–145
Wigner's transition state dynamics:
hydrogen cyanide isomerization, **A:**202–205
saddle region energy landscapes, **A:**180–183
theoretical background, **A:**179

Statistical law, recursive cell production and evolution, catalytic reaction network, autocatalytic network, **B:**588–595
deviation from universal statistics, **B:**593–595
universal statistics, **B:**593

STIRAP scheme, quantum chaos, **A:**436

Stochasticity, chaotic transition, regularity, two-basin landscapes, saddle crossings, **A:**165–166

Stochastic parameter:
fluctuation-dissipation theorem, excess heat production, adiabatic invariant, **B:**365–367
recursive cell production and evolution, catalytic reaction networks, minority control hypothesis, **B:**561–573

Stokes phenomenon, fringed tunneling models:
complex semiclassical techniques, **A:**410
global structure of branches, **A:**416–417

Stretched exponential function, globally-coupled Hamiltonian systems, relaxation and diffusion, equilibrium diffusion, **B:**489–490

SUBJECT INDEX

Strong stochasticity threshold (SST), slow relaxation, internal degrees of freedom, Hamiltonian systems, nearly integrable systems, **B:**394–398
Structural analysis, onset dynamics, argon clusters, **B:**136
Structural stability, phase-space transition states, skelton structures, chaotic itinerancy, **A:**386–389
Superstatistical equilibrium distributions, fluctuation-dissipation theorem, excess heat production, **B:**359–361
Surface diffusion, rapid alloying, microcluster dynamics, **B:**167–170
Survival probability, dissociation dynamics, unimolecular reactions, Hamiltonian equations, **A:**118–123
Survival time distribution, Hamiltonian system multiergodicity, clustering motions, **B:**471–474
Switching mechanism, recursive cell production and evolution, catalytic reaction network, **B:**583–585
  evolution models, **B:**587–588
  itinerant dynamics, **B:**596
Symmetrized autocorrelation function, vibrational energy relaxation, **B:**184–185
Symmetry breaking, catalytic reaction network, recursive cell production and evolution, minority control hypothesis, **B:**573

Takens' theorem, multibasin landscapes, chaotic transition, regularity:
  delay coordinates, **B:**306–309
  state-space structural embedding, **B:**288
  temperature dependence in dimensionality of folding dynamics, **B:**297–299
Tangency, phase-space transition states:
  hyperbolicity breakdown, **A:**391–392
  Melnikov integral, **A:**340, **A:**358–371
  normally hyperbolic invariant manifolds:
    branching, **A:**377–385
    connections, **A:**339–340
Tannor-Rice scheme, quantum chaos, **A:**436
Taylor series expansion:
  intramolecular dynamics, floppy molecules, canonical perturbation theory, **A:**270–278
  phase-space transition states:

Lie perturbation theory, **A:**353–358
normally hyperbolic invariant manifolds, **A:**348–352
vibrational energy relaxation:
  cytochrome c, CD stretching mode, coupling constants, **B:**192–195
  force-force-correlation function approximations, **B:**186–187
Wigner's transition state dynamics:
  hydrogen cyanide isomerization, **A:**205–207
rank-one saddle phase-space structure, "apt" coordinates, **A:**184–186
Temperature dependence:
  multibasin landscapes, chaotic transition, regularity, folding dynamics dimensionality, **B:**293–299
  multichannel isomerization:
    linear surprisal theory, **B:**70–71
    microcanonical states, **B:**53–70
    Arrhenius-like relation, **B:**60–62
    canonical temperature, **B:**70
    definition, **B:**57–60
    density of states evaluation, **B:**55–56
    exponential relation, average lifetimes, **B:**62–70
    liquid-like ergodicity and nonergodicity, **B:**67–69
    M_7 single exponential form case study, **B:**65–67
    multiexponential form, **B:**62–63
    single exponential form, **B:**64–65
    lifetime averaging law, **B:**53–55
    local temperatures, **B:**60
  onset dynamics, argon clusters, **B:**135–136
  phase-space transition states:
    angular momentum, astrophysics applications, **A:**256–261
    Melnikov integral, **A:**360–371
  rapid alloying, microcluster dynamics:
    nano-sized clusters, **B:**158
    radial and surface diffusion, **B:**168–170
Thermal bath, chaotic transition, regularity, two-basin landscapes, Kramers-Grote-Hynes theory, **A:**164–165
Thermal diffusivity, heat transfer, quantum energy flow, clusters and macromolecules, **B:**221–248

Thermally averaged constant, unimolecular reaction, quantized ARRKM theory vs. rigorous quantum rate, **A:**111–114
Thermodynamic laws, fluctuation-dissipation theorem, excess heat production, **B:**355–356
second law from, **B:**369–370
Thomas-Fermi statistics, onset dynamics, argon clusters, configuration entropy, **B:**140
Three-atom clusters:
celestial mechanics, **A:**309–312
collinear eZe case, mass ratio effect, **A:**319–330
antiproton-proton-antiproton system, **A:**330
triple collision manifold, **A:**319–323
triple collision orbits, **A:**323–329
free-fall case, **A:**330–332
molecular internal space, metric force collective coordinates, **B:**94–106
democratic centrifugal force, mass-balance asymmetry and trapping trajectories, **B:**103–106
internal motion equations, **B:**97–99
principal-axis hyperspherical coordinates, **B:**94–97
topographical mapping, PAHC, **B:**99–103
phase-space transition states, Arnold web model and tangency in, **A:**378–385
theoretical background, **A:**305–309
two-dimensional case, zero angular momentum, **A:**312–319
Three degrees of freedom structures, phase-space transition state geometry, Hénon-Heiles potential, **A:**237–244
Three-dimensional wavefunctions, intramolecular dynamics, adiabatic vs. nonadiabatic delocalization, **A:**279–286
"Three threes," Wigner's formulation of, **A:**172–175
Time autocorrelation function:
dissociation dynamics, unimolecular reactions, Hamiltonian equations, **A:**118–123
unimolecular reaction rate, wave packet dynamics, **A:**124–128
Time-continuous systems, multidimensional barrier tunneling:
global dynamics, **A:**403–406
Time delay coordinate system:

multibasin landscapes, chaotic transition, regularity:
average mutual information, **B:**293–294
state-space structural embedding, **B:**286–288
Whitney embedding theorem, **B:**305–309
slow relaxation, internal degrees of freedom, Hamiltonian systems, nearly integrable systems, **B:**395–398
Time-dependent equations:
coarse-grained representation, quantum chaos systems, **A:**443, **A:**446
Rabi state and frequency, **A:**446–448
multidimensional barrier tunneling, low-frequency approximation, **A:**421–422
phase-space transition states:
Arnold web model, **A:**372–377
tangency and, **A:**379–385
Time-evolving wave packet, unimolecular reaction rate, **A:**123–128
Time scales:
multibasin landscapes, chaotic transition, regularity, nonstationarity in energy fluctuations, **B:**278–282
multichannel isomerization, inter-basin mixing, **B:**50–51
phase-space transition states:
global reaction dynamics, **A:**341–342
singular perturbation theory, **A:**342–345
tangency and, **A:**383–385
Time-series analysis, multichannel chemical isomerization, multi-basin potential, **B:**32–34
TIP3 potential, heat transfer, quantum energy flow, **B:**223–227
Top-down technique, Wigner's transition state dynamics, research background, **A:**173–175
Topographical mapping, molecular internal space, principal-axis hyperspherical coordinates (PAHC), **B:**99–103
Tori structures:
chaotic transition, regularity, two-basin landscapes, reactive island theory, **A:**155–163
Coulomb three-body problem:
theoretical background, **A:**307–309
triple collision orbits, **A:**324–329
free-fall three-body problem, **A:**330–332

SUBJECT INDEX

globally coupled Hamiltonian systems,
    B:478–480
multidimensional Hamiltonian systems,
    resonance and transport:
    diffusion coefficient, B:440–442
    fast transition pathways, B:454–457
multidimensional phase space slow dynamics:
    Arnold model, B:429–430
    global motion, B:425–427
phase-space transition states:
    Arnold model, A:371–377
    chaotic itinerancy, A:386–389
    Melnikov integral, A:362–371
    n degrees of freedom structures,
        A:236–237
slow relaxation, internal degrees of freedom,
    Hamiltonian systems:
    mixed-phase space, B:381–387
    nearly integrable systems, B:396–398
unimolecular reaction rate theory:
    isomerization, cyclobutanone, A:102–104
    KAM theorem, A:13–14
Torsional angle energy fluctuation, multibasin
    landscapes, chaotic transition, regularity,
    nonstationarity in, B:282 285
Total momentum, globally-coupled Hamiltonian
    systems, relaxation and diffusion,
    Hamilton mean field (HMF) model,
    B:481
Transition chain crossings, multidimensional
    phase space, slow dynamics, Arnold
    model, B:428–430
Transition diagram, multidimensional
    Hamiltonian resonance and transport
    structures, B:452–454
Transition state theory (TST):
    chaotic transition, regularity, two-basin
        landscapes, A:144–147
    Kramers-Grote-Hynes theory and,
        A:163–165
    historical background, A:176–179
    molecular internal space, democratic
        centrifugal force, B:104–106
    motion trapping, B:121–123
    multibasin landscapes, chaotic transition,
        regularity, B:258–260
    multichannel isomerization, liquid-like state,
        B:34–35
    phase-space transition state geometry:
        angular momentum, A:247–261

astrophysics applications, A:256–261
inelastic scattering, A:257–261
rotating frame dynamics, A:248–256
relative equilibrium, A:249–251
van der Waals complex, A:251–256
Hamiltonian dynamics, A:219–223
dimensions, A:221–223
general equations, A:219–221
n degrees of freedom, A:234–247
Hénon-Heiles potential, A:237–244
normally hyperbolic invariant manifolds,
    dimensions and, A:234–237
geometric characteristics, A:236–237
linear regime, A:235–236
tri-body dynamics, zero angular
    momentum, A:244–247
one degree of freedom, A:223–228
linear case, linearization, A:223–227
nonlinearities, A:227
theoretical background, A:218–219
two degrees of freedom, A:228–234
linear theory, A:229–230
periodic orbit dividing serfaces,
    A:230–234
unimolecular reaction, double well system
    isomerization, A:87–88
Wigner's formulation:
    hydrogen cyanide isomerization,
        A:198–212
    Hamiltonian equation, A:200–202
    model system, A:199–200
    nonreactive degrees of freedom
        quantization, A:211–212
    normal form transformation, A:205–207
    stationary flow points, A:202–205
    visualization techniques, A:207–210
Lie transformation normalization,
    A:194–198
    basic principles, A:194–197
    normal-form coordinates, dynamics,
        A:197–198
rank-one saddle phase-space structure,
    A:183–194
"apt" coordinates, Normal Form theory,
    A:184–186
n-degree-of-freedom Hamiltonian,
    A:183–184
normally hyperbolic invariant manifolds
    (NHIMs), stable/unstable manifolds,
    A:186–190

Transition state theory (TST) (Continued)
transition state:
defined, **A:**190–191
flux, **A:**193–194
search technique, **A:**191–193
research background, **A:**172–175
saddles in energy landscapes,
A:180–183
phase space energy landscapes,
A:182–183
phase space vs. coordinate space,
A:180–181
stability, **A:**181–182
Transmission coefficient:
atomic clusters, nonlinear canonical
transformation, **B:**21–22
chaotic transition, regularity, two-basin
landscapes:
phase space geometrics, **A:**152–163
transition state theory, **A:**145–147
Transporting regular island structures,
faster-than-classical anomalous
diffusion, **A:**135–137
Transport structure, multidimensional
Hamiltonian dynamical systems:
basic principles, **B:**450–452
deflected diffusion, **B:**460–462
diffusion coefficient, **B:**440–442
fast transition pathway, **B:**454–457
model components, **B:**438–439
rotation number, **B:**439–440
transition diagram, **B:**452–454
Trans-stilbene photoisomerization, unimolecular
reaction kinetics, **B:**220–221
Trapped motion:
Hamiltonian system multiergodicity, survival
time distribution, **B:**472–474
molecular internal space, democratic
centrifugal force, **B:**104–106
multichannel isomerization, **B:**83
Triatomic structures, phase-space transition state
geometry:
n degrees of freedom, **A:**244–247
nonzero angular momentum, **A:**247–248
Tricapped tetrahedron (IST) structures,
multichannel isomerization:
inter-basin mixing, **B:**46–47
inter-basin mixing, time scales, **B:**50–51
nonstatistical behavior, low energy phase,
B:44–45

Trigonal pyramidal structures, atomic clusters,
regularity, chaos, and ergodicity, **B:**14–20
Triple collision manifold (TCM):
collinear eZe configuration:
mass ratio effect, **A:**320–323
theoretical background, **A:**308–309
Coulomb three-body problem:
future research issues, **A:**333–334
zero angular momentum, **A:**312–319
n-body problem in celestial mechanics,
A:310–312
Triple collision orbits, Coulomb three-body
problem:
collinear eZe configuration, mass ratio effect,
A:323–329
zero angular momentum, **A:**317–319
Triple-well systems, unimolecular reaction,
isomerization, **A:**88–91
True dynamics, molecular internal space, gauge
field reaction rates, **B:**109–110
Truncation-diagonalization analysis,
intramolecular dynamics, adiabatic *vs.*
nonadiabatic delocalization,
A:280–286
Tsallis distribution, fluctuation-dissipation
theorem, excess heat production,
B:354–355
superstatistical equilibrium distributions,
B:361
Tsallis nonadditive statistical mechanics,
multibasin landscapes, chaotic transition,
regularity, **B:**300–302
Two-basin landscapes, chaotic transition,
regularity:
Kramers-Grote-Hynes theory, **A:**163–165
phase space geometrics, **A:**151–163
reactive island theory, **A:**153–163
saddle crossing stochasticity, **A:**165–166
saddle regions, dynamical regularity,
A:147–151
Two degrees of freedom models, phase-space
transition state geometry, **A:**228–234
linear theory, **A:**229–230
periodic orbit dividing surfaces, **A:**230–234
Two-dimensional barrier systems:
fringed tunneling models, **A:**428–431
multidimensional barrier tunneling, global
dynamics, **A:**404–406
Two-dimensional egg-crate potential, globally-
coupled Hamiltonian systems, relaxation

and diffusion, nonstationary state,
  B:494–496
Two-dimensional free particle:
  Coulomb three-body problem, free-fall
    problem, **A:**330–332
  unimolecular fragmentation mapping,
    Morse-like kicking field, **A:**22–27
Two-electron structures, theoretical background,
  A:306–309

Unimolecular reaction kinetics:
  fragmentation mapping models, **A:**22–30
    Morse-like kicking field:
      four-dimensional free rotor, **A:**27–30
      two-dimensional free particles,
        A:22–27
  future research issues, **A:**137–140
  isomerization theory, **A:**66–104
    cyclobutanone, **A:**100–104
    double-well system, **A:**84–88
    Gray-Rice theory, **A:**66–70
      reactive island theory vs., **A:**80–84
    HCN → CNH, **A:**96–100
    3-phospholene, **A:**91–96
    reactive island theory, **A:**75–80
      Gray-Rice theory vs., **A:**80–84
    triple-well system, **A:**88–91
    Zhao-Rice approximation, **A:**70–75
  multichannel isomerization, liquid-like state,
    B:34–35
  phase-space structure, molecular dynamics,
    A:9–22
    action/angle variables, **A:**11–12
    canonical transformation, **A:**9–10
    few-dimensional system bottlenecks,
      A:18–19
    invariant measure, **A:**10–11
    KAM theorem, **A:**12–14
    many-dimensional system bottlenecks,
      A:19–20
    normally hyperbolic invariant manifold,
      A:20–22
    Poincaré surface of section, **A:**14–16
    stability analysis, **A:**17–18
  predissociation theory, **A:**30–66
    Davis-Gray analysis, **A:**30–39
    Gray-Rice-Davis ARRKM theory,
      A:39–41
    HeCl$_2$ and NeCl$_2$, **A:**63–66
    HeICl and NeICl, **A:**66

reaction path analysis, **A:**54–59
separatrix rate crossing constant, **A:**50–54
van der Waals molecules, **A:**59–63
  HeI$_2$, **A:**59–61
  NeI$_2$ and ArI$_2$, **A:**61–63
Zhao-Rice approximation (MRRKM),
  A:41–54
  intramolecular bottleneck:
    approximate dividing surface, **A:**46–48
    zeroth-order rate constant calculation
      for crossing, **A:**48–50
  separatrix approximate dividing surface,
    A:45–46
quantum energy flow:
  cyclohexane ring inversion, **B:**216–221
  heat transfer in clusters and
    macromolecules, **B:**221–248
    energy diffusion, **B:**222–223
    proteins, **B:**241–248
    protein vibrational energy:
      anharmonic decay, **B:**237–241
      anomalous subdiffusion, **B:**227–237
    water clusters, **B:**223–227
  localization and rate influence, **B:**208–221
    dynamical corrections to RRKM from
      LRMT, **B:**215–216
    local random matrix theory, **B:**209–214
    Rice-Ramsperger-Kassel-Marcus theory,
      B:214–215
  theoretical background, **B:**206
quantum/semiclassical approaches,
  A:104–128
  ARRKM theory, **A:**108–111
  Hamiltonian approach, **A:**116–123
  phase space quantum scars, **A:**106–108
  rigorous quantum rate semiclassical
    approximation, **A:**114–116
  rigorous quantum rate vs. quantized
    ARRKM, **A:**111–114
  wave packet dynamics, **A:**123–128
  Wigner function and Weyl's rule,
    A:104–106
quantum transport, classically chaotic
  systems, **A:**128–137
  Arnold diffusion suppression,
    A:131–134
  Cantori model, **A:**129–131
  faster-than-classical anomalous diffusion,
    A:134–137
research background, **A:**4–8

Unitary transformations, intramolecular
 dynamics, floppy molecules,
 canonical perturbation theory,
 A:272–278
Universality conjecture, Hamiltonian system
 multiergodicity, **B:**469–471
Universal statistics, recursive cell production
 and evolution, catalytic reaction
 networks, **B:**593
 fluctuation control, **B:**596–597
Unstable periodic orbits:
 chaotic transition, regularity, two-basin
 landscapes, reactive island theory,
 A:161–163
 unimolecular reaction rate, reactive island
 theory (RIT), **A:**76–80

Validated computation, multidimensional phase
 space, slow dynamics, multi-precision
 numerical method, **B:**430–435
Van der Waals molecules:
 molecular internal space, democratic
 centrifugal force, **B:**106
 multibasin landscapes, chaotic transition,
 regularity:
 minimalistic 46-bead protein models,
 B:266–270
 nonstationarity in energy fluctuations,
 B:285
 phase-space transition state geometry, angular
 momentum, rotating frame dynamics,
 A:251–256
 equilibria, **A:**251–252
 transport, **A:**252–256
 phase-space transition states, tangency in,
 A:378–385
 unimolecular reaction rate theory:
 Davis-Gray predissociation analysis,
 A:34–39
 Gray-Rice-Davis ARRKM theory,
 A:39–41
 predissociation, **A:**59–63
 helium-iodine, **A:**59–61
 neon-iodine/argon iodine, **A:**61–63
 Zhao-Rice approximation, **A:**41–54
 separatrix dividing surface,
 A:45–46
 zeroth-order rate constant,
 intramolecular bottleneck crossing,
 A:48–50

Van Vleck perturbation:
 chaotic transition, regularity, two-basin
 landscapes, **A:**146–147
 saddle regions, dynamical regularity,
 A:148–151
 intramolecular dynamics, floppy molecules,
 canonical perturbation theory,
 A:274–278
"Velocity vector," ARRKM rate quantization,
 A:109–111
Vibrational energy relaxation (VER):
 cytochrome c CD stretching, **B:**190–200
 carbon monoxide myoglobin (MbCO),
 B:200
 classical calculation, **B:**197
 coupling constants calculation, **B:**192–195
 full width at half maximum spectra,
 B:199–200
 future research applications, **B:**200–201
 lifetime parameter assignment, **B:**195–196
 quantum calculation, **B:**197–199
 system and bath characteristics,
 B:190–192
 force-force correlation function
 approximations, **B:**186–190
 first term contribution, **B:**187–190
 Taylor expansion, **B:**186–187
 general formula, **B:**182–184
 heat transfer, quantum energy flow,
 anomalous subdiffusion, **B:**228–238
 heat transport, theoretical background,
 B:206–207
 perturbation expansion, **B:**181–182
 quantum correction factor, **B:**185–186
 research background, **B:**180–181
 symmetrized autocorrelation function,
 B:184–185
Vibrational frequency:
 phase-space transition states:
 hyperbolicity breakdown, **A:**391–392
 Lie perturbation theory, **A:**353–358
 Melnikov integral, **A:**359–371
 unimolecular reaction rate theory, Davis-Gray
 predissociation analysis, **A:**34–39
Vibrational self-consistent field (VSCF)
 formula, vibrational energy relaxation,
 quantum correction factor and,
 B:186
Vibrational wave packets, quantum energy flow,
 theoretical background, **B:**208

SUBJECT INDEX 681

Vielbein formalism, molecular internal space, four-body systems, PAHC equations of motion, **B:**117–118
Viscosity measurements, singular perturbation theory, **A:**345
Visualization techniques, Wigner's transition state dynamics, hydrogen cyanide isomerization, **A:**207–210
Vlasov-Poisson equation, globally-coupled Hamiltonian systems, relaxation and diffusion:
 Hamilton mean field (HMF) model, **B:**481
 quasi-stationary state, **B:**491–493
Von Neumann equation, vibrational energy relaxation, perturbation expansion, **B:**181–182

Wannier ridge configuration:
 Coulomb three-body problem:
  free-fall problem, **A:**332
  zero angular momentum, **A:**313–319
 three-body problem, theoretical background, **A:**307–309
Water clusters:
 heat transfer, quantum energy flow, **B:**223–227
 slow relaxation, internal degrees of freedom, Hamiltonian systems, nearly integrable systems, **B:**393–398
Watershed region, onset dynamics, argon clusters, **B:**146–148
Wavefunctions:
 heat transfer, quantum energy flow, **B:**224–227
 intramolecular dynamics, adiabatic *vs.* nonadiabatic delocalization, **A:**282–286
 resonantly coupled isomerizing/dissociating systems, polyad folding and saddle-node bifurcation, **A:**289–296
Wave packet dynamics:
 optimal control theory, kicked rotor system, **A:**443
 unimolecular reaction rate, **A:**123–128
Weibull distribution, Hamiltonian system multiergodicity, **B:**472–474
 complex behavior, **B:**474–475
Weyl's rule, unimolecular reaction rate, **A:**104–106
 semiclassical approximation, rigorous quantum rate, **A:**115–116

Whisker mapping:
 phase-space transition states:
  chaotic itinerancy, **A:**386–389
  tori structures, Melnikov integral, **A:**365–371
 unimolecular reaction rate, quantum transport, classically chaotic systems, **A:**129–131
Whitney's embedding theorem, multibasin landscapes, chaotic transition, regularity, **B:**302–309
Width parameter, vibrational energy relaxation, force-force-correlation function approximations, **B:**190
Wiener-Khinchin theorem, multibasin protein landscapes, chaotic transition, regularity, nonstationarity in energy fluctuations, **B:**270–285
Wigner distribution, atomic clusters, energy level distribution, **B:**4–5
Wigner function, unimolecular reaction rate, **A:**104–106
 phase-space quantum scars, **A:**106–108
 wave packet dynamics, **A:**124–128
Wigner's dynamical perspective, transition state theory:
 hydrogen cyanide isomerization, **A:**198–212
  Hamiltonian equation, **A:**200–202
  model system, **A:**199–200
  nonreactive degrees of freedom quantization, **A:**211–212
  normal form transformation, **A:**205–207
  stationary flow points, **A:**202–205
  visualization techniques, **A:**207–210
 Lie transformation normalization, **A:**194–198
  basic principles, **A:**194–197
  normal-form coordinates, dynamics, **A:**197–198
 rank-one saddle phase-space structure, **A:**183–194
  "apt" coordinates, Normal Form theory, **A:**184–186
  n-degree-of-freedom Hamiltonian, **A:**183–184
  normally hyperbolic invariant manifolds (NHIMs), stable/unstable manifolds, **A:**186–190
  transition state defined, **A:**190–191
  transition state flux, **A:**193–194
  transition state search technique, **A:**191–193

Wigner's dynamical perspective, transition state theory (*Continued*)
saddles in energy landscapes, **A:**180–183
phase space energy landscapes, **A:**182–183
phase space *vs.* coordinate space, **A:**180–181
stability, **A:**181–182
Wilson G matrix, vibrational energy relaxation, force-force-correlation function approximations, **B:**187–190
Window function, vibrational energy relaxation, cytochrome *c*, CD stretching mode, **B:**197

Zero angular momentum:
Coulomb three-body problem, **A:**312–319
free-fall problem, **A:**330–332
triatomic dynamics, **A:**244–247
Zero-order vibrational state space, unimolecular reaction kinetics, quantum energy flow, **B:**210–214
Zeroth-order rate constant:
dissociation dynamics, unimolecular reactions, Hamiltonian equations, **A:**116–123
unimolecular reaction rate theory:
isomerization, cyclobutanone, **A:**102–104
Zhao-Rice approximation, intramolecular bottlenecks, **A:**48–50
Zhao-Rice approximation (MRRKM), unimolecular reaction rate:
isomerization theory, **A:**70–75
cyclobutanone, **A:**102–104
double-well potentials, **A:**85–88
HCN to CHN, **A:**97–100
3-phospholene, **A:**91–96
triple-well systems, **A:**88–91
predissociation theory, **A:**41–54
helium chloride/neon-chloride molecules, **A:**63–66
helium-iodine, **A:**60–61
intramolecular bottleneck:
approximate dividing surface, **A:**46–48
zeroth-order rate constant calculation for crossing, **A:**48–50
neon-iodine/argon-iodine molecules, **A:**61–63
separatrix approximate dividing surface, **A:**45–46
wave packet dynamics, **A:**125–128
Zhu-Botina-Rabitz formula, optimal control theory:
coarse-grained transition element, **A:**450–453
kicked rotor system, **A:**442–446
perfect control solution, **A:**453–456
quantum chaos systems, **A:**437–438
random matrix, **A:**439–441
Ziglin analysis, three-body problem, theoretical background, **A:**307–309